D1297170

Electroactive Polymer (EAP) Actuators as Artificial Muscles

Reality, Potential, and Challenges

SECOND EDITION

Electroactive Polymer (EAP) Actuators as Artificial Muscles

Reality, Potential, and Challenges

SECOND EDITION

Yoseph Bar-Cohen
Editor

SPIE PRESS
A Publication of SPIE—The International Society for Optical Engineering
Bellingham, Washington USA

Library of Congress Cataloging-in-Publication Data

Electroactive polymer (EAP) actuators as artificial muscles : reality, potential, and challenges / Yoseph Bar-Cohen, editor. – 2nd ed.
p. cm. – (SPIE Press monograph)
Includes bibliographical references and index.
ISBN 0-8194-5297-1
1. Polymers in medicine. 2. Conducting polymers. 3. Muscles. I. Bar-Cohen, Yoseph. II. Series.

R857.P6E43 2004
610'.28–dc022 2004002977

Published by

SPIE—The International Society for Optical Engineering
P.O. Box 10
Bellingham, Washington 98227-0010 USA
Phone: +1 360 676 3290
Fax: +1 360 647 1445
Email: spie@spie.org
Web: http://spie.org

Printed in the United States of America.

Contents

Preface

This book reviews the state of the art of the field of electroactive polymers (EAPs), which are also known as artificial muscles for their functional similarity to natural muscles. This book covers EAP from all its key aspects, i.e., its full infrastructure, including the available materials, analytical models, processing techniques, and characterization methods. This book is intended to serve as reference tool, a technology users' guide, and a tutorial resource, and to create a vision for the field's future direction. In preparing this second edition, efforts were made to update the chapters with topics that have sustained major advances since the first edition was prepared three years ago. Following the reported progress and milestones that were reached in this field has been quite heartwarming. These advances are bringing the field significantly closer to the point where engineers consider EAPs to be actuators of choice. In December 2002, the Japanese company Eamex produced a robot fish that swims in a water tank without batteries or a motor. For power, the robot fish uses an inductive coil that is energized from the top and bottom of the tank. Making a floating robot fish may not be an exciting event, but this is the first commercial product to use an EAP actuator.

EAPs are plastic materials that change shape and size when given some voltage or current. They always had enormous potential, but only now is this potential starting to materialize. Advances reported in this second edition include an improved understanding of these materials' behavior, better analytical modeling, as well as more effective characterization, processing, and fabrication techniques. The advances were not only marked with the first commercial product; there has also been the announcement by the SRI International scientists who are confident they have reached the point that they can now meet the challenge posed by this book's editor of building a robot arm with artificial muscles that could win an arm wrestling match against a human. This match may occur in the coming years, and the success of a robot against a human opponent will lead to a new era in both making realistic biomimetic robots and implementing engineering designs that are currently considered science fiction.

For many years the field of EAP has received relatively little attention because the number of available materials and their actuation capability were limited. The change in this view occurred in the early 1990s, as a result of the development of new EAP materials that exhibit a large displacement in response to electrical stimulation. This characteristic is a valuable attribute, which enabled myriad potential applications, and it has evolved to offer operational similarity to biological muscles. The similarity includes resilient, damage tolerant, and large actuation strains (stretching, contracting, or bending). Therefore, it is natural to consider EAP materials for applications that require actuators to drive biologically inspired mechanisms of manipulation, and mobility. However, before these materials can be applied as actuators of practical devices their actuation force and robustness will need to be increased significantly from the levels that are currently exhibited by the available materials. On the positive side, there has already been a series of reported successes in demonstrating miniature manipulation devices, including a catheter steering element, robotic arm, gripper, loudspeaker, active diaphragm, dust-wiper, and many others. The editor is hoping that the information documented in this book will continue to stimulate the development of niche applications for EAP and the emergence of related commercial devices. Such applications are anticipated to promote EAP materials to become actuators of choice in spite of the technology challenges and limitations they present.

Chapter 1 of this book provides an overview and background to the various EAP materials and their potential. Since biological muscles are used as a model for the development of EAP actuators, Chapter 2 describes the mechanism of muscles operation and their behavior as actuators. Chapter 3 covers the leading EAP materials and the principles that are responsible for their electroactivity. Chapter 4 covers such fundamental topics as computational chemistry and nonlinear electromechanical analysis to predict their behavior, as well as a design guide for the application of an example EAP material. Modeling the behavior of EAP materials requires the use of complex analytical tools, which is one of the major challenges to the design and control of related mechanisms and devices. The efforts currently underway to model their nonlinear electromechanical behavior and develop novel experimental techniques to measure and characterize EAP material properties are discussed in Chapter 6. Such efforts are leading to a better understanding of the origin of the electroactivity of various EAP materials, which, in turn, can help improve and possibly optimize their performance. Chapter 5 examines the processing methods of fabricating, shaping, electroding, and integrating techniques for the production of fibers, films, and other shapes of EAP actuators. Generally, EAP actuators are highly agile, lightweight, low power, mass producible, inexpensive, and possess an inherent capability to host embedded sensors and microelectromechanical systems (MEMS). Their many unique characteristics can make them a valuable alternative to current actuators such as electroactive ceramics and shape memory alloys. The making of miniature insectlike robots that can crawl, swim and/or fly may become a reality as this technology evolves as discussed in Chapters 7 and 8. Processing techniques, such as ink-jet printing, may potentially be employed to make complete devices that are driven by EAP actuators. A device may be fully produced in 3D detail, thereby allowing rapid prototyping and subsequent mass production possibilities. Thus, polymer-based EAP-actuated devices may be fully produced by an ink-jet printing process enabling the rapid implementation of science-fiction ideas (e.g., insectlike robots that become remotely operational as soon as they emerge from the production line) into engineering models and commercial products. Potential beneficiaries of EAP capabilities include commercial, medical, space, and military that can impact our life greatly.

In order to exploit the greatest benefit that EAP materials can offer, researchers worldwide are now exploring the various aspects of this field. The effort is multidisciplinary and cooperation among scientists, engineers, and other experts (e.g., medical doctors) are underway. Experts in chemistry, materials science, electro-mechanics, robotics, computer science, electronics, and others are working together to develop improved EAP materials, processing techniques, and applications. Methods of effective control are addressing the unique and challenging aspects of EAP actuators. EAP materials have a significant potential to improving our lives. If EAP materials can be developed to the level that they can compete with natural muscles, drive prosthetics, serve as artificial muscle implants into a human body, and become actuators of various commercial products, the developers of EAP would make tremendously positive impact in many aspects of human life.

Yoseph Bar-Cohen
February 2, 2004

ACKNOWLEDGMENTS

The research at Jet Propulsion Laboratory (JPL), California Institute of Technology, was carried out under a contract with National Aeronautics and Space Administration (NASA) and Defense Advanced Research Projects Agency (DARPA). The editor would like to thank everyone who contributed to his efforts, both as part of his team advancing the technology as well as those who helped with the preparation of this book. The editor would like to thank the team members of his task LoMMAs that was funded by the Telerobotic and the Cross-Enterprise Programs of the NASA Code S. The team members included Dr. Sean Leary, Mark Schulman, Dr. Tianji Xu, and Andre H Yavrouian, JPL, Dr. Joycelyn Harrison, Dr. Joseph Smith, and Ji Su, NASA LaRC; Prof. Richard Claus, VT; Prof. Constantinos Mavroidis and Charles Pfeiffer, Rutgers University. Also, the editor would like to thank Dr. Timothy Knowles, ESLI, for his assistance with the development of the EAP dust-wiper mechanism using IPMC. Thanks to Marlene Turner, Harry Mashhoudy, Brian Lucky, and Cinkiat Abidin, former graduate students of the Integrated Manufacturing Engineering (IME) Program at UCLA, for helping to construct the EAP gripper and robotic arm. A special thanks to Dr. Keisuke Oguro, Osaka National Research Institute, Japan, for providing his most recent IPMC materials; and to Prof. Satoshi Tadokoro, Kobe University, Japan, for his analytical modeling effort. Prof. Mohsen Shahinpoor is acknowledged for providing IPMC samples as part of the early phase of the LoMMAs task. The editor would like also to thank his NDEAA team members Dr. Virginia Olazábal, Dr. José-María Sansiñena, Dr. Shyh-Shiuh Lih, and Dr. Stewart Sherrit for their help. In addition, he would like to thank Prof. P. Calvert, University of Arizona, for the information about biological muscles. Thanks to Dr. Steve Wax, DARPA's EAP program Manager, and Dr. Carlos Sanday, NRL, for their helpful comments related to the development of EAP characterization methods. A specific acknowledgement is made for the courteous contribution of graphics for the various chapters of this book; and these individuals and organizations are specified next to the related graphics or tables.

The editor would like to acknowledge and express his appreciation to the following individuals, who took the time to review various book chapters of the first edition and particularly those who where not coauthors of this book. Their contribution is highly appreciated and it helped to make this book of significantly greater value to the readers:

Aluru Narayan, University of Illinois/Urbana-Champaign

Paul Calvert, University of Arizona,

Sia Nemat-Nasser, Dr. Chris Thomas and Mr. Jeff McGee, University of California, San Diego

Christopher Jenkins, and Dr. Aleksandra Vinogradov, South Dakota School of Mines and Technology

Kenneth Johnson, Brett Kennedy, Shyh-Shiuh Lih, Virginia Olazabal, José-María Sansiñena, and Stewart Sherrit, Jet Propulsion Laboratory

Giovanni Pioggia and Danilo de Rossi, University of Pisa, Italy
Elizabeth Smela, University of Maryland
Gerald Pollack, Washington University
Hugh Brown, Paul Keller, Geoff Spinks, and Gordon Wallace, University of Wollongong, Australia
Jeffrey Hinkley, Ji Su, and Kristopher Wise, NASA LaRC
Jiangyu Li, Caltech
Michael Marsella, University of California, Riverside
Miklos Zrinyi, Budapest University of Technology and Economics, Hungary
Mourad Bouzit and Dinos Mavroidis, Rutgers University
Peter Sommer-Larsen, Risoe Research Center, Denmark
Mary Frecker and Qiming Zhang, Pennsylvania State University
Robert Full, University of California, Berkeley,
Roy Kornbluh and Ron Pelrine, SRI International
Satoshi Tadokoro, Kobe University, Japan,

Also, the editor would like to acknowledge and express his appreciation to the following individuals who took the time to review the 12 chapters that were revised in this second edition and particularly those who where not coauthors of this book. Their contribution is highly appreciated and it helped making this book of significantly greater value to the readers:

Ahmed Al-Jumaily, Wellesley Campus, Auckland, New Zealand
Narayan Aluru, University of Illinois/Urbana-Champaign Andy Adamatzky, University of the West of England, Bristol, England
Mary Frecker, Penn State University
Kaneto Keiichi, Kyushu Institute of Technology, Japan
David Hanson, University of Texas at Dallas and Human Emulation Robotics, LLC
Roy Kornbluh, SRI International, CA
Don Leo, Virginia Polytechnic Institute and State University
John Madden, University of British Columbia, Canada
Chris Melhuish, University of the West of England, UK
Walied Moussa, University of Alberta, Canada
José-María Sansiñena, Los Alamos National Laboratory (LANL)
Stewart Sherrit, Jet Propulsion Laboratory, CA
Peter Sommer-Larsen, Risoe Research Center, Denmark
Geoffery Spinks, University of Wollongong, Australia
Ji Su, NASA Langley Research Center (LaRC), VA
Minuro Taya, University of Washington
Aleksandra Vinogradov, Montana State University
Eui-Hyeok Yang, Jet Propulsion Laboratory, CA

Electroactive Polymer (EAP) Actuators as Artificial Muscles

Reality, Potential, and Challenges

SECOND EDITION

Topic 1

Introduction

CHAPTER 1

EAP History, Current Status, and Infrastructure

Yoseph Bar-Cohen
Jet Propulsion Lab./California Institute of Technology

1.1 Introduction

Polymers have many attractive characteristics; they are generally lightweight, inexpensive, fracture tolerant, and pliable. Further, they can be configured into almost any conceivable shape and their properties can be tailored to suit a broad range of requirements. Since the early 1990s, new polymers have emerged that respond to electrical stimulation with a significant shape or size change, and this progress has added an important capability to these materials. This capability of the electroactive polymer (EAP) materials is attracting the attention of engineers and scientists from many different disciplines. Since they behave very similarly to biological muscles, EAPs have acquired the moniker "artificial muscles." Practitioners in biomimetics (a field where robotic mechanisms are developed based on biologically inspired models) are particularly excited about these materials since the artificial muscle aspect of EAPs can be applied to mimic the movements of animals and insects [Bar-Cohen and Breazeal, 2003]. In the foreseeable future, robotic mechanisms actuated by EAPs will enable engineers to create devices previously imaginable only in science fiction. One such commercial product has already emerged in December 2002: a form of fish robot (Eamex, Japan). An example of this fish robot is shown in Fig. 1. It swims without batteries or a motor and it uses EAP materials that simply bend upon stimulation. For power it uses inductive coils that are energized from the top and bottom of the fish tank. This fish represents a major milestone for the field, as it is the first reported commercial product to use electroactive polymer actuators.

For several decades, it has been known that certain types of polymers can change shape in response to electrical stimulation. Initially, these EAP materials

Figure 1: The first commercial EAP product—a fish robot that is made by Eamex. (Courtesy of EAMEX Corporation, Japan.)

were capable of inducing only a relatively small strain. However, since the beginning of the 1990s, a series of new EAP materials has been developed that can induce large strains, which leads to a great change in the view of these materials' capability and potential. Generally, EAPs can induce strains that are as high as two orders of magnitude greater than the striction-limited, rigid, and fragile electroactive ceramics (EAC). Further, EAP materials are superior to shape memory alloys (SMA) in higher response speed, lower density, and greater resilience [Hollerbach et al. 1992; Hunter and Lafontaine, 1992; Huber et al. 1997; and Madden et al. in press]. In Tables 1 and 2, comparisons are given between EAP, EAC, and SMA for general and specific material categories. These tables show the superiority of EAP in density, actuation strain, and power. The current limitations of actuators that are made of EAP materials, including low actuation force and robustness, are limiting the scope of their practical applications.

Table 1: Comparison of the properties of EAP, SMA, and EAC.

Property	EAP	SMA	EAC
Actuation strain	Over 300%	<8% short fatigue life	Typically 0.1–0.3 %
Force (MPa)	0.1–40	200	30–40
Reaction speed	μsec to min	msec to min	μsec to sec
Density	1–2.5 g/cc	5–6 g/cc	6–8 g/cc
Drive voltage	1–7 V for ionic EAP and 10–150 V/μm for electronic EAP (Section 1.5)	5-Volt	50–800 V
Consumed power *	m-Watts	Watts	Watts
Fracture behavior	Resilient, elastic	Resilient, elastic	Fragile

*Note: The power consumption was estimated for macro devices that are driven by such actuators.

In recognition of the need for international cooperation among the developers, users, and potential sponsors, the author organized the first EAP conference on March 1–2, 1999, through The International Society for Optical Engineering (SPIE) as part of the Smart Structures and Materials Symposium [Bar-Cohen, 1999]. This conference was held in Newport Beach, California, USA, and was the largest ever on this subject, marking an important milestone and turning the spotlight onto these emerging materials and their potential. Following this success, a Materials Research Society (MRS) conference was initiated to address fundamental issues related to the material science of EAP [Zhang et al., 1999]. In addition, other conferences are increasingly including sessions on EAP as part of their program. The ACTUATORS biannual meeting

Table 2: Comparison of typical actuation materials.

Property	Electrostatic silicone elastomer*	Polymer electrostrictor†	Single crystal electrostrictor‡	Single crystal magnetostrictor§
Actuation strain	100%	4%	1.7%	2%
Blocking force/area	0.2 MPa	0.8 MPa	65 MPa	100 MPa
Reaction speed	msec	μsec	μsec	μsec
Density	1.5 g/cc	3 g/cc	7.5 g/cc	9.2 g/cc
Drive field	144 V/μm	150 V/μm	12 V/μm	2500 Oe
Fracture toughness	large	large	low	large

* Pelrine et al., 1998. †Zhang et al, 1998. ‡ Park and Shrout, 1997. §Hathaway and Clark, 1993. ††Note: This is the force that is required to block the actuator and is normalized by its surface area. Values were calculated assuming the elastic properties were independent of applied field and are therefore approximated.

in Breman, Germany, is an example [Borgmann, 2002]. The SPIE conferences are now organized annually and have been steadily growing in number of presentations and attendees. Currently, there is a website called the *WorldWide EAP (WW-EAP) Webhub*∞ that archives related information and links to homepages of EAP research and development facilities worldwide. Since June 1999, the semi-annual WW-EAP newsletter has been published with short synopses from authors worldwide to provide a snapshot of the advances in the field as it progresses. This Newsletter is published electronically and its issues are accessible via the website mentioned above. Finally, this book provides a comprehensive documented reference, technology-users guide, and tutorial resource with a vision for the future direction of this field. The book covers the field of EAP from all its key aspects, i.e., its full infrastructure, including the available materials, analytical models, processing techniques, and characterization methods.

The increased resources, the growing number of investigators conducting research related to EAP, and the improved collaboration among developers, users, and sponsors are leading to rapid progress in this field. In March 1999, the author challenged the worldwide community of EAP experts to develop a robotic arm actuated by artificial muscles capable of winning an arm-wrestling match with a human opponent (Fig. 2). At the rate of development reported in recent

∞ http://ndeaa.jpl.nasa.gov/nasa-nde/lommas/eap/EAP-web.htm

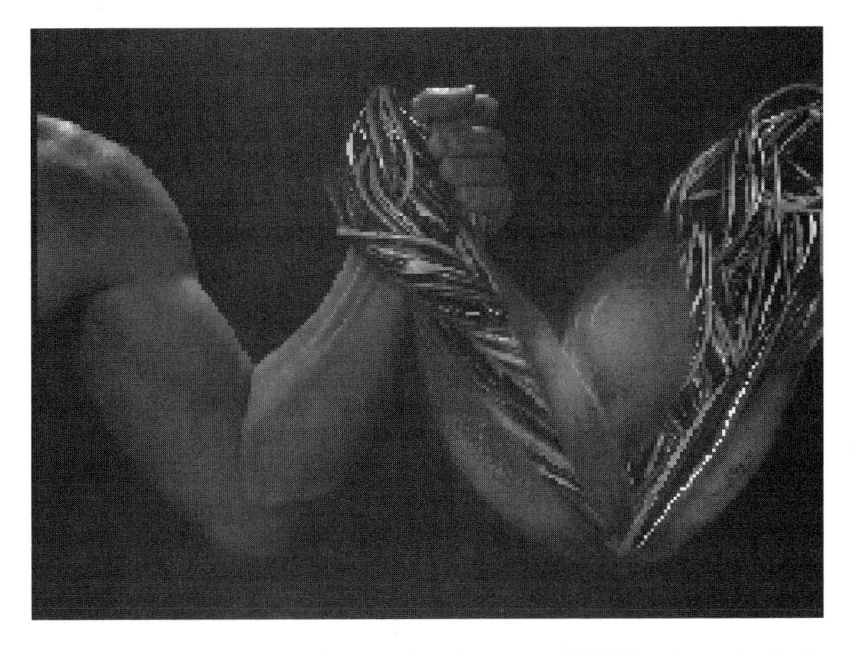

Figure 2: Grand challenge for the development of EAP actuated robotics.

international EAP conferences it is believed that this challenge may be met in the coming years rather than decades from now. Progress toward this goal will lead to great benefits, particularly in the medical area, including effective prosthetics. Decades from now, EAP may be used to replace damaged human muscles, actuate prosthetics, and potentially lead to a "bionic human." A remarkable milestone for the EAP field would be when a disabled person could use this technology to jog to the grocery store.

Various aspects of EAP material properties and applications have been reported in many technical publications. These reports were scattered over a large number of journals and handbooks across a broad range of disciplines. This book brings the accumulated information about the field together into one volume in an effort to provide the reader with a state-of-the-art overview of all aspects of current EAP materials, as well as their existing and conceptual applications. It is intended to serve as a reference book, an educational tool, and an information resource to inspire the development of improved EAP formulations and an expanded range of applications. The book describes the biological muscle as a model for EAP actuators, discusses EAP materials currently under development including processing and characterization techniques and presents some of the challenges that are facing scientists and engineers who are now engaged in developing EAP materials and their practical applications.

This book reviews the various aspects from modeling to applications and it is divided into topics that follow the field infrastructure roadmap as viewed by the author. In Topic 2, Chapter 2, biological muscles are described as a model for the development of EAP, whereas Topic 2, Chapter 3 describes both the biological model and its operation as an electromechanical actuator. Topic 3 reviews the various EAP materials, their mechanism of actuation, and identifies some of the deficiencies and challenges to their application. Topic 4 provides analytical discussions that include computational chemistry, electrochemical, and electromechanical models. Also, an example is given describing an electromechanical design approach to one type of EAP material. Topic 5 covers various processing EAP materials including fabrication, electroding, and shaping. Methods of testing and characterizing EAP are reviewed in Topic 6, providing foundations for establishing a properties database for these materials and metrics for comparing their performance with other electroactive materials. Finally, Topic 7 describes applications of EAP actuators with an emphasis on examples for dielectric EAPs as actuators, biologically inspired robotics, application to entertainment industry, haptic interfaces, and membrane structures in the form of gossamers. Topic 8 provides a lesson learned from using EAP materials to develop a planetary mechanism, a summary of the leading EAP materials, as well as a review of current and potential applications of EAP materials.

1.2 Biological Muscles

Muscles are considered highly optimized systems since they are fundamentally the same for all animals and the differences between species are small. As described in Topic 2, natural muscles are driven by a complex mechanism and are capable of lifting large loads with short response time (milliseconds). The operation of muscles depends on chemically driven reversible hydrogen bonding between two polymers, actin and myosin. Muscle cells are roughly cylindrical in shape, with diameters between 10 and 100 μm, having a length of up to several centimeters. It is difficult to determine the performance of muscles; most measurements have been made on large shell-closing muscles of scallops [Marsh et al., 1992]. A peak stress of 150–300 kPa is developed at a strain of about 25%, while the maximum power output is 150–225 W/kg. The average power is about 50 W/kg with an energy density of 20–70 J/kg that decreases with increasing speed. Although muscles produce linear forces, all motions at joints are rotary. Therefore, the strength of an animal is not just muscle force, but muscle force modified by the mechanical advantage of the joint [Alexander, 1988], which usually varies with joint rotation. The mechanical energy is provided by a chemical-free energy of a reaction involving adenosine triphosphate (ATP) hydrolysis. The release of Ca^{2+} ions is responsible for turning on and off the conformational changes associated with muscle striction.

1.3 Historical Review and Currently Available Active Polymers

There are many types of polymers that have controllable properties and this behavior can be triggered by a variety of stimulators. Some of these polymers sustain a permanent change, while others exhibit reversible responses. Generally, polymers may be passive, but by embedding active materials can be made as smart structures. Such structures allow shape control and self-sensing, taking advantage of the resilience and toughness characteristics of the host polymers.

Generally, electrical excitation is only one type of stimulator that can induce elastic deformation in polymers [Sec. 1.5; Topic 3]. Other activation mechanisms include chemical [Sec. 1.4.3.1; Kuhn et al., 1950; Steinberg et al., 1966; Herod and Schlenoff, 1993; Otero et al., 1995], thermal [Sec. 1.4.3.2 and 1.4.3.6; Tobushi et al., 1992; Li et al., 1999], pneumatic [Sec. 1.4.3.3], optical [Sec. 1.4.3.4; van der Veen and Prins, 1971], and magnetic [Sec. 1.4.3.5; Zrinyi et al., 1997]. Polymers that are chemically stimulated were discovered more than half a century ago when collagen filaments were demonstrated to reversibly contract or expand when dipped in acid or alkali aqueous solutions, respectively [Katchalsky, 1949]. Even though relatively little has since been done to exploit such "chemomechanical" actuators, this early work pioneered the development of synthetic polymers that mimic biological muscles [Steinberg et al., 1966]. However, the convenience and the practicality of electrical stimulation, as well as the technical progress, led to a growing interest in EAP materials.

EAP's beginning can be traced back to an 1880 experiment that was conducted by Roentgen using 16 × 100 cm strips of natural rubber that were charged and discharged with a fixed end and a mass attached to the free end [Roentgen, 1880]. It is interesting to note that Roentgen attributed the observed volume change to thermal effects of the electric field and the interaction with the rubber strip that is made of a dielectric material. Later, Sacerdote [1899] formulated the strain response to electric field activation. These milestones were followed in 1925 with the discovery of a piezoelectric polymer called electret, when carnauba wax, rosin, and beeswax were solidified by cooling while they were subjected to a dc bias field [Eguchi, 1925].

Following the 1969 observation of a substantial piezoelectric activity in Poly(vinylidene fluoride) (PVF2) [Bar-Cohen et al., 1996; Zhang et al., 1998], investigators started to examine other polymer systems, and a series of effective materials have emerged. The largest progress in EAP materials development has occurred in the last fifteen years where effective materials that can induce strains exceeding 100% have emerged [Pelrine et al., 1998].

Generally, EAPs can be divided into two major categories based on their activation mechanism: electronic and ionic (see further information in Topic 3). The electronic polymers (electrostrictive, electrostatic, piezoelectric, and ferroelectric) require high activation fields (>150 V/μm) close to the breakdown level. However, they can be made to hold the induced displacement under activation of a dc voltage, allowing them to be considered for robotic applications. Also, these materials have a faster response and they can be operated in air with no major constraints. In contrast, ionic EAP materials (gels, polymer-metal composites, conductive polymers, and carbon nanotubes) require drive voltages as low as 1–5 V. However, there is a need to maintain their wetness, and except for conductive polymers it is difficult to sustain dc-induced displacements. The induced displacement of both the electronic and ionic EAP can be designed geometrically to bend, stretch, or contract. Any of the existing EAP materials can be made to bend with a significant curving response, offering actuators with an easy-to-see reaction and an appealing response. However, bending actuators have relatively limited applications for mechanically demanding tasks due to the low force or torque that can be induced.

1.4 Polymers with Controllable Properties or Shape

The emphasis of this book is on the electrically activated polymers that exhibit a shape change in response to an electric field or current. However, there is a large group of polymers capable of responding to a stimulation that is not electric, and/or can respond with a change that is not necessarily mechanical or elastic. For the sake of completeness, this section reviews the active polymers that can be stimulated by nonelectrical signals.

1.4.1 Smart Structures and Materials

Generally, polymers including their configuration as composites are widely used as passive materials. However, attaching or embedding actuators and/or sensors allows them to change their shape while sensing their environment, respectively [Brebbia and Samartin, 2000]. Structures and materials that sense external stimuli and respond accordingly in real or near-real time are called "smart." The broad and strongly interdisciplinary field of smart materials applies multifunctional capabilities to existing and new structures. Current activities in this field involve design, fabrication, and testing of fully integrated structural systems enabling capabilities in various discipline areas (e.g., materials, sensing and actuation techniques, control algorithms, and architectures). Smart structures are typically produced by embedding or attaching fiber optics and piezoelectric ceramics to composite structural materials. Using embedded sensors in discrete or distributed locations allows real-time assessment of structural integrity, both during the processing of the composite material and the service operation as a system. Actuators can be used either dynamically in such applications as vibration suppression, or (quasi-) statically for shape control. Actuators for smart structures include shape memory alloys (SMA), piezoelectric and electrostrictive ceramics, magnetostrictive materials, and electro- and magneto-rheological fluids and elastomers. Embedding a sensor, actuator, signal processing network, and appropriate control system enables changing or adapting structural performance (e.g., to compensate for damage) to meet various operational criteria (e.g., change wing lift).

Another aspect of smart structures is the ability to determine their present state [Brebbia and Samartin, 2000]. Based on a set of actions, the system can make a decision to change the structural state to a more desirable one while carrying out the decision in a controlled manner and in a short period of time. Theoretically, such structures can accommodate unpredictable environmental changes, meet exacting performance requirements, and compensate for failure of minor system components. With the evolution of smart structure technology, a new generation of intelligent materials has evolved that reacts to external stimuli in a nonconventional manner. These materials are supported by effective computational methods that allow modeling, controlling, and managing the material and structural behavior. They require effective sensors and actuators, adaptive materials, physical systems modeling and analysis, active and passive control strategies, testing and verification, and analysis tools. The design of smart structures and materials involves many disciplines, and the technology is applicable to civil, mechanical, aerospace, ocean, and biomedical engineering, as well as to the microstructure and acoustic communities. Some of the possibilities of smart structures include the formation of a switchable multitude of stable mechanical configurations or shapes, which can be enabled by adjusting the direction and properties of the different layers of graphite composite panels while they are produced. Examples of such stable shapes of a composite structure are shown in Fig. 3.

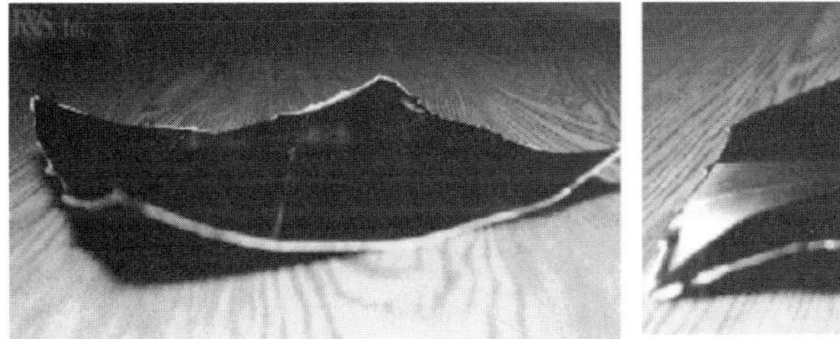
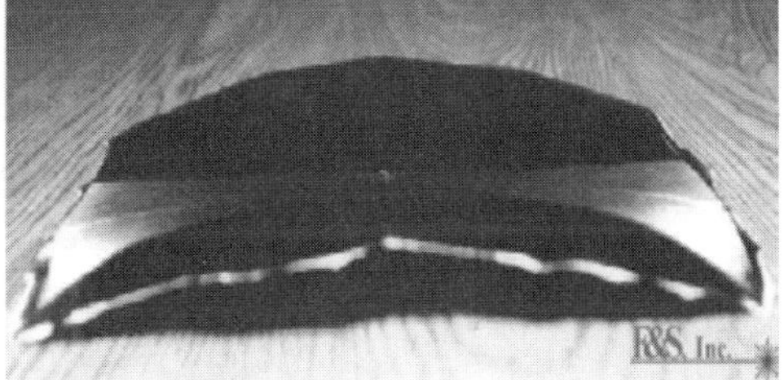

Figure 3: Stable shapes of a laminate with embedded sensors. (Courtesy of Luna Innovations.)

Figure 4: Fiber optic sensors embedded into a patch on a freeway overpass. (Courtesy of Steve E. Watkins, Applied Optics Laboratory, University of Missouri-Rolla.)

By integrating optical-fiber sensors into a structure, one can monitor the deformation of modular composite structures, such as bridge decks, over time. The sensor provides real-time measurements of periodic interrogation over a number of years. It can withstand the process of impregnation and cure of vinyl ester resins used in the manufacture of composite bridge decks. Such bridge structures are being increasingly monitored while in service. An example of such an instrumented bridge is the concrete structure of an overpass on Interstate 44 in south central Missouri, which was damaged in a traffic accident. The University of Missouri-Rolla replaced a column and patched a cap using composite sheets to strengthen the patch. In order to conduct long-term monitoring of the overall repair, a set of five fiber optic EFPI (external Fabry-Pérot interferometer)-type strain sensors was integrated into the concrete and the composite sheets. A view of the repair and the attached sensors is shown in Fig. 4.

Studies are currently under way to design and construct materials and structures in order to realize performance gains in aerodynamic and hydrodynamic control. Some of the applications that are being considered include shape control of aircraft wings, engine inlets and marine propulsion systems, wake control of underwater vehicle vortices, noise and vibration control for helicopter rotor blades, and torpedo quieting. Other applications that are under consideration include aircraft with dramatically improved range, maneuverability, and enhanced survivability, as well as marine vehicle turbo-machinery that operates with reduced acoustic signatures. This capability can be achieved by shape changes that affect the flow fields in the surroundings. Other applications of smart materials and structures include vortex wake control, as shown in Fig. 5. The ability to control the vortex wake structure downstream from lifting surfaces is a topic of continuing importance for a variety of applications. For example, there is a significant safety hazard associated with jet transports while they approach landing due to encountered wakes produced by preceding aircraft. The current standards of conservative separation between aircraft (driven chiefly by the time required for vortex wakes to dissipate) limit the capacity of all major U.S. airports. SMA technology provides an actuation mechanism with a suitable combination of force and deflection capability, and compactness.

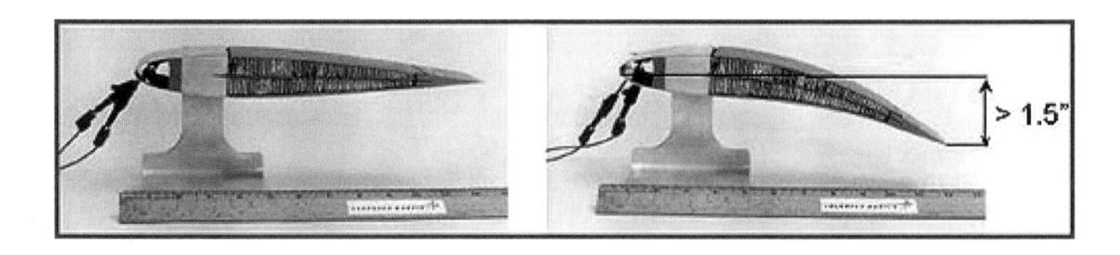

Figure 5: Shape control using smart structures. (Courtesy of T. R. Quackenbush, Continuum Dynamics, Inc.)

1.4.2 Electrically Conductive and Photonic Polymers

Over the past 20 years, investigators have addressed the challenge of developing polymers with high conductivity and photonic characteristics. The success of these studies has led to the emergence of conductive polymers that exhibit electronic properties approaching the levels in metals and semiconductors [Cao et al., 1991] and offering the processing advantages and mechanical properties of polymers. In 2000, the work of the pioneers of this field, Heeger, MacDiarmid, and Shirakawa, earned them the Nobel Prize in Chemistry [McGehee et al., 1999]. The physics and chemistry of conducting polymers has matured to the point that these novel materials are becoming ready for commercialization in a variety of electronic and electrochemical applications. High-performance devices have been demonstrated, including light-emitting diodes, light-emitting electrochemical cells, photodiodes, and lasers. The brightness of light-emitting diodes that are based on photonic polymers is as high as a fluorescent lamp using only a few volts. The performance of photonic polymers has been improved to a level that makes them comparable to or better than their inorganic counterparts. Polyaniline is currently being used by Uniax, Raytheon Computational Sensors Corporation and SFST to develop an analogue image processor [McAlvin et al., 1999; Wax et al., 1999]. Compared to conventional charge-coupled devices (CCDs), electroactive polymer-based systems are expected to respond about 10 times faster, provide larger pixel arrays at low power, and effectively operate at low light levels.

Through optimization of a polymer's chemical structure, the University of Michigan, Ann Arbor, fabricated organic polymer light-emitting heterostructure devices on both glass and flexible plastic substrates [He et al., 1999]. These devices showed a brightness of ~10,000 cd/m^2, an external quantum efficiency of 3.8%, an emission efficiency of 14.5 cd/A, and a luminous efficiency of 2.26 lm/W. Using a CCD calibration method developed by this group, the spectral distribution of the luminance and the photon emission were measured, as shown in Fig. 6. Current studies are seeking to develop active-matrix organic polymer light-emitting displays on both glass and plastic substrates. Such technology has the potential of making graphic displays that may replace hardcopy documents such as books and newspapers. Combining such a display with wireless communication will have even further ramifications.

In recent years, organic light emitting devices (OLED) have emerged as commercial products (e.g., IBM, Kodak, Uniax/Dupont, Universal Display Corp., etc.), where a series of organic thin films are placed between two conductors. When an electrical current is applied, a bright light is emitted. OLEDs are lightweight, durable, power efficient and ideal for portable applications, their processing involves fewer steps and both fewer and lower-cost materials than LCD displays. There is an increasing confidence that OLEDs will replace the current technology in many applications due to the following performance advantages over LCDs: greater brightness, faster response time for full-motion

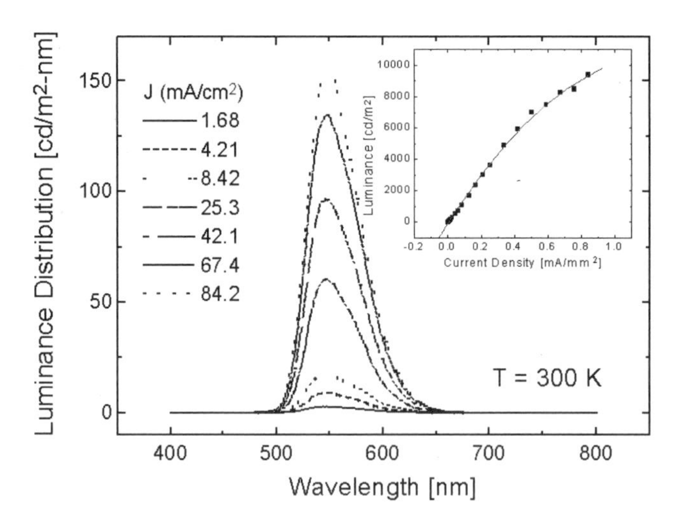

Figure 6: Spectral distribution of an organic polymer light-emitting device under different operating current densities. The inset shows the brightness vs. injection current. (Courtesy of J. Kanicki, University of Michigan.)

video, fuller viewing angles, lighter weight, greater environmental durability, more power efficiency, broader operating temperature ranges, and greater cost-effectiveness.

1.4.3 Nonelectrically Deformable Polymers

There are many polymers that exhibit volume or shape change in response to perturbation of the balance between repulsive intermolecular forces that act to expand the polymer network, and attractive forces that act to shrink it. Repulsive forces are usually electrostatic or hydrophobic in nature, whereas attraction is mediated by hydrogen bonding or van der Waals interactions. The competition between these counteracting forces, and hence the volume or shape change, can thus be controlled by subtle changes in parameters such as solvent or gel composition, temperature, pH, light, etc. This section reviews the type of polymers that can be activated by nonelectrical means.

1.4.3.1 Chemically Activated Polymers

Polymers can interact with chemicals causing them to change dimensions in a relatively slow process. For example, placing a ping-pong ball in kerosene leads to volume expansion that can cause the ball to reach several times its original size. One of the notable studies of chemomechanically active polymer gels is attributed to the pioneering work of Katchalsky in the late 1940s with his discovery of a contractile polymer gel [Katchalsky, 1949; Sperling, 1992]. His research indicated that certain polymers are very sensitive to the pH in their aqueous environments and will swell or contract. The addition of acid causes contraction of the polymer, due to an increase of the concentration of hydrogen ions, whereas alkaline solutions result in expansion due to removal of the hydrogen in the form of water [Brock, 1991; Glass, 1989; Tanaka et al., 1982; Dusěk, 1993; Osada et al, 1992]. Generally, polymer gels consist of macromolecules that are cross-linked in a three-dimensional network with an interstitial spacing that is occupied by fluid. The properties of the gel are directly dependent on the interaction with ionic species that are in contact with the gel, including the pH environment or adjacent solvent [Katchalsky, 1949; Forsterling and Kuhn, 1991]. One polymer that has been used to produce pH-activated materials is a poly(vinyl alcohol) poly(acrylic acid) (PVA-PAA) derivative. This polymer, which is readily available commercially, expands or contracts as the pH gradually increases or decreases, respectively [Woojin, 1996]. Unfortunately, PVA-PAA is not durable and is far from being useful for practical applications; poly(acrylonitrate) (PAN) offers a tougher alternative. To develop a more robust alternative, combinations of PVA-PAA and PAN as well as other gel polymers are currently being explored. Investigators at MIT, University of Arizona, and University of New Mexico [Schreyer et al., 2000] are studying methods to control this behavior electrically as discussed in Sec. 1.5.2.1 and Topic 3.2.

1.4.3.2 Shape-Memory Polymers

Shape-memory polymers (SMP), which can be switched between two phases with different dimensions, have been studied by a number of researchers [Osada and Matsuda, 1995; Liang et al., 1997]. These researchers used co-acrylic acid-n-stearyl acrylate hydrogels and polyurethane SMP, respectively. Using polyurethane in open cellular (foam) structures Sokolowski and his research team [Sokolowski, 1999] formed "cold hibernated elastic memory" (CHEM) structures. Such structures are self-deployable via foam's elastic recovery and shape memory, allowing the erection of structures that have various shapes and sizes. In practice, these foam materials are compacted to small volumes at temperatures above their softening (glass transition) temperature, Tg, and then they can be stored without constraint at temperatures below this Tg level. Heating to a temperature above the Tg level restores the original shape and dimensions. A schematic diagram of the packing and shape restoration processes is shown in Fig. 7. The advantage of this technology is that structures, when compressed and stored below Tg, are a small fraction of their original size and are lightweight. Examples of stowed and deployed CHEM structures are shown in Fig. 8. For commercial applications, these materials may be applied to shelters, hangars, camping tents, rafts or outdoor furniture. Such parts can be transported and stored in small packages, and heating them when needed at the outdoor site can expand them. After expansion, the parts are cooled to an ambient temperature below the Tg level and they become rigid as needed for use. CHEM foam materials are currently under development by the Jet Propulsion Laboratory (JPL) and Mitsubishi Heavy Industries (MHI). This material provides a stowed volume ratio of up to 40, where the cold hibernation allows long stowage, the material is impact and radiation resistant, and has very good thermal and electrical insulation properties. The disadvantage of these materials is that packing requires a pressure mechanism, and such a capability may not be readily available for outdoor users of the technology.

A shape-memory polymer has recently become commercially available (mnemoScience, http://www.mnemoscience.de) in either linear, thermoplastic multiblock copolymers, or covalently cross-linked polymer networks. In chemically cross-linked shape-memory networks the permanent shape is stabilized by the covalent net-points. In analogy to linear block copolymers with shape memory, the temporary shape of covalently cross-linked shape-memory networks can be either fixed by crystallized segment chains or by glass transition of the segment chains, which is in the temperature range of interest.

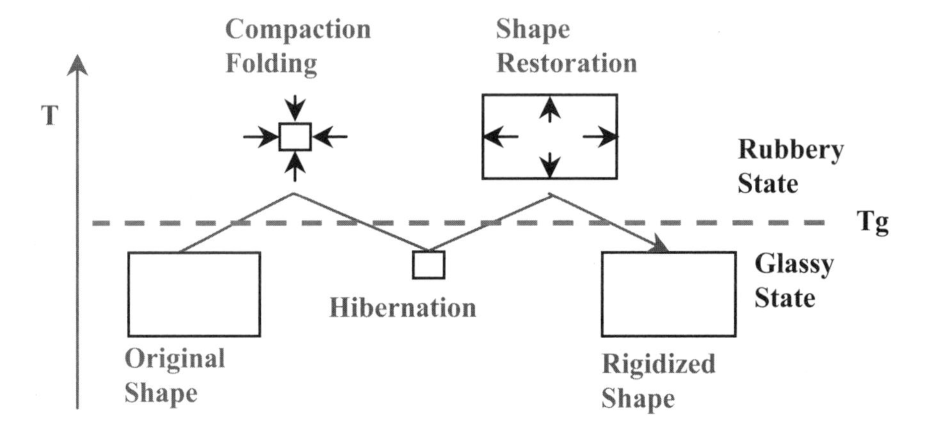

Figure 7: Shape-change process for shape memory polymers [Sokolowski, 1999]. (Courtesy of Witold Sokolowski, JPL.)

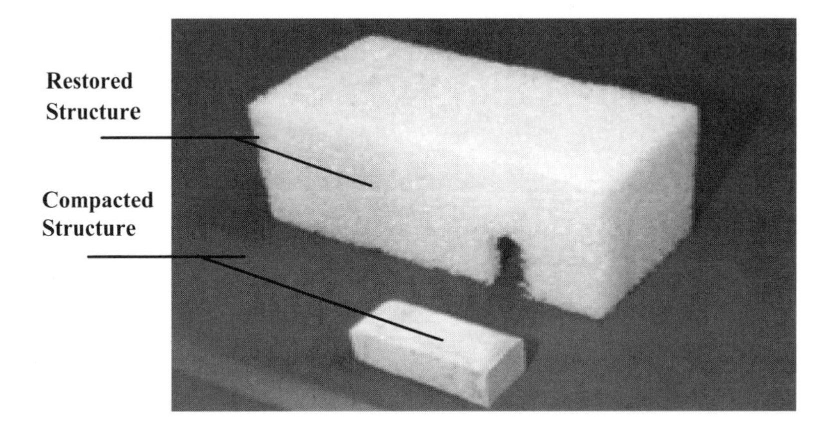

Figure 8: Stowed and deployed CHEM structures [Sokolowski, 1999]. (Courtesy of Witold Sokolowski, JPL.)

1.4.3.3 Inflatable Structures

One of the most common ways to achieve a large structural volume change is to use inflatables. Through the use of inflatables one can also create an actuator in the form of a McKibben muscle that induces significant forces approaching biological muscle levels.

1.4.3.3.1 Aerobots and balloons

Prior to the emergence of aircraft as the flight vehicle of choice, lighter-than-air balloons and other inflatables were the leading form of air transportation. To obtain the necessary flexibility, elasticity, and strength, balloons are made from polymers. The capability of inflatable structures has evolved to a level such that in 1999 a balloon was successfully used to fully circulate the earth. Further, under a NASA task, researchers at the Jet Propulsion Laboratory are currently studying various forms of aerobot mechanisms for planetary exploration, including inflatable rovers. One rover that has been developed (shown in Fig. 9) is a large-wheeled, lightweight, remote-controlled vehicle for transport of instrument payloads on distant planets and moons. The use of inflatables to increase speed and range is a critical enabling technology that will allow robotic outpost development, transportation of astronauts, and long-distance transfer of heavy equipment and *in-situ* resources such as water and ice from the martian north pole.

Generally, there are five basic mechanisms of inflating and mobilizing balloons flown in the earth's atmosphere. The simplest mechanism is known as a *zero-pressure helium balloon* consisting of a balloon partially filled with helium gas.

Figure 9: Inflatable rover under development at JPL. (Courtesy of Jack Jones, JPL.)

This balloon is, at best, metastable at any altitude and eventually it goes up or down depending on the amount of helium in the balloon. Multiple day flotation with this balloon type is possible only by venting helium during the day and dropping ballast at night. Another type of helium balloon is known as a *superpressure helium balloon.* Because it has an internal pressure slightly higher than ambient, it floats at an altitude where the mass of ambient gas it displaces is equal to the entire mass of the balloon system. This type of balloon has a higher pressure during the day than at night, and it floats at a nearly constant altitude.

Another common type of commercial balloon is a *hot air balloon,* where ambient air is burned with propane and the hot exhaust gas creates buoyancy. A variation of this is a *solar-heated balloon*, wherein a solar-absorptive black balloon creates buoyancy by heating the contained ambient air. The fifth balloon type is known as an *infrared Montgolfiere* and it flies in the earth's cold stratosphere by heat from the sun during the day and by trapped lower planetary radiation at night. These balloons generally have fixed daytime and nighttime altitudes.

1.4.3.3.2 McKibben artificial muscle

The McKibben artificial muscle is a pneumatic actuator that exhibits performance characteristics that have some similarities to biological muscles [Chou and Hannaford, 1994 and 1996]. Its springlike characteristics, physical flexibility, and lightweight make it applicable to powered prosthetics, and its high force-to-weight ratio makes it effective for mobile robots. The actuator consists of an expandable internal bladder (an elastic tube) surrounded by a braided shell. When the internal bladder is pressurized, it expands in a balloonlike manner against the braided shell, which acts to constrain the expansion in order to maintain a cylindrical shape. As the volume of the internal bladder increases due to the increase in pressure, the actuator shortens and/or produces tension if coupled to a mechanical load. These actuators can be easily constructed in a variety of sizes, and the first device was developed in the 1950s for use in artificial limbs. More recently, Bridgestone Rubber Company of Japan commercialized it in the 1980s for robotic applications, followed by Shadow Robot Group of England in the 1990s. Using a finite element modeling approach, the interior stresses and strains of McKibben actuators were estimated [Chou and Hannaford, 1996] and the results were used to improve their actuator design (see Fig. 10).

A typical application of McKibben muscle requires a significant number of contractions and extensions of the actuator. This cycling operation leads to fatigue and failure of the actuator, yielding a life span in the range of 1–10 Kcycle, which is an important design issue. To some degree this is a materials science challenge, and studies have shown that natural latex bladders have a fatigue limit that is 24 times greater than synthetic silicone rubber bladders. The force generated by a McKibben artificial muscle depends on the weave characteristics of the braided shell, the material properties of the elastic tube, the actuation pressure, and the muscle's length. The actuation properties of the McKibben actuator are reasonably close to biological muscle; however, they are deficient with regard to the speed of response (fraction of a second to several seconds). An example of a simulated human arm driven by such muscles is shown in Fig. 11.

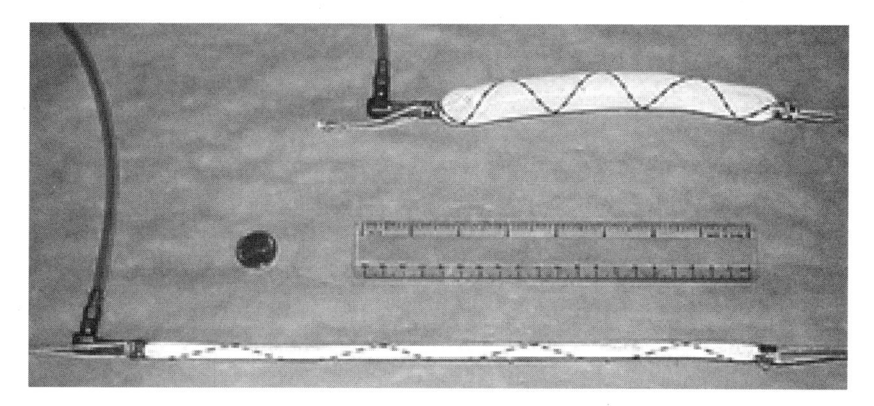

Figure 10: View of a McKibben muscle in activated and relaxed shapes. (Courtesy of B. Hannaford and D. P. Ferris, Univ. of Washington.)

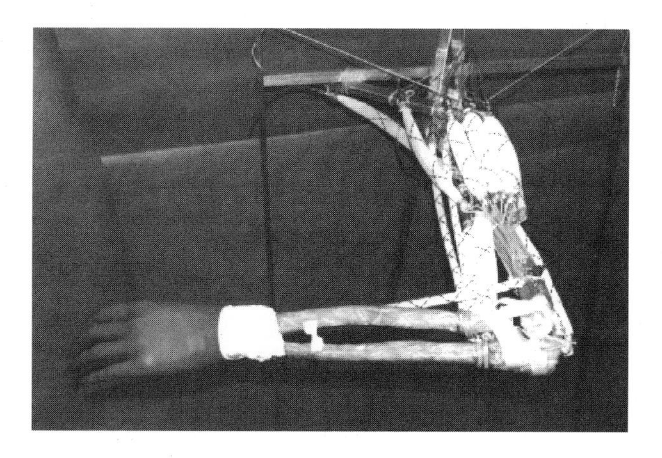

Figure 11: View of a simulated human arm with McKibben muscles. (Courtesy of B. Hannaford, Univ. of Washington.)

1.4.3.4 Light-Activated Polymers

Studies of polymers that exhibit volume and shape change when subjected to light were reported in the 1970s. Aviram [1978] reported that UV illumination induces an expansion of about 35% in a polymer gel made of poly(N,N-dimethyl glutamanilide). Such a photoinduced ionization was also demonstrated by Smets

[1995] who used a cross-linked poly(ethyl methacrylate) derivative with spirobenzopyranes. Later, Holme et al. [1999] induced surface corrugation in liquid-crystalline polymers on a micron scale using polarized laser light. In 2000, a study took place at the SPAWAR System Center (San Diego, CA) to develop polymers that change volume in response to light activation. Efforts were made to identify polymers called "jump molecules" that exhibit such expansion. Results showed a polymer gel that contract by about 20% when exposed to 455-nm of light radiation and was shown to return to its original shape when the light source was turned off [Becker and Glad, 2000]. In another study, Juodkazis et al. [2000] reported the use of a focused laser beam on the center of an experimental cylinder made of a polymer-water gel based on N-isopropylacrylamide (see Fig. 12). The beam excites polymer molecules locally, causing functional groups on nearby molecules to temporarily attract each other; this causes the polymer gel to shrink at the point where the beam is focused. Relaxing the beam releases these attraction forces and the gel returns to its original size. Experiments have shown that the shrinkage of the polymer is not related to heating of the water in the gel; but it is possibly caused by laser-induced phase transitions. This exposure causes shear-relaxation processes and gel shrinkage of up to several tens of micrometers away from the irradiation spot.

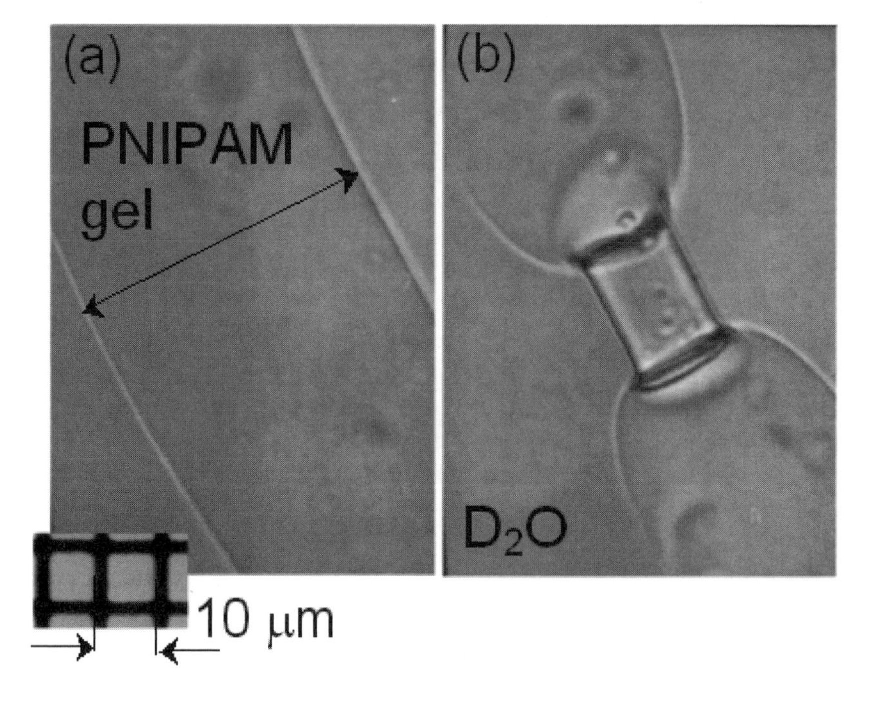

Figure 12: Micro-video image of poly(*N*-isopropylacrylamide) (PNIPAM) gel rod in D_2O before (a) and after (b) illumination by 0.75 W power laser illumination at λ = 1064 nm wavelength. (Courtesy of Prof. Hiroaki Misawa, Hokkaido University, Japan).

1.4.3.5 Magnetically Activated Polymers

Magnetically activated gels, so-called ferrogels, are chemically cross-linked polymer networks that are swollen by the presence of a magnetic field [Zrinyi et al., 1999]. Such a gel is a colloidal dispersion of monodomain magnetic particles with a typical size of about 10 nm. Subjecting the ferrogel material to a spatially nonuniform magnetic field causes its particles to move in reaction to the field. Changes in molecular conformation can accumulate and lead to shape change, where a 3-g ferrogel was demonstrated to release mechanical work at the level of about 5 mJ. The magnetic field drives and controls the displacement of the individual particles, and a balance between the magnetic and elastic interactions dictates the final shape. The ferrogel material can be made to bend, straighten, elongate, or contract repeatedly, and it has a response time of less than one tenth of a second, which seems to be independent of the particle size. Generally, the material is incompressible and does not change volume under activation. An example of a ferrogel subjected to the effect of two different magnetic fields is shown in Fig. 13.

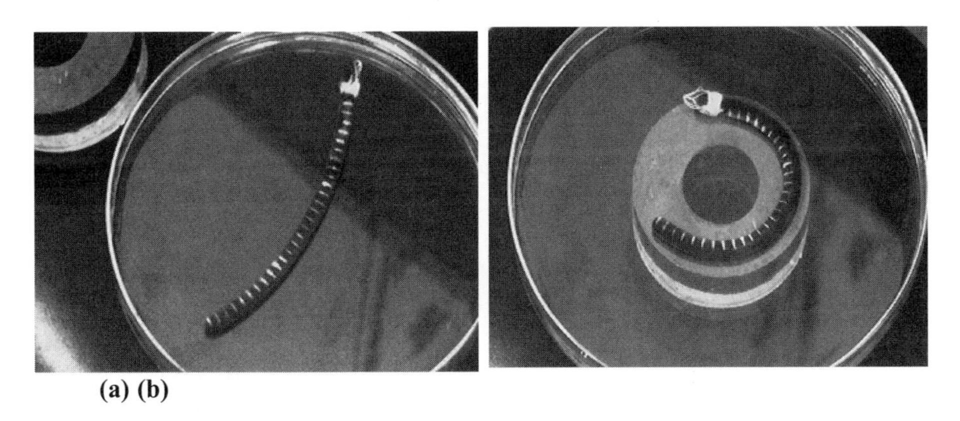

(a) (b)

Figure 13: Shape change of ferrogel induced by a nonuniform magnetic field of a permanent magnet where (a) the ferrogel is located 15 cm from the magnet; and (b) the magnet is placed under the ferrogel. (Courtesy of Miklos Zrinyi, Budapest University of Technology and Economics, Hungary.)

1.4.3.6 Thermally Activated Gels

Certain polymer gels undergo a thermal phase transition that involves volume change. This phenomenon was first reported by Hirokawa and Tanaka [1984]. The phase transition occurs in the range of 20 to 40°C, exhibiting a contractile force that can reach 100 kPa with a response time of 20–90 sec. Such thermally activated gels were the subject of studies using mostly N-substitute polyacrylamide derivatives and polypeptides. Ito [1989 and 1990, respectively] synthesized polyacrylamide derivatives with alkyl and alkoxy groups using radical polymerization. Thermally activated gels precipitate from their solutions above the phase transition temperature, which depends on the polymer concentration. Their structure contains hydrophilic; its moieties, and their hydrogen bonds are weakened at the higher temperature accompanied by release of hydration water and phase separation. The material can be configured into a fiber shape, allowing formation of a musclelike actuator. The application of this type of gel has been considered for various fields, including drug delivery systems and agriculture [Ichijo et al., 1995], automatic valves that respond to cold and hot water [Ichijo et al., 1995], biomolecular machines [Urry, 1995; Urry et al., 1996], and actuators that simulate muscles [de Rossi et al., 1992; Ichijo et al., 1995; Kishi, 1993; Suzuki and Hirasa, 1993].

1.5 Electroactive Polymers (EAP)

Polymers that exhibit shape change in response to electrical stimulation can be divided into two distinct groups: electronic (driven by electric field or Coulomb forces) and ionic (involving mobility or diffusion of ions). This section briefly reviews the currently documented electroactive polymers in each of the two groups, and the significant materials are discussed in depth in Topic 3 of this book.

1.5.1 Electronic EAP

1.5.1.1 Ferroelectric Polymers

Piezoelectricity was discovered in 1880 by Pierre and Paul-Jacques Curie, who found that when certain types of crystals are compressed (e.g., quartz, tourmaline, and Rochelle salt) along certain axes, a voltage is produced on the surface of the crystal. The year afterward, they observed the reverse effect—that upon the application of an electric current, these crystals sustain an elongation. The phenomenon is described by a third-rank tensor that is defined by

$$d = \left(\frac{\partial D}{\partial T}\right)_E = \left(\frac{\partial S}{\partial E}\right)_T, \tag{1}$$

where D is the dielectric displacement, E is the electric field, T is the mechanical stress; and S is the mechanical strain.

Piezoelectricity is found only in noncentro-symmetric materials and the phenomenon is called ferroelectricity when a nonconducting crystal or dielectric material exhibits spontaneous electric polarization. Poly(vinylidene fluoride) (also known as PVDF or PVF_2) and its copolymers are the most widely exploited ferroelectric polymers [Bar-Cohen et al., 1996; Topic 3.1, Chapter 4]. These polymers are partly crystalline, with an inactive amorphous phase, and having a Young's modulus near 1–10 GPa. This relatively high elastic modulus offers high mechanical energy density. A large applied ac field (~200 MV/m) can induce electrostrictive (nonlinear) strains of nearly 2%. Unfortunately, this level of field is dangerously close to dielectric breakdown, and the dielectric hysteresis (loss, heating) is very large. Sen et al. [1984] investigated the effect of heavy plasticization (~65% wt.) of ferroelectric polymers hoping to achieve large strains at reasonable applied fields. However, the plasticizer is also amorphous and inactive, resulting in decreased Young's modulus, permittivity, and electrostrictive strains. In 1998, Zhang and his coinvestigators. [Zhang et al, 1998] introduced defects into the crystalline structure using electron radiation to reduce the dielectric loss dramatically in a P(VDF-TrFE) copolymer. This permits ac switching with much less generated heat. Extensive structural investigations indicate that high-electron irradiation breaks up the coherence polarization domains and transforms the polymer into a nanomaterial consisting of local nanopolar regions in a nonpolar matrix. It is the electric-field-induced change between polar and nonpolar regions that is responsible for the large electrostriction observed in this polymer. As large as 5% electrostrictive strains can be achieved at low-frequency drive fields having amplitudes of about 150 V/μm. Furthermore, the polymer has a high elastic modulus (~1 GPa), and the field-induced strain can operate at frequencies higher than 100 kHz, resulting in a very high elastic power density compared with current electroactive polymers. Ferroelectric EAP polymer actuators can be operated in air, vacuum, or water, and in a wide temperature range. Figure 14 shows a bimorph actuator based on this EAP both at rest and in activated configurations. To reduce the level of voltage that is needed to activate the ferroelectric EAP, Zhang et al. [2002] used an all-organic composite with high-dielectric-constant organic particulates ($K > 10{,}000$). These researchers used a blend of the particulates in a polymer matrix to increase the dielectric constant from single digits to the range of 300 to 1000 (at 1 Hz). This approach of producing a composite EAP was conceived in the first edition of this book in Chapter 4. It led to an EAP that requires significantly lower voltage. For example, a strain of about 2% is generated by a field of 13 V/μm for a CuPc-PVDF based terpolymer composite that has an elastic modulus of 0.75 GPa (the strain is proportional to the square of the applied field). Further details are discussed in Chapter 4.

As with ceramic ferroelectrics, electrostriction can be considered as the origin of piezoelectricity in ferroelectric polymers [Furukawa, 1990]. A dc-bias polarization can be present either via a *poling* process introduced before making a device, in which case a remnant polarization persists, or via a large dc electric field applied during operation of the material as a device. In the latter case, no remnant polarization is observed when the bias is removed and, as a ferroelectric material, a very small hysteresis is observed in the polarization-electric field loop. Unlike electrostriction, piezoelectricity is a linear effect, where not only will the material be strained when voltage is applied, a voltage signal will be induced when a stress is applied. This enables them to be used as sensors and transducers. Care must be taken to avoid applying too large voltage, mechanical stress, or temperature for fear of depoling the material.

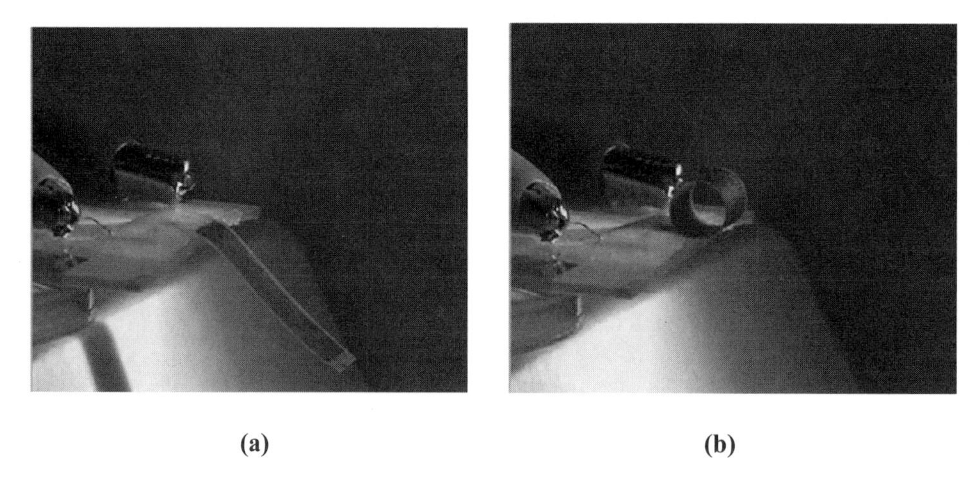

(a) (b)

Figure 14: (a) Piezolectric bimorph with no voltage applied to it. (b) The same bimorph after the voltage was turned on [Chang et al., 2000]. (Courtesy of Q. Zhang, Penn State University.)

1.5.1.1.1 Electrets

Piezoelectric behavior in polymers also appears in electrets, which are materials that consist of a geometrical combination of hard and soft layers with nonconventional routes for symmetry breaking [Sessler and Hillenbrand, 1999]. Discovered in 1925, electrets are materials that retain their electric polarization after being subjected to a strong electric field [Eguchi, 1925]. The positive and negative charges within the material are permanently displaced along and against the direction of the field, respectively, making a polarized material with a net zero charge. Electrets can be prepared from polymers, ceramics, and certain waxes, in which the individual molecules are initially randomly arranged, and under an electric field they sustain a permanent polarization. Some electret materials are made by subjecting a molten material to a strong electric field while it solidifies. Current applications of electrets include electrostatic microphones.

1.5.1.2 Dielectric EAP

Polymers with low elastic stiffness and high dielectric constant can be used to induce large actuation strain by subjecting them to an electrostatic field. In Chapter 16 this EAP is described in detail and the applications that were considered are reviewed. This dielectric EAP, also known as electrostatically stricted polymer (ESSP) actuators, can be represented by a parallel plate capacitor [Pelrine et al., 1998]. In Fig. 15 a schematics shows a dielectric EAP that is in reference and activated conditions (top). The bottom left section of the figure shows a silicone film that is in a reference (top) and activated (bottom) condition. The bottom right section of the figure shows an EAP actuator that was made using a silicone film that was scrolled to the shape of rope. The rope, which is about 2 mm in diameter and 3 cm long, was demonstrated to lift and drop a 17-g rock using about 2.5 KV.

The induced strain is proportional to the square of the electric field square, multiplied by the dielectric constant and it is inversely proportional to the elastic modulus. The use of polymers with high dielectric constants and the application of high electric fields lead to large forces and strains. To reach the required electric field levels, one needs to either use high voltage and/or employ thin films. Under an electric field, the film is squeezed in the thickness direction, causing expansion in the transverse direction. For a pair of electrodes with a circular shape, the diameter and thickness changes can be determined using the volume conservation of the film

$$D^2 t = D_0^2 t_0 , \qquad (2)$$

where D and t are the diameter and thickness of the film when the electric field is applied, and D_0 and t_0 are the corresponding original dimensions of the film without the application of the field.

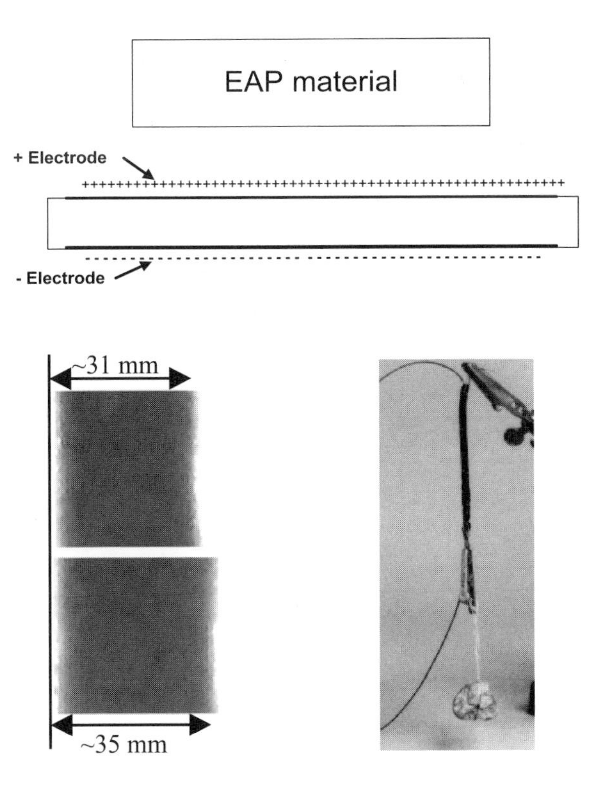

Figure 15: Under electroactivation, a dielectric elastomer with electrodes on both surfaces expands laterally and can be made to operate longitudinally.

The above characteristic allows production of longitudinal actuators using dielectric elastomer films with flexible electrodes that appear to act similarly to biological muscles. For this purpose, SRI International scientists constructed actuators using two silicone layers with carbon electrodes on both sides of one of the layers, where the layers were wrapped to form an actuator that is shaped like a rope. The rope-shaped actuator expanded longitudinally as a result of the Poisson effect of transverse expansion of the film caused by its being squeezed by Maxwell Forces. Thereby, the lateral expansion leads to a lateral stretch of the EAP rope. Dielectric EAP actuators require large electric fields (~100 V/μm) and can induce significant levels of strain (10 to 200%). Overall, the associated voltages are close to the breakdown strength of the material, and a safety factor that lowers the potential is needed. Moreover, the relatively small breakdown strength of air (2–3 V/μm) presents an additional challenge. The Young's modulus is fairly temperature independent until the glass transition temperature is reached, at which point a sharp increase in the modulus occurs, making the material too stiff to be used as an actuator at low temperatures.

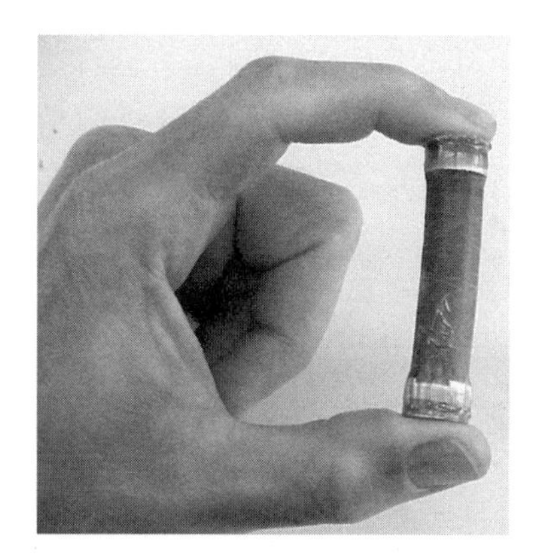

Figure 16: Multifunctional Electroelastomer Roll (MER). (Courtesy of SRI International.)

In 2000, SRI International researchers identified a new class of polymers that exhibits an extremely high strain response [Pelrine et al., 2000]. These acrylic elastomers (such as 3M VHB tapes) have produced planar strains of more than 300% for biaxially symmetric constraints and linear strains of up to 215% for uniaxial constraints. Recently, researchers from this team designed a multifunctional electroelastomer roll (MER) in which highly prestrained dielectric EAP electroelastomer films were rolled around a compression spring to form an actuator [Pei et al, 2002]. When released, the spring holds the polymer films in high circumferential and axial prestrain while allowing for axial actuation. Actuator configuration in one-degree-of-freedom (1-DOF), 2-DOF, and 3-DOF were demonstrated wherein the compliant electrodes were made either with no pattern or patterned on two and four circumferential spans, respectively. An example of this MER actuator is shown in Fig. 16 and it represents advancement in making practical EAP based actuators with a standard configuration.

1.5.1.3 Electrostrictive Graft Elastomers

In 1998, a graft-elastomer EAP was developed at NASA Langley Research Center that exhibits a large electric-field-induced strain due to electrostriction [Su et al., 1999 and Chapter 4]. This electrostrictive polymer consists of two components, a flexible backbone macromolecule and a grafted polymer that can form a crystalline structure. The grafted crystalline polar phase provides moieties in response to an applied electric field and cross-linking sites for the elastomer system. This material offers high electric-field-induced strain (~4%), relatively high electromechanical power density, and excellent processability. Combination of the electrostrictive-grafted elastomer with a piezoelectric poly(vinylidene fluoride-trifluoroethylene) copolymer yields several compositions of a ferroelectric-electrostrictive molecular composite system. Such combination can be operated both as a piezoelectric sensor and as an electrostrictive actuator.

Figure 17: An electrostrictive grafted elastomer-based bimorph actuator in an unexcited state (middle), one direction excited state (left), and opposite direction excited state (right). (Courtesy of Ji Su, NASA.)

Careful selection of the composition allows the creation and optimization of the molecular composite system with respect to its electrical, mechanical, and electromechanical properties. An example of an electroactivated graft EAP is shown in Fig. 17, where a bimorph actuator is fabricated using a backing with a passive grafted elastomer. The actuator can bend in both directions under controlled electric field excitation.

1.5.1.4 Electrostrictive Paper

The use of paper as an electrostrictive EAP actuator was demonstrated by Inha University, Korea [Kim et al., 2000]. Paper is composed of a multitude of discrete particles, mainly of a fibrous nature, which form a network structure (see Fig. 18). Since paper is produced in various mechanical processes with chemical additives, it is possible to prepare a paper that has electroactive properties. Such an EAP actuator has been prepared by bonding two silver laminated papers with silver electrodes placed on the ouside surface. When an electric voltage is applied to the electrodes the actuator produces bending displacement, and its performance depends on the excitation voltages, frequencies, type of adhesive,

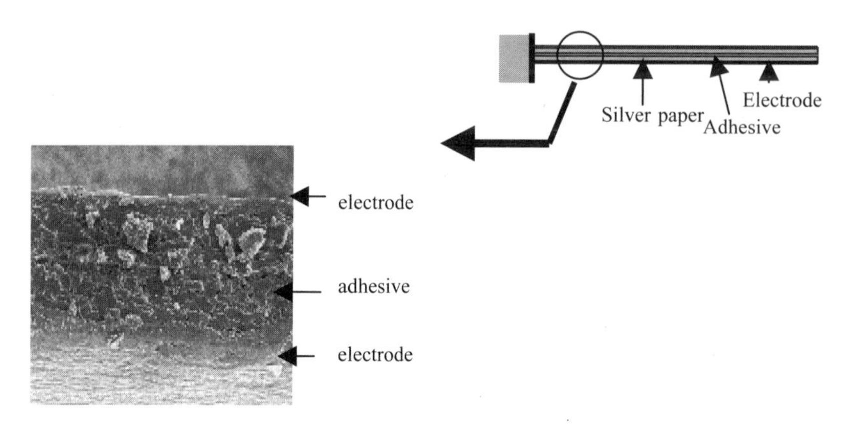

Figure 18: A view of a paper EAP, where a cross-section micrograph is shown on the left and schematic view of the actuator on the right. (Courtesy of J. Kim, Inha University, Korea.)

and the host paper. Studies indicate that the electrostriction effect that is associated with this actuator is the result of electrostatic forces and an intermolecular interaction of the adhesive. The demonstrated actuator is lightweight and simple to fabricate. Various applications currently being considered include active sound absorbing materials, flexible speakers, and smart shape-control devices.

1.5.1.5 Electrovisco-elastic Elastomers

Electrovisco-elastic elastomers represent a family of electroactive polymers that are composites of silicone elastomer and a polar phase. Before cross-linking, in the uncured state they behave as electrorheological fluids. An electric field is applied during curing to orient and fix the position of the polar phase in the elastomeric matrix. These materials then remain in the "solid" state but have a shear modulus (both real and imaginary parts) that changes with an applied electric field (<6 V/μm) [Shiga, 1997]. A stronger magnetorheological effect can also be introduced in an analogous manner and as much as a 50% change in the shear modulus can be induced [Klapcin′ski et al., 1995; Davis, 1999; Zrinyi et al., 1999]. These materials may be used as alternatives to electrorheological fluids for active damping applications. Further, they may be used to perform active damping in support of precision control of robotic arms in a closed-loop system.

1.5.1.6 Liquid-Crystal Elastomer (LCE) Materials

Liquid-crystal elastomers were pioneered at Albert-Ludwigs University (Freiburg, Germany) [Finkelmann et al., 1981]. These materials can be used to form an EAP actuator that has piezoelectric characteristics and can be electrically activated by inducing Joule heating. The actuation mechanism of these materials involves phase transition between nematic and isotropic phases over a period of less than a second. The reverse process is slower, taking about 10 sec, and it requires cooling, which causes expansion of the elastomer to its original length. The mechanical properties of LCE materials can be controlled and optimized by effective selection of the liquid crystalline phase, density of cross-linking, flexibility of the polymer backbone, coupling between the backbone and liquid crystal group, and the coupling between the liquid crystal group and the external stimuli.

Generally, liquid crystals are supramolecular ordered assemblies and as such have an excellent framework for incorporating anisotropy and functionalities that include response to external stimuli. Slight cross-linking of polymeric liquid crystals show elasticity similar to that of conventional elastomers but with some special properties [Warner and Terentjev, 2003]. In such a cross-linked polymer, it is possible to create deformations on macroscopic length scales by changing the orientational order of the mesogenic group by external stimuli (Fig. 19). The uniqueness of the LCE lies in its uniaxial response to external stimuli such as temperature [Thomsen et al., 2001; Wermter and Finkelmann, 2001; Naciri et al., 2003], light [Finkelmann et al., 2001; Shenoy et al.; 2002, Li et al., 2003] or electric field [Lehmann, 2001]. In recent years, researchers have explored the possibility of developing liquid-crystal nematic-elastomer actuator materials that can mimic muscle performance [Thomsen et al., 2001; Wermter and Finkelmann, 2001; Naciri et al., 2003]. In the nematic phase, the orientational order of the mesogen forces the polymer backbone to be elongated along the average direction of orientation of the mesogens. However, upon heating the isotropic phase, the nematic order is lost thereby allowing the polymer backbone to relax to its coiled conformation. LCEs provide a number of advantages including the possibility of using them in the dry state and the ease to introduce multifunctionality into them. By their very chemical nature, LCEs have a number of handles, such as the liquid-crystalline phase, density of cross-linking, flexibility of the polymer backbone, coupling between the backbone and liquid-crystal group, and coupling between the liquid-crystal group and the external stimuli. All these handles can be easily tuned to optimize the mechanical properties of the material. The elastomers that were most frequently studied have been those based on side-chain liquid-crystalline polymers or a combination of side-chain and main-chain polymers. LC elastomers with monodomain orientation of liquid crystalline groups with highly anisotropic mechanical properties have been produced in the form of films [Thomsen et al., 2001], fibers [Naciri et al., 2003], and rods [Kupfer and Finkelmann, 1981].

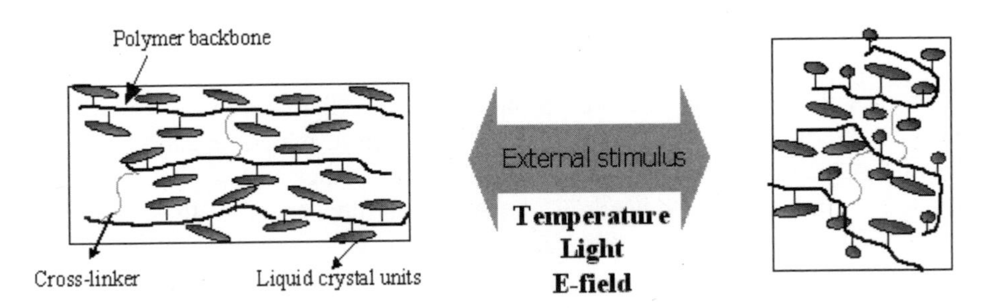

Figure 19: Schematic diagram of the concept of actuation in a liquid crystal elastomer [Naciri et al., 2003a]. (Courtesy of Autonomous Undersea Systems Institute.)

At the Naval Research Laboratory it has been demonstrated that liquid-crystal elastomers can exhibit elastic properties comparable to muscles. Anisotropic nematic films [Thomsen et al., 2001] and fibers [Naciri et al., 2003] have been evaluated for their mechanical properties. The operating temperature, strain, and force generation can be tuned by varying the composition of the liquid crystal monomers, cross-linking density, and cross-linking method [Naciri et al, 2003a]. Figure 20 shows a typical thermoelastic curve having a sharp induction of strain at the nematic to isotropic phase transition. Further, Fig. 21 shows a demonstration of the actuation in a 300-μm-diameter fiber in response to an electric current through a surrounding nichrome coil. LCEs have shown strain responses of 45 percent and blocked stress values of 460 kPa and in some systems, strain values as large as 400% has been achieved [Kupfer and Finkelmann, 1981]. It is interesting to note that the stress-strain loops exhibit negligible viscoelastic losses. Further, carbon coating on LCE film has been shown to be an effective approach to enhance the response to an external optical stimulus such as an IR diode laser [Shenoy et al., 2002]. Photothermal actuation has also been demonstrated by doping the film with a blue dye for maximum absorption at 635 nm for which a diode laser is available (Fig. 22).

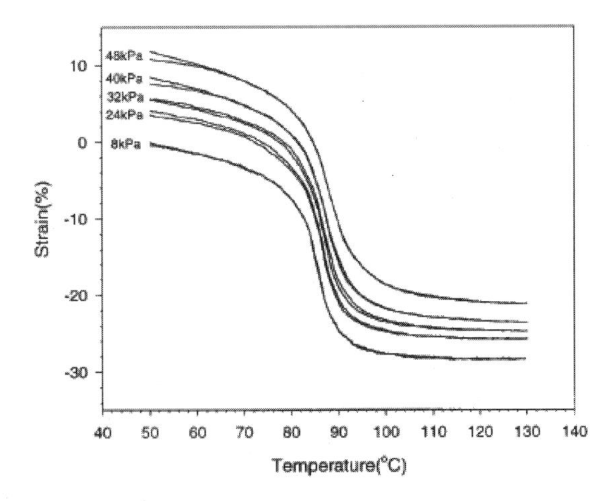

Figure 20: Thermoelastic curves for an elastomer film at different loads. (Reprinted with permission from Naciri et al., 2003. Copyright 2003, American Chemical Society.)

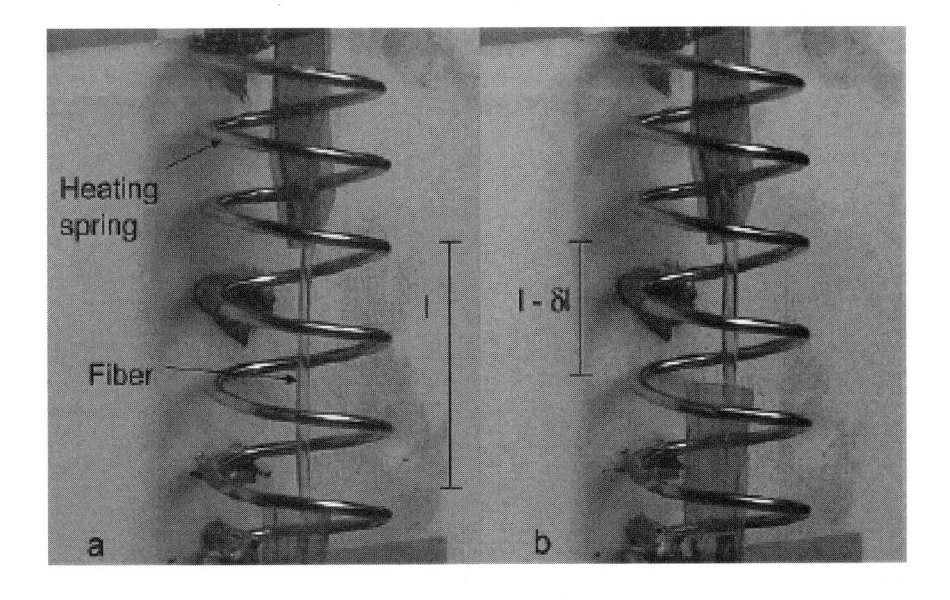

Figure 21: Actuation of a fiber. (a) Extended state in the nematic phase, and (b) contracted state in the isotropic phase. (Reprinted with permission from Naciri et al., 2003. Copyright 2003, American Chemical Society.)

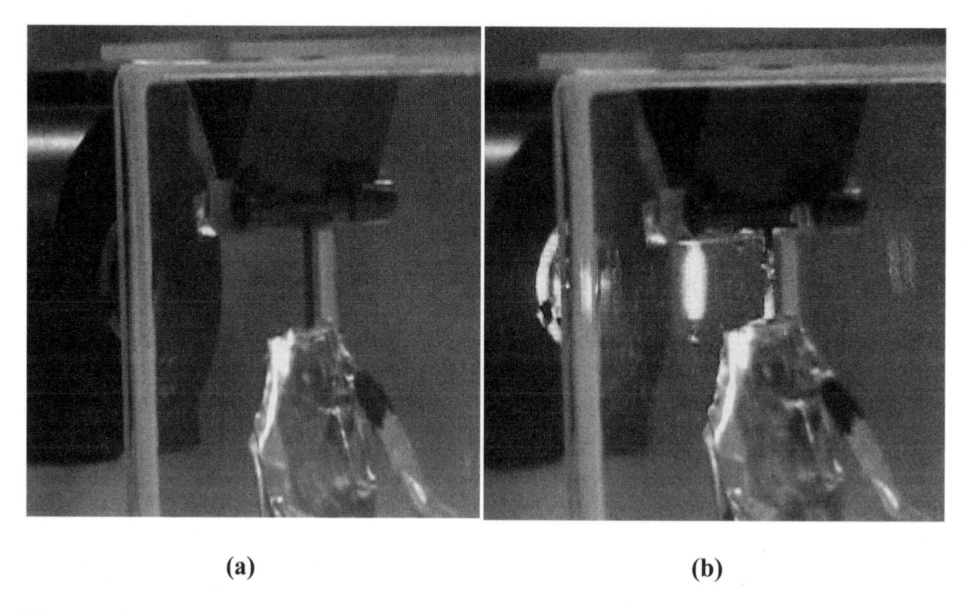

(a) (b)

Figure 22: Digital images taken from a video. Elastomer film (a) before actuation, irradiating with laser and (b) actuated/contracted, during illumination with 10mW 635 nm diode laser. (Courtesy of B. R. Ratna, Naval Research Laboratory.)

1.5.2 Ionic EAP

1.5.2.1 Ionic Polymer Gels (IPG)

Polymer gels can be synthesized to produce strong actuators having the potential of matching the force and energy density of biological muscles [Chapter 5]. These materials (e.g., polyacrylonitrile) are generally activated by a chemical reaction [see Sec. 1.4.3.1], changing from an acid to an alkaline environment causing the gel to become dense or swollen, respectively. This reaction can be stimulated electrically, as was shown by researchers at the University of Arizona, USA [Calvert et al., 1998; Liu and Calvert, 2000]. When activated, these gels bend as the cathode side becomes more alkaline and the anode side more acidic. However, the response of this multilayered gel structure is relatively slow because of the need to diffuse ions through the gel. Expanding and shrinking a layered gel from 6 × 6 cm to 3 × 3 cm (see Fig. 23) occurred over a period of about 20 min, which is far from practical. Also, the induced large displacement caused damage to the electrodes, leading to failure of the actuator after 2 to 3 activation cycles. Current efforts are directed toward the development of thin layers and more robust electroding techniques. Progress in making such materials with fiber shape electrodes was reported by researchers at the University of New Mexico using a mix of conductive and PAN fibers [Schreyer et al., 2000].

Figure 23: Shape change by electrical stimulation emulating pH modification [Liu and Calvert, 2000]. (Courtesy of Wiley-VCH.).

Nonionic polymer gels containing a dielectric solvent can be made to swell under a dc electric field with a significant strain. At Shinshu University, Japan, a study is under way to create bending and crawling EAP using a poly(vinyl alcohol) gel with dimethyl sulfoxide [Hirai et al., 1999]. A 10 × 3 × 2-mm actuator gel was subjected to an electrical field and exhibited bending at angles that were greater than 90 deg at a speed of 60 msec. This phenomenon is attributed to charge injection into the gel and a flow of solvated charges that induce an asymmetric pressure distribution in the gel. Another nonionic gel is poly(vinyl chloride) (PVC), which is generally inactive when subjected to electric fields. However, if PVC is plasticized with dioctyl phthalate (DOP), a typical plasticizer, it can maintain its shape and behave as an elastic nonionic gel. When the gel is placed between a pair of electrodes and an electric field is applied, the gel curves toward the anode. This deformation can be maintained as long as the field is held on, and the original shape is restored once the electric field is turned off [Hirai et al., 1999].

The mechanism that is responsible for the chemomechanical behavior of ionic gels under electrical excitation is described by Osada and Ross-Murphy [1993] and a model for hydrogel behavior as contractile EAP is described in Gong et al., [1994]. A significant amount of research and development was conducted at the Hokkaido University, Japan, and applications using ionic gel polymers were explored. These include electrically induced bending of gels [Osada and Hasebe, 1985; Osada et al., 1992] and electrically induced reversible volume change of gel particles [Osada and Kishi, 1989]. Fundamentals and modeling issues that are related to the chemo-electro-mechanical aspect of polymers such as ionic gels are described in Chapter 11.

1.5.2.2 Ionomeric Polymer-Metal Composites (IPMC)

Ionomeric polymer-metal composite (IPMC) is an EAP that bends in response to an electrical activation (Fig. 24) as a result of mobility of cations in the polymer network (see also Chapters 6 and 13). In 1992, IPMC was realized to have this electroactive characteristic by three groups of researchers: Oguro et al. [1992] in

Figure 24: IPMC in reference and activated state.

Japan, Shahinpoor [1992] and Sadeghipour et al. [1992] in the United States. The operation as actuators is the reverse process of the charge storage mechanism associated with fuel cells [Heitner-Wirguin, 1996]. A relatively low voltage is required to stimulate bending in IPMC, where the base polymer provides channels for mobility of positive ions in a fixed network of negative ions on interconnected clusters. Two types of base polymers are used to form IPMC: Nafion (perfluorosulfonate made by DuPont) and Flemion (perfluorocarboxylate, made by Asahi Glass, Japan). The chemical structure of Nafion is shown in Fig. 25. Prior to using these base polymers as EAP, they were widely employed in fuel cells for production of hydrogen (hydrolysis) [Holze and Ahn, 1992]. In order to chemically electrode the polymer films, metal ions (platinum, gold or others) are dispersed throughout the hydrophilic regions of the polymer surface

```
[-(CF2-CF2)n-CF-CF2-]m
                |
                O-CF-CF2-O-CF2-SO3-M
                   |
                  CF3
```

Figure 25: Nafion chemical structure (where $n \sim 6.5$, $100 < m < 1000$, and M^+ is the counter ion (H^+, Li^+, Na^+, or others).

and are subsequently reduced to the corresponding zero-valence metal atoms. Generally, the ionic content of the IPMC is an important factor in the electromechanical response of these materials [Nemat-Nasser and Li, 2000; Li and Nemat-Nasser, 2000; Bar-Cohen et al., 1999]. Examining the bending response shows that using low voltage (1–10 V) induces a large bending at frequencies below 1 Hz, and the displacement significantly decreases with the increase in frequency. The bending response of IPMC was enhanced using Li+ cations that are small and have higher mobility or large tetra-n-butylammonium cations that transport water in a process that is still being studied (see Chapter 6). The actuation displacement of IPMC was further increased using gold metallization as a result of the higher electrode conductivity [Abe et al., 1998; and Oguro et al., 1999].

1.5.2.3 Conductive Polymers (CP)

Conductive polymers typically function via the reversible counter-ion insertion and expulsion that occurs during redox cycling [Chapter 7, Otero et al., 1995; and Gandhi et al., 1995]. Oxidation and reduction occur at the electrodes, inducing a considerable volume change due mainly to the exchange of ions with an electrolyte. A sandwich of two conductive polymer electrodes (e.g., polypyrrole, or polyaniline, or PAN doped in HCl) with an electrolyte between them forms an EAP actuator (see Fig. 26). When a voltage is applied between the electrodes, oxidation occurs at the anode and reduction at the cathode. Ions (H^+) migrate between the electrolyte and the electrodes to balance the electric charge. Addition

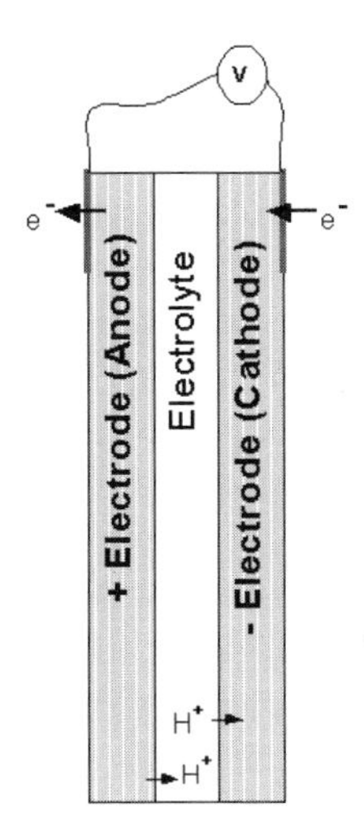

Figure 26: Schematic view of the electro-chemo-mechanical mechanism of actuation in conductive polymers.

of the ions causes swelling of the polymer and conversely their removal results in shrinkage. As a result, the sandwich bends. One of the parameters that affects the response is the thickness of the layers; thinner layers are faster (as fast as 40 Hz) but induce lower force. Since strong shear forces act on the electrolyte layer, attention is needed to protect the material from premature failure. Conductive polymer actuators generally require voltages in the range of 1–5 V, and the speed increases with the voltage having relatively high mechanical energy densities of over 20 J/cm^3 but with low efficiencies at the level of 1%. An example of the response of a conductive polymer is shown in Fig. 27 [Otero and Sansiñena, 1998]. In this figure, a bilayer strip made of 3-mg polypyrrole film (PPy) (area = 3 cm^2 and thickness = 6 μm) and an adhesive tape (no volume-change film, NVC) is shown to lift 1-g steel ball (about 300 times heavier than the PPy). The strip was activated by a 30-mA electric current and is shown after partial oxidation (left), natural state (middle), and partial reduction (right).

(a) Partial oxidation

(b) Neutral

(c) Partial reduction

Figure 27: Photographic views of a multilayered conductive polymer actuator lifting 1-g steel ball using 30 mA (right) [Otero and Sansiñena, 1998]. (Courtesy of Wiley-VCH.)

Various approaches are being considered to operate this EAP in dry environments where the use of quasi-solid electrolytes and the application of a protective coating are some of the alternatives. In recent years, several conductive polymers were reported, including polypyrrole, polyethylenedioxythiophene, poly(p-phenylene vinylene)s, polyanilines, and polythiophenes. Complexes between polypyrrole and sulfonated detergents offer relatively good stability in aqueous media, but they are relatively soft compared to other conjugated polymers. Most actuators that use conductive polymers exploit voltage-controlled swelling to induce bending. Conjugated polymer microactuators were first fabricated at Linköpings Universitet, Sweden. Among the devices that were demonstrated include miniature boxes that can be opened and closed [Smela et al., 1995]. Using this technology, a microrobot was developed; and other applications may evolve, including surgical tools or robots to assemble other microdevices in a factory-on-a-desk [Jager et al., 2000].

Operation of conductive polymers as actuators at the single-molecule level is currently being studied taking advantage of the intrinsic electroactive property of individual polymer chains. Researchers at the University of California, Riverside, [see Chapter 9, Marsella and Reid, 1999] used {4n}annulene chemistry as a method of "encoding" actuation into individual polymer chains. It is well known that cyclooctatetraene (a tube-shaped {8}annulene) becomes planar upon oxidation or reduction. During such a process, the distance between carbons 1 and 4 is altered. By utilizing an {8} annulene as a polymer repeat unit, such redox-induced conformational changes can be translated into a change in the effective monomer length. The effective polymer chain length is ultimately altered, thereby causing the alteration of the effective length of the monomer. Thus, a single-molecule electromechanical actuator can be enabled.

1.5.2.4 Carbon Nanotubes (CNT)

In 1999, carbon nanotubes with diamondlike mechanical properties emerged as formal EAP [Chapter 3.2.4; Baughman et al, 1999]. The carbon-carbon bond in nanotubes (NT) that are suspended in an electrolyte and the change in bond length are responsible for the actuation mechanism. A network of conjugated bonds connects all carbons and provides a path for the flow of electrons along the bonds. The electrolyte forms an electric double layer with the nanotubes and allows injection of large charges that affect the ionic charge balance between the NT and the electrolyte. The more charges are injected into the bond the larger the dimension changes. Removal of electrons causes the nanotubes to carry a net positive charge, which is spread across all the carbon nuclei causing repulsion between adjacent carbon nuclei and increasing the C-C bond length. Injection of electrons into the bond also causes lengthening of the bond resulting in an increase in nanotube diameter and length. These dimension changes are translated into macroscopic movement in the network element of entangled nanotubes and the net result is extension of the CNT.

Considering the mechanical strength and modulus of the individual CNTs and the achievable actuator displacements, this actuator has the potential of producing higher work per cycle than that of any previously reported actuator technologies and of generating much higher mechanical stress. Further, since carbon offers high thermal stability, carbon nanotubes may eventually be used at temperatures greater than 1000°C, which far exceeds the capabilities of alternative high-performance actuator materials. The material consists of nanometer-size tubes and was shown to induce strains at the range of 1% along the length. The key obstacle to the commercialization of this EAP is its high cost and the difficulty in mass production of the material. A carbon nanotube actuator can be constructed by laminating two narrow strips that are cut from a carbon nanotube sheet using an intermediate adhesive layer, which is electronically insulated. The resulting "cantilever device" is immersed in an electrolyte such as a sodium chloride solution, and an electrical connection is made in the form of two nanotube strips. Application of about 1.0 V is sufficient to cause bending, and the direction depends on the polarity of the field.

1.5.2.5 Electrorheological Fluids (ERF)

Electrorheological fluids (ERFs) experience dramatic changes in their viscosity when subjected to an electric field (see Fig. 28). These fluids are made from suspensions of an insulating base fluid and particles 0.1–100 μm in size. Winslow first explained the electrorheological effect in the 1940s, using the oil dispersions of fine powders [Winslow, 1949]. The electrorheological effect, sometimes called the *Winslow* effect, is thought to arise from the difference in the dielectric constants of the fluid and particles. In the presence of an electric field, the particles, due to an induced dipole moment, will form chains along the field lines. This induced structure changes the ERF's viscosity, yield stress, and other

Figure 28: Electrorheological fluid at reference (left) and activated states (right). (Courtesy of Smart Technology Limited, UK.)

properties, allowing the ERF to change consistency from that of a liquid to something that is viscoelastic, such as a gel, with response times to changes in electric fields on the order of milliseconds. A good review of the ERF phenomenon and the theoretical basis for their behavior can be found in Block and Kelly [1988]. Control over a fluid's rheological properties offers the promise of new possibilities in engineering for actuation and control of mechanical motion. The ability to control the viscosity and the rapid response of ERF allows simplification of the mechanisms that are involved with hydraulics. Their solidlike properties in the presence of a field can be used to transmit forces over a large range and have found a large number of applications including shock absorbers, active dampers, clutches, adaptive gripping devices, and variable flow pumps. A detailed description of ERF and its various applications, including haptic interfaces, is given in Chapter 19 of this book.

1.6 The EAP Roadmap—Need For An Established EAP Technology Infrastructure

As polymers, EAP materials can be easily formed into various shapes, their properties can be engineered, and they can potentially be integrated with MEMS sensors to produce smart actuators. As mentioned earlier, their most attractive feature is their ability to emulate the operation of biological muscles with high fracture toughness, large actuation strain and inherent vibration damping. Construction of mobile or articulation systems that are actuated by EAP materials requires the components shown in the block diagram of Fig. 29. While each of the listed components is at various research and development stages, EAP actuator research is the least developed, and extensive efforts are required to bring it to a mature stage.

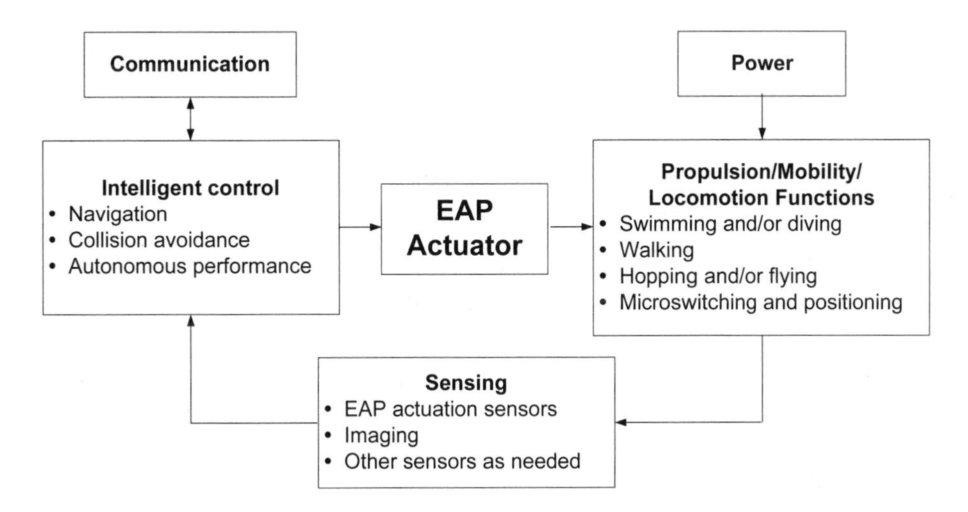

Figure 29: A schematic diagram of the basic components of an EAP-driven system.

Unfortunately, the EAP based actuators that have been developed so far still exhibit low force, they are far below their efficiency limits, are not robust, and there is no standard commercial material available for consideration in practical applications. As described in Sec. 1.5 and Topic 3 of this book, the documented EAP materials that induce large strains are driven by many different phenomena [Bar-Cohen, 1999a, 2000, 2001, 2002, 2003; Bar-Cohen, et al, 2003, Zhang et al., 1999]. Each of these materials requires adequate attention to their unique properties and constraints. In order to be able to take these materials from the development phase to use as effective actuators, there is a need to have an establish EAP infrastructure. Establishing the infrastructure should be part of the technology development roadmap for the field. The author's view of this infrastructure and the areas that need development are shown schematically in

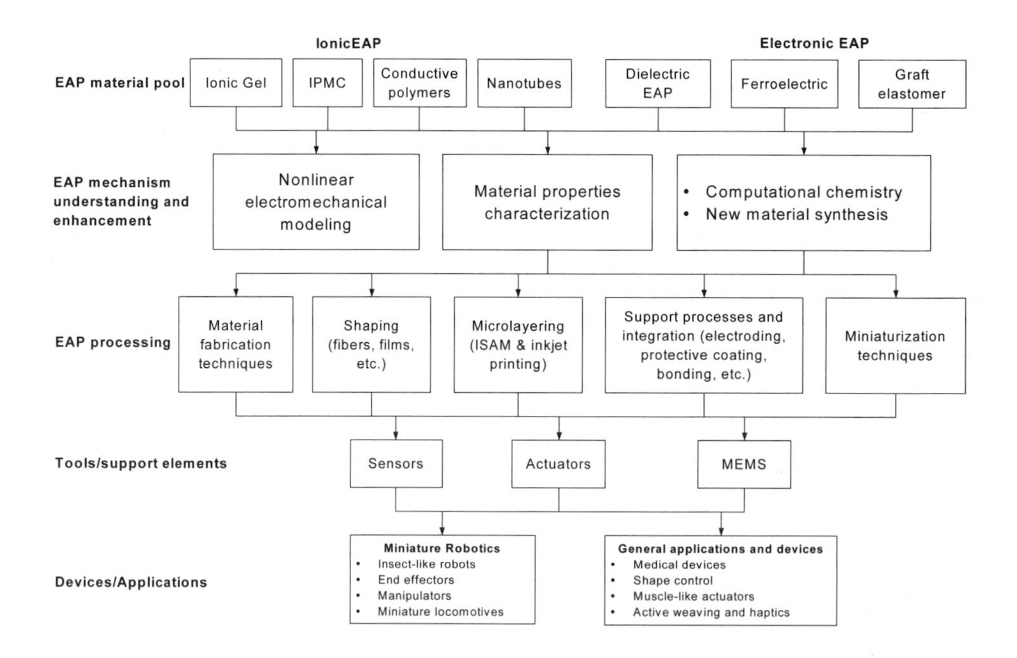

Figure 30: EAP infrastructure and areas needing attention.

Fig. 30. Effectively addressing the requirements of the EAP infrastructure involves developing adequate understanding of EAP materials' behavior, as well as processing and characterization techniques. Enhancement of the actuation force requires an understanding of the basic principles using computational chemistry models, comprehensive material science, electromechanics analytical tools and improved material processing techniques. Efforts are needed to gain a better understanding of the parameters that control the EAP electro-activation force and deformation. The processes of synthesizing, fabricating, electroding, shaping, and handling will need to be refined to maximize the EAP materials actuation capability and robustness. Methods of reliably characterizing the response of these materials are required to establish databases with documented material properties in order to support design engineers considering use of these materials and toward making EAP as actuators of choice. Various configurations of EAP actuators and sensors will need to be studied and modeled to produce an arsenal of effective smart EAP-driven systems. The roadmap for the development of the infrastructure is multidisciplinary and requires international collaboration. This book is structured to address the infrastructure requirements and efforts have been made to cover the state of the art in each of the areas that are identified in Fig. 30.

1.6.1 Mechanical Modeling of Large Strain Actuation

To allow effective design of EAP-actuated mechanisms with large induced strains, it is necessary to have adequate analytical tools for predicting the behavior of EAP actuators as well as simulating their response as part of rapid prototyping methodologies. The topic of modeling EAP is discussed in detail in Topic 4 of this book where the current modeling capabilities, including computational chemistry, electrochemical and nonlinear electromechanical analysis, and design considerations for IPMC are covered. The model of piezoceramic actuators can be used to establish constitutive relations for EAP materials. The coefficient matrices and tensors of the equations that describe the material behavior are material constants that need to be determined through laboratory tests, and such tests are described in Topic 5. The mechanical response of an active structure that is subjected to a given electrical field can be determined from a suitable equation of motion supplemented by constitutive relations. A distinguishing feature of electrostrictive materials is the presence of the quadratic terms related to the electric field (particularly relevant to electrostrictive EAP materials). Since one of the major advantages of the EAP materials is their ability to produce large displacements, their strains are not "infinitesimal," and the quasi-linear constitutive relations may not be adequate to predict the electromechanical behavior of the associated structures. Therefore, there is a need for constitutive relations that include the *nonlinear terms in the strain* and the *electric fields*. Chapters 11 and 12 of this book describe approaches to addressing the nonlinearity behavior of EAP. As the field evolves, it would be necessary to determine the expected actuation efficiency of the various EAPs in order to establish goals toward reaching these predicted efficiency limits

1.7 Potential

Electroactive polymers have emerged with great potential and may enable the development of unique devices that can emulate biological equivalences, and some commercial products are starting to emerge. The various EAP materials, processing techniques and capabilities are reviewed in this book with a structure that follows the infrastructure block diagram described in Fig. 30. The development of an effective infrastructure for this field is critical to the commercial availability of robust EAP actuators and the emergence of practical applications. The challenges are enormous, but the recent trend of international cooperation, the greater visibility of the field, and the surge in funding of related research offer great hope for the future of these exciting new materials. The potential to operate biologically inspired mechanisms driven by EAP as artificial muscles enable capabilities that are currently considered science fiction. Topics 7 and 8 cover some examples of EAP-driven mechanisms that have been developed or are being considered. The arm-wrestling challenge of a match between an EAP-actuated robot and a human opponent highlights the potential of EAP. Progress toward this goal will lead to great benefits to mankind, particularly in the area of medical prosthetics. The author believes that an emergence of a niche application that addresses a critical need will significantly accelerate the transition of EAP from novelty to actuators of choice. In niche cases, these materials will be used in spite of their current limitations, taking advantage of their uniqueness.

1.8 Acknowledgments

The research at Jet Propulsion Laboratory (JPL), California Institute of Technology, was carried out under a contract with National Aeronautics and Space Agency (NASA) and Defense Advanced Research Projects Agency (DARPA). The author would like to thank everyone who contributed to the material that was used in this chapter including Dr. Ji Su, NASA LaRC, for contributing the section on Graft EAP. Also, the author would like to thank Elisabeth Smela, University of Maryland, Peter Sommer-Larsen, Risoe National Laboratory, Denmark, and Geoff Spinks, University of Wollongong, Australia, for their valuable contributions to this chapter. The author would like to acknowledge the contribution Dr. B. R. Ratna, Naval Research Laboratory. She wrote the paragraph that describes her research and recent progress in Liquid Crystals as EAP materials. Special thanks to the team members of the Telerobotic task LoMMAs led by the author that was funded by the NASA Code S program. The author would like to acknowledge the contribution of Dr. Sean Leary, Mark Schulman, Dr. Tianji Xu, and Andre H Yavrouian, JPL, Dr. Joycelyn Harrison, Dr. Joseph Smith, and Ji Su, NASA LaRC, Prof. Richard Claus, Virginia Tech., Prof. Constantinos Mavroidis, and Charles Pfeiffer, Rutgers University. Also, the author would like to thank Dr. Timothy Knowles, ESLI, for his assistance with the development of the EAP dust-wiper mechanism using IPMC, and Marlene Turner, Harry Mashhoudy, Brian Lucky, and Cinkiat Abidin, former graduate students of the Integrated Manufacturing Engineering (IME) Program at UCLA, for helping to construct the EAP gripper and robotic arm. The author would like to express a special thanks to Dr. Keisuke Oguro, Osaka National Research Institute, Japan, for providing his most latest IPMC materials; and to Prof. Satoshi Tadokoro, Kobe University, Japan, for his analytical modeling effort. Prof. Mohsen Shahinpoor is acknowledged for providing IPMC samples as part of the early phase of the LoMMAs task. The author would also like to thank his NDEAA team members Dr. Xiaoqi Bao, Dr. Shyh-Shiuh Lih, and Dr. Stewart Sherrit and the former members Dr. Virginia Olazábal, Dr. José-María Sansiñena, for their help. Also, the author would like to thank Dr. Anita Flynt, Micro Propulsion, for raising the need to confirm the attribution to Roentgen (1880) for being the first to experiment with EAP and to Prof. Rainer W. Gülch Universitat Tubingen, Germany, and Leon Rogson, TRW, USA, for confirming the content of Roentgen's publication on this subject and to Prof. Carlos F. Lange, U. of Alberta, Canada for the confirmation of the details of this important reference. A specific acknowledgment is made for the courteous contribution of graphics from individuals and organizations as specified next to the related graphics or tables.

1.9 References

Abe Y., A. Mochizuki, T. Kawashima, S. Tamashita, K. Asaka and K. Oguro, "Effect on bending behavior of counter cation species in perfluorinated sulfonate membrane-platinum composite," *Polymers for Advanced Technologies*, Vol. 9 (1998), pp. 520-526.

Alexander R. M., *Elastic Mechanisms in Animal Movement*, The Cambridge University Press: Cambridge, 1988.

Aviram A., "Mechanophotochemistry," *Macromolecules*, Vol. 11, (1978) pp. 1275.

Bar-Cohen, Y., (Ed.), *Proceedings of the SPIE's Electroactive Polymer Actuators and Devices Conf.*, 6th Smart Structures and Materials Symposium, SPIE Proc. Vol. 3669, (1999), pp. 1-414.

Bar-Cohen Y., (Ed.), *Proceedings of the SPIE's Electroactive Polymer Actuators and Devices Conf.*, 7th Smart Structures and Materials Symposium, SPIE Proc. Vol. 3987, (2000) pp. 1-360.

Bar-Cohen Y. (Ed.), *Proceedings of the SPIE's Electroactive Polymer Actuators and Devices,* 8th Smart Structures and Materials Symposium, Vol. 4329, (2001), pp. 1-524.

Bar-Cohen Y. (Ed.), *Proceedings of the SPIE's Electroactive Polymer Actuators and Devices,* 9th Smart Structures and Materials Symposium, Vol. 4695, (2002), pp. 1-506.

Bar-Cohen Y. (Ed.), *Proceedings of the SPIE's Electroactive Polymer Actuators and Devices,* 10h Smart Structures and Materials Symposium, Vol. 5051, (2003).

Bar-Cohen Y., Q.M. Zhang, E. Fukada, S. Bauer, D. B. Chrisey, and S. C. Danorth (Eds.), "Electroactive Polymers (EAP) and Rapid Prototyping," 2001 MRS Symposium Proceedings, Vol. 698, Warrendale, PA, (2003), pp. 1-359.

Bar-Cohen Y., "Electroactive polymers as artificial muscles - capabilities, potentials and challenges," Keynote Presentation, Robotics 2000 and Space 2000, Albuquerque, NM, USA, February 28 - March 2, (2000), pp. 188-196.

Bar-Cohen Y., and C. Breazeal (Eds), "Biologically-Inspired Intelligent Robots," SPIE Press, Vol. PM122, (May 2003), pp. 1-393.

Bar-Cohen Y., S. Leary, K. Oguro, S. Tadokoro, J. Harrison, J. Smith and J. Su, "Challenges to the transition of IPMC artificial muscle actuators to practical application," *Proceedings of the 1999 Fall MRS Symposium*, Vol. 600, (1999).

Bar-Cohen Y., T. Xue and S.-S., Lih, "Polymer piezoelectric transducers for ultrasonic NDE," First International Internet Workshop on Ultrasonic NDE, Subject: Transducers, organized by R. Diederichs, UTonline Journal, Germany, http://www.ndt.net/article/yosi/yosi.htm (Sept. 1996).

Baughman R.H., C. Cui, A. A. Zakhidov, Z. Iqbal, J. N. Basrisci, G. M. Spinks, G. G. Wallace, A. Mazzoldi, D. de Rossi, A. G. Rinzler, O. Jaschinski, S.

Roth and M. Kertesz, "Carbon Nanotune Actuators," *Science*, Vol. 284, (1999) pp. 1340-1344.

Becker C. and W. Glad, "Light activated EAP materials," JPL's NDEAA Technologies, WW-EAP Newsletter, Vol. 2, No. 1 (2000), pp. 11.

Biggs A.G., K. M. Blackwood, A. Bowles, S. Dailey and A. May, "A study of liquid crystalline elastomer as piezoelectric devices," MRS Symposium Proceedings, Vol. 600, Warrendale, PA, (1999) pp. 159-164.

Block, H. and Kelly, J. P., "Electro-rheology", *Journal of Physics, D: Applied Physics,* Vol. 21, (1988) pp. 1661-1677.

Borgmann, H. (ed.), Actuator 2002, International conference on new actuators, 10-12 June 2002, Conference proceedings, Messe Bremen GmbH, Bremen, (2002).

Brebbia C.A., and A. Samartin, "Computational methods for smart structures and materials II," http://www.witpress.com/c400.html, Series: Structures and Materials, Vol. 7, (2000), pp. 1-184.

Brock, D. L., "Review of artificial muscle based on contractile polymers," MIT AI Memo No. 1330 (1991).

Calvert P., J. O'Kelly, C. Souvignier, "Solid freeform fabrication of organic-inorganic hybrid Materials," *Materials Science and Engineering*, Vol. C 6, (1998), pp. 167-174.

Cao Y., P. Smith and A.J. Heeger, *Polymer*, Vol. 32 (1991), pp. 1210.

Cheng Z.-Y., T.-B.Xu, V. Bharti, T.X. Mai, Q.M. Zhang, T. Ramotowski and R. Y. Ting, "Characterization of electrostrictive P(VDF-TrFE) copolymer films for high-frequency and high-load applications," Proceedings of SPIE, 3987 (2000), pp. 73-80.

Chou C.P., B. Hannaford, "Measurement and modeling of Mckibben pneumatic artificial muscles," *IEEE Transactions on Robotics and Automation*, Vol. 12, (Feb. 1996), pp. 90-102.

Chou, C. P. and B. Hannaford, "Static and dynamic characteristics of Mckibben pneumatic artificial muscles," Proceedings of the IEEE International Conference on Robotics and Automation, San Diego, CA, May 8-13, 1994, 1:281-286.

Davis, L.C., "Model of magnetorheological elastomers," *Journal of Applied Physics*, Vol. 85, No.6 (1999), pp.3348-3351.

de Rossi D., "Realization, characterization and modeling of electroactive polymer actuators," WW-EAP Newsletter, Vol. 1, No. 2 (1999), pp. 4-5.

de Rossi, D., M. Suzuki, Y. Osada, and P. Morasso, "Pseudomuscular gel actuators for advanced robotics," *J. of Intelligent Material Systems and Structures*, Vol. 3, (1992), pp. 75-95.

Dusěk K. (Ed.), Advances in Polymer Science, 109 and 110, *Responsive Gels, Volume Transitions I and II*, Springer-Verlag Berlin, Heidelberg 1993.

Eguchi M., "On the Permanent Electret," Philosophical Magazine, Vol. 49 (1925) pp. 178.

Finkelmann H., H.J. Kock, and G. Rehage, "Investigations on liquid crystalline silozanes: 3. liquid crystalline elastomer – a new type of liquid crystalline material," *Makromolecular Chemistry*, Rapid Communications, Vol. 2 (1981), pp. 317.

Finkelmann H., E. Nishikawa, G.G. Pereira, and M. Warner, "A New Opto-Mechanical Effect in Solids," *Phys. Rev. Lett.,* Vol. 87, No. 1, (2001) pp. 015501-1 to -4.

Forsterling H. -D., and H. Kuhn, Praxis der Physikalische Chemie, Section 6.3, "Polyacryl-sauremiskel," VCH, (1991).

Furukawa, T. and Seo, N., "Electrostriction as the origin of piezoelectricity in ferroelectric polymers," *Japan J. of Applied Physics,* Vol. 29, No. 4 (1990), pp. 675-680.

Gandhi M.R., P. Murray, G.M. Spinks, and G. G. Wallace, "Mechanisms of electromechanical actuation in polypyrrole," *Synthetic Metals*, Vol. 75 (1995), pp. 247-256.

Glass, J. E. (Ed.), *Polymers in aqueous media,* American Chemical Society, Washington D.C. (1989).

Gong J. P., T. Nitta and Y. Osada, "Electrokinetic modeling of the contractile phenomena of polyelectrolyte gels-one dimensional capillary model," *J. Phys. Chem.*, 98, (1994), pp. 9583-9587.

Hathaway K.B., Clark A.E., "Magnetostrictive materials," MRS Bulletin, Vol.18, No. 4 (April 1993), pp. 34-41.

He Y., S. Gong and J. Kanicki, "Organic polymer light-emitting devices on the plastic substrates," Y. Bar-Cohen (Ed.), Proceedings of the SPIE Conference on EAPAD, SPIE Proc. Vol. 3669, (1999) pp. 330-335.

Heitner-Wirguin, C. "Recent advances in perfluorinated ionomer membranes: Structure, properties and applications," *Journal of Membrane Science*, V 120, No. 1, (1996), pp. 1-33.

Herod T. E., and J. B., Schlenoff, Doping induced strain in polyaniline: stretchoelectrochemistry *Chemistry of Materials*, vol. 5, pp. 951-955, 1993.

Hirai T, M. Watanabe, M. Yamaguchi, "PVC gel deforms like a tongue by applying an electric field," WW-EAP Newsletter, Vol. 1, No. 2, 1999, pp. 7-8.

Hirai T., J. Zheng, M. Watanabe, "Solvent-drag bending motion of polymer gel induced by an electric field," Y. Bar-Cohen, (Ed.), Proceedings of the SPIE's 6th Annual International Symposium on Smart Structures and Materials, Vol. 3669, (1999) pp. 209-217.

Hirokawa Y. and T. Tanaka, "Volume transition in nonionic gel," *J. Chem. Physics*, Vol. 81, (1984), pp. 6379.

Hollerbach J., I. Hunter & J. Ballantyne, "A comparative analysis of actuator technologies for robotics", O. Khatib, J. Craig & Lozano-Perez (Eds), *The Robotics Review* 2, MIT Press, Cambridge MA (1992), pp. 299-342.

Holme N.C.R., L. Nikolava, S. Hvilsted, P.H. Rasmussen, R.H. Berg, P.S. Ramanujam, "Optically induced surface relief phenomena in azobenzene polymers," *Appled Physics Letters*, Vol. 74, (1999), pp. 519.

Holze R., and J. C. Ahn, "Advances in the use of perfluorinated cation-exchange membranes in integrated water electrolysis and hydrogen - oxygen fuel systems," *J. Membrane Sci.*, Vol. 73, No. 1, (1992), pp. 87-97.

Huber J., N.A Fleck, and M.F. Ashby, 'The selection of mechanical actuators based on performance indices' in *Proc. R. Soc. Lond. A*, **453**, (1997), pp. 2185-2205.

Hunter I. & S. Lafontaine, "A comparison of muscle with artificial actuators", *Technical Digest IEEE Solid State Sensors & Actuators Workshop (*1992), pp. 178-185.

Ichijo, H., O. Hirasa, R., Kishi, M. Oowada, K. Sahara, E. Kokufuta, and S. Kohno, "Thermo-responsive gels," *Radiation Physics and Chemistry*, Vol. 46, (1995), pp. 185-190.

Ito S., "Phase transition of aqueous solution of poly(N-alkoxyacrylamide) derivatives-effect of side chain structure," *Koubunshu*, Vol. 46, No. 7 (1989), pp. 437.

Ito S., "Phase transition of aqueous solution of poly(N-alkoxyacrylamide) derivatives-effect of side chain structure," *Koubunshu*, Vol. 47, No. 6 (1990), pp. 467.

Jager E. W. H., O. Inganäs, and I. Lundström, "Microrobots for micrometer-size objects in aqueous media: potential tools for single cell manipulation," *Science*, Vol. 288 (2000) pp. 2335-2338.

Jones J. and J. J. Wu, "Inflatable rovers for planetary applications," Proceedings of the SPIE International Symposium on Intelligent Systems and Advanced Manufacturing (1999).

Juodkazis S., N. Mukai, R. Wakaki, A. Yamaguchi, S. Matsuo and H. Misawa, "Reversible phase transitions in polymer gels induced by radiation forces," *Nature*, Vol. 408, No. 6809 (2000), pp. 178-180.

Katchalsky, A., "Rapid swelling and deswelling of reversible gels of polymeric acids by ionization," *Experientia*, Vol. V, (1949), pp. 319-320.

Kim J., J.-Y. Kim and S.-J. Choe, "Electro-active papers: its possibility as actuators," Y. Bar-Cohen Y., (Ed.), Proceedings of the SPIE's EAPAD Conf., part of the 7th Annual International Symposium on Smart Structures and Materials, SPIE Proc. Vol. 3987, (2000) pp. 203-209.

Kishi R., H. Ichijo, and O. Hirasa, "Thermo-responsive devices using poly(vinyl methyl ether) hydrogels," *J. Intell. Mater. Syst. Struct.* Vol. **4**, (1993) pp. 533-537.

Klapcin´ski T., A. Galeski and M. Kryszewski, "Polyacrylamide gels filled with ferromagnetic anisotropic powder: a model of a magnetochemical device," *J. Appl. Polymer Science*, Vol. 58, (1995) pp. 1007.

Kuhn W., B. Hargitay, A. Katchalsky, and H. Eisenburg, "Reversible dilatation and contraction by changing the state of ionization of high-polymer acid networks," *Nature*, Vol. 165 (1950), pp. 514-516.

Kupfer, J. and H. Finkelmann, "Nematic liquid single crystal elastomers," *Macromol. Chem. Rapid Commun.*, Vol. 12, pp. 717 (1991).

Lehmann, W, L. Hartmann, F. Kremer, P. Stein, H. Finkelmann, H. Kruth, and S. Diele, "Direct and inverse electromechanical effect in ferroelectric liquid crystalline Elastomers," J. Appl. Phys., 86, (1999) pp. 1647.

Li M., P. Keller, B. Li, X.Wang, and M. Brunet, *Adv. Mater.*, Vol. 15 (2003) pp. 569.

Li F. K., W. Zhu, X. Zhang, C. T. Zhao, and M. Xu, "Shape memory effect of ethylene-vinyl acetate copolymers," *J. Appl. Polym. Sci.*, Vol. 71, No. 7 (1999), pp. 1063-1070.

Li, J. Y. and S. Nemat-Nasser, "Micromechanical analysis of ionic clustering in Nafion perfluorinated membrane," *Mechanics of Materials*, Vol. 32, No. 5, (2000), pp. 303-314.

Liang C., C.A. Rogers, E. Malafeew, "Investigation of shape memory polymers and their hybrid composites," *J. Intelligent Material Systems and Structures*, Vol. 8, (1997) pp. 380.

Liu Z. and P. Calvert, "Multilayer hydrogels and musclelike actuators," *Advanced Materials*, Vol. 12, No. 4 (2000), pp. 288-291.

Madden J. D., N. Vandesteeg, P. G. Madden, A. Takshi, R.l Zimet, P A. Anquetil, S R. Lafontaine, P A. Wierenga and I W. Hunter, "Artificial Muscle Technology: Physical Principles and Naval Prospects," Special Issue on Biorobotics, IEEE Journal of Ocean Engineering, in press.

Marsella M. J., and R. J Reid, *Macromolecules*, Vol. 32, (1999), pp. 5982-5984.

Marsh R. L., J. M. Olson, and S. K. Guzik, "Mechanical performance of scallop adductor muscle during swimming," *Nature*, 357: (6377) (1992), pp. 411-413.

McAlvin J., J.D. Langan, R. Behm, M. Costolo and A. J. Heeger, "Spatial frequency filtering using hybrid polymer/VLSI technology," Proceedings of the SPIE's 6th Annual International Symposium on Smart Structures and Materials, SPIE Proc. Vol. 3669, (1999), pp. 336-344.

McGehee M.D., E. K. Miller, D. Moses and A.J. Heeger, "Twenty years of conductive polymers: from fundamental science to applications," published in *Advances in Synthetic Metals: Twenty Years of Progress in Science and Technology*, P. Barnier, S. Lefrant and G. Bidan, Eds., Elsevier (1999), pp 98-203.

Naciri J., A. Srinivasan, H. Jeon, N. Nikolov, P. Keller, and B.R. Ratna, "Nematic elastomer fiber actuator," *Macromolecules,* Vol. 36 (2003), pp. 8499.

Naciri J., A. Srinivasan, W. Sandberg, R. Ramamurti, and B.R. Ratna, "Nematic liquid crystal elastomers as artificial muscle materials," *13th International Symposium on Unmanned Untethered Submersible Technology*, New Hampshire (2003a).

Nemat-Nasser, S., and J. Y. Li, "Electromechanical response of ionic polymer-metal composites," *J. Applied Physics*, Vol. 87, No. 7, (2000), pp. 3321-3331.

Oguro K., N. Fujiwara, K. Asaka, K. Onishi, and S. Sewa, "Polymer electrolyte actuator with gold electrodes," Proceedings of the SPIE's 6th Annual

International Symposium on Smart Structures and Materials, SPIE Proc. Vol. 3669, (1999), pp. 64-71.

Oguro, K., Y. Kawami and H. Takenaka, "Bending of an ion-conducting polymer film-electrode composite by an electric stimulus at low voltage," *Trans. Journal of Micromachine Society*, Vol. 5, (1992) pp. 27-30.

Osada Y., A. Matsuda, "Shape memory in hydrogels," *Nature*, Vol. 376 (1995), p. 219.

Osada Y., and M. Hasebe, "Electrically activated mechanochemical devices using polyelectrolyte gels," *Chemistry Letters*, (1985), pp.1285-1288.

Osada Y., and R. Kishi, "Reversible volume change of microparticles in an electric field," *J. Chem. Soc.*, 85, (1989), pp. 665-662.

Osada Y., and S. Ross-Murphy, "Intelligent gels," *Scientific American*, 268, (1993), pp. 82-87.

Osada, Y., H. Okuzaki, and H. Hori, "A polymer gel with electrically driven motility," *Nature,* Vol. 355, (1992), pp. 242-244.

Otero T. F., and Sansiñena, J. M., "Soft and wet conducting polymers for artificial muscles," *Advanced Materials* 10 (6), (1998) pp. 491-494.

Otero T. F., H. Grande, J. Rodriguez, "A new model for electrochemical oxidation of polypyrrole under conformational relaxation control," *J. Electroanal. Chem.*, Vol. 394 (1995), pp. 211-216.

Park S.E., Shrout T. R., "Relaxor based ferroelectric single crystals for electro-mechanical actuators," Materials Research Innovations, Vol.1, No.1 (1997), pp.20-25.

Pei, Q., M. Rosenthal, R. Pelrine, S. Stanford, and R. Kornbluh. 2002. "3-D Multifunctional Electroelastomer Roll Actuators and Their Application for Biomimetic Walking Robots," *Smart Structures and Materials 2002: Industrial and Commercial Applications of Smart Structures Technology*, ed. A. McGowan, *Proc. SPIE*, Vol. 4698, pp. 246–253.

Pelrine R., R. Kornbluh, and J. P. Joseph, "Electrostriction of polymer dielectrics with compliant electrodes as a means of actuation," *Sensor Actuat. A*, Vol. 64 (1998), p. 77-85.

Pelrine R., R. Kornbluh, Q. Pei, and J. Joseph, "High speed electrically actuated elastomers with strain greater than 100%," *Science*, Vol. 287, (2000) pp. 836-839.

Ratna B. R., "Liquid crystalline elastomers as artificial muscles: role of side chain – backbone coupling," To be published in the Proceedings of the SPIE's 8th Annual International Symposium on Smart Structures and Materials, EAPAD Conf. (2001).

Roentgen, W.C., "About the changes in shape and volume of dielectrics caused by electricity," Section III in G. Wiedemann (Ed.), *Annual Physics and Chemistry Series,* Vol. 11, John Ambrosius Barth Publisher, Leipzig, German (1880) pp. 771-786 (in German).

Sacerdote M. P.,"On the electrical deformation of isotropic dielectric solids," *J. Physics*, 3 Series, t, VIII, 31 (1899), 282-285 (in French).

Sadeghipour, K., R. Salomon, and S. Neogi, "Development of a novel electrochemically active membrane and 'smart' material based vibration sensor/damper," *Smart Materials and Structures*, (1992) pp. 172-179.

Shenoy D. K., D.L. Thomsen III, A. Srinivasan, P. Keller, and B.R. Ratna, *Sensor and Actuators A,* "Carbon coated liquid crystal elastomer film for artificial muscle applications," Vol. 96, (2002) p. 184.

Schreyer,H. B., N. Gebhart, K. J. Kim, and M. Shahinpoor, "Electric activation of artificial muscles containing polyacrylonitrile gel fibers," *Biomacromolecules J.*, ACS Publications, Vol. 1, (2000), pp. 642-647.

Sen A., Scheinbeim, J.I., and Newman, B.A., "The effect of platicizer on the polarization of poly(vinylidene fluoride) films," *J. Appl. Phys.,* Vol.56, No.9, (1984), pp.2433-2439.

Sessler G. M., and J. Hillenbrand, "Novel polymer electrets," MRS Symposium Proceedings on Electroactive Polymers (EAP), Vol. 600, Warrendale, PA, (1999) pp. 143-158.

Shahinpoor M., "Elastically-activated artificial muscles made with liquid crystal elastomers," Y. Bar-Cohen, (Ed.), Proceedings of the SPIE's 7th Annual International Symposium on Smart Structures and Materials, EAPAD Conf. Vol. 3987, (2000) pp. 187-192.

Shahinpoor, M., "Conceptual design, kinematics and dynamics of swimming robotic structures using ionic polymeric gel muscles," *Smart Materials and Structures*, Vol. 1, No. 1 (1992) pp. 91-94.

Shiga, T. "Deformation and viscoelastic behavior of polymer gels in electric fields," *Adv. Polym. Sci.,* Vol. 134, (1997), pp. 131-163.

Smela E., O. Inganäs, and I. Lundström, "Controlled folding of micrometer-size structures," *Science*, Vol. 268 (1995) pp. 1735-1738.

Smets G., "New developments in photochromic polymers," *J. Polymers Science*, Polymers Chemistry, Vol. 13 (1975) p. 2223.

Sokolowski W. M., A. B. Chmielewski, and S. Hayashi, "Cold hibernated elastic memory (CHEM) self-deployable structures," Y. Bar-Cohen, (Ed.), Proceedings of the SPIE's 6th Annual International Symposium on Smart Structures and Materials, SPIE Proc. Vol. 3669, (1999) pp. 179-185.

Sperling, L. H., *Introduction to Physical Polymer Science, 2nd Ed.,* John Wiley & Sons, New York (1992).

Steinberg I. Z., A. Oplatka, and A. Katchalsky, "Mechanochemical engines," *Nature*, Vol. 210, (1966) pp. 568-571.

Su J., J. S. Harrison, T. St. Clair, Y. Bar-Cohen, and S. Leary, "Electrostrictive graft elastomers and applications," MRS Symposium Proceedings, Vol. 600, Warrendale, PA, (1999) pp. 131-136.

Suzuki M. and O. Hirasa, "An approach to artificial muscle using polymer gels formed by microphase separation," *Advanced Polymers Science,* Vol. 110, (1993) pp. 241-261.

Tanaka T., I. Nishio, S.T. Sun, and S. U. Nishio, "Collapse of gels in an electric field," *Science*, vol. 218, (1982), pp. 467-469.

Thomsen III D.L., P. Keller, J. Naciri, R. Pink, H. Jeon, D. Shenoy and B.R. Ratna, "Liquid crystal elastomers with mechanical properties of a muscle," *Macromolecules,* Vol. 34 (2001) pp. 5868.

Tobushi H., S. Hayashi, S. Kojima, "Mechanical properties of shape memory polymer of polyurethane series," *JSME Int. J.*, Ser. I. Vol. 35, No 3 (1992), pp. 296-302.

Urry D. W., "Plastic biomolecular machines," *Scientific America*, (Jan. 1995) pp. 44-49.

Urry D.W., L.C. Hayes, and S.Q. Peng, "Design for advanced materials by the ΔT_t-mechanism," in Smart Structures and Materials 1996: Smart Materials Technologies and Biomimetics, Andrew Crowson, Ed., SPIE Proc. Vol. 2716, (1996) pp. 343-346.

van der Veen G. and W. Prins, *Phys. Sci.*, Vol. 230 (1971), pp. 70.

Wang, T.T., J.M. Herbert and A.M. Glass, Ed., *The Applications of Ferroelectric Polymers*, Chapman and Hall, New York (1988).

Warner M., and E.M.Terentjev, *Liquid Crystal Elastomers*, Clarendon Press, Oxford (2003).

Wax S.G., R.R. Sands, and L.J. Buckley, "Compliant actuators based on electroactive polymers," Q.M. Zhang, T. Furukawa, Y. Bar-Cohen, and J. Scheinbeim, Eds., Electroactive Polymers (EAP), MRS Symposium Proceedings, Vol. 600, Warrendale, PA, (1999) pp 3-11.

Wermter H. and H. Finkelmann, "Liquid crystalline elastomers as artificial muscles," *e-Polymers,* no. 013 (2001) pp. 1.

Winslow, W. M., "Induced fibrillation of suspensions," *Journal of Applied Physics*, Vol. 20, (1949), pp. 1137.

Woojin L, "Polymer gel based actuator: dynamic model of gel for real time control," Ph.D. Thesis, Massachusetts Institute of Technology (1996).

Zhang Q. M., V. Bharti, and X. Zhao, "Giant electrostriction and relaxor ferroelectric behavior in electron-irradiated poly(vinylidene fluoride-trifluorethylene) copolymer," *Science*, Vol. 280, pp. 2101-2104 (1998).

Zhang Q., "Electroactive polymers with high electrostrictive strain and elastic power density," WW-EAP Newsletter, Vol. 1, No. 2 (1999) p. 12.

Zhang, Q. M., Hengfeng Li, Martin Poh, Haisheng Xu, Z.-Y. Cheng, Feng Xia, and Cheng Huang. "An all-organic composite actuator material with high dielectric constant," *Nature*, 419, 284 (2002).

Zhang Q.M., T. Furukawa, Y. Bar-Cohen, and J. Scheinbeim, Eds., Proceedings of the Fall MRS Symposium on electroactive polymers (EAP), Vol. 600, Warrendale, PA, (1999) pp. 1-336.

Zrinyi M., D. Szabo and J. Feher, "Comparative studies of electro- and magnetic field sensitive polymer gels," Y. Bar-Cohen, (Ed.), Proceedings of the SPIE's 6th Annual International Symposium on Smart Structures and Materials, EAPAD Conf., SPIE Proc. Vol. 3669, (1999) pp. 406-413.

Zrinyi M., L. Barsi, D. Szabo, and H. G. Kilian, "Direct observation of abrupt shape transition in ferrogels induced by nonuniform magnetic field," *J. Chem. Phys.,* Vol. 106, No. 13 (1997), pp. 5685-5692.

TOPIC 2

Natural Muscles

CHAPTER 2

Natural Muscle as a Biological System

Gerald H. Pollack, Felix A. Blyakhman, Frederick B. Reitz,
Olga V. Yakovenko, and Dwayne L. Dunaway
University of Washington

Muscle is the natural contractile system that artificial systems attempt to emulate. For proper emulation it is evidently necessary to understand the basic underlying mechanism. In the material that follows, we consider the evidence that muscle is a polymer gel, and that the gel's polymeric filaments contract by a polymer-gel phase-transition. This is an unorthodox conclusion, and we therefore begin by considering the relevant background.

2.1 Conceptual Background

Until the mid-1950s muscle contraction was held to occur by a protein-folding mechanism [for review, cf. A. F. Huxley, 1957]. The folding mechanism is similar enough to the one that will be suggested here that it may be considered a progenitor.

With the discovery of interdigitating thick and thin filaments in the mid-1950s, it was tempting to discard the notion of folding, and suppose instead that contraction arose out of pure filament sliding. This supposition led H. E. Huxley and Sir Andrew Huxley to examine independently whether the interdigitating filaments (Fig. 1) remained at constant length during contraction. Their optical microscopic studies of muscle-striation patterns were published back-to-back in *Nature* [Huxley and Hanson, 1954; Huxley and Niedergerke, 1954]. Together with more detailed follow up [Huxley and Niedergerke, 1958], these studies became the field's pivotal works.

These studies examined relaxed and activated specimens alike. Relaxed specimens were manually stretched and released. During these maneuvers, the width of the A-band remained absolutely invariant. Since A-band width corresponds to the length of the thick filament (Fig. 1), the result implied that thick filaments did not change length. This strengthened the emerging notion of filament sliding.

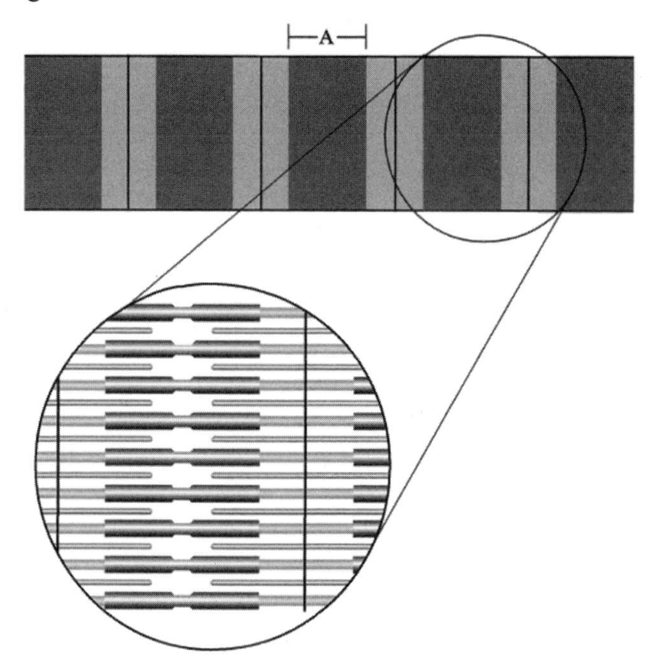

Figure 1: Molecular structure of the sarcomere. Cartoon of a microscopic view (above) corresponds to set of interdigitating filaments. Thick filaments correspond to dark A-band. Corresponding molecular structure (below) shows connecting filaments in series with thick filaments. Thin filaments lie in parallel with thick and connecting filaments.

Less conclusive, however, were the results obtained when muscle length changed actively, for in this case sarcomere shortening was sometimes accompanied by A-band narrowing. The narrowing was dismissed as quantitatively inconsequential in the Huxley–Hanson study, and in the Huxley–Niedergerke study it was relegated to the microscope's limited resolution. With these dismissals, all results could remain interpretable within the framework of constant-length filaments.

These back-to-back papers by the Huxleys dispatched the protein-folding paradigm to the heap of forgotten notions. The constant filament-length paradigm took hold, and has held remarkably firm ever since—notwithstanding subsequent findings of thick filament or A-band shortening in more than 30 reports [for review, see Pollack, 1983; Pollack, 1990].

With the emerging notion of sliding filaments, the central issue became the nature of the driving force. Impressive numbers of researchers sought out a mechanism, and soon settled on the one schematized in Fig. 2. In this mechanism, the cross-bridges attach to the thin filament, swing, and then detach in the presence of ATP, readying themselves for the next cycle. Bridge swinging drives the thin filament toward the center of the sarcomere, thereby shortening the sarcomere and the muscle. The mechanism explains many known features of contraction and has therefore become broadly accepted [for review, cf. Spudich, 1994; H. E. Huxley, 1996; Block, 1996; Howard, 1997; Cooke, 1997].

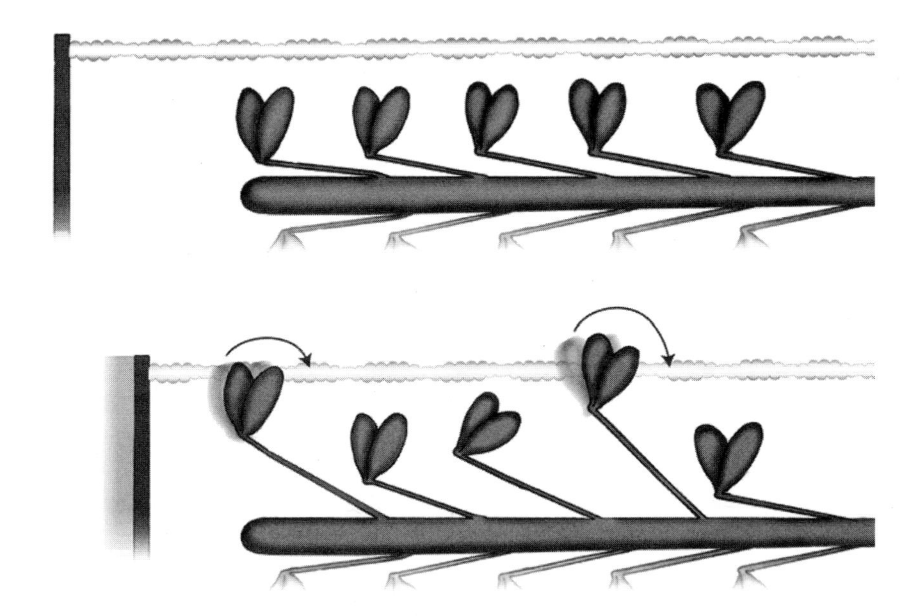

Figure 2: Cross-bridge model of muscle contraction. In the relaxed state (top) bridges projecting from thick filament do not interact appreciably with the thin filament. In the activated state (bottom), bridges attach and rotate, thereby driving the thin filament rightward, toward the center of the sarcomere. Z-line is at left.

Apart from the concern of building on the uncertain foundation of constant-length filaments, a serious concern is the absence of evidence for cross-bridge swinging. To test for bridge-angle changes, numerous molecular scale approaches have been applied [for review, see Thomas, 1987]. Electron-spin resonance, x-ray diffraction, and fluorescence polarization have produced largely negative results, as has high-resolution electron microscopy [Katayama, 1998]. Of all these results, the most supportive has been an angle change of 3 deg measured on a cross-bridge (myosin-light chain) appendage [Irving et al., 1995]. This 3 deg change is still far short of the anticipated 45 deg.

Absence of swinging is only one of several areas of concern. Concerns run the gamut from instability [Iwazumi, 1970] to mechanics [Pollack, 1983], structure [Schutt and Lindberg, 1993, 1998], and chemistry [Oplatka, 1996, 1997]. Many of these concerns are detailed in my 1990 book [Pollack, 1990]. But the full flavor is best gleaned by reading the original reviews, particularly those by Schutt and Lindberg and by Oplatka, whose spicy, no-holds-barred bluntness entertains as it educates. I also commend an insightful work by the late Graham Hoyle [1983], in which some understanding of the field's continued focus on seemingly unproductive paradigms is offered in a chapter labeled "Why Do Muscle Scientists 'Lose' Knowledge?"

In the material that follows, we pick up where the pioneers of a half-century ago who championed the protein folding mechanism, including Nobelist Albert Szent-Györgyi, left off. We first deal with several relevant structural considerations, and then with possible mechanisms, keeping a sharp eye toward relevance for artificial muscles.

2.2 Structural Considerations

The functional unit of contraction is the myofibril, which comprises several hundred protein filaments (Fig. 3). Many parallel myofibrils make up the muscle fiber, or cell; and many muscle fibers make up the muscle. The arrangement is hierarchical.

The myofibrillar sarcomere comprises three types of polymeric filament, two of which were illustrated in Fig. 2. The central one is the thick filament, which is built of multiple repeats of the protein myosin. The thick filament is flanked at either end by a connecting filament, which in turn connects to the Z-line. In vertebrate muscle the connecting filament is built largely of the protein titin—a polymer made up principally of tandem repeats of discrete immunoglobulinlike domains. The thin filament is also polymeric. Along with various regulatory proteins, it consists mainly of multiple repeats of the protein actin. Thus, all three of the sarcomere's longitudinal structures are biological polymers.

The myofibril is striking in the almost crystalline precision with which filaments align with one another in parallel, conferring sharpness on the A-I boundaries. Such alignment is realized through extensive cross-linking throughout the sarcomere, both in the A-band and in the I-band.

I-band cross-links are illustrated in Fig. 4. The top panel is a freeze-fracture image, which reveals the cross-links in lifelike form. The bottom panel is a conventional thin section. Cross-linking is a general rule among diverse actin filaments, and therefore it is no surprise that it occurs in the sarcomere.

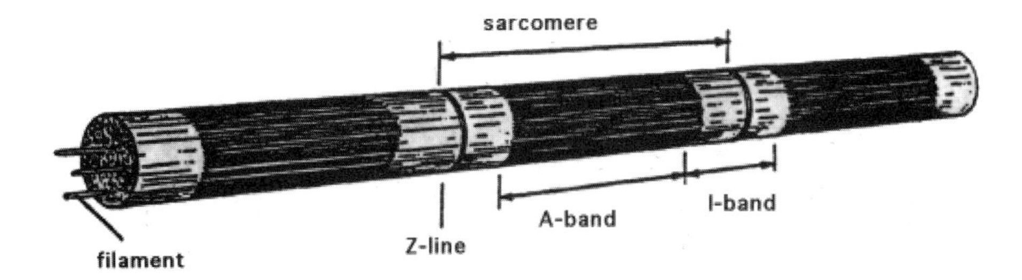

Figure 3: Structure of an intact myofibril. The A-band contains a parallel array of thick filaments. Connecting filaments lie in the I-band. Thin filaments originate at the Z-line and partially overlap thick filaments in the A-band.

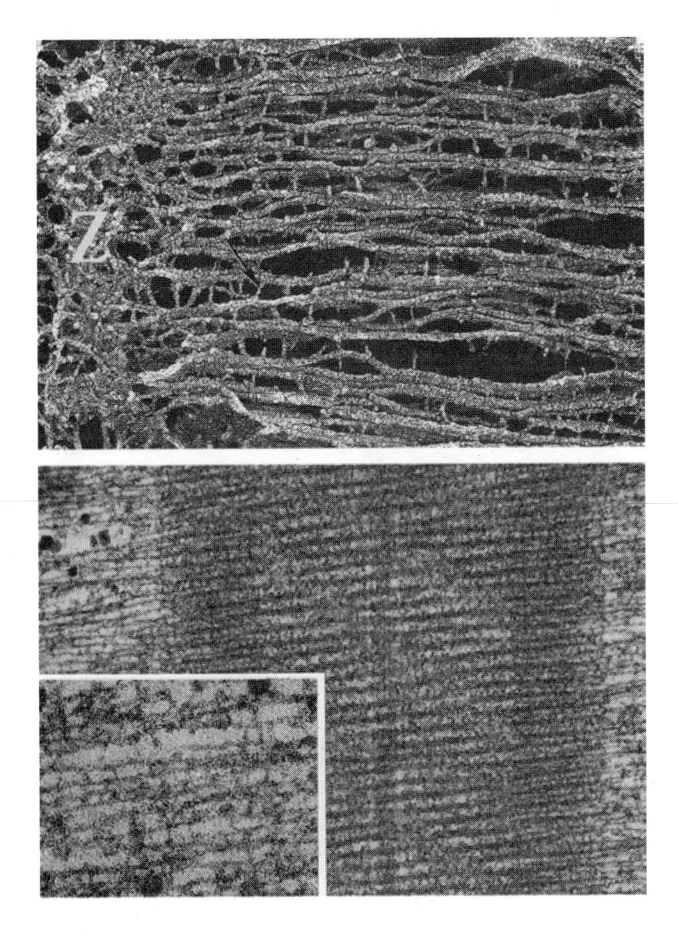

Figure 4: Top: Freeze fracture image of I-band filaments. Vertically oriented mesh at left is the Z-line. Arrow indicates interconnection. Bottom: Ultrathin section showing regular repeat of I-band interconnections. View from shallow angle, below. (Reprinted from Baatsen et al. 1988. Copyright 1988, with permission from Elsevier.)

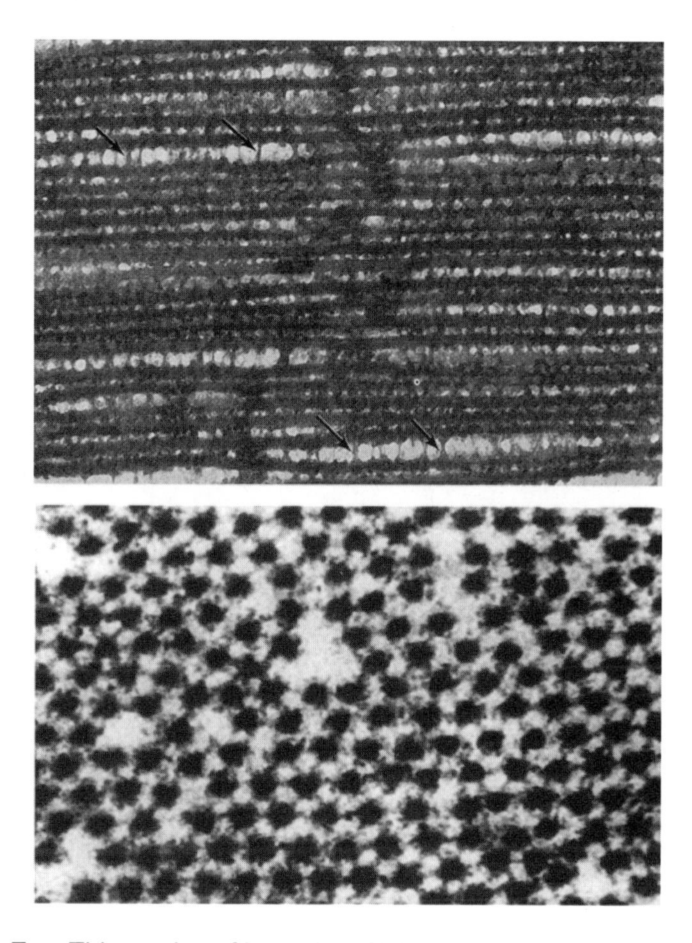

Figure 5: Top: Thin section of honeybee flight muscle in which thin filaments had been withdrawn to increase clarity of thick filament zone. Arrows denote interconnections. Bottom: cross-section through A-band of moderately stretched sarcomere that grazed tips of thin filaments, the latter seen as occasional small dots. Interconnections between thick filaments are evident.

A-band interconnections are shown in Fig. 5. The top panel shows a specimen that had been stretched by about 30% to withdraw thin filaments from the lattice in order to reduce visual congestion. Stretch distorts the lattice somewhat, but cross-links between thick filaments nevertheless remain clear.

The thick-to-thick interconnections are almost certainly built of myosin, for myosin is the only A-band protein in sufficient quantity to account for structures so abundant. Links of appropriate length can be created if cross-bridges extending from adjacent filaments bind to one another at their tips [Pollack, 1990]. The cross-links visible in these micrographs are not widely recognized in the muscle field, where free, swingable cross-bridges are required for the theory.

On the other hand, these essentially static elements may explain why attempts to detect myosin-head tilting have failed; rungs cannot tilt.

Cross-links align the filaments and thereby confer regularity on the lattice; that is why the A–I junctions are so distinct. They also maintain lattice integrity by limiting swelling. Highly charged polymers such as those in the sarcomere can imbibe water up to thousands of times their volume [Osada and Gong, 1993], and may ultimately dissolve if not cross-linked. The muscle-filament lattice, then, is essentially a highly cross-linked, water-filled polymer gel.

2.3 Does Contraction Involve a Phase Transition?

This gel is evidently designed to contract, and certain features of its behavior imply a phase transition. Triggers of contraction, for one, are the same as those for transitions in ordinary polymer-gels [Hoffman, 1991]. In contractile protein bundles of demembranated cells, for example, contraction can be initiated by increased salt, pH change, temperature jump—even electrical current will trigger contraction of an isolated myofibril [Garamvolgyi, 1959]. And like polymer-gel transitions, triggering is critical, or "razor-edge" (see below). Nothing happens until a threshold is crossed, whereupon contractile action is massive.

Ironically, such critical behavior had been evident in model studies carried out more than a half-century ago, and that is perhaps the reason why contraction was presumed to involve something akin to a phase transition. Vanguard experiments of the era focused on suspensions of actin and myosin. Such suspensions could form a gel, the condensation of which was considered a working model of muscle contraction. The gel remained stable until ambient conditions were edged just past a critical threshold; then it contracted massively. As it contracted, the matrix folded and water was released—much the same as it is released in the polymer-gel (Fig. 6).

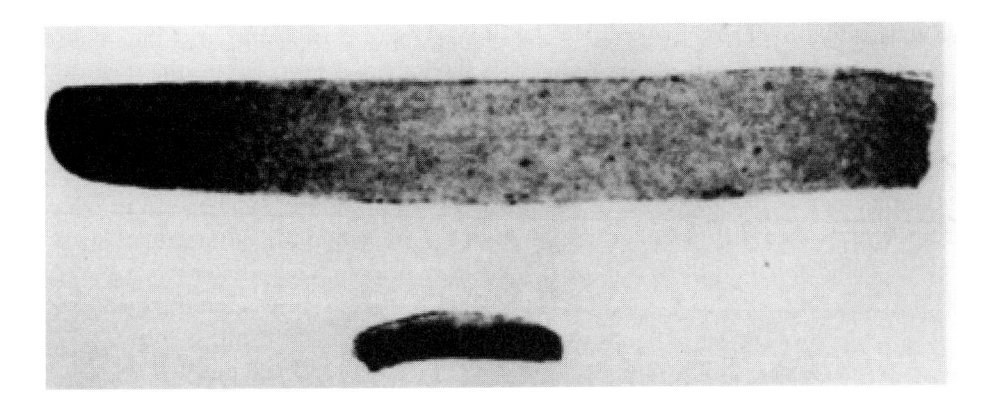

Figure 6: The actomyosin gel undergoes massive volume change in response to a slight increase of ATP concentration. (Reprinted from Szent-Györgyi, 1951. Copyright 1951 with permission from Elsevier.)

Like the polymer gel, the cross-linked polymer network of the sarcomere is invested with solvent—an aqueous salt solution. The solvent remains trapped within the network; it does not leak out. This can be seen in specimens whose membranes have been removed (skinned fibers): such specimens do not easily lose solvent. Again, this phenomenon is similar to that of the gel, the solvent presumably held by strong hydrophilic forces that retain the water within the sarcomere's polymeric framework [Pollack, 2001]. Thus, muscle exhibits the essential features of polymer gels—except that its polymers are well organized into a regular framework and not randomly dispersed.

If muscle contraction were to involve a phase transition, the resulting dimensional change would be axial because polymers are naturally aligned along the axis of the muscle; hence, polymer shortening would produce shortening along the axis of the muscle. On the other hand, polymeric fragments of actin and myosin can be used to construct random or semi-random gels, in which case contraction should be isotropic, as in the gel of Fig. 6, above. The behavior illustrated in Fig. 6 is rather similar to the behavior of polyacrylamide gels exposed to varying ratios of organic/inorganic solvents [Tanaka, 1981]. Not only is contraction fairly isotropic, but it also has a critical nature: no change at all until the stimulus reaches a threshold, and then massive condensation.

Critical behavior is also seen when muscle polymers are naturally oriented, as they are in striated muscle. In Fig. 7 the concentration of calcium, the physiological trigger, is progressively increased. Stripped of its membrane, the contractile specimen is immersed in a physiological solution in which the level of free calcium is progressively elevated. Once the concentration crosses threshold, full tension is observed within a narrow concentration window. Similar behavior is observed when calcium is replaced by other divalent cations such as barium or strontium. Contraction is essentially all-or-none.

Critical behavior is also seen when an organic solvent surrounding the specimen is progressively replaced by an aqueous solvent; see Fig. 8. Again, the contraction is largely all-or-none, within a narrow window of organic/aqueous solvent ratio. Such behavior is a classical signature of a polymer gel phase-transition. Thus, critical "razor-edge" behavior is preserved in the naturally oriented polymeric system just as it is in the random muscle gel.

The evidence above implies that muscle contracts in the same way as an ordinary polymer gel contracts. How, then, might such behavior be manifested at the molecular level? The evidence is most consistent with a mechanism in which the polymers themselves can contract. In fact, evidence in the literature, as shown below, implies that all three polymers can shorten.

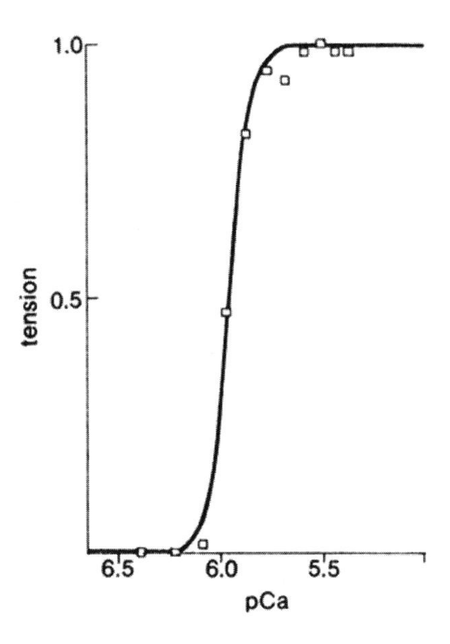

Figure 7: Effect of increase of calcium concentration on relative isometric tension development in single rabbit muscle cells [Pollack, 1990]. (Courtesy of Ebner & Sons.)

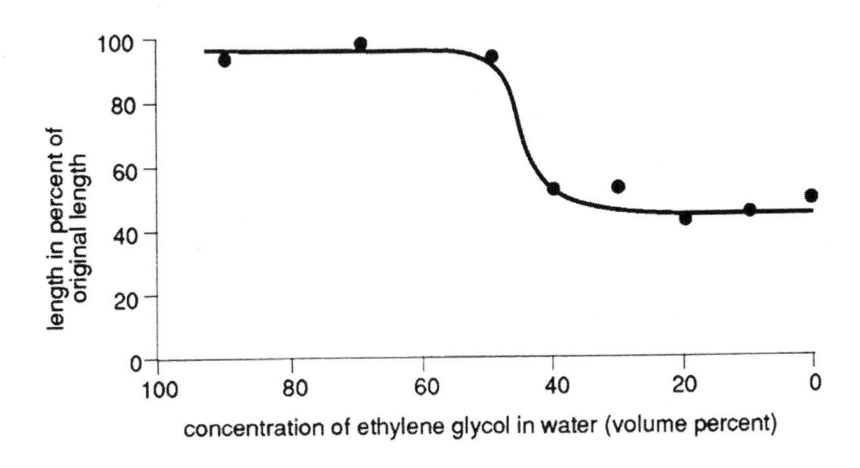

Figure 8: Effect of solvent variation on contraction in rabbit muscle [Pollack, 1990]. (Courtesy of Ebner & Sons.)

2.4 Molecular Basis of the Phase Transition

In the proposed paradigm, all three filaments shorten by a phase transition, and this shortens the sarcomere. I will briefly sketch some of the evidence that leads to this kind of mechanistic view, considering filaments one at a time. The paradigm suggested here is outlined in Pollack [1990], with some newer elements considered in a more recent review article [Pollack, 1996].

2.4.1 Connecting Filaments

Connecting filaments are springlike polymers that retract when stretched. When unactivated muscles are forcibly stretched, the connecting filaments are strained. If the muscle is then released, connecting filaments retract and their tension is relieved. Connecting filaments are thereby responsible for the muscle's so-called resting tension. Resting tension ensures that the stretched sarcomere returns to its natural length, and also keeps the thick filament centered in the sarcomere. All of these considerations refer to muscles that are not stimulated to contract actively; thus, the force range under consideration is relatively modest.

Dynamics of the connecting filament have been revealed in elegant experiments on isolated molecules of titin [Rief et al., 1997; Tskhovrebova et al., 1997]. In the Rief et al. study, ramp-length changes imposed on single titin molecules elicited sawtoothlike tension changes. This implied a one-by-one unfolding of the filament's tandem immunoglobulinlike domains—each unfolding event giving rise to one "tooth" in the sawtooth-tension waveform. The length change occurs as each immunoglobulin "beta-barrel" domain gives way to an extended random structure (Fig. 9). Many such unfoldings lengthen the filament in steplike fashion.

In a conceptually similar experiment [Tskhovrebova et al., 1997], single titin molecules were stretched and held. During the holding phase, the decline of tension (stress-relaxation) occurred in stepwise fashion, not smoothly. Again, the implication is a progression of discrete unfolding events, each one dropping the tension by an increment.

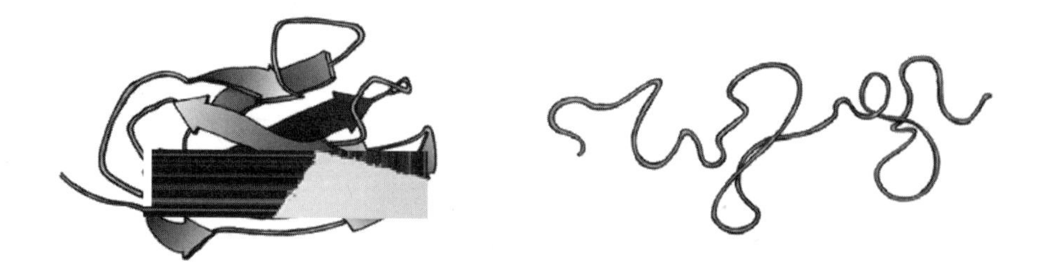

Figure 9: Unfolding of immunoglobulin domain (left to right) results in discrete length change. Unfolding of successive domains yields a large-scale length change of the connecting filament.

Stepwise length changes are also observed under similar conditions in sarcomeres [Yang et al., 1998]. These experiments interrogated the single myofibril, the smallest functional unit that retains muscle's natural structure (see Fig. 3). When the unactivated myofibril is stretched or released by a linear ramp, the sarcomere-length change is stepwise. An example is shown in Fig. 10.

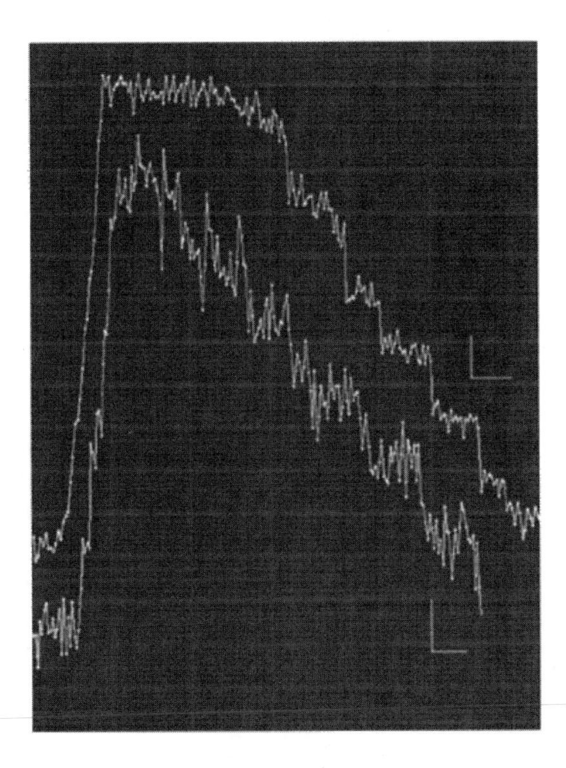

Figure 10: Time course of sarcomere-length change during smooth trapezoidal length change imposed on single isolated myofibril. Calibration bars: 20 nm; 1 sec.

Myofibril sarcomere steps reveal that discrete behavior persists at levels of organization considerably higher than that of the single molecule, for the myofibrillar sarcomere contains hundreds of filaments in parallel. In fact, discrete behavior remains detectable all the way up to the cellular level [Granzier et al., 1985], reinforcing the highly cooperative nature of the step.

The step's character implies a phase transition. First, the onset and termination are abrupt—distinct states appear to be involved. Second, the step is not an isolated event: the fact that it occurs over many different molecules at the same time implies a quasi-all-or-none behavior, which again is characteristic of a phase transition. The simplest interpretation is that the phase-transition progresses collectively along all parallel connecting filaments, shortening or lengthening by one domain at a time.

2.4.2 Thick Filaments

The thick filament is built of successive repeats of the myosin molecule. Myosin contains a long alpha-helical rodlike tail, most of which lies in the thick filament backbone (extensions of the angled lines in Fig. 2); it also contains two globular heads, or cross-bridges, which radiate from the filament backbone at nominally right angles. The rodlike tails overlap in a staggered, helical pattern. The filament is mirror-symmetrical about its midpoint.

Although it is generally accepted that thick filaments do not change length during contraction, the evidence is mixed. Classical studies imply a constancy of filament length under most conditions. But some thirty-plus papers published since those classical studies report substantial shortening during contraction [for review, cf. Pollack, 1983]. These studies were conducted on vertebrate heart and skeletal muscles as well as invertebrate muscles; they employed techniques ranging from electron microscopy through various types of optical microscopy, and they included isolated thick filaments extracted from the muscle and exposed to physiological activating agents. The evidence for thick-filament shortening is therefore appreciable.

A way in which thick filaments might shorten is if myosin rods were able to slide past one another. The filament could then "telescope" to a shorter length. The driving force for such telescopic action would lie in the filament itself, presumably in the myosin rods that comprise the backbone. In this context it has been shown in many experiments that the alpha-helical rod is able to shorten by a helix-coil transition [for review, cf. Pollack, 1990]. The helix-coil transition is a phase transition in which the molecular structure undergoes radical change. It is a force-producing process: because the equilibrium length of the random coil is near zero, the coil will always wants to shorten from its extended length. The retractive force depends on the degree of extension, much like a spring. Thus, the transition shortens the myosin molecule in much the same way as a wool sweater is shortened by excess heat. Here, however, the transition is reversible.

A model illustrating how localized shortening in a myosin rod could generate rod sliding is shown in Fig. 11. Adjacent molecules are held together by cohesive forces derived from molecular surface charges that alternate semi-regularly along the surface. The unstable zone of one molecule lying near the filament's midpoint undergoes a helix-coil transition. (The transitions would occur symmetrically about the filament's midpoint.) The transition results in local shortening, which draws the remainder of the filament toward the mid-zone. Then the same shortening event occurs in the next molecule along the filament, and the next, etc. In such a way the filament shortens step by step, on either side of the midpoint. And because molecules are cross-linked to respective molecules on the next filament in parallel, the transition is cooperative over the muscle cross section.

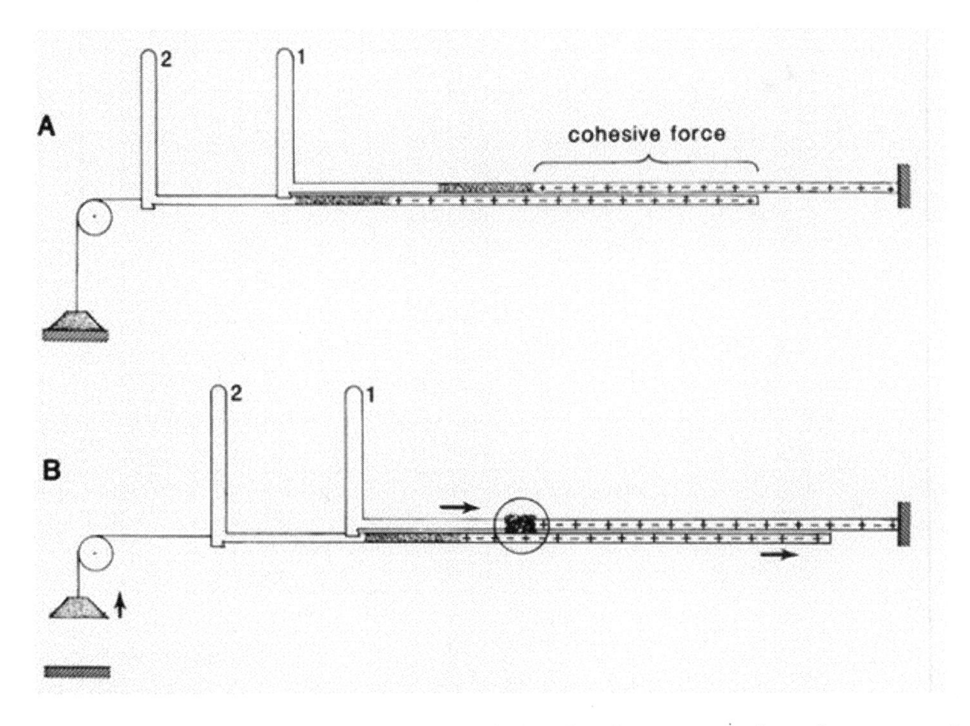

Figure 11: Mechanism by which local shortening in one myosin rod can propel the adjacent rod to slide. Only two of the many molecules are shown.

This mechanism is in good agreement with structural evidence from x-ray diffraction [Huxley and Brown, 1967; Yagi and Matsubara, 1984]. The x-ray patterns show that during active sarcomere shortening the spacing between molecules along the thick filament does not change; only the x-ray intensity changes. This is precisely what is anticipated. The x-ray pattern is dominated by contributions from those molecules along the thick filament that remain regularly arrayed and have not yet shortened; others have shortened by variable amounts and do not contribute significantly. As the number of transitioned molecules increases, their intensity contribution therefore diminishes, explaining the observed x-ray intensity diminution. The model is therefore in good agreement with ultrastructural and x-ray diffraction evidence.

If thick filaments shorten during contraction, the likelihood, then, is that such shortening occurs in steps, one molecule at a time in each half-filament. If the thin filament is bound to the cross-bridges during these thick filament-shortening steps, the sarcomere will likewise shorten in steps. In such a way the myosin phase transition contributes to the shortening of the sarcomere.

2.4.3 Thin Filaments

Thin filaments may bring about stepwise length changes as well. Unlike the thick and connecting filaments, whose shortening can directly shorten the sarcomere, thin filaments are differently situated (Fig. 1). Because of their arrangement, they would need to facilitate contraction in a different way. One potential mechanism is an inchwormlike process, similar to that reported in PAMPS gels [Osada and Gong, 1993]. In that mechanism the cylindrical gel undergoes repeated cycles of curling and straightening. The ends of the gel are hooked to a ratchet in such a way that each bending cycle results in a step advance. Repeated cycling advances the gel by a significant magnitude.

The same principle could apply in the actin filament. If the filament were to pass through cycles of curling and straightening, such action could be harnessed to propel the filament toward the center of the sarcomere. More realistically, any such curling would arise from a local phase-change, which would then propagate along the filament to produce a wormlike reptation. Each cycle of propagation would result in a step of filament translation. The extent of translation would be a function of the number of cycles. A schematic illustration of such an inchworm process is shown in Fig. 12.

The wavelike motion that would be anticipated in the actin filament is broadly observed [for review, cf. Pollack, 1996]. The evidence draws from as early as four decades ago when undulations were directly observed to propagate along actin-filament bundles responsible for active streaming. It also follows from modern studies of single actin filaments. Such motion could be generated by a local molecular structural change, which is observed in the electron microscope [Menetret et al., 1991] and in actin crystals [Schutt and Lindberg, 1992]. That structural changes can propagate along the filament is supported by the observation that the binding of a ligand to one end of the actin filament affects the physical and mechanical properties of the entire filament [Prochniewicz et al., 1996]. Indeed, direct microscopic visualization of actin filaments shows a translation pattern quite strikingly characteristic of snakelike motion [deBeer et al., 1997].

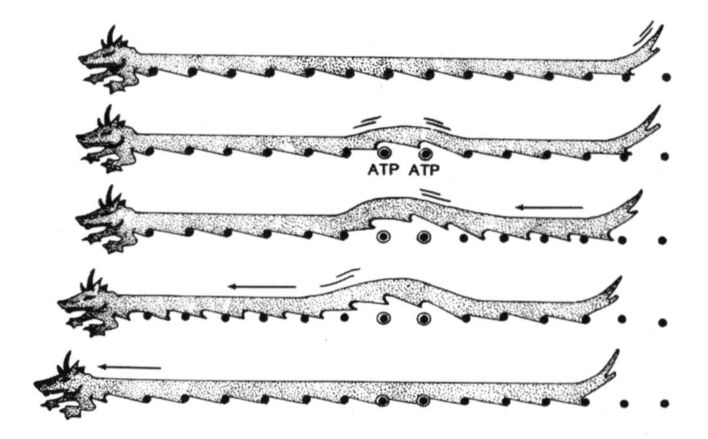

Figure 12: Reptation model explaining translation of thin filament over thick filament. Black dots represent myosin cross-bridges. In this model a phase-transition propagates from tail to head, advancing the thin filament by one notch.

Evidence for inchwormlike behavior is also inferred from dynamics of the intact sarcomere, as shown below. During active sliding of thin filaments past thick, the pattern of sarcomere shortening is stepwise. The measurement is made by projecting the striated image of the myofibril onto a photodiode array, and scanning the array regularly to track individual sarcomere length. Step sizes were measured initially in Blyakhman et al. [1999], and an updated but similar result is shown in the histogram of Fig. 13. The histogram shows multiple peaks, indicating steps of discrete size. Although interpeak spacing is not precisely uniform, there is a strong tendency for peaks to repeat at regular intervals. The best-fit spacing between peaks is 2.7 nm, indicating a step size that is an integer multiple of 2.7 nm. This value is equal to the linear advance of actin monomers along the thin filament.

Such "quantal" behavior is inevitable if an actin monomer is bound to the (immobile) cross-bridge between translation steps; filament translation steps must then be $n \times$ the actin-monomer advance (Fig. 14). That actin and myosin bind to one another is well recognized, and it is presumed that such binding is responsible for the sustenance of active tension. Successive bindings give rise to the inevitable quantum step. The size of the quantum, 2.7 nm, follows the expectation of the inchwormlike mechanism.

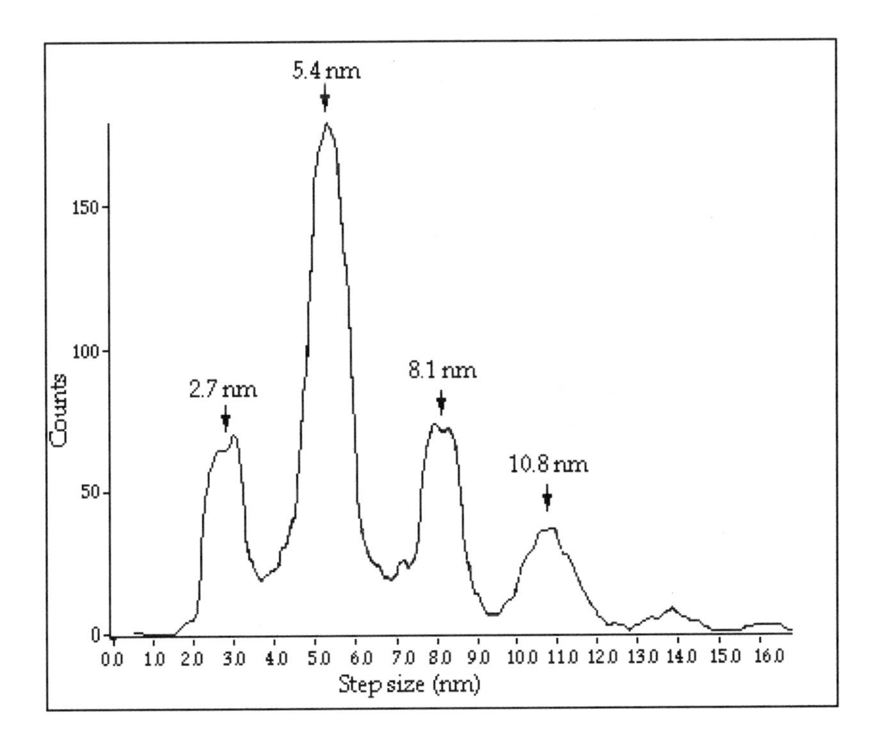

Figure 13: Histogram of step size measured during active contraction of single isolated myofibrils. Spacing between major peaks is 2.7 nm [Yakovenko et al., 2002]. (Courtesy of the American Physiological Society.)

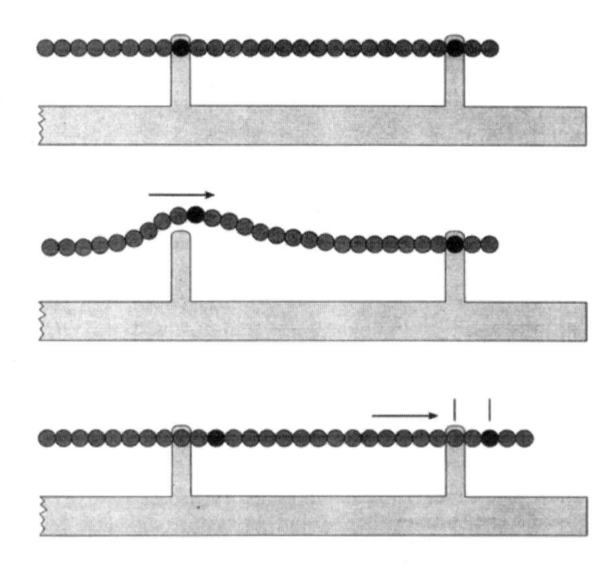

Figure 14: Inchworm advance. The model predicts that the size of the step advance will be an integer multiple of the actin monomer spacing.

Of particular interest is the fact that these active contraction steps are preserved at higher levels of organization. Not only do the steps appear in the single sarcomere, they appear as well at the level of the single cell [Pollack et al., 1977]. If the process is considered as a phase transition, the transition is, again, highly cooperative.

2.5 Lessons from the Natural Muscle System That May Be Useful for the Design of Polymer Actuators

There is ample evidence that the filamentous polymers of the sarcomere—all three filamentous polymers—shorten in a discrete, cooperative manner. If so, muscle contraction very much resembles phase transitions that occur in synthetic polymeric systems. The biological transition may, however, be a lot "smarter" because of the sophistication of its responsiveness. Thus, the connecting filament shortens in the absence of any activation; it behaves as a discrete elastic band. The actin filament undergoes transition after the level of activation crosses threshold. Because the actin filament, but not the myosin filament, is found in relatively simple motile cells, the actin mechanism may be more primitive. Like other primitive mechanisms, it is limited in its capacity and unable to function under too high a load—just as a weight hung on a caterpillar's tail can inhibit upward crawling, even though the caterpillar may still cling. Beyond this critical load, the operative agent is the thick filament, which can shorten under the highest of loads. Indeed, the helix-coil transition has been demonstrated to produce a force that accounts quantitatively for the maximum force muscle can produce [Harrington, 1979]. Thus, the muscle polymer gel capitalizes on the strengths of all three proteins. This is perhaps why muscle, a highly evolved organelle, is as versatile as it is.

One upshot of these considerations is that the mechanism in muscle contraction is not as different from the mechanisms used to construct artificial muscles as had been thought. Both employ polymer-gel phase transitions. Like the polymer gel, the muscle can be triggered by any variety of stimuli ranging from a change of pH, salt, temperature, solvent type, and even electrical current. Therefore, it may be possible to construct artificial muscles with greater speed and efficacy by looking more closely at how nature accomplishes the task of biological contraction.

This can be approached in several ways. One way is to use gels constructed of polymers having some of the same features as muscle proteins. It may be possible, for example, to construct simple gels of only one of the three protein filaments. These polymers could be arranged randomly, or perhaps cross-linked to one another along their length in a parallel arrangement. The parallel, cross-linked arrangement may be particularly advantageous for achieving high speed: cooperativity may be enhanced because the linked units must act in unison, or not at all. Muscle can contract very rapidly. A common difference between natural and artificial muscle is that natural muscle filaments (as well as other biological filaments) are generally parallel and cross-linked, whereas artificial gels are typically random. It may be that the parallel arrangement is a critical factor for high speed.

Cross-linking may also be critical. Muscle filaments are hydrophilic and tend to adsorb water. It is common experience that muscle water is difficult to remove, even by centrifugation. In the absence of cross-linking, muscle-filament polymers will swell, much like the gel of Fig. 6. Cross-links keep filaments closely packed and unable to swell appreciably. The cross-linked matrix therefore has more potential energy than the uncross-linked matrix, and may use some of this chemical potential to contract more powerfully.

A related consideration is size. One conspicuous feature of muscle is that it is built of many parallel units, or myofibrils (Fig. 1). Myofibril diameter is consistently 1–2 μm, rarely more. This diminutive structure is surrounded by its own activating system, called the sarcoplasmic reticulum, which supplies the activating signal. Thus, diffusion distances are only 1–2 μm in natural muscle, which presumably contributes to high speed. Emulation of this kind of arrangement could result in higher contraction speed in artificial actuators.

As for the type of polymer that might be most effective, again, copying nature's design could lead to materials more responsive than those used to date. For high force applications, the choice may be a myosinlike polymer. The myosin melt force, as mentioned, is sufficient to account for full muscle tension. For small-scale snakelike movements, the actin-based gel may be a better choice. For bending, nature often employs microtubules, built of the protein tubulin. The axostyle, for example, consists of a bundle of parallel microtubules cross-linked along their length. Bending of the axostyle is frequently associated with shortening of microtubules on one edge relative to the other [McIntosh, 1973], like a bimetal strip. The challenge is to synthesize polymers that have many of the features of these proteins—or, to use the proteins themselves.

These are a few directions implied by the success of Mother Nature. It may be that artificial actuators could ultimately produce movement comparable to what nature has had almost four billion years to produce.

2.6 References

Baatsen, P. H. W. W., K. Trombitas, and G. H. Pollack, "Thick filaments of striated muscle are laterally interconnected," *J. Ultras. & Mol. Str. Res. 97*, 39-49, 1988.

Blyakhman, F., T. Shklyar, G. H. Pollack, "Quantal length changes in single contracting sarcomeres," *J. Mus Res. Cell Motility, 20*, 529-538, 1999.

Cooke, R., "Actomyosin interaction in striated muscle," *Physiol. Rev.* 77(3): 671-697, 1997.

deBeer, E. L., A. M. A T. A. Sontrop, M. S. Z. Kellermayer, G. H. Pollack, "Actin-filament motion in the in vitro motility assay has a periodic component," *Cell Motil. Cytoskel. 38*, 341-50, 1997.

Granzier, H. L. M., and G. H. Pollack, "Stepwise shortening in unstimulated frog skeletal muscle fibres," *J. Physiol. 362*, 173-88, 1985.

Harrington, W. F., "On the origin of the contractile force in skeletal muscle," *Proc. Nat'l. Acad. Sci. 76*, 5066-70, 1979.

Hoffman, A. S., "Conventional and environmentally-sensitive hydrogels for medical and industrial uses: a review paper," *Polymer Gels* 268(5): 82-87, 1991.

Hoyle, G., *Muscles and their Neural Control*, Wiley, N. Y. 1983.

Huxley, A. F., "Muscle structure and theories of contraction," *Prog. Biophys. Biophys. Chem.* 7: 255-318, 1957.

Huxley, A. F., and R. Niedergerke, "Structural changes in muscle during contraction: Interference microscopy of living muscle fibres," *Nature* 173: 971-973, 1954.

Huxley, A. F., and R. Niedergerke, "Measurement of the striations of isolated muscle fibres with the interference microscope," *J. Physiol.* 144: 403-425, 1958.

Huxley, H. E., "A personal view of muscle and motility mechanisms," *Ann. Rev. Physiol.* 58: 1-19, 1996.

Huxley, H. E., and J. Hanson, "Changes in the cross-striations of muscle during contraction and stretch and their structural interpretation," *Nature* 173: 973-976, 1954.

Huxley, H. E., and W. Brown, "The low angle X-ray diagram of vertebrate striated muscle and its behaviour during contraction and rigor," *J. Mol. Biol. 39*, 383-434, 1967.

Iwazumi, T, "A new field theory of muscle contraction," Ph. D. Dissertation, Univ. of Pennsylvania, 1970.

Katayama, E., "Quick-freeze deep-etch electron microscopy of the actin - heavy meromyosin complex during the *in vitro* motility assay," *J. Mol. Biol.* 278: 349-367, 1998.

McIntosh, J. R., "The axostyle of Saccinobaculus. II. Motion of the microtubule bundle and a structural comparison of straight and bent axostyles," *J. Cell Biol.* 56: 324-339, 1973.

Menetret, J. F., W. Hoffman, R. R. Schroder, R. Gapp, R. S. Goody, "Time-resolved cryo-electron microscopic study of the dissociation of actomyosin induced by photolysis of photolabile nucleotides," *J. Mol. Biol. 219*, 139-44, 1991.

Oplatka, A., "The rise, decline, and fall of the swinging crossbridge dogma," *Chemtracts Bioch. Mol. Biol. 6*, 18-60, 1996.

Oplatka, A, "Critical review of the swinging cross-bridge theory and of the cardinal active role of water in muscle contraction," *Crit. Rev. Biochem. Mol. Biol. 32*, 307-60, 1997.

Osada, Y. and J. Gong, "Stimuli-responsive polymer gels and their application to chemomechanical systems," *Prog. Polym. Sci. 18*, 187, 1993.

Pollack, G. H., T. Iwazumi, H. E. D. J. ter Keurs, and E. F. Shibata, "Sarcomere shortening in striated muscle occurs in stepwise fashion," *Nature 268*, 757-9, 1977.

Pollack, Gerald H., *Cells, Gels, and the Engines of Life*, Ebner and Sons, Seattle, 2001.

Pollack, G. H., *Muscle and Molecules: Uncovering the Principles of Biological Motion,* Ebner and Sons, Seattle, 1990.

Pollack, G. H., "Phase transitions and the molecular mechanism of contraction," *Biophys. Chem. 59*, 315-28, 1996.

Pollack, G. H., "The cross-bridge theory," *Physiol. Reviews 63*, 1049-113, 1983.

Prochniewicz, E., Q. Zhang, P. A. Janmey, D. D. Thomas, "Cooperativity in F-actin: Binding of gelsolin at the barbed end affects structure and dynamics of the whole filament," *J. Mol. Biol. 260*, 756-66, 1996.

Rief, M., M. Gautel, F. Oesterhelt, J. M. Fernandez, and H. E. Gaub, "Reversible unfolding of individual titin immunoglobulin domains by AFM," *Science 276*, 1109-12, 1997.

Schutt, C. E. and U. Lindberg, "A new perspective on muscle contraction," *FEBS Lett.* 325: 59-62, 1993.

Schutt, C. E. and U. Lindberg, "Actin as a generator of tension during muscle contraction," *Proc. Nat'l. Acad. Sci. 89*, 319-23, 1992.

Schutt, C. E. and U. Lindberg, "Muscle contraction as a Markov process I:enertetics of the process," *Acta Physiol. Scan.* 163: 307-324, 1998.

Spudich, J. A., "How molecular motors work," *Nature* 372: 515-518, 1994.

Tanaka, T., "Gels," *Sci. Amer.* 244, 110, 1981.

Thomas, D. D, "Spectroscopic probes of muscle cross-bridge rotation," *Ann Rev. Physiol.* 49: 641-709, 1987.

Trombitas, K. P. H. W. W. Baatsen, and G. H. Pollack, "I-bands of striated muscle contain lateral struts," *J. Ultras. & Mol. Str. Res. 100*, 13-30, 1988.

Tskhovrebova, L., J. Trinick, J. A. Sleep, and R. M. Simmons, "Elasticity and unfolding of single molucules of the giant muscle protein titin," *Nature 387*, 308-12, 1997.

Yagi, N. and I. Matsubara, "Cross-bridge movements during slow length change of active muscle," *Biophys. J. 45*, 611-4, 1984.

Yakovenko, O., Blyakhman, F. and Pollack, G. H., "Fundamental step size in single cardiac and skeletal sarcomeres," *Am J. Physiol (Cell)* 283(3): C735-C743, 2002.

Yang, P., T. Tameyasu, and G. H. Pollack, "Stepwise dynamics of connecting filaments measured in single myofibrillar sarcomeres," *Biophys. J. 74*, 1473-83, 1998.

CHAPTER 3

Metrics of Natural Muscle Function

Robert J. Full
University of California at Berkeley

Kenneth Meijer
Technische Universiteit Eindhoven/Universiteit Maastricht (The Netherlands)

3.1 Caution about Copying and Comparisons

Natural muscle is a spectacular actuator. Why? After millions of years, nature has evolved actuators that allow breathtaking performances. Cheetahs can run, dolphins can swim, and flies can fly like no artificial technology can. It is often argued that if human technology could mimic muscle, then biologicallike performance would follow. Unfortunately, the blind *copying* or mimicking of a part of nature [Ritzmann et al., 2000] does not often lead to the best design, for a host of reasons [Vogel, 1998]. Evolution works on the "just good enough" principle. Optimal designs are not the necessary end product of evolution. Multiple satisfactory solutions can result in similar performances. Animals do bring to our attention amazing designs, but these designs carry with them the baggage of their history. Why should these historical vestiges be incorporated into an artificial technology? Moreover, muscle design is constrained by factors that may have no relationship to human-engineered designs. Muscles must be able to grow over time, but still function along the way. Muscles remain plastic in adulthood and can self-repair. Muscles are intimately tied to pressure in the fluid system that supports them. Muscles are involved in metabolic regulation and can even serve as a source of fuel in starvation. Finally, muscles are obviously not the only part of an animal that makes spectacular performances possible. We must understand what muscle uniquely contributes to an integrated, tuned system that includes multiple muscles, joints and sensors, a transport system for fuel delivery, and a complex control system, all of which functions through skeletal scaffolding.

To design an artificial muscle is a worthy endeavor. However, we strongly urge that nature's technologies provide biological inspiration for artificial technologies. Biological inspiration should involve the transfer of principles or lessons discovered in a diversity of animals. Our knowledge of biological muscle should be able to assist us in the construction of an actuator with desired performance capacities only observed in animals. However, the performance of biological actuators should not be and has not been the single design by which we measure our success. We have and will continue to design human-made actuators that exceed natural muscle in performance in particular metrics and for specific tasks.

If we are to call a human-made actuator an artificial muscle, we must detail precisely the tasks that uniquely define what muscles do. Metrics can best be *compared* under common conditions. To develop these appropriate tests is an ongoing challenge, because we are still discovering how muscles work in animals. Moreover, engineers have a multitude of metrics that have made relevant, direct comparisons nearly impossible. The design of an artificial muscle will require novel interdisciplinary collaborations between muscle biologists and engineers. Biologists can provide inspiration and detail about what is known at present, but engineers can reciprocate with quantitative hypotheses and novel instrumentation that will lead to new tests and discoveries of muscle function.

With these important caveats declared, this chapter serves as a guide through some of the classical metrics of muscle function that can allow comparisons.

3.2 Common Characterizations—Partial Picture

3.2.1 Maximum Isometric Force Production Depends on the Level of Neural Activation

The maximum force an active muscle can generate is most often measured at a set length when muscles are not allowed to shorten. This type of contraction is termed isometric. The force generated by muscle during an isometric contraction is a function of neural activation (Fig. 1). Single neural stimuli produce small, transitory increases in force called twitches. Consecutive stimuli in the form of a train of neural spikes lead to a summation in force. Maximum, sustained isometric force can be attained at the greatest stimulation frequencies when the muscle is in tetanus. Maximum force increases with the cross-sectional area of the muscle stimulated. Values of maximum isometric stress vary by over 100-fold and range from 0.7 to 80 Ncm^{-2} [Full, 1997; Josephson, 1993]. In general, force development tends to be the greatest in the slowest muscles.

3.2.2 Rate of Force Production and Relaxation Varies Among Muscles

The rates of muscle force generation and relaxation are important because they reveal possible limitations to duty cycles required for rhythmic activity. Slow kinetics can limit behavior if antagonistic muscles co-contract. Muscles operating in escape or producing small strain vibrations such as the insect flight muscle and the rattlesnake shaker muscle demonstrate the fastest kinetics [Rome, 1998]. Reported contraction times to peak force vary by over 200-fold. Values range from 0.004 to 0.79 sec, but the lack of standardization of conditions makes direct comparison difficult [Full, 1997]. Time to 50% relaxation varies by over 100-fold and ranges from 0.009 to as long as 1.1 sec.

3.2.3 Maximum Isometric Force Depends on Muscle Length

The magnitude of a muscle's isometric force development depends on the length at which it is set (Fig. 2). Active muscle force is reduced at the shortest and longest muscle lengths because of the filamentous nature of the contractile structure. Maximum, active isometric force is attained at intermediate muscle lengths. Animals tend to operate on the ascending part of the active force-length curve. It is important to note that passive muscle force increases curvilinearly at longer muscle lengths where active force declines. As a consequence, total muscle force can increase at longer muscle lengths.

Active force-length curves vary in the degree to which force decreases with changes in length (Fig. 3). The largest differences are seen when flight muscle is compared to the body-wall muscle of animals that crawl using hydrostatic skeletons, like insect larvae [Full, 1997]. Flight muscle can only generate maximum force over a very narrow range of strain (2 to 4%) as these muscles appear to simply vibrate. By contrast, body-wall musculature of soft-bodied animals operates over a large range of length changes (200%) corresponding with the considerable shape changes in these species.

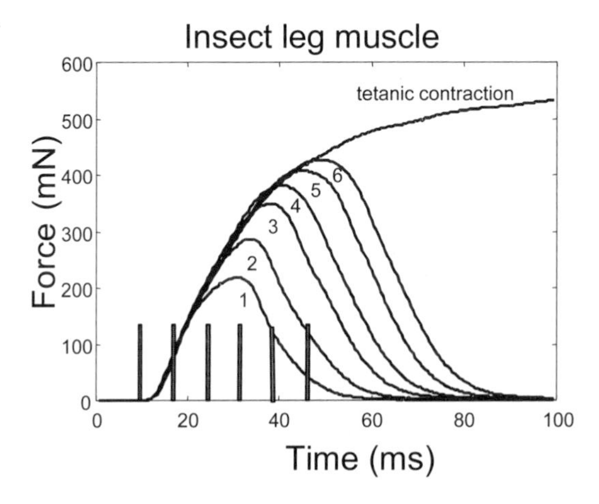

Figure 1: Force produced by activation of a muscle. Force production of a cockroach muscle (muscle #177c) increases as a function of the number of input pulses.

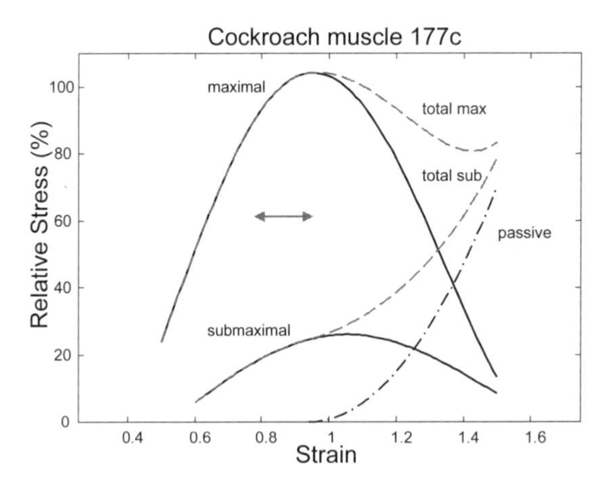

Figure 2: Strain-stress characteristics of maximally and submaximally activated cockroach muscle (#177c; represented by continuous lines) as well as the passive length-force characteristics (dashed-dotted line). Total muscle force is equal to the sum of active and passive force (represented by gray dashed lines). Stress is normalized to the peak isometric tension. Strain is normalized as a fraction of the length that gives peak isometric stress. The arrow indicates range of strains where muscle functions under natural conditions.

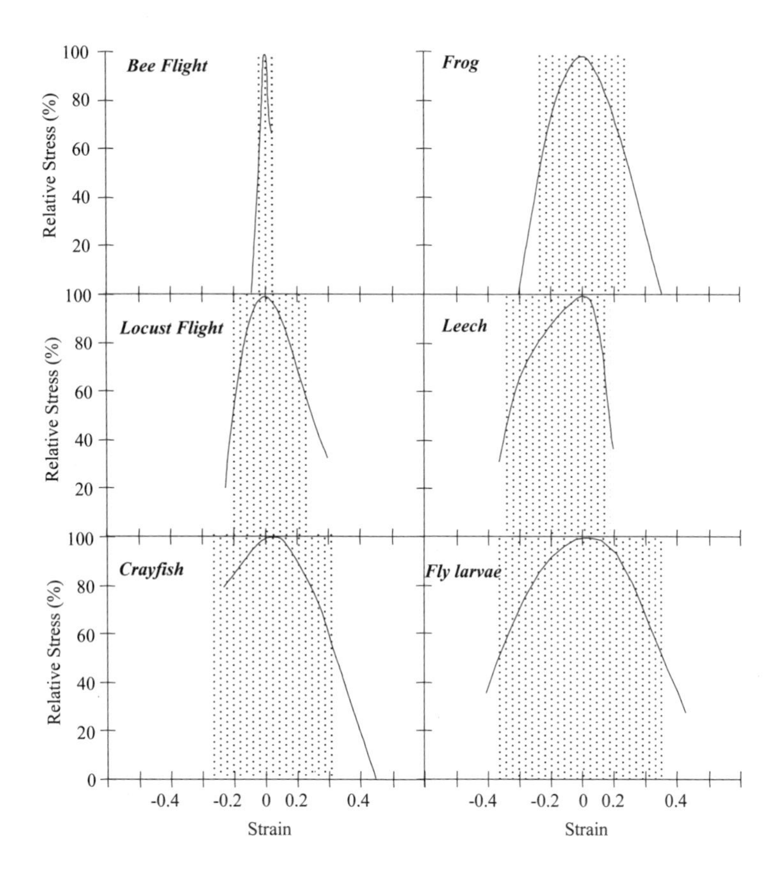

Figure 3: Active length-force or strain-stress curves. Stress is normalized to the peak isometric tension. Strain is normalized as a fraction of the length that gives peak isometric stress. Shaded areas represent strains that correspond to stresses above 50% maximum isometric stress. Data include bee flight, locust flight, crayfish, frog, leech, and fly larvae muscle [Full, 1997].

3.2.4 Force Development Decreases with an Increase in Shortening Velocity

The faster a muscle shortens, the less force it can develop (Fig. 4). Said in another way, when muscles have larger loads, they contract more slowly. Single contractions of maximally activated muscle are studied by measuring either force or velocity, while holding the other constant. These are referred to as isotonic or isovelocity contractions. Data from frog, rat, turtle, fish, locust, beetle, and scallop muscle all show the characteristic rectangular hyperbolic shape with force decreasing with an increase in velocity [Full, 1997; Rome and Lindstedt, 1997; Josephson, 1993]. The maximum shortening velocity of muscles varies both within an individual and among species. Maximum shortening velocity varies by over 50-fold, ranging from 0.3 to 17.0 lengths/sec depending on the slowest muscle of a hydrostat selected for inclusion [Full, 1997; Josephson, 1993]. The force-velocity relationship for lengthening contractions is less well defined [Josephson, 1999] (Fig. 4). However, when muscle is stretched, the force developed often exceeds that of maximum isometric force by nearly twofold.

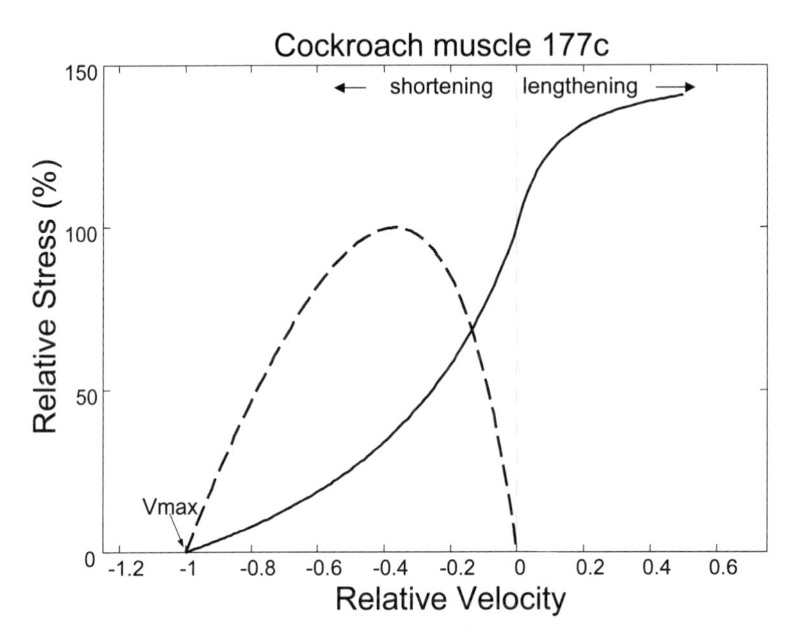

Figure 4: The force-velocity curve for active cockroach muscle (continuous line). Stress is normalized to the peak isometric tension. Velocity is normalized as a fraction of the maximal shortening velocity of the muscle (V_{max} = the highest velocity at which the muscle can still generate force). Force is larger than the isometric force if the muscle is lengthening and lower than the isomeric force when the muscle is shortening. The instantaneous power (force $\times$ velocity) is represented by the dashed line.

3.2.5 "Blocked Stress" Calculations Overestimate a Muscle's Capacity to do Work

Work and power output of a muscle can be bounded from the metrics thus far mentioned. The product of maximum force and length change can yield an estimate of maximum work output [Wax and Sands, 1999]. This blocked stress value can be misleading for several reasons. First, the maximum force produced by muscle does not necessarily occur in the same muscle that produces the largest changes in length. Second, because force is a function of velocity, maximum work and power output for single contractions occur at approximately one-third of maximum velocity. Third, although work and power output for single contractions can be estimated from the force-velocity relationship for a single contraction, they overestimate the work and power of muscles operating in animals [Josephson, 1989]. Muscle physiologists sometimes refer to these estimates as instantaneous work and power. Because most activities involve cyclic contractions, muscle is inactive for nearly one half a cycle. Instantaneous work and power estimates can exceed what muscles are capable of by twofold.

Next we discuss the method now used by muscle physiologists interested in understanding how muscles function during activity in animals. Examining rhythmic activities has shown not only that we need to be cautious about our comparisons of work and power, but that discoveries have revealed that muscle has more diverse roles.

3.3 Work-Loop Method Reveals Diverse Roles of Muscle Function during Rhythmic Activity

Muscles are used by animals in rhythmic activities such as running, swimming, flying, chewing, and communicating. Cycle frequency, level and pattern of neural stimulation, phase of neural stimulation, strain pattern, and magnitude are required to define the workspace for muscles undergoing rhythmic oscillations. A technique referred to as the "work-loop method" allows a muscle physiologist to control each of these parameters. Using this technique, muscle is attached to a moveable lever, subjected to cyclic length changes, and stimulated at a distinct phase in its motion, while muscle force is measured [Josephson, 1985] [Fig. 5(a)]. From muscle force and length changes, a work-loop is created. Net work equals the work done during shortening minus the work done during lengthening [Fig. 5(b)]. If the shortening or positive work exceeds the lengthening or negative work, then energy is generated by the muscle (positive, counterclockwise loop for stress vs. strain). If the positive work is less than the negative work, then energy is absorbed by the muscle (negative, clockwise loop).

Muscle physiologists, thus far, have used the work-loop technique in at least two ways. First, by varying the controlled parameters, we have searched for a muscle's maximum capacity to do work and generate power. Second, we have input the parameters actually used by an animal during an activity. An advantage of the work-loop method is that the stimulation pattern used by the animal can be better approximated [Dickinson et al., 2000]. The stimulation pattern, magnitude and phase selected by an animal during an activity can be determined from electromyograms (EMGs), and then later played back into the motorneuron, stimulating the muscle attached to the lever. In addition, muscle strain can be estimated from joint kinematics or measured directly in a freely moving animal by sonomicrometry [Biewener et al., 1998a, b]. Strain patterns can be programmed into the lever attached to the muscle to mimic length changes measured in the animal during a behavior in nature.

We first discuss the results from work-loop experiments conducted by muscle physiologists searching for maximum power output and the parameters that produce it. Following this, we show examples of how muscles function when we use the parameter values that animals select in nature.

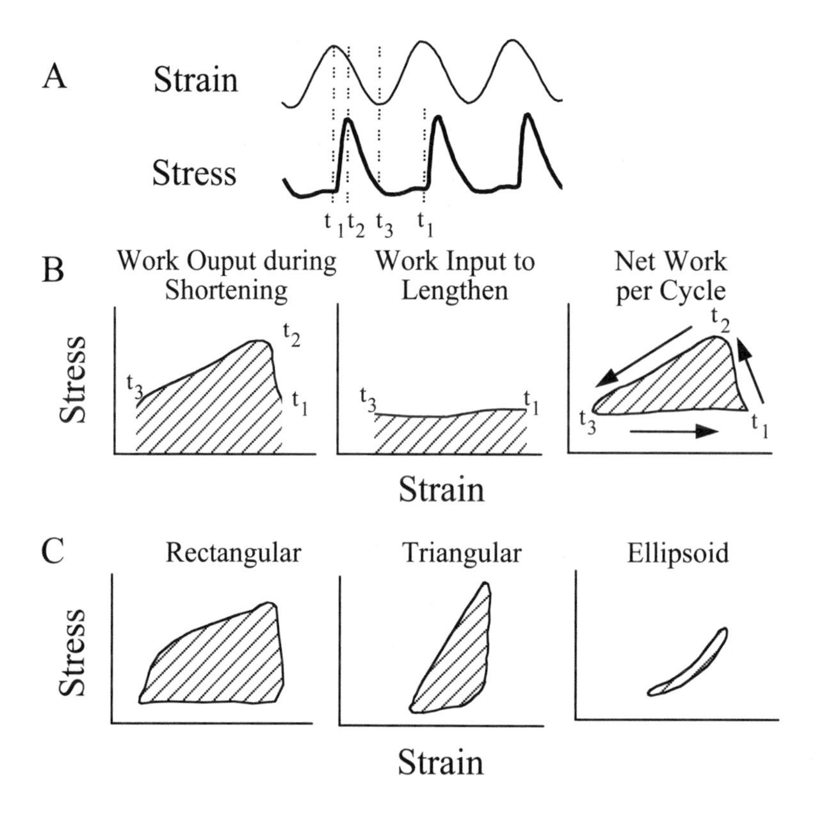

Figure 5: Schematic representation of the work-loop technique. (a) Muscle stress and strain patterns for three cycles. (b) From left to right, the work done by the muscle on its environment, the work done on the muscle and the resulting net work output. (c) Different kind of work loops that are found for different muscles [after Josephson, 1985]. (Courtesy of The Company of Biologists Ltd.)

3.3.1 Work-Loop Method Provides Better Estimate of Maximum Power Output during Rhythmic Activity

Single isotonic or isovelocity contractions do not mimic muscle function in most animals. Propulsive units most often function rhythmically. Instantaneous power output estimates do not consider the frequency of cyclic movement nor do they mimic the neural stimulation pattern in animals. As a result, instantaneous muscle power output (the product of force and velocity from the force-velocity relationship) overestimates power output of a propulsor by two- to threefold because muscles in an animal are usually not generating power for half or more of the cycle. Muscles rarely shorten at a constant force or velocity during activity, but lengthen and shorten in a more sinusoidal fashion [Johnston, 1991; Josephson, 1985]. Muscle force or stress varies with phase of activation, strain, and frequency. Imposing sinusoidal length and velocity changes on muscle reduces maximum power output estimates by as much as 20% compared to estimates assuming linear or constant velocity shortening. A survey of work-loops yielding maximum power output shows that work-loop shape depends on the frequency of operation and seldom results in a shape comparable to the perfect rectangle assumed by a "blocked stress" estimation [Fig. 5(b)].

3.3.2 Mass-Specific Work Output Decreases with Cycle Frequency

From data on oscillatory contractions, muscle mass-specific, work output ranges from 9 to 284 W/kg [Altringham et al., 1993; Askew and Marsh, 1997; Biewener et al., 1998b; Couglin, 2000; Ettema, 1996; Girgenrath and Marsh 1999; Full, 1997; Hammond et al., 1998; James et al., 1995, 1996; Prilutsky et al., 1996; Rome et al., 1999; Swoap et al., 1993]. Mass-specific muscle work per cycle that yields maximum power output decreases with an increase in frequency (Fig. 6). The decrease in work results from a decrease in stress (five-fold for a 100-fold change in frequency) and strain (six-fold decrease; Fig. 7). Strain rate at maximum power output increases by 16 fold with a 100-fold increase in muscle cycle frequency. In addition, the shape of the work-loop changes [Fig. 5(c)]. Work-loops tend to be more rectangular at low frequencies (< 30 Hz). Higher forces are attained by using multiple stimuli. Work-loops are more triangular shaped at intermediate frequencies (~30–60 Hz). The forces measured often result from twitches generated by one or two stimuli per cycle. At the highest frequencies (~60–180 Hz) work-loops are ellipsoid and more springlike.

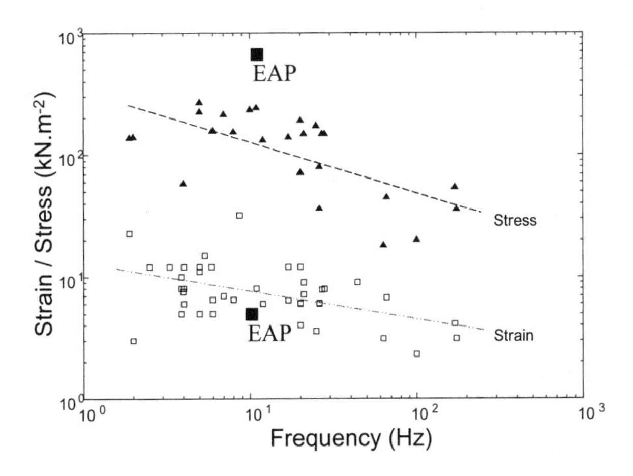

Figure 6: Mass-specific muscle work per cycle as a function of frequency for vertebrate and invertebrate muscles (open circles). Data were obtained using the workloop method [Josephson, 1985]. (Courtesy of The Company of Biologists). Preliminary results show that EAPs fall within the range of values for natural muscle [Full, 1997]. (Courtesy of Oxford University Press.)

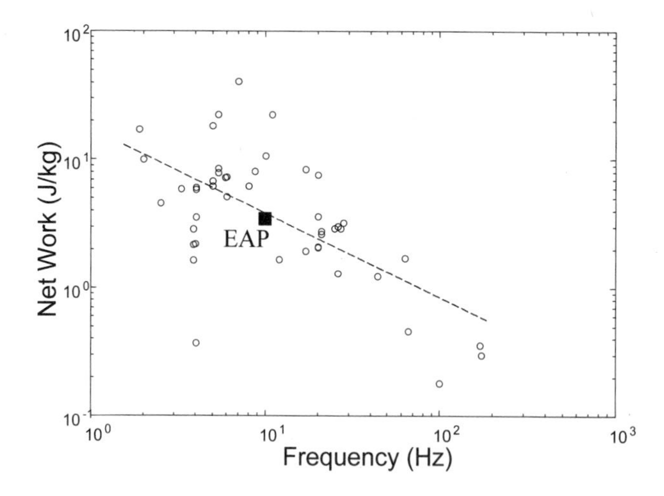

Figure 7: Stress (filled triangles) and strain (open squares) of vertebrate and invertebrate muscle as a function of frequency. Data for EAP muscle is represented by filled square [Full, 1997]. (Courtesy of Oxford University Press.)

3.3.3 Mass-Specific Power Output is Independent of Cycle Frequency

Maximum, mass-specific, power output ranges from 9 to 284 W/kg in animals as diverse as birds, cats, rats, mice, bees, fish, crabs, and scallops [Full, 1997]. Maximal, mass-specific power appears to change little with frequency or body mass (Fig. 8). The decrease in stress, strain and work observed as frequency increases is balanced by the increased frequency, since power output is the product of stress, strain, and frequency.

3.3.4 Work-Loop Method Shows Muscles have a Role in Energy Management and Control

Although a muscle may be capable of generating substantial amounts of power under certain conditions, muscles can play different roles for particular behaviors in nature [Dickinson et al., 2000]. For example, an insect leg muscle is capable of generating power at low strains if stimulated slightly before leg extension. During wedging or pushing underneath an object, the muscle does indeed operate in this space [Full and Ahn 1995]. However, during running, the same muscle is subjected to large strains, produces clockwise work-loops, and therefore, only absorbs energy to control the swing of the leg [Full et al., 1998] (Fig. 9). Similarly, control muscles of flies generate little or no mechanical power, because they function as tunable springs to direct the forces of much larger

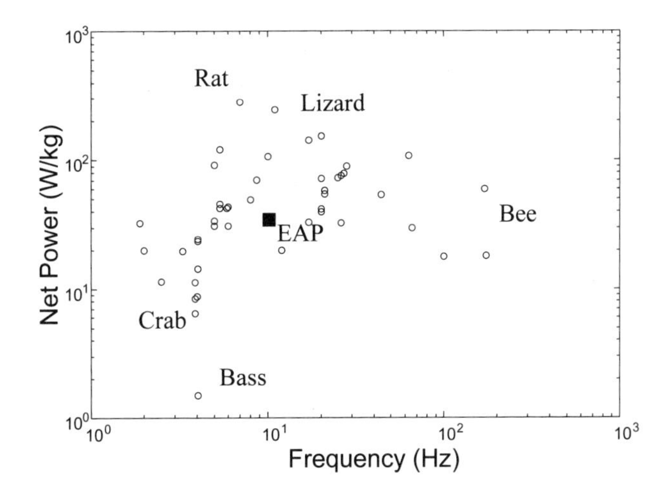

Figure 8: Mass-specific power output of muscles [Full, 1997; Josephson, 1993] (circles) and an electroactive polymer (filled square) as a function of the frequency of oscillation. Data were obtained using the work-loop method [Josephson, 1985]. Preliminary results show that EAPs fall within the range of values for natural muscle. (Courtesy of Oxford University Press, Annual Review of Physiology, and The Company of Biologists.)

power muscles, thus allowing the nervous system to rapidly change wing kinematics by varying the timing of activation. Whereas the large pectoralis muscle of a bird powers flight, another controls joint stiffness, and may modulate wing shape during takeoffs and turns [Dial, 1992]. The axial muscles of fish can serve as either force-generators or energy-transmitters, depending on when they are activated with respect to the undulatory wave that passes along the body. In eels, muscles all along the trunk may contribute to the generation of mechanical power [Gillis, 1998]. In fish that generate the bulk of their hydrodynamic forces using their tail fin, anterior muscles generate energy that is transmitted to the fin through the stiffening action of more posterior muscles.

By measuring both the length and force of a muscle in an active animal at the same time, it is even possible to calculate in vivo work-loops. Direct measurements of muscle length changes in behaving animals have been obtained using sonomicrometry, whereas forces have been measured with strain gauges attached to tendon buckles or to the wing bones of flying birds [Biewener et al., 1998]. In running turkeys and hopping wallabies, muscle fiber length remains nearly constant or even shortens while the tendon stretches [Roberts et al., 1997; Biewener et al., 1998]. Under these conditions, muscles act as struts, permitting the elastic tendons to store and release energy. Examples of multifunctionality and division of labor are likely to become increasingly common as more muscles are measured.

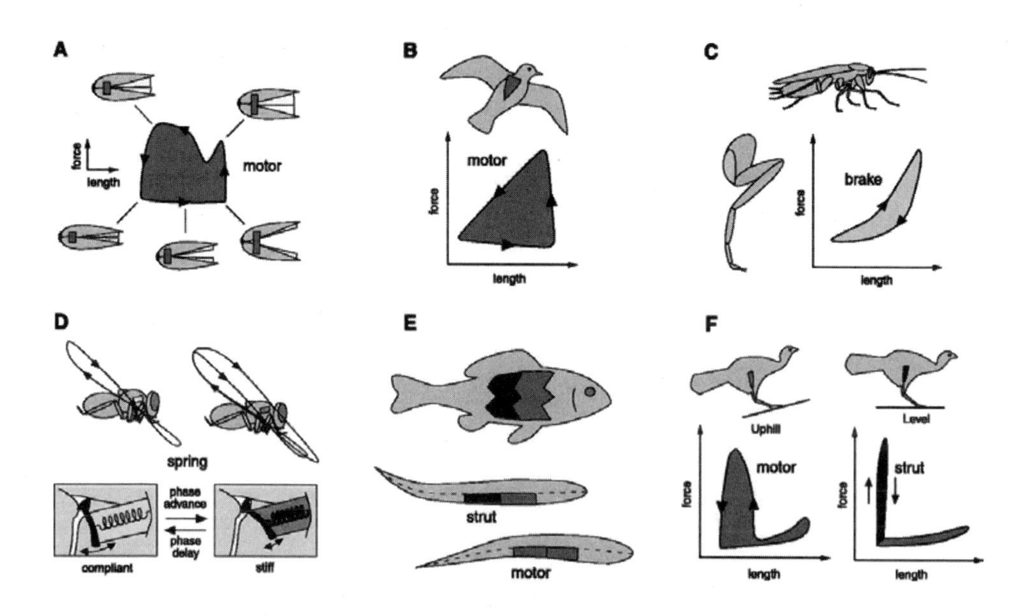

Figure 9: Muscles can act as motors, brakes, springs, and struts. Muscles that generate positive power (motors) during locomotion, and the area within associated work loops, are indicated in red. Muscles that absorb power during locomotion (brakes), and the area within associated work loops, are indicated in blue. Muscles that act as springs of variable stiffness are indicated in green. Muscles that act to transmit the forces (struts) are shown in black. (a) Scallop swimming provides a simple example of a muscle generating positive work to act as a motor [Marsh and Olson, 1994]. (b) The pectoralis muscle of birds generates the positive power required to fly [Biewener et al., 1998b]. (c) In running cockroaches, some muscles that anatomically appear to be suited for shortening and producing power, instead act as brakes and absorb energy because of their large strains [Full et al., 1998]. (d) In flies, an intrinsic wing muscle acts to steer and direct the power produced by the primary flight muscles [Tu and Dickinson, 1996]. (e) In swimming fish, the function of muscles varies within a tail-beat cycle. In some fish designs, early in a beat, the cranial muscle fibers shorten and produce power that is transmitted by more caudal muscle fibers acting as struts. As the beat continues, the fibers that were previously acting as struts change their role to power producing motors. The image at the top shows a fish from the side. Beneath it are shown views from above the fish at two points in the tail-beat cycle [Altringham et al., 1993]. (f) In vivo muscle force and length measurements in running turkeys indicate a dual role for the gastrocnemius muscle. It generates positive power during uphill running, but acts as a strut during level running, which allows the springlike tendons to store and recover energy [Roberts et al., 1997]. (Reproduced with permission from Dickinson et al. [2000]. Courtesy of the American Association of the Advancement of Science.)

3.4 Direct Comparisons of Muscle with Human-Made Actuators

A wide range of materials with musclelike properties are currently available or being developed [Wax and Sands, 1999]. Their properties vary considerably, particularly with respect to the stress, strain, operation frequency, and power output that they can attain. Each material appears to have particular advantages for certain types of activities. Appropriate evaluations of these actuators are most often made with a particular application in mind. In many cases, these evaluations have made a direct comparison to biological muscle difficult. We propose to begin mapping the properties of presumed artificial muscle to the workspace of natural muscle. One way to achieve this goal is to test the proposed artificial muscles in the *same apparatus* with the *same methodology* with which natural muscle is most effectively evaluated. Few studies have compared the mechanical properties of artificial and natural muscles directly [Gonzalez et al., 1997; Klute et al., 1999] and to our knowledge no study has addressed such a comparison from the perspective of muscle function within an animal.

3.4.1 Evaluating Musclelike Properties of Electroactive Polymers (EAPs) Using the Work-Loop Technique

In collaboration with SRI International, we are currently measuring the musclelike properties of electroactive polymer (EAP) actuators [Meijer et al., 1999]. We have examined EAP actuators in the very same experimental apparatus in which we test natural muscle (Fig. 10). The muscle lever system can simultaneously vary actuator length; control the intensity and phase of stimulation; and record position and force. With this system we used the work-loop technique to determine power output.

Figure 10: Acrylic dielectric elastomer actuator (black square area in middle of picture) from SRI International in lever setup used for muscle experiments. The black arrow indicates the length changes that were imposed on the actuator by the muscle lever. Force was recorded at the left tip of the actuator [Meijer et al., 1999]. (Courtesy of Knowledge Press.)

3.4.2 Activation, Stress, and Strain during Cyclic Activity

As in natural muscle, the force production of EAPs varied with the level of stimulation. EAPs, however, require stimulation for relaxation. We have not yet determined the ramifications of this difference with respect to energy consumption nor have we explored the necessary difference in control or connections to skeletal structures. Preliminary results show that the maximum stress attained by EAPs (~80 Ncm^{-2}) falls above the range of values measured for natural muscle. The strain at which the EAPs produced maximal power (2.5%) is at the lower end of the range of values measured for natural muscle (Fig. 7). The EAPs are capable of much higher (> 100%) unloaded strains [Pelrine et al., 2000].

3.4.3 Mass-Specific Power Output of an EAP can Fall within the Range of Natural Muscle

The EAP actuator was capable of both generating and absorbing energy much like natural muscle. When the actuator was stimulated during the lengthening phase of the cycle it overcame the viscoelastic losses of the material and generated power. We obtained a maximal power output of 40 W/kg at a cycle frequency of 10 Hz, a strain of 2.5% and a stimulation voltage of 6 kV. Our preliminary results show that the power output of EAPs falls within the boundary of values for natural muscle when near maximal activation.

3.5 Future Reciprocal Interdisciplinary Collaborations

Natural muscles differ greatly in capacity among animals that crawl, climb, burrow, walk, run, jump, glide, fly, swim, skate, and sing [Full, 1997; Rome and Lindstedt, 1997; Johnston, 1991; Rome, 1998]. Muscles differ both in their fundamental properties as well as how animals use the muscles in nature. Muscles perform in a variety of ways during animal activities. In some cases, muscles and their attachments are primarily force generators used for stabilization and support of limbs and skeletons, allowing for the possibility of springlike function [Biewener 1998]. In other cases, muscles function to produce the rapid movement of limbs or body parts. Most often muscles both generate and absorb energy. These diverse roles of muscle function were discovered using the work-loop technique, but required a directed effort toward understanding how muscles function in the whole animal. To best understand the capabilities of an artificial muscle, it may require building an array of devices within which the artificial muscle may function.

We strongly urge a common standard for direct comparisons between natural and artificial muscle. To call an actuator musclelike, its capabilities should fall within the functional space of natural muscle. No matter what the standard or evaluation technique used for human-made actuators, we recommend more communication between muscle physiologists and engineers in the future.

3.6 Acknowledgments

We thank R. Kornbluh, R. Pelrine, J. Eckerle, and S. V. Shastri from SRI for supplying the EAPs and the expertise to use them. Development of the EAP actuators was done for the U.S. Naval Explosive Ordinance Technology Division supported by the Office of Naval Research under contract N0014-97-C-0352, supported by ONR MURI contract N00014-98-0747 and DARPA-ONR contract N00014-98-1-0669 to RJF. Thanks to Anna Ahn for the use of preliminary data on insect muscle.

3.7 References

Abbot, B. C. and Aubert X. M., "The force exerted by active striated muscle during and after change of length," *J. Physiol.* **117**, pp. 77-86, 1952.

Altringham, J. D., Wardle, C. S., Smith, C. I., "Myotomal muscle function at different locations in the body of a swimming fish," *J. Exp. Biol.* **182** (0), pp. 191-206, 1993.

Altringham, J. D., Young, I S., "Power output and the frequency of oscillatory work in mammalian diaphragm muscle: The effects of animal size," *J. Exp. Biol.* **157**, pp. 381-390, 1991.

Askew, G. N. and Marsh, R. L., "The effects of length trajectory on the mechanical power output of mouse skeletal muscles," *J. Exp. Biol.* **200**, pp. 3119-3131, 1997.

Askew, G. N. and Marsh, R L., "Optimal shortening velocity V/Vmax of skeletal muscle during cyclical contractions: Length-force effects and velocity-dependent activation and deactivation," *J. Exp. Biol.* **201** (10), pp. 1527-1540, 1998.

Biewener, A. A., D. D. Konieczynski, R. V. Baudinette, "In vivo muscle force-length behavior during steady-speed hopping in tammar wallabies," *J. Exp. Biol.* **201** (11), pp. 1681-1694, 1998a.

Biewener, A., "Muscle function in vivo: A comparison of muscles used for elastic energy savings versus muscles used to generate mechanical power," *American Zoologist* **38**, pp. 703-717, 1998.

Biewener, A., W. R. Corning, and B. W. Tobalske, "In vivo pectoralis muscle force-length behavior during level flight in pigeons (Columba livia), *J. Exp. Biol.* **201**, pp. 3293-3307, 1998b.

Coughlin, D. J., "Power production during steady swimming in largemouth bass and rainbow trout," *J. Exp. Biol.* **203**, pp. 617-629, 2000.

Dial, K. P., "Activity patterns of the wing muscles of the pigeon (Columba livia) during different modes of flight, *J. Exp. Zool.*, **262**, pp. 357-373, 1992.

Dickinson, M.H., Farley, C.T., Full, R.J., Koehl, M. A. R., Kram R., and Lehman, S., "How animals move: An integrative view," *Science* **288**, pp. 100-106, 2000.

Ettema, G. J. C., "Mechanical efficiency and efficiency of storage and release of series elastic energy in skeletal muscle during stretch-shorten cycles," *J. Exp. Biol.* **199**, pp. 1983-1997 1996.

Full R. J. and A. N. Ahn, "Static forces and moments generated in the insect leg – comparison of a three-dimensional musculo-skeletal computer model with experimental measurements," *J. Exp. Biol.* **198**, pp. 1285-1298, 1995.

Full, R. J., D. R. Stokes, A. N. Ahn, and R. K. Josephson, "Energy absorption during running by leg muscles in a cockroach," *J. Exp. Biol.* **201**, pp. 997-1012, 1998.

Full, R.J., "Invertebrate locomotor systems," in *The Handbook of Comparative Physiology*, W. Dantzler, ed., pp. 853-930, Oxford University Press, Oxford, 1997.

Gillis, G. B., "Neuromuscular control of anguilliform locomotion: Patterns of red and white muscle activity during swimming in the American eel Anguilla rostrata," *J. Exp. Biol.* **201** (23), pp. 3245-3256, 1998.

Girgenrath M. and Marsh R.L., "Power output of sound-producing muscles in the three frogs *Hyla Versicolor* and *Hyla Chrysoscelis*," *J. Exp. Biol.* **202**, pp. 3225-3257, 1999.

Gonzalez, R., H. Bock, G. Collison, C. Lee, C. Smokowicz and J. Thielman, "Artificial muscles across the human elbow: a performance study of polyacrylonitrile fibers," *Proc. 21st Annual ASB Meeting*, South Carolina, USA, 1997.

Hammond, L., J.D. Altringham and C.S. Wardle, "Myotomal slow muscle function of rainbow trout *Oncorhynchus Mykiss* during steady swimming," *J. Exp. Biol.* **201**, pp. 1659-1671, 1998.

James, R. S., J.D. Altringham, and D.F. Goldspink, "The mechanical properties of fast and slow skeletal muscles of the mouse in relation to their locomotory function," *J. Exp. Biol.* **198**, pp. 491-502, 1995.

James, R.S., I.S. Young, V.M. Cox, D.F. Goldspink and J.D. Altringham, "Isometric and isotonic muscle properties as determinants of work loop power output," *Pflug. Arch.- Eur. J. Physiol.* **432**, pp. 767-774, 1996.

Johnston.I.A., "Muscle action during locomotion - a comparative perspective," *J. Exp. Biol*, **160**, pp. 167-185, 1991.

Josephson, R. K, "Contraction dynamics and power output of skeletal muscle," *Ann. Rev. Physiol.* **55**, pp. 527-546, 1993.

Josephson, R. K., "Dissecting muscle power output," *J. Exp. Biol.* **202**, pp. 3369-3375, 1999.

Josephson, R. K., "Mechanical power output from striated muscle during cyclic contraction," *J. Exp. Biol.* **117**, pp. 493-512, 1985.

Josephson, R. K., "Power output from skeletal muscle during linear and sinusoidal shortening," *J. Exp. Biol.* **147**, pp. 533-537, 1989.

Klute, G. K., J. M. Czerniecki, and B. Hannaford, "Mckibben artificial muscles: pneumatic actuators with biomechanical intelligence," *Proceedings of the IEEE/ASME International Conference on Advanced Intelligent Mechatronics*, Atlanta, USA, 1999.

Marsh, R. L. and J. M. Olson, "Power output of scallop adductor muscle during contractions replicating the in vivo mechanical cycle," *J. Exp. Biol.* **193** (0), pp. 139-156, 1994.

Marsh, R.L., "How muscles deal with real-world loads: the influence of length trajectory on muscle performance," *J. Exp. Biol.* **202**, pp. 3377-3385, 1999.

Meijer, K., S. V. Shastri, and R. J. Full, "Evaluating the musclelike properties of an ElectroActive Polymer actuator," in *Utilization of Electroactive Polymers: Advanced Technologies for the Commercial Development of Novel Applications*, Coronado, CA, USA, 1999.

Pelrine, R., R. Kornbluh, Q. Pei, and J. Joseph, "High-Speed electrically actuated elastomers with strain greater than 100%," *Science* **287**, pp. 836-839, 2000.

Prilutsky, B. I., W. Herzog, and T.L. Allinger," "Mechanical power and work of cat soleus, gastrocnemius and plantaris muscles during locomotion: possible functional significance of muscle design and force patterns," *J. Exp. Biol.* **199**, pp. 801-814, 1996.

Ritzmann, R E., R. D. Quinn, J. T. Watson, and S. N. Zil, "Insect walking and biorobotics: A relationship with mutual benefits," *BioScience* **50** No. 1, pp. 23-33, 2000.

Roberts, T. J., R. L. Marsh, P. G. Weyand, and C. R. Taylor, "Muscular force in running turkeys: The economy of minimizing work," *Science* **275**, pp. 1113-1115, 1997.

Rome, L. C., D.M. Swank, and D.K. Coughlin, "The influence of temperature on power production during swimming. II. Mechanics of red muscle fibers *in vivo*," *J. Exp. Biol.* **202**, pp. 333-345, 1999.

Rome, L.C. and S. L. Lindstedt, "Mechanical and metabolic design of the muscular system in vertebrates," in *The Handbook of Comparative Physiology*, W. Dantzler, ed., Oxford University Press, Oxford, pp. 1587-1652, 1997.

Rome. L. C., "Some advances in integrative muscle physiology," *Comparative Biochemistry and Physiology B* **120**, pp. 51-72, 1998.

Swoap, S. J., Johnson, T. P., Josephson, R. K. and Bennett, A. F., "Temperature, muscle power output and limitations on burst locomotor performance of the lizard *Dipsosaurus Dorsalis*,"*J Exp. Biol.* **174**, pp. 185-197, 1993.

Tu, M. S., Dickinson, M. H., "The control of wing kinematics by two steering muscles of the blowfly (Calliphora vicina)," *J. Comp. Physiol.* A **178** (6), pp. 813-830, 1996.

Vogel S., *Cats' Paws and Catapults*, Norton, New York, p. 382, 1998.

Wax, S. G. and Sands, R.R., "Electroactive polymer actuators and devices," in *Smart Structures and Materials: Electroactive Polymer Actuators and Devices,* Yoseph Bar-Cohen, ed., Proc. SPIE Vol. 3669, pp. 2-10, 1999

Zajac, F.E., "Muscle and tendon: properties, models, scaling, and application to biomechanics and motor control," *Critical Reviews in Biomedical Engineering*, **17**, pp. 359-411, 1989.

Topic 3

EAP Materials

Topic 3.1

Electric EAP

CHAPTER 4

Electric EAP

Qiming Zhang, Cheng Huang, and Feng Xia
The Pennsylvania State University

Ji Su
NASA Langley Research Center

4.1 Introduction

This chapter covers the main features and material examples of the electric field activated electroactive polymers. This class of electroactive polymers (EAPs) is very attractive in performing the energy conversion between the electric and mechanical form and hence can be utilized as both solid-state electromechanical actuators and motion sensors. As will be discussed in the chapter, the electromechanical response in this class of polymers can be linear such as in typical piezoelectric polymers or electrets, or nonlinear such as the electrostrictive polymers and Maxwell stress effect induced response.

Most of the piezoelectric polymers under investigation and in commercial use are based on poled ferroelectric polymers including PVDF-based and nylon-based ferroelectric polymers. This chapter will discuss in detail the properties of these ferroelectric polymers. In comparison with the electromechanical responses in inorganic materials, the electromechanical activity in these polymers is relatively low. In order to significantly improve the electromechanical properties in electric field activated EAPs, new avenues or approaches have to be explored. From the basic material consideration, these approaches include the strain change accompanied with the molecular conformation change, due to the polar vector reorientation, and from the Maxwell stress effect in soft polymer elastomers. This chapter will discuss the recent results based on those approaches, which have produced remarkable improvements in terms of the electric-field-induced strain level, elastic energy density, and electromechanical conversion efficiency in the electric field activated EAPs.

4.2 General Terminology of Electromechanical Effects in Electric EAP

4.2.1 Piezoelectric and Electrostriction Effects

The piezoelectric effect is a linear electromechanical effect where the mechanical strain (*S*) and stress (*T*) are coupled to the electric field (*E*) and displacement (or charge density *D*) linearly, i.e.,

$$S = d\,E, \tag{1a}$$

$$D = d\,T. \tag{1b}$$

In the literature, the effect in Eq. (1a) is often known as the converse piezoelectric effect and in Eq. (1b) as the direct piezoelectric effect. The variable *d* in Eq. (1) is the piezoelectric coefficient. Adding the linear elastic (Hook's law) and dielectric relations to Eq. (1) and writing it out in the full tensor form, we will have the piezoelectric constitutive equations [Nye, 1987; *IEEE Standard on Piezoelectricity*, 1988]:

$$S_{ij} = d_{kij}E_k + s^E_{ijkl}T_{kl} \tag{2a}$$

$$D_i = \varepsilon^T_{ik}E_k + d_{ikl}T_{kl} \tag{2b}$$

where s^E_{ijkl} is the elastic compliance, ε^T_{ik} is the dielectric permittivity, and i, j, k, l = 1–3. In the equation, the convention in which repeated indices are summed is used. The superscripts E and T refer to the condition under which these quantities are measured. That is, compliance is measured under a constant electric field (short circuit condition) and dielectric constant under a constant stress. Due to the electromechanical coupling in a piezoelectric material, the elastic compliance under a constant electric field can be very different from that under constant charge. In order to write the elastic and piezoelectric tensors in the form of a matrix array, a compressed matrix notation is introduced. In the matrix notation, ij or kl is replaced by p or q according to [Nye, 1987; *IEEE Standard on Piezoelectricity*, 1988], so

$$11 \to 1,\ 22 \to 2,\ 33 \to 3,\ 23 \text{ or } 32 \to 4,\ 31 \text{ or } 13 \to 5, \text{ and } 12 \text{ or } 21 \to 6.$$

Equation (2) is the complete constitutive equation for a piezoelectric material. For a given piezoelectric material, the number of independent parameters can be reduced using symmetry relations in the material. For instance, for an unstretched and poled P(VDF-TrFE) copolymer which has a point group ∞m, the piezoelectric coefficient, the dielectric permittivity ($\varepsilon_{ij}=K_{ij}\varepsilon_0$, where K_{ij} is the relative permittivity and ε_0 is the vacuum permittivity), and elastic compliance matrices are

$$\begin{pmatrix} 0 & 0 & 0 & 0 & d_{15} & 0 \\ 0 & 0 & 0 & d_{15} & 0 & 0 \\ d_{31} & d_{31} & d_{33} & 0 & 0 & 0 \end{pmatrix}, \begin{pmatrix} K_{11} & 0 & 0 \\ 0 & K_{11} & 0 \\ 0 & 0 & K_{33} \end{pmatrix},$$

$$\begin{pmatrix} s_{11} & s_{12} & s_{13} & 0 & 0 & 0 \\ s_{12} & s_{11} & s_{13} & 0 & 0 & 0 \\ s_{13} & s_{13} & s_{33} & 0 & 0 & 0 \\ 0 & 0 & 0 & s_{44} & 0 & 0 \\ 0 & 0 & 0 & 0 & s_{44} & 0 \\ 0 & 0 & 0 & 0 & 0 & s_{66}[=2(s_{11}-s_{12})] \end{pmatrix},$$

where the 3-direction is the polymer poling direction.

In addition to Eq. (2) where the stress T and electric field E are the independent variables, the piezoelectric constitutive equations have three other forms [Nye, 1987; *IEEE Standard on Piezoelectricity*, 1988]:

$$T_\lambda = -e_{j\lambda} E_j + c^E_{\lambda\mu} S_\mu \tag{3a}$$

$$D_i = \varepsilon^S_{ij} E_j + e_{i\mu} S_\mu \tag{3b}$$

and

$$E_i = \beta^T_{ij} D_j - g_{i\mu} T_\mu \tag{4a}$$

$$S_\lambda = g_{i\lambda} D_j + s^D_{\lambda\mu} T_\mu \tag{4b}$$

and

$$T_\lambda = -h_{j\lambda} D_j + c^D_{\lambda\mu} S_\mu \tag{5a}$$

$$E_i = \beta^S_{ij} D_j - h_{i\mu} S_\mu \ . \tag{5b}$$

These four equivalent pairs of constitutive equations differ in their combinations of independent and dependent variables, and the coefficients used in these sets are interrelated.

The piezoelectric effect cannot exist in a polymer that possesses the inversion symmetry. In fact, among the 32 classes of crystal point symmetry groups, only 20 can exhibit the piezoelectric effect [Nye, 1987]. On the other hand, the electrostrictive effect that is a quadratic dependence of strain or stress on the polarization P exists in all polymers,

$$S_{ij} = Q_{ijkl} P_k P_l \ , \tag{6}$$

where Q_{ijkl} is the charge related electrostrictive coefficient. For an isotropic polymer,

$$S_3 = Q_{33} P^2 \text{ and } S_1 = Q_{13} P^2, \tag{7}$$

where S_3 and S_1 are the strains along and perpendicular to the polarization direction, known as the longitudinal and transverse strains, respectively. In Eq. (7), the compressed matrix notation is used. For an isotropic polymer, experimental evidence and theoretical consideration indicate that $Q_{33} < 0$ and $Q_{13} > 0$ [Kinase et al., 1955; Shkel et al., 1998]. Hence, for a polymer, an increase in polarization will result in a contraction along the polarization direction.

For a linear dielectric polymer, the polarization is related to the dielectric permittivity as

$$P = (\varepsilon - \varepsilon_0) E \ , \tag{8}$$

where ε_0 is the vacuum dielectric permittivity (= 8.85×10^{-12} F/m). Hence, Eq. (6) can be converted into

$$\begin{aligned} S &= Q\,(\varepsilon - \varepsilon_0)^2\,E^2 \\ &= M E^2\,, \end{aligned} \tag{9}$$

where M is known as the electric-field-related electrostriction coefficient. For an isotropic solid, the longitudinal strain is $S_3 = M_{33}E^2$ and the transverse strain $S_1 = M_{13}E^2$. Therefore, for an isotropic polymer, $M_{33} < 0$ and $M_{13} > 0$. That is, it will contract along the thickness direction and expand along the film direction when an electric field is applied across the thickness.

It should be noted that most polymers exhibit nonlinear dielectric behavior. The strain vs. electric field plots do not exhibit a quadratic nature and at high electric field, the field-induced strain exhibits saturation rather than follow Eq. (9).

For an electrostrictive polymer, a dc electric bias field can be applied and a smaller ac field superimposed on it as shown schematically in Fig. 1. When the dc field is much larger than the ac field, the dominating ac strain response term is [from Eq. (6)]

$$\Delta S = 2\,Q\,P_D\,\Delta P, \tag{10}$$

where P_D is the dc bias field induced polarization and ΔP is the polarization change induced by the small ac field as shown schematically in Fig. 1. Under weak fields, the polarization response is predominantly linear, $\Delta P = (\varepsilon - \varepsilon_0)\Delta E$, which leads to

$$\Delta\Sigma = 2\,\Theta\,\Pi_\Delta\,(\varepsilon - \varepsilon_0)\Delta E\,. \tag{11}$$

Equation (11) is valid for both linear and nonlinear dielectric polymers as long as ΔE is small. The linear relationship between the strain and field in Eq. (11) resembles that in Eq. (1) and therefore, one can introduce an effective piezoelectric coefficient here for dc field-biased electrostrictive polymers:

$$\delta = 2\,\Theta\,\Pi_\Delta\,(\varepsilon - \varepsilon_0)\,. \tag{12}$$

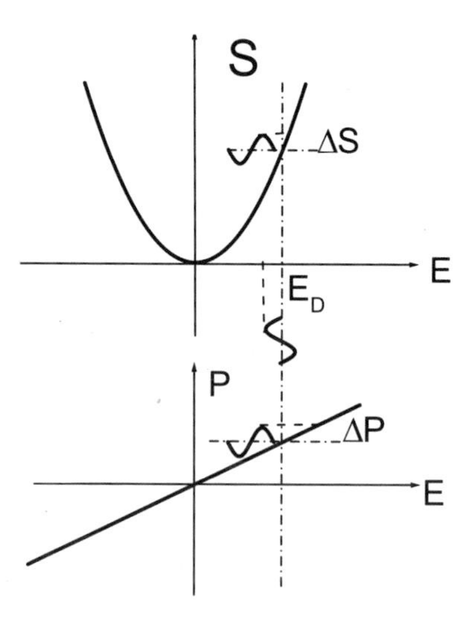

Figure 1: A schematic view of an electrostrictive polymer, with a small ac electric field superimposed on a dc bias field.

As will be discussed below, the piezoelectric state in a poled ferroelectric polymer can be regarded as the remanent polarization-biased response as described in Eq. (12) where P_D is replaced by the remanent polarization. Using the phenomenologic theory, it can be readily shown that the d coefficient in Eq. (12) describes both the converse and direct piezoelectric effects. For an isotropic polymer, Eq. (12) leads to

$$d_{33}=2Q_{33}\,P_D\,(\varepsilon-\varepsilon_0) \text{ and } d_{31}=2\,Q_{13}\,P_D\,(\varepsilon-\varepsilon_0)\,, \qquad (13)$$

where d_{33} and d_{31} are the piezoelectric coefficients along and perpendicular to the induced polarization direction (or the dc bias field direction). For an isotropic polymer, there is only one independent dielectric permittivity ε.

For electromechanical applications, electromechanical coupling factor k, which measures the efficiency of the material in converting energy between the electric and mechanical forms, is one of the most important parameters [*IEEE Standard on Piezoelectricity*, 1988],

$$k^2 = \text{converted mechanical energy / input electric energy} \qquad (14a)$$

or

$$k^2 = \text{converted electrical energy / input mechanic energy} \qquad (14b)$$

In electromechanical applications, an electric field is applied along a certain direction [Eq. (2)] and the electromechanical actuation along the same direction or another direction is used. The coupling factor depends on the direction of the electric field and the mechanical strain (or stress) direction. Hence, for an electromechanical EAP, there are many electromechanical coupling factors, depending on the combination of the directions of electric and mechanical variables. For example, a polymer actuator can be made with the electric field along the 3-direction [Eq. (2)], while the actuation is along the same direction. In this case, the coupling factor is the longitudinal electromechanical coupling factor k_{33} which is related to the parameters in Eq. (2) as

$$k_{33}^2 = d_{33}^2/(\varepsilon_{33}^T s_{33}^E)\,. \tag{15}$$

On the other hand, if the mechanical actuation is along the direction perpendicular to the applied electric field (the 1-direction, for instance), the coupling factor is k_{31} and

$$k_{31}^2 = d_{31}^2 /(\varepsilon_{33}^T s_{11}^E).$$

The coupling factor can also be related to the material coefficient measured under different conditions. For example, the elastic compliance s_{33}^E is related to s_{33}^D (measured under constant charge or open circuit condition) as

$$s_{33}^D = (1 - k_{33}^2)\, s_{33}^E\,. \tag{16}$$

Therefore, a polymer with a large coupling factor will see a large difference in the elastic compliance when used under different external electric boundary conditions, which may be utilized to tune the elastic modulus of the polymeric material by varying the electric conditions.

4.2.2 Maxwell Stress Effect

Analogous to the electrostriction where the strain is approximately proportional to the square of the electric field strength [Eq. (9)], there is another electromechanical actuation mechanism whose strain response is also approximately proportional to the square of the electric field strength. That is the electrostatic force (Maxwell stress)-induced strain, which can be quite significant in soft polymer elastomers.

The general expression for the Maxwell stress for a dielectric material has been derived as [Landau et al., 1970; Kloos, 1995]

$$T_{ij}^M = -E_i D_j + \frac{1}{2}\delta_{ij}\, D_k E_k\,, \tag{17}$$

where $\delta_{ij} = 1$ when $i = j$, and $\delta_{ij} = 0$ otherwise. Using $D_i = \varepsilon_{ij} E_j$, Eq. (17) can be rewritten as

$$T_{ij}^M = -\,\varepsilon_{jm}\, E_i\, E_m + \frac{1}{2}\delta_{ij}\, \varepsilon_{kl} E_l\, E_k\,. \tag{18}$$

Therefore, the strain due to the Maxwell stress in a dielectric material is

$$S_{ij} = -(s_{ijkm}\varepsilon_{ml} - \frac{1}{2}\delta_{ij}s_{ijmn}\varepsilon_{kl})E_k E_l \,, \tag{19}$$

which has the same form as the electrostriction [Eq. (9)]. For an isotropic solid without mechanical constraints, the longitudinal strain induced by the Maxwell stress is

$$S_3 = S_{33} = -\frac{1}{2}\varepsilon(s_{11} - 2s_{12})E_3^2 \,, \tag{20}$$

and the transverse strain is

$$S_1 = S_{11} = \frac{1}{2}\varepsilon s_{11} E_3^2 \,. \tag{21}$$

In Eqs. (20) and (21), we have made use of the fact that for an isotropic polymer, the dielectric tensor is reduced to one constant: ε. For a polymer with Poisson's ratio equal to one half, Eq. (20) becomes

$$S_3 = -\,\varepsilon\, s_{11}\, E_3^2 \,. \tag{22}$$

Analogous to the electrostriction in an isotropic polymer, the Maxwell stress causes contraction along the field direction; and for a soft polymer (a large elastic compliance), the Maxwell stress induced strain (here it is called Maxwell strain) can be quite significant. For large strains, Eq. (22) should be modified to the form

$$Y\,\mathrm{Ln}(\frac{d_0}{d}) = \varepsilon(V/d)^2 \,, \tag{23}$$

where d_0 is the initial film thickness, d is the final thickness, Y is the Young's modulus (Y=1/s_{11}) and V is the applied voltage. These expressions show that for a soft polymer elastomer under high field, the Maxwell stress induced strain can be quite high. For an isotropic material, it has been shown that a thickness strain of 40% can be induced by the Maxwell stress effect before the mechanical instability occurs [Stark et al., 1955]. However, for anisotropic materials or materials with nonlinear elastic behavior, the analysis of the mechanical instability should be modified, which may lead to a strain level higher than 40% before the mechanical instability occurs.

4.3 PVDF-Based Ferroelectric Polymers

A ferroelectric polymer is a polymer possessing spontaneous polarization that can be reoriented between possible equilibrium directions by a realizable electric field. A ferroelectric polymer can be in a single crystal form or, as in most cases encountered, a semicrystalline form in which the ferroelectric crystallites are embedded in an amorphous matrix. Examples of ferroelectric polymers include polyvinyl fluoride (PVF), polyvinylidene fluoride (PVDF), copolymers of PVDF with trifluoroethylene (TrFE) or tetrafluoroethylene (TFE), and odd-numbered nylons. This section will discuss P(VDF-TrFE) copolymers in some detail. Although almost all of the ferroelectric polymers available are in the semicrystalline form, recently there have been reports of research work on single crystalline P(VDF-TrFE) copolymers, which will also be discussed.

4.3.1 Phases and Phase Transitions

PVDF is polymorphic and has at least four major crystalline phases [Lovinger, 1982]. Two of them, the form I (β-phase) and form II (α-phase), are the most relevant phases for practical ferroelectric and piezoelectric applications. In form I, which is also known as the β phase, two chains in an all-*trans* planar zigzag conformation are packed into individual orthorhombic unit cells having lattice dimensions of a = 8.58 Å, b = 4.90 Å, and the chain direction or fiber axis c = 2.56 Å (Fig. 2) [Hasegawa et al., 1972]. It is noted from Fig. 2 that in the all-*transconformation*, the fluorine atoms are positioned on one-side of the unit cell, resulting in a net dipole moment. The form I unit cell is quite polar, having a net dipole of 2.1 D (debye). As the structure of the unit cell of the form I crystal satisfies the symmetry requirement of a piezoelectric crystal, i.e., that the crystal belongs to a noncentrosymmetric class, this is the form of PVDF that is responsible for its piezoelectric properties.

In form II, or the α-phase, the chain conformations are represented as a sequence of alternating *trans* and *gauche* sequences, or $TGT\overline{G}$ (Fig. 3) [Takahashi et al., 1983]. Each unit cell containing two chains is orthorhombic with lattice parameters a = 4.96 Å, b = 9.64 Å, and c = 4.62 Å. In the α-phase, adjacent chains are packed such that the dipole moments of the individual carbon-fluorine bonds are aligned perpendicular to the chain direction, canceling one another out. The directions of the chains consist of a statistical average of up-up and up-down orientations.

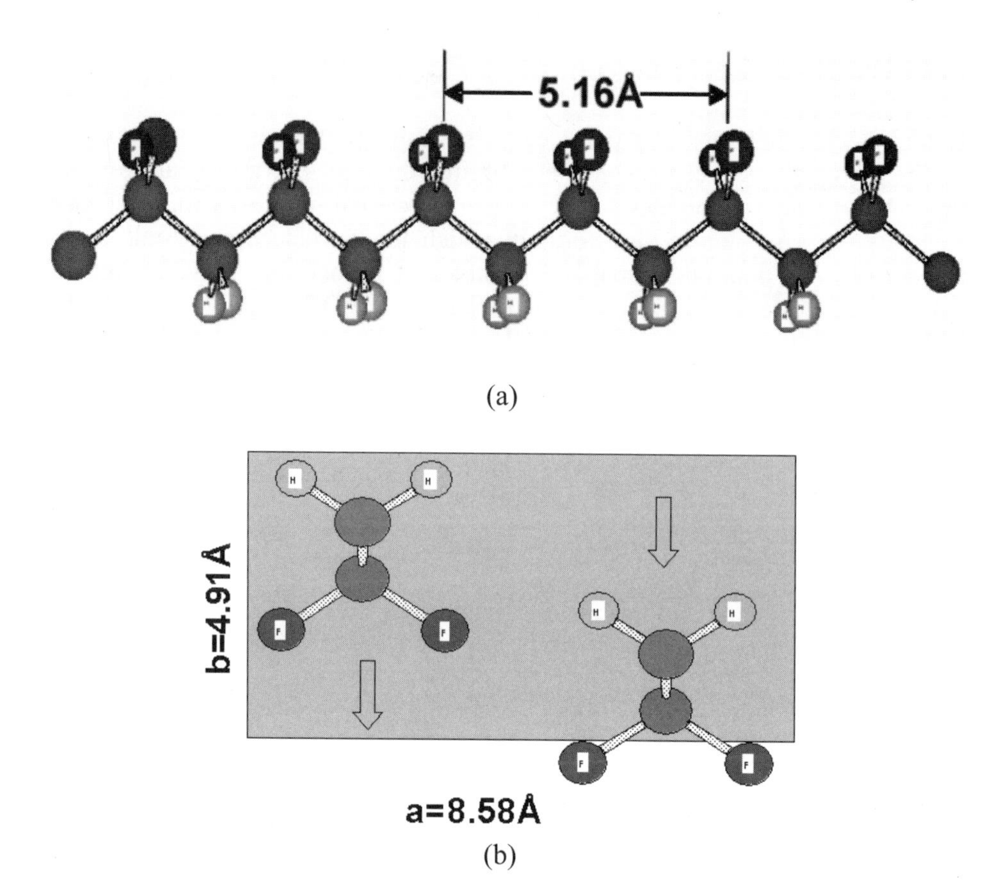

Figure 2: A view of form I, i.e., β phase, having lattice dimensions of a = 8.58 Å, b = 4.90 Å, and the chain direction or fiber axis c = 2.56 Å.

When prepared using the melt crystallization or solution cast, in most cases PVDF will form α-phase, which is not a polar phase. Mechanical stretching is often used in order to convert the α-phase to the ferroelectric β-phase. On the other hand, for P(VDF-TrFE) copolymers with VDF content less than approximately 85 mol%, the β-phase will be formed directly. In the β-phase crystallites there are ferroelectric domains, which are polar but nonetheless orientated in all crystallographically allowed directions. Furthermore, in the semicrystalline polymer, these crystallites are randomly oriented within the sample. This accounts for the absence of any piezoelectric activity unless the sample is poled. To be made piezoelectric, the domains must be oriented in a strong electric field called the "poling field." Poling can be accomplished by electroding the polymer surfaces with a metal, followed by application of a strong electric field to orient the crystallites. An alternative method of poling is the use of a corona discharge where a corona charge is injected into the polymer from a needle electrode placed a centimeter or two from the polymer film. In corona poling, no electroding is required as in the case of direct field poling.

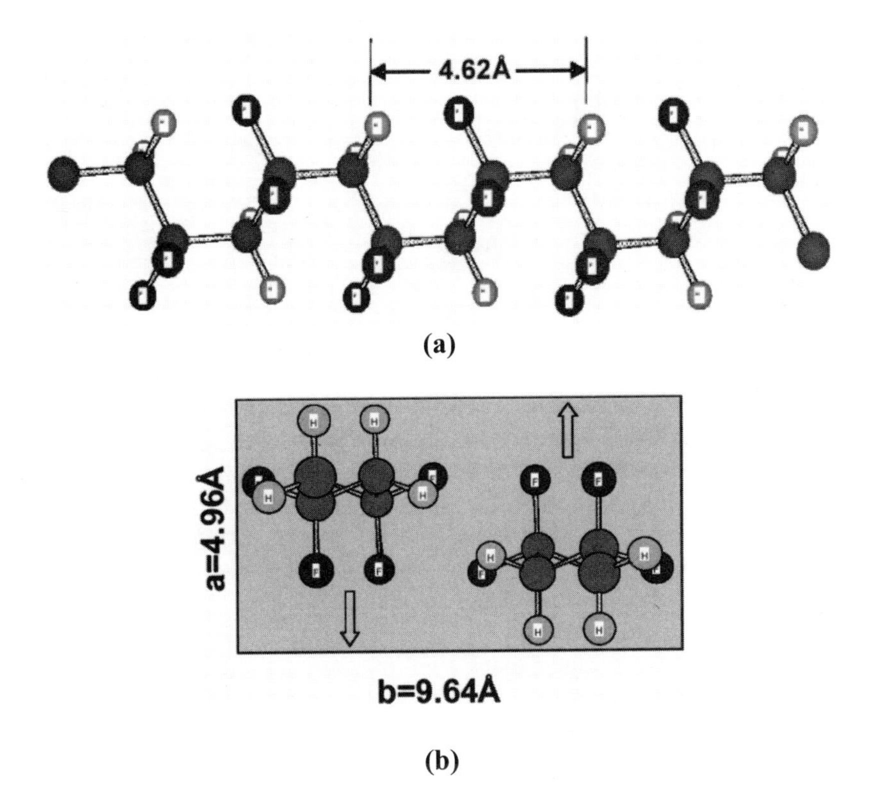

Figure 3: A schematic view of form II, i.e., α-phase, having lattice dimensions of a = 4.96 Å, b = 9.64 Å, and the chain direction or fiber axis c = 4.62 Å [Takahashi et al., 1983]. (Courtesy of the American Chemical Society.)

The phase diagram of PVDF and P(VDF-TrFE) polymers shows a ferroelectric-paraelectric transition that signals a change from a ferroelectric (polar) phase to a paraelectric (nonpolar) phase (Fig. 4). The ferroelectric-paraelectric (F-P) transition temperature increases with vinylidene fluoride mole fraction content. Below the F-P transition, the crystal is best represented as an ordered form I (β) structure with long sequences of all-*trans* bonds. As the temperature of the crystals rises and goes through the F-P transition, an increasing number of *gauche* bonds are introduced into the ordered all-*trans* structure. As a result, the polarization in the crystal regions tends toward disorder, leading to the formation of the *paraelectric* phase containing a random mixing of TG, T$\overline{G}$, TTTG, and TTT$\overline{G}$. Eventually, at higher temperatures the paraelectric phase passes through the melt transition [Lovinger, 1981; Takahashi et al., 1980].

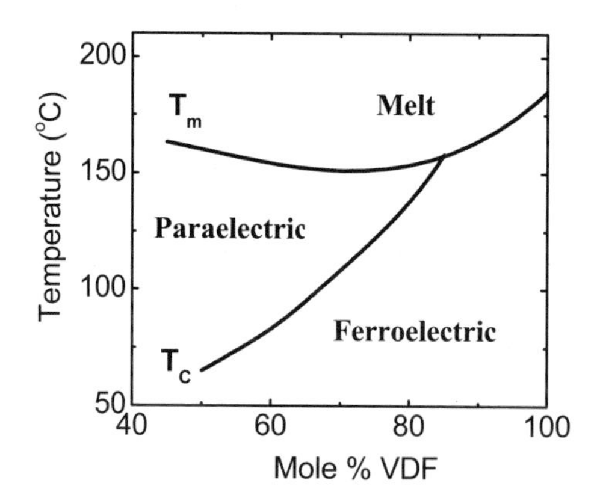

Figure 4: Phase diagram of PVDF and P(VDF-TrFE) polymers showing a ferroelectric-paraelectric transition that signals a change from a ferroelectric (polar) phase to a paraelectric (nonpolar) phase [Lovinger, 1983]. (Courtesy Taylor and Francis.)

One should note from Fig. 4 that PVDF as well as P(VDF-TrFE) copolymers with high VDF concentrations do not appear to possess distinct F-P transitions; rather, it appears that melting takes place before a F-P transition. However, it must be mentioned that even in the ferroelectric phase, conformational defects can be introduced as the temperature of the polymers is raised. These defects are introduced so subtly that they may not be apparent in thermal studies such as differential scanning calorimetry (DSC). Moreover, in P(VDF-TrFE), in addition to a low-temperature (LT) phase where the chain conformation is predominantly all-*trans*, a cooled (CL) phase has been identified. Structural analysis has indicated that the most probable structure of the so-called CL phase is a mixture of two disordered crystalline phases, one transplanar and the other 3/1 helical (frozen-in high temperature phase) [Davis et al., 1982; Lovinger et al., 1982; Fernandez et al., 1987; Tashiro et al., 1986]. Because of this "frozen-in" disorder, copolymers with VDF contents below 50 mol% lose ferroelectricity. Accordingly, there is no clear phase transition signal as shown in the phase diagram.

4.3.2 Ferroelectric Responses

As ferroelectric materials, PVDF and its copolymers with TrFE and TFE exhibit well-defined polarization hysteresis loops. Figure 5(a) shows the polarization loop measured on a P(VDF-TrFE) 68/32 mol% copolymer stretched film for which, at low cyclic electric fields (< 55 MV/m), the polarization loop is nearly linear with very little hysteresis. At field amplitudes exceeding 55 MV/m, a well-defined polarization loop appears. The coercive field E_c (the field when $P = 0$)

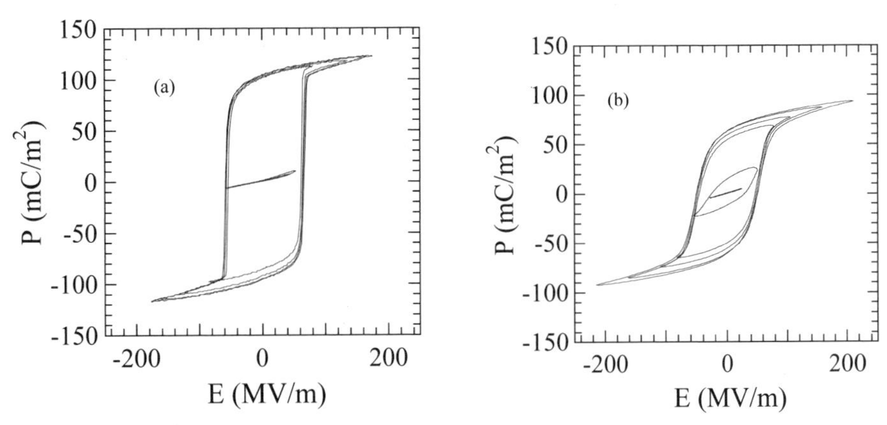

Figure 5: Polarization loops measured under different electric fields on a P(VDF-TrFE) 68/32 mol% copolymer stretched film (left) and a 50/50 mol% film (right).

and remanent polarization P_r (the polarization when $E = 0$) do not change appreciably with the applied field amplitude. In contrast, unstretched 50/50 mol% copolymer films display very little hysteresis at fields below 25 MV/m. As the field amplitude increases beyond that, the polarization loop gradually expands and both E_c and P_r increase with the applied cyclic field amplitude, as shown in Fig. 5(b). Eventually, when the field amplitude exceeds 100 MV/m, E_c and P_r, which are now saturated, define the maximum coercive field and remanent polarization. Moreover, even at fields far below E_c measured from the saturated loop, polarization is often switchable. Such behavior is related to the nucleation process in the polarization switching [Lines et al., 1977; Jona et al., 1993].

Because of the semicrystalline nature of the polymer, the magnitudes of P_r and E_c are particularly sensitive to sample preparation conditions. For instance, Fig. 6 illustrates that both crystallinity and remanent polarization for stretched 68/32 mol% copolymer increase with the sample annealing temperature.

The crystallization process in the copolymer is also influenced by the mobility as well as the stereoirregularity of the polymer chains [Sharples, 1966]. For PVDF homopolymer, the crystallinity is at about 50% [Scheinbeim et al., 1979]. With increased TrFE content, because of the larger size of the TrFE monomer unit, the lattice spacing between the polymer chain expands and thus facilitates and enhances the crystallization process (Fig. 6). Hence, in the copolymer of 75/25 mol%, the degree of crystallinity can reach 90%. In contrast, in polymer with higher TrFE content, the stereoirregularity introduced by TrFE reduces the stability of the crystalline phase with respect to the amorphous and leads to a lowering of the crystallinity. The dependence of the crystallinity with VDF content is depicted in Fig. 7(a). Similarly, the dependence of the remanent polarization and coercive field on VDF content is illustrated in Fig. 7(b).

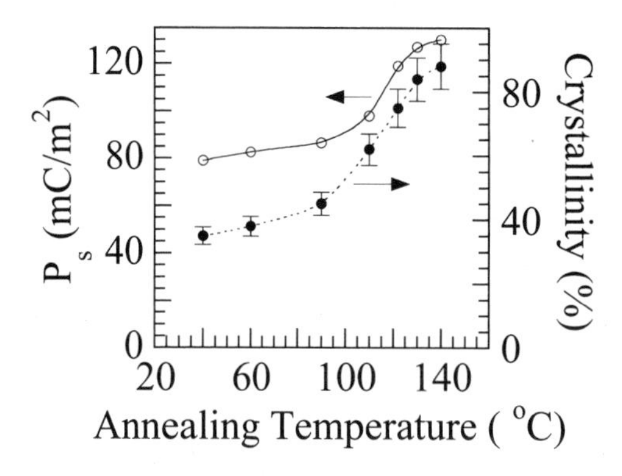

Figure 6: The effect of the annealing temperature on the polarization and crystallinity.

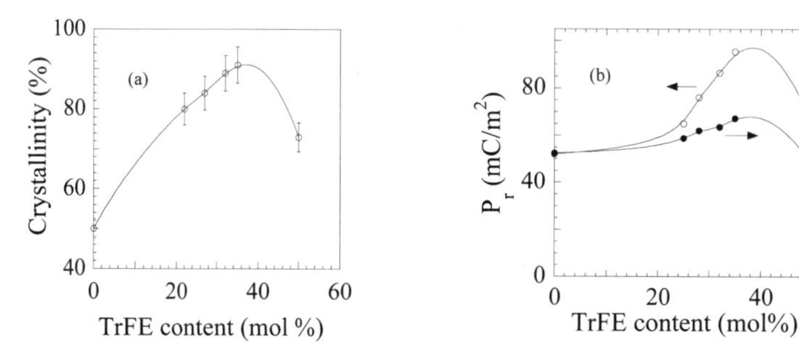

Figure 7: The dependence of the crystallinity with VDF content (left) and the dependence of the remanent polarization and coercive field on VDF content (right).

Besides the crystallinity and remanent polarization, the F-P transition temperature for a given P(VDF-TrFE) copolymer can also be varied by carefully selecting conditions for sample processing [Stack and Ting, 1988; Green et al., 1986; Tanaka et al., 1999]. For instance, it has been demonstrated that for P(VDF-TrFE) 60/40 mol% copolymer, the F-P transition can occur in the temperature range from 65°C to 85°C [Green et al., 1986]. The sample processing methods used in this investigation included melt crystallization as well as crystallization and recrystallization from cast solutions of dimethylacetamide (DMA), dimethylformamide (DMF), and cyclohexanone. This behavior is attributed to variations in the TrFE concentration in the crystalline phase for samples prepared under different conditions. Such variations lead to different F-P transition temperatures. But, it is also quite possible that the polar ordering in the crystalline region is affected by these differing processing conditions.

In the ferroelectric phase, the copolymer is in the all-*transconformation*, resulting in a unit cell with a large lattice constant along the polymer chain direction and a smaller unit cell dimension perpendicular to the chain. In the paraelectric phase, however, where the crystallites adopt conformations containing a mixture of *trans* and *gauche* bonding, the unit cell dimension along the chain direction is significantly shortened, while the cell dimension perpendicular to the chain expands (see Figs. 2 and 3). As a result, when the copolymer goes through the F-P transition, there is a large lattice constant change for the 65/35 mol% copolymer as depicted in Fig. 8 [Tashiro et al., 1984]. In fact, this phenomenon has been observed for copolymers exhibiting F-P transitions. The corresponding lattice strain change through the transition perpendicular to the polymer chain is presented in Fig. 9 with VDF content.

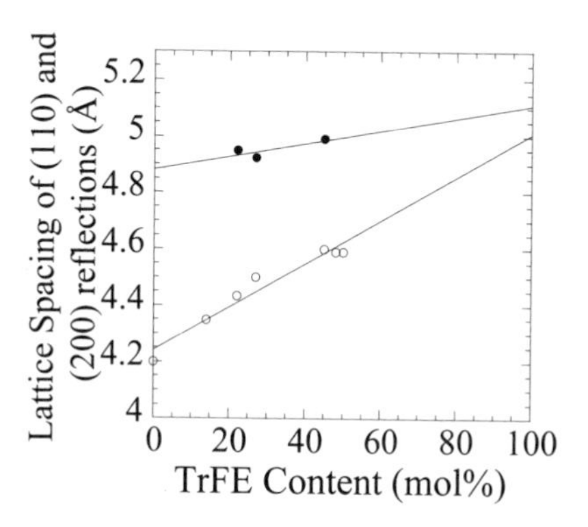

Figure 8: As a result of the copolymer going through the F-P transition, there is a large lattice constant change for the 65/35 mol% copolymer.

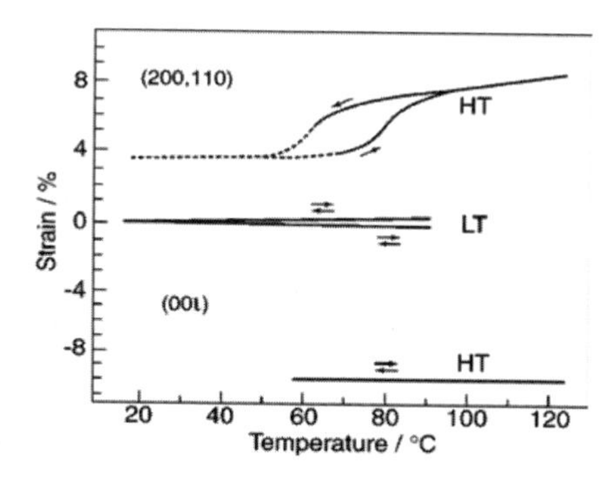

Figure 9: The lattice strain change through the transition perpendicular to the polymer chain is presented for VDF content.

The results presented indicate a strong coupling between the polarization and strain in the P(VDF-TrFE) copolymer, which is the most attractive feature of this polymer. For PVDF and its copolymers, this coupling is electrostrictive in nature [Sundar and Newnham, 1992].

$$S = Q\,P^2. \tag{24}$$

However, in the polarization switching process, it has been observed that Eq. (24) does not adequately describe the strain-polarization relationship. This is primarily due to the fact that the polarization switching is through the domain wall motion. One notes, for example, the polarization-switching loop for 65/35 mol% copolymer and the corresponding strain changes in Fig. 10. As can be seen, there is very little strain change as the polarization switches from "A" to "B" [Furukawa et al., 1990]. Early FT-IR and x-ray studies demonstrated that the switching is primarily through successive 60° domain wall motions. Owing to the

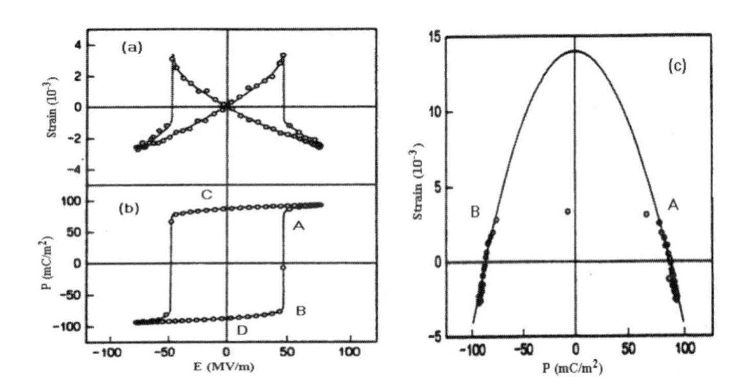

Figure 10: A linear relationship of strain versus P^2 is observed in the nonswitching part of the polarization response [Furukawa et al., 1990]. (Courtesy of the Japanese Journal of Applied Physics.)

pseudo-hexagonal symmetry of the unit cell in the directions perpendicular to the polymer chain, 60° domain wall motions are not expected to generate very high strains [Furukawa et al., 1990; Dvey-Aharon et al., 1980; Takahashi et al., 1984; Takahashi et al., 1980; Guy et al., 1988]. Therefore, to achieve a high strain response in ferroelectric materials through the polarization switching mechanism, efforts should be expended to suppress the domain wall motions that do not generate much strain response. As also shown in Fig. 10, a linear relationship of strain versus P^2 is observed in the nonswitching part of the polarization response.

Most of the electromechanical applications that use PVDF and its copolymers exploit their piezoelectric properties. For ferroelectric PVDF and its copolymers, the samples as prepared usually have negligible piezoelectric responses due to the fact that the polarization at each domain orientates randomly, implying that the net sample polarization is zero. In order to establish the piezoelectric state, the polymers can be poled under a field several times higher than the coercive field to induce a stable remanent polarization (at point "C" or "D" of the polarization loop in Fig. 10). In fact, the piezoelectric state in a poled ferroelectric polymer or any poled ferroelectric can be regarded as the remanent polarization-biased electrostriction.

4.3.3 Piezoelectric and Other Related Electromechanical Properties

Various coefficients in the piezoelectric constitutive equation for PVDF and P(VDF-TrFE) 75/25 copolymer are listed in Table 1, where the data have been taken from several different sources [Wang et al., 1993; Schewe et al., 1982]. In most polymers, the response to an external stimulus often shows strong relaxation behavior [McCrum et al., 1991]; hence, the coefficients in the piezoelectric constitutive equation are frequency dependent and complex [Holland et al., 1967], i.e.,

$$d_{ij}^{*} = d_{ij}^{'} - jd_{ij}^{''},\ \ s_{ij}^{*} = s_{ij}^{'} - js_{ij}^{''},\ \text{and}\ \varepsilon_{ij}^{*} = \varepsilon_{ij}^{'} - j\varepsilon_{ij}^{''}. \tag{25}$$

In Table 1, both the real and imaginary part of these coefficients for P(VDF-TrFE) 75/25 mol% copolymer are listed, which suggests that the imaginary part for the copolymer cannot be neglected.

Table 1: Piezoelectric, dielectric, and elastic properties of PVDF and P(VDF-TrFE) 75/25 mol%.

Material Parameter	PVDF	P(VDF-TrFE) 75/25	
		real	imaginary
d_{31} (pC/N)	28	10.7	0.18
d_{32} (pC/N)	4	10.1	0.19
d_{33} (pC/N)	-35	-33.5	-0.65
d_{15} (pC/N)		-36.3	-0.32
s_{11} (10^{-10} Pa^{-1})	3.65	3.32	0.10
s_{22} (10^{-10} Pa^{-1})	4.24	3.24	0.07
s_{33} (10^{-10} Pa^{-1})	4.72	3.00	0.07
s_{12} (10^{-10} Pa^{-1})	-1.10	-1.44	-0.036
s_{13} (10^{-10} Pa^{-1})	-2.09	-0.89	-0.022
s_{23} (10^{-10} Pa^{-1})	-1.92	-0.86	-0.022
$\varepsilon_{33}/\ \varepsilon_0$	15	7.9	0.09
k_{33}		0.23	
k_{13}	0.13	0.07	
k_t	0.144	0.196	

The temperature dependence of the piezoelectric and dielectric constant of 75/25 copolymer is presented in Figs. 11 and 12. Such temperature dependence behavior, where the piezoelectric coefficient and dielectric constant increase as the temperature is increased toward the F-P transition temperature, is a common phenomenon for all ferroelectric-based piezoelectric materials, including many commercially available piezoceramics [Jona et al., 1993; Devonshire, 1954; Berlincourt et al., 1964].

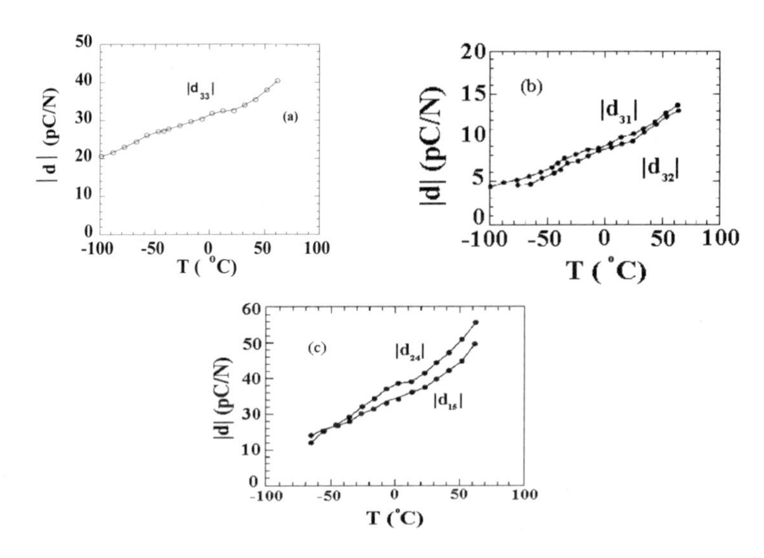

Figure 11: The temperature dependence of the piezoelectric constant of 75/25 copolymer.

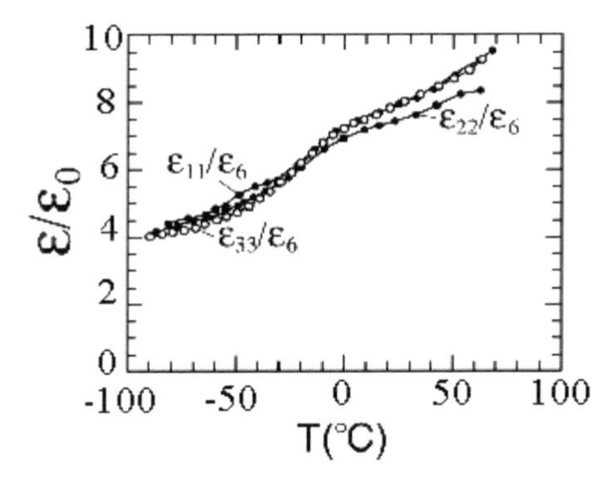

Figure 12: The temperature dependence of the dielectric constant of 75/25 copolymer.

In semicrystalline copolymers, the piezoelectric properties can often be improved by raising the degree of crystallinity, given that the piezoelectric response is mainly from the crystalline region. For example, using high-pressure crystallization to increase the crystallinity and improve the quality of the crystalline phase, Ohigashi et al. reported an improved thickness-coupling factor for 80/20 copolymer to near 0.3 (k_t=0.3) [Ohigashi et al., 1995]. More recently, a relatively large-sized single crystal P(VDF-TrFE) 75/25 mol% copolymer has been grown that has a room temperature d_{33} = –38 pm/V and coupling factor

$k_{33} = 0.33$, values which up to now represent the best piezoelectric performance of known piezoelectric and ferroelectric polymers [Omete et al., 1997]. Figures 13 through 15 show the temperature dependence of the dielectric and piezoelectric properties and the electromechanical coupling factors for a copolymer single crystal.

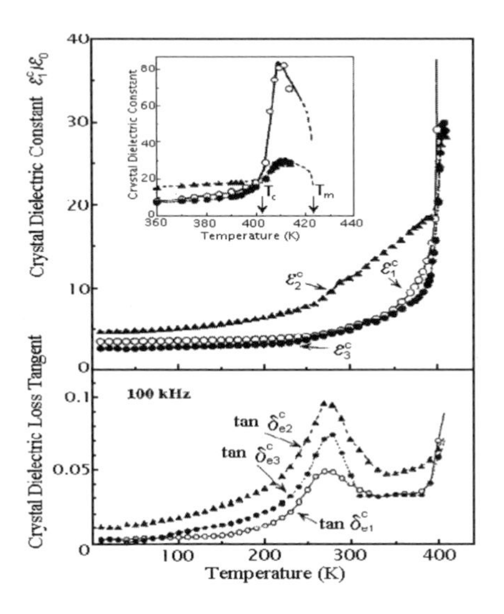

Figure 13: Temperature dependence of the dielectric properties for a copolymer single crystal [Ohigashi et al., 1995]. (Courtesy of the American Institute of Physics.)

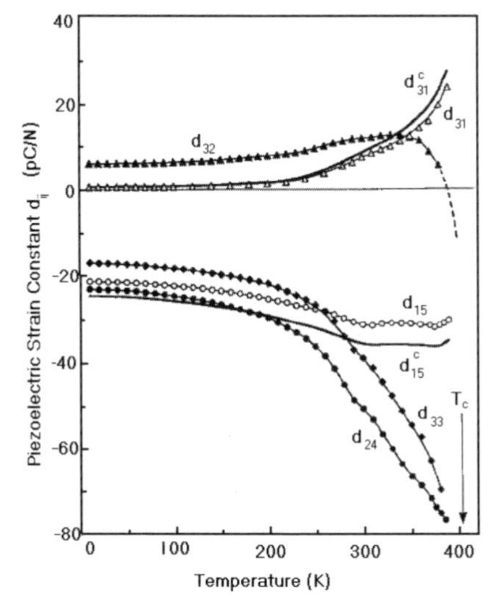

Figure 14: Temperature dependence of the piezoelectric properties for a copolymer single crystal [Ohigashi et al., 1995]. (Copyright 1995, American Institute of Physics.)

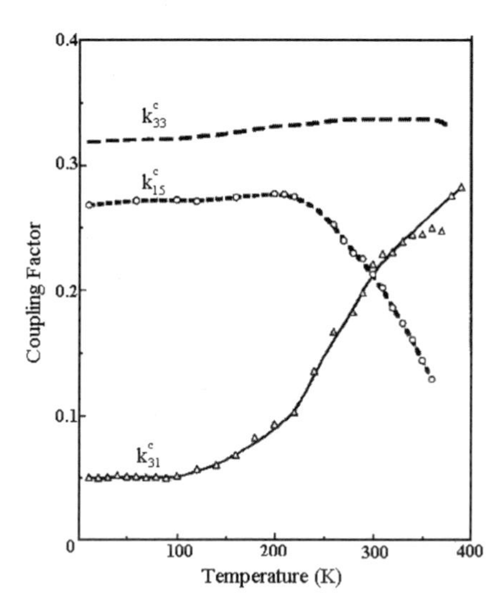

Figure 15: Temperature dependence of the electromechanical coupling factors for a copolymer single crystal [Ohigashi et al., 1995]. (Copyright 1995, American Institute of Physics.)

4.4 Ferroelectric Odd-Numbered Polyamides (Nylons)

Kawai et al. first reported very low piezoelectricity in polyamides in 1970. In 1980, Scheinbeim's group reported significantly high piezoelectricity in nylon 11. The synopsis of ferroelectricity in nylons was not confirmed until 1990 when Scheinbeim et al. reported a typical ferroelectric hysteresis loop from the electric displacement *D* versus the applied electric field *E* test of nylon 11 films, which was produced by a melt-quenched and cold stretching method. This result was achieved after about a persistent 10-year investigation on nylon 11 in the laboratory. The melt-quenched and cold-stretched nylon 11 does not only exhibit the ferroelectric switching mechanism but also increases its piezoelectric property significantly compared to that previously reported, especially at an elevated temperature. The discovery has led to a series of achievements in the research on the ferroelectricity and piezoelectricity of odd numbered nylons including odd-numbered nylon-contained polymeric composites (bilaminates and blends with polyvinylidene fluoride, PVDF or PVF_2) in the group in the following years.

4.4.1 Molecular Structures and Ferroelectricity in Nylons

The alpha-phase crystal structure of nylon 11 suggested by Slichter in 1959 is polar. The molecular conformation of the proposed structure is all trans, and for an odd-numbered nylon, this entails a net dipole moment per chain. Figure 16(a) shows the molecular structures of odd-numbered nylons and even numbered nylons). As can be seen, in odd-numbered nylons, the electric dipoles formed by amide groups (H-N-C=O) with a dipole moment of 3.7 Debyes are sequenced in a way that all the dipoles are in the same direction synergistically. Therefore, the net dipole moment can be formed in one direction. While in even-numbered nylons, the amide group dipoles are in two ways: if one is in one direction, the next one will be in the opposite direction. This results in an intrinsic cancellation of the dipole moments, as demonstrated schematically in Fig. 16(a). The ferroelectricity of nylon 11 and nylon 7 was discovered by Lee et al. in 1991 when typical electric displacement *D* versus electric field *E* hysteresis loops were obtained. The ferroelectric characteristics of the nylons were reported as a function of test temperature. The results are shown in Fig. 16(b).

Lee et al. also proposed a 90-deg-then-180-deg mechanism for the ferroelectricity of nylon 11 (see Fig. 17). The mechanism was confirmed by x-ray diffraction (XRD) and Fourier-transform infrared (FTIR) tests. It was also noticed that the cold stretching following the melt quenching is a very critical step in obtaining the polyamide chains in the parallel form needed for the ferroelectric polarization switching and piezoelectricity.

The remanent polarization of the ferroelectric odd-numbered nylons is a function of the number, which is decisive to the electric dipole density: the smaller the number, the higher the dipole density. This is because the number represents how many carbons are contained in a repeat unit of polyamide molecules; therefore, the concentration of amide groups and hydrogen bonds in the polymer increases when the number goes down since there is one amide group in a repeat unit. The inter-relationship between the number and the remanent polarization is tabulated in Table 2, as reported by Mei et al. in 1993.

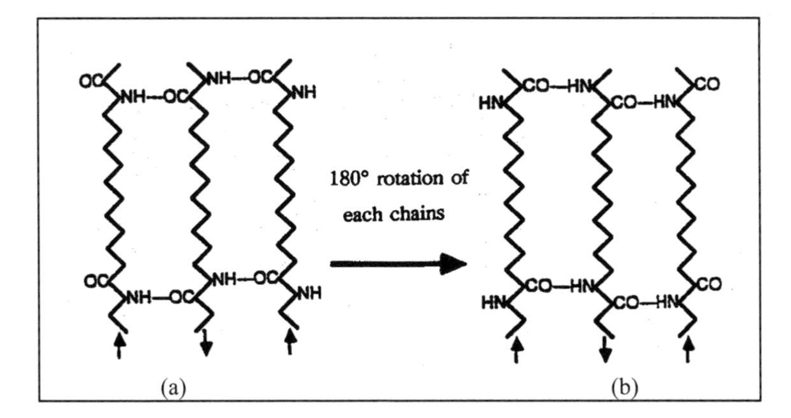

Figure 16: (a) Schematics of molecular structures of odd-numbered nylons and even-numbered nylons and (b) the electric displacement, *D*, versus electric field, E, showing the D-E hysteresis behavior of nylon 11 at different temperatures. (Courtesy of Ji Su, J. Lee, and John Wiley & Sons.)

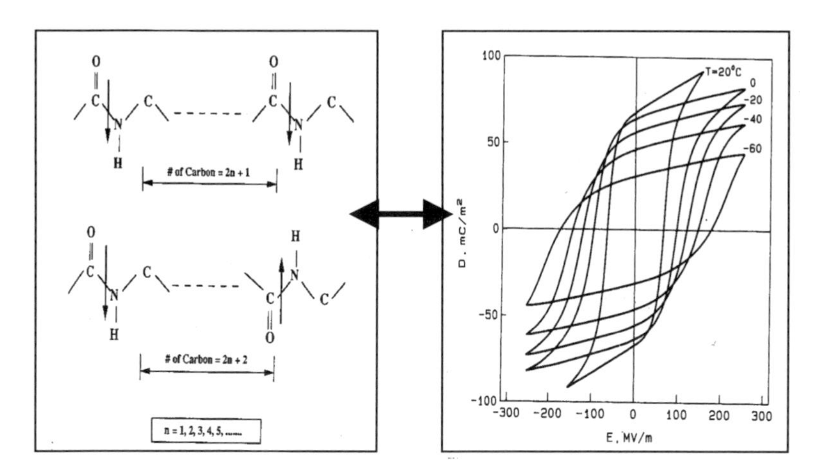

Figure 17: A schematic representative of the proposed 90-deg-then-180-deg mechanism for the ferroelectricity of the nylon 11 [Sheinbeim et al., 1992]. (Courtesy of the American Chemical Society.)

Table 2: Unit molecular weight, dipole density and remanent polarization of odd-numbered nylons.

	3-nylon	5-nylon	7-nylon	9-nylon	11-nylon
Molecular weight of repeat unit	71.08	99.13	127.18	155.23	183.29
Dipole density, D/100A^3	4.30	2.92	2.12	1.65	1.40
Remanent Polarization, mC/m^2	180	125	86	68	56

4.4.2 Piezoelectricity and Pyroelectricity

The piezoelectricity of the odd-numbered nylons poled by an electric field depends on the temperature. At room temperature, the piezoelectricity is about one tenth of that in PVDF. However, when the temperature is higher than its glass transition temperature, the piezoelectricity of odd-numbered nylons show a transitional increase, which makes their piezoelectric response superior over that of PVDF. In 1991, Takase et al. reported the temperature dependence of piezoelectric strain coefficient, d_{31} of nylon 11 and nylon 7, and compared the results with that of PVDF, as shown in Fig. 18. Takase et al. also reported that annealing treatment may improve the thermal stability of the piezoelectricity in the nylons.

Esayan et al. reported that the poled nylon 11 and nylon 7 exhibit significant pyroelectric response. In the experiments, the poled nylon 11 demonstrated 57% of the pyroelectric coefficient, P_y, compared to that of poled PVDF and the poled nylon 7 showed 80% of the coefficient of the poled PVDF.

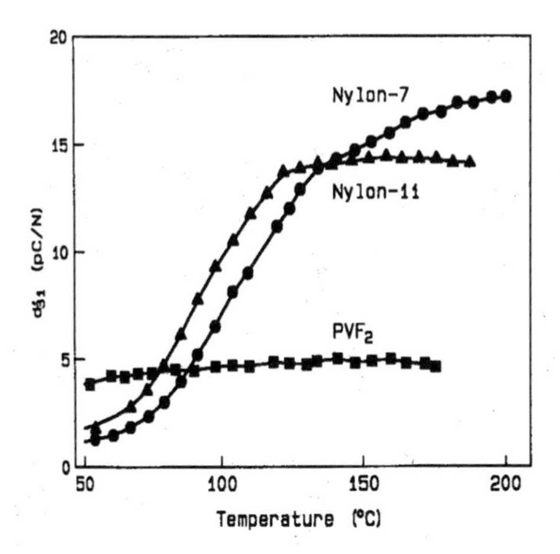

Figure 18: Temperature dependence of piezoelectric strain coefficient, d_{31}, of the poled nylon 11 and nylon 7 with a comparison to that of poled PVDF (PVF_2) [Takase et al., 1991]. (Courtesy of the American Chemical Society.)

4.4.3 Ferroelectricity and Piezoelectricity of Odd-Nylon-Contained Composites

4.4.3.1 PVDF-Nylon 11 bilaminates

Su et al. developed nylon 11-poly(vinylidene fluoride) (PVDF or PVF_2) bilaminates by a co-melt-pressed-stretched process in 1992. The bilaminate exhibits a typical ferroelectric *D-E* hysteresis loop with significantly enhanced remanent polarization, P_r, of 75 mC/m^2, which is 44 % higher than those of the individual institutes, nylon 11 or PVDF (Fig. 19a). The piezoelectric strain coefficients including strain coefficient, d_{31}, stress coefficient, e_{31} and hydrostatic coefficient, d_h) also show significant enhancement. The enhancement in the piezoelectricity becomes more obvious at the elevated temperature over the glass transition temperature of nylon 11. Figure 19b shows the temperature dependence of the piezoelectric strain coefficient, d_{31}, of nylon 11-PVDF bilaminate having a 50-50 fraction in comparison to those of individual nylon 11 or PVDF.

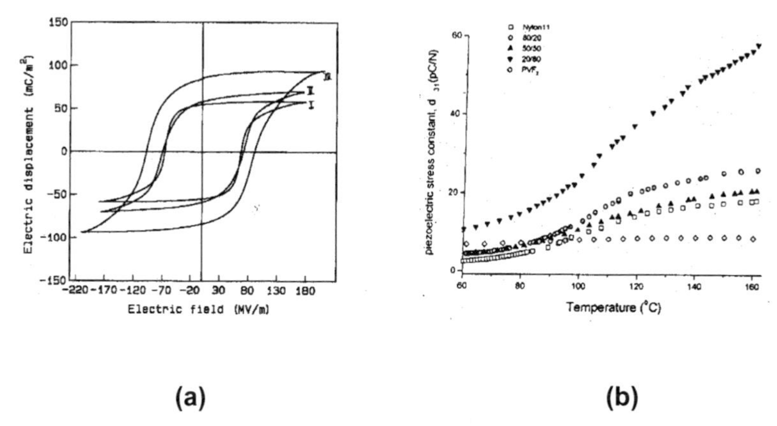

(a) **(b)**

Figure 19: (a) Curves of electric field displacement, D, versus applied electric field, E, (D-E) and (b) temperature dependence of the piezoelectric strain coefficient, d_{31}, for -a-: nylon 11/PVDF bilaminate, -b-: PVDF, and -c-: nylon 11 films [Su, 1995]. (Courtesy of John Wiley & Sons.)

The mechanism resulting in the enhancement was attributed to the interfacial space charge accumulation and the asymmetric distribution of the accumulated space charges along the direction across the interface between the two constitutes [Su, 1995]. The remanent polarization and piezoelectric coefficients of the bilaminates are a function of the fraction of the two constitutes.

4.4.3.2 Nylon 11-PVDF blends

In 1999, Gao et al. reported the development of nylon 11-PVDF blends which also exhibit a significantly enhanced remanent polarization, P_r. The P_r of the blend with the 50-50 composition gives the P_r of 85 mC/m^2, which is more than 60% higher than those of individual nylon 11 or PVDF with the Pr of 52 mC/m^2. The curves of electric field displacement, D, versus applied electric field, E, are shown in Fig. 20(a). The same paper also reported the dependence of the ferroelectricity of the blends is dependent on the fraction of the two constitutes, and the enhancement might also be attributed to the space charge accumulation and distribution. The piezoelectric strain coefficient, d_{31}, of nylon 11-PVDF blend films also shows a significant enhancement when compared with the individual nylon 11 or PVDF films (Gao et al., 2000). The coefficient is also a function of the fraction of the two constitutes. Figure 20(b) shows the temperature dependence of the piezoelectric strain coefficient, d_{31} of the nylon 11-PVDF blends compared with those of individual nylon 11 and individual PVDF.

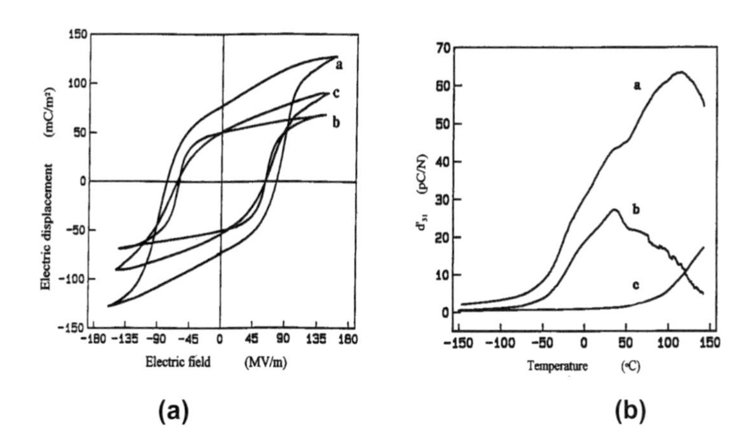

Figure 20: (a) Curves of electric field displacement, *D*, versus applied electric field, *E*, (*D-E*) [(a) nylon 11, (b) PVDF films and for (c) nylon 11/PVDF blend] [Gao et al., 1999]. (Courtesy of John Wiley & Sons.) (b) The temperature dependence of the piezoelectric strain coefficient, d_{31},of nylon 11-PVDF blends with different compositions compared with those of individual nylon 11 and individual PVDF [Gao et al., 2000]. (Courtesy of the American Chemical Society.)

4.4.4 Remarks

Because of the strong hydrogen bond effects, the odd numbered nylons are sensitive to humidity. The humidity sensitivity is a drawback for the polymers to be utilized in many applications even though this may be an advantage for applications of the materials in high-temperature humidity-sensing technologies.

4.5 Electrostriction

Electrostriction is the most fundamental electromechanical coupling effect in all-insulating materials. In most insulating polymers, the electrostriction is relatively weak. One of the reasons for the small electrostrictive strain in polymers is the low dielectric constant. A recent study of electrostriction in polymers shows that the electrostriction coefficient M ($S = ME^2$) for most polymers with the elastic modulus 1 GPa is below 2×10^{-20} m^2/V^2. Hence, even under a field of 100 MV/m, the electrostrictive strain is below 2×10^{-4}. On the other hand, for polymers with low elastic modulus, as pointed out by Newnham et al., the electrostrictive coefficient may become quite large. This will result in a large electrostrictive strain although the force level of the polymers is relatively low compared with the piezoelectric polymers [Newnham et al., 1998].

4.5.1 Electrostrictive Responses in High-Energy Electron-Irradiated P(VDF-TrFE) Copolymers

One of the approaches to significantly increase the electrostrictive response with a high force level and high elastic energy density in polymers is to work with polymers that have a high dielectric constant. Recently, it has been shown that in high-energy electron-irradiated P(VDF-TrFE) copolymers, the dielectric constant of almost 60 can be achieved at room temperature and, as a consequence, a high electrostrictive strain of about 5% under 150 MV/m field has been demonstrated, as shown in Fig. 21(a), [Zhang et al., 1998; Zhao et al., 1998; Cheng et al., 1999]. The plot of strain versus P^2 yields a straight line, indicating the response is electrostrictive in nature [($S_3 = Q_{33}P_3^2$, Fig. 21(b)]. For the irradiated copolymer, Q_{33} is found in the range of – 4 to –15 m^2/C (the electrostrictive *M* coefficient is greater than 2×10^{-18}), depending on the sample processing conditions. The strain response does not change appreciably with temperature, as suggested by the plot in Fig. 21(c).

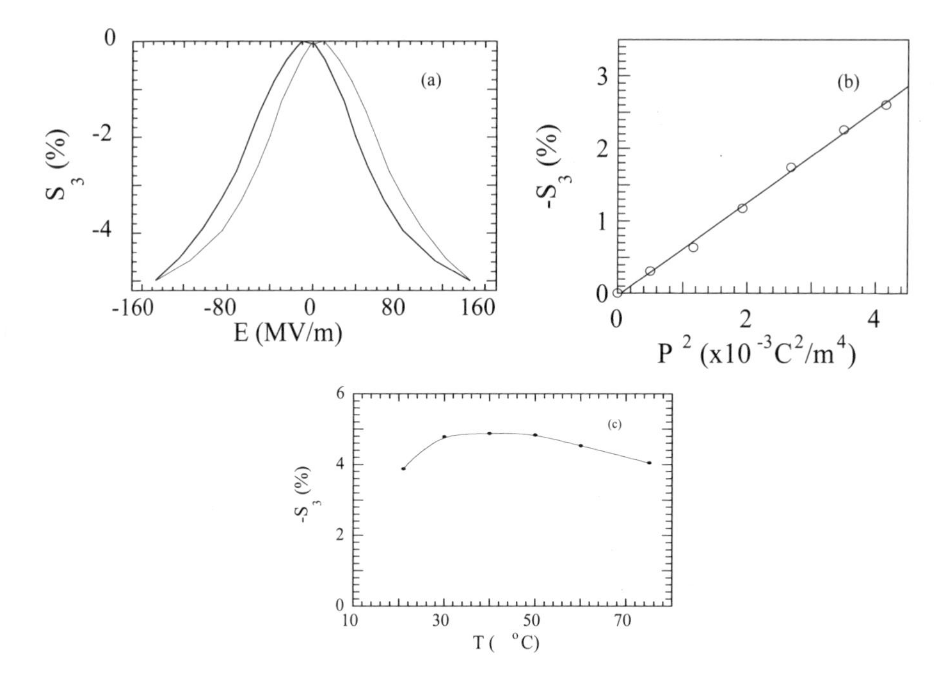

Figure 21: A high electrostrictive strain, about -5% under 150 MV/m field.

The basis for such a large electrostriction in the irradiated P(VDF-TrFE) copolymers is the large change in the lattice strain as the polymer goes through the transformation between the ferroelectric (F) and paraelectric (P) phase. Besides the large strain response at F-P transition, there is another interesting feature associated with the F-P transition—the possibility that a very large electromechanical coupling factor (*k*~1) can be obtained near a first-order F-P transition temperature [Zhang et al., 1995]. Here, the irradiation effect is to introduce defect structures in the polymer to move the transformation process to near room temperature, eliminate the hysteresis in the polarization response to the applied field, and broaden the transformation temperature region so that a high electrostriction can be obtained over a relatively broad temperature range.

Of special interest is the discovery that in P(VDF-TrFE) copolymers, large anisotropy in the strain responses exists along and perpendicular to the chain direction, as can be deduced from the change in the lattice parameters between the polar and nonpolar phases. Therefore, the transverse strain can be tuned over a large range by varying the film processing conditions. For unstretched films, the transverse strain is relatively small (~+1% at ~100 MV/m) while the amplitude ratio between the transverse strain and longitudinal strain is less than 0.33 [Cheng et al., 1999]. This feature is attractive for devices using the longitudinal strain such as ultrasonic transducers in the thickness mode, and actuators and sensors making use of the longitudinal electromechanical responses of the material. For example, with a very weak transverse electromechanical response in comparison with the longitudinal, one can significantly reduce the influence of the lateral modes on the thickness resonance and improve the performance of the thickness transducer. On the other hand, for stretched films, a large transverse strain (S_1) along the stretching direction can be achieved as shown in Fig. 22, where a transverse strain of about +4.5% was observed in the irradiated copolymer under an electric field of 85 MV/m [Xia et al., 2002].

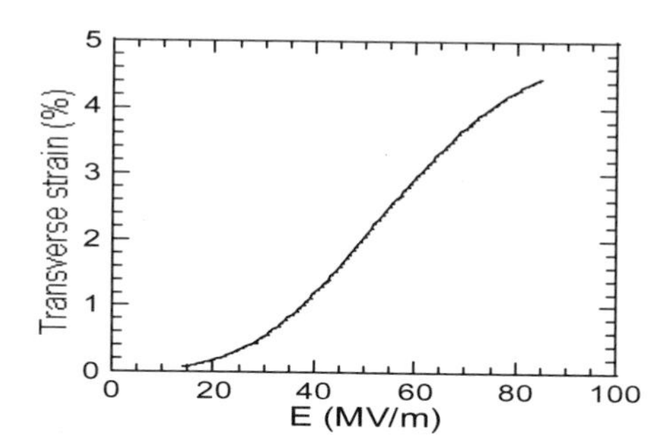

Figure 22: For stretched films, a large transverse strain (S_1) along the stretching direction can be achieved.

It is also interesting that the strain along the thickness direction (parallel to the electric field) is always negative for P(VDF-TrFE) copolymers regardless of the sample processing conditions. In fact, this is a general feature for a system in which the polarization response originates from the dipolar interaction and is true for all polymeric piezoelectric and electrostrictive responses [Kinase et al., 1955; Shkel et al., 1998]. The sign of the strains perpendicular to the applied field direction will depend on the sample processing conditions. For theanisotropically stretched films discussed here, the electric-induced strain along the stretching direction, which is perpendicular to the applied field, is positive, whereas in the direction perpendicular to both stretching and applied field directions, the strain is negative. For unstretched samples that are isotropic in the plane perpendicular to the applied field, the strain component in the plane is an average of the strains along the chain (positive) and perpendicular to the chain (negative) and is, in general, positive.

For electrostrictive materials, the electromechanical coupling factor (k_{ij}) has been derived by Hom et al. [1994] based on the consideration of electrical and mechanical energies generated in the material under the external field:

$$k_{3i}^2 = \frac{kS_i^2}{s_{ii}^D \left\{ P_E \ln\left(\frac{P_S + P_E}{P_S - P_E}\right) + P_S \ln\left[1 - \left(\frac{P_E}{P_S}\right)\right] \right\}^2} , \quad (26)$$

where i=1 or 3 correspond to the transverse or longitudinal direction (for example, k_{31} is the transverse coupling factor), s_{ii}^D is the elastic compliance under constant polarization, and S_i and P_E are the strain and polarization responses, respectively, for the material under an electric field E. The coupling factor depends on E, the electric field level. In Eq. (26), it is assumed that the polarization-field (P-E) relationship follows approximately

$$|P_E| = P_S \tanh\left(k|E|\right) , \quad (27)$$

where P_S is the saturation polarization and k is a constant.

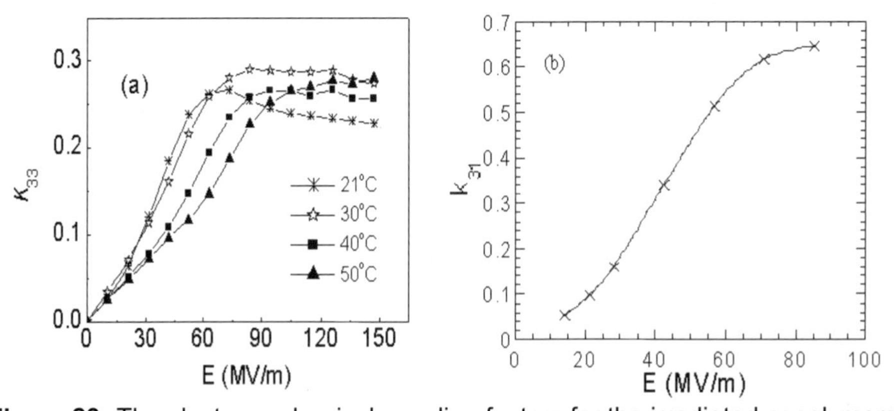

Figure 23: The electromechanical coupling factors for the irradiated copolymers.

The electromechanical coupling factors for the irradiated copolymers calculated based on Eq. (26) are shown in Fig. 23. Near room temperature and under an electric field of 80 MV/m, k_{33} can reach more than 0.3, which is comparable to that obtained in a single-crystal P(VDF-TrFE) copolymer. More interestingly, k_{31} of 0.65 can be obtained in a stretched copolymer, which is much higher than values measured in unirradiated P(VDF-TrFE) copolymers [Omote et al., 1997]. These results are also verified by recent resonance studies in these polymers.

In a polymer, there is always a concern about the electromechanical response under high mechanical load; that is, whether the material can maintain high strain levels when subject to high external stresses. Figure 24(a) depicts the transverse strain of a stretched and irradiated 65/35 copolymer under a tensile stress along the stretching direction and the longitudinal strain of unstretched and irradiated 65/35 copolymer under hydrostatic pressure [Bharti et al., 1999; Gross et al., 1999]. As can be seen from the figure, under a constant electric field the transverse strain increases initially with the load and reaches a maximum at the tensile stress of about 20 MPa. Upon a further increase of the load, the field-induced strain is reduced. One important feature revealed by the data is that even under a tensile stress of 45 MPa, the strain generated is still nearly the same as that without load, indicating that the material has a very high load capability. Shown in Fig. 24(b) is the longitudinal strain under hydrostatic pressure. At low electric fields, the strain does not change much with pressure, while at high fields it shows increase with pressure.

Figure 25 presents the wide-angle x-ray scattering data obtained on the irradiated copolymer when subject to different external fields. It shows that the x-

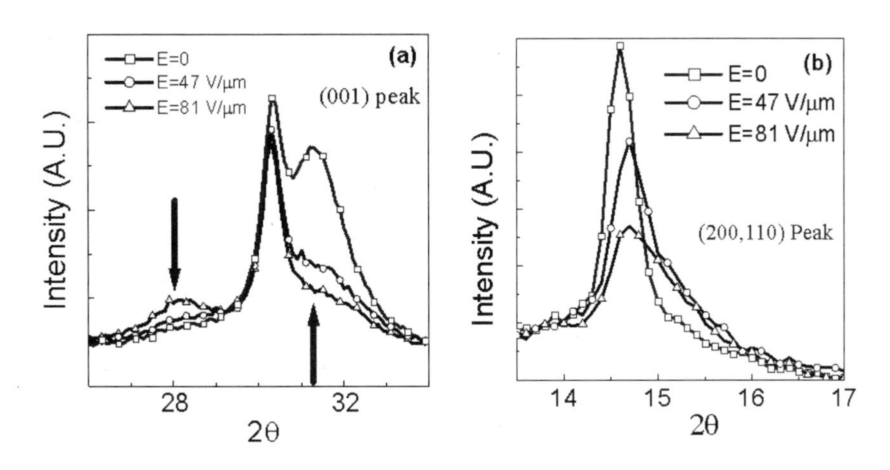

Figure 24: (a) The transverse strain (amplitude) of stretched and irradiated 65/35 copolymer under a tensile stress along the stretching direction. (b) The longitudinal strain of unstretched and irradiated 65/35 copolymer under hydrostatic pressure. [Bharti et al. 1999]. (Copyright 1999, American Institute of Physics.)

ray peaks change with external field, reflecting that there is a local field induced conformation change in the polymer. The data confirm that it is the reversible field induced transformation between the polar and nonpolar conformations that is responsible for the observed large electromechanical responses in these electrostrictive polymers.

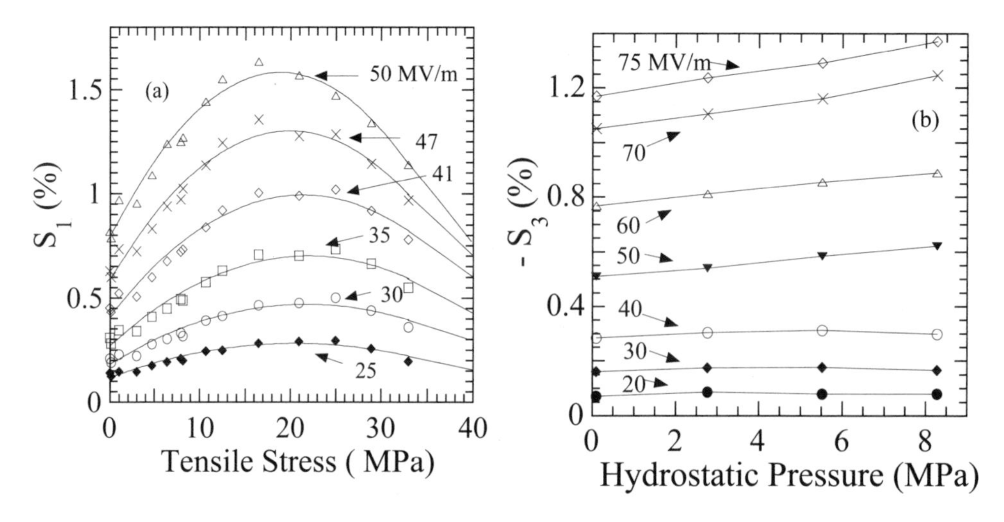

Figure 25: X-ray diffraction pattern measured at room temperature as a function of applied electric field for the irradiated film. (a) The (001) diffraction observed using transmission scan (the peak at $2\theta \sim 30.2^o$ is from the Au electrodes) and the arrows indicate the polar and nonpolar phase x-ray peaks, and (b) the (110, 200) diffraction peak obtained using reflection scan [Huang et al., 2003c]. (Copyright, IEEE.)

4.5.2 Electrostrictive Responses in P(VDF-TrFE) Terpolymers

Although high energy irradiations can be used to convert the normal ferroelectric P(VDF-TrFE) into a relaxor ferroelectric with high electrostriction, the irradiation also introduces many undesirable defects to the copolymer, such as the formation of crosslinkings, radicals, and chain scission [Mabboux and Gleason, 2002; Bharti et al., 2000]. From the basic ferroelectric response point of view, the defects modification of the ferroelectric properties can also be realized by introducing randomly in the polymer chain a third monomer, which is bulkier than VDF and TrFE. Furthermore, by a proper molecular design which enhances the degree of molecular level conformational changes in the polymer, the terpolymer can exhibit a higher electromechanical response than the high energy electron irradiated copolymer, as will be shown by the terpolymer containing chlorofluoroethylene (CFE, -CH_2-CFCl-) as the termonomer. To facilitate the discussion and comparison with the irradiated P(VDF-TrFE) copolymer, the composition of the terpolymer is labeled as VDF_x-$TrFE_{1-x}$-CFE_y, where the mole ratio of VDF/TrFE is $x/1-x$ and y is the mol% of CFE in the terpolymer.

Presented in Fig. 26(a) is the field-induced strains along the thickness direction (S_3) as a function of the applied electrical field for the terpolymer P(VDF-TrFE-CFE) 62/38/4 mol%. Under a field of 130 MV/m, a thickness strain of – 4.5% can be achieved, which is comparable to that observed in the irradiated copolymers. The high elastic modulus in this P(VDF-TrFE-CFE) terpolymer (Y~1.1 GPa for unstretched films) results in a high elastic energy density, $YS^2/2$, ~1.1 J/cm^3. The longitudinal coupling factor k_{33} is presented in Fig. 26(b). It is interesting to note that the coupling factor k_{33} can reach more than 0.55 [Xia et al., 2002].

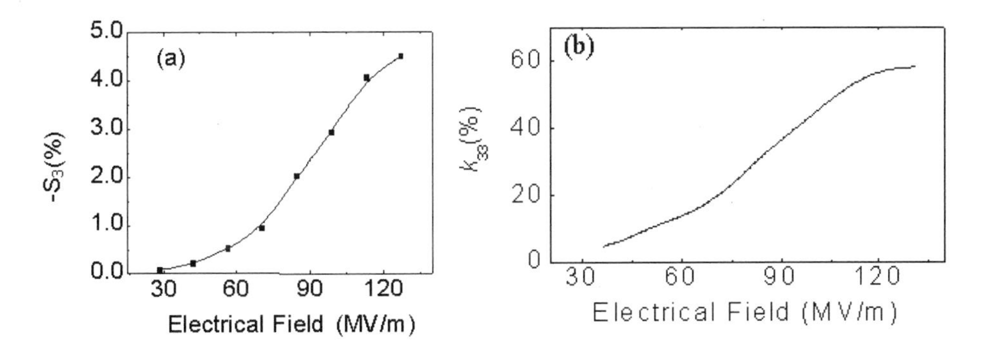

Figure 26: (a) The field induced longitudinal strain (S_3) as a function of the applied field and (b) The longitudinal coupling factor k_{33} for the P(VDF-TrFE-CFE) 62/38/4 mol% terpolymer versus the applied field amplitude.

By increasing the ratio of VDF/TrFE in the terpolymer, the field induced strain level can be raised due to the fact that the lattice strain between the polar conformation and nonpolar conformations increases with the VDF/TrFE ratio. Presented in Fig. 27(a) is the thickness strain of the P(VDF-TrFE-CFE) terpolymer at composition of 68/32/9 mol% and a thickness strain of more than 7% can be reached, which is by far the highest among all the HIEEPs and terpolymers investigated [Cheng et al., 2001; Xia et al., 2002; Huang et al., 2003c; Xu et al., 2001]. The results here demonstrate the potential of the terpolymers in achieving very high electromechanical responses through optimized composition. The transverse strain response of this terpolymer was also characterized. Presented in Fig. 27(b) is the transverse strain S_1 from a stretched film.

Table 3 summarizes the electromechanical properties of the terpolymers and high-energy electron irradiated copolymer (HEEIP), where the typical strain level, the elastic energy density, and coupling factors are listed.

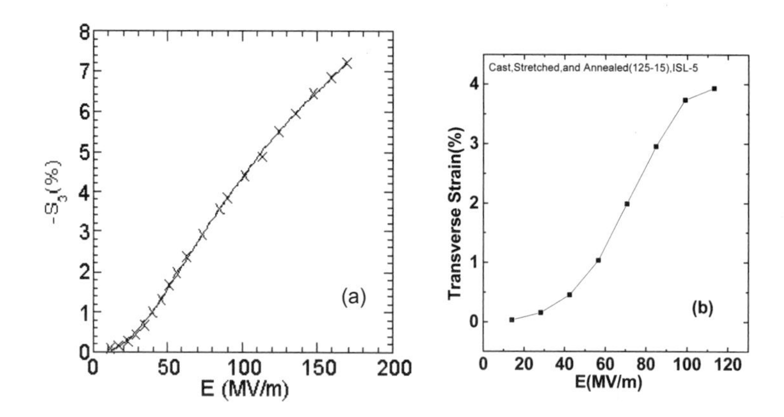

Figure 27: (a) The field induced longitudinal strain (S_3) and (b) transverse strain as a function of the applied field for the unstretched terpolymer P(VDF-TrFE-CFE) 68/32/9 mol%.

Table 3: Comparison of the electromechanical properties of the modified PVDF based polymers.

Polymer		$*S_M$(%)	Y (GPa)	$YS_M^2/2$ (J/cm^3)	k_{33}	k_{31}
**HEEIP	S3	–5	0.4	0.5	0.30	
	S1	4.5	1.0	1.0		0.65
P(VDF-TrFE-CFE)						
62/38/4 mol%	S3	–4.5	1.2	1.2	0.55	
68/32/9 mol%	S3	–7	0.3	0.73		

$*S_M$ is the strain and $YS_M^2/2$ is the volumetric elastic energy density, Y is the elastic modulus along the actuation direction. **HEEIP: high-energy electron irradiated copolymer.

4.5.3 Electrostrictive Graft Elastomers

Su et al. [2000a] recently reported the observation of a high strain (thickness strain ~4%) in a grafted elastomer. The graft elastomer consists of two components, a flexible backbone polymer and grafted crystalline polar groups. The flexible backbone polymer provides the amorphous chains for the formation of the three-dimensional network that is physically cross-linked by the grafted crystalline moieties. The grafted polar crystalline moieties also provide the electric-field responsive mechanism for the electric-field-induced dimensional change, or electromechanical properties. The schematics in Figs. 28(a) and (b) show the structure and molecular morphology, respectively, of the grafted elastomers.

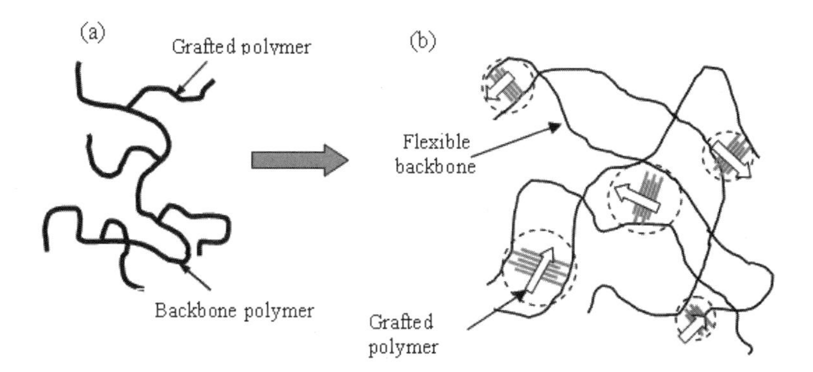

Figure 28: Schematics of (a) structure the electrostrictive graft elastomers. (b) Morphology of the electrostrictive graft elastomers [Su et al., 2000b]. (Copyright IEEE.)

In the electrostrictive graft elastomer studied, the flexible backbone polymer is a copolymer of chlorotriflouroethylene and triflouroethylene, while the grafted polar crystalline polymer is a copolymer of vinylidene-flouride and triflouroethelene, P(VDF-TrFE). The electric-field-induced-longitudinal strain, S_{33}, was measured at room temperature. As shown in Fig. 29, the induced strain in the electrostrictive graft elastomer exhibits a quadratic dependence with the applied electric field. In the case of polymeric elastomers exhibiting a large electric-field-induced strain, two intrinsic mechanisms are considered as primary contributors: electrostriction and the Maxwell stress effect. The contributions from both mechanisms exhibit a quadratic dependence with an applied electric field. For the newly developed electrostrictive graft elastomers, more than 95% of the electric-field-induced strain response is contributed by the electrostriction mechanism, while the contribution from the Maxwell stress effect is less than 5%. Due to the relatively high elastic modulus (~0.55 GPa), the grafted elastomer also exhibits a high elastic energy density of 0.44 MJ/m^3.

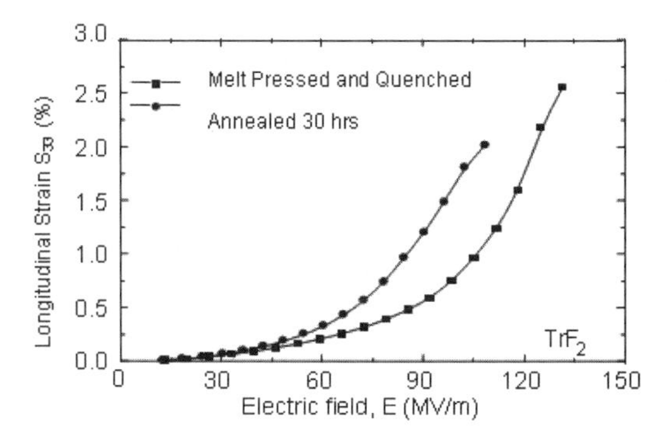

Figure 29: The relationship between the electric-field-induced longitudinal strain, *S*, and the applied electric field at room temperature [Su et al., 2000b]. (Copyright IEEE).

4.5.3.1 Composition and Morphology Control

As an electroactive polymer containing two components: the flexible backbone polymer and the grafted polymer crystalline domains that are polar and electric field responsive, it offers an advantage to tailoring the electromechanical properties by controlling the chemical composition of the components and the molecular morphology of the graft elastomeric system. Su et al. (2003) reported that the electric-field-induced strain of the elastomer can be controlled by thermal annealing treatment, which can increase the crystallinity of the grafted electric field responsive domains. The higher crystallinity is achieved, the higher electric field-induced strain is obtained. Since the mechanism of the electromechanical response in an electrostrictive graft elastomer is hypothesized to involve rotation of the polar crystal domains under an applied electric field, the overall electric-field-induced strain should critically depend on the number or size of the polar domains; and therefore, the degree of crystallinity. Accordingly, the electric-field-induced strain of the elastomers significantly increases with annealing treatment. These results are shown in Figs. 30(a) and 30(b), respectively.

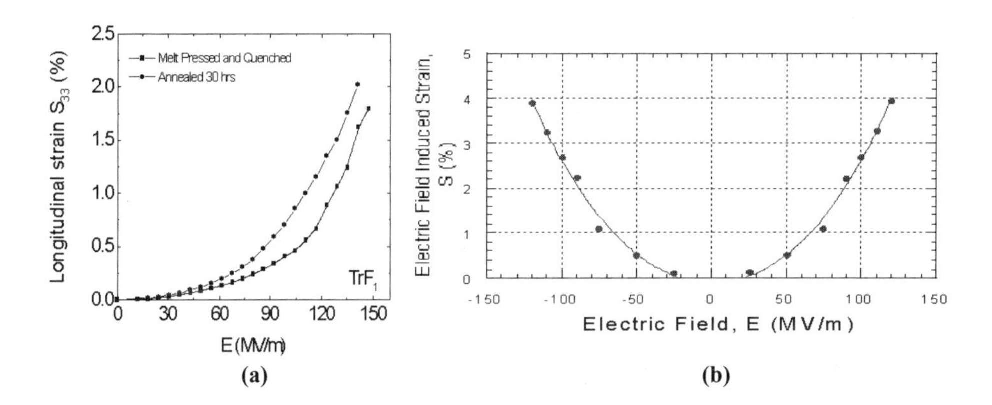

Figure 30: Effects of the composition and the annealing treatment on the electric-field-induced strain longitudinally, S_{33}, of the electrostrictive graft elastomer (a) containing 30 mol.% of grafted VDF-TrFE copolymer (TrF_1) and (b) containing 50 mol.% of grafted VDF-TrFE copolymer (TrF_2).

These trends are summarized in Fig. 31. In general, the strain responses increase with total crystallinity. At a given crystallinity, however, the elastomer containing more grafted polar domains gives a higher response. It is thought that the crystal size and distribution may also affect the overall electromechanical response.

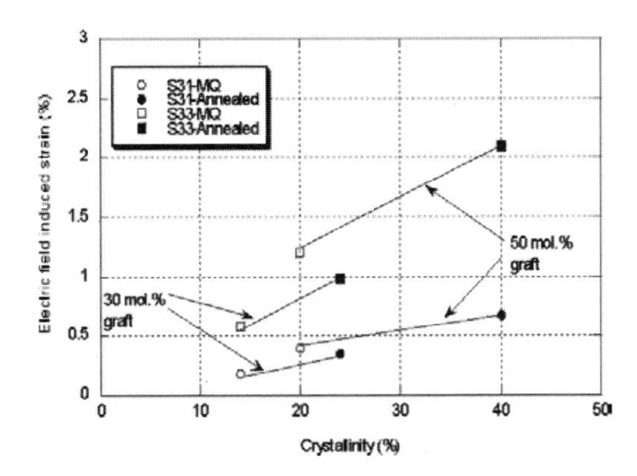

Figure 31: Comparison of the effects of the crystallinity in the elastomer and the thermal treatment on the electric-field-induced strain, S_{31} and S_{33}, of the electrostrictive graft elastomers. (MQ represents "melt-pressed-quenched" and Annealed represents "annealed for 30 hours at 120°C"; 31 represents "the transverse strain" and 33 represents "the longitudinal strain.")

4.5.3.2 Blends for Sensing-Actuating Dual Functionality

Recently a sensing-actuating dually functional polymeric material was developed at NASA Langley Research Center (Su et al., 2001b and 2001c) using electroactive graft elasomer (g-elastomer) and piezoelectric poly(vinylidene fluoride-trifluoroethylene) [P(VDF-TrFE)] copolymer. The blend contains both piezoelectric and electrostrictive components after electric field poling (EFP). The advantage of this kind of sensing-actuating dually functional material has functions for both piezoelectric response for sensors and electrostrictive response for actuators at the same time if the driving field for actuation is not beyond the depolarization field for the piezoelectric component. The dual functionality can make the process of integrated multifunctional systems, such as micro-electro-mechanical systems (MEMS), much easier since a single film can be used for both sensor and actuator elements. Therefore, the occupied space, the weight, and the processing cost can be reduced while the device's density can be increased.

4.5.3.3 Piezoelectricity and Electrostriction

The piezoelectric strain coefficient in the blend shows the dependence on the copolymer content since the piezoelectricity is essentially controlled by the remanant polarization that is proportional to the amount of the copolymer. Figure 32 shows the effects of the poling field on the piezoelectric coefficient, d_{31}, and the electric-field-induced longitudinal strain, S_{31}, of the films. The unpoled films have low piezoelectric coefficients while the electric-field-induced strain is relatively high. The piezoelectric coefficient increases with the increase of the

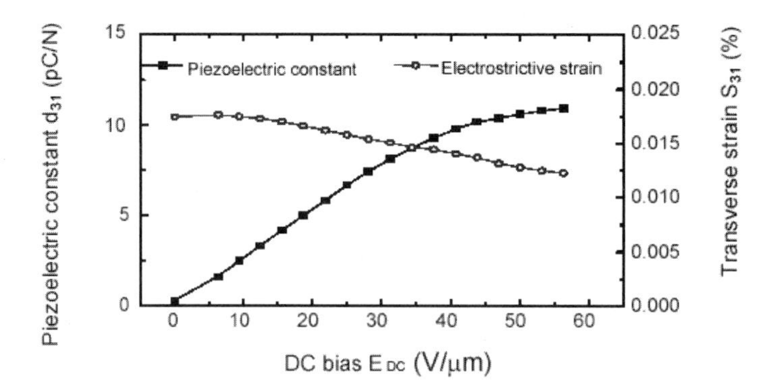

Figure 32: Effects of the poling field and the driving field on the piezoelectric strain coefficient (solid symbols and right axis) and the electrostrictive strain (open symbols and lift axis) of the piezo-electrostrictive of the 50/50 blend.

poling field while the poling treatment decreases the electric-field-induced strain of the films. The increase of the piezoelectric coefficient with the increase of the poling field is due to the fact that the higher poling field can align the piezoelectric copolymer component more; therefore higher remanent polarization, which is significant for the piezoelectric response. However, the electric-field-induced strain, or electrostriction, is a response of the polar domains in the film to the applied electric field. The unpoled films contain polar domains in a random state while the poled films have the piezoelectric copolymer component permanently aligned. The system containing aligned piezoelectric copolymers may reduce the amount of the contributors to electrostrictive response and increase the barrier to the electrostrictive response. Therefore, the highest induced electrostrictive strain is given by the unpoled blend system.

Figure 33 shows the effects of a dc bias electric field on the piezoelectric strain coefficient, d_{31} and the electrostrictive strain, S_{31}. The piezoelectric strain coefficient increases when the dc bias field increases while the electrostrictive strain decreases with the increased dc bias field. The increase in the piezoelectric response occurs because the bias field can enhance the alignment of the piezoelectric copolymer in the blend resulting in an increase in the remanent polarization. When the bias field is high enough, the enhancement of the bias field on the piezoelectric response becomes less important because the alignment is close to a saturated state. This is reflected when the bias field is higher than about 40 V/μm. On the other hand, the increase of the bias field generates higher restricting force on the electrostrictive polar domains, preventing their response, thus decreasing the electrostrictive strain.

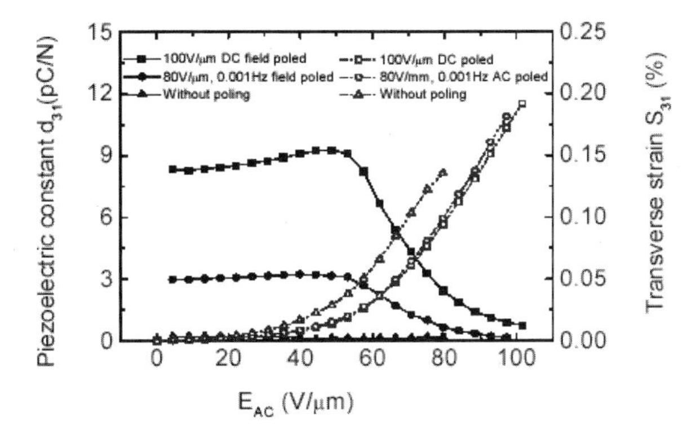

Figure 33: The effects of the dc bias field on the piezoelectric response (solid symbols and right axis) and the electrostrictive strain (open symbols and left axis) of the piezo-electrostrictive polymer blend.

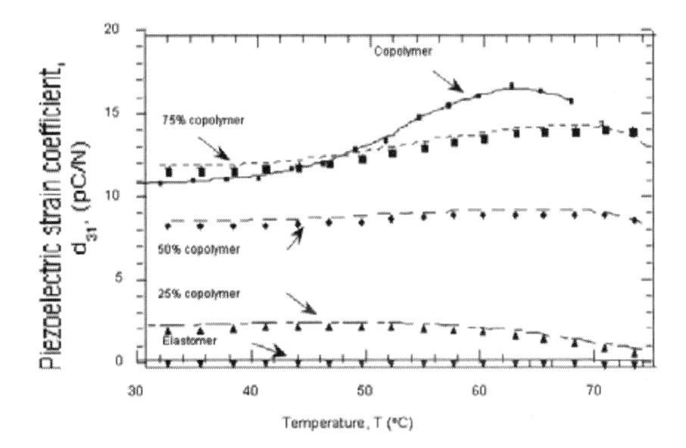

Figure 34: Temperature dependence of the piezoelectric strain coefficient, d_{31}, of the film blends (1 Hz) as a function of the relative composition [Su et al. 200b]. (Copyright IEEE.)

4.5.3.4 Temperature Dependence of the Piezoelectric Property

The piezoelectric strain coefficient, d_{31}, of the poled films was measured as a function of temperature. The results are shown in Fig. 34. The piezoelectric strain coefficient, d_{31}, increases with copolymer content. However, the film blend can offer the highest d_{31} (75 wt.% copolymer) due to the combinational effects of the electric polarization and mechanical modulus on the piezoelectric strain response. For the same reason, a particular composition (50 wt.% piezoelectric copolymer) can result in almost temperature independence of the piezoelectric strain coefficient, d_{31}, from the room temperature to the temperature of 70°C. The temperature independence of the piezoelectric strain coefficient simplifies device design and offers advantages for applications of the electroactive polymers.

4.5.3.6 Computational Study on Mechanisms of Electrostrictive Graft Elastomers

The mechanisms of electrostriction in the graft elastomers are explained in a 2D model recently established by Wang et al. (2003). The model indicates that the mechanisms of the electric-field-induced strain in the electrostrictive graft elastomer are attributed to the rotation of grafted polar crystal units and the reorientation of flexible backbone chains when an electric field is applied on the material. Figure 35 shows the element of the model and the response of the element to an electric field.

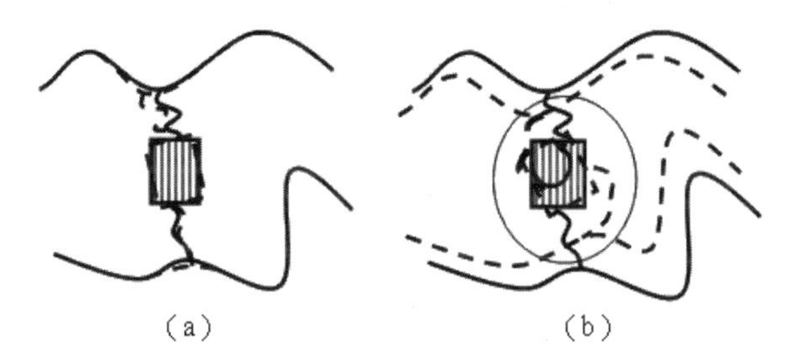

Figure 35: Schematics of (a) the element of the model and (b) the element responding to an electric field [Wang et al., 2003].

4.6 Field-Induced Strain Due to Maxwell Stress Effect

Recently, the Maxwell stress-induced strain in polymers has gained a great deal of attention. In 1994, Zhenyi et al. observed that in certain polyurethane elastomers, a high elastic strain (>3%) can be induced by an external electric field (Zhenyi et al., 1994). Since then, SRI International has investigated the field-induced strain response in soft polymer elastomers extensively, and it was demonstrated recently that a thickness strain of more than 60% and lateral strain of more than 150% can be achieved in some very soft polymer elastomers (Young's modulus ~ 1 MPa) after being mechanically stretched [Kornbluh et al., 2000; Pelrine et al., 2000]. Such a high strain level in the elastomers results in a high elastic energy density (>1 MJ/m^3). Table 4 summarizes some of the results reported by the research group at SRI.

Table 4: Electromechanical responses of some elastomers.

	HS3	CF19-2186	VHB 4910
Prestrain (*x*,*y*) (%)	(280)	(100)	(540.75)
Thickness Strain (%)	-54	-39	-68
Lateral Strain (%)	117	63	215
Field applied (MV/m)	128	181	239
Elastic energy density (MJ/m^3)	0.16	0.2	1.36
Stress (MPa)	0.4	0.8	2.4

It is interesting to note that due to the E^2 dependence of the Maxwell stress, any nonuniform field distribution across the thickness direction can result in enhancement of the strain response. That is, the strain can be much larger than that predicted from Eqs. (26) or (27), which was observed in a polyurethane elastomer thin film where a strain of –3.5% can be induced under an external field of 20 MV/m (Young's modulus ~ 20 MPa and dielectric constant $\varepsilon/\varepsilon_0 \sim 7$) [Zhenyi et al., 1994]. Furthermore, in the same polyurethane elastomer, it was also observed that the field-induced strain strongly depends on the film thickness [Su et al., 1997]. Although the strain response in a thick sample (~ 1-mm thickness) can be accounted for mostly by the Maxwell strain, the strain in a 0.1-mm-thick film is much higher than that in thick samples. For example, under a field of 7 MV/m, a strain of –0.85% can be induced in a 0.1-mm thick polyurethane film coated with Au [Su et al., 1997]. Such a high strain response is caused by the charge injection in the polyurethane elastomer, which results in a surface-charged region and strongly affects the field distribution in thin film samples. This effect may also be interesting and useful for electrostrictive polymers since the electrostrictive strain also depends on the square of the local electric field.

For the Maxwell stress-induced strain, under a dc bias field, an effective piezoelectric state can be induced with the effective piezoelectric coefficient $d \sim 2\ \varepsilon s_{11} E_{DC}$ [from Eq. (22)]. From Eq. (13), the coupling factor at this dc field-biased state can be deduced:

$$k_{33}^2 = 4\varepsilon\ s_{11} E_{DC}^2 . \tag{28}$$

Therefore, a large dielectric constant and elastic compliance will result in a large coupling factor. In addition, the Maxwell strain can be increased significantly by raising the elastic compliance and dielectric constant of the polymer. However, increasing the elastic compliance to a very high level may not be desirable since it will significantly reduce the response frequency of the system, which is inversely proportional to the square root of the elastic compliance. It may also cause mechanical instability of the polymer when subjected to external mechanical stresses. In comparison, increasing the dielectric constant of the polymer is a more favorable approach, which will be discussed in the next section.

4.7 High Dielectric Constant Polymeric Materials as Actuator Materials

One severe drawback for the newly developed field type EAPs, as has been shown in the proceeding sections, is the high electric field required in order to generate high strain with high elastic energy density. For actuator materials, elastic energy density is a key parameter measuring the strain and stress generation capability of an electromechanical material. From an energy conservation point of view, the output elastic energy density U_s cannot exceed the input electric energy density, which is equal to $K\varepsilon_0E^2/2$. Therefore, a high-input electric energy density is required in order to have a high elastic energy density output. In almost all the current electric-field-activated EAP, the dielectric constant is not very high. Even in the high-energy irradiated P(VDF-TrFE) copolymers that possess the highest room temperature dielectric constant among the known electric EAPs, the dielectric constant (~ 60 at 100 Hz) is still far below that in ceramic systems where the dielectric constant in many cases is much higher than 5000. As a result, a high electric field is required to make up for the low dielectric constant. For example, to generate a U_s of 0.1 J/cm^3, which corresponds to the elastic energy density of the best performed piezoelectric and electrostrictive ceramics, in a polymer with a dielectric constant 10, assuming a 50% energy conversion efficiency, which is very high for the current field type EAPs, the field required is 67 V/μm. All of these considerations point to the need to search for materials and mechanisms that will enhance the dielectric responses significantly in electric-field-activated EAPs, which can improve the coupling factor and induced strain response, and reduce the driving electric field. Recently, two approaches have been investigated along this direction and have shown promising results.

4.7.1 An All-Organic High Dielectric Constant Composite

One of the approaches is based on the dielectric composite concept in which high dielectric constant particulates are blended into a polymer matrix to raise the dielectric constant of the composite. In these new dielectric composites, high dielectric organic solids were used for the fillers, which elastic modulus is not very much different from that of the polymer matrix. As a result, the elastic modulus of the new composites is not very much different from that of the polymer matrix. Furthermore, the P(VDF-TrFE) based electrostrictive polymers were chosen for the matrix, which possess a room-temperature dielectric constant near 60 and a high electrostrictive strain. Both features are highly desirable for a composite to achieve a high dielectric constant and high field-induced strain [Zhang et al., 2002].

In the study reported, copper-phthalocyanine (CuPc) was the high dielectric constant filler (dielectric constant > 10,000), whose molecular structure is shown schematically in Fig. 36. The high dielectric constant of CuPc can be explained in terms of the electron delocalization within CuPc molecules. Figure 37(a) presents the dielectric constant and loss of the composites containing 40 wt% of CuPc as a function of the applied field amplitude. As can be seen, under a field of 15 V/μm, the dielectric constant of the composite can reach more than 400 while the dielectric loss is nearly the same as that of the polymer matrix. More interestingly, the composite exhibits a high field induced strain under a much reduced electric field (~2% strain under a field of 13 V/μm), as shown in Fig. 37(b). In the figure, the electrostrictive strain from the polymer matrix is also

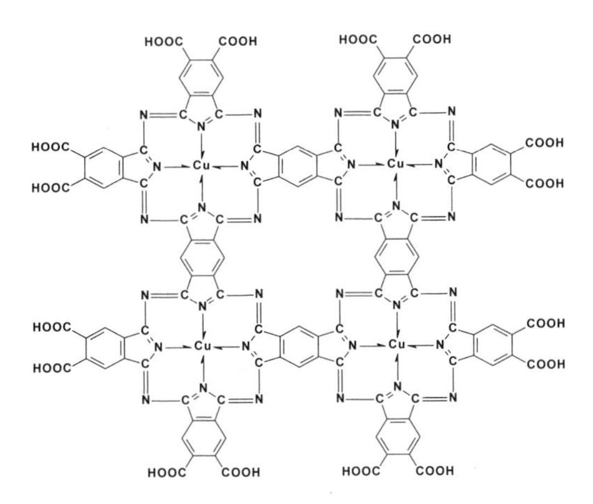

Figure 36: Schematic drawing of CuPc oligomer. (Reprinted by permission from Nature, Zhang et al. copyright 2002, Macmillan Publishers Ltd.)

shown for the comparison (as the dashed line). From the elastic modulus (0.75 GPa), the elastic energy density has also be estimated, which is 0.13 J/cm^3.

From the dielectric and elastic data, the contribution to the strain response from the Maxwell stress and electrostriction can also be estimated, which shows that the strain obtained is about one order of magnitude higher than the estimated strain response assuming a uniform local field throughout the composites. Apparently, in a dielectric composite, there exists a nonuniform electric field distribution in the polymer matrix due to the high dielectric constant filler. This nonuniform field distribution can significantly enhance the strain response if thestrain is proportional to the square of the local field. Furthermore, in a recent theoretical modeling on composites consisting of a high-dielectric constant and a low-dielectric constant media, it was shown that the exchange coupling at the interface between a high dielectric and a low dielectric media can result in a significant change in the local polarization level at the interface region. This can result in a large enhancement in both the dielectric constant and strain response of the composite compared with the case when there is no such effect if there is a large difference in the dielectric constant between the two media [Li, 2003], which is the case for the CuPc filler in the electrostrictive P(VDF-TrFE) matrix. Because the exchange coupling exists only in the near interface region, by working with a smaller dielectric filler, one can enhance the effect of this exchange coupling to achieve a high electromechanical response under a low applied field in a polymericlike material [Zhang et al., 2002].

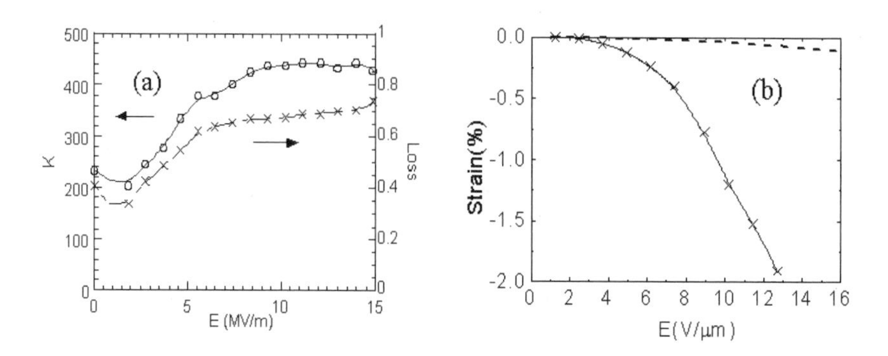

Figure 37: (a) The real part of the dielectric constant (K′) and dielectric loss (D) as a function of the applied field amplitude and (b) the strain amplitude as a function of the applied field amplitude measured at room temperature for the composite with 40 wt% CuPc. For the comparison, the strain from the electrostrictive P(VDF-TrFE) copolymer at the same field range is also shown (the dashed curve) in (b). (Reprinted by permission from Nature, Zhang et al. copyright 2002, Macmillan Publishers Ltd.)

4.7.2 High-Dielectric-Constant All-Polymer Percolative Composites

It has been known that in a composite containing metal (conductive) particulates when the metal concentration f is close to the percolation concentration f_c, there is a marked increased in the dielectric constant, i.e.,

$$K = K_m \left\{ \frac{f_c - f}{f_c} \right\}^{-q}, \tag{29}$$

where K_m is the dielectric constant of the insulation matrix, and q is a critical index (~1 for a three dimensional composite) [Efros and Shklovskii, 1976]. Compared with the approach of high dielectric constant particles in a polymer matrix, the percolative approach requires much lower-volume concentration f of the filler to raise the dielectric constant.

In the all-polymer percolative composites reported, a conductive polymer, polyaniline (PANI) was used as the conductive filler. Because it has a relatively low elastic modulus (~ 2.3 GPa) in comparison to metal particles, it will not change the modulus of the composite very much from that of the polymer matrix. A P(VDF-TrFE-CTFE) terpolymer was used for the matrix. As shown in Fig. 38, a composite with f = 23% exhibits a dielectric constant near 2000, a significant increase compared with the electrostrictive terpolymer matrix. For the same composite, a thickness strain of more than 2.6% can be induced under a field of less than 16 V/μm, a significant reduction compared with the field required for the electrostrictive terpolymers [Huang et al., 2003a]. The elastic modulus of the composite is 0.535 GPa and the corresponding elastic energy density is 0.18

J/cm^3. The data analysis indicates that the induced strain is at least 3 times higher than that estimated from the combined contribution of the Maxwell stress and electrostriction. Apparently, the nonuniform field distribution in the composites and the interface effects, as have been discussed in the proceeding section, may contribute to the observed strain-enhancement effect.

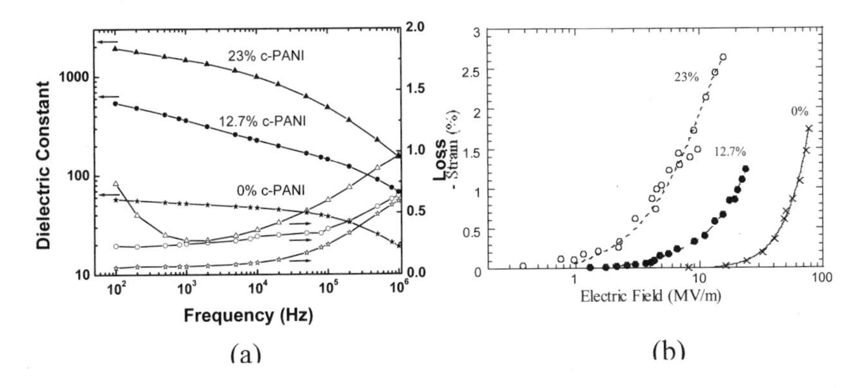

Figure 38: (a) The dielectric constant and (b) the strain amplitude as a function of the applied-field measured at room temperature for the composites with 12.7% and 23% *c*-PANI. For comparison, the dielectric constant and strain from the electrostrictive P(VDF-TrFE-CTFE) terpolymer are also shown (0%) [Huang et al., 2003a]. (Copyright 2003, American Institute of Physics.)

4.8 Electrets

A polymer electret is a polymeric dielectric, which, when subjected to an applied electric field, can retain a state of polarization after the field is removed. Placing conducting electrodes on opposing sides of a polymer film and applying the required voltage to obtain the devised electric field usually performs this process, called direct or dc poling.

Electrets can also be formed by the injection of real charges in a process known as corona poling or by electron beam charging. In the former case, a high dc voltage applied to a wire creates a corona charge that can be directed toward the surface of a polymer film by use of a fine metal mesh screen held at the appropriate potential. In the latter case, an electron beam is directed toward the surface of a polymer film with the injected electrons forming a negatively charged space layer.

The polarization of an electret may result from the freezing-in of the orientation of real dipoles into the applied field direction. This is usually done by heating a polar polymer above its glass transition temperature, T_g, applying the electric field, and then cooling the polymer below T_g before removing the field. It may also result from the injection of real charges into the polymer film, which become trapped as temperature is decreased, or trapped at heterogeneities within the polymer. The interfaces between crystalline and amorphous regions or at the interface between phase separated block copolymers are examples of such heterogeneities.

In the years since the initial work of Eguchi [1922] on the solidification of wax electrets, polymer electrets have become commercially important in applications including microphones, sensors, transducers, and filters. They can exhibit stable piezoelectric properties comparable or exceeding those found in ferroelectric polymers such as uniaxially oriented PVDF or its copolymers with trifluoroethylene.

More recently, a new type of electrets based on charged microporous foam that exhibited piezoelectric coefficient much higher than those observed in the ferroelectric PVDF and its copolymers was first introduced by VTT Chemical Technology [Paajanen et al., 2000; Sessler and Hillenbrand, 1999; and Neugschwandtner et al., 2000]. The quasi-static piezoelectric coefficient d_{33} measured from foamed polypropylene (PP) foils, and, as can be seen in Fig. 39, a d_{33} higher than 300 pC/N was achieved [Hillenbrand et al., 2002]. One of the

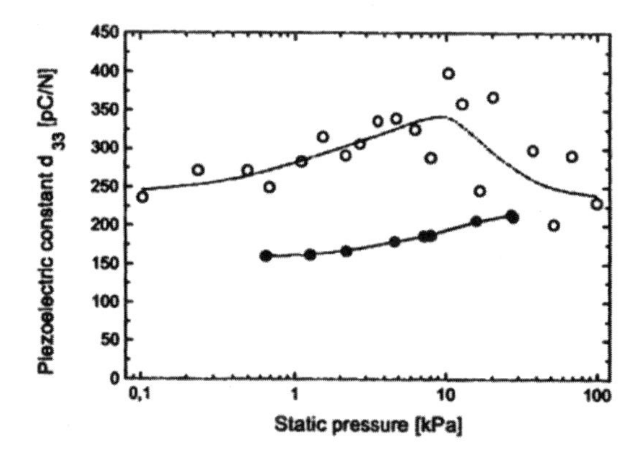

Figure 39: Quasi-static d_{33} -constant of a nonstretched (closed circles) and a stretched cellular PP foil (open circles) [Hillenbrand et al., 2002]. (Copyright IEEE.)

interesting features of this new class of electret is its microporous structure in which macroscopic dipoles form in the voids, as illustrated in Fig. 40 [Neugschwandtner et al., 2000]. During corona poling, the large electric fields in the voids cause Paschen breakdown of the gas. The charges generated during the plasma discharge are trapped in the surface states where oriented macroscopic dipoles are formed. The large piezoelectric response originates from the low elastic modulus of the cellular structure and a small stress can induce large dimensional change of the voids. As a result, a large change of the macroscopic dipole density occurs.

Both the quasi-static and dynamic piezoelectric responses of the microporous foam PP electrets have been systematically investigated. The electrets show a quasi-static elastic modulus of 0.95 MPa and a dynamic elastic modulus of 2.2 MPa. At 600 kHz, the piezoelectric d_{33} coefficient can still reach 140 pC/N.

Typical piezoelectric resonance spectra from the foam PP electrets are shown in Fig. 41 [Neugschwandtner et al., 2000].

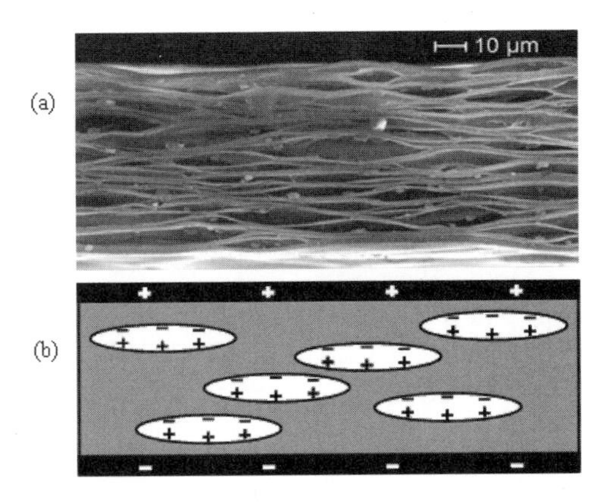

Figure 40: (a) Scanning electron micrograph of the cross section of a charged piezoelectric-polymer foam. (b) Schematic representation of the nonsymmetric charge distribution in the foam [Neugschwandtner et al., 2000]. (Copyright 2000, American Institute of Physics.)

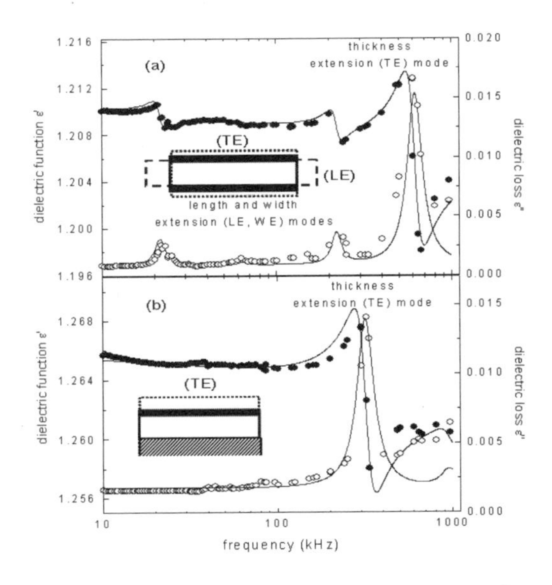

Figure 41: (a) Dielectric function ε′ (full circles) and loss ε″(open circles) of a mechanically free, charged polypropylene foam, together with fit curves. (b) ε′ and ε″ of the same sample in clamped conditions with fit curves. Mechanical clamping suppresses length and width extension modes and reduces the antiresonance frequency of the thickness extension mode [Neugschwandtner et al., 2000]. (Copyright 2000, American Institute of Physics.)

The thermal stability of the foam PP electrets was also studied and it was found that the electrets were safely operated at temperatures below 50°C. Above that, the stored charges became unstable [Neugschwandtner et al., 2001]. To address this issue, other porous or cellular polymers with strong piezoelectric activity and higher thermal stability have been investigated. It was found that another microporous electret, corona-charged sandwich films of porous and nonporous amorphous fluoropolymer (PTFE), exhibit both strong piezoelectricity (d_{33} ~600 pC/N) and thermal stability of the piezoelectric response to above 130°C. Figure 42 shows the dynamic elastic stiffness, piezoelectric coefficient d_{33}, and the electromechanical coupling factor k_{33} as a function of temperature for the porous/nonporous PTFE sandwich electrets film. The coupling factor k_{33} of the sandwich film is about 0.1 and elastic stiffness is less than 0.2 Mpa [Mellinger et al., 2001]. The strong piezoelectric activity, low elastic modulus, and acoustic impedance make this class of electrets very attractive for acoustic transducers used in air. The drawback of this porous/nonporous sandwich structure is the cumbersome manufacture process in comparison with the cellular PP electrets [Gerhard-Multhaupt, 2002]. For porous PTFE electrets, it was found that the pores are usually open and often extend from one film surface to the other. Therefore, it is very difficult to prepare single microporous film with high charge levels in the bulk and with two electrodes on the outer surfaces. A porous/nonporous sandwich film structure is needed and effective to establish stable and high piezoelectric responses.

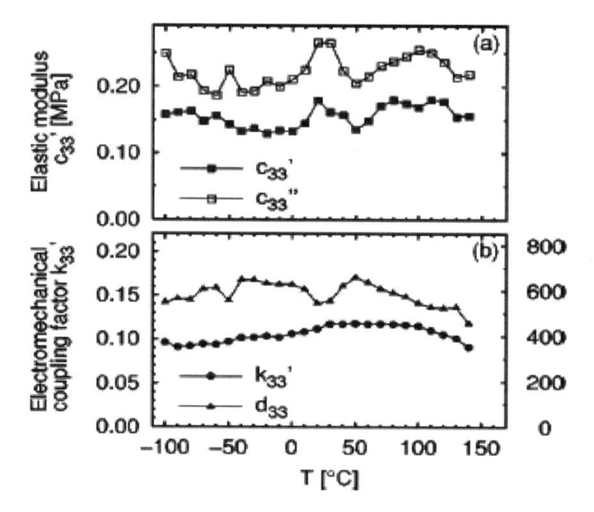

Figure 42: Temperature dependence of the real and imaginary elastic modulus c'_{33} and c''_{33}, the real part of the electromechanical-coupling factor k'_{33}, and the piezoelectric coefficient d_{33} for the porous/nonporous Teflon AF sandwich film [Mellinger et al., 2001]. (Copyright 2001, American Institute of Physics.)

4.9 Liquid-Crystal Polymers

Liquid crystals (LC) are unique molecular materials because of the large anisotropies in many of their properties, such as the molecular level dielectric, optical, and mechanical anisotropies [Demus et al., 1998; Huang et al., 2003b]. The large shape anisotropy of the liquid crystal molecules is especially attractive for actuator applications. As schematically shown in Fig. 43, a reorientation of the liquid crystal molecules can generate large dimensional change. However, because of the liquid elastic properties, liquid crystals can neither sustain nor exert mechanical stress. Hence, pure liquid crystals are not suitable for practical actuator applications. It is the liquid-crystal polymers, in which the rigid mesogenic units are either in the main chain where they alternate with flexible fragments or in the side chains where they are joined to the main chain by flexible spacers, which are interesting for practical actuator device applications. The polymer networks establish the solid elastic properties and couple the local mesogenic reorientation to generate macroscopic strain. Furthermore, in properly designed liquid-crystal polymers, the mesogenic units can still be arranged in various self-organized liquid crystalline phases and the polymer network can be used to preserve such liquid crystal ordering on a macroscopic length scale [Plate, 1993]. For example, extensive works by Finklemann's group have demonstrated that in cross-linked side-chain liquid-crystal elastomers with a monodomain structure, the thermally induced transition between a nematic and isotropic phase can generate a strain of more than 50% [Kundler and Finkelmann, 1998].

Combining the large shape anisotropy with the dielectric anisotropy, liquid-crystal polymers can also exhibit large electromechanical responses (see Fig. 43) [Kremer et al., 2000] Recently, Lehmann et al. [2001] reported that in a ferroelectric liquid crystal elastomer, an electrostrictive strain of up to 4% can be induced under a field of 1.5 MV/m. The reported electrostriction originated from the electroclinic effect in a chiral-smectic phase. The elastic modulus of the elastomer reported is below 3 MPa and hence, the elastic energy density is below 0.002 J/cm^3.

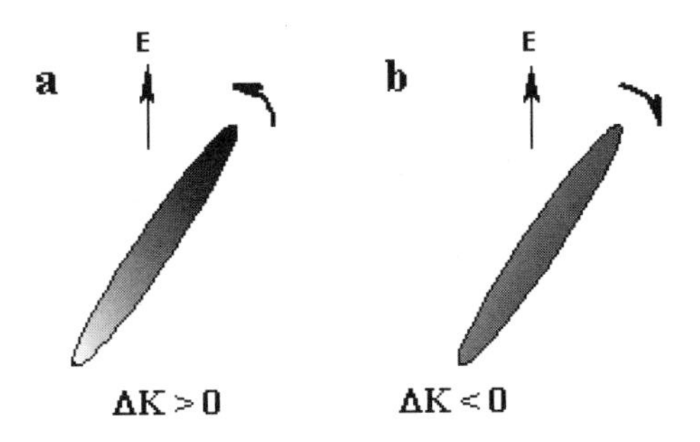

Figure 43: Schematic of the reorientation of a mesogenic molecule under an applied electric field for (a) positive dielectric anisotropy and (b) negative dielectric anisotropy [Huang et al., 2003b]. (Courtesy of Wiley-VCH.)

To raise the elastic energy, another liquid crystal polymer, anisotropic liquid crystal gels (ALCGs) have recently been investigated [Huang et al., 2003b]. ALCGs have exhibited many attractive features. For example, their elastic properties can be easily tailored by varying the volume ration between the mesogenic polymer network and the low-molecular-weight liquid crystals. Compared with other electroactive polymer gels in which actuation is driven by ionic diffusion in the gel network, ALCGs offer a much faster actuation speed due to the direct coupling of the molecular orientation to external fields. As presented in Fig. 44, for a homeotropically aligned ALCG in its nematic phase, a thickness strain of near 2% can be induced under an external field of 25 MV/m.

In addition, the ALCG exhibits a much higher elastic modulus (100 MPa) in comparison with the liquid-crystal elastomers that yield higher elastic energy density. Furthermore, the ALCG also shows a high electromechanical conversion efficiency (75%) under an electric field of 25 MV/m. The elastic energy density of the ALCG is about 0.02 J/cm^3.

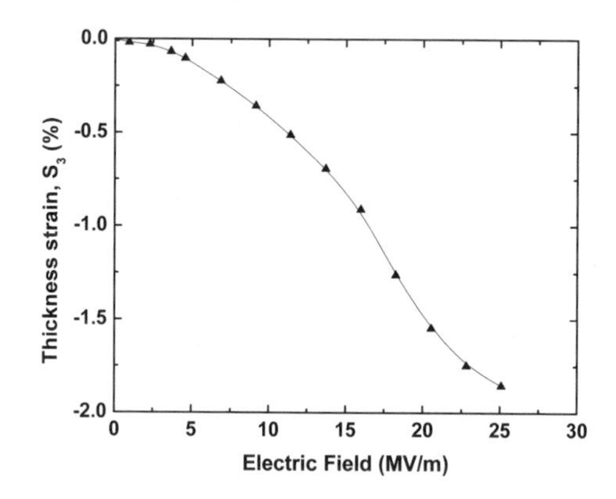

Figure 44: The thickness strain S_3 amplitude as a function of the applied field amplitude measured at room temperature under 1 Hz electric field for the ALCG in homeotropic alignment [Huang et al., 2003b]. (Courtesy of John Wiley & Sons.)

4.10 Acknowledgments

This work was supported by NIH under Grant 8R01-EB002078-04, ONR under Grant N000140210418, DARPA under contract N00173-99-C-2003, and NASA NAG-1-03008. We also acknowledge the assistance of Dr. J. Scheinbeim, Dr. Tian-bing Xu, and Mr. T. Wilkers.

4.11 References

Berlincourt, D.A., D.R. Curran, and H. Jaffe, "Piezoelectric and piezomagnetic materials and their function in transducers in physical acoustics," edited by W. P. Mason, Vol. I, Part A, Academic Press, New York, (1964).

Bharti, V., Z.-Y. Cheng, S. Gross, T.-B. Xu, and Q.M. Zhang, "High electrostrictive strain under high mechanical stress in electron irradiated poly(vinylidene fluoride-trifluorethylene) copolymer," *Appl. Phys. Lett.* 75, 2653 (1999).

Bharti, V., H. Xu, G. Shanthi, Q. M. Zhang, and K. Liang, "Polarization and structural properties of high energy electron irradiated P(VDF-TrFE) copolymer films," *J. Appl. Phys.* 87, 452-461 (2000).

Chen, G.C., J. Su, and L.J. Fina, "FTIR-ATR studies of drawing and poling in polymer bilaminate films," *J. Polym. Sci. Part B,* 32, 2065 (1994).

Cheng, Z.-Y., T.-B. Xu, V. Bharti, S. Wang, and Q.M. Zhang, "Transverse strain responses in the electrostrictive poly(vinylidene fluoride-trifluoroethylene) copolymer," *Appl. Phys. Lett.* 74, 1901-1903 (1999).

Cheng, Z.-Y., V. Bharti, T.-B. Xu, H. Xu, T. Mai, and Q.M. Zhang, "Electrostrictive poly(vinylidene fluoride-trifluoroethylene) copolymers," *Sensors and Actuators A. Phys.* 90, 138 (2001).

Davis, G.T, T. Furukawa, A.J. Lovinger, and M.G. Broadhurst, "Structural and dielectric investigation on the nature of the transition in a copolymer of vinylidene fluoride and trifluoroethylene (52/48 mol%)," *Macromolecules* 15, 329 (1982).

Demus, D., J.W. Goodby, G.W. Gray, H.-W. Spiess, and V. Vill, *Handbook of Liquid Crystals,* Wiley-VCH, Weinheim (1998).

Devonshire, A.F., "Theory of ferroelectric," *Philosophical Magazine* 3 (10), 86 (1954).

Dvey-Aharon, H., T.J. Sluckin, and P.L. Taylor, "Kink propagation as a model for poling in poly(vinylidene fluoride)," *Phys. Rev. B* 21, 3700 (1980).

Efros, A. L. and B. I. Shklovskii, "Critical behavior of conductivity and dielectric constant near the metal-nonmetal transition threshold," *Phys. Status Solidi B* 76, 475 (1976).

Eguchi, M., "On the permanent electret," *Phil. Mag.*, 49, 178, (1925).

Esayan, S. Scheinbeim, J. I. and Newman, B. A., "Pyroelectricity in nylon 7 and nylon 11 ferroelectric polymers," *Appl. Phys. Lett.*, 67, 623 (1995).

Fernandez, M.V., A. Suzuki, and A. Chiba, "Study of annealing effects on the structure of vinylidene fluoride-trifluoroethylene copolymers using WAXS and SAXS," *Macromolecules* 20, 1806 (1987).

Furukawa, T. and N. Seo, "Electrostriction as the origin of piezoelectricity in ferroelectric polymers," *Jpn. J. Appl. Phys.* 29 (4), 675 (1990).

Gao, Q., Scheinbeim, J. I. and Newman, B. A., "Ferroelectric properties of nylon 11 and poly(vinylidene fluoride) blends," *J. Polym. Sci. Part B*, 37, 3217 (1999).

Gao, Q. and Scheinbeim, J. I., "Dipolar intermolecular interactions, structure development, and electromechanical properties in ferroelectric polymer blends of Nylon-11 and poly(vinylidene fluoride)," *Macromolecules* 33, 7564. (2000).

Kawai, H. and Heiji, I., "Electrostriction and piezoelectricity of elongated polymer films," *Oyo Butsuri* 39, 413 (1970).

Gerhard-Multhaupt, R., "Voided polymer electrets—new materials, new challenges, new chances," *Proc. 11th Intern. Symp. Electrets,* 36 (2002).

Green, J.S., B.I. Farmer, and J.F. Rabolt, "Effect of thermal and solution history on the curie point of VF2-TrFE random copolymers," *J. Appl. Phys.* 60 (8), 2690 (1986).

Gross, S.J., Z.-Y. Cheng, V. Bharti, and Q.M. Zhang, "Mechanical load effects on the electrostrictive strain of P(VDF-TrFE) copolymer and the development of a high-resolution hydrostatic-pressure dilatometer," *Proc. IEEE 1999 Int. Symp. Ultrasonics,* 1019-1024 (1999).

Guy, I.L. and J. Unworth, "Observation of a change in the form of polarization reversal in a vinylidene fluoride/trifluoroethylene copolymer," *Appl. Phys. Lett.* 52, 532 (1988).

Hasegawa, R., Y. Takahashi, Y. Chatani, and H. Tadokoro, "Crystal structures of three crystalline forms of poly(vinylidene fluoride)," *Polymer J.* 3, 600 (1972).

Hillenbrand, J., Z. Xia, X. Zhang, and G. M. Sessler, "Piezoelectricity of cellular and porous polymer electrets," *Proc. 11th Intern. Symp. Electrets,* 46 (2002).

Holland, R., "Representation of dielectric, elastic, and piezoelectric losses by complex coefficients," *IEEE Transactions on Sonics and Ultrasonics*, Su-14 (1), 18 (1967).

Hom, C., S. Pilgrim, N. Shankar, K. Bridger, M. Masuda, and S. Winzer, "Calculation of quasi-static electromechanical coupling coefficients for electrostrictive ceramic materials," *IEEE Trans. Ultrason. Ferro. Freq. Cntr.* 41, 542-551 (1994).

Huang, Cheng, Q.M. Zhang, and Ji Su, "High-dielectric-constant all-polymer percolative composites," *Appl. Phys. Lett.* 82, 3502 (2003a).

Huang, Cheng, Q.M. Zhang, and Antal Jákli, "Nematic anisotropic liquid crystal gels: self-assembled nanocomposites with high electromechanical response," *Adv. Funct. Mater.* 13, 525 (2003b).

Huang Cheng, R. Klein, Feng Xia, Hengfeng Li, Q. M. Zhang, Francois Bauer, and Z.-Y. Cheng, "Poly(vinylidene fluoride-trifluoroethylene) based high performance electroactive polymers," Submitted to *IEEE Trans. Diel. & Eelc. Insu.* (2003c).

IEEE Standard on Piezoelectricity, *ANSI/IEEE Std 176-1987*, IEEE New York, (1988).

Jona, F. and G. Shirane, *Ferroelectric Crystals*, Dover Publications, New York, p. 138 (1993).

Kawai, H. and Heiji. I, Oyo Butsuri 39, 413 (1970).

Kinase, W. and H. Takahashi, *J. Phys. Soc. Jpn.* 10, 942 (1955).

Kloos, G., "The correction of interferometric measurements of quadratic electrostriction for cross effects," *J. Phys.* D 28, 939 (1995).

Kornbluh, R., R. Pelrine, Q. Pei, S. Oh, and J. Joseph, "Ultrahigh strain response of field-activated elastomeric polymers," *Smart Structures and Materials 2000*, Ed., Yoseph Bar-Cohen, Proc. SPIE Vol. 3987, pp. 51-64 (2000).

Kremer, F., H. Skupin, W. Lehmann, L. Hartmann, P. Stein, and H. Finkelmann, "Structure, mobility, and piezoelectricity in ferroelectric liquid crystalline elastomers," *Adv. Chem. Phys.* 113, 183 (2000).

Kundler, I, and H. Finkelmann, "Director reorientation via stripe-domains in nematic elastomers: influence of cross-link density, anisotropy of the network and smectic clusters," *Macromol. Chem. Phys.* 199, 677 (1998).

Landau, L. D. and E. M. Lifshitz, *Electrodynamics of Continuous Media*, Pergamon, Oxford (1970).

Lee, J. W. Takase, Y. Newman, B. A. and Scheinbeim, J. I., "Ferroelectric polarization switching in Nylon-11," *J. Polym. Sci.Part B* 29, 273 (1991).

Lehmann, W., H. Skupin, C. Tolksdorf, E. Gebhard, R. Zentel, P. Krüger, M. Lösche, and F. Kremer, "Giant lateral electrostriction in ferroelectric liquid-crystalline elastomers," *Nature* 410, 447, (2001).

Li, Jiangyu, "Exchange coupling in P(VDF-TrFE) copolymer based all-organic composites with giant electrostriction," *Phys. Rev. Lett.* 90, 217601 (2003).

Lines, M.E. and A.M. Glass, *Principles and Applications of Ferroelectrics and Related Materials*, Clarendon Press, Oxford (1977).

Lovinger, A. J., "Unit cell of the β phase of poly(vinylidene fluoride)," *Macromolecules* 14, 322 (1981).

Lovinger, A.J., G.T. Davis, T. Furukawa, and M.G. Broadhurst, "Crystalline forms in a copolymer of vinylidene fluoride and trifluoroethylene (52/48 mol %)," *Macromolecules* 15, 323 (1982).

Lovinger, A.J., "Poly(vinylidene fluoride), in developments in crystalline polymers-1," Ed. D. C. Bassett, Applied Science Publishers, London and New Jersey, p. 195 (1982).

Lovinger, A. J. and T. Furukawa, *Ferroelectrics* 50, 227 (1983).

Mabboux, P., and K. Gleason, "F-19 NMR characterization of electron beam irradiated vinylidene fluoride-trifluoroethylene copolymers," *J. Fluorine Chem.* 113, 27 (2002).

Mason, W. P., *Physical Acoustics: Principles and Methods*, Academic Press, New York, Vol. I, Part A, Chapter 3 (1964).

McCrum, N.G., B.E. Read, and G. Williams, *Anelastic and Dielectric Effects in Polymeric Solids*, Dover Publications Inc., New York, Chapter 4 (1991).

Mei, B. Z. Scheinbeim, J. I.and Newman, B. A., "The ferroelectric behavior of odd-numbered Nylons," *Ferroelectrics* 144, 51 (1993).

Mellinger, A., M. Wegener, W. Wirges, and R. Gerhard-Multhaupt, "Thermally stable dynamic piezoelectricity in sandwich films of porous and nonporous amorphous fluoropolymers," *Appl. Phys. Lett.* 79, 1852 (2001).

Nalwa, H. S., L. Dalton, and P. Vasudevan, "Dielectric properties of copper-phthalocyanine polymer," *Eur. Polym. J.* 21, 943-947 (1985).

Newman, B. A. Chen, P. Pae, K. D. and Scheinbeim, J. I., "Piezoelectricity in Nylon 11," *J. Appl. Phys.* 51, 5161 (1980).

Neugschwandtner, G. S., et al., "Large and broadband piezoelectricity in smart polymer-foam space-charge electrets," *Appl. Phys. Lett.* 77, 3827 (2000).

Neugschwandtner G. S., R. Schwodiauer, S. bauer-Gogonea, S. Bauer, M. Paajanen, and J. Lekkela, "Piezo- and pyroelectricity of a polymer-foam space-charge electrets," *J. Appl. Phys.* 89, 4503 (2001).

Newnham, R.E., V. Sundar, R. Yimmirun, J. Su, and Q.M. Zhang, "Electrostriction in dielectric materials," *Ceramic Trans.* Vol. 88, 15-39 (1998).

Nye, J.F., *Physical Properties of Crystals*, Clarendon Press, Oxford (1987).

Ohigashi, H. and T. Hattori, "Improvement of piezoelectric properties of PVDF and its copolymers by crystallization under high pressure," *Ferro.* 171, 11 (1995).

Omote, K., H. Ohigashi, and K. Koga, "Temperature dependence of elastic, dielectric, and piezoelectric properties of "single crystalline" films of vinylidene fluoride trifluoroethylene copolymer," *J. Appl. Phys.* 81 (6), 2760 (1997).

Paajanen, M., J. Lekkala, and K. Kirjavanen, "Electromechanical films (EMFi)-a new multipurpose electret materials," *Sen. Actuat.* 84, 95 (2000).

Pelrine, R., R. Kornbluh, Q. Pei, and J. Joseph, "High-speed electrically actuated elastomers with strain greater than 100%," *Science*, 287, 836 (2000).

Plate, N. A., (editor), *Liquid-Crystal Polymers*, Plenum Press, NY (1993).

Scheinbeim, J., C. Nakafuku, B.A. Newman, and K.D. Pae, "High-pressure crystallization of poly(vinylidene fluoride)," *J. Appl. Phys.*, 50, 4399 (1979).

Scheinbeim, J. I. Lee, J. W. and Newman, B. A., "Ferroelectric polarization mechanisms in Nylon 11," *Macromolecules* 25, 3729 (1992).

Scheinbeim, J. I. and Newman, B. A., "Electric field induced changes in odd-numbered Nylons," *Trends in Polym. Sci.*, 1, 384 (1993).

Schewe, H., "Piezoelectricity of uniaxially oriented polyvinylidene fluoride," *Ultrasonics Symposium Proceedings,* Vol. 1, IEEE, New York (1982).

Sessler, G. M. and J. Hillenbrand, "Electromechanical response of cellular electret films," *Appl. Phys. Lett.* 75, 3405 (1999).

Sharples, A., *Introduction to Polymer Crystallization*, St. Martin's Press, New York (1966).

Shkel, Y.M. and D.J. Klingenberg, "Electrostriction of polarizable materials: comparison of models with experimental data," *J. Appl. Phys.* 83, 415 (1998).

Slichter, W. P., "Crystal structures in polyamides made from ω-amino acids," *J. Polym. Sci.* 36, 259 (1959).

Stack, G.M. and R.Y. Ting, "Thermodynamic and morphological-studies of the solid state transition in copolymers of vinylidene fluoride and trifluoroethylene," *J. Polym. Sci. Part B* 26, 55 (1988).

Stark, K. H. and C. G. Garton, "Electric strength of irradiated polythene," *Nature* 176, 1225 (1955).

Su, J., "Ferroelectric and piezoelectric properties of ferroelectric polymer composite systems," Ph.D Thesis, Rutgers University (1995).

Su, J. Hales, K. and Xu, T. B. "Composition and annealing effects on the response of electrostrictive graft elastomers," *Proceedings of SPIE Smart Structures and Materials*, Vol.5051, 191 (2003).

Su, J, Harrison, J. S. and St. Clair, T. *"Electrostrictive Graft Elastomers,"* U. S. Patent No.6,515,077.

Su, J., J. S. Harrison, T. Clark, Y. Bar-Cohen, and S. Leary, "Electrostrictive graft elastomer and applications," *Mat. Res. Soc. Symp. Proc.* Vol. 600, 131 (2000a).

Su, J. Ma, Z. Y. Scheinbeim, J. I. and Newman, B. A., "Ferroelectric and piezoelectric properties of Nylon 11/poly(vinylidene fluoride) bilaminate films," *J. Polym. Sci. Part B*, 33, 85 (1995).

Su, J., Q. M. Zhang, and R. Y. Ting, "Space charge enhanced electromechanical response in thin film polyurethane elastomers," *Appl. Phys. Lett.* 71, 386 (1997).

Su, J. Harrison, J. S. and St. Clair, T., "Novel polymeric elastomers for actuation," *Proceedings of IEEE International Symposium on Application of Ferroelectrics,* 811, (2000b).

Su, J. Ounaies, Z. Harrison, J. S. Bar-Cohen, Y. and Leary, S., "Electromechanically active polymer blends for actuation," *Proceedings of SPIE-Smart Structures and Materials*, 3987, 65 (2000c).

Su, J. Xu, T. B. and Harrison, J. S., *Proceedings of The First World Congress on Biomemitics and Artificial Muscles*, CD version, (2003).

Sundar, V. and R.E. Newnham, "Electrostriction and polarization," *Ferroelectrics* 135, 431 (1992).

Takahashi, N. and A. Odajima, "On the structure of poly(vinylidene fluoride) under a high electric field," *Ferroelectrics* 57, 221 (1984).

Takahashi, T., M. Dale, and E. Fukada, "Dielectric hysteresis and rotation of dipoles in polyvinylidene fluoride," *Appl. Phys. Lett.* 37 (9), 791 (1980).

Takahashi, Y. and Tadokoro, H., "Crystal structure of form III of poly(vinylidene fluoride)," *Macromolecules* 13, 1317 (1980).

Takahashi, Y., Y. Matsubara, and H. Tadokoro, "Crystal Structure of Form II of PVDF," *Macromolecules* 16, 1588 (1983).

Takahashi, Y., Y. Nakagawa, H. Miyaji, and K. Asai, "Direct evidence for ferroelectric switching in poly(vinylidene fluoride) and poly(vinylidene fluoride-trifluoroethylene) crystals," *J. Polym. Sci.* Part C 25, 153 (1987).

Takase, Y. Lee, J. W. Scheinbeim, J. I. and Newman, B. A., *Macromolecules*, 24, 6644 (1991).

Tanaka, R., K. Tashiro, and M. Kobayashi, "Annealing effect on the ferroelectric phase transition behavior and domain structure of vinylidene fluoride (VDF)-trifluoroethylene copolymers: a comparison between uniaxially oriented VDF 73 and 65% copolymers," *Polymer* 40, 3855 (1999).

Tashiro, K. and M. Kobayashi, "Ferroelectric phase transition and specific volume change in vinylidene fluoride-trifluoroethylene copolymers," *Rep. Progr. Polymer Phys. Jpn.* 29, 169 (1986).

Tashiro, K., K. Takano, M. Kobayashi, Y. Chatani, and H. Tadokoro, "Structural study on ferroelectric phase transition of vinylidene fluoride-trifluoroethylene copolymers (III) dependence of transitional behavior on VdF molar content," *Ferroelectrics* 57, 297-326 (1984).

Wang, H., Q.M. Zhang, L.E. Cross, and A.O. Sykes, "Piezoelectric dielectric, and elastic properties of polyvinylidene fluoride/trifluoroethylene," *J. Appl. Phys.* 74, 3394 (1993).

Wang, Y. Q. Sun, X. K. Sun, C. J. and Su, J. "Two-dimensional computational model for electrostrictive graft elastomer," *Proceedings of SPIE Smart Structures and Materials*, 5051, pp. 100-111 (2003).

Xia, F., Z.-Y. Cheng, H. Xu, H. Li, Q. M. Zhang, G. Kavarnos, R. Ting, G. Abdul-Sedat, and K. D. Belfield, "High electromechanical responses in a poly(vinylidenefluoride-trifluoroethylene-chlorofluoroethylene terpolymer," *Adv. Mater.* 14, 1574 (2002).

Xia, Feng, Y. K. Wang, Hengfeng Li, Cheng Huang, Y. Ma, Q. M. Zhang, Z.-Y. Cheng, Fred B. Bateman. "Influence of the annealing conditions on the polarization and electromechanical response of high energy electron irradiated poly(vinylidene fluoride-trifluoroethylene) copolymer," *J. Poly. Sci. A. Poly. Phys.* 41, 797-806 (2003).

Xu, H., Z.-Y. Cheng, D. Olson, T. Mai, Q. M. Zhang, and G. Kavarnos. "Ferroelectric and electromechanical properties of poly(vinylidene-fluoride-trifluoroethylene-chlorotrifluoroethylene) terpolymer," *Appl. Phys. Lett.*, Vol.78, 2360 (2001).

Zhang, Q.M., J. Zhao, T. Shrout, N. Kim, L.E. Cross, A. Amin, and B.M. Kulwicki, "Characteristics of the electromechanical response and polarization of electric field biased ferroelectrics," *J. Appl. Phys.* 77, 2549 (1995).

Zhang, Q.M., V. Bharti, and X. Zhao, "Giant electrostriction and relaxor ferroelectric behavior in electron irradiated poly(vinylidene fluoride – trifluoroethylene) copolymer," *Science* 280, 2101 (1998).

Zhang, Q. M., Hengfeng Li, Martin Poh, Haisheng Xu, Z.-Y. Cheng, Feng Xia, and Cheng Huang. "An all-organic composite actuator material with high dielectric constant," *Nature*, 419, 284-287 (2002).

Zhao, X., V. Bharti, Q.M. Zhang, T. Ramotowski, F. Tito, and R. Ting, "Electromechanical properties of electrostrictive poly(vinylidene fluoride-trifluoroethylene) copolymer," *Appl. Phys. Lett.* 73, 2054 (1998).

Zhenyi, M., J. Scheinbein, J. W. Lee, and B. Newman, "High field electrostrictive response of polymers," *J. Polym, Sci. B: Poly. Phys.* 32, 2721 (1994).

TOPIC 3.2

Ionic EAP

CHAPTER 5

Electroactive Polymer Gels

Paul Calvert
University of Arizona

5.1 Introduction—the Gel State

The term *gel* resists any precise definition. What is generally meant is a solution or colloidal suspension that undergoes a physical or chemical change to a solid while retaining much of the solvent within the structure. The usual test for gelation of a liquid is that it shows a yield point and a measurable elastic modulus, as opposed to a viscosity. In many cases the yield point is sufficiently low so that the gel may also be considered a structured liquid, with properties closer to mayonnaise than to Jell-O. This puts gel more in the regime of the rheologist than the mechanical engineer. There is a recent, excellent review of the rheology of complex fluids in Larson [1999].

For the purposes of understanding gel muscles it is important to keep the word *complex* in mind. These materials contain several components with quite small interaction energies, so that the structure and properties are very sensitive to the details of composition and structure. Hence, the question "what are the properties of polyacrylic acid gel?" is essentially meaningless without a detailed specification of the state of the gel. Water-swollen polymer structures, such as the Nafion used in IPMCs, (see Chapter 6), can be thought of as gels with a low solvent content and so have much of the complexity to be discussed here.

5.2 Physical Gels

In many cases of suspension of colloidal particles, aggregation results in the formation of a gel. Thus, in clays the presence of positive ions on the flat surfaces of the platelike particles and negative ions on the edges results in the formation of open aggregates with edge-to-surface bonding. These aggregates will break up and reform on mechanical mixing. Such physical gels are not well understood but are used in a wide range of pharmaceuticals, cosmetics, and foods. They might be applicable to actuators but have not really been explored except in the context of electrorheological fluids.

A second class of physical gels includes the familiar gelatin desserts. In this case a solution of long-chain polymer undergoes a phase transition on cooling, which results in aggregation of individual chains into clusters, through either helix formation or crystallization. For short chains, the result would simply be a suspension of aggregates, but long chains may take part in several separate clusters with intervening sections of chain remaining surrounded by solvent. The three-dimensional mesh that is formed gives rise to the gel properties.

The familiar gelatin gel is the product of the dissolution of collagen in water at high temperature. During tissue growth, collagen triple helices aggregate outside the cell to form the strong fibers that make up tendons and are reinforcing elements in skin, cartilage, and bone. After dissolution of collagen at high temperatures, the less organized gelatin gel forms as a result of the rapid aggregation that occurs upon cooling.

Collagen can also be induced to denature in concentrated salt solutions. A collagen-gel engine was constructed by Steinberg et al. [1966], where a belt of the polymer ran over pulleys through concentrated lithium bromide or urea solution and water. The contraction on denaturation drove the system to turn the pulleys, the direction being determined by a difference in radii. The system could be quite efficient on a thermodynamic basis but the power/weight ratio was poor.

Similar gelation processes occur in other proteins and in polysaccharides. Agarose (agar) forms strong gels on cooling dilute solutions. Alginate solutions gel by the addition of calcium ions. Many synthetic polymers form gels from organic solvents. These materials have not been exploited recently for gel muscles, but may be more suitable than many chemical gels in that the clustered structure gives superior strength, as will be discussed below.

5.3 Chemical Gels

The archetypal chemical gel is polyacrylamide, which has been used for many years as a matrix for gel electrophoresis of proteins and has been heavily studied in that context. A standard recipe [Sandler and Karo, 1992] starts from a dilute

(1–10%) aqueous solution of acrylamide monomer, methylene bisacrylamide cross-linking agent (at 5–20% of the acrylamide), a persulfate free-radical catalyst, and a tertiary amine (usually TEMED, tetramethylethylenediamine) as an activator. The activator causes decomposition of the catalyst to produce free radicals, which initiate the polymerization. As the polyacrylamide chains grow by addition of acrylamide monomer, occasional bisacrylamide units are incorporated (see Fig. 1). Where two separate growing chains incorporate opposite ends of the same bisacrylamide, the chains become linked and so a network is built with a cross-link density that depends on the ratio of bisacrylamide to acrylamide.

Acrylamide, but not the polymer polyacrylamide, is a known neurotoxin and causes central nervous system paralysis. In factory workers with continuous exposure to acrylamide, symptoms of intoxication have been reported. Serious problems for laboratory workers have apparently not resulted despite many years of use in molecular biology and biochemistry.

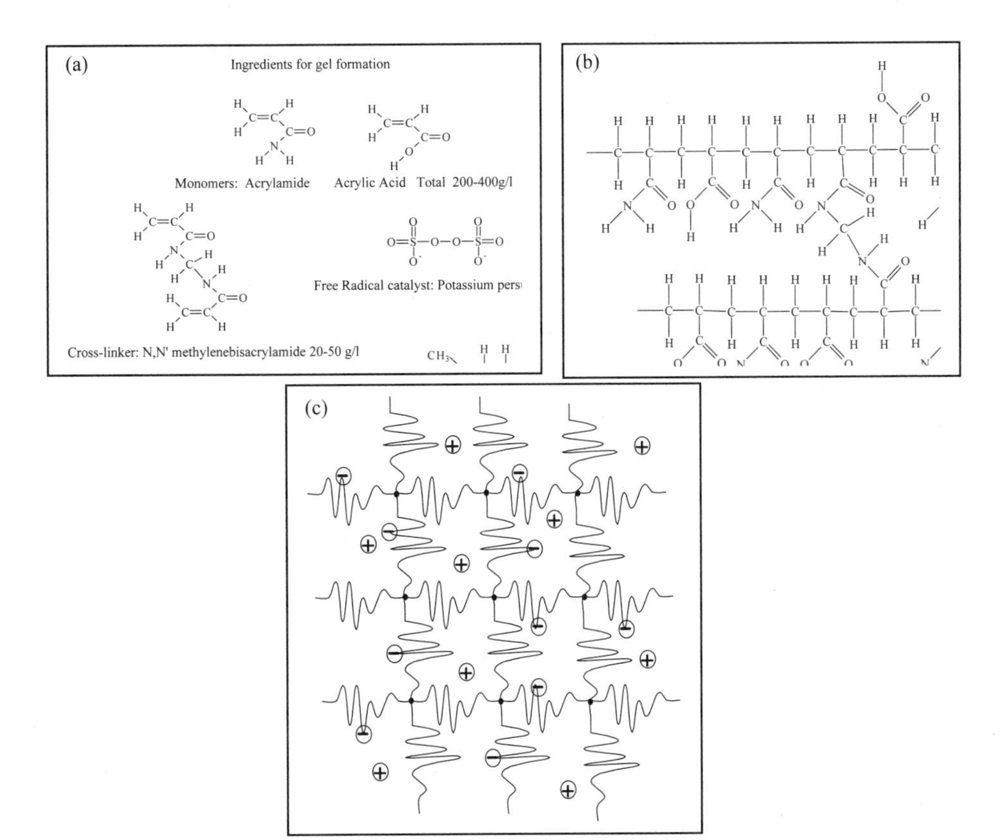

Figure 1: Schematic of gel formation by cross-linking polymerization of acrylamide. (a) Chemical components for gel synthesis. (b) Chemical structure of gel. (c) Schematic of partly charged gel in partly swollen state.

Since the efficient separation of proteins in electrophoresis depends on the gel structure, there has been much interest in defining whether there is some clustering of polyacrylamide chains in the gel. The gels do tend to be slightly cloudy, which implies that there may be some density fluctuations. Electron microscopy is not very informative because most specimen preparation methods would be expected to result in some clustering. Polyacrylamide is not highly water soluble (see thermodynamics of gels, below), so clustering is quite likely. There have been many studies on the degree of inhomogeneity in gels [Candau et al., 1992; Suzuki et al., 1992].

Almost any water-soluble polymer can be prepared as a gel by carrying out the polymerization in the presence of a cross-linking agent or by carrying out a cross-linking reaction on a solution of the polymer. The properties of the gel will be very dependent on the degree of cross-linking. In addition to polyacrylamide gels, which are widely used, polyacrylic acid gels are used as water-absorbers in diapers, and various gel compositions have been studied for soft contact lenses, including polymers of hydroxyethylmethacrylate, some silicones, and vinylpyrrolidone. A range of similar gels has been studied in the context of searching for a synthetic replacement for cartilage [Corkhill et al., 1993].

5.4 Thermodynamic Properties of Gels

Gel actuators change volume by taking up or expelling water. This process is the equivalent of a dissolved polymer chain expanding or contracting as the solvent quality is changed. For instance, polystyrene dissolves well in toluene, and the chain takes on the form of a highly expanded random walk in the solvent (see Fig. 2). As ethanol, a nonsolvent, is added, the polymer chain contracts to increase the fraction of polymer-polymer neighbors. Many gels are simply cross-linked polymer solutions. A cross-linked polystyrene gel swollen with toluene would likewise contract and expel liquid if the gel was exposed to ethanol vapor. Swollen gels of ionized polyacrylic acid in water (a solvent) contract when immersed in acetone (a nonsolvent) [Shiga et al., 1992].

The standard simple theory of polymer solutions is the Flory-Huggins theory [Flory, 1953], in which the key parameter is χ, the polymer-solvent interaction parameter. In brief, if χ is less than one-half, the polymer is quite soluble, and if χ is greater than 0.5, it is insoluble. A swollen gel will expand or contract strongly as χ is shifted around 0.5. χ can be related to the solubility parameter difference between polymer and solvent and so can be estimated from tables and used to design a gel with actuator properties in any solvent combination.

One key polymer property in the Flory-Huggins theory is the molecular weight of the polymer. A lightly cross-linked gel effectively has a molecular weight of infinity. If the polymer is heavily cross-linked, there is an additional complexity in that the polymer structure is itself changed and this can shift the thermodynamics of dissolution. Physical polymer gels, arising from chain clustering or helix formation, would be expected to follow roughly the same

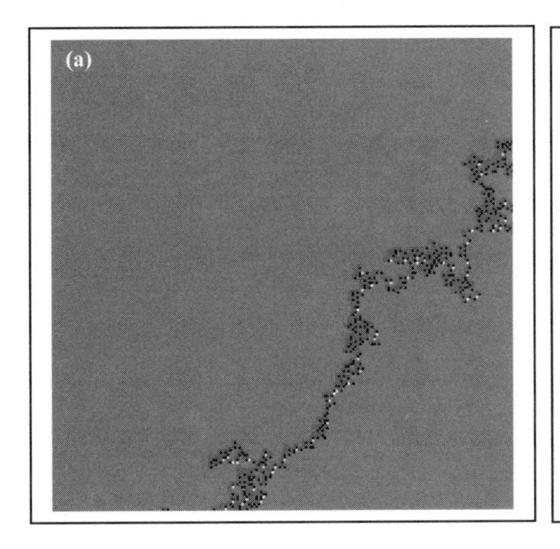

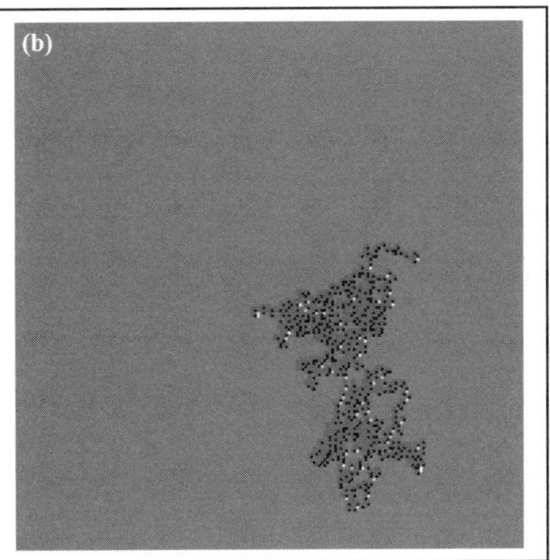

Figure 2: (a) Expanded random walk simulation for a polymer chain in a good solvent and (b) contracted random walk for a polymer chain in a poor solvent. Gels expand and contract similarly.

rules, but we generally do not know the polymer-solvent interactions or the extent of cross-linking.

5.5 Transport Properties of Gels

The expansion and contraction of gels depends on the diffusion of water or solvent in and out of the matrix. For small changes in the concentration of solutes, the process can be treated as simple diffusion, obeying Fick's laws. The diffusion coefficients of solutes in gels have been measured and do not differ dramatically from those in solution [Pavesi and Rigamonti, 1995]. Diffusion processes in gels can also be conveniently studied by conductivity [Sheppard et al., 1997]. The response time of a sample to a change of external conditions will then be given roughly by $x^2/2D$, where x is the characteristic distance, for instance half the thickness of a tube or the radius of a rod, and D is the diffusion coefficient, which will be of the order of 10^{-6} cm^2/sec. A 1-mm thick sheet sample would then be expected to respond in about 20 minutes. Changes due to a pH shift may be faster in that the diffusion coefficient for hydrogen ions in water is much faster than that of most solutes. A thorough discussion of the mathematics of such diffusion processes is given by Crank [1975].

Larger changes in volume and the expulsion of significant amounts of water will be more complex. As a result of a pH, temperature, or other change, pressure gradients will be set up that produce a flow of liquid out of the gel, and densification of the gel will result in a slowing of the flow. The theory of large volume changes in gels has been studied by Tanaka's group [Shibayama and Tanaka, 1993] and others [Khokhlov et al., 1993]. The process, by which liquid is extruded from a gel or is taken up during a change of state, is known as syneresis.

The response time of any gel actuator will be very dependent on its size and the resulting distance for fluid flow. Any practical actuator would need to be formed on the scale of a few microns if the response is to be rapid. A recent paper presents a numerical solution to the swelling of a gel in salt solution [Achilleos et al., 2001].

5.6 Polyelectrolyte Gels

A special class of polymer solutions is that in which the polymer is ionizable. Included in this group are such anionic polymers as the salts of polyacrylic acid, polymethacrylic acid, polystyrenesulfonic acid, and polyvinylphosphonic acid. There is also a family of cationic polymers including polydimethyldiallylammonium chloride, polylysine hydrochloride, and polyvinylpyridine hydrochloride.

These would be expected to share the acid-base characteristics of the equivalent salts. The pK_a of acrylic acid is 4.25. Hence, it will ionize as pH is increased from pH 3 to pH 5, by slowly adding a base. Polyacrylic acid differs in its ionization behavior from the monomer, however, because there are interactions between ions along the chain. This broadens the titration curve for the polymer to higher pH values in aqueous solutions of low ionic strength. High ionic strengths suppress the ion-ion repulsion, and the titration curve thus becomes sharper. As the chain becomes ionized, the ion-ion repulsions also cause a great expansion of the polymer coil to something closer to a straight chain. The sulfonated polymers will behave similarly but will shift to a much lower pH range. The cationic polymers will show a similar effect as the pH is decreased from 14. The physical chemistry of polyelectrolytes is discussed well by Tanford [1961].

Most gel actuator materials are based on cross-linked polyacrylic acid where the expansion caused by increasing pH results in the gel taking in more water and increasing in volume. As explained above, the extent of this response to pH will depend on the ionic strength and the cross-link density [Skouri et al., 1995].

5.7 Mechanical Properties of Gels

As with conventional materials, a gel can be characterized by an elastic modulus and yield point or fracture strength. For a chemically cross-linked gel in a good solvent for the polymer, the elastic modulus would be expected to be less than that of a rubbery polymer—below 10 Mpa—and to be roughly proportional to concentration. The strength of gels is seldom studied because they are rarely called upon to support significant loads. For a cross-linked unstructured rubbery polymer, a typical strength is 2 MPa. The strength of a gel might then be expected to be a similar value multiplied by the volume fraction of the solid phase. Thus, a 10% gel would have a strength of about 0.2 MPa. This suggests that gel muscles will not be capable of delivering stresses in excess of about 0.2 MPa.

When a gel is immersed in excess liquid and a stress is applied, there will be a slow viscoelastic response associated with the flow of liquid into or out of the system. Thus, articular cartilage, which is essentially a gel, has an initial modulus of 2.5 MPa, but has an equilibrium modulus, allowing for fluid flow, of about 0.7 MPa [Wainwright et al., 1986]. It is for this reason that humans must fidget in order to avoid excessive compression of the joint cartilage. Scherer has measured time-dependent elastic properties of silica gels [Scherer, 1994]. Chiarelli et al. [1992] have discussed the response in terms of water flow in a porous medium using Darcy's law. One peculiar consequence of the water loss is that gels can show a negative Poisson's ratio [Hirotsu, 1991; Li et al., 1993].

These low moduli mean that the extension or contraction of a gel actuator will be very sensitive to load in the stress range above about 10 kPa, since a structural extension may be cancelled out by simple elastic deformation if the opposing load is high.

For dense polymers, glassy materials are typically stronger than crystalline polymers (in their normal use range between the glass transition and melting point) because crystalline polymers yield and thus avoid undergoing brittle fracture. Crystalline polymers, however, are much tougher as slip of chains in the crystal adds greatly to the energy absorbed during fracture. Unstructured and unfilled rubbers tend to be very weak, and this weakness becomes extreme in simple gels. Natural rubber gains strength by crystallizing under stress, and noncrystallizable rubbers, such as silicone and styrene-butadiene tire rubber, are generally reinforced with silica or carbon black. Similarly, it is clear that structure can greatly enhance the strength of a gel by forming a mixture of energy-absorbing clusters and intervening flexible regions. The structure will also increase the modulus of the gel. Thus agarose gels are clearly stronger than the equivalent polyacrylamide gels. Likewise, cartilage owes its strength of about 1.5 MPa to a combination of collagen fibrils and soluble cross-linked polysaccharides. Developers of active gels have exploited such two-phase structures for greater strength.

5.8 Chemical Actuation of Gels

Early studies of gel muscles [Osada, 1987] were based on cross-linked polyacrylic acid gels. These materials show a large expansion on being moved from an acid solution to a base. However, the gels are quite weak and the response is slow. During the decade between 1985 and 1995, a number of new gel systems were developed with stronger responses.

One such system is a composite of polyvinylalcohol blended with polyacrylic acid and polyallylamine hydrochloride [Suzuki, 1991]. Nanbu had discovered in 1983 that polyvinylalcohol gels could be produced in a strong porous form by repeated freezing and thawing cycles. The ice formed during these cycles drives the polymer into a concentrated region around the crystals where it becomes strongly hydrogen-bonded. The resulting gel has an elastic modulus of about 0.3 MPa and a strength of about 0.6 MPa. The condensed polymer regions provide high strength, while the pores allow rapid exchange of reagents [Hickey and Peppas, 1997; Mori et al., 1997]. This dense, hydrogen-bonded gel swells strongly in water, without dissolution but contracts in organic solvents where the polymer complex has no solubility. On being shifted from water to acetone, a 10-μm film with 87% water content would respond in about 0.2 sec with a 30% contraction at a stress of 200 kPa.

A second approach to strong gels is to saponify polyacrylonitrile (PAN) fibers [Umemoto et al., 1991; Solari, 1994]. PAN is a textile fiber that is used in place of wool for many knitted garments. It is also a precursor for the manufacture of carbon fiber. On heating to about 220°C in air, while being clamped to constant length, the fiber partly oxidizes and cross-links. If it is then boiled in 1 molar sodium hydroxide, the remaining nitrile ($-C \equiv N$) units are converted to carboxylate ($-COO^-$). The result is a densely cross-linked, oriented polyacrylic acid fiber. On moving from acid to base, these fibers will change in length by about 70% in about 5 seconds. A force of 1.2 MPa can be generated in highly cross-linked fibers but greater cross-linking results in smaller length changes. Various workers have used these fibers to build chemically driven muscles [Brock et al., 1994; Salehpoor et al., 1997]. A recent paper describes similar studies on cross-linked composite fibers [Sun et al., 2000].

Chemically driven bending can be achieved by building a graded gel or by attaching the gel to a flexible unresponsive substrate [Park et al., 1999]. Temperature can also be regarded as a chemical input. A number of gels have been developed that go through a volume transition as temperature is raised near room temperature. One, which has been widely studied, is poly(N-isopropylacrylamide), which contracts when heated to about 35°C [Shibayama and Tanaka, 1993]. The gels are quite weak, having a strength of 3–12 kPa at gel concentrations of about 10% [Cicek and Tuncel, 1998]. Typical response times for 3-mm gels are hours, being governed by the diffusion of water out of the gel. Faster responses can be achieved with porous structures [Wu et al., 1992]. There is much interest in the use of these gels for drug delivery, triggered by body temperature.

5.9 Electrically Actuated Gels

Natural muscle is chemically actuated but chemical actuation of synthetic gels is not really desirable, except possibly for underwater applications. It would be much more desirable to use an electrically driven actuator for most applications.

When considering the electrical response of gels, the electrochemistry of the system must be taken into account. An acid or alkali solution at 1 molar will release hydrogen at the negative electrode and oxygen at the positive electrode at a potential difference of 1.2 V. The rate of this process will depend on the catalytic activity of the electrodes, platinum being very active. This or other electrolytic processes will be superimposed on the electrical response of most gel actuators. Accompanying the electrolysis, the region near the negative electrode will become more basic as hydrogen is released and OH^- ions are formed in solution. Likewise, the region near the positive electrode will become more acidic.

In the absence of ion flow, a simple, uniform gel sample will not be expected to respond to an electric field since, unlike a piezoelectric material, it has a center of symmetry. Once current flows, any asymmetry between the two sides of the sample would be expected to induce bending. As discussed by Doi et al. [1992]; and Shibuya et al. [1995], polyacrylic acid gel samples are expected to bend with the positive side concave if the gel is in contact with the anode, but will bend with the negative side concave if the gel does not touch either electrode. In the first case the bending is driven by the buildup of acidity, in the second by the differences in ion diffusion rates in the gel and in the electrolyte. Thus the bending response is not a material property but depends on the dynamics of the electrolytic reactions in the system, and hence on the geometry of the system, the composition of the electrolyte and the electrodes, and on the previous history of the gel. One can argue that an electroactive gel should be regarded not as a material but as a battery, which undergoes a shape change as it charges or discharges. In addition, a bending effect, which is driven by an electrochemically induced compositional change, will be less forceful than an effect which is due to a uniform change in composition, such as that arising from a change in pH or solvent.

Shiga and coworkers have shown that the bending of strips of polyelectrolyte gel can be fitted to the osmotic pressure difference predicted by the Flory-Huggins theory for polymer solutions [Shiga and Kurauchi, 1990] (see Fig. 3). The volume-pH curve for a polyacrylic acid gel shows a sharp increase in the region of pH 4 as the gel becomes partly ionized. Full ionization does little to increase the volume. Thus, the electrical response of a gel will be greatest if it is initially in this most-responsive pH region. Likewise, Shiga et al. have shown that the response of a gel can be tuned by varying the pH and acetone concentration such that the gel is just on the shrunken side of the shrinking-swelling transition. As with chemical gels, polyacrylic acid-polyvinylalcohol gels that have good properties and respond rapidly can be made by the freeze-thaw technique [Kim et al., 1999]. In an oscillating field at about 5 Hz, strips of gel with one end clamped show a wavelike response that can be fitted to models based on a vibrating beam [Shiga et al., 1993 and 1994; Shiga, 1997]. Presumably this reflects some coupling between ion displacements and the elasticity of the gel.

Other theories of the bending response attribute the effect to transport of water of hydration by the cation flow [Osada et al., 1991; Budtova et al., 1995] or to ion displacement and bilayer formation [Shahinpoor, 1995]. It is not clear whether there is an absolute distinction between the outward transport of water by cations and the deionization of the carboxylate to carboxylic acid followed by expulsion of water.

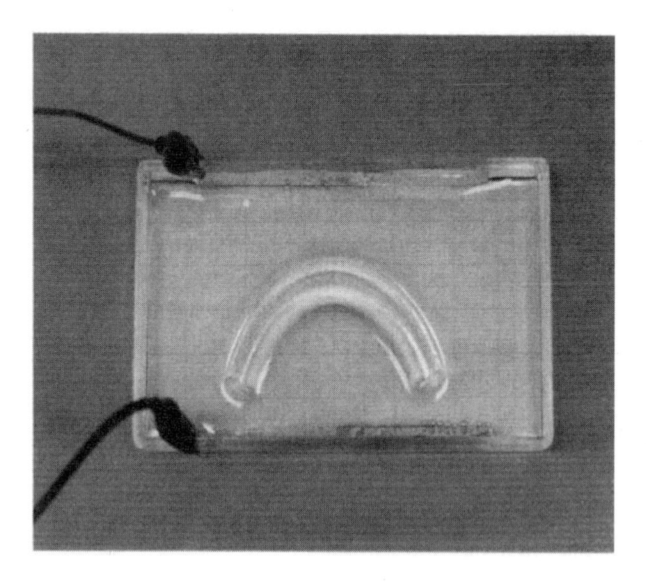

Figure 3: Bending of a polyacrylic acid gel rod sodium hydroxide. DC applied field, cathode (negative) at bottom. Gel swells on the anode side and bends toward the cathode [Shiga, 1997]. (Copyright Springer.)

A separate response mechanism has been seen in electrically activated gels in surfactant solution [Okuzaki and Osada, 1994 and 1995]. Driving counter-ions out of the gel causes a complex to form between the polymer and surfactant with a resulting shrinkage. The association of surfactant with anionic polymers has been studied in detail [Yoshida et al., 1998].

Rather than gels that respond when immersed in water, it is of potentially greater interest to devise systems that could run without external water. One step is to incorporate one or more electrodes into the gel. Chiarelli et al. [1991] made a gel with entrapped conducting polypyrrole as an internal electrode. There is also a report of highly anisotropic conductivity induced in a phthalocyanine-gel complex [Osada and Ohnishi, 1991].

Reports describe a polyvinylalcohol gel swollen with dimethylsulfoxide in place of water, which bends rapidly in a field [Hirai et al., 1994 and 1995; Zheng et al., 2000]. The use of nonionic gels with this or other nonaqueous electrolytes could solve problems with evaporation and electrolysis in gel actuators but in this case the fields are high and the response may be electrostrictive, like that of elastomer fibers in a high field (see Chapters 4 and 7).

Recent work by Calvert and Liu [2000] has shown that a linear contraction can be induced in an asymmetric stack of gel layers arranged such that one side responds to pH change while the other does not (see Fig. 4). This does demonstrate that more useful electrical responses can be obtained from complex designs, but this particular example does not overcome the soft response or low speed of simple polyacrylic acid gels or the concurrent electrolysis and gas

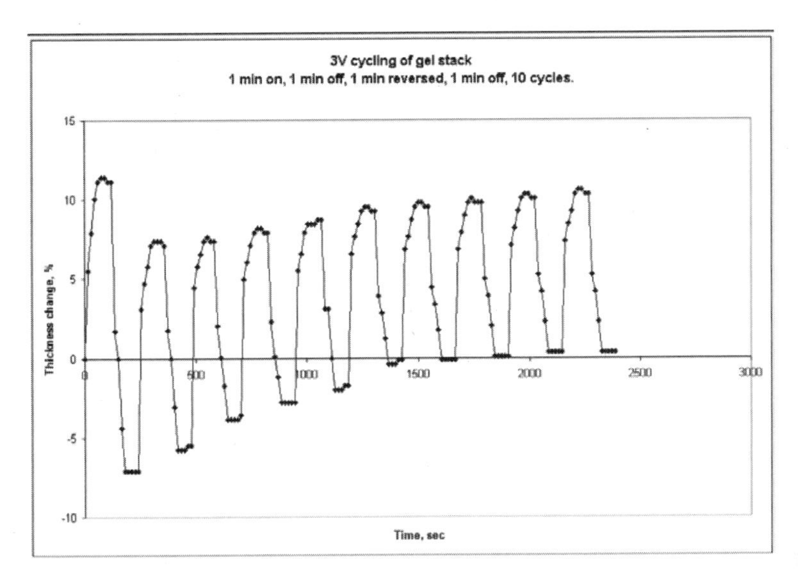

Figure 4: Response of an asymmetric gel stack to a cyclic field [Liu and Calvert 2000]. (Courtesy of Wiley-VCH.)

release. Since this composite relies on a shape change, without expulsion of water it could be used as a sealed system.

Much faster responses can be obtained by miniaturizing the system [Iwamoto et al., 1997]. A recent paper describes 0.5-mm gel actuators formed by lithography as valves for microfluidic systems [Beebe et al., 2000].

A model of ion distribution in gel actuators suggests that the contraction only occurs at the gel-electrode interface and suggests a number of strategies for enhancing the performance of actuators [Tamagawa and Taya, 2000]. It has been reported that graded gels can be formed by polymerization in an electric field [Shan et al., 1999].

The properties of various gel muscle systems are summarized in Table 1. Recent reviews of other gel actuators are given by Gong and Osada [1998 and 2000] and by Shiga [1997].

Table 1: The properties of various gel muscle systems [T. Shiga, personal communication].

Stimulus	**Material**	**Deformation mode**	**Force MPa**	**Response time, sec**	**Power density W/cm^3**
Electric field	PVA gel film thickness 120 μm	Shrinking	0.01–0.1	1000	
Electric field	PVA-PAA gel, 7-mm diam, 70-mm long	Bending	0.001	1–10	<0.001
Acetone/water change	PVA-PAA gel film thickness 10 μm	Shrinking	0.2	0.4–1	0.01–0.1
	Muscle	Sliding	0.2–1	<0.1	0.1–0.5

5.10 Recent Progress

As yet, there is no working, electrically driven gel muscle in practical use. This section reviews recent progress to improve upon the various drawbacks of gel systems and take them closer to a practical artificial muscle. Progress is needed on speed, strength, processability and force developed.

5.10.1 Strength

As was mentioned in Sec. 5.7 above, gel strength is not well understood but most gels fail by crack propagation like a brittle plastic. Natural gels, such as cartilage or the bodies of soft marine animals, tend to be composites with reinforcing fibers. Since there are many actual or potential medical applications for gels, including contact lenses and soft tissue prostheses, there is considerable interest in improving their properties. Most of this work involves preparation of composites, but also explores promising combinations of materials rather than the detailed design of a system.

If we want a gel that delivers a maximum stress, similar to muscle, of about 300 kPa, we would hope for a gel with a breaking strength of 500 kPa or more. An N-isopropylacrylamide gel responding to electrical heating has been reported to have a strength of 10 kPa [Kato et al., 1998]. A cross-linked copolymer gel of hydrophilic and hydrophobic segments was reported to have a strength of 200–500 kPa [Isayeva et al., 2002]. The strength is very dependent on water content and many modifications will increase strength by reducing swelling. For example, contact lenses have water contents of 30–50% and strengths of 2–4 MPa [Willis et al., 2001]. There have been many recent studies of composite gels made by irradiation of mixed solutions of polymers and increases in strength have been reported when compared to single polymer gels (Relleve et al., 1999). One would expect that the properties of these disordered systems would primarily follow the water content.

Likewise freeze-thaw modified polyvinylalcohol gels with water contents of around 80% were found to fail in compression at a few MPa [Stammen et al., 2001]. As was found by earlier workers [Suzuki, 1991], this freeze-thaw process produces a two-phase composite structure that is stronger and has recently been studied in more detail [Matsumura et al., 1998; Shapiro, 1999]. Composite gels with polyvinylalcohol reinforcing water-soluble polymers have also been studied [Nho and Park, 2002].

A number of studies have considered reinforcement of gels with inorganic fibers or plates both added before gelation and grown *in situ* in the gel and a significant increase in modulus is certainly seen [Gao et al., 1999; Xia et al., 2003; Yamaguchi et al., 2003]. One very attractive approach, based on the analogy to collagen-reinforced biological gels, is to reinforce gels with fibrils of rigid-rod polymers [Philippova et al., 1998]. This particular system does show a significant increase in modulus but from low values and a lack of strength, data is given. Thus the full potential of composites of this type has yet to be explored. A simple version of this approach is to reinforce a gel with a textile, such as nonwoven polypropylene [Wu et al., 1999; Lopergolo et al., 2002].

5.10.2 Speed

Unlike piezoelectric or dielectric polymers, gel muscles change volume by a local uptake or loss of water. Since this is a diffusional process, the response time will increase roughly with the square of the diameter of the muscle fiber. There will be additional dependencies on the rate of generation of ions electrically, on the relationship between gel charge and equilibrium swelling and on the chemical kinetics of ionization of the gel. The complex inter-relationship of these factors means that a complete muscle system must be modeled in order to predict a response rate. Such a model has been presented for polyacrylic acid-polyvinylalcohol gels [Marra et al., 2002].

In general porous composites will respond more rapidly than monolithic gels (Kato et al., 1998; Chen and Park 2000; Gemeinhart et al., 2000) and fibers (Schreyer et al., 2000) Recently, 30-μm contractile bodies have been found in plants [Knoblauch et al., 2003]. These "forisomes" contract in response to calcium ions in about 50 msec and develop a force of 11 kPa.

5.10.3 Processability

Many of the electro-active polymers are unable to be processed by conventional polymer methods, such as extrusion and molding. A successful artificial gel muscle is likely to be an assembly of fine fibers or films and metal electrodes. Cross-linked gels may be formed by irradiation of existing shapes or by casting and polymerization of monomer solutions. Liu and Calvert (2000) did apply free-form fabrication methods to making more complex shapes, but there is a need for new gel chemistries that might be more readily processed. In particular, the free-radical polymerization used to make polyacrylic acid is very air and impurity sensitive, so that fine-scale structures cannot be made without elaborate precautions.

With this in mind, Liu and Calvert explored gels based on epoxy chemistry, which is relatively robust and reproducible. A range of gels was made based on water-soluble di-epoxides of glycerol and glycols coupled with aminated polyethers [Yoshioka and Calvert, 2002]. These materials could be readily formed by extruding toothpastelike physical gels and warming them to drive the cross-linking reaction. The gels were highly swollen in acid and contracted in base. Small cylinders, about 100 μm in diameter, responded to pH change or electrical stimulus in about 5 minutes.

Extrudable blends of colloidal conducting polymer and hydrogels have been described [Kim et al., 2000]. Photo cross-linkable gels have been developed based on a nitrocinnamate photosensitive group attached to peg [Micic et al., 2003]. Self-assembly methods have been discussed for artificial muscles and other responsive polymers (Zhang, 2002). A method has been described for making microvalves from polyaniline-hydrogel on micromachined silicon[Low et al., 2000]. A blend of polyaniline, polyhydroxyethylmethacrylate and polyvinylpyrrolidone was shown to have a large electrically driven swelling response.

5.10.4 Force

There is no standard method for comparing forces developed by proposed artificial-muscle systems. A sample of gel immersed in an aqueous solution will respond to a change in pH or other conditions by a volume change, taking up or losing water. Generally, gel systems have a low elastic modulus and extend under a static load but this modulus will depend on the degree of swelling. As a result there is a complex interaction between the active and passive responses of the gel. While the contraction of an unloaded system maybe characterized, there are few systematic studies of the response under load.

In general the force developed increases with the solids loading of the gel, while larger and faster volume changes will be seen with more dilute gels. Denser gels are also stronger and therefore can exert a greater maximum stress. Thus, the response of any gel system will also depend on the details of the chemical structure of the sample. The response to an electrical stimulus adds further complexity. At present most samples are tested by measuring a rate of bending for a sample between electrodes, thus there is no figure of merit for comparing different gel materials. More detailed studies of conducting polymer (Skaarup et al., 2000), conducting polymer gel [Irvin, et al. 2001] and IPMC [de Gennes et al., 2000; Nemat-Nasser, 2002] actuators also shed light on the expected response of gels.

5.11 Future Directions

At this point it seems clear that the combination of properties required from a musclelike gel system can only be provided by a composite structure. Mechanical strength requires some dense phase, probably in fibrous form, while rapid response requires a porous structure that allows water and solutes to flow rapidly. This porosity needs to be on multiple length scales to allow rapid volume change. We do have evidence that suitable forces and speeds are possible from very small systems, the problem is to combine these into a large mass of muscle.

Electrical drive energy for gel muscles functions by generating a local pH change. This must be intrinsically inefficient as hydrogen ions are generated at the anode hydroxyl ions are generated at the cathode. Diffusion tends to erase the pH gradient. Also, the concentration of hydrogen ions needed to change the state of how a gel corresponds to large current flows, which are hard to generate except in systems of high ionic strength. This suggests that alternative energy inputs should be considered in more detail. This includes optical inputs or electrically driven thermal inputs. Biological muscle depends on an electrically triggered release of stored ATP or calcium ions (Pollack, 2001 and Chapter 2) and it might be possible to devise some similar store-and-release system for artificial muscles.

The complex thermodynamics and kinetics of force development have been studied for polyacrylic acid gels. We need to know more, but there is little point in more detailed studies before there is a combination of material and energy source that approaches the performance of natural muscle.

5.12 References

Achilleos, E., Christodoulou, K. and Kevrekidis, I. (2001), "A transport model for swelling of polyelectrolyte gels in simple and complex geometries," *Comput. Theor. Polym. Sci.* 11, pp. 63-80.

Beebe, D. J., Moore, J. S., Bauer, J. M., Yu, Q., Liu, R. H., Devadoss, C. and Jo, B-H. (2000), "Functional hydrogel structures for autonomous flow control inside microfluidic channels," *Nature* 404, pp. 588-590.

Brock, D., Lee, W., Segalman, D. and Witkowski, W. (1994), "A dynamic model of a linear actuator based on polymer hydrogel," *Journal of Intelligent Material Systems and Structures* 5, pp. 764-771.

Budtova, T., Suleimenov, I. and Frenkel, S. (1995), "Electrokinetics of the contraction of a polyelectrolyte hydrogel under the influence of constant electric-current," *Polym. Gels Netw.* 3, pp. 387-393.

Candau, S., Moussaid, A., Munch, J. and Schosseler, F. (1992), "Dynamic light-scattering by weakly ionized gels—effects of inhomogeneities and nonergodicity," *Makromolekulare chemie-macromolecular symposia* 62, pp. 183-189.

Chen J and Park K (2000), "Synthesis and characterization of superporous hydrogel composites." *J. Control. Release* 65, pp.73-82.

Chiarelli, P., Basser, P., Derossi, D. and Goldstein, S. (1992), "The dynamics of a hydrogel strip," *Biorheology* 29, pp. 383-398.

Chiarelli, P., Umezawa, K. and DeRossi, D. (1991), "A polymer composite showing electrocontractile response," *Polymer Gels*, D. DeRossi, K. Kajiwara, Y. Osada and A. Yamauchi, Plenum, NY, pp. 195-203.

Cicek, H. and Tuncel, A. (1998), "Preparation and characterization of thermoresponsive isopropylacrylamide-hydroxyethylmethacrylate copolymer gels," *J. Polymer Sci. A, Polymer Chemistry* 36, pp. 527-541.

Corkhill, P., Fitton, J. and Tighe, B. (1993), "Towards a synthetic articular cartilage," *J. Biomater. Sci.-Polym. Ed.* 4, pp. 615-630.

Crank, J. (1975), *The Mathematics of Diffusion*, Oxford University Press, Oxford.

de Gennes, P., Okumura, K., Shahinpoor, M. and Kim, K. (2000), "Mechanoelectric effects in ionic gels." *Europhys. Lett.* 50, pp. 513-518.

Doi, M., Matsumoto, M. and Hirose, Y. (1992), "Deformation of ionic polymer gels by electric-fields," *Macromolecules* 25, pp. 5504-5511.

Flory, P. J. (1953), *Principles of Polymer Chemistry*, Cornell University Press, Ithaca, NY.

Gao D., Heimann, R., Williams, M., Wardhaugh, L., and Muhammad, M. (1999), "Rheological properties of poly(acrylamide)-bentonite composite hydrogels." *J. Mater. Sci.* 34, pp. 1543-1552.

Gemeinhart, R., Chen, J., Park, H., and Park, K. (2000), "pH-sensitivity of fast responsive superporous hydrogels." *J. Biomater. Sci.-Polym. Ed.* 11, pp. 1371-1380.

Gong, J. P. and Osada, Y. (2000), "Gel actuators," *Polym. Sens. Actuators*, Y. Osada and D. E. DeRossi, eds., Springer-Verlag, Berlin, Germany, pp. 273-294.

Hickey, A. and Peppas, N. (1997), "Solute diffusion in poly(vinyl alcohol) poly(acrylic acid) composite membranes prepared by freezing/thawing techniques," *Polymer* 38, pp. 5931-5936.

Hirai, M., Hirai, T., Sukumoda, A., Nemoto, H., Amemiya, Y., Kobayashi, K. and Ueki, T. (1995), "Electrically-induced reversible structural-change of a highly swollen polymer gel network," *J. Chem. Soc.-Faraday Trans.* 91, pp. 473-477.

Hirai, T., Nemoto, H., Hirai, M. and Hayashi, S. (1994), "Electrostriction of highly swollen polymer gel - possible application for cell actuator," *J. Appl. Polym. Sci.* 53, pp. 79-84.

Hirotsu, S. (1991), "Softening of bulk modulus and negative Poisson's ratio near the volume phase transition of polymer gels," *J. Chem. Phys.* 94, pp. 3949-3957.

Irvin D, Goods S and Whinnery L (2001), "Direct measurement of extension and force in conductive polymer gel actuators." *Chem. Mat.* 13, pp. 1143-145.

Isayeva I, Gent A and Kennedy J (2002). "Amphiphilic networks part XVIII - Synthesis and burst strength of water-swollen immunoisolatory tubules." *J. Polym. Sci. Pol. Chem.* 40, pp. 2075-2084.

Iwamoto, K., Kurashima, F., Ozawa, M. and Yamazaki, T. (1997), "Preparation of an ionic polymer gel microactuator and measurement of its periodic motions," *Nippon Kagaku Kaishi*, pp. 609-614.

Kato, N., Morito, T., and Takahashi, F., (1998), "An energy-saving method of heating for a chemomechanical poly(N-isopropylacrylamide) gel with Joule's heat." *Mater. Sci. Eng. C-Biomimetic Mater. Sens. Syst.* 6, pp. 27-31.

Kato, N., Morito, T., and Takahashi, F. (1998), "An energy-saving method of heating for a chemomechanical poly(N-isopropylacrylamide) gel with Joule's heat." *Mater. Sci. Eng. C-Biomimetic Mater. Sens. Syst.* 6, pp. 27-31.

Khokhlov, A., Starodubtzev, S. and Vasilevskaya, V. (1993), "Conformational transitions in polymer gels - theory and experiment," *Adv. Polymer Sci.* 109, pp. 123-175.

Kim, B., Spinks, G., Too, C., Wallace, G., and Bae, Y. (2000), "Preparation and characterisation of processable conducting polymer-hydrogel composites." *React. Funct. Polym.* 44, pp. 31-40.

Kim, S., Shin, H., Lee, Y. and Jeong, C. (1999), "Properties of electroresponsive poly(vinylalcohol)/poly(acrylic acid) IPN hydrogels under an electric stimulus," *J. Appl. Polym. Sci.* 73, pp. 675-683.

Knoblauch, M., Noll, G. A., Müller, T., Prüfer, D., Schneider-Hüther, I., Scharner, D., Bel, A. J. E. V., and Peters, W. S. (2003). "ATP-Independent Contractile Proteins From Plants." *Nature Materials* 2, pp. 600-603.

Larson, R. G. (1999), *The Structure and Rheology of Complex Fluids*, Oxford University Press, Oxford.

Li, C., Hu, Z. and Li, Y. (1993), "Poisson's ratio in polymer gels near the phase-transition point," *Phys. Rev. E* 48, pp. 603-606.

Liu, Z. and Calvert, P. (2000), "Multilayer hydrogels as musclelike actuators," *Adv. Mater.* 12, pp. 288-291.

Lopergolo, L., Lugao, A., and Catalaini, L., (2002), "Development of a poly(N-vinyl-2-pyrrolidone)/poly (ethylene glycol) hydrogel membrane reinforced with methyl methacrylate-grafted polypropylene fibers for possible use as wound dressing." *J. Appl. Polym. Sci.* 86, pp. 662-666.

Low, L., Seetharaman, S., He, K., and Madou, M., (2000), "Microactuators toward microvalves for responsive controlled drug delivery." *Sens. Actuator B-Chem.* 67, pp. 149-160.

Marra, S., Ramesh, K., and Douglas, A., (2002), "The actuation of a biomimetic poly(vinyl alcohol)-poly(acrylic acid) gel." *Philos. Trans. R. Soc. Lond. Ser. A-Math. Phys. Eng. Sci.* 360, pp. 175-198.

Matsumura, K., Hyon, S., Oka, M., Ushio, K., and Tsutsumi, S. (1998), "Scanning electron microscopy and atomic force microscopy observations of surface morphology for articular cartilages of dog's knee and poly(vinyl alcohol) hydrogels." *Kobunshi Ronbunshu* 55, pp. 786-790.

Micic, M., Zheng, Y., Moy, V., Zhang, X., Andreopoulos, F., and Leblanc, R. (2003), "Comparative studies of surface topography and mechanical properties of a new, photo-switchable PEG-based hydrogel." *Colloid Surf. B-Biointerfaces* 27, pp. 147-158.

Mori, Y., Tokura, H. and Yoshikawa, M. (1997), "Properties of hydrogels synthesized by freezing and thawing aqueous polyvinyl alcohol solutions and their applications," *J. Mater. Sci.* 32, pp. 491-496.

Nemat-Nasser, S. (2002), "Micromechanics of actuation of ionic polymer-metal composites." *J. Applied Phys.* 92, pp. 2899-2915.

Nho, Y., and Park, K. (2002), "Preparation and properties of PVA/PVP hydrogels containing chitosan by radiation." *J. Appl. Polym. Sci.* 85, pp.1787-1794.

Okuzaki, H. and Osada, Y. (1994), "Electro-driven polymer gels with biomimetic motility," *Polym. Gels Networks* 2, pp. 267-277.

Okuzaki, H. and Osada, Y. (1995), "Electro-driven polyelectrolyte gel with biomimetic motility," *Electrochim. Acta* 40, pp. 2229-2232.

Osada, Y. (1987), "Conversion of chemical into mechanical energy by synthetic-polymers (chemomechanical systems)," *Advances In Polymer Science* 82, pp. 1-46.

Osada, Y. and Gong, J. (1998), "Soft and wet materials: Polymer gels," *Adv. Mater.* 10, pp. 827-837.

Osada, Y. and Ohnishi, S. (1991), "Electrophoretic orientation of phthalocyanine molecules in polymer gel," *Macromolecules* 24, pp. 3020-3022.

Osada, Y., Gong, J. P., Ohnishi, S., Sawahata, K. and Hori, H. (1991), "Synthesis, mechanism, and application of an electro-driven chemomechanical system using polymer gels," *J. Macromol. Sci., Chem.* A28, pp. 1189-1205.

Park, G. B., Kagami, Y., Gong, J. P., Lee, D. C. and Osada, Y. (1999), "Chemomechanical bending behaviors of ionizable thin films with gradient network-size," *Thin Solid Films* 350, pp. 289-294.

Pavesi, L. and Rigamonti, A. (1995), "Diffusion constants in polyacrylamide gels," *Physical Review E* 51, pp. 3318-3323.

Philippova, O., Rulkens, R., Kovtunenko, B., Abramchuk, S., Khokhlov, A., and Wegner, G. (1998), "Polyacrylamide hydrogels with trapped polyelectrolyte rods." *Macromolecules* 31, pp. 1168-1179.

Pollack, G. H. (2001), *Cells, Gels and the Engines of Life*, Ebner & Sons.

Relleve, L., Yoshii, F., dela Rosa, A., and Kume, T. (1999), "Radiation-modified hydrogel based on poly(N-vinyl-2-pyrrolidone) and carrageenan." *Angew. Makromol. Chem.* 273, pp. 63.

Salehpoor, K., Shahinpoor, M. and Mojarrad, M. (1997), "Some experimental results on the dyamic performance of PAN muscles," SPIE Proc.Vol. 3040, pp. 169-173.

Sandler, S. R. and Karo, W. (1992), *Polymer Syntheses, Vol. 1* , Academic Press, Boston.

Scherer, G. W. (1994), "Relaxation of a viscoelastic gel bar I," *J. Sol-Gel Sci & Tech.* 1, pp. 169-175.

Scherer, G. W. (1994), "Relaxation of a viscoelastic gel bar II," *J. Sol-Gel Sci & Tech.* 2, pp. 199-204.

Schreyer, H., Gebhart, N., Kim, K., and Shahinpoor, M. (2000), "Electrical activation of artificial muscles containing polyacrylonitrile gel fibers." Biomacromolecules 1, pp. 642-647.

Shahinpoor, M. (1995), "Microelectromechanics of ionic polymeric gels as electrically controllable artificial muscles," *J. Intell. Mater. Syst. Struct.* 6, pp. 307-314.

Shan, J., Chen, J., Shen, X. and Chen, R. (1999), "Mechanical properties of poly(acrylamide-sodium acrylate) hydrogels synthesized under direct current electric fields with high voltages," *Chem. Lett.*, pp. 427-428.

Shapiro, Y. (1999), "H-1 NMR self-diffusion study of morphology and structure of polyvinyl alcohol cryogels." *J. Colloid Interface Sci.* 212, pp. 453-465.

Sheppard, N., Lesho, M., Tucker, R. and SalehiHad, S. (1997), "Electrical conductivity of pH-responsive hydrogels," *J. Biomater. Sci.-Polym. Ed.* 8, pp. 349-362.

Shibayama, M. and Tanaka, T. (1993), "Volume phase-transition and related phenomena of polymer gels," *Adv. Polymer Sci.* 109, pp. 1-62.

Shibuya, T., Yasunaga, H., Kurosu, H. and Ando, I. (1995), "Spatial information on a polymer gel as studied by h-1-nmr imaging .2. Shrinkage by the application of an electric-field to a polymer gel," *Macromolecules* 28, pp. 4377-4382.

Shiga, T. (1997), "Deformation and viscoelastic behavior of polymer gels in electric fields," *Adv. Polymer Sci.* 134, pp. 131-163.

Shiga, T. and Kurauchi, T. (1990), "Deformation of polyelectrolyte gels under the influence of an electric field," *J. Appl. Polymer Sci.* 39, pp. 2305-2320.

Shiga, T., Hirose, Y., Okada, A. and Kurauchi, T. (1992), "Electric field-associated deformation of polyelectrolyte gel near a phase transition point," *Journal of Applied Polymer Science* 46, pp. 635-640.

Shiga, T., Hirose, Y., Okada, A. and Kurauchi, T. (1993), "Bending of ionic polymer gel caused by swelling under sinusoidally varying electric fields," *Journal of Applied Polymer Science* 47, pp. 113-119.

Shiga, T., Hirose, Y., Okada, A. and Kurauchi, T. (1994), "Deformation of ionic polymer gel films in electric-fields," *J. Mater. Sci.* 29, pp. 5715-5718.

Skaarup, S., West, K., Gunaratne, L., Vidanapathirana, K., and Careem, M. (2000), "Determination of ionic carriers in polypyrrole." *Solid State Ion.* 136, pp. 577-582.

Skouri, R., Schosseler, F., Munch, J. P. and Candau, S. J. (1995), "Swelling and elastic properties of polyelectrolyte gels," *Macromolecules* 28, pp. 197-210.

Solari, M. (1994), "Evaluation of the mechanical-properties of a hydrogel fiber in the development of a polymeric actuator," *J. Intell. Mater. Syst. Struct.* 5, pp. 295-304.

Stammen, J., Williams, S., Ku, D., and Guldberg, R. (2001), "Mechanical properties of a novel PVA hydrogel in shear and unconfined compression." *Biomaterials* 22, pp. 799-806.

Steinberg, I. Z., Oplatka, A. and Katchalsky, A. (1966), "Mechanochemical engines" *Nature* 210, pp. 568–571.

Sun, S., Wong, Y., Yao, K. and Mak, A. (2000), "A study on mechano-electro-chemical behavior of chitosan/poly(propylene glycol) composite fibers," *J. Appl. Polym. Sci.* 76, pp. 542-551.

Suzuki, M. (1991), "Amphoteric poly(vinyl alcohol) hydrogel and electrodynamic control method for artificial muscles," *Polymer Gels*, D. deRossi, New York, Plenum, pp. 221-236.

Suzuki, Y., Nozaki, K., Yamamoto, T., Itoh, K. and Nishio, I. (1992), "Quasi-elastic light-scattering study of the formation of inhomogeneities in gels," *Journal of Chemical Physics* 97, pp. 3808-3812.

Tamagawa, H. and Taya, M. (2000), "A theoretical prediction of the ions distribution in an amphoteric polymer gel," *Materials Science and Engineering A* 285, pp. 314-325.

Tanford, C. (1961), *Physical Chemistry of Macromolecules*, Wiley, New York.

Umemoto, S., Okui, N. and Sakai, T. (1991), "Contraction behavior of polyacrylonitrile gel fibers," *Polymer Gels*, D. DeRossi, K.Kajiwara, Y. Osada and A. Yamauchi, Plenum, NY, pp. 257-270.

Wainwright, S. A., Biggs, W. D., Currey, J. D. and Gosline, J. M. (1986), *Mechanical Design in Organisms*, Princeton University Press, Princeton, NY.

Willis, S., Court, J., Redman, R., Wang, J., Leppard, S., O'Byrne, V., Small, S., Lewis, A., Jones, S., and Stratford, P., (2001) "A novel phosphorylcholine-coated contact lens for extended wear use." *Biomaterials* 22, pp. 3261-3272.

Wu, M., Bao, B., Chen, J., Xu, Y., Zhao, S., and Ma, Z. (1999), "Preparation of thermosensitive hydrogel (PP-g-NIPAAm) with one-off switching for controlled release of drugs." *Radiat. Phys. Chem.* 56, pp. 341-346.

Wu, X., Hoffman, A. and Yager, P. (1992), "Synthesis and characterization of thermally reversible macroporous poly(n-isopropylacrylamide) hydrogels," *J. Polym. Sci. Pol. Chem.* 30, pp. 2121-2129.

Xia, X., Yih, J., D'Souza, N, and Hu, Z. (2003) "Swelling and mechanical behavior of poly (N-isopropylacrylamide)/Na-montmorillonite layered silicates composite gels." *Polymer* 44, pp. 3389-3393.

Yamaguchi, I., Itoh, S., Suzuki, M., Osaka, A., and Tanaka, J. (2003), "The chitosan prepared from crab tendons: II. The chitosan/apatite composites and their application to nerve regeneration." *Biomaterials* 24, pp. 3285-3292.

Yoshida, K., Sokhakian, S. and Dubin, P. (1998), "Binding of polycarboxylic acids to cationic mixed micelles: Effects of polymer counterion binding and polyion charge distribution," *J. Colloid Interface Sci.* 205, pp. 257-264.

Yoshioka, Y. and Calvert, P. (2002), "Epoxy-based electroactive polymer gels." *Experimental Mechanics* 42, pp. 404-408.

Zhang, S. (2002), "Emerging biological materials through molecular self-assembly." *Biotechnol. Adv.* 20, pp. 321-339.

Zheng, J., Watanabe, M., Shirai, H. and Hirai, T. (2000), "Electrically induced rapid deformation of nonionic gel," *Chem. Lett.*, pp. 500-501.

CHAPTER 6

Ionomeric Polymer-Metal Composites

Sia Nemat-Nasser and Chris W. Thomas
University of California, San Diego

6.1 Introduction

Ionomeric polymer-metal composites (IPMCs) as bending actuators and sensors are sometimes referred to as "soft actuators-sensors" or "artificial muscles." A typical IPMC consists of a thin (200 μm) polymer membrane with metal electrodes (5–10-μm thick) plated on both faces; see Fig. 1. The polyelectrolyte matrix is neutralized with an amount of counter-ions, balancing the charge of anions covalently fixed to the membrane. When an IPMC in the solvated (i.e., hydrated) state is stimulated with a suddenly applied small (1–3 V, depending on the solvent) step-potential, both the fixed anions and mobile counter-ions are subjected to an electric field, with the counter-ions being able to diffuse toward one of the electrodes. As a result, the composite undergoes an initial fast bending deformation, followed by a slow relaxation, either in the same or in the opposite direction, depending on the composition of the backbone ionomers and the nature of the counter-ion. The magnitude and speed of the initial fast deflection also depend on the same factors, as well as on the structure of the electrodes, and other conditions (e.g., the time-variation of the imposed voltage). IPMCs that are made from Nafion and are neutralized with alkali metals or with alkyl-ammonium cations (except for tetrabutylammonium,TBA^+), invariably first bend towards the anode under a step direct current (dc), and then relax towards the cathode, while the applied voltage is being maintained, often moving beyond their starting position. In this case, the motion towards the anode can be eliminated by slowly increasing the applied potential at a suitable rate. For Flemion-based IPMCs, on the other hand, the initial fast bending and the subsequent relaxation are both towards the anode, for all counter-ions that have been considered. With TBA^+ as the counter-ion, no noticeable relaxation towards the cathode has been recorded for either Nafion- or Flemion-based IPMCs. When an IPMC membrane is suddenly bent, a small voltage of the order of millivolts is produced across its faces. Hence, IPMCs of this kind can serve as soft actuators and sensors.

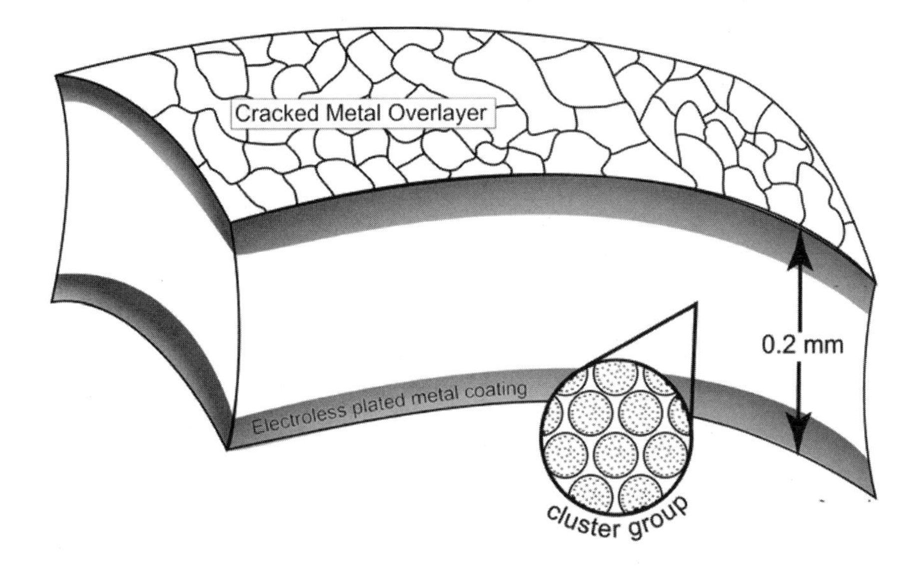

Figure 1: IPMC schematics indicating surface morphology, interface, and cluster structure.

In certain applications, IPMC materials may offer advantages over conventional mechanical, hydraulic, or pneumatic actuators, in that IPMCs lack moving parts and require only modest operating voltages for actuation. Investigations into the mechanism of IPMC actuation and sensing seek to foster development of actuators that generate greater displacement magnitudes and forces, and sensors that are more sensitive to imposed deformations. Developers seek to exploit these materials in a number of applications including medical, space, robotic, soft microelectronic machine (MEMS), and entertainment devices. This chapter addresses the known properties of IPMC materials, their manufacture, and the methods for their characterization. In addition, we provide a summary of models proposed to account for the mechanisms of IPMC actuation and sensing, experimental results to support or invalidate these models, and concluding with a hybrid model that integrates electrical, chemical, and mechanical forces to produce results in accord with experimental observations. Finally, we highlight a few applications that have been proposed for these materials.

6.2 Brief History of IPMC Materials

IPMCs represent one class of electroactive polymeric materials, a recent entry into a field of shape-modifying polymers that dates back more than 50 years. While a total survey of these materials is beyond the scope of this chapter, they have been addressed elsewhere [Segalman et al., 1992; Shahinpoor et al., 1998]. Recent developments in electroactive polymers are addressed throughout this book, and it is the intention of this chapter to highlight the elements that distinguish IPMC materials from other electroactive polymers. One of the most significant aspects of IPMCs is their actuation based on the active (anisotropic, directional) motion of ions and solvent molecules within the membrane under applied stimulus, and other materials whose actuation derives from the passive (isotropic, volumetric) response to stimuli.

Polymer-metal composites were developed as early as 1939 via the precipitation of colloidal silver on prepared substrates [*Modern Plastics*, 1938; Feynman, 1985]. These early materials suffered from delamination of the metal overlayer, and were little more than decorative curiosities. Recently, sputtering methods have provided routes to polymer-metal composites, but these were prone to delamination as well [Bergman, 1970; McCallum and Pletcher, 1975]. Indeed, it was not until the late 1960s when researchers at Dow Chemical showed that the permselective properties of ionomeric resins could be used to facilitate selective reduction of metal salts at the surface of an ion exchange membrane using chemical reductants such as sodium borohydride ($NaBH_4$) or hydrazine (N_2H_4) [Levine and Prevost, 1968; Bartrum, 1969; Wiechen, 1971]. Later, these methods were applied to Nafion-type membranes by many Japanese groups, including workers at Hitachi [Hitachi, 1983; Sakai et al., 1985a; 1985b; Takenaka et al., 1982, 1985] in the early 1980s. Millet and coworkers further developed the technique of IPMC formation, characterizing the plating mechanism to improve the morphology of the metal electrodes in IPMCs [Millet et al., 1989, 1990, 1992, 1993, 1995]. Figure 2 displays the chemical structure of three perfluorinated ionomers used to produce IPMCs, varying in the length and number of side chains and in the nature of ionic side group—usually sulfonate or carboxylate anions. As will be discussed, the acidity strength of the ionic side group has a profound effect on the actuation of the resulting IPMC [Nemat-Nasser and Wu, 2003a], and, indeed, lies at the root of the back-relaxation mechanism in Nafion-based IPMCs.

IPMC materials were developed as solid polymer electrolyte fuel cell membranes [Kordesch and Simader, 1995]. During investigations into the use of IPMCs as hydrogen pressure transducers, Sadeghipour et al. [1992] found that IPMC materials could act as vibration sensors. They reported that a platinum-IPMC in the Na^+ form generated over 12 mV/g of acceleration. The maximum volt response of their sensor occurred at 2000 Hz oscillation. These authors point out that the dynamic behavior of their "cell" filters and amplifies vibratory input to the IPMC, and that the signal response of the IPMC is not strongly frequency

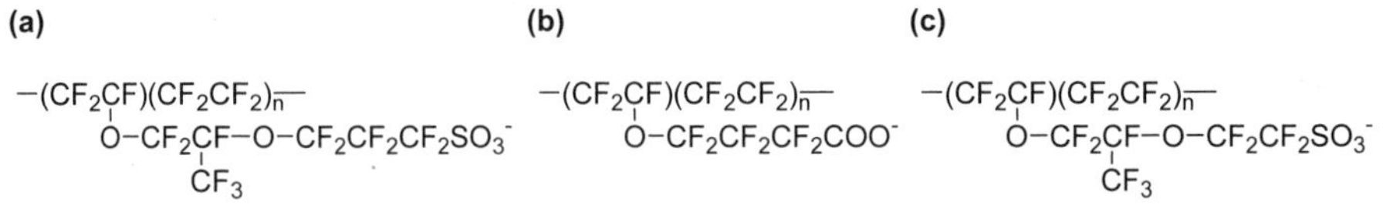

(a)

$-(CF_2CF)(CF_2CF_2)_n-$
$O-CF_2CF-O-CF_2CF_2CF_2SO_3^-$
CF_3

(b)

$-(CF_2CF)(CF_2CF_2)_n-$
$O-CF_2CF_2CF_2COO^-$

(c)

$-(CF_2CF)(CF_2CF_2)_n-$
$O-CF_2CF-O-CF_2CF_2SO_3^-$
CF_3

Figure 2: Perfluorinated ionomers used in IPMC manufacture include (a) Aciplex, (b) Flemion, and (c) Nafion.

dependent. In noting that the transduction of an IPMC generates a voltage, Sadeghipour and coworkers should be credited for proposing this material as a soft sensor. It should be noted that they were tantalizingly close to discovering IPMC's use as an actuator. Instead, in Japan, that same year, Oguro et al. described the ability of an IPMC material to bend under an applied voltage [1992]. In the USA, researchers, such as Mojarrad and Shahinpoor [1996], have since sought to improve the performance of Nafion-based IPMC actuators through optimization of the method of manufacture, including sample size, dimensions, and particularly the electrode morphology. More recently, Nemat-Nasser [2002], Nemat-Nasser and Wu [2003a], and Nemat-Nasser and Zamani [2003] have performed thorough and systematic studies of the properties and actuation of both Nafion- and Flemion-based IPMCs, in various cation forms and with different solvents, seeking to identify the underpinning mechanisms of actuations, model the phenomena of the IPMC actuation and sensing in order to predict properties quantitatively and relate these to the composition, processing, and microstructure of IPMC materials.

6.3 Materials and Manufacture

6.3.1 Ionomer (Nafion) Structure

For convenience, to date all IPMC materials known to us have used commercially available perfluorinated ionomeric polymer membranes. Several vendors offer ionomer membranes, including DuPont (Nafion), Asahi Glass (Flemion), Asahi Chemical (Aciplex), and others. The general method [Millet, 1999] for preparing IPMCs consists of three basic steps, as shown in Fig. 3. After cleaning with strong acids such as HNO_3, a clean sample of the

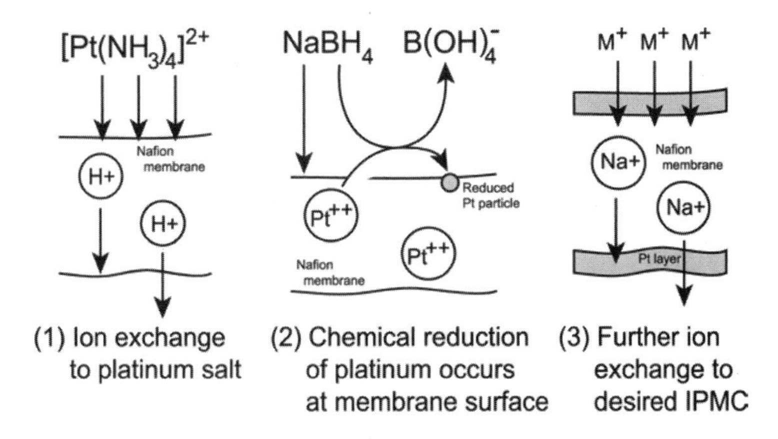

Figure 3: Scheme for IPMC fabrication: (1) ion exchange with noble metal salt; (2) reduction of metal at surface; (3) ion exchange with desired cation.

*poly*perfluoroethylenesulfonate membrane is soaked in a solution of an appropriate metal salt [e.g., $Pt(NH_3)_4Cl_2$] to populate each ionomer exchange site with reducible metal. The prepared sample is immersed in a solution of a suitable chemical reductant, such as sodium borohydride, which cannot penetrate into the ion exchange membrane. The metal salt diffuses out of the membrane and is reduced when it encounters the reducing agent at the surface of the membrane. The reduction process creates metal particles between 3 and 10 nm in diameter. The distribution of the metal plate is greatest at the surface of the membrane, and decreases significantly through the first 10–20-μm depth into the membrane (Fig. 4), although some particles are found throughout the membrane sample, even at its center. Optimizations of the techniques for IPMC manufacture have been reported by Liu et al. [1992], Homma and Nakano [1999], and Rashid and Shahinpoor [1999].

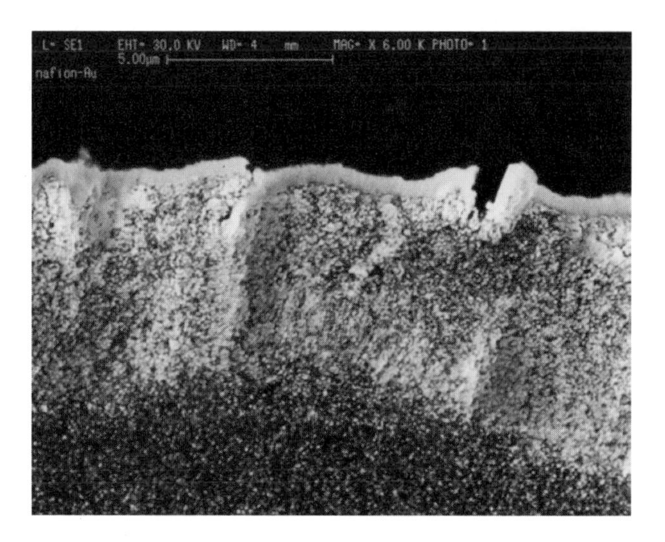

Figure 4: Transmission electron micrograph (TEM) of Pt/Au-IPMC. A cross section at the surface indicates a gold overcoating and the decay in metal (Pt) concentration deeper within the Nafion ionomer.

Variations of IPMC manufacture have been reported: Fujita and Muto describe a method in which platinum is precipitated in the presence of a solution of Nafion125, and the resulting mixture of Nafion and colloidal platinum is then coated onto the surface of a Nafion membrane [Fujita and Muto, 1986]. IPMC materials have been made using a number of different metals including Ni, Pb, Cu, Ag [Chen and Chou, 1993], Au [Oguro et al., 1999; Fujiwara et al., 2000] and Ir [Millet et al., 1989]. Formation of cadmium selenide (CdS) clusters within Nafion membranes has been reported [Nandakumar et al., 1999]. Metal reduction has been accomplished using gamma radiation, as reported by Platzer et al. [1989], and Belloni et al. [1998]. As well, Salehpoor et al. [1998] have prepared IPMCs by chemical vapor deposition. For IPMCs manufactured from anion exchange membranes, see Dube [1998]. Si et al. [1998] have reported on possible mechanisms for platinum particle synthesis, and have noted preferential growth of Pt(111) crystal faces in IPMC materials. Bennett and Leo [2003] have manufactured and characterized IPMCs with non-precious metal electrodes.

Optical micrographs and SEM images of the surfaces of plated IPMCs indicate a two-part construction of these materials. Beyond the surface of the membrane, a thicker overlayer of metal is deposited, usually less than 10-μm thick. Optimization of this layer is crucial, as greater thicknesses yield greater surface conductivity—essential in charging the membrane and generating actuative bending. At the same time, greater metal thicknesses increase the composite's stiffness, increasing the force required for the same displacement.

In a recent work, Nemat-Nasser and Wu [2003a] have examined the microstructure and actuation of Flemion-based IPMCs, having gold electrodes of fine dendritic structure, comparing the response of this composite with that of Nafion-based IPMCs in various cation forms. Figure 5 shows the electrode morphology of a Flemion-based IPMC.

Figure 5: Cross section of a typical Au-plated Flemion-1.44, showing dendritic structure of gold electrodes.

The metal surface in an IPMC usually appears fractured, displaying discrete islands of metal deposition between 5–20 μm across, as shown in Fig. 6. It is presumed that water swelling and electroactive bending generate these islands when the resulting strains exceed the tensile strength of the thin metal layer. This process is not well studied, and anecdotal evidence suggests that IPMC actuators

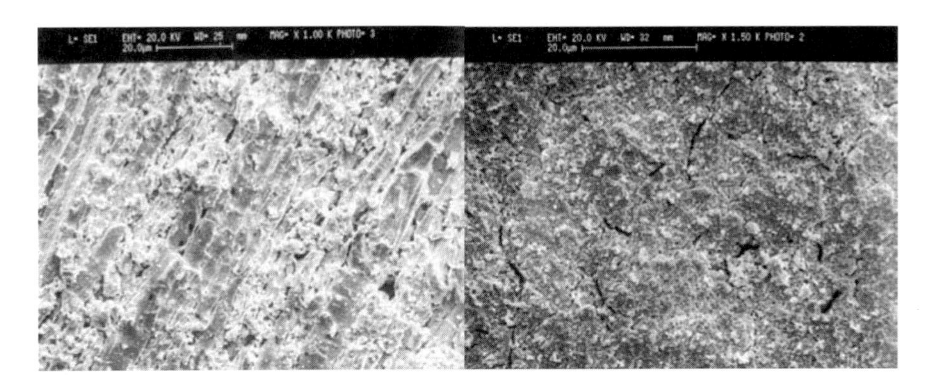

Figure 6: Scanning electron microscopic photos of IPMC surfaces, showing microcracks on Nafion- (left) and on Flemion- (right) based IPMCs.

require an initial "working" of the sample to develop sufficient surface fractionation. Regardless of the presence of fractionation, surface conductivity must be sufficiently large to produce maximum actuation, as conduction occurs through the metallic underlayer or via contact of individual metal islands at the surface of the membrane.

6.4 Properties and Characterization

As stated before, if an IPMC sample in the solvated (i.e., hydrated) state is suddenly bent, a voltage is produced across its faces. If, on the other hand, the same strip has an alternating voltage imposed across its faces, the sample exhibits an oscillatory bending motion. When the same strip is subjected to a suddenly imposed and sustained constant voltage (dc) across its faces, the initial relatively "fast" displacement is generally followed by a slower relaxation, in the reverse direction for Nafion-based IPMCs and in the same direction for Flemion-based IPMCs, in aqueous or other environments. Finally, when the two faces of the strip are shorted during this slow relaxation, a sudden fast motion, in the same direction for Nafion-based and in the opposite direction for Flemion-based IPMCs, occurs. This is then followed by slow relaxation in the opposite (Nafion-based) or the same (Flemion-based) direction. By imposing a ramp voltage of a suitable constant rate, the initial motion towards the anode can be eliminated for Nafion-based IPMCs in all tested cation forms (except for TBA^+). Improvements in the sensing and actuating attributes of IPMC materials have involved measurement of the actuation and sensing properties of the material, as well as modeling to understand the underpinning actuation and sensing mechanisms.

The first improvements in IPMCs' bending response were obtained by optimizing the method of IPMC fabrication. Building on the work of Millet et al., Shahinpoor and coworkers developed a number of improved forms by varying the metal ion used, the chemistry of metal reduction, and the surface treatments of IPMC samples [Rashid and Shahinpoor, 1999]. On a different tack, Oguro et al. [1992] drew from the work of Takenaka et al. [1982] and researchers at

Hitachi [Hitachi Patent, 1983], adopting a method that includes the use of hydrazine during chemical reduction.

A second round of investigation indicated the strong influence of countercation species on the actuation and sensing behavior of IPMCs [Abé et al., 1998]. These studies were limited to monovalent cations such as alkali metal salts or tetraalkylammonium salts. Some early results from these investigations showed that lithium salts produced the greatest displacement of the tip of a cantilever strip over other alkali metals; this may be due to the effect of the countercation on the bending stiffness of the strip, on the resulting internal forces, or on both, as is discussed in Sec. 6.4.7. Other alkali metals showed different degrees of tip displacement and large variation in the extent of slow reverse relaxation. Tetrabutylammonium (TBA^+) samples exhibit the interesting property that dc actuation produces a slow deformation and no observed back relaxation in either Nafion- or Flemion-based IPMCs. On the other hand, lithium-containing Nafion-based samples have been observed to exhibit fast actuation followed by some small reverse relaxation [Nemat-Nasser and Li, 2000; Nemat-Nasser, 2002], while extensive back relaxation is observed for sodium-, potassium, rubidium-, cesium-, thallium-containing Nafion-based IPMCs [Nemat-Nasser and Thomas, 2001; Nemat-Nasser and Wu, 2003a; Nemat-Nasser and Zamani, 2003]. Combined Na^+ and TBA^+ cations have been used to tailor the actuation of Nafion-based IPMCs, thereby controlling the initial speed of bending towards the anode and the subsequent relaxation in the opposite direction [Nemat-Nasser and Wu, 2003b].

Throughout our investigations, focus has been placed on understanding the mechanisms of actuation and sensing, in order to understand these materials and improve their performance. We accomplish this by examining the effect of the nature of the backbone ionomer, the morphology of the metal electrode, the countercation, or the solvent used to produce the IPMC, as well as the optimization of the procedures of electrode deposition and sample preparation.

6.4.1 Polymer Properties

Nafion is a perfluorinated copolymer of polytetrafluoroethylene (PTFE) and a perfluorinated vinyl ether sulfonate. Varying the nature of the vinyl ether side chain gives routes to alternative species, usually through the addition of single CF_2 groups (Aciplex) or modification of the length of the side chain (Flemion). Within a limited range, the ratio of monomers can be used to tailor the polymer equivalent weight (grams dry polymer per mole of ion), although large variation can significantly influence the strength, flexibility, and stability of the resulting membranes. Primary physical data for fluorinated ionomers have been summarized [Fernandez, 1999].

Small-angle x-ray scattering [Roche et al., 1981] and TEM analysis [Xue et al.1989] of Nafion (hydrated and dry H^+ and Na^+ membranes) indicate an essentially constant spacing at about 50 Å with an intensity dependent on water content. This spacing has been attributed to the spacing between individual ionic cluster domains within the Nafion structure [Gierke et al., 1981]. These domains are a result of the particular structure of Nafion polymers. Presumably, the hydrophobic PTFE backbones aggregate to form a semicrystalline matrix and the individual side chains orient their hydrophilic-end groups to form water-filled clusters [Heitner-Wirguin, 1996]. James et al. [2000] have used atomic force microscopy (AFM) to probe the surface of Nafion membranes under a variety of hydration conditions. They observe small clusters (2–5-nm diameter) gathering into super-clusters up to 30-nm diameter. Upon hydration, they noted that the total number of clusters decreases while the size of each cluster increases, suggesting a cluster-consolidation model for Nafion hydration.

6.4.2 Ion-Exchange Capacity

The ion-exchange capacity of an IPMC material indicates the number of sulfonate (Nafion) or carboxylate (Flemion) groups within a fixed volume of material. This number should correspond to the number of substituted (non-H^+) charge-balancing cations within the IPMC. During preparation, sheets of bare Nafion are surface roughened by sanding or sandblasting. Ablation of the ionomeric polymer reduces the mass of the ionomer and, thus, the total moles of ion per cm^2 of sheet. (Note that the measurement of IPMCs in cm^2 derives from the fact that these materials are prepared and handled in sheet form.) Moreover, the addition of metal to the membrane increases the density of the IPMC. Since a sample of IPMC weighs more than bare Nafion and contains fewer ionic sites per unit mass, the equivalent weights of plated samples is larger than the bare Nafion. The required scaling factor can be found by dissolving the metal over layer in aqua regia and determining the mass of remaining IPMC. We find for some early IPMC samples provided to us by Shahinpoor [Shahinpoor, M., personal communication; Rashid and Shahinpoor, 1999] that they are composed of Nafion117, approximately 88% by volume (58% by weight). More recently, similar results (about 59 to 61% by weight) were reported by Nemat-Nasser and Wu [2003a] for various samples that have been plated by Shahinpoor and coworkers.

Because of these modifications to the sample composition during plating, a sample's equivalent weight changes. Equivalent weight is an operational quantity, representing the mass of dry IPMC that contains one mole of ion. For IPMCs containing ions other than protons, the true equivalent weight is given by

$$EW_{\text{Ion}} = \frac{EW_{H+} - 1.008 + FW_{\text{Ion}}}{SF},$$

where EW_{H+} is the equivalent weight of the dry ionomer in proton form (1100 for Nafion117), 1.008 is the formula weight of the proton in grams per mole, FW_{Ion} is the formula weight of the cation used, and SF is a scaling factor which accounts for any added electrode mass. For an unplated membrane, $SF = 1$; for the plated

IPMCs described above, the scaling factor is the weight fraction of Nafion within the sample ($SF = 0.58$ to 0.61).

The ion capacity of prepared samples can be determined from the weight difference of dry IPMCs containing different metal ions or by measuring the pH change of a salt solution after immersion of an H^+-form IPMC sample. A summary of ion capacity and hydration of Nafion versus IPMC samples is presented in Table 1 for an IPMC provided by the University of New Mexico's, Artificial Muscle Research Institute [Rashid and Shahinpoor, 1999]. Samples in the H^+-form contain considerably more water than the values given in Table 1, since a fully dried state cannot be easily attained [Gierke et al., 1981]. The data in Table 1 should be viewed as illustrative rather than definitive. Indeed, specific (e.g., per liter dry) ion content should be independent of the (monovalent, Table 1) cation form. Hence, the variation in the values of n_{ion} in Table 1, for example, is mainly due to experimental error as well as to possible mass changes during cation exchange.

6.4.3 Solvent Content and Swelling

The uptake of solvent from a dried IPMC sample is dependent on the method of solvation and the cation activity. The weight difference between dry and solvated samples indicates the mass of solvent taken up by each ion form. The ion exchange process is selective [Miyoshi et al., 1990; Iyer et al., 1992] and correlates with the extent of solvation.

Table 1: Illustrative results: ion exchange capacity of IPMC vs. Nafion117.

	Membrane	H^+	Rb^+	Cs^+	Tl^+	TMA^+	TBA^+
Eq. Wt. (W_e) *g* dry IPMC/mole ion	Nafion 117	1100	1184	1232	1303	1173	1341
	Metal	1929	2013	2061	2132	2002	2170
Hydration Volume Change (%)	Nafion 117	57%	29%	25%	31%	35%	28%
	Metal	53%	27%	24%	31%	33%	38%
One liter dry material hydrated to N-form has:							
n_{H2O} (moles)	Nafion 117	31.7	16.1	13.9	17.2	19.4	15.6
	IPMC	29.4	15.0	13.3	17.2	18.3	21.1
n_{ion} (moles)	Nafion 117	1.74	1.78	1.78	1.81	1.59	1.38
	IPMC	1.50	1.59	1.47	1.41	1.37	1.32
x_{H2O} (10^{-1})	Nafion 117	9.01	8.19	7.96	8.26	8.60	8.50
	IPMC	9.07	8.25	8.19	8.60	8.70	8.89
x_{M+} (10^{-2})	Nafion 117	4.95	9.05	10.2	8.70	7.01	7.52
	IPMC	4.63	8.74	9.04	7.02	6.49	5.57

Gebel and Pineri have also investigated the hydrative swelling of Nafion117 membranes [Gebel et al., 1993]. As well, Nafion hydration has been measured by quartz crystal gravimetric analysis [Shi and Anson, 1996] and infrared spectroscopy [Blanchard and Nuzzo, 2000; Ludvigsson et al., 2000]. We propose that the balance of osmotic pressures, electrostatic forces, and elastically induced interfacial stresses within the membrane determines the extent of solvation expansion. The elastic resistance (stiffness) of the backbone polymer is, in turn, dependent on the nature of the neutralizing cation. Dimensional measurements of IPMC samples indicate that solvent-induced swelling strains are smaller than in bare (nonplated) ionomer samples. Presumably, the stiffness of metal-coated layers restricts the expansion of IPMC samples and results in a smaller observed solvation.

It should be noted that hydration studies of Nafion-type polymers usually refer to three common states: the normal (N) form, in which dried samples have been hydrated at 25°C, the prepared (P) form, in which nondried hydrated H^+ samples have been ion exchanged at 25°C, or the expanded (E) form, in which dry ion-exchanged samples are hydrated at 100°C [Yeager et al., 1982]. Samples prepared in the expanded form can retain their swollen characteristics for long periods, often more than several weeks (Jeff McGee, personal communication; and McGee, 2002). Presumably, the hydration state of IPMC samples is affected by the hydration of H^+ samples near their glass transition temperature (104°C). This treatment should provide for the optimized water-swollen state for the H^+ form of the IPMC. The polymeric structure may thus become biased toward the ideal H^+ swollen state; subsequent ion exchanged states are then subject to the constraints imposed by this prepared state.

6.4.4 Ion Migration Rates

Xue et al. [1991] have shown that the diffusion of alkali metal ions through Nafion membranes varies with the ion used. At high concentration, it has been shown that the rate of diffusion is in the order $Li^+ < Cs^+ < Rb^+ < Na^+ < K^+$. Under applied voltages, cations within the membrane migrate toward the cathode. Ion motion occurs in tandem with water migration, either through primary solvation-shell migration of water or other solvent molecules, or through secondary electro-osmotic migration of water. These processes are determined by several factors, including the diffusion of water [Zelsmann et al., 1990] and migration of metal ions (e.g., sodium, cesium, and zinc) through Nafion membranes [Sodaye et al., 1996; Rollet et al., 2000].

6.4.5 Metal Content and Distribution

The distribution and morphology of deposited metal at the surface of IPMC materials has a significant effect on their actuation behavior. Many studies of metal deposition in IPMCs, including TEM analysis [Millet et al., 1989], x-ray microprobe analysis [Millet et al., 1992], and analysis of the surface area of implanted electrodes have been performed [Chen and Chou, 1993]. Performance of IPMCs as actuators correlates with their surface morphology, with the best electrodes for IPMC actuators being those that exhibit the largest available surface area and greatest surface conductivity. Large surface areas can be attributed to the development of a highly dispersed particle distribution or the small size of individual metal particles (Fig. 4) or fine dendritic structure, as seen in Fig. 5, for gold plated Flemion. An indication of an electrode's surface area can be determined through measurement of an IPMC's behavior as a parallel plate capacitor. Moreover, surface conductivity can be measured directly (Nemat-Nasser and Wu, 2003a].

6.4.6 Modeling of Cluster Size

Hsu and Gierke [1982] proposed a model for ionic clustering in Nafion that describes the experimental data well. This model is based on elastic interaction between the fluorocarbon matrix and the ionic cluster. It ignores the electrostatic dipole interaction that is believed to be the driving force for the clustering [Eisenberg, 1970]. The electrostatic dipole interaction results in an increase in the stretching of the polymer chains, and thus an increase in the elastic energy of the matrix [Forsman, 1982]. Using a computer simulation, Datye et al. [1984] have shown that the electrostatic and elastic forces acting on the pendant ionic groups and their neutralizing counter-ions produce a dipole layer at the surface of an ionic cluster. Monte Carlo simulation [Datye and Taylor, 1985] has further revealed that the electrostatic energy of an ionic cluster in an ionomer is not very sensitive to the variation of the cluster shape. More recently, Li and Nemat-Nasser [2000] have sought to determine the cluster size and shape from free-energy minimization, taking into account the electrostatic dipole interaction energy, the elastic energy of the polymer chain reorganization during clustering, the cluster surface energy, and the electro-elastic interaction energy of the ionic clusters and the fluorocarbon polymer matrix. They also have examined the effect of the cluster morphology on the macroscopic electro-elastic and transport properties of the hydrated Nafion, using a micromechanical multi-inclusion model proposed by Nemat-Nasser and Hori [1993, 1999]. In what follows, the main results of this investigation that correspond to the average cluster size are summarized.

Consider a cluster of fixed radius containing a fixed number of dipoles. These dipoles are arranged on the cluster surface so as to minimize the free energy of the cluster. In such a situation, the spacing of the dipole pairs will be proportional to the cluster radius, while the corresponding orientation will be independent of the cluster radius. Thus the electrostatic energy of a cluster with N dipoles may be expressed as

$$U_{ele} = -g\frac{N^2}{4\pi\kappa_e}\frac{m^2}{r_c^3} = -\frac{N^2}{4\pi\kappa_e}\frac{m^{*2}}{r_c^3}, \tag{1}$$

where g is a geometric factor, depending only on the detailed arrangement of the dipoles on the cluster surface, r_c is the radius of the cluster, $m^* = \sqrt{g}m$ is the effective dipole moment, and κ_e is the effective electric permittivity of the water-swollen Nafion.

Datye et al. [1984] use a simple model of rubber elasticity to obtain the per-cluster elastic energy associated with clustering. The result can be expressed as

$$U_{ela} = \frac{3NkT}{<h^2>}\left(\sqrt[3]{\frac{N\,EW}{\rho^* N_A}} - r_c\right)^2, \tag{2}$$

where k is Boltzmann's constant, T is the absolute temperature, $<h^2>$ is the mean square end-to-end chain length, EW is the equivalent-weight of Nafion (i.e., the weight in grams of dry polymer per mole of ion exchange sites), ρ^* is the effective density of the water-swollen Nafion membrane, and N_A is Avogadro's constant. In deriving Eq. (2) it is assumed that half of the pendant side-chain ions terminate on the same cluster while the remaining chains terminate on a nearest-neighbor cluster.

Finally, the surface energy of the cluster is expressed as

$$U_{sur} = 4\pi r_c^2 \gamma, \tag{3}$$

where γ is the surface energy density. Since the surface energy is composed of the hydrophilic energy between the water and the ion pairs, and the hydrophobic energy between the water and the fluorocarbon matrix, a small decrease in the surface energy density γ with an increase in the volume fraction of water is expected.

The number of clusters, n, per unit volume is given by

$$n = \frac{3(c_w + c_i)}{4\pi r_c^3}, \tag{4}$$

where $c_w = w/(1 + w)$ and $c_i = w'/(1 + w')$ are the volume fractions of the water and the ion exchange sites in the membrane, respectively; w denotes the water volume per unit dry polymer, and $w' = N_A V_i /(EW/\rho_d)$, with V_i being the volume of a single ion exchange site. Now, the total energy per unit volume can be calculated from the sum of the electrostatic dipole interaction energy, the elastic energy of the polymer chain reorganization, and the cluster surface energy. Minimizing this expression with respect to the cluster radius, Li and Nemat-Nasser [2000] obtained the following result for the cluster radius:

$$r_c^3 = \frac{(w+w')}{\rho_d}\left(1-\sqrt[3]{\frac{4\pi\rho_d}{3\rho^*(w+w')}}\right)^{-2}, \quad (5)$$

where $\rho^* = (\rho_d + w\rho_w)/(1 + w)$ is the effective density. Assuming $<h^2> = \beta\ EW$ [Forsman, 1986], Eq. (5) suggests that a plot of r_c^3 versus

$$\Phi = \frac{EW^2(w+w')}{\rho_d}\left(1-\sqrt[3]{\frac{4\pi\rho_d}{3\rho^*(w+w')}}\right)^{-2}$$

should be a straight line crossing the origin for all membranes of different equivalent weights, different cations, or water intake, with the slope given by $\gamma\beta/2N_AkT$. A typical value for Φ/EW^2 is 0.19 for $w = 0.443$. Using data from Gierke et al. [1981], Li and Nemat-Nasser show that a linear relation generally holds. Table 2 provides sample results reported by Gierke et al. [1981]. For more information and for the calculation of the cluster shape, see Li and Nemat-Nasser [2000].

Table 2: Cluster size of 1200 equivalent weight Nafion with different cations [Gierke et al., 1981].

Cation	H	Li	Na	K	Rb	Cs
Dry density (g/cm^3)	2.075	2.078	2.113	2.141	2.221	2.304
Volume gain (%)	69.7	61.7	44.3	18.7	17.9	13.6
Cluster diameter (nm)	4.74	4.49	4.21	3.45	3.56	3.5

6.4.7 Stiffness

Nafion polymers are viscoelastic, exhibiting first-order tensile moduli from 50–1500 MPa and greater. The large variance in stiffness is primarily due to the presence or absence of solvent, but the nature of the countercation present is a significant stiffness factor, especially for dry samples. The variation in stiffness is attributed to the action of ions within the membrane increasing stiffness by acting as cross-linking agents [Eisenberg et al., 1996]. Ionomeric polymers containing polyvalent cations may act as noncovalent cross-linking agents, binding multiple anionic side-chains by ligand coordination. Similarly, monovalent cations such as alkali metals or tetraalkylammonium salts may serve to cross-link ionomers via dipolar interactions between individual salt pairs.

The stiffness of dry IPMC samples correlates with the radius of the alkali metal countercation (Li^+, Na^+, K^+, Rb^+, Cs^+, and Tl^+). Smaller cations may more closely approach the sulfonate anion, yielding a smaller ion pair dipole [Gejji et al., 1997]. Larger dipoles in interaction will require a greater energy for reorganization. This can result in a larger observed stiffness of the material.

The Young's moduli of IPMC materials for a variety of cations have been determined. These values were compared to the same-cation unplated ionomer samples. Measurements were obtained using a mini-load frame developed in the author's lab by Jon Isaacs (Fig. 7). Some typical results of these experiments are summarized in Table 3. In these tests, the same sample is sequentially neutralized by the indicated countercations, and tested. Therefore, the differences in the measured stiffness are due to the effect of the cations. Similar measurements are performed on several other IPMCs that have been produced by Shahinpoor and coworkers using new processing techniques, as well as on Flemion and Flemion-based IPMCs that were provided by Dr. Kenji Asaka. Additional results are given later on in this section; see also Nemat-Nasser [2002], Nemat-Nasser and Wu [2003a,b], Nemat-Nasser and Zamani [2003].

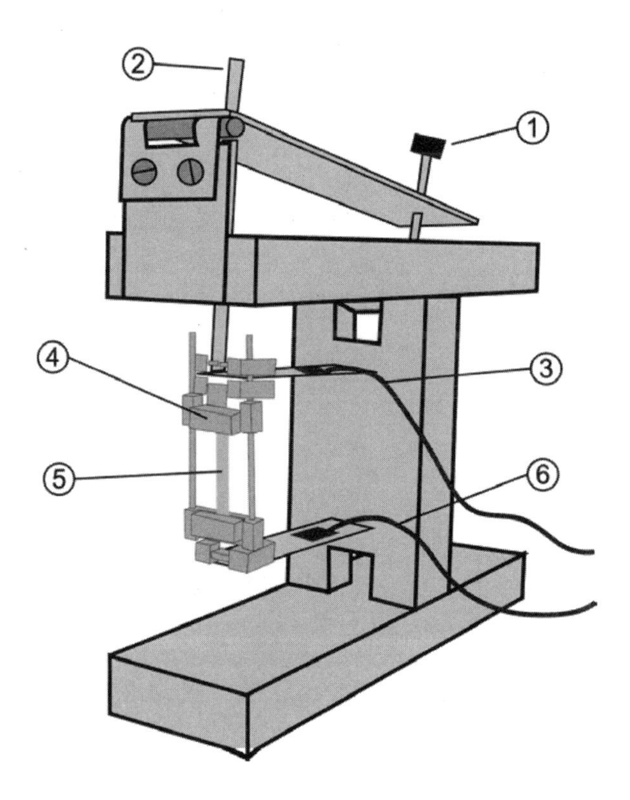

Figure 7: Diagram of mini-load frame. (1) Force adjuster, motorized; (2) force arm; (3) displacement gauge; (4) sample clamps; (5) IPMC sample; (6) load-displacement gauge.

Table 3: Stiffness of bare Nafion and IPMC samples in dry and hydrated form for indicated cations.

	Nafion 117		IPMC	
Ion	wet	dry	wet	dry
H			140	340
Li	70	300	90	650
Na	80	500	90	
K	120	1010	170	–
Rb	–	850		–
Cs	210	1200	190	1270
Tl		750	190	1300
TMA	110	760	140	830
TBA	–	–	130	1400

The measured axial stiffness of Nafion and IPMC samples in dry and solvated states allows estimation of the corresponding bending stiffness. An expression for the extensional stiffness of the IPMC in terms of the Young's moduli of bare ionomer and the surrounding thin surface electrodes is presented. The result is used to estimate the bending stiffness of the IPMC. Since the extensional stiffness is also directly measured, the model estimate of the extensional stiffness of the IPMC provides an assessment of the effective stiffness of the plating region. The thin metal electrodes contain numerous microcracks, and a diffuse metal particle distribution or a fine dendritic metal structure provides for an electrode with large surface area, as shown in Figs. 4 and 5. Hence, only an effective extensional modulus can be assigned to these regions within the IPMC. By direct measurement of the extensional stiffness of the bare membrane and the IPMC, the effective stiffness may be calculated and used as a measure of the effectiveness of the plating procedure.

In a recent work, Nemat-Nasser [2002] provides a micromechanical model to estimate the extensional stiffness of both the membrane and the corresponding IPMC as functions of the solvation level for each neutralizing countercation. The model is based on the observation that a dry sample of a bare polymer or an IPMC immersed in a solvent absorbs the solvent until the resulting pressure within its clusters is balanced by the elastic stresses that are consequently developed within its backbone polymer membrane. From this observation then the stiffness of the membrane is calculated as a function of the solvent uptake for various cations. In this calculation, first the balance of the cluster pressure and the elastic stresses for the bare polymer (no metal plating) is considered, and then the results are used to calculate the stiffness of the corresponding IPMC by including the effect of the added metal electrodes. The procedure also provides a way of estimating many of the microstructural parameters that are needed for the modeling of the actuation of the IPMCs. Since, for both the Nafion- and Flemion-based IPMCs, the overall stiffness of both bare membrane and the corresponding IPMCs has been measured directly as a function of the hydration (or solvation by other solvents), the basic assumptions and the results can be subjected to experimental verification [Nemat-Nasser, 2002; Nemat-Nasser and Wu 2003a; Nemat-Nasser and Zamani, 2003].

6.4.7.1 Extensional Stiffness of Bare Polymer

The stresses within the bare polymer may be estimated by modeling the polymer matrix as an incompressible elastic material [Treolar, 1958; Atkin and Fox, 1980]. It will prove adequate to consider a neo-Hookean model for the matrix material, where the principal stresses σ_I are related to the principal stretches λ_I by

$$\sigma_I = -p_0 + K\lambda_I^2; \tag{6}$$

here, $p_0(w)$ is an undetermined parameter (pressure) to be calculated from the boundary data; in spherical coordinates, $I = r, \theta, \varphi$, for the radial and the two hoop components; and $K = K(w)$ is an *effective stiffness*, which depends on the cation type and its concentration, and on the solvent uptake, w.

The aim is to calculate K and p_0 as functions of w for various ion-form membranes. For this, examine the deformation of a unit cell of the solvated polyelectrolyte by considering a spherical cavity of initial (dry state) radius a_0 (representing a cluster), embedded at the center of a spherical matrix of initial radius R_0, and placed in a *homogenized solvated membrane,* referred to as the *matrix*. Assume that the stiffness of both the spherical shell and the homogenized matrix is the same as that of the (yet unknown) overall effective stiffness of the solvated membrane. For an isotropic expansion of a typical cluster, the two hoop stretches are equal, $\lambda_\varphi = \lambda_\theta$, and incompressibility yields $\lambda_r \lambda_\theta^2 = 1$, leading to

$$\sigma_r(r_0) = -p_0 + K\lambda_\theta^{-4}(r_0), \quad \sigma_\theta(r_0) = -p_0 + K\lambda_\theta^2(r_0), \tag{7}$$

where r_0 measures the initial radial distance from the center of the cluster. The basic observation is that the *effective elastic resistance* of the (homogenized solvated) membrane balances the cluster's pressure, p_c, which is produced by the combined osmotic and electrostatic forces within the cluster.

Upon solvation from an initial state with w_0 solvent uptake, material points initially at r_0 move to r,

$$r^3 = r_0^3 + a_0^3(w/w_0 - 1), \quad n_0 = (a_0/R_0)^3, \quad w_0 = n_0/(1-n_0), \tag{8}$$

where n_0 is the initial porosity (volume of saturated or dry void divided by total volume). The radial and hoop stresses at an initial distance of r_0 from the cluster center then become

$$\begin{aligned} \sigma_r(r_0) &= -p_0 + K[(r_0/a_0)^{-3}(w/w_0 - 1) + 1]^{-4/3}, \\ \sigma_\theta(r_0) &= -p_0 + K[(r_0/a_0)^{-3}(w/w_0 - 1) + 1]^{2/3}. \end{aligned} \tag{9}$$

The radial stress, σ_r, must equal the pressure, p_c, in the cluster, at $r_0 = a_0$, i.e., $\sigma_r(a_0) = -p_c$. In addition, the volume average of the stress tensor, taken over the entire membrane, must vanish in the absence of any externally applied loads, i.e.,

$$\frac{1}{V_{dry}} \int_{V_{dry}} \frac{1}{3}(\sigma_r + 2\sigma_\theta)\, dV_{dry} - w p_c = 0. \tag{10}$$

This is a *consistency condition* that to a degree accounts for the interaction among clusters. These conditions are sufficient to yield the undetermined pressure p_0, and the stiffness, K, in terms of w and w_0, for each ion-form bare membrane.

To estimate the cluster pressure, p_c, note that in the absence of an applied electric field, this pressure consists of an osmotic part, $\Pi(M^+)$, and an electrostatic (dipole-dipole interaction) part, p_{DD}, i.e., $p_c = \Pi(M^+) + p_{DD}$, where M^+ stands for the considered cation. Detailed calculations are given by Nemat-Nasser [2002]. The final expression is

$$p_c = \frac{2\rho_B RT\phi}{EW_{Ion} w} + \frac{1}{3\kappa_e}\left(\frac{\rho_B F}{EW_{Ion}}\right)^2 \frac{\pm\alpha^2}{w^2}, \tag{11}$$

where F is Faraday's constant (96,487 C/mol), ρ_B is the dry density of the bare membrane, R = 8.31 J/mol/K is the gas constant, T (= 300 K) is the test temperature, EW_{Ion} is the equivalent weight of bare membrane, $\kappa_e = \kappa_e(w)$ is the effective electric permittivity in the cluster, $\alpha = \alpha(w)$ is an effective dipole length, and ϕ is the osmotic coefficient. From $\sigma_r(a_0) = -p_c$ [Eqs. (10), and (11)], it follows that

$$K(w)=p_c\frac{(1+w)}{w_0I_n-\left(\frac{w_0}{w}\right)^{4/3}},\quad I_n=\frac{1+2An_0}{n_0(1+An_0)^{1/3}}-\frac{1+2A}{(1+A)^{1/3}}$$

$$p_0(w)=K\left(\frac{w}{w_0}\right)^{-4/3}+p_c,\quad A=\frac{w}{w_0}-1. \tag{12}$$

It turns out that the electrostatic forces are a dominating element in this calculation, with the dielectric properties of the solvent playing an important role. Consider for illustration, water as the solvent. As part of the hydration shell of an ion, water has a dielectric constant of 6, whereas as free molecules, its room-temperature dielectric constant is about 78. The number of mole water per mole ion within a cluster is

$$m_w=\frac{EW_{\text{Ion}}w}{36\rho_B}. \tag{13}$$

Hence, when the water uptake is less than CN moles per mole of ion within a cluster, set $\kappa_e=6\kappa_0$, where $\kappa_0=8.85\times10^{-12}$ F/m is the electric permittivity of the free space and CN is the coordination number (number of water molecules per ion in bulk). On the other hand, when more water is available in a cluster, i.e., when $m_w > CN$, calculate κ_e as follows:

$$\kappa_e=\frac{7+6f}{7-6f}6\kappa_0,\quad f=\frac{m_w-CN}{m_w}, \tag{14}$$

where $(14)_1$ is obtained as a special case of Eq. (60) which is discussed in Sec. 6.5.7.9. As a first-order approximation, assume α^2 in Eq. (11) is linear in w for $m_w\le CN$,

$$\pm\alpha^2=a_1w+a_2, \tag{15}$$

and estimate a_1 and a_2 from the experimental data. For $m_w>CN$, assume that the distance between the two charges forming a pseudo-dipole is controlled by the effective electric permittivity of their environment (i.e., water molecules), and set (measured in meters)

$$\alpha=10^{-10}\frac{7+6f}{7-6f}(a_1w+a_2)^{1/2}, \tag{16}$$

which is obtained from Eq. (15)$_1$ by setting $\kappa_e = 10^{-10}(a_1 w + a_2)^{1/2}$.

Figure 8 shows the experimentally measured Young modulus of the bare Nafion-117 and the corresponding IPMC, in Cs$^+$-form. The Young's modulus Y_B of the hydrated strip of bare polymer relates to the stiffness K, by $Y_B = 3K$, based on incompressibility. The lower solid curve (bare Nafion) is obtained from Eq. (13), using the following parameters: FW_{Cs+} = 132.91 g/mol, ρ_B = 2.16 g/cm^3 (measured dry density), $n_0 = 0.01$, and $\phi = 1$. The values of a_1 and a_2 in Eq. (A9) are obtained as 1.6383 × 10^{-20} and -0.0807 × 10^{-20} (in m^2), respectively, by setting Y_B = 1130 MPa for w = 0.02 and Y_B = 158 MPa for w = 0.42. Figure 9 shows the experimentally measured Young's modulus of the bare Flemion-1.44 and the corresponding IPMC in Cs$^+$-form. The model results (lower solid curve), are obtained based on: EW_{H+} = 694.4 g/mol, ρ_B = 2.19 g/cm^3, and a_1 = 0.8157 × 10^{-20} and a_2 = –0.0606 × 10^{-20}, obtained by setting Y_B = 1006 MPa for w = 0.036 and Y_B = 147 MPa for w = 0.54.

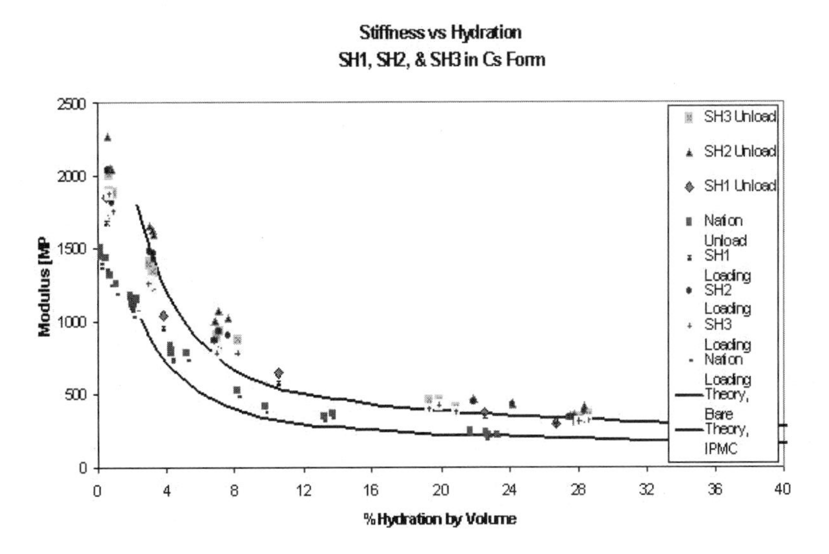

Figure 8: Stiffness versus hydration of Nafion ionomer (lower data points and the solid curve) and IPMCs (upper data points and the solid curve) in Cs+-form.

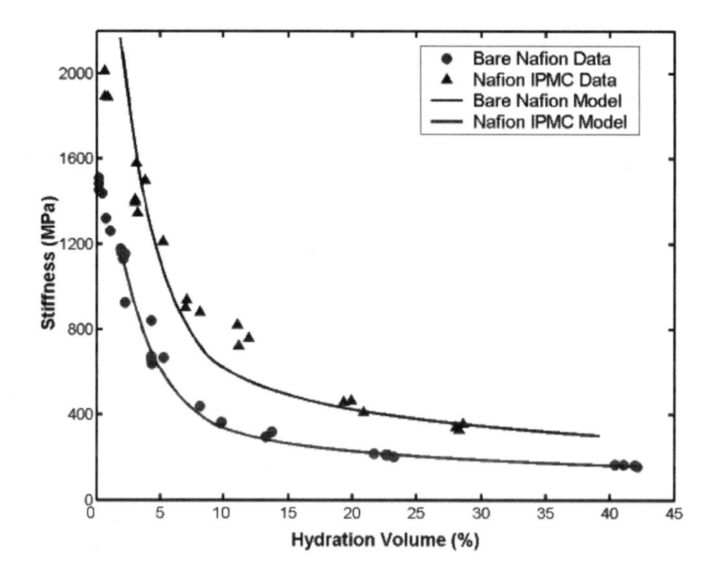

Figure 9: Stiffness versus hydration of Flemion ionomer (lower data points and the solid curve) and IPMCs (upper data points and the solid curve) in Cs+-form.

6.4.7.2 Extensional Stiffness of IPMC

To include the effect of metal plating, assume a uniaxial stress state and by volume averaging obtain

$$\bar{\varepsilon}_{IPMC} = f_{MH}\bar{\varepsilon}_M + (1 - f_{MH})\bar{\varepsilon}_B,$$
$$\bar{\sigma}_{IPMC} = f_{MH}\bar{\sigma}_M + (1 - f_{MH})\bar{\sigma}_B, \; f_{MH} = \frac{f_M}{1+w}, \tag{17}$$

where the barred quantities are the average uniaxial values of the strain and stress in the IPMC, metal, and bare polymer, respectively, and f_M is the volume fraction of the metal plating in a dry sample, given by

$$f_M = \frac{(1-SF)\rho_B}{(1-SF)\rho_B + SF\rho_M}, \tag{18}$$

where ρ_M is the mass density of the metal plating and *SF* is the scaling factor, representing the weight fraction of dry polymer in the IPMC. The average stress in the bare polymer and in the metal are assumed to relate to the overall average stress of the IPMC, by

$$\bar{\sigma}_B = A_B \bar{\sigma}_{IPMC}, \; \bar{\sigma}_M = A_M \bar{\sigma}_{IPMC}, \tag{19}$$

where A_B and A_M are the concentration factors. Setting $\bar{\sigma}_B = Y_B \bar{\varepsilon}_B$, $\bar{\sigma}_M = Y_M \bar{\varepsilon}_M$, and $\bar{\sigma}_{IPMC} = Y_{IPMC} \bar{\varepsilon}_{IPMC}$, obtain

$$\bar{Y}_{IPMC} = \frac{Y_M Y_B}{BA_B Y_M + (1 - BA_B) Y_B},$$

$$B = \frac{(1+\bar{w})(1-f_M)}{1+\bar{w}(1-f_M)}, \; w = \bar{w}(1-f_M). \tag{20}$$

Here, Y_B is evaluated at solvation of $\bar{w}$ when the solvation of the IPMC is w. The latter is measured directly at various hydration levels.

The result for the Nafion-based IPMC is shown in Fig. 8 (the upper solid curve), using a scale factor of 0.6, and ρ_M = 20 g/cm^3 for the combined overall density of gold and platinum, with Y_M = 75 GPa (the results are insensitive to this quantity) and A_B = 0.55. For the Flemion-based IPMC, the same procedure is used with SF = 0.54, ρ_M = 19.3 g/cm^3 (for gold), Y_M = 75 GPa, and A_B = 0.5, leading to the results given in Fig. 9 by the upper solid curve.

6.4.7.3 Bending Stiffness of IPMC

For a given (antisymmetric) axial stress distribution, $\sigma(x)$, over a cross section of an IPMC strip, the bending moment M, acting at that cross section is given by (see Fig. 10)

$$M = \int_{-H}^{H} \sigma(x) x dx = 2 \int_{0}^{H} Y(x) \varepsilon(x) x dx. \tag{21}$$

Since IPMCs are very thin, the Euler beam theory applies. The strain $\varepsilon(x)$ is linear over the cross section of the strip,

$$\varepsilon(x) = \varepsilon_{\max} \frac{x}{H}, \tag{22}$$

and hence Eq. (21) becomes

$$M = \frac{2\varepsilon_{\max}}{H} \int_{0}^{H} Y(x) x^2 dx = \frac{2}{3} \varepsilon_{\max} H^2 (3\bar{Y}_{IPMC} - 2Y_B) + O(t/H)^2, \tag{23}$$

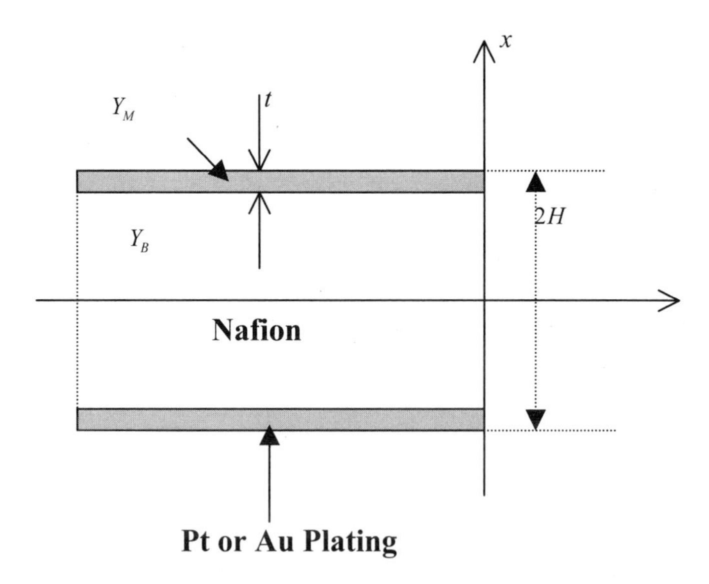

Figure 10: Edge-on diagram of IPMC used in modeling force output.

where $O(t/H)^2$ indicates (negligible) terms of the second order and higher in the small parameter t/H.

If we define the effective bending stiffness $\overline{Y}$ as the stiffness of a uniform strip having the same geometry, to the first order in (t/H) the effective Young's modulus corresponding to bending is given by $\overline{Y} \equiv 3\overline{Y}_{IPMC} - 2Y_B$.

6.4.8 Internal Force in IPMCs

The bending moment M can be used as a measure of the internal forces that produce a corresponding displacement. The displacement of a cantilever strip subjected to an electric potential can be measured directly. From this, ε_{max} and the corresponding bending moment along the length of the strip can be calculated.

Let the radius of curvature at a typical point along the strip be denoted by R_0 before the application of a voltage, and by R afterwards (see Fig. 11). For an element of the strip initially subtended by angle $d\theta_0$, which upon pure bending changes to $d\theta$, the maximum strain at the top (tension, +) and at the bottom (compression, –) are equal to

$$\varepsilon_{max} = \frac{R \pm H}{R_0 \pm H} \frac{d\theta}{d\theta_0} - 1 = \pm H \left(\frac{1}{R} - \frac{1}{R_0} \right) + O\left(\frac{H}{R} \right)^2, \tag{24}$$

where the plus sign corresponds to the upper, and the minus sign to the lower edge of the cross section.

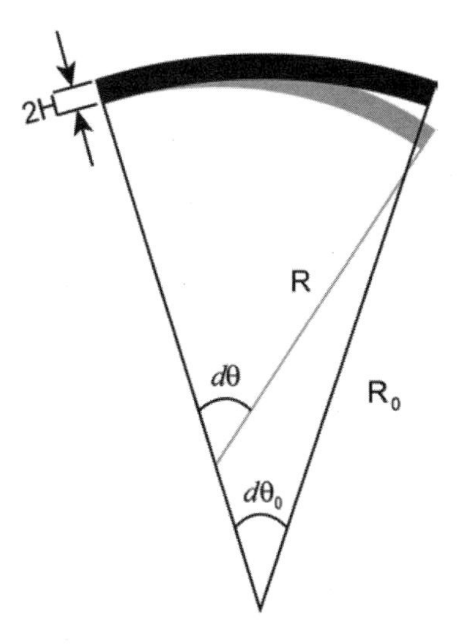

Figure 11: Pure bending of an element of an IPMC strip corresponds to a change in the local radius of curvature.

For small deflection of the strip, observed when the applied voltage is below that required for water electrolysis, we may use the following approximation:

$$\frac{1}{R} \approx \frac{d^2 u}{dz^2}, \quad \frac{1}{R_0} \approx \frac{d^2 u_0}{dz^2}, \tag{25}$$

where u and u_0 are the deformed and initial deflections of the strip, respectively. From Eqs. (24) and (25) it follows that, for a cantilevered strip,

$$\varepsilon_{\max} = \pm \frac{2H}{z^2}(u - u_0), \tag{26}$$

where z measures length along the strip from its fixed end.

Assume that the internal forces generated in the strip are represented by equivalent forces that are concentrated at a distance $H - t$ from the centerline of the strip, as shown in Fig. 7. We may then represent these forces by

$$\begin{aligned} F &= \frac{M}{2(H-t)} = \frac{M}{2H}\left(1 + \frac{t}{H}\right) + O\left(\frac{t}{H}\right)^2 \\ &= \frac{1}{3}\varepsilon_{\max} H\left(1 + \frac{t}{H}\right)(3\bar{Y} - 2Y_B) + O\left(\frac{t}{H}\right)^2, \end{aligned} \tag{27}$$

where ε_{max} is given by Eq. (26), and F is the resultant internal force producing the bending moment M necessary to induce the displacement $u - u_0$.

As has been pointed out before, the stiffness of the IPMC strip depends on the nature of the countercation and the level of solvation. Considerably greater internal forces are required for Cs^+-form IPMCs to yield the same induced displacement $u - u_0$, as compared with Li^+ or Na^+.

Other methods have been used to estimate the force output. These methods usually involve the measurement of total tip deflection and determination of the required restoring force. This method has been used by Asaka and Oguro [2000], Bar-Cohen et al. [1999b], McGee [2002], Mojarrad and Shahinpoor [1996, 1997], and Shahinpoor [1999]. These materials can be actuated at tens of Hz, depending on the sample geometry and boundary conditions. This limiting bandwidth factor will determine the ultimate power that can be generated by IPMC materials [Shahinpoor et al., 1997].

6.5 Actuation Mechanism

Several models for the mechanism of IPMC actuation have been presented. De Gennes et al. [2000] have suggested a model that incorporates water pressure gradients and the overall electric field as thermodynamic forces that induce ion/water fluxes as primary mechanisms for actuation. Shahinpoor et al. offer an electromechanical model for IPMC motion [Shahinpoor, 1995; Shahinpoor and Thompson, 1995; Shahinpoor, 1999], and Asaka and Oguro [2000] have proffered a model by which water flow, induced by pressure gradients and electro-osmotic flow, may generate swelling stresses to drive actuation. Nemat-Nasser and Li [2000] presented a model that includes ion and water transport, electric field, and elastic deformation, emphasizing that ion transport may dominate the initial fast motion of IPMC materials. More recently, Nemat-Nasser [2002] has examined in some detail hydraulic, osmotic, electrostatic, and elastic forces that may affect actuation of fully hydrated Nafion-based IPMCs in various cation-forms, concluding that the electrostatic and osmotic forces within the clusters and the elastic resisting force of the backbone ionomer basically control the actuation, with water flowing into or out of the clusters as a response to these forces. Direct measurement of cation charge accumulation in the cathode region of an IPMC strip has decidedly shown that this continued charge accumulation is accompanied by cathode contraction which produces bending towards the cathode [Nemat-Nasser and Wu, 2003a]. This is illustrated in Fig. 12, from Nemat-Nasser and Wu [2003a]. As is seen, back relaxation occurs soon after actuation has begun while cations are migrating into the cathode side, presumably carrying with them their hydration water molecules.

In this section, we present a detailed description of the phenomenon of actuation under a 1 to 3V applied dc signal. Following this, we present a summary of the requisite points of the Nemat-Nasser and Li [2000] and Nemat-Nasser [2002] model for IPMC sensing and actuation, including comments as appropriate regarding the relationship of this model to other proposed models. We discuss both the initial fast motion as well as the subsequent relaxation aspects of the actuation of Nafion-based IPMCs. First, we provide an outline of the primary physical components of the model. Then, we present specific details supporting the model as have been worked out to date, including new observations and results. Numerical results and comparison with experimental data are reported elsewhere [Nemat-Nasser, 2002].

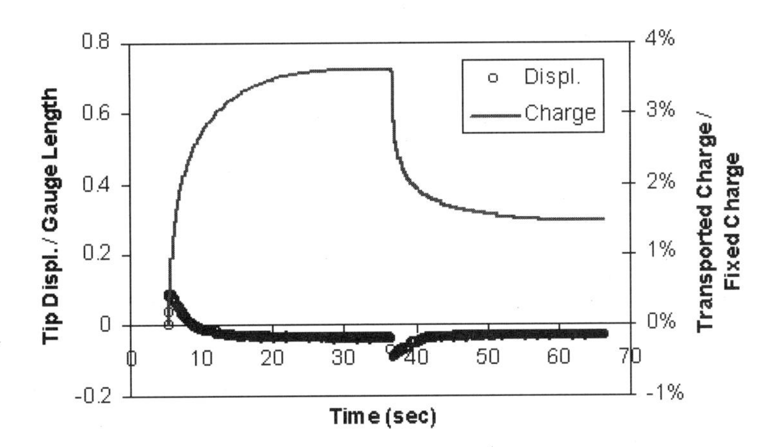

Figure 12: Tip displacement and accumulated charge versus time for a Nafion-based IPMC in Na^+-form, with 1.5V DC applied and then shorted.

6.5.1 Model Summary

Consider a model of an IPMC, where the hydrophobic backbone fluorocarbon polymer is separated from the hydrophilic clusters, as schematically represented in Fig. 13. The anions within the clusters are attached to the fluorocarbon matrix, while under suitable conditions the associated unbound cations may move within the water that permeates the interconnected clusters. Under an electric field, cations redistribute and migrate toward the cathode. This redistribution produces several significant changes in the local properties of the composite, specific to the anode and cathode. These changes form the basis for our model of IPMC actuation, and are suitable for explaining the observed actuation and subsequent relaxation of IPMC materials. Our research is aimed at quantifying each effect and determining which are best suited to improving the actuating properties of IPMCs.

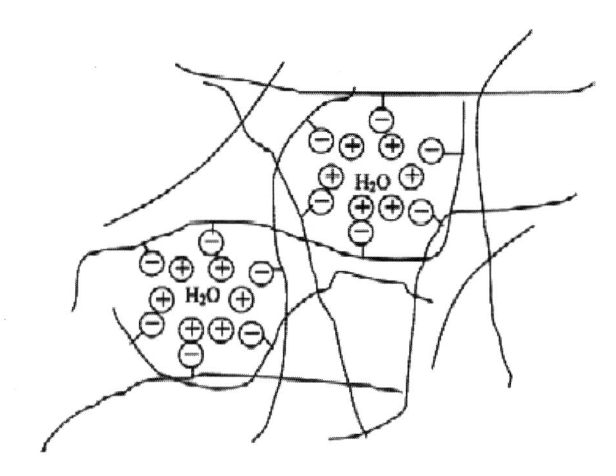

Figure 13: Schematic depiction of polymer domains within IPMC materials.

At the *anode*, the application of the electric field results in depletion of cations from their clusters. We refer to this thin layer as the *anode boundary layer*. The reduction in cation concentration may produce any of the following effects within the anode layer:

(1) A decrease in the effective stiffness of the polymer.

(2) A repulsive electrostatic force among the fixed anions within each cluster, which tends to increase the average cluster volume and also relax the prestretched polymer chains between adjacent clusters. This leads to an increase in the associated entropy and a decrease in the elastic energy of the polymer; polymer chains coil back to lower energy states.

(3) Reorientation of the remaining water in the clusters, which tends to increase the effective electric permittivity of the clusters and hence reduce the electrostatic repulsive forces among the fixed charges within each cluster.

(4) A decrease in the osmotic pressure inside the clusters, due to the reduced ion concentration.

(5) Removal or addition of water (solvent), in response to reduction or increase in the average volume of the clusters.

At the *cathode*, another boundary layer is formed in which local clusters contain an excess in the cation concentration. We refer to this thin layer as the *cathode boundary layer*. The excess cations may produce any of the following effects within the cathode boundary layer:

(1) An increase in the effective stiffness of the polymer.

(2) A change in the attractive electrostatic force within each cluster, which tends to hold the fixed anions more tightly within the cluster, decreasing the associated entropy and increasing the elastic energy of the polymer.

(3) A reduction in the effective electric permittivity of the clusters, which tends to increase the electrostatic forces between cations and the fixed anions within each cluster.
(4) An increase in the osmotic pressure within the clusters, due to the increased ion concentration.
(5) A reorganization of ions in cation-rich clusters, through interaction with fixed anions, depending on the nature of the cations (whether soft, as Tl^+, or hard, as H^+) and the anions (whether strongly acidic, as sulfonate, or weakly acidic, as carboxylate), leading to changes in the corresponding electrostatic interaction forces.
(6) Removal or addition of water (solvent), in response to reduction or increase in the average volume of the clusters.

As is shown in Sec. 6.4.7, the stiffness of the IPMC membrane is critically affected by the nature of the bound cations. Effects (1) and (2) pertain to this fact: the greater the interaction forces between the cations and the fixed anions within the clusters, the greater the resulting stiffness of both Nafion and IPMCs in both solvated and dry forms. Therefore, it is reasonable to expect that removal of all the cations from the clusters would lead to a decrease in the corresponding stiffness, and that their addition would tend to have a reverse effect. Furthermore, the anion-cation coupling within the clusters provides a pseudo cross-linking and a structured arrangement with concomitant increase in the internal energy and decrease in the entropy of the system. Effects (1) and (2) are therefore closely connected. In addition, reorganization of ions in cation-rich clusters, may occur depending on the nature of the cations (whether soft, as Tl^+, or hard, as H^+) and that of the fixed anions (whether strongly acidic, as sulfonate, or weakly acidic, as carboxylate), leading to changes in the corresponding electrostatic interaction forces. This reorganization appears to be at the heart of the mechanism responsible for the observed large and slow relaxation motion of the Nafion-based IPMCs that contain alkali-metal cations, as well as soft cations such as Tl^+; see Fig. 14.

Effect (3) stems from the fact that free water exhibits a dielectric constant of 78 at room temperature. Such a high dielectric constant reflects the ability of water, a polar molecule, to reorient to oppose applied electric fields. Waters of hydration, bound to ions in the solvent, are restricted by this association and exhibit a reduced effective dielectric constant, on the order of 6 at room temperature [Bockris and Reddy, 1998]. Similar comments apply to other polar solvents, such as ethylene glycol, glycerol, and various crown ethers [Nemat-Nasser and Zamani, 2003].

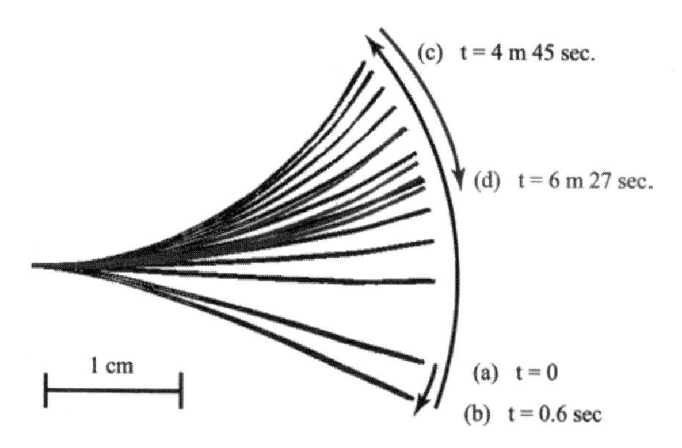

Figure 14: Representation of IPMC actuation. Sample is in the thallium (I) ion form, and a 1-V dc signal is suddenly applied and maintained during the first five minutes, after which the voltage is removed and the two electrodes are shorted. Initial fast bend toward the bottom (anode) occurs during the first 0.6 sec, followed by a long relaxation to upward (cathode) over 4.75 min. Upon shorting, the sample displays a fast bend in the same upward direction (not shown), followed by slow downward relaxation during the next 1.75 min.

Some of the primary solvation molecules may be carried away from or into the clusters by the cations, producing an osmotic imbalance and hence some solvent exchange with the surrounding bath. This principle is not without exception, as tetrabutylammonium cations carry no hydration waters, and H^+ migrates via the *Grotthuss mechanism*, an entirely different mechanism of ion migration that stems from the hydrogen-containing nature of water. This mechanism does not involve diffusion, but rather hopping of H^+ from one water molecule to another at the speed of about 100 μm per μs. In this event, H^+ does not transport any water with it, and the *transient* process is so fast that no significant diffusion can accompany it. (A steady-state proton transport through a membrane may involve some water transport, but not in the transient Grotthuss mechanism). In addition, the proton has a large number of tightly held (via electrostriction) hydration water molecules, about 10 H_2O per each H^+. As an H^+ leaves a cluster under the action of a suddenly applied electric field, its tightly held hydration water molecules are left behind, expanding up to possibly 10% in volume. Nevertheless, experiments show an initial fast response (albeit limited) for protonated Nafion-based IPMCs, similar to that for other metal cations (i.e., towards the anode), followed by a similar reverse slow relaxation. This suggests that effects (1) to (3) [rather than the often-assumed hydraulic water-pressure effect] are dominant in producing the initial fast response of the IPMC for certain cations, as well as its subsequent back relaxation. In general, all listed effects are coupled and occur in tandem.

Experimental observations suggest that the slow relaxation is dominated by effect (5) within the cathode boundary layer, where a redistribution of the excess cations occurs at a slower rate than the rate of cation supply. Recent experiments by Nemat-Nasser and Wu have shown that with a suitably slow voltage buildup, the fast initial motion towards the anode of Nafion-based IPMCs can be precluded for cations that otherwise display substantial initial quick motion in that direction, e.g., sodium. The slow potential buildup allows for reconfiguration of the cations within the cathode clusters, as fewer cations move into these clusters, creating contractive forces that reduce the cluster size and actually expel solvent from these clusters. This process is accompanied by the solvent diffusion into the anode. Indeed, experiments using glycerol as the solvent show that, while the anode surface is drying during bending relaxation towards the cathode, droplets are appearing on the cathode surface.

The two boundary layers, together with their corresponding charged electrodes, form two *double layers,* one on the anode side and the other on the cathode side. These double layers provide *shielding* of the remaining part of the IPMC from the effects of the applied electric potential. This suggests that essentially all critical processes occur at the two boundary layers, especially the cathode boundary layer. Each boundary layer has its own distinct characteristics.

In passing, we mention that a fast motion in the same (cathode) direction follows the slow reverse motion of Nafion-based IPMCs, once the two electrodes are shorted. This fast motion is then followed by a slow relaxation in the opposite direction, i.e., towards the initial state of the sample. Here again, this fast motion towards the cathode is due to the rather quick redistribution of the cations, which may or may not (e.g., for H^+) involve water transport. The subsequent slow relaxation towards the anode is caused by the reconfiguration of the cations within the clusters, and the osmotic water transport due to stiffness and ion concentration changes.

6.5.2 Actuation Process

Asaka et al. [1995] describe the actuation of a Pt-IPMC sample in the sodium form. This sample "quickly bends to the anode side and bends back to the cathode side gradually." The initial deflection observed was sufficient to move the tip horizontally a distance approximately 7% of the overall length of the sample. This initial deflection is fast, achieving full displacement within 100 ms. We have made similar observations for several cations, including Tl^+, as shown in Fig. 14.

Substitution of alkali metal cations, as well as tetramethylammonium (TMA) and tetrabutylammonium (TBA) cations yields different deflection speeds and displacements for the same IPMC. The relaxation phenomenon varies widely by ion, with samples such as thallium displaying relaxation displacements greater than five times the initial actuating displacement (Fig. 14).

In the following section, we present the basic concepts and equations necessary for modeling the actuation phenomena observed in IPMCs, starting with an estimate of the internal forces induced by the hydration of a dry IPMC.

6.5.3 Osmosis-Induced Pretension

Consider a *dry* strip of IPMC neutralized by a given cation, M^+. When this strip is immersed in water, osmotic pressures develop in order to reduce the concentration of anions and cations within the clusters. This pressure induces hydration within the sample. As water diffuses into the clusters, the clusters dilate until the pressure induced by the elasticity of the surrounding hydrophobic matrix material balances the osmotic pressure that tends to add water to each cluster and further dilute the salt concentration within each cluster. The pressure in the clusters produces tension within the surrounding backbone polymer matrix. In the absence of external forces other than the surrounding water pressure, the *average* (equilibrium) hydrostatic tension in the fluorocarbon-backbone matrix material can be calculated, as follows, where, to be specific, Nafion-based IPMCs with water as solvent are considered.

Assuming a pure water bath, denote the chemical potential of the water external to the IPMC, by $\mu_w(p)$, where p is the water pressure used as the reference pressure. Let the volume fraction of the IPMC's hydration water at *equilibrium* be denoted by w and denote by $M_{SO_3^-}$ the concentration of sulfonate ions in mole per cubic meter of dry IPMC; here w is the (total) volume of the intake water per unit volume of the dry IPMC. The total ion concentration (measured per unit volume of water) is $n(\nu M_{SO_3^-}) = \nu M_{SO_3^-} / w$, where ν is the number of ions formed when one molecule of salt dissolves (ν = 2 for alkali sulfonates). Since the molar concentration of pure water is $10^6/18$ mol/m^3, the mole fraction of water in the clusters is given by

$$x(H_2O) = (10^6/18)/(10^6/18 + \nu M_{SO_3^-}/w). \tag{28}$$

The corresponding osmotic pressure in MPa, $\Pi(M^+)$, associated with the cation M^+ is now given by

$$\Pi(M^+) = -\frac{RT}{18} g \ln[x(H_2O)] = \frac{RT}{1000} \phi \nu m, \tag{29}$$

where ϕ is the osmotic coefficient of the metal sulfonate salt at the molality m in the cluster environment, (i.e., m is the ion concentration in mole per kilogram solvent); and g is the "rational" osmotic coefficient [Robinson and Stokes, 1965]. The rational osmotic coefficient, g, relates to ϕ approximately by $g = \phi(1 + 0.009\,\nu m)$. Except for the value of these coefficients, Eq. (29) for a given IPMC depends on the volume fraction of water, w, only. Metal salts of trifluoromethanesulfonic acid (CF_3SO_3H, triflate) can be used as model compounds for ionomer-bound sulfonic acid groups [Bonner, 1981]. Table 4 gives estimates of the osmotic pressure $\Pi(M^+)$ for several cations, assuming the ideal situation where $\phi = 1$. As is seen, the smaller the volume fraction of water, the greater is the resulting osmotic pressure. However, since for $\phi = 1$, Eq. (29) becomes independent of the nature of the cations, it may be concluded that the substantial differences in the osmotic pressures reported in Table 4 are basically due to the (elastic) resistance provided by the fluorocarbon matrix material. This observation is in line with the differences in the measured stiffness of the composite reported in Table 3 for various cations. Similar differences are observed for plain Nafion. Hence, cations associated with smaller hydration volumes provide greater pseudo-cross-link forces that result in stiffer composites. It is, therefore, this stiffness that defines the level of hydration associated with each cation. The stiffer membrane would thus be under greater nominal tension when hydrated. The average and the nominal stresses carried by the backbone polymer are computed, as follows; see also Sec. 6.5.7.9.

The stress in the clusters is hydrostatic compression, given by $\boldsymbol{\sigma}_0 = -\Pi(M^+)\,\mathbf{1}$, where $\mathbf{1}$ is the second-order identity tensor, and compression is viewed *negative*. Denote the *variable* stress tensor in the composite by $\boldsymbol{\sigma}(x)$, where x is the position vector of a typical point in the composite. When measured relative to the surrounding water pressure p, the average stress of the composite in the absence of any externally applied loads must be zero, i.e., $\int_V \boldsymbol{\sigma}(\boldsymbol{x})\,d\boldsymbol{V} = \mathbf{0}$, where V is the total volume of the composite [Nemat-Nasser and Hori, 1993]. This yields,

$$\bar{\boldsymbol{\sigma}}_M = w\Pi(M^+)\,\mathbf{1} = \phi\frac{2\rho_d RT}{EW}\mathbf{1}, \tag{30}$$

where, as before, ρ_d is the dry density and EW is the equivalent weight, and where $\bar{\boldsymbol{\sigma}}_M$ is the stress $\boldsymbol{\sigma}(x)$ averaged over the volume of the matrix material only, i.e., excluding the clusters. Hence, for a fully hydrated composite, the *average* pretension in the backbone polymer matrix due to hydration is hydrostatic tension and independent of the water intake, w (except through the dependency of ϕ on w). Indeed, since the value of EW/ρ_d is independent of the nature of the cations, only the osmotic coefficient ϕ in Eq. (30) depends on the cation and its concentration. For Nafion 1200 with $\phi = 1$, for example, the average pretension is about 8.6 MPa. On the other hand, the *nominal* cross-sectional stress (this is the force carried by the polymer matrix, measured per unit total area of the cross section of the hydrated membrane including the area of the clusters) due to the average stress in the polymer, is a function of the amount of the hydration and hence varies with the cations. By considering a spherical cluster in a cubic unit cell, it is readily shown that this nominal tension (due to the average stress only) may be estimated by

$$t_n = (2\phi\rho_d RT / EW)\left[1 - \pi\left(\frac{3}{4\pi}\frac{w}{1+w}\right)^{2/3}\right],$$

where the quantity inside the brackets is the matrix area fraction, i.e., the area of the matrix material divided by the total cross-sectional area; for a discussion of the full stress field, see Sec. 6.5.7.9.

Table 4: Calculated osmotic pressures developed during hydration of bare (Nafion1200) samples. Ion molalities are derived from Gierke [1981]; osmotic coefficients used are for the ideal case (ϕ = 1).

Ion	**H^+**	**Li^+**	**Na^+**	**K^+**	**Rb^+**	**Cs^+**
ρ_{dry}	2.075	2.078	2.113	2.141	2.221	2.304
EquivalentWeight, EW	1200	1206	1222	1238	1284	1332
Water Content, *w*	0.697	0.617	0.443	0.187	0.179	0.136
1 L 'dry' Nafion 1200 upon hydration contains:						
n(H_2O), moles	38.72	34.28	24.61	10.39	9.94	7.56
n(ion), moles	1.73	1.72	1.73	1.73	1.73	1.73
ions per water	0.04	0.05	0.07	0.17	0.17	0.23
x(H_2O)	0.92	0.91	0.88	0.75	0.74	0.69
x(ion)	0.08	0.09	0.12	0.25	0.26	0.31
ν m(ion) molality	4.96	5.59	7.81	18.49	19.32	25.44
Π (MPa), Osmotic Pr.	12.35	13.91	19.44	46.05	48.11	63.34

6.5.4 Basic Model Assumptions

Consider a model in which the anions are permanently attached to the fluorocarbon backbone, while the countercations are free to move through the water-saturated interstitial regions. Under the action of an electric potential applied to the electrodes of an IPMC, the cations are redistributed, resulting in a microscopic, *locally* imbalanced net charge density that produces internal stresses acting on the backbone polymer. The redistribution of the charges is generally accompanied by changes in the volume of the clusters caused by the electrostatic, osmotic, and elastic interaction forces that develop within and outside the clusters. The change in the cluster volume is accompanied by the commensurate change in its water content, as well as in relaxation or further stretching of the polymer chains between the clusters.

In general, all of these effects (i.e., electrostatic, osmotic, and elastic) are significant in producing the initial fast response of the IPMC, depending on the cation. Models are proposed which emphasize various possible effects in describing the IPMC's initial fast motion. The *hydraulic model* asserts that ions in solution are hydrated to a greater or lesser extent, and the ions migrating within IPMC materials carry water molecules whose bulk contributes to volumetric swelling stresses within the membrane. The *electrostatic model* asserts that the locally imbalanced net charge density produces internal stresses acting on the backbone polymer, relaxing the polymer chains in the anion-rich regions and further extending them in the cation-rich regions. This model is thus based on the observation by Forsman that the clustering of ionic groups in sulfonate ionomers is accompanied by an increase in the mean length of the polymer chains [Forsman, 1982]. This prestretching of the polymer is in addition to the hydration-induced pretension discussed earlier. The *hybrid model* asserts that all the electrostatic, osmotic, and elastic effects are important, although one or the other may dominate, depending on the case considered, and that water (or other solvents) migration in and out of the clusters *is a response to rather than the cause of the cluster swelling.*

For the slow relaxation motion, it appears that the electro-osmotic forces, as well as the change in the polymer stiffness, play important roles, and that *the reorganization of cations within the cation-rich clusters is the most significant controlling factor*. It is expected that viscous effects would mediate the flow of water during ion migration via electro-osmosis. Solute concentration and matrix stiffness strongly influence the magnitude of electro-osmotic flow. Ions migrating through inhomogeneous structures may exhibit a *solvation number* (SN) smaller than the bulk, owing to the tortuous route through which the ion must travel.

6.5.5 Hydraulic Model

Hydraulic models for IPMC actuation assume that water migration between ionomer clusters causes differential swelling, yielding bending actuation. The mechanism of water migration has been attributed to primary and secondary hydraulic motions. Applied voltages cause migration of ionic species, and the primary hydraulic effect is assigned to the motion of water molecules that are closely associated to ions by solvation. The secondary hydraulic effect includes forces that derive from the initial migration of ions—electro-osmotic effects.

6.5.5.1 Primary Hydraulic Effects

Ions in aqueous environments are closely associated with water molecules (*solvation water*). It has been proposed that the migration of ions through IPMC materials involves the concomitant migration of water, either through solvation water or electro-osmotic migration. According to this model, the increase in cluster volume caused by water migration is sufficient to cause volumetric strain within the IPMC matrix, yielding the observed bending strain in the material and actuation towards the anode. As pointed out before, experimental observations do not support the basic assumptions of this model, as further discussed in the sequel.

The instantaneous contact of water molecules with an ion is a *static* description of the ion's local environment. This is known as the ion's *coordination number* (CN) or the *secondary hydration number* (SN). Ion CN is determined from small-angle x-ray and neutron diffraction studies.

Under dynamic conditions, ions in aqueous environments are not freely moving species. Instead, ions are solvated by a shell of strongly interacting water molecules known as water of solvation, or *primary hydration water*. The SN is a dynamic value, representing the number of water molecules that migrate with ions during diffusion and motion. Generally, water molecules that remain in association with the ion for longer periods than the diffusive lifetime (τ) are considered solvation waters. Ion SNs are derived from ultrasonic vibration potentials [Zana and Yeager, 1966]. Solvation numbers may be deduced from the apparent molar volume of an ion, as derived from the electro-osmotic flow of water between two electrodes. Strong local electric fields surrounding an ion have the effect of compressing the volume of surrounding waters. As a result, the apparent molar volume is not a direct representation of the absolute solvation number of an ion.

Protons provide an interesting exception to this case, as the proton has a large number of tightly held secondary waters of hydration, as pointed out before. Its motion through the water, though, does not include these hydration waters. Its migration is not by diffusion, but rather by a *hopping* mechanism that is an order of magnitude faster than diffusion—the *Grotthuss mechanism* [Agmon, 1995, 1999]. In this event, H^+ does not transport any water with it, and the process is so fast that no significant diffusion can accompany it. As discussed in Sec. 6.5.1, *the cluster volume actually increases with proton depletion and decreases with excess proton concentration.* The hydraulic model without inclusion of other effects cannot account for the resulting motion of the associated IPMC, which is principally similar to that of other metallic cation composites. Furthermore, the hydraulic model does not have any convincing explanation for the observed large back relaxation (see Fig. 13) that occurs under a suddenly applied and sustained step voltage (dc). As mentioned already, *through a slow voltage buildup, a slow bending towards the cathode only (i.e., without any preceding bending towards the anode) can be produced, contradicting the hydraulic model.* We thus conclude that the hydraulic model does not have any basis in experiments and should be abandoned.

6.5.6 Hybrid Models

We expect models that incorporate each effect to prove the most useful. The scale of each effect must be determined, as well as the time-scale contribution of each term to the overall displacement of IPMC materials (cf. dc actuation). Each effect may be investigated by variations of known material parameters such as ion charge, radius, and solvation (i.e., hydration); solvent viscosity and dielectric constant; ion "hydrophobicity" or "hydrophilicity"; and polymer-matrix cluster size, equivalent weight, branch length, and end group anion species. Correlation between mechanical and micromechanical observables and molecular interactions will prove the most useful in improving the force yield of IPMC materials. Nemat-Nasser [2002] proposes a model that incorporates electrostatic, osmotic, and elastic effects in the actuation of IPMCs, and leads to results in good accord with experiments, including initial fast motion, subsequent back relaxation, and the response upon shorting. Various essential features of this model are summarized in what follows.

6.5.7 Basic Equation of Hybrid Model

The basic field equations, which include the electrostatic and the osmotic effects, are now summarized and discussed. To be specific, the solvent is assumed to be water, with actuation occurring in an aqueous environment.

6.5.7.1 Electrostatic and Transport Equations

The distribution of cations under the action of an imposed electric potential must first be established. With *D, E,* and φ, respectively denoting the electric displacement, the electric field, and the electric potential, the charge distribution is governed by the following field equations:

$$\frac{\partial D}{\partial x} = \rho, \; D = \kappa_e E, \; E = -\frac{\partial \varphi}{\partial x}, \; \rho = (C^+ - C^-)F, \tag{31}$$

where only the one-dimensional case is considered, with x measuring length along the cross section; ρ is the charge density (C/m^3), C^+ and C^- are the positive and negative ion densities (mol/m^3), respectively, F is Faraday's constant (96,487 C/mol), and κ_e is the effective electric permittivity of the polymer. The distribution of the counter-ions and the associated solvent molecules are governed by coupled continuity equations,

$$\frac{\partial}{\partial t}\ln(1+w) + \frac{\partial v}{\partial x} = 0, \quad \frac{\partial C^+}{\partial t} + \frac{\partial J^+}{\partial x} = 0, \tag{32}$$

where $v = v(x, t)$ is water velocity (flux) and $J^+ = J^+(x,t)$ is the cation flux, both in the thickness (i.e., x) direction. These fluxes are usually expressed as linear functions of the gradients $\partial\mu^w/\partial x$ and $\partial\mu^+/\partial x$ (i.e., forces), through empirical coefficients. The electro-chemical potentials of the solvent (water, here) and the cations are given by

$$\mu^{w}=\mu_{0}^{w}+18(10^{-6}p-10^{-3}RT\phi\nu m),\ \mu^{+}=\mu_{0}^{+}+RT\ln(\gamma^{+}C^{+})+F\varphi, \tag{33}$$

where the first term on the right side of these equations include the reference and the normalizing values of the associated potential; see also comments after expression (29).

As noted before, some water molecules are carried by hydrophilic cations (except for H^+). If SN is the *dynamic solvation number*, i.e., mole water molecules actually transported per mole migrating cations, the resulting volumetric strain then would be

$$\varepsilon_{v}^{*}=18\times10^{-6}(C^{+}-C^{-})SN. \tag{34}$$

However, recent current-flow measurements (Nemat-Nasser and Wu, 2003a) show continued cation addition to the cathode clusters while these clusters are actually in contact with each other (Nafion-based IPMCs in alkali cation forms), suggesting that the cations rather than their solvation water molecules are of primary importance in the IPMC actuation. Therefore, in what follows, SN is set equal to zero, and the volume strain is calculated directly by considering cation transport and the resulting forces that affect the volume of the clusters, driving water into or out of the clusters.

6.5.7.2 Estimate of Length and Time Scales

Consider now the problem of cation redistribution and the *resulting induced* water diffusion. The diffusion-controlled water migration is generally a slow process. Thus, only the second continuity equation in Eq. (32) is expected to be of importance in describing the initial cation redistribution, while the first continuity equation applies to the subsequent resulting water diffusion. The corresponding cation flux can be expressed as

$$J^{+}=-\frac{D^{++}C^{+}}{RT}\frac{\partial\mu^{+}}{\partial x}+\frac{D^{+w}C^{+}}{RT}\frac{\partial\mu^{w}}{\partial x}, \tag{35}$$

where D^{++} and D^{+w} are empirical coefficients. It is very difficult to estimate or measure these coefficients in the present case. Therefore, instead, use the well-known Nernst equation [Lakshminarayanaiah, 1969],

$$J^{+}=-D^{+}\left(\frac{\partial C^{+}}{\partial x}+\frac{C^{+}F}{RT}\frac{\partial\varphi}{\partial x}+\frac{C^{+}V^{+w}}{RT}\frac{\partial p}{\partial x}\right)+C^{+}v, \tag{36}$$

with D^{+} being the ionic diffusivity coefficient. Upon linearization, obtain [Nemat-Nasser and Li, 2000; and Nemat-Nasser and Thomas, 2001],

$$\frac{\partial}{\partial x}\left[\frac{\partial(\kappa E)}{\partial t}-D^{+}\left(\frac{\partial^{2}(\kappa E)}{\partial x^{2}}-\frac{C^{-}F^{2}}{\kappa RT}(\kappa E)\right)\right]=0. \tag{37}$$

This equation includes a natural length scale, ℓ, and a natural time scale, τ, respectively characterizing the length of the boundary layers and the relative speed of cation redistribution,

$$\ell=\left(\frac{\bar{\kappa}RT}{C^{-}F^{2}}\right)^{1/2}, \quad \tau=\frac{\ell^{2}}{D^{+}}, \tag{38}$$

where $\bar{\kappa}$ is the overall electric permittivity of the hydrated IPMC strip, which can be estimated from its measured capacitance (if *Cap* is the measured overall capacitance, then $\bar{\kappa}=2H\,Cap$). Therefore, ℓ can be estimated directly from Eq. $(38)_1$, whereas the estimate of τ will require an estimate for D^{+}. It turns out that the capacitance of the Nafion-based IPMCs (about 200- to 225-μm thick) that have been examined in the first author's laboratories ranges from 1 to 60 mF/cm^2, depending on the sample and the cation form. This suggests that ℓ may be 0.5 to $6\,\mu m$, for the Nafion-based IPMCs. Thus, ℓ^2 is of the order of $10^{-12}\,m^2$, and for the relaxation time τ to be of the order of seconds, D^{+} must be of the order of $10^{-12}\,m^2/s$. Our most recent experimental results (Nemat-Nasser and Wu, 2003a) suggest this to be the case. Indeed, direct measurement of current flow through Nafion-based IPMC strips under a constant voltage shows $\tau=O(1)$ s in an aqueous environment. Since ℓ is linear in $\sqrt{\bar{\kappa}}$ and $\bar{\kappa}$ is proportional to the capacitance, it follows that ℓ is proportional to the square root of the capacitance.

6.5.7.3 Equilibrium Solution

To calculate the ion redistribution caused by the application of a step voltage across the faces of a hydrated strip of IPMC, first examine the time-independent *equilibrium* case with $J^{+}=0$. In the cation-depleted (anode) boundary layer the charge density is $-C^{-}F$, whereas in the remaining part of the membrane the charge density is $(C^{+}-C^{-})F$. Let the thickness of the cation-depleted zone be denoted by ℓ', and set

$$Q(x,t)=\frac{C^{+}-C^{-}}{C^{-}}, \quad Q_{0}(x)=\lim_{t\to\infty}Q(x,t). \tag{39}$$

Then, it can be shown that [Nemat-Nasser, 2002] the equilibrium distribution is given by

$$Q_0(x) \equiv \begin{cases} -1 & \ldots for\ x \leq -h + \ell' \\ \frac{F}{RT}[B_0 \exp(x/\ell) - B_1 \exp(-x/\ell)] \ , & \\ & \ldots for - h + \ell' < x < h \end{cases} \tag{40}$$

where the following notation is used:

$$B_1 = K_0 \exp(-a'), \quad \frac{\ell'}{\ell} = \sqrt{\frac{2\phi_0 F}{RT}} - 2, \quad B_2 = \frac{\phi_0}{2} - \frac{1}{2} K_0 \left[\left(\frac{\ell'}{\ell} + 1 \right)^2 + 1 \right],$$

$$B_0 = \exp(-a) \left[\phi_0 / 2 + B_1 \exp(-a) + B_2 \right], \quad K_0 = \frac{F}{RT}, \tag{41}$$

ϕ_0 is the applied potential, $a \equiv h/\ell$, and $a' \equiv (h - \ell')/\ell$. Since ℓ is only 0.5 to 3 μm, $a \equiv h/\ell, a' \equiv h'/\ell) >> 1$, and hence $\exp(-a) \approx 0$ and $\exp(-a') \approx 0$. The constants B_0 and B_1 are very small, of the order of 10^{-17} or even smaller, depending on the value of the capacitance. Therefore, the approximation used to arrive at Eq. (40) does not compromise the accuracy of the results. Remarkably, the estimated length of the anode boundary layer with constant negative charge density of $-C^- F$, i.e., $\ell' = \left(\sqrt{2\phi_0 F/RT} - 2\right)\ell$, depends only on the applied potential and the characteristic length ℓ. For $\phi_0 = 1$ V, for example, $\ell' \approx 6.8\ell$.

6.5.7.4 Temporal Variation of Cation Distribution

Since $a = h/\ell \approx 50$, it can be shown that

$$Q(x,t) \approx g(t) Q_0(x), \quad g(t) = 1 - \exp(-t/\tau). \tag{42}$$

Thus, the spatial variation of the charge distribution can be separately analyzed and then modified to include the temporal effects.

6.5.7.5 Clusters in Anode Boundary Layer

Consider the anode boundary layer, and note from

$$\int_{-h'}^{0} Q_0(x) dx = \ell \tag{43}$$

that the *effective* total length of the anode boundary layer, L_A, can be taken as

$$L_A \equiv \ell' + \ell = \left(\sqrt{\frac{2\phi_0 F}{RT}} - 1 \right) \ell. \tag{44}$$

Thus, the equilibrium charge density is $Q_0(x) = -1$, in $-h \le x < -h + L_A$, and zero in $-h + L_A \le x < 0$. In the anode boundary layer, $-h \le x < -h + L_A$, the cation density is $C^+ = C^- \exp(-t/\tau)$ and the total ion (cation and anion) concentration is $C^- + C^+ = C^-[2 - g(t)]$. These expressions are then used to estimate the osmotic pressure, $\Pi_A(t)$, the effective electric permittivity, $\kappa_A(t)$, and the electrostatic forces within the anode boundary layer. The details are found in Nemat-Nasser [2002]. The electrostatic effects include anion-anion interaction, $p_{AA}(t)$, and dipole-dipole interaction, $p_{ADD}(t)$, forces. The former increases in time while the latter decreases as cations migrate out of the anode boundary layer. The decrease in ion concentration reduces the osmotic pressure, increases the dielectric parameter, both of which tend to decrease the average cluster size in this zone, contributing to the initial bending towards the anode. This also shows that the back relaxation must necessarily be the result of cation activity in the cathode boundary layer, as discussed below. Finally, the resulting pressure within a typical cluster in the anode boundary layer then becomes

$$\begin{aligned}
&t_A = \Pi_A(M^+, t) + p_{AA}(t) + p_{ADD}(t) + \sigma_r(a_0, t), \\
&\Pi_A(t) = \frac{\phi Q_B^- K_0}{w_A(t)}[2 - g(t)], \quad Q_B^- = \frac{\rho_B F}{EW_{ion}}, \\
&p_{AA}(t) = \frac{g(t)}{18\kappa_A(t)} Q_B^{-2} \frac{R_0^2}{[w_A(t)]^{4/3}}, \\
&p_{ADD}(t) = \frac{1 - g(t)}{3\kappa_A(t)} Q_B^{-2} \frac{\pm[\alpha_A(t)]^2}{[w_A(t)]^2}, \\
&\sigma_r(a_0, t) = -p_0(t) + K(t)(w_A(t)/w_0)^{-4/3},
\end{aligned} \tag{45}$$

where the last term in Eq. $(45)_1$ is the elastic resistance of the backbone polymer, R_0 is the dry cluster size, and other terms have been defined before; see Nemat-Nasser [2002] for further comments. As shown, all interaction forces are calculated using the corresponding time-dependent water uptake, $w_A(t)$. This water uptake is computed incrementally, using the diffusion equation and the initial and boundary conditions. Since the boundary layer is rather thin, assume uniform w_A and t_A in the boundary layer, and replace the diffusion equation with

$$\frac{\dot{w}_A}{1+w_A} = D_{H_2O}\frac{t_A}{(L_A/2)^2} = D_A t_A, \tag{46}$$

where D_{H_2O} is the hydraulic permeability coefficient. Note that coefficient D_A here includes the effect of the thickness of the boundary layer. We assume that D_A is constant.

6.5.7.6 Clusters in the Cathode Boundary Layer

Consider now the clusters within the *cathode boundary layer*. Unlike in the anode boundary layer, the ion concentration in the cathode boundary layer is sharply variable, and we have

$$\nu_C(x,t) = 2 + Q(x,t) = 2 + \frac{B_0}{K_0}\exp(x/\ell)g(t), \tag{47}$$

based on which the osmotic pressure and dielectric parameter must be computed. In this boundary layer, there are two forms of electrostatic interaction forces. One is repulsion due to the cation-anion pseudo-dipoles already present in the clusters, and the other is due to the extra cations that migrate into the clusters and interact with the existing pseudo-dipoles. The additional stresses produced by this latter effect may tend to expand or contract the clusters, depending on the distribution of cations relative to the fixed anions. Each effect may be modeled separately, although in actuality they are coupled. The dipole-dipole interaction pressure in the clusters may be estimated as

$$p_{CDD}(x,t) = \frac{Q_B^{-2}}{3\kappa_C(x,t)}\frac{\pm[\alpha_C(x,t)]^2}{[w_C(x,t)]^2}[1-g(t)], \tag{48}$$

where the subscript C denotes the corresponding quantity in the cathode boundary layer; $\alpha_C(x,t)$ is the dipole arm that can evolve in time, as the cations reconfigure under the action of the strong sulfonates (but not necessarily under the action of the weak carboxylates). We represent the interaction between the pre-existing dipoles and the additional cations that move into a cluster under the action of an applied voltage, by dipole-cation interaction stresses defined by

$$p_{DC}(x,t) = \frac{2Q_B^{-2}}{9\kappa_C(x,t)}\frac{a_C(x,t)\alpha_C(x,t)}{[w_C(x,t)]^2}g(t) \approx \frac{2Q_B^{-2}}{9\kappa_C(x,t)}\frac{R_0\alpha_C(x,t)}{[w_C(x,t)]^{5/3}}g(t). \tag{49}$$

This equation is obtained by placing the extra cations at the center of a sphere of (current) radius $a_C(x,t)$, which contains uniformly distributed radial dipoles-of-moment arm $\alpha_C(x,t)$ on its surface, and then multiplying the result by $g(t)=1-\exp(-t/\tau)$.

6.5.7.7 Reverse Relaxation of IPMC under Sustained Voltage

For sulfonates in a Nafion-based IPMC, extensive restructuring and redistribution of the extra cations, appear to underpin the observed reverse relaxation of the Nafion-based IPMC strip. To represent this, modify Eq. (49) by a *relaxation factor*,

$$\begin{aligned} p_{DC}(x,t) &\approx \frac{2Q_B^{-2}}{9\kappa_C(x,t)} \frac{R_0\alpha_C(x,t)}{[w_C(x,t)]^{5/3}} g(t) g_1(t), \\ g_1(t) &= [r_0 + (1-r_0)\exp(-t/\tau_1)], \quad r_0 < 1, \end{aligned} \tag{50}$$

where τ_1 is the relaxation time, and r_0 is the equilibrium fraction of the dipole-cation interaction forces.

The total stress in clusters within the cathode boundary layer is now approximated by

$$t_C = \sigma_r(a_0,t) + \Pi_C(x,t) + p_{CDD}(x,t) + p_{DC}(x,t). \tag{51}$$

The rate of change of water uptake in the cathode boundary layer is governed by the diffusion equation,

$$\frac{\dot{w}_C}{1+w_C} = D_{H_2O} \frac{\partial^2 t_C}{\partial x^2}, \tag{52}$$

subject to the boundary and initial conditions

$$\begin{aligned} &t_C(h,t) = 0, \quad w_C(0,t) = w_0, \quad t > 0, \\ &w_C(x,0) = w_0, \quad 0 < x < h, \end{aligned} \tag{53}$$

where w_0 is the uniform water uptake just prior to the cation redistribution. This is a nonlinear initial-boundary value problem, whose complete solution would require a numerical approach. To reveal the essential micro-mechanisms of the actuation, use the thinness of the cathode boundary layer and replace the spatial gradients by the corresponding difference expression, to obtain an average value of the water uptake in this boundary layer.

The overall charge neutrality requires that

$$\int_0^h Q(x,t)dx = L_A g(t). \tag{54}$$

Define an *effective* length and ion density for an equivalent cathode boundary layer of uniform cation distribution, respectively by

$$L_C = 2\left(h - \frac{1}{L_A}\int_0^h xQ_0(x)dx\right), \quad \overline{\upsilon}_C = 2 + \frac{L_A}{L_C}g(t), \tag{55}$$

and calculate the osmotic pressure and all other quantities using this *equivalent* boundary layer with *uniform ion and water distribution.* In particular, the average water uptake, $\overline{w}_C(t)$, is obtained using

$$\frac{\dot{w}_C}{1+w_C} = D_C\overline{t}_C, \quad \overline{t}_C(t) = \overline{\sigma}_r(a_0,t) + \overline{\Pi}_C(t) + \overline{p}_{CDD}(t) + \overline{p}_{DC}(t), \tag{56}$$

where the barred quantities denote the average values. The anode and the equivalent cathode boundary layer thicknesses are related by $D_C = (L_A / L_C)^2 D_A$.

6.5.7.8 Tip Displacement

The tip displacement of the cantilever is now calculated incrementally, using the rate version of Eqs. (22) and (23), which can be expressed as

$$\frac{\dot{u}}{L} = \frac{LY_{BL}}{4H^3(3\overline{Y}_{\text{IPMC}} - 2Y_B)}\int_{-h}^{h} x\frac{\dot{w}(x,t)}{1+w(x,t)}dx. \tag{57}$$

For the equivalent uniform boundary layers, the integral in the right-hand side of this equation can be computed in closed form, yielding

$$\frac{\dot{u}}{L} = \frac{Y_{BL}}{(3\overline{Y}_{\text{IPMC}} - 2Y_B)}\frac{hL}{4H^2}\left(\frac{\dot{w}_A}{1+w_A}\frac{L_A}{H} - \frac{\dot{w}_C}{1+w_C}\frac{L_C}{H}\right), \tag{58}$$

where we have neglected terms of the order of $O(L_A / H)^2$. Combining this with Eqs. (46) and (56), we now have

$$\frac{\dot{u}}{L} = \frac{Y_{BL}}{(3\bar{Y}_{\text{IPMC}} - 2Y_B)} \frac{hLL_A}{4H^3} D_A \left(t_A - \bar{t}_C \frac{L_A}{L_C} \right), \tag{59}$$

where t_A and $\bar{t}_C$ are given by Eqs. (51) and (56), respectively. This equation may now be integrated incrementally.

6.5.7.9 Estimate of κ_e for Hydrated Nafion

Using a double inclusion model, Nemat-Nasser and Li [2000] show that the effective electric permittivity of the membrane can be evaluated as

$$\kappa_e = \frac{\kappa_p + \kappa_w + c_w(\kappa_p - \kappa_w)}{\kappa_p + \kappa_w - c_w(\kappa_p - \kappa_w)} \kappa_p, \tag{60}$$

where the subscripts p and w refer to the electric permittivity of water (solvent) and polymer, respectively. This expression has been used to obtain the estimate of Eq. (14) as well as Eq. (16).

6.5.7.10 Example of Actuation

To check whether or not the proposed model captures the essential features of the observed actuation response of the Nafion-based IPMCs, consider a cantilevered strip of fully hydrated IPMC in Na^+-form. Figure 15 shows (geometric symbols) the measured tip displacement of a 15mm long cantilever strip that is actuated by applying a 1.5V step potential across its faces, maintaining the voltage for about 32 seconds and then removing the voltage while the two faces are shorted. The initial water uptake is $w_{\text{IPMC}} = 0.46$, and the volume fraction of metal plating is 0.0625. Hence, the initial volume fraction of water in the Nafion part of the IPMC is given by $w_0 = w_{\text{IPMC}}/(1 - f_M) = 0.49$. The formula weight of sodium is 23, and the dry density of the bare membrane is 2.02g/cm^3. The equivalent weight for the bare Nafion (and *not* the IPMC) then becomes EW_{Na^+} = 1122 g/mol. The initial value of C^- for the bare Nafion then becomes

$$C^- = 10^6 \frac{\rho_B}{EW_{Na^+}(1 + w_0)} = 1{,}208\,\text{mol/m}^3.$$

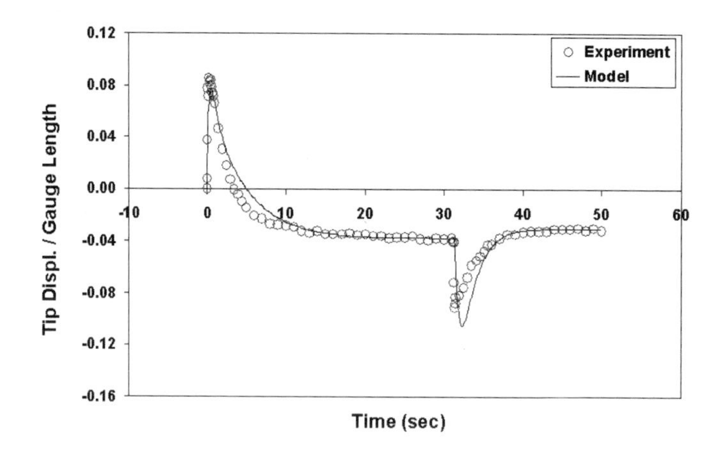

Figure 15: Tip displacement of a 15mm cantilevered strip of a Nafion-based IPMC in Na^+-form, subjected to 1.5 V step potential for about 32 seconds, then shorted; the heavy solid curve is the model and the geometric symbols are experimental points.

The thickness of the hydrated strip is measured to be $2H = 224\,\mu\text{m}$, and based on inspection of the microstructure of the electrodes, set $2h \approx 212\,\mu\text{m}$. The effective length of the anode boundary layer for $\phi_0 = 1.5\,\text{V}$ is $L_A = 9.78\ell$, from Eq. (55) the thickness of the equivalent uniform cathode boundary layer becomes $L_C = 2.84\ell$, and $\ell = 0.862\mu\text{m}$ for $\phi_0 = 1.5$ V. The electric permittivity and the dipole length are calculated using $a_1 = 1.5234\times10^{-20}$ and $a_2 = -0.0703\times10^{-20}$. The measured capacitance ranges from 10 to 20 mF/cm^2, and we use Cap = 15 mF/cm^2.

The measured stiffness of bare Nafion and the corresponding IPMC in Na-form are shown in Fig. 16, as function of the hydration, for loading and unloading. The solid curves are the corresponding model predictions, calculated using expression (12), with the following values of the parameters: $n_0 = 0.01$, $\phi = 1$, CN = 4.5. Other actuation-model parameters are: $D_A = 10^{-2}$ (when pressure is measured in MPa), $\tau = 1/4$ and $\tau_1 = 4$ (both in seconds), SN = 0, and $r_0 = 0.25$. In line with the observation that the cations continue to move into the cathode boundary layer long after the back relaxation has started, we have set SN = 0. Since $w_0 \approx (a/R_0)^3$, where a is the cluster size at water uptake w_0, we have set $R_0 \approx w_0^{-1/3}a$, and adjust a to fit the experimental data; here, a = 1.65 nm, or an average cluster size of 3.3 nm, prior to the application of the potential. The fraction of cations that are left after shorting is adjusted to $r = 0.03$.

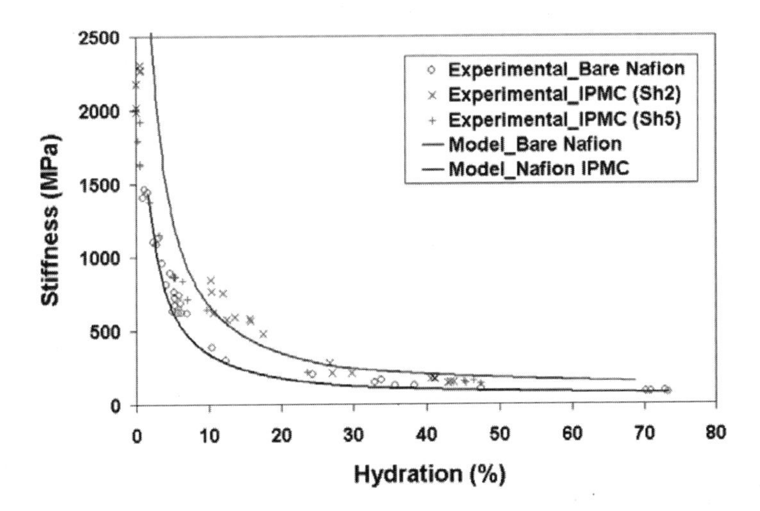

Figure 16: Uniaxial stiffness (Young's modulus) of bare Nafion 117 (lower data points and the solid curve, model) and an IPMC (upper data points and light curve) in the Na+-form versus hydration water.

The solid line in Fig. 15 shows the model result. While the values of the parameters that are used are reasonable, they are chosen to give good comparison with specific experimental data. These data are from one test only. The variation of the response from sample to sample, or even for the same sample tested at various times, is often so great that only a qualitative correspondence between the theoretical predictions and the experimental result in general can be expected, or reasonably required. In examining the influence of various competing factors, it has become clear that the electrostatic forces are most dominant, as has also been observed by Nemat-Nasser and Li (2000) who used a different approach. Although the osmotic effects are also relevant, they have less impact in defining the initial actuation and subsequent relaxation of the Nafion-based IPMCs.

6.5.7.11 Hydrated IPMC Strip as Sensor

Assume the IPMC strip is suddenly bent. An electric potential will be generated across the composite. Nemat-Nasser and Li [2000] assume that this is due to the differential displacement of the effective centers of the anions and cations within each cluster, producing an effective dipole. Since this differential displacement is less than second-order in magnitude, the resulting electric potential will also be of the same order of magnitude. The displacement along the x-axis of the membrane subjected to an applied bending curvature is given by Love [1944]:

$$u=(z^2+\nu x^2-\nu y^2)/2R_c, \tag{61}$$

where ν is the Poisson ratio of the membrane, and R_c is the imposed radius of curvature. This imposed displacement field distorts the ionic clusters, creating an effective dipole within each cluster. To estimate the value of this dipole, consider a spherical cluster of radius r_c, with the center at $(x_\alpha, 0, 0)$. Assume that the fixed anions are uniformly distributed on the surface of this sphere, while the cations are uniformly distributed over a sphere of the same center but of radius $r_c - r_i$, where r_i is the distance between the anion and cation in an ion pair [see Fig. 17(a)].

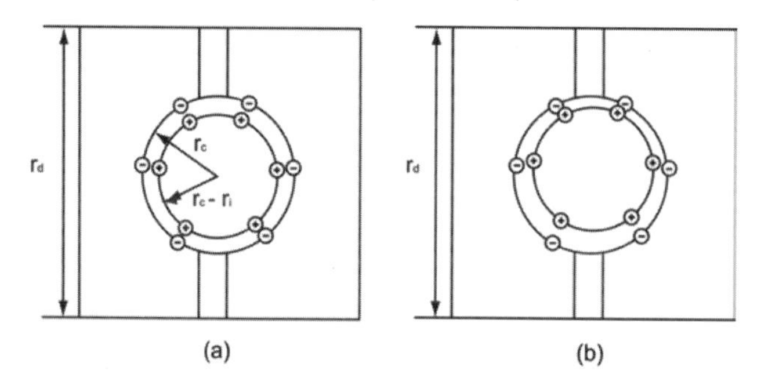

Figure 17: Schematic representation of IPMC clusters: (a) an arrangement of ions into dipoles within cluster, and (b) a dipole induced by imposed bending curvature.

The distributed charges are equivalent to an effective total charge located at the common center of the two spheres prior to the applied distortion, producing zero total charge and dipole. The imposed deformation displaces the effective anion and cation charge centers by different amounts, producing an effective dipole [see Fig. 17(b)]. The displacement of the effective charge center is estimated by averaging the resulting surface displacement over each surface. The separation of the effective centers of anions and cations in a cluster, due to the imposed bending curvature, denoted by $\Delta d(x_\alpha)$ for the α-th cluster, is then given by

$$\Delta d(x_\alpha) = [r_c^2 - (r_c - r_i)^2]/6R_c \approx r_c r_i / 3R_c, \tag{62}$$

which is independent of x_α. The separation of the effective charge centers introduces a dipole $\boldsymbol{\mu}$ in the cluster, which induces a potential field. The magnitudes of the dipole, μ, and the resulting potential, ϕ_α, are given by

$$\mu(x_\alpha) = q_{total}\Delta d = C^- F r_d^3 r_c r_i / 3R_c, \quad \phi_\alpha = (q_{total} r^2 + \boldsymbol{\mu} \cdot \boldsymbol{r})/(4\pi\kappa_w r^3), \tag{63}$$

where q_{total} is the total positive charge inside the cluster, $\boldsymbol{r}$ is the position vector of magnitude r with respect to the induced dipole, and κ_w is the electric permittivity of water ($78\kappa_0$). Now we have a series of clusters of water embedded in the polymer backbone, each with a dipole at its center. The electric field in the polymer, induced by the dipole located at the cluster center, is given by

$$E = 12\sum_{n=0}^{n=\infty} \mu/[(2n+1)^3 4\pi\kappa_w (2\kappa_p/\kappa_w + 1)(r_d/2)^3], \tag{64}$$

where E is the magnitude of the resulting electric field. This leads to a potential difference across the thickness, given by

$$\Delta\varphi = 2hE. \tag{65}$$

For reasonable estimates of the parameters, this equation yields potential values of tens of mV for a sudden deformation of an IPMC. Figure 17, from Nemat-Nasser and Li [2000], illustrates this.

6.6 Development of IPMC Applications

6.6.1 Proposed Applications

The ultimate success of IPMC materials depends on whether or not suitable applications can be found. Whereas IPMC materials cannot offer the force output (of some materials) or the driving frequency (of others), a large number of applications have been proposed which take advantage of IPMCs' bending actuation, low voltage/power requirements, small and compact design, lack of moving parts, and relative insensitivity to damage. In this section, a number of suggested and prototype applications are described. As well, applications of IPMC materials are presented elsewhere within this book (Oguro, Chapter 13).

Osada and colleagues have described many applications for IPMCs, including catheters [Sewa et al., 1998; Oguro et al., 1999], elliptic friction drive elements [Tadokoro et al., 1997], and ratchet-and pawl-based motile species [Osada et al., 1992]. IPMCs have been suggested in applications to mimic biological muscles; Caldwell has investigated artificial muscle actuators [Caldwell, 1990; Caldwell and Tsagarakis, 2000], and Shahinpoor has suggested applications ranging from peristaltic pumps [Segalman et al., 1992] and devices for augmenting human muscles [Shahinpoor, 1996], to robotic fish [Shahinpoor, 1992]. These materials have been suggested for use as fabrics for use in theatrical costuming and special effects (A. Lauer, Alley Theater, Houston, TX, personal communication).

Bar-Cohen and others have discussed the use of IPMC actuators in space-based applications [Bar-Cohen et al., 1998, 1999a], as their lack of multiple moving parts is ideal for any environment where maintenance is difficult. As described above, IPMCs offer application in vibration sensing applications [Sadeghipour et al., 1992]. Besides these novel applications, IPMCs find applications in other disciplines, including fuel cell membranes, electrochemical sensing [DeWulf and Bard, 1988], and electrosynthesis [Potente, 1988]. Much of the interest in these materials has stemmed from these already established applications for Nafion-type materials. The commercial availability of Nafion has allowed significant opportunity to find alternate uses for these materials. Growth and maturity in the field of IPMC actuators will require exploration beyond these commercially available polymer membranes, as the applications for which they have been designed are not necessarily optimal for IPMC actuators.

6.7 Discussion: Advantages/Disadvantages

6.7.1 Force Generation

IPMC materials do not offer an appreciable power output (estimated at 10 W/kg [Wax and Sands, 1999]), as compared with natural muscle (1000 W/kg), pneumatic, piezoelectric, or hydraulic components. Indeed, it must be recognized that each method of force generation may prove optimal in specific applications. IPMC materials will show similar niche applications based on those properties unique to IPMC actuators. Some of these properties are addressed below.

6.7.2 Low Power Requirements

Unlike other EAP materials [Pelrine et al., 2000], IPMC materials can actuate under applied potentials as low as 1 V. Larger displacements can be obtained using larger voltages, but operation at voltages above 1.23 V is prohibitive due to concomitant electrolysis of water. The electrolytic breakdown of water significantly increases the operating currents required for IPMC actuators, and the production of oxygen and hydrogen gases can contaminate an operating platform. Nonplatinum electrodes, on the other hand, are more easily polarized and develop an overpotential that increases the voltage required for water electrolysis. Usually this is disadvantageous for efficient electrolysis, but Oguro has taken advantage of this phenomenon by creating IPMCs containing all-gold electrodes that can be driven at 2.0 V without a noticeable increase in current or gas evolution [Oguro et al., 1999].

6.7.3 Hydration Requirements

IPMC materials require the presence of water for operation. Without water as a solvent, migration of cation species is curtailed, reducing or eliminating power output. Strategies that seek to contain water within the IPMC must adhere strictly to the 1.23-V electrolysis limit, as the splitting of water has the effect of dehydrating each sample.

Nafion ionomers are compatible with a large number of solvents having higher electrolytic voltages. This would allow operation above the 1.23-V limit for water. Unfortunately, it is expected that the lower conductivity of these organic solvents will have a detrimental effect on IPMCs' overall force generation.

6.7.4 Sample Contamination

Due to the equilibrium nature of ion exchange in ionomeric polymers, care must be taken to prevent sample contamination when IPMCs are operated in the presence of any competing cationic species (such as salt water). For example, a sample prepared in the lithium form will rapidly exchange with sodium ions if available. Since power output in sodium IPMCs is less than in lithium samples, sample composition is an important parameter that must be considered. Simply handling IPMCs in the lithium form with one's bare hands is sufficient to contaminate the sample with sodium ions. It is recommended that all IPMC materials be handled using clean forceps or protective gloves. Treatment of prepared samples with a cladding layer to prevent further ion exchange has been proposed, but protective strategies used have not been sufficient to prevent ion contamination in IPMC materials.

Regardless of an IPMC's ion exchange equilibrium constant [Steck and Yeager, 1980; Cwirko and Carbonell, 1992], it must be realized that ion exchange is a statistical process. One piece of sodium-form IPMC in a solution of lithium electrolyte will experience ion exchange, given enough time and a large enough reservoir of electrolyte solution. This may limit the use of IPMC actuators in open environments (seawater, blood, or beneath the surface of Europa).

6.7.5 Manufacturing Cost

Current methods for IPMC manufacture rely on expensive noble metals (Pt $20/g, Au $10/g) for the plating process. The method of manufacture includes steeping in hot nitric or hydrochloric acids for cleaning and ion exchange—processes that demand use of unreactive noble metals. In addition, use of IPMC materials at potentials greater than 1.23 V can create transient oxygen and hydrogen species that can corrode non-noble metals. It is anticipated that changes in processing and operating conditions may be able to reduce reliance upon these metals.

In addition, IPMC materials have heretofore been manufactured using Nafion perfluorosulfonate membrane materials. Since these materials are somewhat expensive ($8/g), some consideration should be made for alternative matrix materials. Of course, the degree to which IPMC behavior is dependent on the unique properties of perfluorinated sulfonic acid ionomers may limit these opportunities.

6.7.6 Soft Materials—Consistency of Processing

To some degree, IPMC materials suffer from their lack of rigidity. Processing through multiple steps yields materials whose properties may not be consistent from batch to batch. Variations including surface morphology, small bends, ripples, and other processing imperfections can yield large variations in actuator performance. Our solutions to these inconsistencies include working with individual batches and repeating experiments on large numbers of samples. Additionally, our models for force generation anticipate variation in each sample's shape. Bulk processing methods can be addressed after force and power output of IPMC materials are known, and applications are found that justify further optimization of processing conditions.

6.7.7 Bending Mode of Actuation

Description of IPMC materials as "artificial muscles" is problematic, as natural muscles exhibit contractile forces, and IPMC actuation occurs by bending motion. Applications requiring linear actuators may not be ideally suited for IPMC materials; instead, IPMC actuators can find use as positioning actuators or other specialized applications. However, Oguro and colleagues demonstrated several unique applications utilizing bending actuators, including an elliptical friction drive, a ratchet-based crawler, and a two-axis catheter positioner.

IPMC actuators might better be described as plantlike in their motion rather than as animal muscles [Shahinpoor and Thompson, 1995]. Motion in vascular plants is caused by the differential swelling and contraction of waterlogged tissues at either end of a plant's stem. It may prove useful to study the mechanics of plant movement while searching for suitable applications for IPMC actuators.

6.7.8 Scalability

Natural muscles consist of millions of strands of individual muscle fibers. IPMC materials studied to date are large (1–10-mm wide) and are operated as single slabs. A combination of multiple pieces to form IPMC bundles could be used to scale the performance of these materials. This engineering challenge requires electrical connections between IPMC slabs that prevent short-circuiting between slabs and lubricate the contact between neighboring slabs.

In his studies of IPMCs based on a Nafion117 matrix, Zawodzinski [2000] observed that competition between ion conduction and electro-osmotic migration would limit the minimum effective thickness of IPMC actuators to ~100 μm. He suggests that this is based on matrix topography, solvent, and ion properties. Understanding the required properties of IPMC matrix materials will assist in the development of replacements for Nafion117.

6.8 Acknowledgments

We wish to thank Drs. Steve Wax (DARPA), Len Buckley (NRL), Carlos Sanday (NRL), and Randy Sands for their continued encouragement and many stimulating discussions while the original version of this chapter was being prepared. We thank Prof. Mohsen Shahinpoor and Dr. K. J. Kim for providing the Nafion-based IPMC materials and many informative discussions, and Dr. Kinji Asaka for providing the Flemion-based IPMC materials; also Jon Isaacs (who designed the mini-load frame) and David Lischer for assistance with mini-load frame testing, Jeff McGee for helpful discussions, and Nicolas Daza and Jean-Charles Duchêne, visiting summer students from the Université de Toulon et du Var. This work has been supported by DARPA grant number MDA972-00-1-0004 to the University of California, San Diego.

6.9 References

Abé, Y., Mochizuki, A., Kawashima, T., Yamashita, S., Asaka, K., and Oguro, K., "Effect on bending behavior of counter cation species in perfluorinated sulfonate membrane-platinum composite," *Polym. Adv. Technol.*, v. 9, no. 8, 1998, pp. 520-526.

Agmon, N., "The Grotthuss Mechanism," *Chem. Phys. Lett.*, v. 244, no. 5-6, 1995, pp. 456-462.

Agmon, N., "Proton solvation and proton mobility," *Israel J. Chem.*, v. 39, no. 3-4, 1999, pp. 493-502.

Asaka, K., Oguro, K., Nishimura, Y., Mizuhata, M., and Takenaka, H. "Bending of polyelectrolyte membrane-platinum composites by electric stimuli. Part 1. Response characteristics to various waveforms," *Polymer J.*, v. 27, no. 4, 1995, pp. 436-440.

Asaka, K., and Oguro, K. "Bending of polyelectrolyte membrane platinum composites by electric stimuli, Part II. Response kinetics," *J. Electroanal. Chem.*, v. 480, no. 1-2, 2000, pp. 186-198.

Atkin, R. J., and Fox, N., *An Introduction to the Theory of Elasticity*, London: Longman, 1980.

Bar-Cohen, Y., Xue, T., Shahinpoor, M., Simpson, J. O., and Smith, J., "Low-mass muscle actuators using electroactive polymers (EAP)," Proc. SPIE Vol.3324, pp. 218-223, 1998.

Bar-Cohen, Y., Leary, S., Shahinpoor, M., Harrison, J. O., and Smith, J., "Electro-active polymer (EAP) actuators for planetary applications," *SPIE Conference on Electroactive Polymer Actuators*, Proc. SPIE Vol. 3669, pp. 57-63, 1999a .

Bar-Cohen, Y., Leary, S. P., Shahinpoor, M., Harrison, J. S., and Smith, J., "Flexible low-mass devices and mechanisms actuated by electroactive polymers," Proc. SPIE Vol. 3669, pp. 51-56, 1999b.

Bartrum, B. E. (Dow Chemical Co.), "Process for plating permselective membrane," GB Patent: 1,143,883, 1969.
Belloni, J., Mostafavi, M., Remita, H., Marignier, J. L., and Delcourt, M. O., "Radiation-induced synthesis of mono- and multi-metallic clusters and nanocolloids," *New J. Chem.*, v. 22, no. 11, 1998, pp. 1239-1255.
Bennett, M., and Leo, D.J., "Manufacture and characterization of ionic polymer transducers with non-precious metal electrodes," *Smart Materials and Structures*, vol. 12, no. 3, 2003, pp. 424-436.
Bergman, I. (Minister of Power, London), "Improvements in or relating to membrane electrodes and cells," GB Patent: 1,200,595, 1970.
Blanchard, R. M., and Nuzzo, R. G., "An infrared study of the effects of hydration on cation-loaded nafion thin films," *J. Polym. Sci. B Polym. Phys.*, v. 38, no. 11, 2000, pp. 1512-1520.
Bockris, J. O'M., and Reddy, A. K. N., *Modern Electrochemistry*, v. 2, 1977.
Bockris, J. O'M., and Reddy, A. K. N., *Modern Electrochemistry 1: Ionics*, v. 1, New York: Plenum Press, 1998.
Bonner, O. D., "Study of methanesulfonates and trifluoromethanesulfonates. Evidence for hydrogen bonding to the trifluoro group," *J. Am. Chem. Soc.*, v. 103, no. 12, 1981, p. 3262-3265.
Caldwell, D. G., "Pseudomuscular actuator for use in dextrous manipulation," *Med. Biol. Eng. Comp.*, v. 28, no. 6, 1990, pp. 595.
Caldwell, D. G., and Tsagarakis, N., "'Soft' grasping using a dextrous hand," *Ind. Robot*, v. 27, no. 3, 2000, pp. 194-199.
Chen, Y.-L., and Chou, T.-C., "Metals and alloys bonded on solid polymer electrolyte for electrochemical reduction of pure benzaldehyde without liquid supporting electrolyte," *J. Electroanal. Chem.*, v. 360, no. 1-2, 1993, p. 247-259.
Cwirko, E. H., and Carbonell, R. G., "Ionic equilibria in ion-exchange membranes: a comparison of pore model predictions with experimental results," *J. Membr. Sci.*, v. 67, no. 2-3, 1992, p. 211-226.
Datye, v. K., Taylor, P. L., and Hopfinger, A. J., "Simple model for clustering and ionic transport in ionomer membranes," *Macromolecules*, v. 17, 1984, pp. 1704-1708.
Datye, v. K., and Taylor, P. L., "Electrostatic contributions to the free energy of clustering of an ionomer," *Macromolecules*, v. 18, 1985, pp. 1479-1482.
de Gennes, P. G., Okumura, K., Shahinpoor, M., and Kim, K. J., "Mechanoelectric effects in ionic gels," *Europhys. Lett.*, v. 50, no. 4, 2000,p p. 513-518.
DeWulf, D. W., and Bard, A. J., "Application of Nafion/platinum electrodes (solid polymer electrolyte structures) to voltammetric investigations of highly resistive solutions," *J. Electrochem. Soc.*, v. 135, no. 8, 1988, pp. 1977-1985.
Dube, J. R. (United Technologies Corporation, USA), "Formation of electrode on anion exchange membrane," US Patent: 5,853,798, 1998.
Eisenberg, A., "Clustering of ions in organic polymers, a theoretical approach," *Macromolecules*, v. 3, 1970, pp. 147-154.

Eisenberg, A., and Kim, J.-S., "Ionomers (Overview)" in *Polymeric Materials Encyclopedia* Salamone, J. C., (Ed.) pp. 3435-3454, CRC Press, Boca Raton, FL, 1996.

Fernandez, R. E., "Perfluorinated Ionomers" in *Polymer Data Handbook*, Mark, J. E., (Ed.) p. 233-238, Oxford University Press, New York, 1999.

Feynman, R. P., *Surely You're Joking, Mr. Feynman!*, First Ed., New York: W.W. Norton, 1985.

Forsman, W. A., "Effect of segment-segment association on chain dimensions," *Macromolecules*, v. 15, 1982, pp. 1032-1040.

Forsman, W. A., "Statistical mechanics of ion-pair association in ionomers," *Proceedings of the NATO Advanced Workshop on Structure and Properties of Ionomers*, 1986, pp. 39-50.

Fujita, Y., and Muto, T. (Japan Storage Battery Co., Ltd., Japan), "Manufacture of ion exchange resin membrane-electrode connections," JP Patent: 61,295,387, 1986.

Fujiwara, N., Asaka, K., Nishimura, Y., Oguro, K., and Torikai, E., "Preparation of gold-solid polymer electrolyte composites as electric stimuli-responsive materials," *Chem. Mater.*, v. 12, no. 6, 2000, pp. 1750-1754.

Gebel, G., Aldebert, P., and Pineri, M., "Swelling study of perfluorosulfonated ionomer membranes," *Polymer*, v. 34, no. 2, 1993, pp. 333-339.

Gejji, S. P., Suresh, C. H., Bartolotti, L. J., and Gadre, S. R., "Electrostatic potential as a harbinger of cation coordination: CF3SO3- ion as a model example," *J. Phys. Chem. A*, v. 101, 1997, p. 5678-5686.

Gierke, T. D., Munn, C. E., and Walmsley, P. N., "The morphology in Nafion perfluorinated membrane products, as determined by wide- and small-angle X-ray studies," *J. Polym. Sci., Polym. Phys. Ed.*, v. 19, 1981, pp. 1687-1704.

Heitner-Wirguin, C., "Recent advances in perfluorinated ionomer membranes - structure, properties and applications," *J. Membr. Sci.*, v. 120, no. 1, 1996, pp. 1-33.

Hitachi, "Formation of an electrode film on an ion-exchanging membrane" (Hitachi Shipbuilding and Engineering Co., Ltd., Japan), JP Patent: 58,185,790, 1983.

Homma, M., and Nakano, Y., "Development of electro-driven polymer gel/platinum composite membranes," *Kagaku Kogaku Ronbunshu*, v. 25, no. 6, 1999, pp. 1010-1014.

Hsu, W. Y., and Gierke, T. D., "Elastic theory for ionic clustering in perfluorinated ionomers," *Macromolecules*, v. 13, 1982, pp. 198-200.

Iyer, S. T., Nandan, D., and Iyer, R. M., "Thermodynamic equilibrium constants of alkali metal ion-hydrogen ion exchanges and related swelling free energies in perfluorosulfonate ionomer membrane (Nafion-117) in aqueous medium," *Indian J. Chem., Sect. A: Inorg., Bio-inorg., Phys., Theor. Anal. Chem.*, v. 31A, no. 6, 1992, pp. 317-322.

James, P. J., Elliott, J. A., McMaster, T. J., Newton, J. M., Elliott, A. M. S., Hanna, S., and Miles, M. J., "Hydration of Nafion (R) studied by AFM and X-ray scattering," *J. Mater. Sci.*, v. 35, no. 20, 2000, pp. 5111-5119.

Kordesch, K. v., and Simader, G. R., "Environmental impact of fuel cell technology," *Chem. Re v.,* v. 95, no. 1, 1995, pp. 191-207.

Lakshminarayanaiah, N., *Transport Phenomena in Membranes*, New York: Academic Press, 1969.

Lauer, A., Alley Theater, Houston, TX, personal communication.

Levine, C. A., and Prevost, A. L. (Dow Chemical Co.), "Metal plating permselective membranes," FR Patent: 1,536,414, 1968.

Li, J. Y., and Nemat-Nasser, S., "Micromechanical analysis of ionic clustering in Nafion perfluorinated membrane," *Mechanics of Materials*, v. 32, no. 5, 2000, pp. 303-314.

Liu, R., Her, W. H., and Fedkiw, P. S., "*In situ* electrode formation on a Nafion membrane by chemical platinization," *J. Electrochem. Soc.*, v. 139, no. 1, 1992, pp. 15-23.

Love, A. E. H., A Treatise on the Mathematical Theory of Elasticity, New York: Dover, 1944.

Ludvigsson, M., Lindgren, J., and Tegenfeldt, J., "FTIR study of water in cast Nafion films," *Electrochim. Acta*, v. 45, no. 14, 2000, pp. 2267.

McCallum, C., and Pletcher, D., "Reduction of oxygen at a metallized membrane electrode," *Electrochim. Acta*, v. 20, no. 11, 1975, pp. 811-814.

McGee, J., "Mechano-electrochemical response of ionic polymer-metal composites, " Ph.D Thesis Dissertation, UC San Diego, 2002.

Millet, P., Pineri, M., and Durand, R., "New solid polymer electrolyte composites for water electrolysis," *J. Appl. Electrochem.*, v. 19, no. 2, 1989, pp. 162-166.

Millet, P., Durand, R., and Pineri, M., "Preparation of new solid polymer electrolyte composites for water electrolysis," *Int. J. Hydrogen Energy*, v. 15, no. 4, 1990, pp. 245-253.

Millet, P., Durand, R., Dartyge, E., Tourillon, G., and Fontaine, A., "Study of the precipitation of metallic platinum into ionomer membranes by time-resolved dispersive x-ray spectroscopy," *AIP Conf. Proc.*, 1992, v. 258, pp. 531-538.

Millet, P., Durand, R., Dartyge, E., Tourillon, G., and Fontaine, A., "Precipitation of metallic platinum into Nafion ionomer membranes. I. Experimental results," *J. Electrochem. Soc.*, v. 140, no. 5, 1993, pp. 1373-1380.

Millet, P., Andolfatto, F., and Durand, R., "Preparation of solid polymer electrolyte composites: investigation of the ion exchange process," *J. Appl. Electrochem.*, v. 25, no. 3, 1995, pp. 227-232.

Millet, P., "Noble metal-membrane composites for electrochemical applications," *J. Chem. Ed.*, v. 76, no. 1, 1999, pp. 47-49.

Miyoshi, H., Yamagami, M., and Kataoka, T., "Characteristic coefficients of Nafion membranes," *Chem. Express*, v. 5, no. 10, 1990, pp. 717-720.

Modern Plastics, "Metalizing plastic surfaces," v. 16, Sept., 1938, pp. 30.

Mojarrad, M., and Shahinpoor, M., "Ion exchange membrane-platinum composites as electrically controllable artificial muscles," *Third International Conference on Intelligent Materials*, SPIE Proc. Vol. 2779, 1996, pp. 1012-1017.

Mojarrad, M., and Shahinpoor, M. "Ion-exchange-metal composite artificial muscle actuator load characterization and modeling," SPIE Proc. Vol. 3040, 1997, pp. 294-301.

Nandakumar, P., Vijayan, C., Murti, Y.v.G.S., Dhanalakshmi, K., and Sundararajan, G., "Proton exchange mechanism of synthesizing CdS quantum dots in Nafion," *Indian J. Pure Appl. Phys.*, v. 37, no. 4, 1999, pp. 239-241.

Nemat-Nasser, S., "Micromechanics of Actuation of Ionic Polymer-metal Composites," *J. Appl. Phys*., v. 92, no. 5, 2002, pp. 2899-2915.

Nemat-Nasser, S., and Hori, M., *Micromechanics: Overall Properties of Heterogeneous Materials*, First Ed., Amsterdam: North-Holland, 1993.

Nemat-Nasser, S., and Hori, M., *Micromechanics: Overall Properties of Heterogeneous Materials*, Second Ed., Amsterdam: Elsevier, 1999.

Nemat-Nasser, S., and Li, J. Y. "Electromechanical response of ionic polymer-metal composites," *J. Appl. Phys.*, v. 87, no. 7, 2000, pp. 3321-3331.

Nemat-Nasser, S., and Thomas, C., "Ionomeric polymer-metal composites," *Electroactive Polymer (EAP) Actuators as Artificial Muscles—Reality, Potential, and Challenges*, Yoseph Bar-Cohen, Ed., PM98, SPIE Press, 2001, pp. 139-191.

Nemat-Nasser, S., and Wu, Y., "Comparative Experimental Study of Nafion-and-Flemion-Based Ionic Polymer-metal Composites (IPMC)," *J. Appl. Phys.*, v. 93, no. 9, 2003a, pp. 5255-5267.

Nemat-Nasser, S., and Wu, Y., "Tailoring Actuation of Ionic Polymer-metal Composites Through Cation Combination," *Proceedings of SPIE*, 5051, 2003b, pp. 245-253.

Nemat-Nasser, S., and Zamani, S., "Experimental Study of Nafion-and Flemion-Based Ionic Polymer-Metal Composites (IPMC's) with Ethylene Glycol as Solvent," *Proceedings of SPIE*, 5051, 2003, pp. 233-244.

Oguro, K., Kawami, Y., and Takenaka, H., "An actuator element of polyelectrolyte gel membrane-electrode composite," *Osaka Kogyo Gijutsu Shikensho Kiho*, 43, no. 1, 1992, pp. 21-24.

Oguro, K., Fujiwara, N., Asaka, K., Onishi, K., and Sewa, S., "Polymer electrolyte actuator with gold electrodes," SPIE Proc. Vol. 3669, 1999, pp. 64-71.

Osada, Y., Okuzaki, H., and Hori, H., "A polymer gel with electrically driven motility," *Nature*, v. 355, no. 6357, 1992, pp. 242-244.

Pelrine, R., Kornbluh, R., Pei, Q. B., and Joseph, J., "High-speed electrically actuated elastomers with strain greater than 100%," *Science*, v. 287, no. 5454, 2000, pp. 836-839.

Platzer, O., Amblard, J., Belloni, J., and Marignier, J. L. (Atochem S. A., Fr.), "Ion-exchange fluoropolymers containing metal aggregates and their preparation," EP Patent: 309,337, 1989.

Pomes, R. "Theoretical studies of the grotthuss mechanism in biological proton wires," *Israel J. Chem.*, v. 39, no. 3/4, 1999, pp. 387-395.

Potente, J. M., "Gas-phase electrosynthesis by proton pumping through a metalized Nafion membrane: hydrogen evolution and oxidation, reduction of ethene, and oxidation of ethane and ethene," Ph.D. Thesis., North Carolina State University, Raleigh, North Carolina, 1988.

Rashid, T., and Shahinpoor, M., "Force optimization of ionic polymeric platinum composite artifical muscles by means of an orthogonal array manufacturing method," *SPIE Conference on Electroactive Polymer Actuators and Devices*, SPIE Proc. Vol. 3669, 1999, pp. 289-298.

Robinson, R. A., and Stokes, R. H., *Electrolyte Solutions*, Second (Revised) Ed., London: Butterworths, 1965.

Roche, E. J., Pineri, M., Duplessix, R., and Levelut, A. M., "Small-angle scattering studies of Nafion membranes," *J. Polym. Sci., Polym. Phys. Ed.*, v. 19, no. 1, 1981, pp. 1-11.

Rollet, A. L., Simonin, J. P., and Turq, P., "Study of self-diffusion of alkali metal cations inside a Nafion membrane," *Phys. Chem. Chem. Phys.*, v. 2, no. 5, 2000, pp. 1029.

Sadeghipour, K., Salomon, R., and Neogi, S., "Development of a novel electrochemically active membrane and 'smart' material based vibration sensor/damper," *Smart Mater. Struct.*, v. 1, no. 2, 1992, p. 172-179.

Sakai, T., Takenaka, H., and Torikai, E. (Agency of Industrial Sciences and Technology, Japan), "Metal-containing ion-exchange membranes for gas separation," JP Patent: 60,135,434, 1985a.

Sakai, T., Takenaka, H., Wakabayashi, N., Kawami, Y., and Torikai, E. "Preparation of Nafion-metal fine particle composite membranes," *Osaka Kogyo Gijutsu Shikensho Kiho*, v. 36, no. 1, 1985b, pp. 10-16.

Salehpoor, K., Shahinpoor, M., and Razani, A. "Role of ion transport in actuation of ionic polymeric-platinum composite (IPMC) artificial muscles," SPIE Proc. Vol. 3330, 1998, pp. 50-58.

Segalman, D. J., Witkowski, W. R., Adolf, D. B., and Shahinpoor, M., "Theory and application of electrically controlled polymeric gels," *Smart Mater. Struct.*, v. 1, no. 1, 1992, pp. 95-100.

Sewa, S., Onishi, K., Asaka, K., Fujiwara, N., and Oguro, K., "Polymer actuator driven by ion current at low voltage, applied to catheter system," *Proc. - IEEE Ann. Int. Workshop Micro Electro Mech. Syst., 11th*, 1998, pp. 148-153.

Shahinpoor, M., "Conceptual design, kinematics and dynamics of swimming robotic structures using active polymer gels," *Act. Mater. Adapt. Struct., Proc. ADPA/AIAA/ASME/SPIE Conf.*, 1992, pp. 91-95.

Shahinpoor, M., "Micro-electro-mechanics of ionic polymeric gels as electrically controllable artificial muscles," *J. Intel. Mat. Syst. Struct.*, v. 6, no. 3, 1995, pp. 307-314.

Shahinpoor, M., "Ionic polymeric gels as artificial muscles for robotic and medical applications," *Iran. J. Sci. Technol.*, v. 20, no. 1, 1996, p. 89-136.

Shahinpoor, M., and Thompson, M. S., "The Venus flytrap as a model for a biomimetic material with built-in sensors and actuations," *Mater. Sci. Eng., C*, v. 2, no. 4, 1995, pp. 229-233.

Shahinpoor, M. "Electro-mechanics of iono-elastic beams as electrically-controllable artificial muscles," *SPIE Conference on Electroactive Polymer Actuators and Devices*, SPIE Proc. Vol. 3669, 1999, pp. 109-121.

Shahinpoor, M., Mojarrad, M., and Salehpoor, K., "Electrically induced large amplitude vibration and resonance characteristics of ionic polymeric membrane-metal composites artificial muscles," SPIE Proc. Vol. 3041, 1997, pp. 829-838.

Shahinpoor, M., Bar-Cohen, Y., Simpson, J. O., and Smith, J., "Ionic polymer-metal composites (IPMCs) as biomimetic sensors, actuators and artificial muscles. A review," *Smart Mater. Struct.*, v. 7, no. 6, 1998, pp. R15-R30.

Shi, M., and Anson, F. C., "Effects of hydration on the resistances and electrochemical responses of Nafion coatings on electrodes," *J. Electroanal. Chem.*, v. 415, 1996, p. 41-46.

Si, Y.-C., Han, Z.-Q., and Chen, Y.-X., "Fabrication of Pt/Nafion membranes," *Huaxue Xuebao*, v. 56, no. 10, 1998, p. 1027-1031.

Sodaye, H. S., Pujari, P. K., Goswami, A., and Manohar, S. B., "Diffusion of Cs^+ and Zn^{2+} through Nafion-117 ion exchange membrane," *J. Radioanal. Nucl. Chem.*, v. 214, no. 5, 1996, p. 399-409.

Steck, A., and Yeager, H. L., "Water sorption and cation-exchange selectivity of a perfluorosulfonate ion-exchange polymer," *Anal. Chem.*, v. 52, 1980, p. 1215-1218.

Tadokoro, S., Murakami, T., Fuji, S., Kanno, R., Hattori, M., Takamori, T., and Oguro, K., "An elliptic friction drive element using an ICPF actuator," *IEEE Control Systems Magazine*, v. 17, no. 3, 1997, p. 60-68.

Takenaka, H., Torikai, E., Kawami, Y., and Wakabayashi, N., "Solid polymer electrolyte water electrolysis," *Int. J. Hydrogen Energy*, v. 7, no. 5, 1982, p. 397-403.

Takenaka, H., Torikai, E., Kawami, Y., Wakabayashi, N., and Sakai, T., "Solid polymer electrolyte water electrolysis. II. Preparation methods for membrane-electrocatalyst composite," *Denki Kagaku Oyobi Kogyo Butsuri Kagaku*, v. 53, no. 4, 1985, p. 261-265.

Treolar, L. R. G., *Physics of Rubber Elasticity*, Oxford University Press, 1958.

Wax, S. G., and Sands, R. R., "Electroactive polymer actuators and devices," *SPIE Conference on Electroactive Polymer Actuators and Devices*, SPIE Proc. Vol. 3669, 1999, p. 2-10.

Wiechen, A. (Thiele, Heinrich), "Isopor membranes," DE Patent: 1,303,142, 1971.

Xue, T., Trent, J. S., and Osseo-Asare, K., "Characterization of Nafion membranes by transmission electron microscopy," *J. Membr. Sci.*, v. 45, no. 3, 1989, p. 261-271.

Xue, T., Longwell, R. B., and Osseo-Asare, K., "Mass transfer in Nafion membrane systems: effects of ionic size and charge on selectivity," *J. Membr. Sci.*, v. 58, no. 2, 1991, p. 175-189.

Yeager, H. L., Twardoski, Z., and Clark, L., *J. Electrochem. Soc.*, v. 129, 1982, p. 328.

Zana, R., and Yeager, E., "Determination of ionic partial molar volumes from ionic vibration potentials," *J. Phys. Chem.*, v. 70, no. 3, 1966, p. 954-955.

Zawodzinski, T. A., Jr., *220th ACS National Meeting*, Washington, D.C., 2000, v. PMSE - 270.

Zelsmann, H. R., Pineri, M., Thomas, M., and Escoubes, M., "Water self-diffusion coefficient determination in an ion exchange membrane by optical measurement," *J. Appl. Polym. Sci.*, v. 41, 1990, p. 1673-1684.

CHAPTER 7

Conductive Polymers

José-María Sansiñena and Virginia Olazábal
NASA-Jet Propulsion Lab./California Institute of Technology

7.1 Brief History of Conductive Polymers

It was thought for many decades that all polymers were inherently electrical insulators. This concept began to change in 1973 when Walatka et al. discovered that polysulfurnitride glasses [$(SN)_x$] were metallic [Walatka et al., 1973]. The different studies carried out on polysulfurnitride led to the discovery in 1975 that this polymer behaves as a superconductor material at temperatures below 0.3 K [Greene et al., 1975]. This fact accelerated the study of several methods of the synthesis of similar materials [Street et al., 1977a; Wolmershäuer et al., 1978].

In 1977, it was proven that the conductivity of polysulfurnitride at room temperature could be increased by several orders of magnitude by adding halogen derivatives [Street et al., 1977b; Akhtar et al., 1977; Chiang et al., 1977]. It was also seen that polyacetylene [$(CH)_x$] and its derivatives became conductive when they were partially oxidized or reduced with acceptors or donors of electrons [Shirakawa et al., 1977; MacDiarmid and Heeger, 1980; Pochan et al., 1980].

In 1979, two new conductive polymers (CP) were obtained by two different methods. It was possible to synthesize poly(p-phenylene) chemically modified with AsF_5 or alkali metals [Ivory et al., 1979], and polypyrrole/BF_4 was electrochemically deposited onto a metallic electrode [Díaz et al., 1979; Kanazawa et al., 1979; Kanazawa et al., 1979–80].

Díaz et al. were the first authors who electrochemically synthesized polypyrrole (PPy) in acetonitrile. This polymer was obtained on a platinum electrode as a fragile and insoluble deposit that could be detached from the electrode and could be subjected to electrical and analytical testing. The results demonstrated that it was possible to get a CP film with an electrical conductivity of up to 100 $S \cdot cm^{-1}$ [Díaz, 1981]. That was an important fact compared to the powdered PPy previously synthesized in the presence of sulfuric acid by Dall'Ollio et al. that showed an electric conductivity that was only around 8 $S \cdot cm^{-1}$ [Dall'Ollio et al., 1968].

Conductive polymers based on different aromatic monomers such as pyrrole, thiophene, aniline, azulene, carbazole, etc. and their derivatives have been studied as an improvement of both properties and applications of conductive polymers [Mohilner et al., 1962; Bacon and Adams, 1968; Waters and White, 1968; Bargon et al., 1983a; Bargon et al; 1983b; Gurunathan et al., 1999; Inzelt et al., 2000]. All these conductive polymers are π–conjugated systems where single and double bonds alternate along the polymer chains. In their neutral state they are intrinsically insulating with conductivities in the range of 10^{-10} to 10^{-5} $S \cdot cm^{-1}$. However, in their conductive state their conductivities are in the range of 10^2 to 10^3 $S \cdot cm^{-1}$.

Because of the great energy separation (more than 3 eV) between the valence band (highest occupied band) and the conduction band (lowest unoccupied band), it is necessary to subject the polymers to a transformation process called "doping" to get them to their conductive state. This doping process in conductive polymers can be carried out either chemically (through the employment of oxidizing or reducing agents) or electrochemically. However, the basic process is not dependent on the method used:

Doping *p*: partial oxidation of the neutral polymer chains to generate polycations and simultaneous insertion of compensating anions to neutralize the positive charges created in the polymer:

$$(M)_n - nye^- + nyA^- \rightarrow \left[(M)^{y+} (A^-)_y \right]_n$$

M = monomer unit

A^- = anion responsible for maintaining electroneutrality

Doping *n*: partial reduction of the neutral polymer chains to generate polyanions and simultaneous insertion of compensating cations to neutralize the negative charges created in the polymer:

$$(M)_n + nye^- + nyA^+ \rightarrow \left[(M)^{y-} (A^+)_y \right]_n$$

M = monomer unit

A^+ = cation responsible for maintaining electroneutrality

Most of the final electrical and physical properties of a conductive polymer and its applications will be determined by the type of doping mechanism, the nature of the interchanged ions, and their diffusion during the doping process.

7.2 Applications of Conductive Polymers

The change in the oxidation state of a conductive polymer is accompanied by modifications in its properties such as electrical conductivity, color, volume, stiffness, etc. This fact has been used in many applications, as follows:

Batteries

The possibility of a reversible oxidation/reduction process of conductive polymers allows the use of these materials such as electrodes for rechargeable batteries [Mohammadi et al., 1985; Shimidzu et al., 1986; Jonas and Heywang, 1994; Arbizzani et al., 1997; Schopf and Kossmehl, 1997]. A conductive polymer battery system having the configuration of a dry cell (Leclanche) type, which could be recharged with a cyclability of 100 cycles, would be the preferred choice. Prototypes of commercial lithium batteries with polypyrrole (PPy) or polyaniline (PANi) have been developed and the actual challenge is to get new cathodic materials [Munstedt et al., 1987; Naegele and Bithin, 1988; Akhtar et al., 1988; Genies et al., 1989; Santhanam and Gupta, 1993; Gustafsson et al., 1994; Trivedi, 1997]. Good results were obtained with substituted polythiophenes and poly[1,2-di(2-thienyl)ethylene] but these electrodes showed an insufficient cycle stability compared with inorganic systems, and a high discharge rate [Novak et al., 1997].

Electrochromic Devices

All redox active and electrically conductive polymers are potentially electrochromic in thin-film form [Díaz et al., 1982; Díaz, 1986]. The development and application of electrochromic devices (ECD) based on conductive polymers increased strongly in the 1990s [Jelle et al., 1992; Mastragostino, 1993a; Jelle and Hagen, 1993a; Hyodo, 1994; Panero et al., 1995a]. The electrochromic properties are related to the doping/undoping process that modifies the polymer electronic structure producing both new electronic states in the band gap and a color change. The advantages of these electrochromic materials are their low power requirements, gray scale and low polarization needed to change their color reversibly. Also important is their high transmission in the region of visible light [Díaz and Logan, 1980; Garnier et al., 1983; Duek et al., 1992; Mastragostino, 1993b; Duek et al., 1993; Jelle et al., 1993b; Jelle et al., 1993c; Morita, 1994; Mortimer, 1995].

Electrochromic displays are typically assembled by combining a transparent electrode of indium-tin oxide (ITO) covered with a thin layer of electrochromic materials, a transparent solid polymer electrolyte, and a complementary electrochromic material as a counter-electrode. If the back of the counter-electrode is covered with a reflective material the assembly will act as an electrochromic mirror [Bange and Gamble, 1990; Girotto and De Paoli, 1998].

Selective Membranes

Chemical separations are a major cost component of most chemical, pharmaceutical, and petrochemical processes. Membrane-based separations are potentially important because they can be less energy intensive and more economical than competitive separation technologies. Conductive polymers can be utilized like membranes due to their porosity, the possible switching of their selectivity, and the excellent separation effect for some systems. They have been used for both ion exchange and filtration in electrolytic solutions [Dotson and Woodward, 1982; Martin, 1994; Sata, 1994; Nishizawa et al., 1995; Ehrenbeck and Jüttner, 1996; Schow et al., 1996; Jirage et al., 1997; Liu et al., 1997; Talanova et al., 1998;] and for enantiomer separation (enantioseparation) in many drug, pharmaceutical, and flavor compounds that are racemic mixtures [Aoki et al., 1996; Ogata, 1997; Abe et al., 1997; Lakashima and Martin, 1997; Jirage and Martin, 1999]. Several studies reported also a variable permeability for water, organic liquids, or gases, depending on the redox state of the conductive polymer. For electrochemically formed PANi and PPy on metal or conductive grids, a large increase of water permeability was observed for doped (oxidized) films compared with undoped (reduced) films [Schmidt, 1995]. However, if these types of membranes are to make greater inroads into industrial separation processes, higher flux and greater selectivity will be required.

Anti-Corrosion Protection

Several conductive polymers, such as polyaniline (PANi), polypyrrole (PPy), and polythiophene (PT) (and their derivatives), have been used as corrosion protection layers with interesting results [Naugi et al., 1981; DeBerry, 1985; Wessling, 1994; Roth, 1995; Ahmad, 1996]. These conductive polymers have been applied directly by electrodeposition onto the active material or by coating with formulated solutions of these polymers. The efficiency and mechanism of this anti-corrosion protection are not yet confirmed, but an inhibition of the corrosion process due to a geometric blocking and a reduction of the active surface have been proposed.

Anti-Static Materials

In thin-film technologies, conductive polymers can be used as conductive layers for anti-static protection [De Paoli et al., 1985; Heywang and Jonas, 1991; Jonas and Lerch, 1997] and avoid electromagnetic interferences [Taka, 1991]. CP materials have been incorporated as filler in common polymers to substitute for carbon black. A large-scale technological process has also been developed for the through-hole plating of printed circuit boards [Meyer et al., 1994; Hupe et al., 1995; Schattka et al., 1997].

Super-Capacitors

Conductive polymers have also been proposed as electrode materials in super-capacitors due to their excellent ionic conductivity that allows high discharge rates and capacitance [Arbizzani et al., 1997]. These super-capacitors are expected to work in conjunction with batteries in electrical vehicles to provide necessary peak power performance and possibly to reduce the size and enhance the life expectancy of the battery [Gottesfield et al., 1986; Genies et al., 1988; Panero et al., 1995b; Arbizzani et al., 1995; Kogan et al., 1995, Calberg and Inganäs, 1997]. Also, CP materials are now proposed as electrode material in capacitors due to their high corrosion resistance, low temperature coefficient of resistance, and good stability against breakdown phenomena because of the loss of conductivity at a higher field strength [Jonas and Heywang, 1995; Schopf and Kossmehl, 1997].

Microelectronics

Microelectronic technology is a field where conductive polymers may be applied as conductive resists for delineating device patterns on wafers [Kaneko and Wöhrle, 1981; Potember et al., 1987; Anglelopoulos et al., 1993; Wong et al., 1998]. The use of a CP resist layer in analytical scanning electron microscopy (SEM) will make the observation of resist patterns easier. In ordinary analytical SEM, the insulating nature and high-beam energy of conventional resists can lead to charging, joule heating, and image distortion [Anglelopoulos et al., 1993; Lauchlan et al., 1997]. Also, it was observed that the dry etching rates of small holes and narrow trenches in pattern transfer by reactive ion etching (RIE) were significantly reduced when conductive resists based on conductive polymers were utilized [Arnold and Sawin, 1991; Abdou et al., 1991; Wong and Ingram, 1992; Schiested et al., 1994].

Sensors

Conductive polymers have been applied in sensors to perform an electrode modification that improves its selectivity or to support other sensor molecules. In this way, sensors based on CP-modified electrodes already exist for many different substances such as glucose [Schuhmann et al., 1990; Janda and Weber, 1991; Umana and Waller, 1996], urea [Contractor et al., 1994], hemoglobin [Contractor et al., 1994; Schopf and Kossmehl, 1997], xanthin [Xue and Mu, 1995], NO_2 and SO_2 [Hanawa and Yoneyama, 1989], and humidity [Hwang et al., 1993]. Furthermore, conductive polymers can play either an active role (as a catalytic layer, redox mediator, chemically modulated resistor, etc.) [Schuhmann et al., 1990; Korri-Youssoufi et al., 1997; Kranz et al., 1998;

Cosnier et al., 1998; Wallace et al., 1999] or a passive role (as a matrix) [Bartlett and Cooper, 1993; Trojanowicz and Krawczynski vel Krawcyk, 1995a; Trojanowicz et al., 1995b; Schuhmann, 1995; Cosnier, 1997; Cosnier, 1999; Céspedes and Alegret, 2000].

The advantages associated with these devices are their high selectivity, operation at room temperature, and simplicity of use in complex media, as well as the possibility of developing compact and portable analyzers. In addition, the electrochemical formation of CP layers of controlled thickness constitutes a reproducible and automatic procedure of biosensor fabrication. This approach has received considerable attention due to the increased demand for miniaturized biosensors.

In this way, "Electronic Noses" (ENoses) and "Electronic Tongues" (ETongues) based on conductive polymers have been studied for several applications. ENoses are instruments that include arrays of solid-state gas sensors and a variety of transducers all oriented to the classification and quantification of chemical clusters of volatile compounds, including aromas. ETongues are instruments that include arrays of solid-state ion sensors and different types of transducers all oriented to the classification of liquid samples and the quantification of chemical species presented into them, including flavors [Bartlett and Gardner, 1992; Gardner et al., 1995; Gibson et al., 1997; Albert et al., 2000].

Electromechanical Actuators

Conductive polymers show volume changes during their electrochemical oxidation/reduction process due to the exchange of counter-ions between the polymer and an ionic electrolyte related to doping mechanism. These volume changes are responsible for the electromechanical properties and applications of conductive polymers that are the subject of the following sections in which we present both an explanation of the basic mechanism of CP actuators and a brief review of the CP actuators that have been developed. We also analyze the advantages and disadvantages of actuators based on conductive polymers.

7.3 Basic Mechanism of CP Actuators

The construction of electromechanical devices based on conductive polymers is possible due to the volume changes that take place in these materials during their electrochemical oxidation/reduction [Marque and Roncali, 1990; Chiarelli et al., 1994; Chen and Inganäs, 1995; Takashima et al., 1995a; Takashima et al., 1995b]. These volume changes are due to the incorporation/expulsion of ions and solvent into/from the polymeric structure during its oxidation/reduction process [Okabayashi et al., 1987; Hillman et al., 1989; Bruckenstein et al., 1989; Slama and Taguy, 1989; Hillman et al., 1990; Peres et al., 1992; Christensen et al., 1993; Clarke et al., 1993]. It was reported that this process of charge compensation varies according to the size of the anion used in the polymer electrogeneration:

- If the conductive polymer is grown in the presence of bulky anions (dodecylsulfate, dodecylbenzenesulfonate, etc.) or polyanions (polyvinylsulfonate, polystyrenesulfonate, etc.), mostly cations and solvent molecules are inserted and removed to compensate charge in the polymer. In this case, polymer volume increases during the reduction process and decreases in the oxidation process [Shimidzu et al., 1987; Zhou et al., 1987; Bidan et al., 1988; Naoi et al., 1989; Naoi et al., 1991b; Elliot et al., 1991; Baker and Reynolds, 1991; Dusemund and Schwitzgebel, 1991].

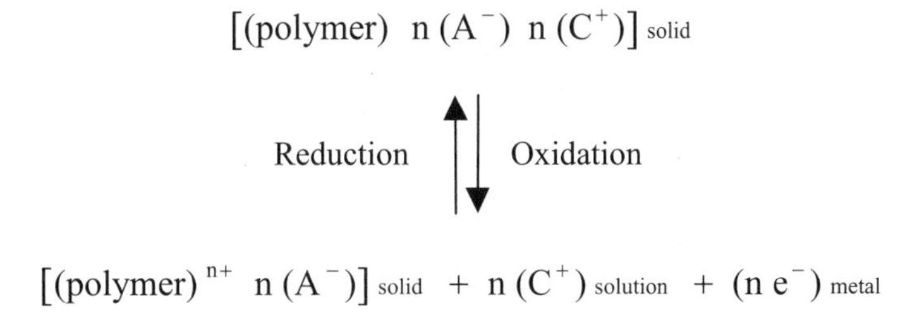

- When conductive polymer is electrogenerated in the presence of small anions (ClO_4^-, BF_4^-, Cl^-, etc.) mostly anion and solvent molecule motion is observed during the redox process. In this case, polymer volume increases during the oxidation and decreases in the reduction. However, apparent cation motion also becomes significant for higher oxidation and/or reduction states [Pei and Qian, 1990; Naoi et al., 1991a; Zhao, 1992; Peres et al., 1992; Reynolds et al., 1993].

$$(\text{polymer})_{\text{solid}} + n\,(A^-)_{\text{solution}}$$

$$\text{Reduction} \;\; \uparrow\downarrow \;\; \text{Oxidation}$$

$$[(\text{polymer})^{n+}\; n\,(A^-)]_{\text{solid}} + (n\,e^-)_{\text{metal}}$$

- There is an intermediate situation that occurs when the conductive polymer is formed in the presence of medium-size anions (p-toluensulfonate, naftalensulfonate, etc.). In this case, simultaneous anion and cation motion are observed during the redox process and the volume change that occurs in the polymer is lower than in the preceding two cases [Genies and Pernault, 1984; Chao et al., 1987; Naoi et al., 1991a; Iseki et al., 1991; Zhao et al., 1992; Reynolds et al., 1993; Beck and Dahlhaus, 1993a; Gandhi et al., 1995].

When the challenge is the construction of a CP actuator, it is necessary to obtain a system that transforms the volume variations that occur inside the conductive polymers (Fig. 1) during their oxidation/reduction process into a macroscopic movement.

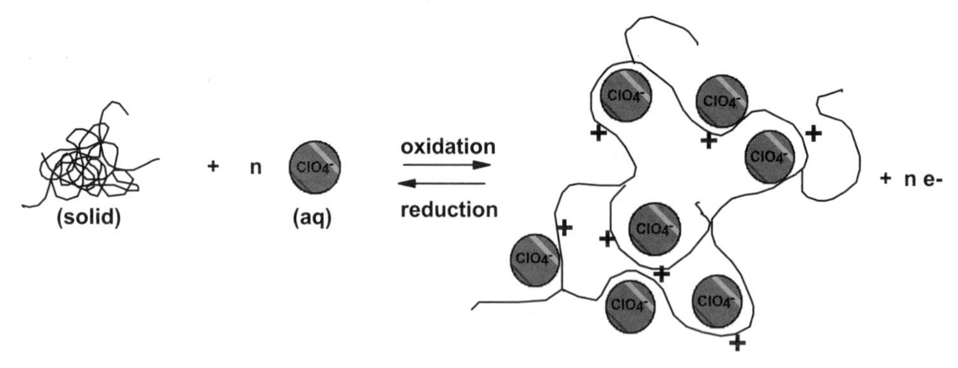

Figure 1: Conformational changes and subsequent volume changes occurring during electrochemical doping in conductive polymers. Reprinted from *Advanced Materials*, 10(6), Otero and Sansiñena, "Soft and wet conducting polymers for artificial muscles," [Otero and Sansiñena, 1998]. (Courtesy of Wiley-VCH.)

CP//NVC Bilayer Devices

The most studied CP actuators are based on a "conductive polymer film//no volume-change film" bilayer structure (CP//NVC), where the conductive polymer is the only electromechanically active material. The conductive polymer film is connected as a working-electrode in an electrochemical cell; an additional counter-electrode and an electrolytic solution are necessary.

When the conductive polymer is electrogenerated in the presence of small anions (ClO_4^-, BF_4^-, Cl^-, etc.), the mechanism of actuation is like that described in Fig. 2. The volume increase of the conductive polymer film during its oxidation due to the entrance of counter-ions (anions) from the electrolytic solution generates an expanding stress gradient at the interface CP//NVC. This stress gradient produces a bending movement of the device so that the CP film is located at the external side of the bilayer [Fig. 2(a)]. In the same way, conductive polymer contraction that occurs during the reduction process due to counter-ions (anions) leaving the polymer toward the electrolytic solution causes a contracting stress gradient at the interface CP//NVC opposite to the one described for the oxidation process. The device bends so that the CP film remains at the internal side of the bilayer [Fig. 2(b)].

When conductive polymer is grown in the presence of bulky anions or polyanions counterion movement and the bending direction are the opposite of those described above. Now counterions are cations that leave the CP film during its oxidation process (contraction) and enter the polymer structure during its electrochemical reduction (expansion).

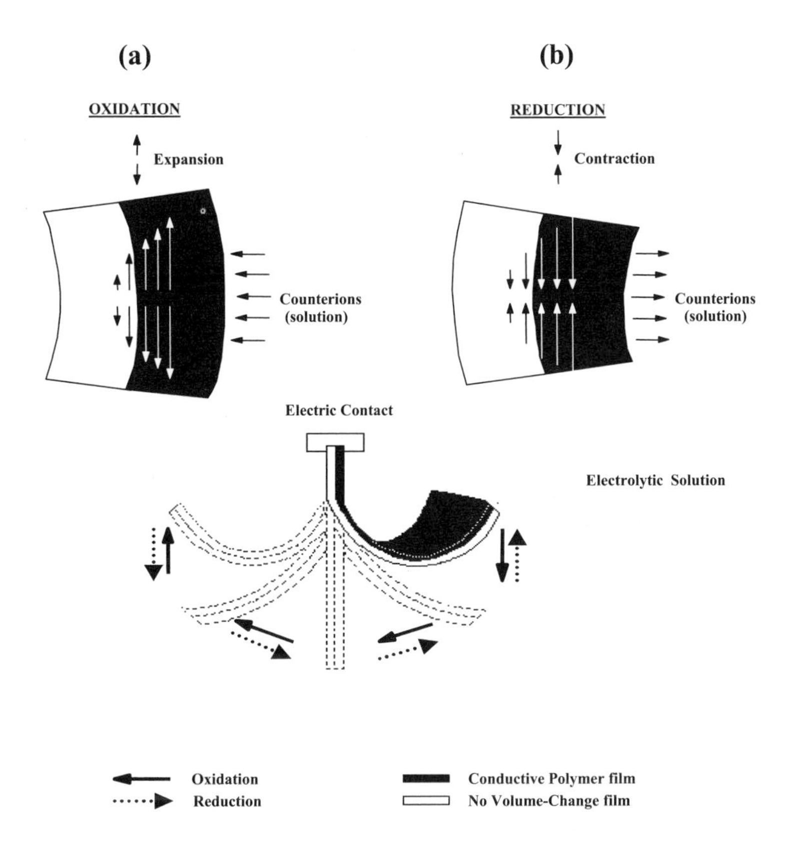

Figure 2: Bending movement of a conductive polymer film/no volume-change film bilayer (CP//NVC) produced by the stress gradient generated at the interface CP//NVC due to volume variations taking place in the CP film during oxidation/reduction processes. Reprinted from *Advanced Materials*, 10(6), Otero and Sansiñena, "Soft and wet conducting polymers for artificial muscles," [Otero and Sansiñena, 1998]. (Courtesy of Wiley-VCH.)

CP//NVC//CP Trilayer Devices

Some studies have been carried out with CP actuators based on a trilayer "conductive polymer film//no volume-change film//conductive polymer film" (CP//NVC//CP), where two CP films are adhered to each side of a NVC film; one of the CP films is connected as working-electrode and the other one as counter-electrode in an electrochemical cell. In this case, the presence of an electrolytic solution is still necessary to make feasible the ionic conductivity between the two CP electrodes, but the use of an additional counter-electrode is not necessary.

Let us suppose that CP films are electrogenerated in the presence of small anions (ClO_4^-, BF_4^-, Cl^-, etc.) and that the CP film on the right side of the trilayer acts as working-electrode while the other one on the left side acts as a counter-electrode (Fig. 3).

When a positive electric potential is applied between the working-electrode and counter-electrode [Fig. 3(a)] an anodic current flows through the CP film that acts as working-electrode (right side), while a simultaneous cathodic current flows through the one that acts as counter-electrode (left side). Therefore, the CP film on the right side is oxidized and the one on the left side is simultaneously reduced.

The oxidation of the CP film placed on the right side causes its expansion, and an expanding stress gradient is generated at its interface with the NVC film. This expanding stress gradient induces a bending movement of the trilayer toward its left side. Simultaneously, the reduction of the CP film placed on the left side causes its contraction, and a contracting stress gradient is generated at its interface with the NVC film that is opposite to that generated at the other CP//NVC interface. This contracting stress gradient induces a bending movement of the trilayer toward its left side. Therefore, both CP films push the trilayer toward the same side and add their contribution to the movement.

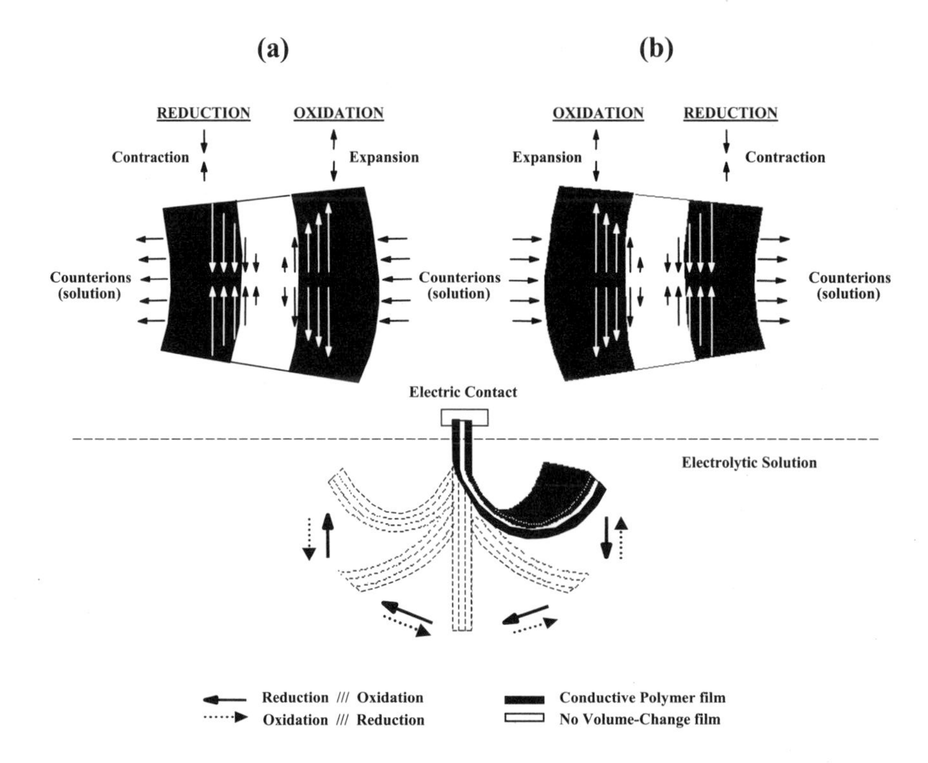

Figure 3: Bending movement of a conductive polymer film/no volume-change film/conductive polymer film trilayer (CP//NVC//CP) produced by the stress gradient generated at the CP//NVC interfaces due to volume variations that take place in both CP films during oxidation/reduction processes [Sansiñena, 1998].

When a negative potential is applied between working-electrode and counter-electrode [Fig. 3(b)] the CP film on the right side is reduced (contraction), the other one on the left side is simultaneously oxidized (expanded) and the trilayer bends toward its right side.

However, if the conductive-polymer films were electrogenerated in the presence of bulky anions counter-ions are cations. The direction of the bending during oxidation/reduction process is the opposite of that described above.

A particular example of a CP//NVC//CP system is the use of either a paper film previously soaked in an electrolytic solution, a solid polymer electrolyte or an ionic polymer gel between two CP films. These electrolytic films allow ionic contact between two CP films while avoiding a direct electric contact that could produce a short-circuit in the system. Therefore, the electrolytic solution is substituted for the electrolytic film that simultaneously acts like a NVC film.

The actuation mechanism of these actuators is similar to that described previously for general CP//NVC//CP trilayers. The only difference is that the counterion exchange occurs through the interfaces between the CP films and the electrolytic film that play the role of the electrolytic solution. Improvement of this type of actuator will allow the development of future CP actuators that will be based only on polymer materials and will avoid the necessity of using an electrolytic solution so they can work in air.

On the other hand, when one of the CP films that are included in a CP//NVC//CP trilayer is grown in the presence of small anions and the other one in the presence of bulky anions, a linear movement in the axial direction of the trilayer is obtained. In this case, the oxidation of CP film with small anions and the simultaneous reduction of CP film with bulky anions promote an expansion in both CP films and the trilayer shows a linear extension. In the same way, the reduction of CP film with small anions and the simultaneous oxidation of CP film with bulky anions promote a contraction in both CP films, and the trilayer shows a linear contraction.

7.4 Development of CP Actuators

An increasing number of laboratories around the world have been studying several systems and materials in the last decade to develop new electromechanical actuators based on conductive polymers and to improve the properties of the existing actuators. In general, CP actuators have been developed by one of the two following procedures:

- A conductive polymer film is deposited onto a flexible plastic film (e.g., polyethylene, polyimide) sputtered with a thin layer of metal (e.g., gold, platinum) (Fig. 4). The CP film can be deposited electrochemically or by casting. Ultimately, the flexible conductive substrate will be the "no volume-change (NVC) film" of the actuator.
- A conductive polymer film is electrochemically deposited onto a metallic electrode and it is later removed from the electrode with a flexible and adhesive polymer film that will work as the actuator NVC film (Fig. 5).

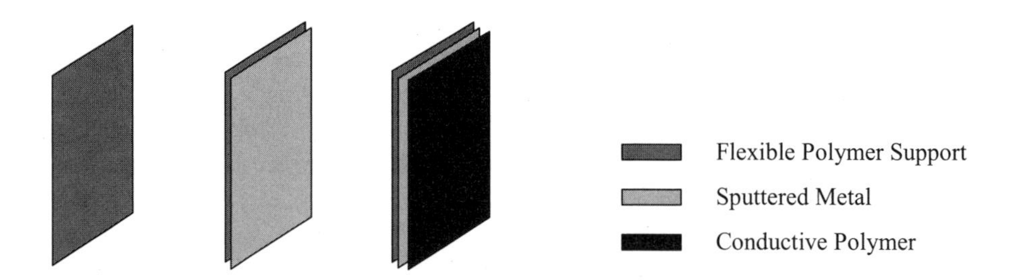

Figure 4: Deposition of a conductive polymer (CP) film onto a flexible plastic film sputtered with a thin layer of metal, which will be the "no volume-change (NVC) film" film of the actuator.

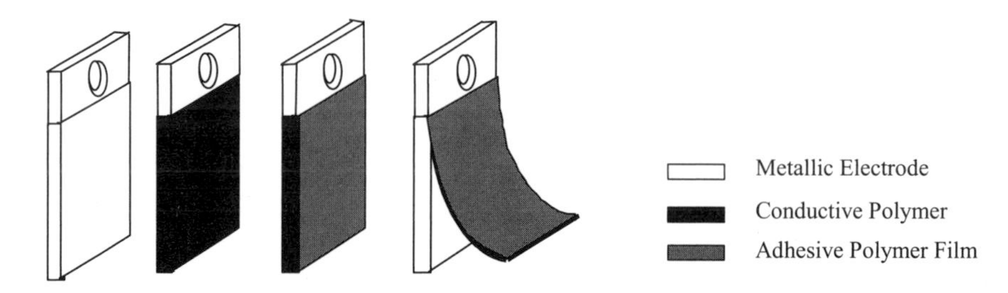

Figure 5: Electrochemical deposition of a conductive polymer (CP) film onto a metallic electrode. The CP is later removed from the electrode with a flexible and adhesive polymer film that will work as the actuator's NVC film [Sansiñena, 1998].

Several CP actuators have been developed since Baughman et al. proposed the first designs of bending electromechanical devices based on conductive polymers in the beginning of the 1990s [Baughman et al., 1990; Baughman, 1991a; Baughman et al., 1991b; Baughman et al., 1992]. These authors constructed a device where the conductive polymer was a poly(3-methylthiophene-co-3-n-octylthiophene) copolymer film. This copolymer was sputter-coated on one side with gold and immersed in $LiClO_4$ propylene carbonate.

The polymer film was connected to a current source as working-electrode and metallic lithium was used as a counter-electrode. This system was denoted "unimorph" by the authors because it had a single electromechanical active electrode (like a CP//NVC bilayer). For comparison, they used the term "bimorph" for systems based on two integrated electromechanical electrodes (working-electrode and counter-electrode) that contribute complementarily to actuator movement (like a CP//NVC//CP trilayer).

Between 1992 and 1995, the Laboratory of Applied Physics at Linköping University (Sweden) constructed three different unimorph systems: polypyrrole//Cr-Au//polyethylene, poly(3-octylthiophene)//polypyrrole and poly(3-octylthiophene)//polyethylene which worked in a liquid electrolytic solution with an additional counter-electrode separate from the device [Pei and Inganäs, 1992; Pei and Inganäs, 1993a; Pei et al., 1993b; Pei and Inganäs, 1993c;

Pei and Inganäs, 1993d; Pei and Inganäs, 1993e; Pei and Inganäs, 1993f; Pei and Inganäs, 1993g; Smela et al., 1993a; Smela et al., 1993b]. They also fabricated microactuators where PPy//Au bilayers showed large bending and could act as hinges to rotate rigid plates [Smela et al., 1995].

Furthermore, this laboratory used the so-called bending beam method (BBM) based on the self-induced bending of a bilayer strip to detect small volume changes in the active layer of a CP actuator [Pei and Inganäs, 1992; Smela et al., 1993a; Smela et al.,1993b]. This BBM has also been used by these authors to study the ion transport mechanism in several conductive polymers [Pei and Inganäs, 1993a; Pei and Inganäs, 1993c; Pei and Inganäs, 1993f; Pei and Inganäs, 1993g; Chen and Inganäs, 1995].

In 1999, the construction of a microactuator with all on-chip was also accomplished in the Linköping Laboratory to study the effect of mechanical stimulation of living cells [Jager et al., 1999a; Jager et al., 1999b]. However, the authors reported that the necessity of using a liquid electrolyte was an important limitation of this actuator, and they said it is a very important challenge to get a dry microactuator integrated on-chip (Fig. 6).

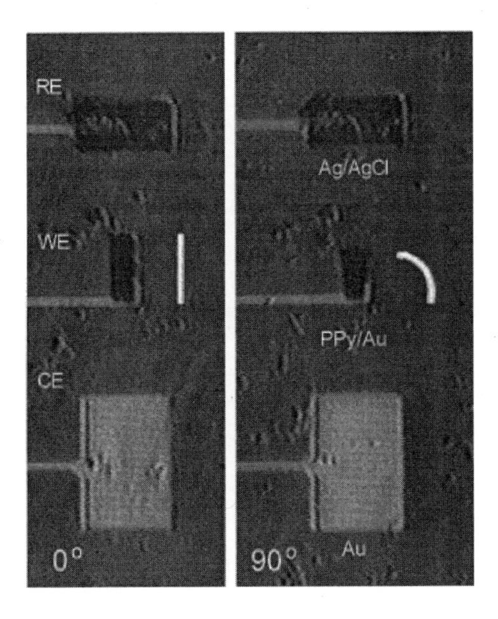

Figure 6: A photograph of the three microelectrodes in an electrolyte solution. The top electrode is the Ag/AgCl quasi-reference electrode (50 × 100 μm^2), the middle electrode is the PPy/Au microactuator or working electrode (20 × 70 μm^2) and the bottom electrode is the Au counter-electrode (100 × 150 μm^2). On the left, the microactuator lies flat on the substrate and, on the right, it is bent perpendicular to the substrate (90 deg). Reprinted from *Sensors and Actuators* B, 56(1-2), Jager et al., "On-chip microelectrodes for electrochemistry with moveable PPy bilayer" [Jager et al., 1999a]. (Copyright 1999, with permission from Elsevier Science.)

In the same year, this group constructed small "fingers" with polypyrrole and gold (PPy//Au) using microfabrication techniques [Smela et al., 1999c]. The authors found that the lifetime of these fingers was limited by the loss of contact between the CP film and the Au due to a delamination process. They also that said once this delamination problem is solved, life will probably be determined by degradation of the PPy because conjugated polymers are not chemically stable, and their charging capacity gradually declines when they are cycled [Beck et al., 1993b]. However, these preliminary studies demonstrated the possibility of using conventional microfabrication techniques to develop microactuators based on conductive polymers.

In 1995, the Department of Computer Science and Electronics of the Kyushu Institute of Technology (Japan), in collaboration with the Department of Chemistry of the University of Pennsylvania (USA) constructed and characterized a trilayer actuator called "backbone-type." This actuator was based on two polyaniline (PANi) films separated by an adhesive. One of the PANi films was connected as working-electrode and the other one as counter-electrode in a 1M HCl aqueous solution. When an electric potential was applied, the actuator covered a bending movement of 360 deg and responded to frequencies up to 44 Hz [Kaneto et al., 1995a; Kaneto et al., 1995b; Takashima et al., 1995a; Takashima et al., 1995b]. Also, a called "shell-type" actuator moving in air was constructed by substituting the adhesive polymer that separated the two PANi films with a paper film previously soaked in HCl (Fig. 7).

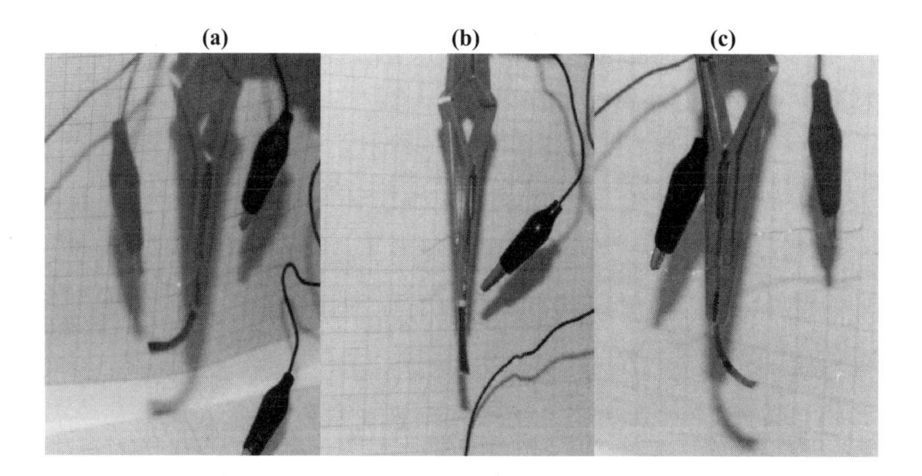

Figure 7: The "shell-type" actuator operated by 1.5 V battery in air atmosphere. No-voltage (a), positive bias applied to the left-hand side of ES films (b), and reversed bias (c). Reprinted from *Synthetic Metals*, 71(1-3), Kaneto et al., "Artificial muscle: Electrochemical actuators using polyaniline films" [Kaneto et al., 1995a]. (Copyright 1995, with permission from Elsevier Science.)

The authors proposed many potential usages for these actuators, for example, medical equipment, artificial muscles of robots, valves for chemicals, and positioning of fine mechanics. However, even though this actuator was an important landmark in the field of CP actuators, the authors reported that the movement was very slow and not very stable:

- The slow diffusion process of HCl from the paper into the PANi films limited the velocity of the actuator movement.
- The actuator lifetime was limited by how long it took the paper to dry up, causing the two PANi films to lose their ionic contact.

In parallel to the actuators mentioned before, in 1992, the Laboratory of Electrochemistry of the Faculty of Chemistry at San Sebastián (Spain) constructed and patented a CP actuator based on a polypyrrole film and a flexible and adhesive tape; it was able to develop a reversible movement of 180 deg in $LiClO_4$ aqueous solution [Otero et al., 1992a; Otero et al., 1992b]. This actuator supplied the first step for further studies carried out by the authors at this laboratory.

In this way, several studies about the synthesis and degradation processes of conductive polymers [Otero et al., 1995a; Otero et al., 1995b; Otero and Olazábal, 1995c; Otero et al., 1996a; Otero et al., 1996b; Otero and Olazábal, 1996c; Otero et al., 1996d] and about the development of CP actuators [Otero et al., 1995d; Otero and Sansiñena, 1995e; Otero and Sansiñena, 1996e; Otero and Sansiñena, 1998] were carried out.

Those studies allowed the group to construct new bilayers based on a PPy/ClO_4 film and an adhesive polymer tape (PPy//NVC//PPY) that showed a reversible and reproducible bending movement of 360 deg in an aqueous $LiClO_4$ solution. Furthermore, these actuators were able to move a weight up to 1,000 times higher than the mass of PPy film included in the CP actuator (Fig. 8).

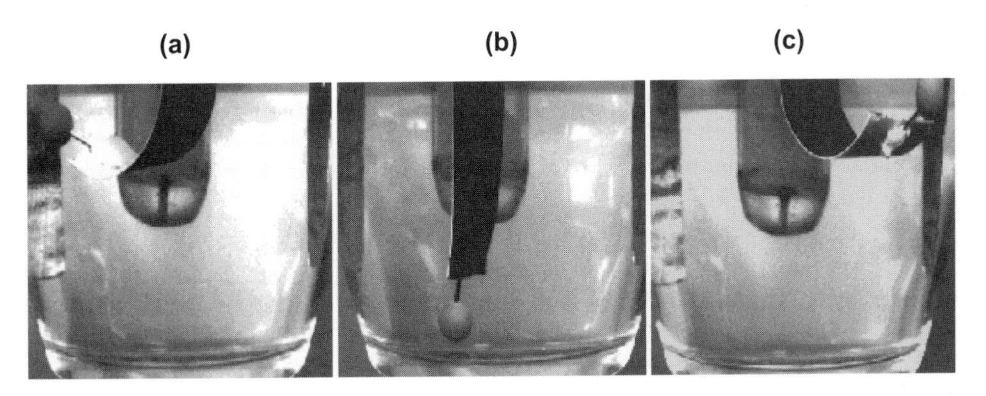

Figure 8: Movement in $LiClO_4$ aqueous solution of a (PPy//NVC) bilayer device constructed using a 3-mg PPy film (black) and an adhesive tape (white) trailing a steel pin of 1 g (≅ 300 times higher than PPy mass) attached at its free end: (a) after partial PPy oxidation, (b) neutral state, (c) after partial PPy reduction [Otero and Sansiñena, 1998]. (Courtesy of Wiley-VCH.)

These CP actuators were systematically studied to determine direct relationships between several variables and bending movement. Based on results obtained in these studies, three general principles for the CP actuator movement were established [Otero and Sansiñena, 1995f; Otero and Sansiñena, 1995g; Otero and Sansiñena, 1997]:

- The angle (rad) covered by the CP actuator during its bending movement is proportional to the electric charge ($\int I\ dt$) consumed per CP mass unit ($mC \cdot mg^{-1}$).
- The angular velocity ($rad \cdot s^{-1}$) of CP actuator bending is proportional to the electric current applied per CP mass unit ($mA \cdot mg^{-1}$).
- The electric energy ($\int I \cdot V\ dt$) consumed during the movement per CP mass unit ($J \cdot g^{-1}$) is proportional to the applied electric current (mA).

The relationships obtained for CP actuators based on a PPy/ClO_4 film and an adhesive tape were the following:

$$\text{Angle (rad)} = 0.04 \cdot Q/m\ (mC \cdot mg^{-1}) \tag{1}$$

$$\text{Angular Velocity}\ (rad \cdot s^{-1}) = 0.04 \cdot I/m\ (mA \cdot mg^{-1}) \tag{2}$$

$$\text{Energy per CP mass}\ (mJ \cdot mg^{-1}) = 1.47 \cdot I\ (mA) \tag{3}$$

These general equations were expressed in terms of CP mass because this parameter is easier to control by electrochemical methods and to measure in a laboratory of chemistry or electrochemistry. However, because mass and volume are directly related by the specific density of the material ($\rho_{(PPy/ClO_4)}$ = Mass / Volume = $1.51\ g \cdot cm^{-3}$), it is possible to express the relationship shown above in terms of volume. That should be the preferable way for physicists and engineers.

These principles are independent of both CP actuator size and mass and allow for prediction of the CP actuator behavior when a change of scale is incorporated into the design. That is a very important point to control the construction and application of macro- and micro-CP actuators.

Also, CP actuators based on a trilayer constructed with two similar PPy/ClO_4 films adhered to a double-sided adhesive tape (PPy//NVC//PPy) were developed (Fig. 9) [Sansiñena, 1998].

The obtained results pointed to the same relationships shown above for CP actuators based on a PPy//NVC bilayer. However, the force developed by a trilayer was the double the force developed by a bilayer based on a similar PPy/ClO_4 film. This result predicted that it would be possible to construct more powerful CP actuators by increasing the number of complementary CP layers.

Also, in 1997, a CP actuator based on a "polypyrrole//solid polymer electrolyte//polypyrrole" (PPy//SPE//PPy) trilayer capable of a reversible bending movement of 360 deg in air was constructed (Fig. 10). This work was carried out by the authors in collaboration with da Fonseca and De Paoli during their stay as visiting researchers at the Laboratory of Conductive Polymers at the UNICAMP University (Brazil) [Sansiñena et al., 1997; Otero et al., 1999].

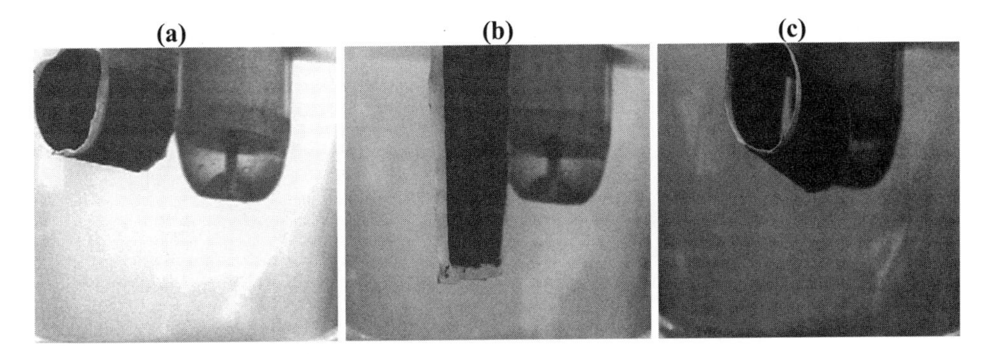

Figure 9: Movement in $LiClO_4$ aqueous solution of a (PPy//NVC//PPy) trilayer device constructed using two PPy films (black) and an adhesive tape (white). One of the PPy films was connected as working-electrode and the other one as counter-electrode [Sansiñena, 1998].

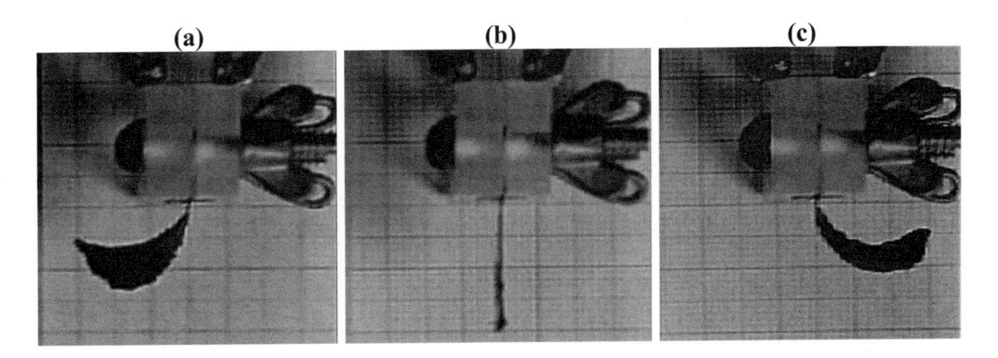

Figure 10: Movement in air of a (PPy//SPE//PPy) trilayer device constructed using two PPy films (black) and a solid polymer electrolyte (SPE) that was placed between both PPy films with a face to face configuration. One of the PPy films was connected as working-electrode and the other one as counter-electrode [Sansiñena et al., 1997]. (Reproduced by permission of The Royal Society of Chemistry.)

This fully polymeric CP actuator was constructed using two PPy film electrodes and a solid polymer electrolyte (SPE) film that was put between both PPy films with a face-to-face configuration. The SPE was based on a film of poly(epichlorohydrin-co-ethylene oxide) with a $LiClO_4$ salt previously dissolved in its matrix [P(ECH-co-EO)]/$LiClO_4$. The electrochemical characterization of the movement of this all-polymer actuator pointed to the same linear relationships "angle vs. electric charge per CP mass unit," "angular velocity vs. current per CP mass unit," and "energy per CP mass vs. current," as those shown in $LiClO_4$ aqueous solution [Sansiñena et al., 1997].

In parallel to the work mentioned above, as the result of their collaboration the Intelligent Polymer Research Institute of the University of Wollongong

(Australia) and the "Enrico Piaggio" Center of the Faculty of Engineering at the University of Pisa (Italy) developed another all-polymer CP actuator bending in air (Fig. 11) [Lewis et al., 1997].

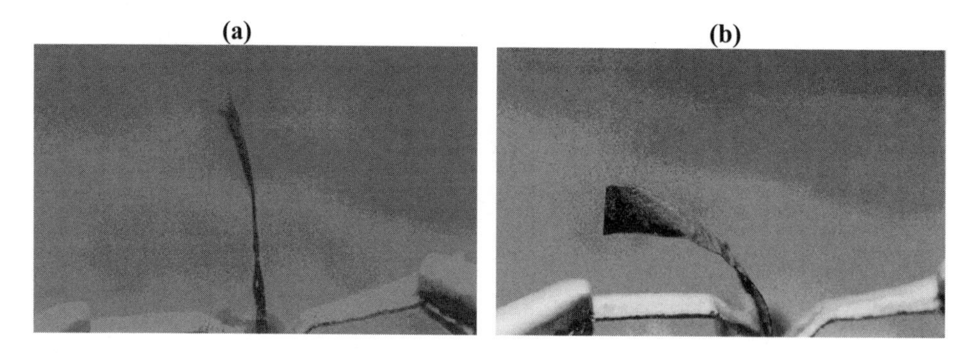

Figure 11: (a) PPy//SPE//PPy sandwich in air, with one film connected as working-electrode and the other one as the auxilliary, no applied potential. (b) As in (a), but with a potential difference of 1.5 V applied [Lewis et al., 1997].

This actuator was built in a (PPy//SPE//PPy) sandwich configuration where the SPE was prepared from a polyacrylonitrile (PAN) film with a $NaClO_4$ salt previously dissolved in its matrix. This actuator could be made to bend in air by 90 deg in each direction when a +/− 1.5 V potential between the two PPy films was applied.

Although electric conductivity and lifetime have to be improved, these CP actuators based on (PPy//SPE//PPy) trilayers opened the possibility of getting all-polymer electromechanical actuators working out of a liquid electrolyte. This fact may involve an important increase of the potential applications of CP actuators in both macro and micro systems.

In spite of the promising results that have been obtained studying CP actuators with bending movement, a simultaneous development of linear CP actuators should be a foreground challenge. Linear CP actuators that exhibit the same properties as biological muscles (high force, power density, and linear displacement with adequate response time and efficiency) are needed by the robotics and bioengineering fields.

In this regard, in 1997 the "Enrico Piaggio" Center of the Faculty of Engineering at the University of Pisa (Italy) measured the length change of a polypyrrole film doped with benzensulfonate anions in a liquid sodium benzenesulfonate bath. They obtained a linear of the "length change vs. exchanged charge" dependence of about 1 $mm \cdot C^{-1}$ and modeled the CP-electrolyte system [Della Santa et al., 1997a; Della Santa et al., 1997b]. This group also reported the construction and characterization of a linear actuator prototype made of PANi fibers, a solid polymer electrolyte (based on polyacrylonitrile and cupric perchlorate), and a spiral-shaped copper wire as counter-electrode (Fig. 12).

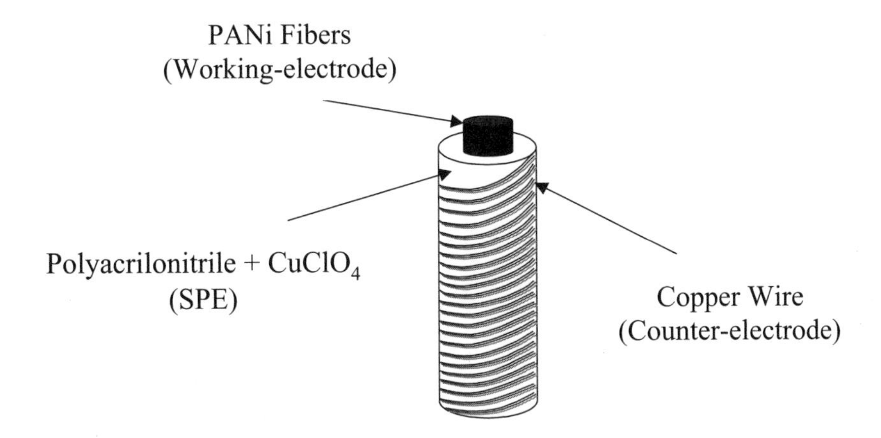

Figure 12: Scheme of the linear actuator made of PANi fibers, a solid polymer electrolyte (SPE) based on polyacrilonitrile and cupric perchlorate, and a spiral-shaped copper wire as counter-electrode [Mazzoldi et al., 1998]. (Copyright 1998, with permission from Elsevier Science.)

Whereas measurements carried out in both 1 M $HClO_4$ aqueous solution and in a dry configuration showed relatively modest strain values (0.2–0.3%), the stress values developed by the dry actuator (2–3 MPa) were very interesting since they were approximately 10 times higher than those corresponding to human skeletal muscle [Mazzoldi et al., 1998; De Rossi and Mazzoldi, 1999]. Although this linear CP actuator and the others that have been constructed indicate a good beginning, both force and displacement need to be much better before they can be applied in robotics and bioengineering [Madden et al., 1999; Lewis et al., 1999; Madden et al, 2000].

Therefore, the development of new linear CP actuators with improved properties working both in wet and dry configuration is a challenge with top priority for the coming years. Consequently, the challenge of the current research we are carrying out in the NDEAA Technologies Group of the NASA-Jet Propulsion Laboratory (JPL) in Pasadena California, USA is the construction of optimized CP actuators with improved force that would allow to develop selectively linear and bending displacements in an actuator. This research is focused on the development of dry CP actuators based on increasing the number of combined series and parallel CP films [Sansiñena et al., 2000].

7.5 Advantages and Disadvantages of CP Actuators

Following the review of the most important CP actuators, the aim of this section is to analyze the general advantages and disadvantages of actuators based on conductive polymers. Although important improvements have been reported during the last ten years, and CP actuators show great potential possibilities, many limitations still have to be overcome before they can be used widely in practical applications.

The challenge is to get new CP actuators with improved properties capable of showing fast response time, high force density, high efficiency in energy conversion, long lifetime and high stability, large linear displacements, etc. All of these properties are desirable to make CP actuators' performance similar to that of natural muscles.

7.5.1 Advantages

Conductive polymers (CP) are both ionic and electronic conductive materials that allow construction of CP actuators with a basic mechanism similar to natural muscles. In both cases an electric pulse triggers electrochemical processes and an ionic exchange is responsible for the maintenance of the electric neutrality. This ion movement causes a volume change in both systems that are used to perform mechanical work.

Also, many conductive polymers are biocompatible and need low electric potentials (a few volts) to trigger redox electrochemical processes. That is an important fact if biomimetic devices are to be incorporated inside the human body.

In addition, the possibility of electrodepositing conductive polymer films on a conductive material is important point for achieve micro and macro actuators with controlled properties and shapes. In this way, conductive polymer is electrodeposited on an electrode with defined favorable or unfavorable areas for polymer deposition. Several patterning techniques that depend on the nature of the electrode and the final application of the CP actuator can be used to define these favorable and unfavorable areas [Smela, 1999a].

The theoretically calculated force densities of some conductive polymers (undrawn polythiophene, drawn polyaniline, and drawn polyacetylene) could get up to 450 MPa [Baughman, 1996]. This value is enormous when compared to the force density of natural muscles (~0.5 MPa) and demonstrates the great potential of conductive polymers as actuators. This potential capability to develop high force density is an essential point for the construction of powerful actuators for practical applications where either small size or light materials are required.

The experimental force density values developed by CP actuators at present (~3 MPa) are still far from those theoretically calculated [Della Santa et al., 1997c; Kaneko et al., 1997; Mazzoldi et al., 1998]. However, these values are very interesting since they are approximately 10 times higher than the value reported for natural muscles. Also, an increase of total force developed by CP actuators is expected to be achieved in the next few years through the construction of new actuators based on a parallel combination of increasing numbers of conductive polymer films working in a complementary way.

The strain or displacement of CP actuators is other important point that needs to be considered. The challenge is to construct actuators capable of developing large forces and displacements at the same time. The displacement of a CP actuator depends on both the volume change that occurs in the CP film and the configuration utilized to construct the actuator. In this way, 30–40% volume change between oxidized and reduced states in the direction perpendicular to the substrate reported by Smela and Gadegaard in 1999 for polypyrrole doped with dodecylbenzenesulfonate (PPy/DBS) using AFM, point to a promising future of CP actuators [Smela and Gadegaard, 1999b]. Also, the authors found that the volume change is highly anisotropic and they proposed the bending-beams method to be studied in conjunction with the AFM technique to obtain the extent of the volume change in the plane of the film. These data will be very important to model appropriate configuration for future CP actuators with larger linear and/or bending displacements.

7.5.2 Disadvantages

The actuation mechanism of a CP actuator is based on an ion exchange process between the conductive polymer film and the electrolytic medium (either electrolytic solution or dry/wet polymer electrolyte). This is the most important factor that controls and limits the response time of a CP actuator. Therefore, the improvement of both ion diffusion inside the CP film and the ionic conductivity of the electrolyte has to be one of the first challenges to increase the performance of any CP actuator.

This improvement should be obtained by the following ways:

- Increasing the porosity of the CP film that is strongly limited by the polymer nature and its synthesis conditions.
- Minimizing the CP film thickness, which is not compatible with an increase in the force of the CP actuator.
- Increasing the ionic conductivity of the electrolytic medium, which is the most controllable parameter.

Taking into account the third way, when a liquid electrolyte is utilized it is possible to increase ionic conductivity by increasing the ion concentration but the actuator is forced to work in a liquid medium. In this case, it is necessary to encapsulate a CP actuator and liquid electrolyte together for movements out of the bulk solution. The problem is both to find a good encapsulating material whose mechanical stiffness does not impede the actuator's movement and to avoid electronic conduction between conductive polymer and the counter-electrode (which can be either a metallic electrode or a conductive polymer).

The use of polymer electrolytes to produce all-polymer CP actuators capable of working out of a liquid environment presents two different cases:

- If the polymer electrolyte is a gel, both a relatively high ionic conductivity and a low mechanical impediment for the actuator movement can be obtained. However, an encapsulating material is necessary, which shows the same problem as that reported before for the use of a liquid electrolyte.
- If a solid polymer electrolyte (SPE) is utilized, then the encapsulating material is not necessary but its low ionic conductivity is an important problem that has to be resolved. Also, a SPE with low mechanical stiffness has to be utilized, which is not an easy matter.

Further, a delamination process that limited the actuator's lifetime was reported in CP actuators where a conductive polymer film was deposited onto a flexible plastic film sputtered with a thin layer of metal [Smela et al., 1993b; Smela et al., 1999c]. This delamination process is due to a weak contact between the metal and conductive polymer, so metal surface treatments and different metallic substrates have been studied [Smela et al., 1993a; Idla et al., 2000].

Also, degradation processes that occur in the conductive polymer limit the lifetimes of CP actuators. These degradation processes depend on both electrochemical overoxidation and the presence of oxygen:

- The only way to avoid overoxidation processes is the control of the applied potential. This process will depend on the electrolytic medium, metallic substrate, nature of the conductive polymer, dopant used during conductive polymer synthesis, etc.
- The degradation of conductive polymer due to the presence of oxygen can be avoided by CP actuator encapsulation.

Finally, the low linear strain of conductive polymer impedes the construction of effective linear CP actuators working in the same way as natural muscles. This fact is an important limitation that has to be overcome looking for either new conductive polymers or new configurations where longitudinal changes should be amplified.

7.6 Acknowledgments

The authors express their gratitude to the Basque Country Government, Caltech President Fund, and the Defense Advanced Research Agency (DARPA) for their financial support. Also, research at Jet Propulsion Laboratory (JPL), California Institute of Technology, was carried out under a contract with the National Aeronautics and Space Agency (NASA).

7.7 References

Abdou, M.S.A., Díaz-Guijada, G.A., Arroyo, M.I., Holscroft, S., 1991. *Chem. Mater.* 3, 1003.

Abe, K., Goto, M., Nakashio, F., 1997. *Sep. Sci. Technol.* 32, 1921.

Ahmad, N., MacDiarmid, A.G., 1996. *Synth. Met.* 78, 103.

Akhtar, M., Kleppinger, J., MacDiarmid, A.G., Milliken, J., Morán, M.J., Chiang, C.-K., Cohen, M.J., Heeger, A.J., Peebles, D.L., 1977. *J. Chem. Soc. Chem. Commun.* 473.

Akhtar, M., Weakliem, H.A., Paiste, R.H., Gaughan, K., 1988. *Synth. Met.* 26, 203.

Albert, K.J., Lewis, N.S., Schauer, C.L., Sotzing, G.A., Stitzel, S.E., Vaid, T.P., Walt, D.R., 2000. *Chem. Rev.* 100, 2595.

Anglelopoulos, M., Patel, N., Shaw, J.M., Labianca, N.C., Rishton, S.A., 1993. *J. Vac. Sci. Technol. B11,* 2794.

Aoki, T., Shinohara, K., Kaneko, T., Oikawa, E., 1996. *Macromolecules* 29, pp. 4192.
Arbizzani, C., Mastragostino, M., Meneghello, L., 1995. *Electrochim. Acta* 40, 2223.
Arbizzani, C., Mastragostino, M., Meneghello, L., 1997. *Electrochim. Acta* 41, 21.
Arnold, J.C., Sawin, H.H. 1991. *J. Appl. Phys.* 70, 5314.
Bacon, J., Adams, R.N., 1968. *J. Am. Chem. Soc.* 90, 6596.
Baker, Ch., Reynolds, J.R., 1991. *J. Phys. Chem.* 95, 4446.
Bange, K., Gamble, T., 1990. *Adv. Mater.* 2, 10.
Bargon, J., Mohamand, S., Waltman, R.J., 1983a. *IBM J. Res. Dev.* 27, pp. 1452.
Bargon, J., Mohamand, S., Waltman, R.J., 1983b. *Mol. Cryst. Liq. Cryst.* 93, 279.
Bartlett, P.N., Cooper, J.M., 1993. *J. Electroanal. Chem.* 363, 1.
Bartlett, P.N., Gardner, J.W., *Sensors and Sensory Systems for an Electronic Nose,* Gardner, J.W. and Bartlett, P.N. (Eds.), Kluwer Academic, *Dordrecht*, p. 31, 1992.
Baughman, R.H., Shacklette, L.W., Elsenbaumer, R.L., Plichta, E.J., Becht, C., 1990. *Conjugated Polymeric Materials: Opportunities in Electronics, Optoelectronics, and Mole-Electronics,* J.L. Brodas and R.R. Chance (Eds.), Kluwer, Dordrecht, pp. 559.
Baughman, R.H., 1991a .*Makromol. Chem., Macromol. Symp.* 51, 193.
Baughman, R.H., Shacklette, L.W., Elsenbaumer, R.L., Plichta, E.J., Becht, C., 1991b. *Topics in Molecular Organization and Engineering: Engineering Electronics,* P.I. Lazarev (Ed.), Kluwer, Dordrecht, Vol. 7, pp. 267.
Baughman, R.H., Murthy, N.S., Eckhardt, H., Kertesz, M., 1992. *Phys. Rev.* B46, 10515.
Baughman, R.H., 1996. *Synth. Met.* 78, 339.
Beck, F., Dahlhaus, M., 1993a. *J. Electroanal. Chem.,* 357, 288.
Beck, F., Barsch, U., Michaelis, R., 1993b. *J. Electroanal. Chem.,* 351, 169.
Bidan, G., Ehui, B., Lapkowski, M.,1988. *J. Phys. D, Appl. Phys. 21, 1043.*
Bruckenstein, S., Wilde, C.P., Shay, M., Hillman, A.R., Loveday, D.C., 1989. *J. Electroanal. Chem.,* 258, 457.
Calberg, J.C., Inganäs, O., 1997. *J. Electrochem. Soc.* 144, L61.
Céspedes, F., Alegret, S., 2000. *Trends in Analytical Chemistry* Vol. 19 No. 4, 276.
Chao, F., Baudoin, J.L., Costa, M., Lang, P., 1987. *Makromol. Chem. Macromol. Symp.*, 8, 173.
Chen, X., Inganäs, O., 1995. *Synth. Met.* 74, 159.
Chiang, C.-K., Cohen, M.J., Peebles, D.L., Heeger, A.J., Akhtar, M., Kleppinger, J., MacDiarmid, A.G., Milliken, J., Morán, M.J., 1977. *Solid State Commun,* 23, 607.
Chiarelli, P., De Rossi, D., Della Santa, A., Mazzoldi, A., 1994. *Polymers Gels and Network* 2, 289.
Christensen, P.A., Hamnett, A., Hillman, A.R., Swann, M.J., Higgins, S.J., 1993. *J. Chem. Soc. Faraday Trans.,* 89 (6), 921.

Clarke, A.P., Vos, J.G., Glidle, A., Hillman, A.R., 1993. *J. Chem. Soc. Faraday Trans.*, 89 (11), 1695.
Contractor, A.Q., Sureshkumar, T.N., Narayanan, R., Sukeerthi, S., Lal, R., Srinivasa, R.S., 1994. *Electrochim. Acta* 39, 1321.
Cosnier, S., 1997. *Electroanalysis,* 9, 894.
Cosnier, S., Galland, B., Gondran, C., Le Pellec, A., 1998. *Electroanalysis* 10, 808.
Cosnier, S., 1999. *Biosensors and Bioelectronics* 14, 443.
Dall'Ollio, A., Dascola, Y., Varacca, V., Bocchi, V., 1968. *Comptes Rendus* C267, 433.
DeBerry, D.W., 1985. *J. Electrochem. Soc.* 132, 1022.
Della Santa, A., De Rossi, D., Mazzoldi, A., 1997a. *Smart Mater. Struct.* 6, 23.
Della Santa, A., Mazzoldi, A., Tonci, C., De Rossi, D., 1997b. *Materials Science and Engineering* C, 5, 101.
Della Santa, A., De Rossi, D., Mazzoldi, A., 1997c. *Synth. Met.* 90, 93.
De Paoli, M.-A., Waltman, R.J., Díaz, A.F., Bargon, J., 1985. *J. Chem. Soc. Chem. Commun.* 1343.
De Rossi, D., Mazzoldi, A., 1998. In: *Smart Structures and Materials,* Y. Bar-Cohen (Ed.), SPIE Vol. 3669, pp. 35.
Díaz, A.F., Kanazawa, K.K., Gardini, G.P., 1979. *J. Chem. Soc. Chem. Commun.* 635.
Diaz, A.F., Logan, J.A., 1980. *J. Electroanal. Chem.* 111, 111.
Díaz, A.F., 1981. *Chemica Scripta* 17, 145.
Díaz, A.F., Castillo, J.I., Logan, J.A., Lee, W.Y., 1982. *J. Electroanal. Chem.* 129, 115.
Díaz, A.F., 1986. In: *Handbook of Conducting Polymers*, T.A. Skotheim (Ed.), Chapter 3.
Dotson, R.L., Woodward, K.E., 1982. *ACS Symp. Ser., Am. Chem. Soc., Washington DC, USA,* Vol. 180, 311.
Duek, E.A.R., De Paoli, M.-A., Mastragostrino, M., 1992. *Adv. Mat.* 4, 287.
Duek, E.A.R., De Paoli, M.-A., Mastragostrino, M., 1993. *Adv. Mat.* 5, 650.
Dusemund, T., Schwitzgebel, G., 1991. *Ber. Bunsenger. Phys. Chem.* 95, 1543.
Ehrenbeck, C., Jüttner, K., 1996. *Electrochim. Acta* 41, 1815.
Elliott, C.M., Kopelove, A., Albery, W.J., Chen, Z., 1991. *J. Phys. Chem.* 95, 1743.
Gandhi, M.R., Murray, P., Spinks, G.M., Wallace, G.G., 1995. *Synth. Metals* 73, 247.
Gardner, J.W., Pearce, T.C., Friel, P.N., Bartlett, P.N., Blair, N., 1995. *Sens. Actuat. B, Chem.,* 18-19, 240.
Garnier, F., Tourillon, G. Gazard, M., Dubois, J.C., 1983. *J. Electroanal. Chem.* 148, 299.
Genies, E.M., Pernault, J.M., 1984. *Synth. Metals* 10, 117.
Genies, E.M., Hanry, P., Santier, C., 1988. *J. Appl. Electrochem.* 18, 751.
Genies, E.M., Hary, P., Sawtier, C., 1989. *Synth. Met.* 28, C647.

Gibson, T.D., Prosser, O., Hulbert, J.N., Marshall, R.W., Corcoran, P., Lowery, P., Ruck-Keene, E.A., Heron, S., 1997. *Sens. Actuat. B: Chem.,* 44, 413.
Girotto, E.M., De Paoli, M-A., 1998. *Adv. Mater.* 10, 790
Gottesfield, S., Redondo, A., Feldberg, S.W., 1986. *Electrochem. Soc. Extended Abstract No. 507.*
Greene, R.L., Street, G.B., Suter, L.J., 1975. *Phys. Rev. Lett.* 34, 577.
Gurunathan, K., Vadivel Murugan, A., Marimuthu, R., Mulik, U.P., Amalnerkar, D.P., 1999. *Mater. Chemistry and Physics* 61, 173.
Gustafsson, J.C., Inganäs, O., Andersson, A.M., 1994. *Synth. Met.* 62, 17.
Hanawa, T., Yoneyama, H., 1989. *Bull. Chem. Soc. Faraday Trans.* I 84, 1710.
Heywang, G., Jonas, F., 1991. *Adv. Mater.,* 4, 116.
Hillman, A.R., Loveday, D.C., Bruckenstein, 1989. *J. Electroanal. Chem.* 274, 157.
Hillman, A.R., Loveday, D.C., Bruckenstein, S., Wilde, C.P., 1990. *J. Chem. Soc. Faraday Trans.* 86 (2), 437.
Hupe, J., Wolf, G-D., Jonas, F., 1995. *Galvanotechnik* 86, 3404.
Hwang, L.S., Ko, J.M., Rhee, H.W., Kim, C.Y., 1993. *Synth. Met.* 55-57, 3665.
Hyodo, K., 1994. *Electrochim. Acta* 39, 265.
Idla, K., Inganäs, O., Strandberg, M., 2000. *Electrochim. Acta* 45, 2121.
Iseki, M., Sato, K., Kuhara, K., Mizukami, A., 1991. *Synth. Metals* 40, 117.
Inzelt, G., Pineri, M., Schultze, J.W., Vorotyntsev, M.A., 2000. *Electrochim. Acta* 45, 2403.
Ivory, D.M., Miller, G.G., Sowa, J.M., Shacklette, L.W., Chance, R.R., Baughman, R.H., 1979. *J. Chem. Phys.* 71, 1506.
Janda, P., Weber, J., 1991. *J. Electroanal. Chem.* 300, 119.
Jager, E.W.H., Smela E., Inganäs, O., 1999a. *Sensors and Actuators B: Chemical,* 56 (1-2), 73.
Jager, E.W.H., Smela E., Inganäs, O., Lundström, I., 1999b: "Polypyrrole microactuators", *Synt. Met.* 102, 1309.
Jelle, B.P., Hagen, G., Hesjevik, S.M., Öedegärd, R., 1992. *Mater. Sci. Eng.* B13, 239.
Jelle, B.P., Hagen, G., 1993a. *J. Electrochem. Soc.* 140, 3560.
Jelle, B.P., Hagen, G., Nodland, S., 1993b. *Electrochim. Acta* 38, 1497.
Jelle, B.P., Hagen, G., Nodland, S., 1993c. *J. Electrochem. Soc.* 140, 3560.
Jirage, K.B., Hulteen, J.C., Martin, C.R., 1997. *Science* 278, 655.
Jirage, K.B., Martin, C.R., 1999. *Trends in Biotechnology Vol.* 17, 198.
Jonas, F., Heywang, G., 1994. *Electrochim. Acta* 39, 1345.
Jonas, F., Lerch, K., 1997. *Kunstoffe* 87, 1401.
Kanazawa, K.K., Díaz, A.F., Geiss, R.H., Gill, W.D., Kwak, J.F., Logan, J.A., Rabolt, J.F., 1979. *J.Chem. Soc., Chem. Commun.* 854.
Kanazawa, K.K., Díaz, A.F., Gill, W.D., Grant, P.M., Street, G.B., Gardini, G.P., Kwak, J.K., 1979/80. *Synth. Met.* 1, 329.
Kaneko, M., Wöhrle, D., 1981. In: *Advances in Polymer Sciences,* H-J. Cantow Springer, Berlin Hudelber (Ed.), New York, pp 141.
Kaneko, M., Masanori, S., Wataru, T., Keiichi, K., 1997. *Synth. Met.* 84, 795.

Kaneto, K., Kaneko, M., Min, Y., MacDiarmid, A.G., 1995a. *Synth. Met.* 71, 2211.
Kaneto, K., Kaneko, M., Takashima, W., 1995b. *J. Appl. Phys.,* 34, 1837.
Kogan, Y.J., Gedrovich, G.V., Rudakova, M.I., Fokeeva, L.S., 1995. *Russ. J. Electrochem.* 31, 689.
Korri-Youssoufi, H., Garnier, F., Srivasta, P., Godillor, P., Yassar, A., 1997. *J. Am. Chem. Soc.* 119, 7388.
Kranz, C., Wohlschläger, H., Schmidt, H.-L., Schuhmann, W., 1998. *Electroanalysis* 10, 546.
Lakashmi, B.B., Martin, C.R., 1997. *Nature* 388, 758.
Lauchlan, L.J., Nyyssonen, D., Sullivan, N., 1997. In: *Handbook of Microlitography, Micromachining and Microfabrication, Vol. 1,* P. Rai-Choudhury (Ed.), SPIE Press, Bellingham WA.
Lewis, T.W., Spinks, G.M., Wallace, G.G., De Rossi, D., Pachetti, M., 1997. *Polym. Prep. (Am. Chem. Soc., Div. Polym. Chem.)* 38, 520.
Lewis, T.W., Kane-Maguire, L.A.P., Hutchinson, A.S., Spinks, G.M., Wallace, G.G., 1999. *Synth. Met.* 102, 1317.
Liu, Y., Zhao, M., Bergbreiter, D.E., Crooks, R.M., 1997. *J. Am. Chem. Soc.* 119, 8720.
MacDiarmid, A.G., Heeger, A.J., 1980. *Synth. Met.* 1, 101.
Madden, J,D., Cush, R.A., Kanigan, T.S., Brenan, C.J., Hunter, I.W., 1999. *Synth. Met.* 105, 61.
Madden, J,D., Cush, R.A., Kanigan, T.S., Hunter, I.W., 2000. *Synth. Met.* 113, 185.
Marque, P., Roncali, J., 1990. *J. Phys. Chem.* 94, 8614.
Martin, C.R., 1994. *Science* 266, 1961.
Mastragostino, M., 1993a. In: *Applications of Electroactive Polymers,* B. Scrosati (Ed.), Chapman & Hall, London, Chapter 8.
Matragostrino, M., 1993b. *In B. Scrosati (Ed.). Applications of Electroactive Polymers, Chapman & Hall, London,* Chap. 7, pp. 244.
Mazzoldi, A., Degl'Innocenti, C., Michelucci, M., De Rossi, D., 1998. *Materials Science and Engineering C,* 6, 65.
Meyer, H., Nichols, R.J., Schroër, D., Stamp, L., 1994. *Electrochim. Acta* 39, 1325.
Mohammadi, A., Inganäs, O., Lundström, I., 1985. *J. Electrochem. Soc.* 135 (5), 947.
Mohilner, D.M., Adams, R.N., Argensinger W.J. Jr., 1962. *J. Am. Chem. Soc.* 84, 3618.
Morita., M., 1994. *J. Appl. Poly. Sci.* 52, 711.
Mortimer, R.J., 1995. *J. Mater. Chem.* 5, 969.
Munstedt, H., Kohler, G., Mohwald, H., Naegde, D., Bitthin, R., Fly, G., Meissner, E., 1987. *Synth. Met.* 18, 259.
Naegele, D., Bithin, R., 1988. *Sol. State Ion.* 28:30, 983.
Naoi, K., Lien, M.M., Smyrl, W.H., 1989. *J. Electroanal. Chem.* 272, 273.
Naoi, K., Lien, M.M., Smyrl, W.H., 1991a. *J. Electrochem. Soc.* 138 (2), 440.

Naoi, K., Lien, M.M., Smyrl, W.H., 1991b. *J. Electrochem. Soc.* 35, 1971.
Naugi, R., Frank, A.J., Nocic, A.J., 1981. *J. Am. Chem. Soc.* 103, 1849.
Nishizawa, M., Menon, V.P., Martin, C.R., 1995. *Science* 268, 700.
Novak, P., Müller, K., Santhanam, K.S.V., Haas, O., 1997. *Chem. Rev.* 97, 202.
Ogata, N., 1997. *Macromol. Symp.* 118, 693.
Okabayashi, K., Goto, F., Abe, K., Yoshida, T., 1987. *Synth. Met.* 18, 365.
Olazábal, V., 1998: Synthesis, characterization and electrochemical degradation of PPy/PVS composite, Ph.D. Thesis, University of The Basque Country.
Otero, T.F., Angulo, E., Rodríguez, J., Santamaría, C., 1992a. *J. Electroanal. Chem.* 341, 369.
Otero, T.F., Rodríguez, J., Angulo, E., Santamaría, C., 1992b. E.P.-9200095, EP-9202628.
Otero, T.F., Bengoechea, M., Olazábal V., 1995a. *Portugaliae Electrochimica Acta* 13, 493.
Otero, T.F., Grande, H., Rodríguez, J., Sansiñena, J.M., 1995b. *Portug. Electrochim. Acta* 13, 409.
Otero, T.F., Olazábal, V., 1995c. *Proceedings book of III Iberian Meeting of Electrochemistry, Algarve, (Portugal).*
Otero, T.F., Olazábal V., Bengoechea, M., 1995d. *Portugaliae Electrochimica Acta* 13, 403.
Otero, T.F., Sansiñena, J.M., 1995e. *Bioelectrochem. and Bioenerg. 38, 411.*
Otero, T.F., Sansiñena, J.M., 1995f. *Proceedings of the III Iberian Meeting of Electrochemistry F*-7, 231.
Otero, T.F., Sansiñena, J.M., Grande, H., Rodríguez, J., 1995g. *Portug. Electrochim. Acta* 13, 499.
Otero. T.F., Cantero, I., Grande, H., Bengoechea, M., Olazábal, V., 1996a. *Proceedings book of V National Meeting of Materials, Cádiz (Spain),* pp. 255.
Otero, T.F., Grande, H., Cantero, I., Olazábal, V., 1996b. *Proceedings book of 8th Congress of Instrumental Analysis (EXPOQUIMIA 96), Barcelona (Spain),* pp. 94.
Otero, T.F., Olazábal, V., 1996c. "Electrogeneration of polypyrrole in presence of polyvinylsulphonate. Kinetic Study," *Electrochimica Acta* 41 (2), 213.
Otero, T.F., Sansiñena, J.M., Cantero, I., Grande, H., Olazábal, V., Villanueva, S., Bengoechea, M., 1996d. *Proceedings book of XII Ibero-American Meeting of Electrochemistry, Mérida (Venezuela),* pp. 220.
Otero, T.F., Sansiñena, J.M., 1996e. In: *Proceedings of the Third International Conference of Intelligent Materials,* P.F. Gobin, J. Tatibouet (Eds.), pp 365.
Otero, T.F., Sansiñena, J.M., 1997. *Bioelectrochem. and Bioener.,* 42, 117.
Otero, T.F., Sansiñena, J.M., 1998. *Adv. Mater.* 10 (6), 491.
Otero, T.F., Cantero, I., Sansiñena, J.M., De Paoli, M.-A., 1999. In: *Smart Structures and Materials,* Y. Bar-Cohen (Ed.), SPIE Vol. 3669, pp.98.
Panero, S., Scrosati, B., Baret, M., Cecchini, B., Masetti, E., 1995a. *Sol. Energy Mater. Sol. Cells* 39, 239.
Panero, S., Spila, E., Scrosati, B., 1995b. *J. Electroanal. Chem.* 396, 385.

Pei, Q., Qian, R., 1990. *Proc. CMRS International* 3, 195.
Pei, Q., Inganäs, O., 1992. *Adv. Materials* 4, 277.
Pei, Q., Inganäs, O., 1993a. *J. Phys. Chem.* 96, 10507.
Pei, Q., Inganäs, O., Lundström, I., 1993b. *Smart Mater. Struct.* 2, 1.
Pei, Q., Inganäs, O., 1993c. *J. Phys. Chem.* 97, 6034.
Pei, Q., Inganäs, O., 1993d. *Synth. Met.* 55-57, 3718.
Pei, Q., Inganäs, O., 1993e. *Synth. Met.* 55-57, 3724.
Pei, Q., Inganäs, O., 1993f. *Synth. Met.* 55-57, 3730.
Pei, Q., Inganäs, O., 1993g. *Solid State Ionics* 60, 161.
Peres, R.C.D., De Paoli, M.-A., Torresi, R.M., 1992. *Synth. Met.* 48, 259.
Pochan, J.M., Gibson, H.W., Bailey, F.C., 1980. *J. Polym. Sci., Polym. Lett.* 18, 447.
Potember, R.S., Holfman, R.C., Hu, H.S., Cocchiaro, J.E., Viands, C.A., Murphy, R.A., Poehler, T.O., 1987. *Polymer* 28, 574.
Reynolds, J.R., Pyo, M., Qiu, Y.-J., 1993. *Synth. Met.* 55, 1388.
Roth, S., 1995. *One Dimentional Met. VCH Press, Weinheim.*
Sansiñena, J.M., Olazábal, V., Otero, T.F., Polo da Fonseca, C.N., De Paoli, M.-A., 1997. *J. Chem. Soc. Chem. Commun.* 2217.
Sansiñena, J.M., 1998. Artificial Muscles: Electrochemomechanical devices based in conducting polymers, *Ph.D. Thesis, University of The Basque Country.*
Sansiñena, J.M., Olazábal, V., Bar-Cohen, Y., 2000. *Worlwide Electroactive Polymer Newsletter,* Vol.2, N° 1, 10.
Santhanam K.S.V., Gupta, N., 1993. *TRIP* 1, 284.
Sata, T., 1994. *J. Membr. Sci.* 93, 117.
Schattka, D., Winkels, S., Schultze, J.W., 1997. *Metalloberfläche* 51, 823.
Schiested, S., Ensinger, W., Wolf, G.K., 1994. *Nucl. Instr. Meth.* B91, 473.
Schmidt, V.M., Tegtmeyer, D., Heitbaum, J., 1995. *J. Electroanal.Chem.* 385, 149.
Schopf, G., Kossmehl, G., 1997. *Adv. Polymer Sci.* 129, 124.
Schow, A.J., Peterson, R.T., Lamb, J.D., 1996. *J. Membr. Sci.* 111, 291.
Schuhmann, W., Lammert, R., Uhe, B., Schmidt, H.-L., 1990. *Sensors and Actuators* B1, 537.
Schuhmann, W., 1995. *Mikrochim. Acta* 121, 1.
Shimidzu, T., Ohtani, A., Iyoda, T., Honda, K., 1986. *J. Chem. Soc. Chem. Commun.* 198, 1415.
Shimidzu, T., Ohtani, A., Honda, K., 1987. *J. Electroanal. Chem.* 224, 123.
Shirakawa, H., Louis, E.J., MacDiarmid, A.G., Chiang, C.-K., Heeger, A.J., 1977. *J. Chem. Soc. Chem. Commun.* 578.
Slama, M., Tanguy, J., 1989. *Synth. Met.,* 28, C171.
Smela, E., Inganäs, O., Lundström, I., 1993a. *J. Micromech. Microeng.* 3, 203.
Smela, E., Inganäs, O., Pei, Q., Lundström, I., 1993b. *Adv. Mater.* 5, 630.
Smela, E., Inganäs, O., Lundström, L., 1995. *Science* 268, 1735.
Smela, E., 1999a. *J. of Micromech. Microeng.* 9, 1.
Smela, E., Gadegaard, N., 1999b. *Adv. Mater.* Vol. 11, Issue 11, 953.

Smela, E., Kallenbach, M., Holdenried, J., 1999c. *J. of Microelectromech. Syst. 8,* Issue 4, 373.
Street, G.B., Greene, R.L., 1977a. *IBM J. Res. Develop.* 21, 99.
Street, G.B., Gill, W.D., Geiss, R.H., Greene, R.L., Mayerle, J.J., 1977b. *J. Chem. Soc. Chem. Commun.* 407.
Taka, T., 1991. *Synth. Met.* 41, 1177.
Takashima, W., Fukui, M., Kaneko, M., Kaneto, K., 1995a. *J. Appl. Phys.* 34, 3786.
Takashima, W., Kaneko, M., Kaneto, K., MacDiarmid, A.G., 1995b. *Synth. Met.* 71, 2265.
Talanova, G.G., Hwang, H., Talanov, V.S., Barrsch, R.A., 1998. *J. Chem. Soc. Chem. Comm.* 3, 419.
Trivedi, D.C., 1997. In: *Handbook of Organic Conductive Molecules and Polymers,* H.S. Nalwa (Ed.), Wiley, New York , Chapter 12, pp. 506.
Trojanowicz, M., Krawczynski vel Krawcyk, T., 1995a. *Mikrochim. Acta* 12, 167.
Trojanowicz, M., Geschke, O., Krawczynski vel Krawcyk, T., Cammann, K., 1995b. *Sensors and Actuators* B28, 191.
Umana, M., Waller, J., 1996. *Anal. Chem.* 58, 2979.
Walatka, V.V. Jr., Labes, M.M., Perlstein, J.H., 1973. *Phys. Rev. Lett.* 31, 1139.
Wallace, G.G., Smyth, M., Zhao, H., 1999. *Trends in Analytical Chemistry,* Vol. 18 No. 4, 245.
Waters, W.A., White, J.E., 1968. *J. Chem. Soc. C.* 740.
Wessling, B., 1994. *Adv. Mater.* 6, 226.
Wolmershäuser, A., Brulet, C.R., Street, G.B., 1978. *Inorg. Chem.* 17, 3586.
Wong, T.K.S., Ingram, S.G., 1992. *J. Vac. Sci. Technol. B10, 2393.*
Wong, T.K.S., Gao, S., Hu, X., Liu, H., Chan, Y.C., Lam, Y.L., 1998. *Mater. Science and Engineering* B55, 71.
Xue, H., Mu, S., 1995. *J. Electroanal. Chem.* 397, 241.
Zhao, H., Price, W.E., Wallace, G.G., 1992. *J. Electroanal. Chem.* 334, 111.
Zhou, Q.-X., Miller, L.L., Valentine, J.R., 1987. *J. Electroanal. Chem.* 223, 283.

CHAPTER 8

Carbon Nanotube Actuators: Synthesis, Properties, and Performance

Geoffrey M. Spinks, Gordon G. Wallace
University of Wollongong (Australia)

Ray H. Baughman
University of Texas at Dallas

Liming Dai
The University of Akron

8.1 Introduction

Carbon nanotubes are the most recent addition to the growing list of electroactive actuator materials. These materials have exciting electrical and mechanical properties derived from their structure that consists of hollow cylinders of covalently bonded carbon just one atomic layer thick. Recent reports have described several different actuation mechanisms for carbon nanotubes, including actuation by double-layer charge injection [Baughman, Cui et al., 1999], electrostatic actuation [Kim and Lieber, 1999], and photo-thermal actuation [Zhang and Iijima, 1999]. Double-layer charge injection is a particularly promising mechanism for nanotube actuation, wherein supercapacitor charging results in relatively large changes in covalent bond lengths. The achievable actuator strain, coupled with the excellent mechanical properties of carbon nanotubes, mean that these materials potentially offer actuator performance exceeding that of all other actuator materials in terms of work-per-cycle and stress generation.

The accessibility of nanoscale structural features in nanotubes also makes these materials very exciting "building blocks" for nanotechnology. The extremely high stiffness, strength, and damage tolerance of carbon nanotubes makes them ideal structural members for nano-devices. The actuation and sensing capabilities of nanotubes [Kong, Franklin et al., 2000] adds multifunctionality, so that the entire structure may become like a living cell: responding in a pre-determined way to changes in the surrounding environment. Breakthroughs in nanotube processing and nano-assembly [Yu et al., 1999] will provide the platform for integration of nanotubes into existing micro-electro-mechanical systems (MEMS) devices and, in the future, into entirely new nano-electro-mechanical systems (NEMS) applications.

Realization of the potential of nanotube actuators requires a thorough understanding of their synthesis, assembly, and properties aspects. The synthesis of and theoretical and experimental characterization of nanotubes are reviewed, with special emphasis on nanotube mechanical properties. Emerging techniques for the controlled assembly of nanotubes into macroscopic structures, such as fibers, are described. The theory of actuation in carbon nanotubes arising from double-layer charge injection is also discussed. Finally, the experimental results obtained for carbon nanotube actuators are summarized, along with suggested future research directions.

8.2 Nanotube Synthesis

Since the early report of carbon nanotubes by Iijima [1991], various types of carbon nanotubes have been described. Single-wall carbon nanotubes (SWNTs) consist of single graphene layers that have been wound into seamless tubes having nanoscale diameters of 0.4 nm to > 3 nm. Multiwall carbon nanotubes (MWNTs), on the other hand, are heuristically obtained by coaxially wrapping additional graphene tubes around the core of a SWNT (diameters can be several tens of nanometer). Several different synthetic methods have been developed for effective preparation of both SWNTs and MWNTs [Journet, 1997], as discussed below. The major challenges for commercial production of NTs remain the purity, control of helicity and defects, and increasing scale of production so as to reduce costs. Currently, high-purity SWNTs are prohibitively expensive for commodity-type applications (of the order of \$500/g); however, less pure MWNTs can be supplied in kilogram quantities for \$25/kg. Anticipated increases in production capacity should drive the cost of all nanotubes down in the future.

Synthesis of SWNTs: The observation of SWNTs in arc-generated soot, in 1993, prompted several groups to prepare SWNTs by the carbon arc discharge technique using a pure carbon cathode and a carbon anode containing a mixture of transition metal(s) (e.g., Fe, Ni, Co, or Cu) [Bethune et al., 1993; Iijima and Ichihashi, 1993; Saito et al., 1993] and graphite powder. Covaporization of carbon and the metal catalyst in the arc generator produced weblike deposits of SWNTs in the fullerene-containing soot. A few years later, Smalley's group [Thess et al., 1996] developed a pulsed laser vaporization technique for efficient synthesis of internally ordered SWNT bundles from a carbon target containing 1–2% (w/w) Ni/Co. Using Ni/Y (4.2/1 atom%) as the catalyst, Journet and coworkers [1997] have described the improved synthesis of SWNTs using a refined carbon-arc method.

While the arc-discharge and laser vaporization techniques can provide high-quality SWNTs, they do not presently provide a low-cost route for industrial quantity production of pure SWNTs. As an alternative, chemical vapor deposition (CVD) of hydrocarbons over metal catalysts has been investigated as a viable path towards the large-scale synthesis of SWNTs at a low cost. In this regard, Cassell et al. [1999] reported the large-scale CVD synthesis of high-quality SWNTs by pyrolysis of methane over Fe/Mo bimetallic species supported on a novel silica-alumina multicomponent material, By optimizing the chemical compositions and textural properties of the catalyst material, these authors have successfully demonstrated large-scale production of SWNTs (ca. 42 wt% purity) up to the gram level. Starting with prepatterned catalysts on a substrate, the CVD method has also been successfully used to grow SWNT wires between controlled surface sites (see, for example: Cassell, 1999a; Franklin, 2000; Gu, 2001; Joselevich, 2002), indicating the possibility of fabricating nanotube devices by growing SWNTs on substrates to directly create a desired architecture.

On the other hand, the earlier discovery [Dai, 1996a] of SWNT formation via disproportion of CO on preformed catalytic particles (e.g. Co or Ni) has recently attracted increasing interest (see, for example: Alvarez, 2001; Resasco, 2002; Zheng, 2002). By introducing metal catalyst(s) in the form of volatile organometallic molecules [e.g., $Fe(CO)_5$] in continuously flowing CO at high pressure and elevated temperature, Smalley and co-workers [Nikolaev, 1999] have demonstrated a continuous process for the large-scale production of high-purity SWNTs; the so-called HiPco process [Bronikowski, 2001]. The HiPco process can produce up to a few hundred grams of high-quality SWNTs per day [Colbert, 2002] in a modest production facility, and the only major impurity is residual catalyst.

Synthesis of MWNTs: The large-scale synthesis of MWNTs was first reported by Ebbesen and Ajayan [1992]. These authors produced carbon nanotube bundles, along with other disordered carbonaceous material by operating a carbon arc-discharge generator in an inert gas (e.g., He) at a discharge temperature above 3000°C. Later, MWNTs were also produced by decomposition of hydrocarbons at approximately 1100°C in the presence of metal catalyst(s) including Fe, Co, and Ni [Endo et al., 1993; Ivanov et al., 1995; Qin, 1997]. Under the same conditions, conventional vapor-grown carbon fibers are simultaneously obtained [Endo et al., 1993]. The pyrolysis technique has some distinct advantages over the arc-discharge method, especially in terms of the simplicity of the process and high yield of carbon nanotubes.

Synthesis of Aligned MWNTs: Carbon nanotubes are produced in a highly entangled state by most of the above-mentioned techniques [Ajayan, 1995]. The preparation of aligned carbon nanotubes is also of interest since these can be effectively incorporated into working devices. Aligned carbon nanotubes have been prepared using a template synthesis technique [Hoyer, 1996; Li et al., 1996; Che et al., 1998], or using approaches that allow post-synthesis alignment [Ajayan et al., 1994; Ajayan, 1995; Deheer et al., 1995a; Deheer et al., 1995]. Pyrolysis of 2-amino-4, 6-dichloro-*s*-triazine on a silica substrate prepatterned with a cobalt catalyst [Terrones et al., 1997; Terrones et al., 1998] has been used to produce carbon nanotubes aligned in the plane parallel to the substrate. Aligned carbon nanotube arrays have been prepared by pyrolysis of ferrocene at approximately 900°C without the use of template pores or laser-etched tracks [Rao et al., 1998]. Others have synthesized large arrays of well-aligned carbon nanotubes by plasma-enhanced hot filament chemical-vapor deposition of acetylene below 666°C [Ren et al., 1998].

The formation of aligned carbon nanotubes from organic-metal complexes, containing both the metal catalyst and carbon source required for the nanotube growth, is of particular interest. Using this approach, one of the authors (LMD) has prepared "forests" of carbon nanotubes aligned perpendicular to the substrate surface. This was accomplished by the pyrolysis of iron (II) phthalocyanine, $FeC_{32}N_8H_{16}$, (designated as FePc hereafter) in a dual temperature furnace assembly (Fig. 1) [Huang et al., 1999; Yang et al., 1999; Dai and Mau, 2000; Huang et al., 2000; Li et al., 2000; Qidao and Dai, 2000]. Briefly, a predetermined amount of FePc and a quartz glass plate were placed in the first and second furnace, respectively. The initial pyrolysis process was then carried out while maintaining the second furnace at 1100°C and the first furnace at 650–750°C, during which Ar/H_2 (1:1 to 1:2 v/v) was flowing through the system at 20–40 cm^3/min. After 10 min at these temperatures, both furnaces were kept at the pyrolysis temperature (1100°C) for an additional 10 min, allowing the deposition of nanotubes to be completed. As seen in Fig. 2, the as-synthesized carbon nanotubes have a fairly uniform tube length and diameter, and are aligned almost precisely normal to the substrate. High-resolution transmission electron microscopy (HR-TEM) shows that most of the nanotubes are well graphitized with approximately 40 layers of graphite sheets and an outer diameter of approximately 50 nm.

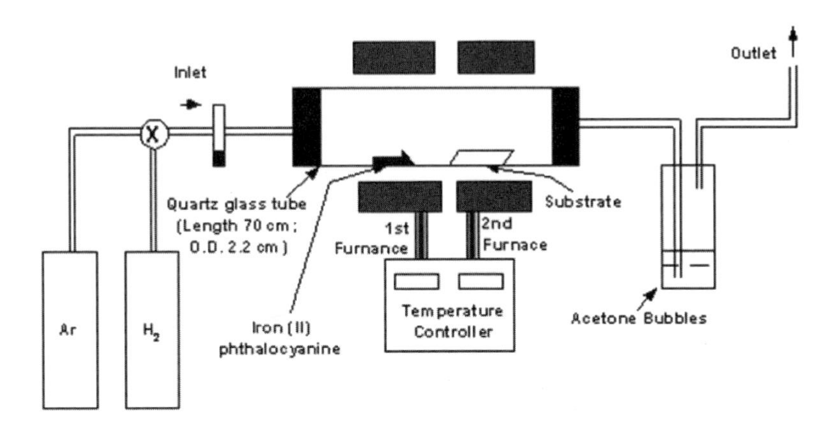

Figure 1: Synthesis facility for preparation of aligned MWNT arrays.

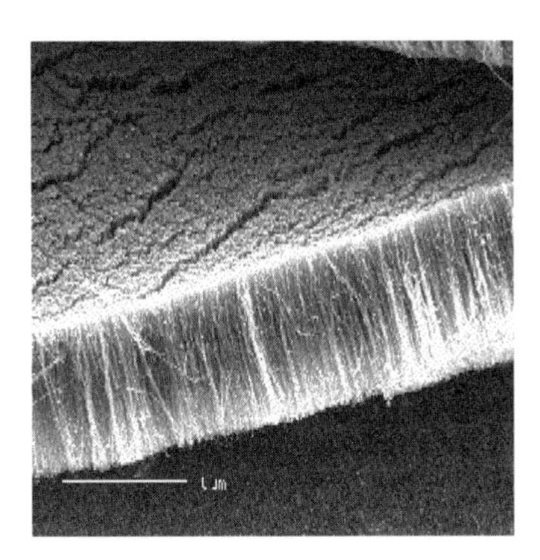

Figure 2: Scanning electron micrograph of MWNTs prepared from the pyrolysis of iron (II) phthalocyanine.

Other methods for preparing aligned nanotubes have also recently been reported. Most prominently, the use of electric fields has been shown to be a viable means for aligning nanotubes while in dispersed form [Senthil, 2003]. This method then allows for the controlled deposition so as to make electrical contact at selective locations. Electric fields are also implicated in the CVD deposition process described above: it has been reported recently that the electric field direction around catalyst particles determines the direction of growth and, therefore, the alignment of the nanotubes [Jang, 2003].

In order to transfer the aligned nanotubes to other surfaces intact, a solution transfer method was developed that involves simply immersing the nanotube-deposited quartz plate into an aqueous hydrofluoric acid solution [Huang et al., 1999; Yang et al., 1999; Dai and Mau, 2000; Huang et al., 2000; Li et al., 2000;

Qidao and Dai, 2000]. This process leads to a free-standing nanotube film floating on the HF/H_2O solution. The floating nanotube film is transferred to a new substrate by sliding this substrate under the floating film, lifting this substrate and the attached nanotube film from the bath, and then drying the film on the new substrate. A dry lift-off technique has also been developed, which involves using Scotch tape to peel the aligned nanotube film from the quartz substrate [Huang et al., 1999; Yang et al., 1999; Dai and Mau, 2000; Huang et al., 2000; Li et al., 2000; Qidao and Dai, 2000]. The availability of these MWNTs having controlled diameters and lengths, together with the methods developed for transferring the aligned nanotube films onto various substrates (e.g. polymer films), provides new opportunities for evaluating carbon-nanotube properties and applications, ranging from electron-field emitters to nanotube actuators.

8.3 Characterization of Carbon Nanotubes

The performance of carbon nanotubes as actuators depends on the ability to efficiently inject and remove charges from these tubes. Consequently, the electrochemical properties of carbon nanotubes are critically important. Controlling these properties, however, are the electronic conductivity, accessible surface area, and wettability. These fundamental material properties, in turn, depend upon the basic nanotube structure and the nature of nanotube aggregation, such as the formation of nanotube bundles.

Tube diameter, tube length, helicity, and the number of concentric shells characterize the basic nanotube structure. Variations in nanotube helicity can be understood by imagining the way in which a graphene sheet may be rolled to form a nanotube cylinder (Fig. 3). If the edge of the graphene sheet is rolled in such a direction that carbon atom O is joined to carbon atom A, then the chiral vector can uniquely define the tube helicity. The chiral vector (n,m) is simply defined by

$$C_h = n\,\mathbf{a}_1 + m\,\mathbf{a}_2\,, \tag{1}$$

where $\mathbf{a}_1$ and $\mathbf{a}_2$ are unit vectors as defined in Fig. 3. There are an infinite number of different types of nanotubes; however all can be classified as one of the following:

- Armchair nanotubes (chiral vector = n,n): metallic
- Zig-zag nanotubes (chiral vector = $n,0$): semiconducting or metallic
- Chiral nanotubes (chiral vector = n,m): semiconducting or metallic.

The diameter of a nanotube (d_t) is given by

$$d_t = \frac{a}{\pi}\sqrt{n^2 + m^2 + nm}\,, \tag{2}$$

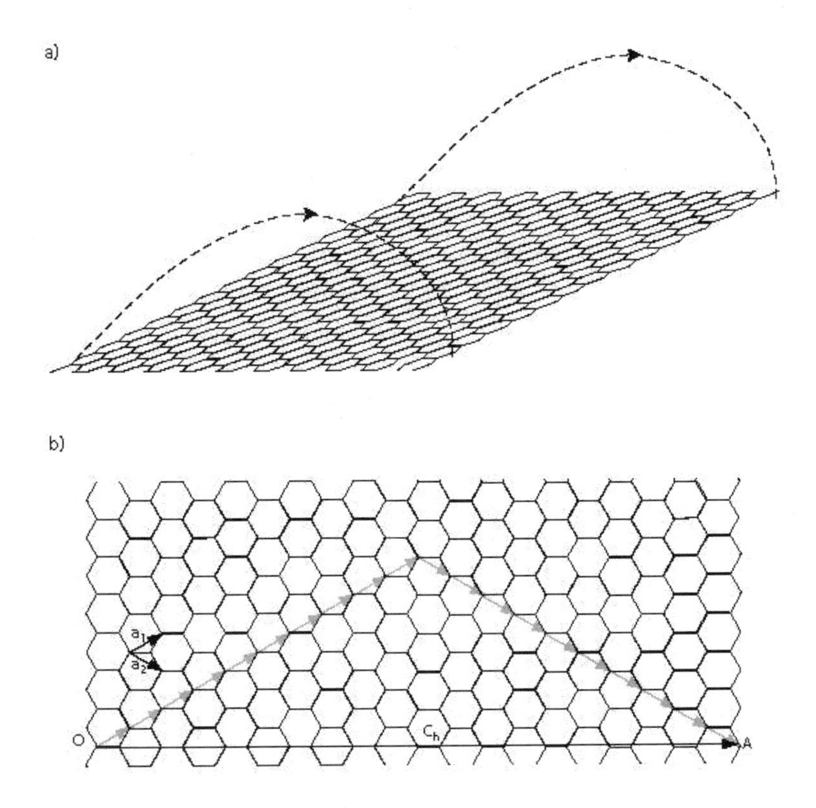

Figure 3: (a) Schematic illustration of the rolling of a single graphene sheet to form a (10,10) nanotube cylinder. (b) Definition of unit vectors and the chiral vector for a (10,10) nanotube.

where a is the lattice constant of 2.49 Å. Thus, a typical (10,10) SWNT has a diameter of 1.37 nm.

Various high-resolution microscopy techniques are used to characterize the structure and morphology of individual carbon nanotubes and carbon nanotube assemblies. As described above, scanning electron microscopy provides details on nanotube alignment and tube diameter. AFM techniques [Tsang et al., 1996] have been used to probe tube helicity and defects in tube structure. High-resolution TEM accurately quantifies tube diameter, tube type, and the number of shells for multiwall nanotubes (Fig. 3). In the case of single-wall nanotubes, HR TEM also enables measurement of the number of CNTs in a nanotube "bundle." The bundles form when the SWNTs spontaneously assemble into micron or longer stringlike structures. The cross section of the bundle may contain tens of nanotubes, and thousands of nanotubes can be included in the entire fiber bundle.

Diffraction methods have provided further information on nanotube structures. Use of nanoelectron diffraction techniques (with an electron beam diameter of 1 nm) has enabled the structure (chirality) of nanotubes to be probed in more detail [Liu and Cowley, 1994]. X-ray diffraction has been useful in studying nanotube bundle structures and has revealed two-dimensional periodicity in the location of carbon nanotubes within bundles, which is enhanced by annealing the nanotubes bundles at 1100°C in an inert atmosphere. This periodicity corresponds to the packing of nanotubes closely into a hexagonal array, like a dense arrangement of straws.

Various electrochemical studies have been carried out on carbon nanotubes. Single-carbon nanotubes have been used as electrodes and shown to exhibit scan-rate behavior typical of ultra-small electrodes [Campbell et al., 1999]. Another interesting aspect of this work was the in-situ electrocoating of the nanotube with the insulator polyphenol.

Liu et al. [1999] have considered the electroactivity of a cast film of single-wall carbon nanotubes and compared it to that observed for fullerenes. Whereas the latter undergoes well-defined oxidation/reduction processes, the charging of the carbon nanotubes appears completely capacitive in nature. Another possibility posed by the authors is that the obtained cyclic voltametry may be an average of many closely spaced peaks representing electron transfer into each nanotube. As pointed out above, the electronic properties of nanotubes are strongly dependent on the diameter, degree of charging, and helicity of the tubes and any suspension or film cast from it will contain a wide variety of tubes with different electronic characteristics. These workers used mass measurements derived from casting the dispersion on a quartz crystal microbalance to determine a specific capacitance value of 283 F/g. For reasons that are unclear, this gravimetric capacitance is higher than reported by other researchers for similar samples of single-wall carbon nanotubes.

In recent investigations of the electrochemical properties of SWNT mats, the mats were made by a process similar to paper making [Rinzler, 1998]. After annealing, the nanotube sheets exhibit predominantly capacitive behavior, which is almost independent of the cation or anion used in the supporting electrolyte [Barisci et al., 2000]. Observed gravimetric capacitances were in the range 18–37 F/g. Investigation of weight changes during charge/discharge processes using a quartz crystal electrochemical microbalance are consistent with the domination of electrochemical processes by double-layer charging [Barisci et al., 2000]. However, in an organic solvent (acetonitrile), incorporation and expulsion of Li^+ ions occurs when the SWNT paper is exposed to reducing/oxidizing conditions. Others (Yang and Wu, 2001) have shown that the degree electrochemical intercalation of Li^+ depends on the use of open or closed tubes.

Li et al. (2002) have reported that the electron transfer rate observed for oxidation/reduction of ferro/ferri cyanide species strongly depends on the number of opened ends and other defects on tubes. This in turn is dependent on the pretreatment/purification steps used in preparing the CNT electrodes, which can also affect the electron transfer rate by modifying the nanotube surface.

Glassy carbon electrodes have been coated with CNT by casting a CNT dispersion in a volatile solvent (Luo et al 2001, Wang et al 2002). These electrodes exhibited electrocatalytic effects towards oxidation of the biomolecules dopamine, epinephrine and ascorbic acid and enabled direct electron transfer to/from cytochrome c. These electrochemical properties have been put to use in a number of areas including sensors (Gao et al 2003), electrocatalysis (Che et al 1998), supercapacitors and electromechanical actuators as discussed below (An et al. 2001 and Niu et al. 1997).

8.4 Macroscopic Assemblies of Nanotubes: Mats and Fibers

One of the most intriguing aspects of carbon nanotube actuators is their potential to operate both at the nanoscale (single nanotubes) and at the macroscale (assemblies of nanotubes). The short-term use of nanotubes for actuators will undoubtedly be at the macroscale and will move toward micro and nano devices as assembly techniques at these levels are further developed. The initial studies on macroscopic assemblies of nanotubes were conducted on porous mats, but more recently improved properties have been achieved by the wet spinning of nanotubes into fibers. The methods for assembling nanotube mats and fibers are briefly reviewed in this section.

Mats of SWNTs are formed by a filtering process that leads to a self-assembled sheet of entangled nanotube ropes [Rinzler, 1998]. The SWNTs are first suspended in a liquid medium (usually water) sometimes with the aid of a dispersing agent and strong sonication. The suspension is then filtered under pressure through a microporous membrane that traps the nanotubes on its surface. The use of fluorinated membranes (e.g., polyvinylidene fluoride) prevents strong adhesion between the membrane and the nanotubes, and allows the washed and dried mats to be peeled from the membrane after filtering. Mats of SWNTs made by the laser-evaporation or HiPco method have adequate handling strength when made to ~20 μm in thickness. The mats are highly porous (Fig. 4) resembling the structure of regular paper and are, hence, known as "bucky paper" [Rinzler, 1998]. The nanotube ropes tend to lie flat in the plane of the mat, but are randomly coiled within the plane. The density of the mats is typically 0.4 g/cm^3 so that the pore volume is about 80%.

Wet spun carbon nanotube fibers are also made from suspensions of nanotubes in water. In the method described in Poulin et al. (2002), and Vigolo et al. (2000 and 2002), the stability of the nanotubes in suspension is controlled by the amount of surfactant (sodium dodecyl sulfate, SDS) and coagulant (polyvinylalcohol, PVA) in the solution. The fiber spinning involves forming a stable suspension using SDS and then injecting this mixture into a coagulation bath of PVA solution. The PVA molecules are strongly attracted to the nanotubes and replace the SDS, causing solidification into a ribbon-shaped continuous fiber. At this point the fibers consist of a SWNT-PVA composite highly swollen with water. Washing these gel fibers in methanol removes much of the water, surfactant and PVA. These as-spun fibers show partial orientation of the nanotubes in the fiber direction and show improved mechanical properties compared with bucky paper, as described in Sec. 8.5.4. Subsequent

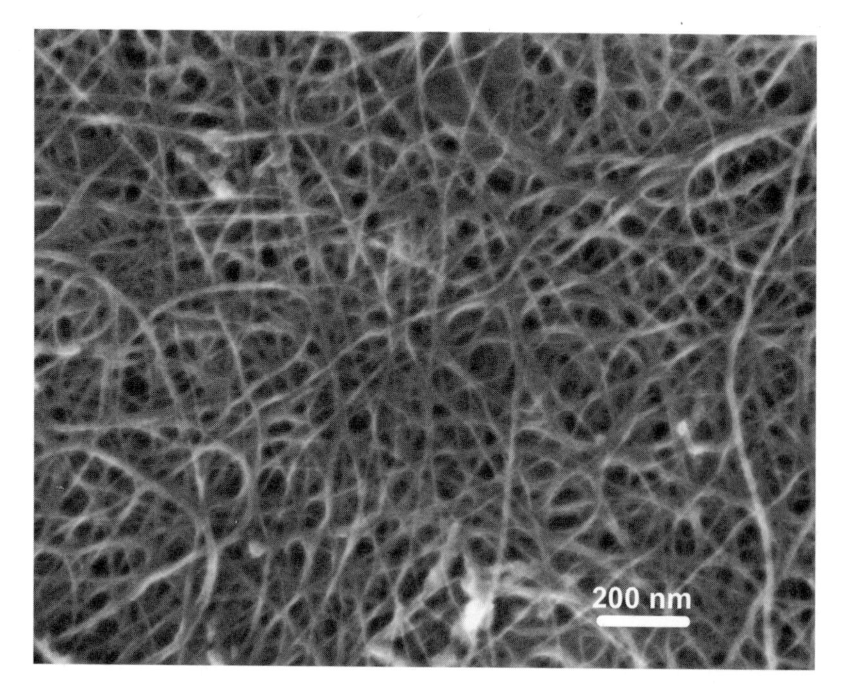

Figure 4: Scanning electron micrograph of filtered mats of SWNTs.

treatments can cause further alignment of the nanotubes by drawing [Poulin, 2002]. Thus, a variety of composite structures can be produced with varying mechanical properties. A modification on the SWNT-PVA wet spinning process has recently been reported [Dalton 2003] in which a continuous process results in effectively unlimited fiber lengths at greatly increased spinning rates, 70 cm/min. (see Sec. 8.5.4). These fibers have about an order of magnitude higher strength than fibers previously spun by the coagulation-based process, and a higher toughness (gravimetric strength-to-break) than previously reported for any material.

8.5 Mechanical Properties of Carbon Nanotubes

8.5.1 Theoretical Predictions

Numerous analytical calculations and numerical modeling studies have been conducted on the mechanical properties of carbon nanotubes in tension, compression, and shear. Much recent focus has been on large strain deformations, since SWNTs are able to reversibly deform to much higher strains without damage or failure compared with other high modulus materials.

Simple analytical estimates of the elastic modulus of carbon nanotubes are based on the known properties of graphite. The in-plane modulus of graphite is about 1 TPa. Salvetat et al. [1999] have used this value to calculate the Young's modulus of a SWNT bundle by determining the cross-sectional area of the bundle occupied by each SWNT. Their calculations give a bundle Young's modulus of 600 GPa. Similarly, these workers used the shear modulus of graphite (4.5 GPa) to estimate a shear modulus of 19.5 GPa for a SWNT bundle. Calculations of the radial compressive modulus of NTs give quite low values. Tersoff and Ruoff [1994], for example, calculated a modulus in radial compression of 9.3 GPa for a bundle of close-packed SWNTs.

SWNTs show remarkable resiliency for large strain deformations (>5%). Yakobson et al. [1996], for example, analyzed the compressive buckling of SWNTs using molecular dynamics simulations. The derived buckling mechanisms are similar to macroscopic beams, and the calculated buckling strains were in good agreement with those obtained from continuum mechanics. The reversible buckling of carbon nanotubes has a practical application in atomic force microscopy (AFM) tips. Here the resiliency means that the nanotube tip resists damage from "crashing" with the surface. The tip simply buckles and recovers, rather than suffering mechanical damage, as traditional silicon nitride AFM tips are apt to do.

8.5.2 Experimental Studies: Multiwall Carbon Nanotubes

Treacy et al. [1996] reported a very high elastic modulus for MWNTs. Their studies involved measuring the amplitude of vibration of MWNT cantilevers in the transmission electron microscope (TEM) as a function of temperature. The mean square amplitude of vibration (σ^2) is related to the thermal energy (kT) and the Young's modulus (Y) by

$$\sigma^2 = \frac{0.4243L^3kT}{Y(a^4 - b^4)}, \tag{3}$$

where L is the NT length and a and b are the outer and inner diameters. The average Young's modulus value found from these studies was 1.8 TPa. This Young's modulus is artificially high, since the authors normalized applied force with respect to the Van der Waals area of the carbon sheets in the nanotubes (rather than to the entire cross-sectional area of the nanotubes).

Pan et al. [1999] measured the tensile properties of MWNT bundles by conventional mechanical tests. The MWNTs were synthesized by the vapor deposition method as aligned arrays and then cleaved from the array to prepare a bundle of > 10 μm in diameter and ~2 mm in length. The measured elastic modulus of the MWNT bundles (10 samples) was 0.45 +/- 0.23 TPa, which is lower than for other measurements. The low modulus value was attributed to the higher number of defects that occur in CVD NTs compared to NTs made by the arc-discharge method. Also, it was recognized that weak binding between different nanotubes and even between layers in a MWNT could result in slippage and, therefore, lead to a decreased modulus.

Shen et al. [2000] studied the radial compression of MWNTs by pressing an AFM tip into a NT lying on a silicon substrate. The radial compressive moduli derived from these studies varied depending upon the maximum force applied, but the values obtained were quite low (around 10 to 80 GPa). The load dependence of the modulus was attributed to the flattening of the tubes at higher loads.

Dai et al. [1996] also comment that the MWNT AFM tips can survive repeated "crashing" into the sample surface. The observation implies that the NT tips can be reversibly buckled at high strain. Falvo et al. [1997] studied such processes by using the AFM tip to deform MWNTs and then to image the deformed NT, which was held in its deformed conformation by the frictional forces between the NT and the substrate. High strain bends induced buckling of the NT. Up to an undetermined strain limit the buckling was reversible, but irreversible NT damage occurred when the nanotubes were bent to very small radii. From their studies of radial compression using an AFM tip, Shen et al. [2000] also reported a high resilience for NTs. These workers could find no evidence of damage or breakage of the indented NTs, and were able to estimate the compressive strength of the NTs to be well above 5.3 GPa.

8.5.3 Experimental Studies: Single-Wall Carbon Nanotubes

There have been a number of recent experimental investigations of the mechanical properties of SWNTs. Chesnokov et al. [1999], for example, have investigated the volume changes induced by pressing SWNT mats (bucky paper). The volume was found to decrease substantially with increasing pressure, and to completely recover upon pressure release. These observations were attributed to flattening of the tubes.

Recently, AFM methods have been used to measure the mechanical properties of individual SWNT bundles. Walters et al. [1999] used elaborate silicon etching techniques to suspend a single SWNT bundle across a 4-μm wide trench, with the SWNT bundle firmly anchored on each side. Next an AFM tip was brought into contact with the suspended NT bundle from the side. The deformations were observed to be fully elastic without any hysteresis. The maximum strain before failure was found to be 5.8%, which is exceptionally large for a high modulus fiber.

Slavetat et al. [1999] have used similar AFM techniques to measure the Young's modulus and shear modulus of SWNT bundles. In their studies, a SWNT bundle was suspended over the pores in a polished membrane (Whatman Anodisc with 200-nm pores). In contrast to the method used by Walters et al. Slavetat and coworkers [1999] made contact to the NT bundle in the vertical direction. The deflection of the NT was analyzed using classical beam theory. The total deflection (δ) is the sum of the deflection due to bending (δ_b) and the deflection due to shear (δ_s):

$$\delta = \delta_b + \delta_s = \frac{FL^3}{192EI} + \frac{f_s FL}{4GA}, \quad (4)$$

where L is the suspended length; Y is the Young's elastic modulus; f_s is the shape factor (10/9 for a cylindrical beam); G is the shear modulus; I is the second moment of area of the beam ($I = \pi D^4/64$ for a filled cylinder); and A is the cross-sectional area. This equation indicates that the shear component is important only when $L/R < 4(Y/G)^{1/2}$, so shear is negligible for long NT bundles having a small radius (R).

The method used by Slavetat et al. [1999] is not capable of independently providing both Y and G. When shear deformations were neglected, the apparent Young's modulus was found to be highly dependent upon the diameter of the NT bundle: decreasing from ~1 TPa to ~0.1 TPa when the diameter increased from 3 nm to 20 nm. The decrease in modulus was attributed to shear processes, since Y depends on the stiffness of the individual nanotubes within the bundle and should, therefore, be independent of bundle diameter (ignoring the effects of tube ends and missing tubes within a bundle). The shear modulus values were then calculated by assuming a Young's modulus of 600 GPa (based on the elastic modulus of graphite and a calculation of the cross-sectional area of a bundle occupied by each tube). This value for Y was in reasonable agreement with the apparent modulus measured for narrow tubes, where shear effects are expected to be smallest.

The thereby derived values for G were low: decreasing from 6.5 GPa to 0.7 GPa as the bundle diameter increased from 4.5 nm to 20 nm. These values are lower than estimated for nanotube bundles from the shear modulus of graphite (19.5 GPa). The lower than expected shear modulus calculated from the AFM data was attributed to structural defects, such as variations in nanotube diameter and/or vacancies in the bundle.

Wood and Wagner [2000] estimated the compressive properties of SWNTs by studying the shift in Raman peak position for the disorder-induced band (2610 cm^{-1} in air) when the NTs were dispersed in various liquid media. The assumption was made that the surrounding media creates an internal pressure that compresses the nanotubes. The internal pressure was estimated from the cohesive energy density (CED) of the liquid and the shift in peak position was found to be sensitive to the CED. Reconstruction of the stress–strain curve showed a linear region of low compressive modulus (100 GPa) up to 1.5% strain, followed by a nonlinear region of increasing modulus up to 2.5% strain. Chesnokov et al. [1999] reported a similar strain-dependent modulus for SWNT bundles from their high-pressure experiments. As the NTs in a bundle are flattened and the minor axis of the tube approaches the interplanar spacing of graphite, further compression becomes extremely difficult due to the high repulsive forces now operating. The shape of the curve obtained by Wood and Wagner is also similar to the compressive stress-strain curves obtained from AFM compression experiments on MWNTs reported by Shen et al. [2000], who attributed the increase in tangential modulus at high strains to increased repulsive forces between the shells in the MWNT as they were further deformed.

8.5.4 Mechanical Properties of Nanotube Assemblies

The mechanical properties of macroscopic mats of SWNTs fall well short of those measured for the individual nanotubes. Modulus values for the porous mats are typically 0.3–5 GPa. The structure and properties resemble that of cellulose-based paper, which consists of cellulose fibers (roughly an order of magnitude larger than nanotube ropes) that are randomly coiled and highly entangled. The modulus of individual cellulose fibers is 130 GPa [Haslach, 2000], while that of paper is only 0.2–10 GPa [Nissan, 1987]. It is possible that the elastic modulus of paper is dominated by the strong hydrogen bonding that occurs at the fiber junctions. Thus, by analogy, the modulus of nanotube mats should also be dominated by the inter-tube bonding. In pure nanotubes, this bonding would be van der Waals interactions, which are quite weak and may account for the relatively low modulus of the nanotube sheets compared with the individual nanotubes.

Similarly, the strength of nanotube sheets is much lower than that predicted for individual nanotubes. The reasons may also be related to the mechanical strength of paper. In a classical study describing the properties of isotropic paper, Page [1969] related the tensile index (σ_T^w) to the fiber and network structure. The tensile index (with units Nm/kg) is the breaking force divided by the sheet width and grammage (mass per unit sheet area) and is related to the tensile strength (σ_T) by density (ρ):

$$\sigma_T^w = \frac{\sigma_T}{\rho}. \tag{5}$$

Page related the tensile index to the inherent fiber strength (the "zero span strength," σ_{ZS}^w) the shear bond strength between fibers (τ_s), fiber length (l) and the relative bonded area per gram of fiber (α):

$$\frac{1}{\sigma_T^w} = \frac{9}{8\sigma_{ZS}^w} + \frac{12}{\tau_s l \alpha}. \tag{6}$$

In the papermaking process, much effort is made to increase paper strength by increasing the bonding between cellulose fibers. The pulping process greatly affects interfiber bonding by altering the fiber surface chemistry, while additives are sometimes included to enhance fiber bonding. When the bond strength is low, the paper fails by fiber pull out and separation. When the bonding is increased (by altering surface interactions or by increasing paper density) the failure mode changes to fiber fracture.

By analogy, these insights may provide the means for increasing the strength of nanotube assemblies. The process is limited at the current time by the lack of quantitative data on the length of nanotube ropes and the various additives/impurities that are incorporated into the nanotube assemblies during their formation. For example, residual surfactant (used to form nanotube dispersions) may reduce the bonding at nanotube junctions and make the bonding susceptible to moisture content due to the hydrophilic nature of the surfactants.

Higher modulus and strength have been obtained for wet-spun nanotube fibers [Vigolo, 2000; Dalton, 2003]. In the work of Poulin et al. [2002], a Young's modulus of 40 GPa has been achieved due to the partial alignment of the nanotubes in the fiber direction. Alignment is achieved by stretching the fibers after immersion in swelling solvents. The strength is 230 MPa, which is nearly twice that of fibers that have not been treated by this solvent-enhanced stretching process. Dalton et al. [2003] have reported substantial improvements in fiber properties. This continuous fiber spinning process produces SWNT-PVA composite fibers that do not neck during drawing, so they can be drawn more than 500%. Incredibly, suitably predrawn fibers (with a strain-to-failure of about 100%) have a toughness exceeding that of any known material, including spider silk, Kevlar, steel, and carbon fiber. After a very high draw, these fibers show an increase in modulus to 80 GPa and a failure of strength of more than 1800 MPa, which is about an order of magnitude higher than obtained by the original coagulation spinning process. Normalized to density, the Young's modulus and tensile strength of these fibers is more than twice that of high-quality steel wire, and they are about 20 times tougher.

8.6 Mechanism of Nanotube Actuation

A review of artificial muscles by Baughman predicted the emergence of carbon nanotubes as electromechanical actuators [Baughman, 1996]. The review described the inherent limitations associated with doping-induced actuation in noncovalent directions for conducting polymers and proposed that actuation in covalently bonded directions would provide superior performance in terms of cycle life, work-density-per-cycle, and operating temperature range. Furthermore, Baughman proposed that non-Faradaic electrochemical charging of high-surface-area electrodes would produce useable dimensional changes for actuation in covalently bonded directions. Materials that combine both high surface area and a covalently bonded network became available with the advent of single-wall nanotubes (ca. 1997). Baughman's group successfully demonstrated carbon nanotube actuation based on double-layer (non-Faradaic) charge injection in 1998 and published the discovery the following year [Baughman et al., 1999].

The inherent limitations of conducting polymer actuators are mainly associated with the doping-induced phase changes that are the primary mechanism for electromechanical actuation in these materials. A series of first-order phase changes accompanies doping and dedoping processes in conducting polymers. The phase changes affect the lateral packing of the polymer chains, and are usually hysteretic and only partially reversible [Baughman, 1996]. Therefore, cycle lifetimes and switching speeds are limited for macroscopic conducting polymer actuators, as well as for large related electrochemical devices (like batteries). The resistance to degradation caused by electrochemical cycling can be dramatically improved by replacing dopant intercalation processes for low-surface-area conjugated materials with double-layer charge injection for high-surface-area conjugated materials in which the covalent structure determines the actuator stroke.

Substantial increases in work-density-per-cycle and high temperature range are further achievable benefits arising from use of the dimensional changes of a covalently bonded network. The volumetric work density is given by

$$W = \frac{1}{2} Y \varepsilon^2 , \tag{7}$$

where Y is the Young's elastic modulus and ε is the strain induced by a change in the level of charge transfer. Highly cross-linked networks (diamond is an extreme example) have a Young's modulus up to 1 TPa, while Y for unoriented linear polymers is typically < 10 GPa. Thus, the work densities achieved using strains in covalently bonded directions of cross-linked networks will be substantially higher than are achievable using unoriented linear polymers, as long as sufficiently large actuator strains result in the covalently bonded directions. An additional benefit is derived by using the strains of graphitic systems for actuation: carbon nanotubes are stable to temperatures in excess of 1400°C (in inert atmosphere), which is far above that of alternative actuator materials.

Further improvements in actuator performance are possible by utilizing a double-layer charge injection mechanism instead of electrochemical doping and dedoping of ions. Previous work has shown that donation of electrons to conjugated polymers, such as polyacetylene, poly(p-phenylene) or poly(p-phenylene vinylene), causes chain length expansion. Similarly, removal of electrons from the polymer chain causes contraction [Baughman et al., 1992]. The same effects have been reported for the in-plane dimension of graphite [Nixon and Perry, 1969]. In fact, the measured strain-charge coefficient for graphite is 0.096 (i.e., 0.96% expansion is produced by injecting 0.1 electrons per carbon atom).

The actuation strain produced by double-layer charge injection is characterized by the strain-voltage coefficient (S_V), which is determined by the product of three factors:

- Interfacial capacity (C_A; $\mu F/cm^2$): the amount of charge that can be stored at the solid/electrolyte interface. Because of the charge compensation that occurs when the ionic double layer forms at a charged surface, a much higher charge can be stored at the surface in an electrolyte than in air or in dielectric capacitors.

- Specific surface area (A_S; m^2/g): by maximizing the surface area, the amount of charge stored is also maximized.
- Strain-charge coefficient ($S_Q=\Delta L/L\Delta y$): strain ($\Delta L/L$) induced per change in the number of injected charges per carbon atom (Δy). Since the charge is stored close to the surface, bond length changes due to charge injection are confined to surface and near-surface atoms. These dimensional changes are constrained by the nondeforming internal atoms. By increasing the proportion of surface atoms to internal atoms, the actuation strain is increased.

Assuming the available data for graphite holds for carbon nanotubes, the actuator strain is predicted to be 0.19% per volt for carbon nanotubes. The calculated value is based on an interfacial capacity for the basal plane of graphite of 10 $\mu F/cm^2$; a specific surface area of 1600 m^2/g for unbundled single wall nanotubes of diameter 14 Å and a strain-charge coefficient for the in-plane direction in graphite of 0.096. Assuming a redox window of 4 V (for organic electrolytes) the maximum achievable strain is predicted to be of the order of 0.8%. This value is much higher than produced by high modulus piezoelectric ceramic actuators (~0.1%), which require high operating voltages.

The predicted actuator strain available from carbon nanotube actuators leads to a very high calculated stress generation capacity under isometric conditions. Since the theoretically derived and experimentally measured elastic modulus of individual nanotubes is about 0.6 TPa, a 0.8% strain would generate 4800 MPa from individual nanotubes. In comparison, the field-induced strain of hard ceramics (like PZT) is limited (by stress-induced grain orientation) to roughly 150 MPa.

Carbon nanotube actuators are also predicted to produce giant work densities per cycle, far exceeding all other available actuator materials (Table 1). The exceptionally high modulus of carbon nanotubes yields predicted work capacity per cycle ~30 times (volumetric) and ~35 times (gravimetric) higher than for the best ferroelectric, magnetic, or electrostrictive material. These huge work densities are a consequence of the giant energy storage capacity of carbon nanotubes: $C_A \times A_S$ = 160 F/g, which is similar to that of the best available supercapacitors. For comparison, ceramic ferroelectrics have an energy storage density similar to ordinary capacitors (~ 10^{-3} F/g for commercial ceramic ferroelectrics).

Since the above predictions are based on the inherent properties of single nanotubes, the predicted performance limits can only be approached for actuators based on single nanotubes or for nanotube assemblies that effectively utilize the mechanical properties of the individual nanotubes. The problem is, presently available nanotube assemblies have a Young's modulus (typically 1 GPa for nanotube sheets or 80 GPa for partially aligned nanotube fibers) and failure strengths (up to 1800 MPa for fibers) that are much lower than that of the inherent properties of the individual nanotubes. As Baughman described in 1996, the challenge for constructing actuators that operate by double-layer charge injection is to make very high-surface-area materials that maintain adequate mechanical strength. An additional requirement is high electrical conductivity, which permits minimization of the IR (current × resistance) potential drop across the sample and minimization of the RC (resistance × capacitance) time for charging and discharge. Fortunately, the recent development of carbon nanotubes provide the ideal material for these non-Faradaic actuators. The nanotubes are readily assembled into large sheets called "bucky paper" (by solution-based self-assembly), as fibers (by solution spinning) or as coatings [by chemical vapor deposition (CVD), or other methods]. Both films and coatings have sufficient mechanical strength for the construction of early prototype macroscopic actuators and partially aligned fibers have substantially improved mechanical properties. In addition, the high electrical conductivity (about 5000 S/cm for bucky paper) is important for efficiency and fast switching. The available data predicts that nanotube actuators operating through non-Faradaic charging should eventually exceed all other actuator materials in terms of stress generation and work-density-per-cycle capabilities. Very fast switching speeds and high cycle life are also expected from non-Faradaic charging. The performance of the first nanotube actuator materials are described below, after the synthesis, characterization, and mechanical properties of these materials are reviewed.

Table 1: Comparison of actuator properties (after Zhang et al., 1998; Hunter and Lafontaine, 1992); PZN-PT is lead-zinc-niobate/lead-titanate single crystal and P(VDF-TrFE) is a random copolymer of vinylidene fluoride and trifluoroethylene irradiated with a 3 MeV electron beam in a nitrogen atmosphere at 120°C).

Material	Elastic Modulus Y (GPa)	Actuator Strain ε_m (%)	Volumetric Work per Cycle $Y\varepsilon_m^2/2$ (J/cm^3)	Gravimetric Work per Cycle $Y\varepsilon_m^2/2\rho$ (J/kg)
Polyaniline	5	2	1.0	670
Piezoceramic	64	0.1	0.13	4
Magnetostrictor	100	0.2	0.2	22
PZN-PT single crystal	7.7	1.7	1.0	130
Polyurethane elastomer	0.02	4	0.016	13
P(VDF-TrFE)	1	4	0.78	420
SWNT (theory)	640	1	32	15,000
SWNT (experiment)	5-80	0.05-0.7	0.04	19

More recent theoretical studies have highlighted the importance of the type of nanotube on the actuation performance [Gartstein, 2002]. By studying a simplified electron lattice model, these workers examined the effect of small amounts of charge on the carbon-carbon bond lengths in carbon nanotubes of different chiral vectors. The results are fascinating and demonstrate that some nanotubes expand while others contract for the same amount (and sign) of charge injected. Also of significance is the predicted value of the expansion, which can exceed that predicted for graphite. These results demonstrate that further improvements in actuation performance from carbon nanotubes can be obtained by controlled synthesis of specific nanotube chirality.

8.7 Experimental Studies of Carbon Nanotube Actuators

The first report of carbon nanotube actuation featuring SWNTs, appeared in 1999, and was closely followed by the demonstration of actuation in MWNTs [Gao et al., 2000]. The following sections discuss the actuator performance of nanotubes and compare the results obtained with the theoretical predictions of Sec. 8.2.

8.7.1 Actuation of SWNT Mats and Fibers

The first actuator experiments on NTs were performed on SWNT mats [Baughman et al., 1999]. The mats (bucky paper) were adhered to opposite sides of double-stick tape and immersed in salt-water electrolyte. By applying a dc voltage to the NT electrodes the bimorph structure was found to bend. Reversing the polarity caused the bimorph to bend in the opposite direction (Fig. 5). Tensile measurements on unsupported bucky paper provided axial strains of up to 0.2% during switching of the electrochemical potential between +0.5 V and –1.0 V (vs. SCE). Figure 6 shows that the actuator strain is approximately parabolic, with a minimum at around +0.5 V. Some hysteresis is observed, but this largely disappeared when the actuator strain was plotted against the charge stored per carbon atom (Fig. 7), rather than electrochemical potential applied. The maximum strain-voltage coefficient of 0.11%/V occurs between –0.2 V and –0.8 V (vs. SCE). This value is close to that predicted (0.096%/V) from the dimensional changes for the in-plane direction of graphite [Baughman, 1996]. The actuator strain sharply decreased with increasing scan rate, although actuator response could be observed at > 1 kHz [Baughman et al., 1999].

The actuator behavior of sheets of SWNTs is quite distinct from that observed for conducting polymers. The mechanism of actuation in polypyrrole, for example, involves oxidation and reduction of the polymer backbone and concomitant anion/cation movement into or out of the polymer [Gandhi et al., 1995]. The ion flows, and consequently the electromechanical actuation behavior (Fig. 8), can be quite complex and are determined by the relative mobility of the polymer counter-ion and the ions in the electrolyte. In contrast, carbon NT actuation appears insensitive to the nature of the electrolyte, so the behavior shown in Fig. 7 is typically observed.

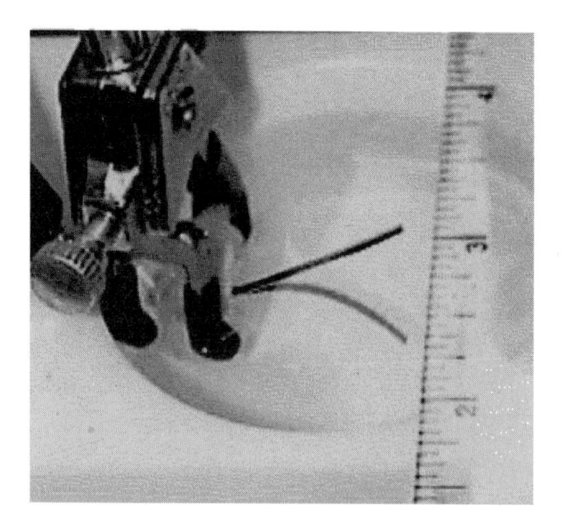

Figure 5: Time-lapse photograph showing bending of carbon NT bimorph structure upon the application of +/- 1.5V. Bending occurred within several seconds.

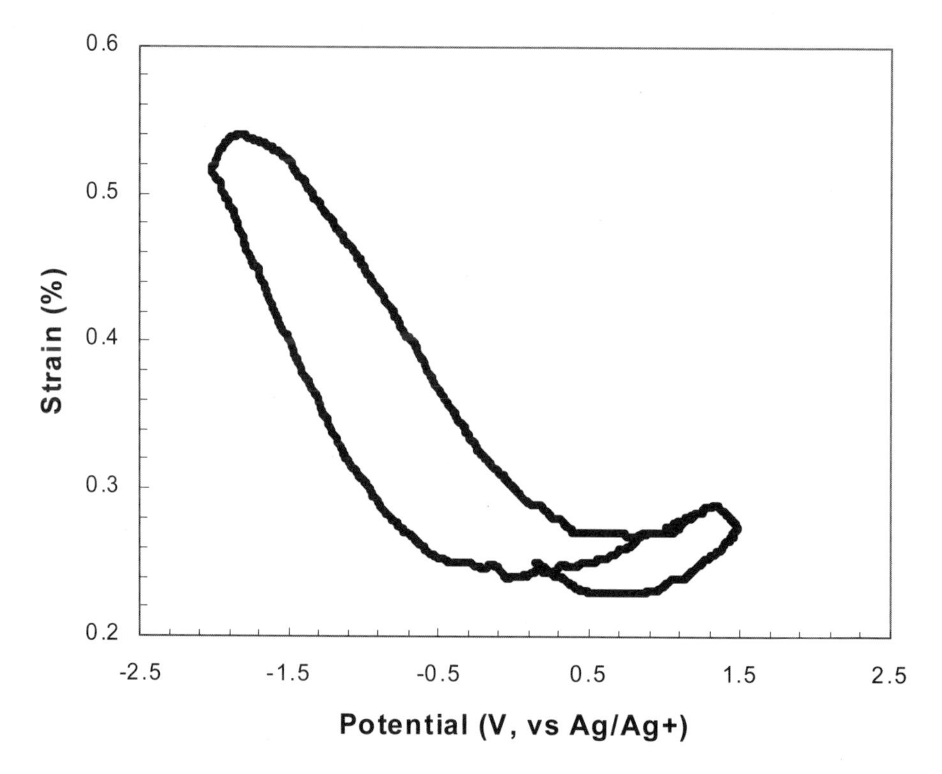

Figure 6: Actuator strain of carbon NT mat as a function of applied electrochemical potential (vs. Ag/Ag^{+}, scan rate = 50 mV/s).

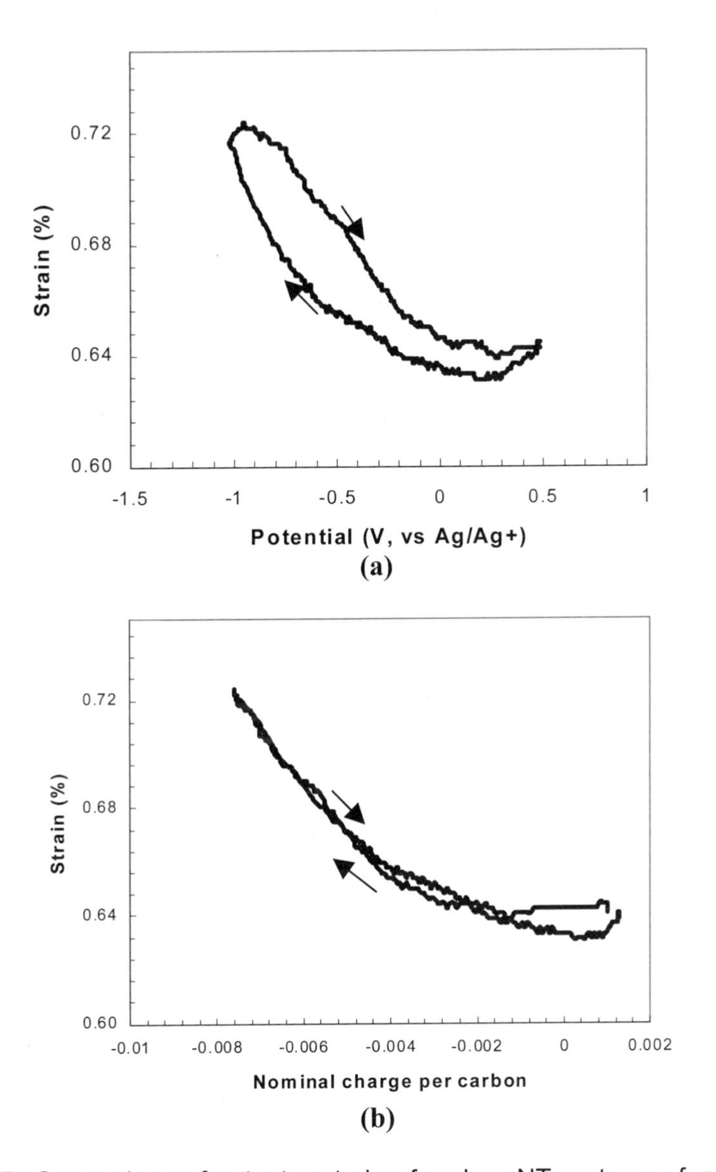

Figure 7: Comparison of actuator strain of carbon NT mat as a function of (a) applied potential and (b) nominal charge stored per carbon atom.

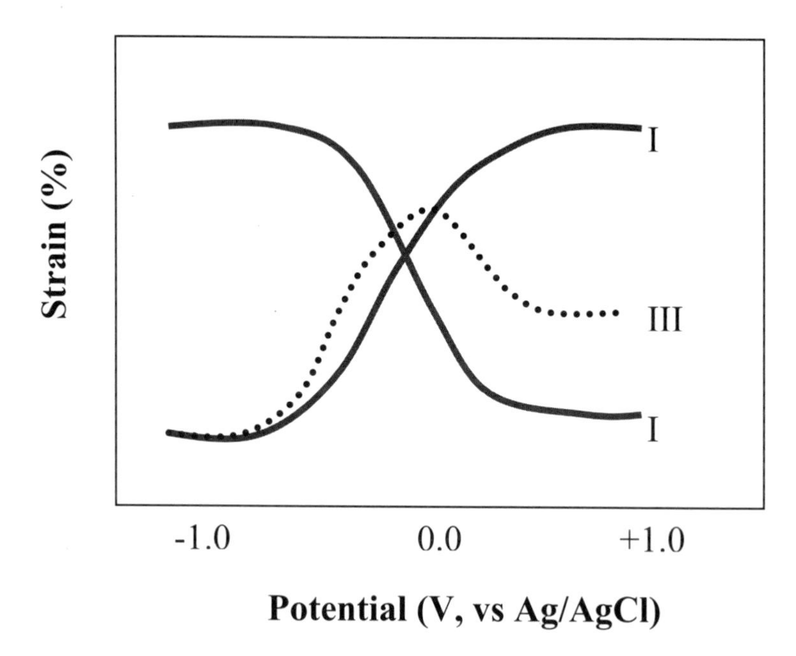

Figure 8: Schematic illustration of the types of actuation behavior observed for polypyrrole: I, large immobile counter-ion; II, small mobile counter-ion and large electrolyte anion; III, counter-ion of intermediate mobility.

Clear differences are also observed in the cyclic voltammograms (CVs) of conducting polymers and carbon nanotubes (Fig. 9). While distinct oxidation and reduction peaks are observed for conducting polymers, the CV for nanotubes is purely capacitive. In contrast with the case for conducting polymers, this actuation for nanotubes at reasonably low applied potentials does not involve Faradaic reactions.

These differences support the concept that the mechanism of nanotube actuation is a new, non-Faradaic process. The nanotube acts as an electrochemical capacitor with charge injected into the nanotube balanced by the electrical double-layer formed by movement of electrolyte ions to the nanotube surface. The charge injection causes quantum chemically based dimensional changes in the carbon-carbon bond length of the surface atoms close to the double layer. In individual single-wall nanotubes the *surface* is also the *bulk*, so expansion or contraction of the surface bonds causes the diameter and length of the entire nanotubes to change. The expansion/contraction of individual nanotubes is transferred to macroscopic dimensions so we also see expansion/contraction of an entire mat of entangled nanotubes. This idealized picture of isolated nanotubes, all in contact with the electrolyte, is not presently realized, since the SWNTs tend to form bundles. As a result of this bundling charge injection is only on the outermost nanotubes in a bundle, and this reduces actuator performance.

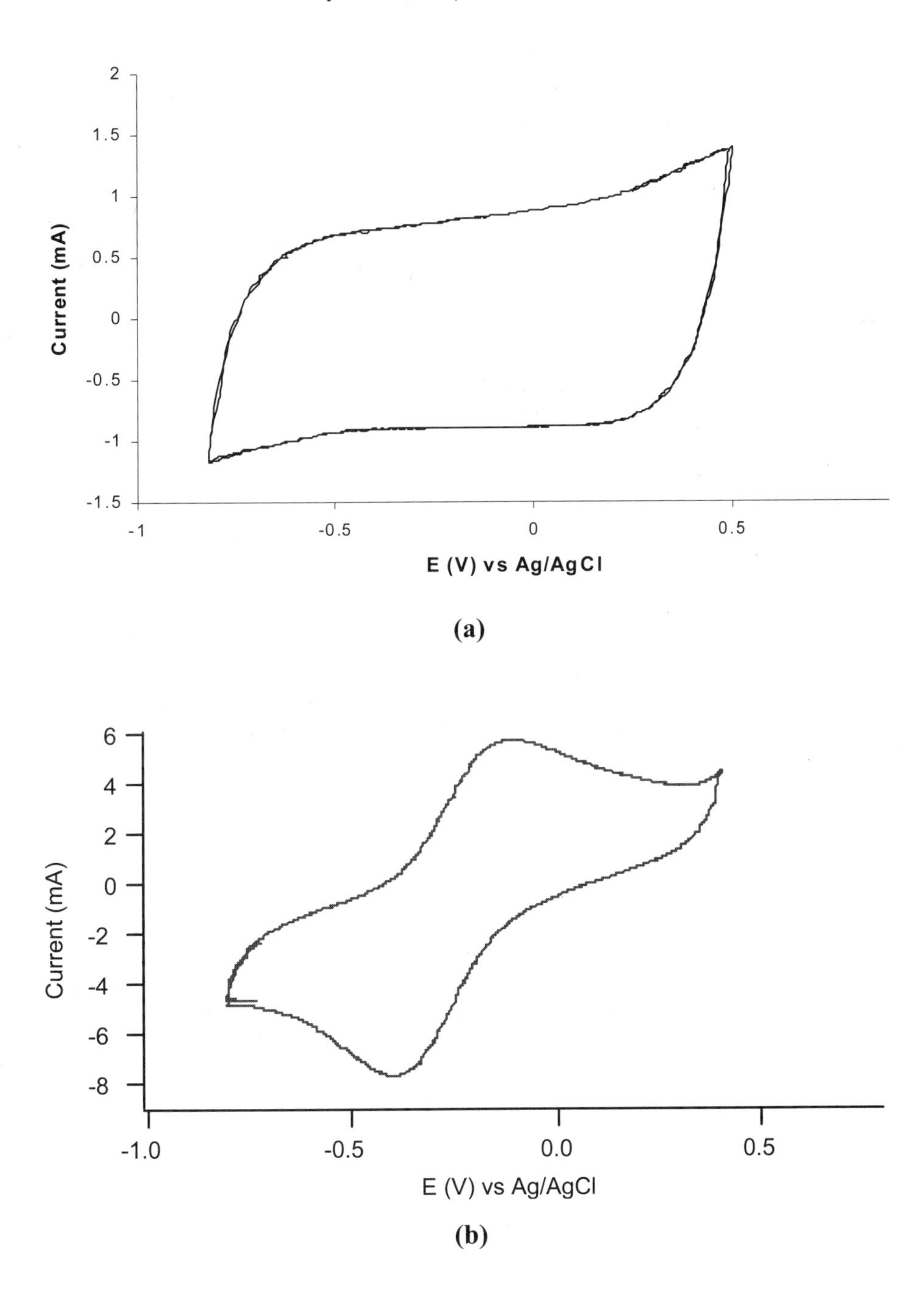

Figure 9: Cyclic voltammograms of (a) single-wall carbon nanotube sheets showing only capacitive behavior; and (b) conducting polymer polypyrrole showing distinct oxidation and reduction peaks.

We have discounted the possibility of major contributions to the actuation from electrostatic effects, at least at low levels of charge injection. It is feasible that the charging of neighboring nanotubes will cause electrostatic repulsion and expansion of the nanotube mats. We see a roughly parabolic dependence of sheet dimension, with a minimum length at about +0.5 V (vs. SCE). If this minimum corresponded to the potential of zero charge (pzc), then these results would be consistent with an electrostatic mechanism for actuation. However, the pzc for graphite has been well established as being around –0.2 V (vs. SCE) for similar salt electrolytes [Radin and Yeager, 1972; Oren et al., 1985; Gerischer et al., 1987] and this value has been indirectly confirmed for carbon nanotubes by electrochemical Raman spectroscopy [Kavan, 2000]. The fact that the experimental minimum occurs 700 mV higher in the positive direction than the pzc, excludes the electrostatic mechanism at low degrees of charge injection and is consistent with the quantum mechanical mechanism for nanotube actuation. However, since band structure related dimensional changes are roughly proportional to injected charge, and electrostatically induced dimensional changes of individual nanotubes are quadratic in the injected charge, the latter effect dominates at high degrees of charge injection.

Nanotube actuation may also arise from ion intercalation into the nanotube bundles. However, our measured capacitance values for the SWNT sheets discount this possibility. We have obtained capacitances in the range of 12 to 30F/g for nanotubes in various electrolytes. Considering a measured surface area (by the BET method) of 300 m^2/g for SWNT sheets, these values equate to a specific capacitance of 4–10 $\mu F/cm^2$. This is consistent with measured values for the basal plane of graphite (~3 to 12 $\mu F/cm^2$) [Radin and Yeager, 1972; Oren et al., 1985; Gerischer et al., 1987], indicating that the rolled up graphene sheets in nanotubes are similar electrochemically to graphite. The measured surface area is equivalent to the calculated surface area of 10-nm-diam cylinders having a crystallographically derived bundle density of 1.33 g/cm^3, as expected for bundled nanotubes [Thess et al., 1996]. In contrast, unbundled nanotubes would have a surface area of as high as 1500 m^2/g.

The reduction in surface area is an adverse consequence of nanotube bundling. Only one fifth of the nanotube surface area is available for double layer charging, thus reducing the capacitance and average degree of charge transfer to the nanotubes. Unless intertube sliding occurs, bundling will reduce the actuator stroke since the expansion of the outer nanotubes is resisted by the inactive inner nanotubes.

The experimentally obtained actuation performance of macroscopic assemblies of nanotubes falls short of that predicted from theoretical considerations (Sec. 8.6). Nanotube bundling is one reason for the lower than expected performance, but another limitation is the much lower modulus available from macroscopic assemblies than predicted for (and measured) in single nanotubes. The lower modulus leads to a much lower work-per-cycle as shown in Table 1. By analogy to ordinary paper, the mechanical properties of porous nanotube assemblies is likely to be dominated by intertube bonding and by alignment of the nanotubes. Already, the fiber spinning processes recently introduced have improved mechanical properties substantially by partial alignment of nanotubes in the fiber direction. Possible further improvements in alignment may come from refinements to the fiber-making process and/or the use of electric fields to assist alignment. Other ways to improve the modulus of nanotube assemblies focus on the intertube bonding. Here, specific chemical functionalization of the nanotubes may facilitate the covalent attachment at junction points. The key requirement for actuation is to increase the bonding without upsetting the electronic/electrochemical properties so that the high specific capacitance and high surface area remains but in a covalently bonded structure.

8.7.2 Pneumatic Actuation in SWNT Mats

A novel high-strain actuation process has recently been reported for SWNT mats [Spinks, 2002]. The process involves electrochemically producing gas bubbles within the porous mats due to the electrochemical breakdown of the electrolyte (producing gases such as oxygen, chlorine and hydrogen, Fig. 10). The pressure of the gas bubbles causes disk-shaped cracks to form and propagate within the plane of the sheet (Fig. 11). Like an accordion, the blistering of the sheet causes a 300% increase in the sheet thickness and a corresponding 3% contraction in in-plane dimensions (Fig. 12). The in-plane contraction occurs over the period of several seconds (although the subsequent expansion caused by the electrochemical removal of the gas bubbles is much faster).

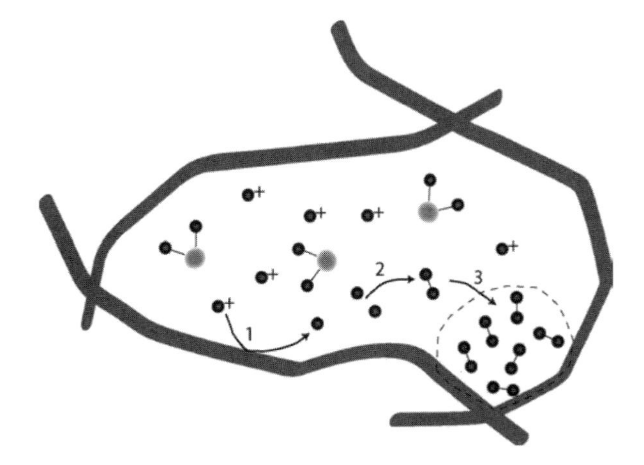

Figure 10: Mechanism of electrochemical formation of gas chambers: (1) Hydrogen ions accept electrons from negatively polarized nanotubes to form atomic hydrogen: (2) when two hydrogen atoms spontaneously form molecular hydrogen, and (3) when the local concentration of molecular hydrogen in the electrolyte becomes too high the hydrogen condenses as a gas bubble. Because of the hydrophobicity of the nanotubes, the hydrogen gas bubble adheres to the nanotubes. Further hydrogen generation causes growth of the gas chambers.

Figure 11: Schematic illustration of nanotube mats formed from layers of randomly oriented nanotube ropes laying flat in each layer and with little entanglement between layers. Internal gas formation due to electrolysis of the electrolyte in the pore space causes delamination and blister formation. The growth of the blisters causes lateral contraction of the nanotubes mats.

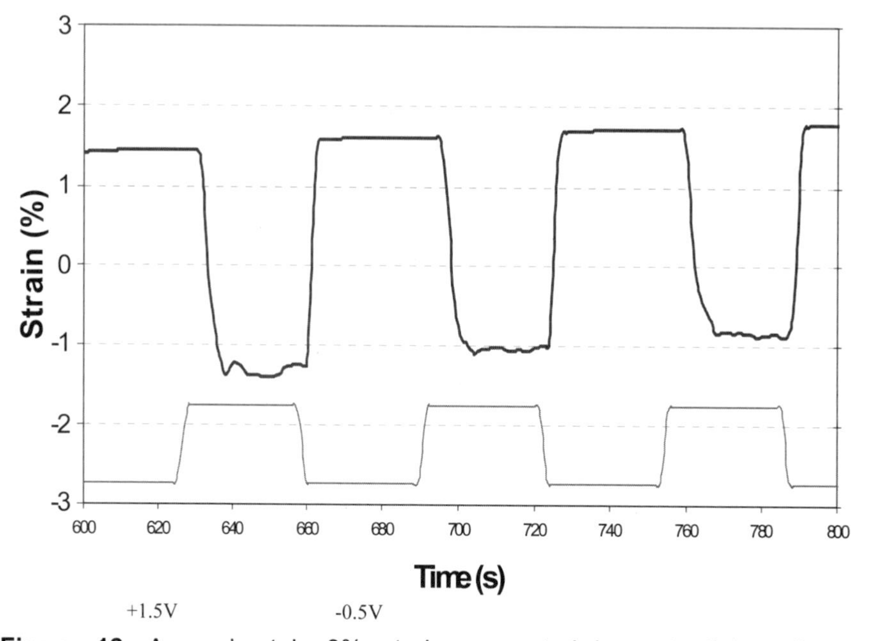

Figure 12: Approximately 3% strain generated by potential cycling carbon nanotube sheets between –0.5V and +1.5V in 5M NaCl (aqueous).

The pneumatic process is energetically inefficient and the number of cycles is limited (due to unstable crack growth leading to mechanical failure). However, the pneumatic process could be used as an emergency "high gear" response for nanotube actuators operating in liquid environments. The cycle lifetime may also be improved by appropriate strengthening of the nanotube mats in the thickness direction to prevent uncontrolled growth of the gas chambers.

8.7.3 Actuation in Aligned MWNT Arrays

To date it has been difficult to study the actuation in MWNTs due to the difficulty in producing macroscopic sheets with adequate mechanical properties. However, the synthesis method giving "forests" of MWNTs and the methods developed to transfer such arrays to other substrates [Huang et al., 1999; Yang et al., 1999; Dai and Mau, 2000; Huang et al., 2000; Li et al., 2000; Qidao and Dai, 2000] have enabled the demonstration of MWNT actuation. MWNT arrays were prepared on a flexible gold foil and immersed in 1 M $NaNO_3$ electrolyte. Application of a potential of +1 V or –1 V vs. SCE produced bending of the unimorph structure in opposite directions. Higher voltages (+/– 2 V produced larger deformations (Fig. 13). Proper patterning of the aligned nanotube array surface with appropriate non-conducting polymers can lead to controllable actuation paths [Soundarrajan 2003], suggesting that the aligned MWNT actuators eventually might be useful for MEMS applications.

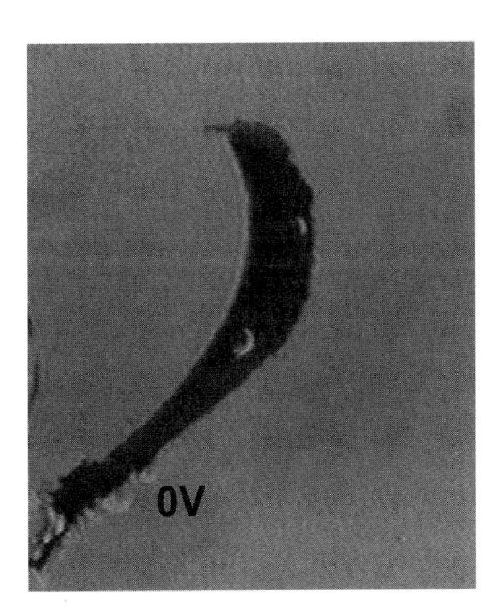

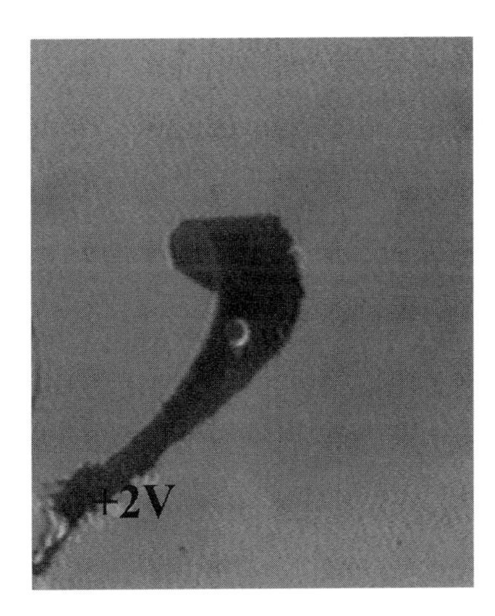

Figure 13: Bending of aligned MWNT arrays on a gold foil by application of +2 V or –2 V (vs. SCE).

The gravimetric capacitance for the forests of MWNTs is about 30 F/g, which is similar to that for the SWNT bucky paper. However, the above cantilever-based actuator response for the forests of SWNTs is believed to be electrostatic. Electrostatic repulsion between different tubes in the nanotube forest would cause repulsion between these tubes. Since the nanotubes are firmly anchored to the substrate, the effect of this repulsion is a bending of the sheet, so that the separation between unanchored nanotube ends is increased. While scientifically interesting, this inter-tube electrostatic actuator mechanism is much less promising for applications than actuation mechanisms that seem to principally involve the direct effect of charge injection on nanotube dimensions.

Moreover, possible pneumatic actuation cannot be ruled out for cantilevered forests of nanotubes, as bubble generation was observed throughout the actuation process.

8.8 Conclusions and Future Developments

The demonstration of carbon nanotube actuation driven by a non-Faradaic charging may eventually lead to game-changing actuator devices. However, major improvements in the structure and mechanical properties of nanotube assemblies (fibers or sheets) are required before applications can be realized. While the observed actuator strains are quite large, the primitively prepared SWNT mats have a Young's modulus (up to 5 GPa) that is well below that determined for individual NT bundles (~600 GPa). Significant improvements in strength and stiffness have been achieved by obtaining indefinitely long nanotube fibers that comprise highly aligned SWNTs in a polymer matrix. The modulus of these assemblies now approaches 100 GPa and the strength reaches about 2 GPa. However, the polymer "glue" that holds the nanotube fibers together in these composite fibers interferes with the charge injection needed for actuation if this polymer is not a good ionic conductor. Alternatively, conversion of the host polymers in the nanotube fibers to ionic conductors (for example, by incorporating phosphoric acid in a polyvinyl alcohol host) degrades the mechanical properties of the fiber. Hence, there is a need for ionic conductor that effectively couples the nanotubes together without interfering with electrochemical charge injection. Recent developments in the covalent attachment of organic groups [Boul et al., 1999] and polymers [Curran et al., 1999] to NTs provide one direction for improving inter-tube bonding in NT mats and fibers. As the understanding of NT synthesis, characterization, and processing further develops, other approaches for improving the bulk mechanical properties of macroscopic NT assemblies are expected.

Nanoscale actuators based on single nanotubes do not have these problems. The challenge here is to develop methods for measuring actuation at the nanoscale, and then to manipulate single nanotubes in the fabrication of practical devices. Recent publications provide some guidance. The AFM techniques, used by Salvetat et al., [1999] and Walters et al., [1999], have enabled mechanical properties measurements (modulus and yield strain) on small-diameter NT bundles. Similar techniques may well demonstrate the true potential of NT actuators in terms of actuator strain, strain rates, cycle life, and stress generation.

8.9 References

Ajayan, P. M., "Aligned carbon nanotube arrays formed by cutting a polymer resin-nanotube composite," *Adv. Mater.*, **7**, p. 489-91 (1995).

Ajayan, P. M., O. Stephan, C. Colliex, and D. Trauth, "Aligned carbon nanotubes in a thin polymer film," *Science*, **265**, p. 1212-14 (1994).

Alvarez, W.E., Kitiyanan, B., Borgna, A., and Resasco, D.E., "Synergism of Co and Mo in the catalytic production of single-wall carbon nanotubes by decomposition of CO," *Carbon*, **39**, p. 547 (2001).

An, K.H., Kim, W.S., Park, Y.S., Choi, Y.C., Lee, S.M., Chung, D.C., Bae, D.J., Lim, S.C., Lee, Y.H., "Supercapacitors using single-walled carbon nanotube electrodes," *Adv. Mat.*, **13**, p. 497 (2001).

Barisci, J.N., Wallace, G.G., Baughman, R.H., "Electrochemical quartz crystal microbalance studies of single-wall carbon nanotubes in aqueous and non-aqueous solutions," *J. Electroanal. Chem.*, **488**, p. 92 (2000).

Barisci, J.N., Wallace, G.G., Baughman, R.H., "Electrochemical studies of single-wall carbon nanotubes in aqueous solutions," *J. Electrochem. Soc.*, **147**, p. 4580 (2000).

Barisci, J.N., Wallace, G.G., Baughman, R.H., "Electrochemical characterization of single-walled carbon nanotube electrodes," *Electrocem. Acta.*, **6**, p. 509 (2000).

Baughman, R.H., "Conducting polymer artificial muscles," *Synth. Met.*, **78**, p. 339-353 (1996).

Baughman, R. H., N. S. Murthy, H. Eckhardt and M. Kertesz, "Charge oscillations and structure for alkali-metal-doped polyacetylene," *Phys. Rev. B*, **46**, Art. No. 10515 (1992).

Baughman, R.H., Cui, C., Zakhidov, A.A., Iqbal, Z., Barisci, J.N., Spinks, G.M., Wallace, G.G., Mazzoldi, A., De Rossi, D., Rinzler, A.G., Jaschinski, O., Roth, S. and Kertesz, M., "Carbon nanotube actuators," *Science*, **284**, p. 1340-1344 (1999).

Bethune, D. S., C. H. Kiang, M. S. de Vries, G. Gorman, R. Savoy, J. Vazquez and R. Beyers, "Cobalt-catalyzed growth of carbon nanotubes with single-atomic-layer walls," *Nature*, **363**, p. 605-7 (1993).

Boul, P. J., J. Liu, E. T. Mickelson, C. B. Huffman, L. M. Ericson, I. W. Chiang, K. A. Smith, D. T. Colbert, R. H. Hauge, J. L. Margrave, et al., "Reversible sidewall functionalization of buckytubes," *Chem. Phys. Lett.*, **310**, p. 367-372 (1999).

Britto, P. J., K. S. V. Santhanam and P. M. Ajayan, "Carbon nanotube electrode for oxidation of dopamine," *Bioelectrochemistry & Bioenergetics*, **41**, p. 121-125 (1996).

Bronikowski, M.J., Willis, P.A., Colbert, D.T., Smith, K.A., Smalley, R.E. "Gas-phase production of carbon single-walled nanotubes from carbon monoxide via the HiPco process: A parametric study," *J. Vac. Sci. Technol. A*, **19**, 1800 (2001).

Campbell, J. K., L. Sun and R. M. Crooks, "Electrochemistry using single carbon nanotubes," *J. Am. Chem. Soc.*, **121**, p. 3779-3780 (1999).

Cassell, A.M., Raymakers, J.A., Kong, J. and Dai, H. "Directed growth of free-standing single-walled carbon nanotubes," *J. Phys. Chem. B*, **103**, p. 6484, (1999).

Cassell, A.M., Franklin, N.R., Tombler, T.W., Chan, E.M., Han, J., Dai, H. "Large scale CVD synthesis of single-walled carbon nanotubes," *J. Am. Chem. Soc.*, **121**, p. 7975 (1999a).

Che, G., B. B. Lakshmi, E. R. Fisher and C. R. Martin, "Chemical vapor deposition based synthesis of carbon nanotubes and nanofibers using a template method," *Nature*, **393**, p. 346-349 (1998).

Chesnokov, S. A., V. A. Nalimova, A. G. Rinzler, R. E. Smalley and J. E. Fischer, "Mechanical energy storage in carbon nanotube springs," *Phys. Rev. Lett.*, **82**, p. 343-346 (1999).

Colbert, D.T. and Smalley, R.E., "Past, present and future of fullerene nanotubes: buckytubes," *Perspectives of Fullerene Nanotechnology*, Osawa, E. (Ed.), Kluwer Academic Publishers: Dordrecht, 2002.

Collins, P. G., A. Zettl, H. Bando, A. Thess and R. E. Smalley, "Nanotube nanodevice," *Science*, **278**, p. 100-103 (1997).

Curran, S., A. P. Davey, J. Coleman, A. Dalton, B. McCarthy, S. Maier, A. Drury, D. Gray, M. Brennan, K. Ryder, et al., "Evolution and evaluation of the polymer/nanotube composite," *Synth. Met.*, **103**, p. 2559-2562 (1999).

Dai, H., J. H. Hafner, A. G. Rinzler, D. T. Colbert and R. E. Smalley, "Nanotubes as nanoprobes in scanning probe microscopy," *Nature*, **384**, p. 147-150 (1996).

Dai, H., Rinzler, A.G., Nikolaev, P., Thess, A., Colbert, D.T., Smalley, R.E. "Single-wall nanotubes produced by metal-catalyzed disproportionation of carbon monoxide," *Chem. Phys. Lett.*, **260**, p. 471 (1996a).

Dai, L. and A. W. H. Mau, "Surface and interface control of polymeric biomaterials, conjugated polymers, and carbon nanotubes," *J.Phys.Chem.B.,* **104**, p. 1891 (2000).

Dalton, A.B., S. Collins, E. Munoz, J.M. Razal, V.H. Ebron, J.P. Ferraris, J.N. Coleman, B.G. Kim, R.H. Baughman. "Super-tough carbon-nanotube fibres," *Nature*, **423**, p. 703 (2003).

Davis, J. J., R. J. Coles and H. A. O. Hill, "Protein electrochemistry at carbon nanotube electrodes," *J. Electroanal. Chem.*, **440**, p. 279-282 (1997).

Deheer, W. A., W. S. Bacsa, A. Chatelain, T. Gerfin, R. Humphreybaker, L. Forro and D. Ugarte, "Aligned Carbon Nanotube Films - Production and Optical and Electronic Properties," *Science*, **268**, p. 845-847 (1995a).

Deheer, W. A., A. Chatelain and D. Ugarte, "A carbon nanotube field-emission electron source," *Science*, **270**, 1179-1180 (1995).

Dujardin, E., T. W. Ebbesen, H. Hiura and K. Tanigaki, "Capillarity and wetting of carbon nanotubes," *Science*, **265**, p. 1850-2 (1994).

Ebbesen, T. W. and P. M. Ajayan, "Large-scale synthesis of carbon nanotubes," *Nature*, **358**, p. 220-2 (1992).

Endo, M., K. T akeuchi, S. Igarashi, K. Kobori, M. Shiraishi and H. W. Kroto, "The production and structure of pyrolytic carbon nanotubes (PCNTs)," *J. Phys. Chem. Solids*, **54**, p. 1841-8 (1993).

Falvo, M. R., G. J. Clary, R. M. Taylor, II, V. Chi, F. P. Brooks, Jr., S. Washburn and R. Superfine, *Nature* "Bending and buckling of carbon nanotubes under large strain," **389**, p. 581-584 (1997).

Frank, S., P. Poncharal, Z. L. Wang and W. A. de Heer, "Carbon nanotube quantum resistors," *Science*, **280**, p. 1744-1746 (1998).

Franklin, N.R. and Dai, H. "An enhanced CVD approach to extensive nanotube networks with directionality," *Adv. Mater.*, **12**, p. 890 (2000).

Gandhi, M. R., P. Murray, G. M. Spinks and G. G. Wallace, *Synth. Met.* "Mechanism of electromechanical actuation in polypyrrole," **73**, p. 247-56 (1995).

Gao, M., L. Dai, R. H. Baughman, G. M. Spinks and G. G. Wallace, "Electrochemical properties of aligned nanotube arrays: basis of new electromechanical actuators," *Proceedings of SPIE*, **3987**, p. 18-24 (2000).

Gao, M., Dai. L., Wallace, G.G., "Biosensors based on aligned carbon nanotubes coated with inherently conducting polymers," *Electroanalysis* ***15***, p. 1089-1094 (2003).

Garstein, Y.N., A.A. Zakhodov, R.H. Baughman, "Charge-induced anisotropic distortions of semiconducting and metallic carbon nanotubes" *Phys. Rev. Lett.*, **89**, Art. No. 045503 (2002).

Gerischer, H., R. McIntyre, D. Scherson and W. Storck,. "Density of the electronic states of graphite: derivation from differential capacitance measurements," *J. Phys. Chem*, **91**, p. 1930-1935 (1987).

Gu, G., Philipp, G., Wu, X., Burghard, M., Bittner, A.M. and Roth, S. "Growth of single-walled carbon nanotubes from microcontact-printed catalyst patterns on thin Si^3N^4 membranes," *Adv. Funct. Mater.*, **11**, p. 295 (2001).

Haslach, H. W., Jr. "The moisture and rate-dependent mechanical properties of paper: A review," *Mechanics of Time-Dependent Materials*, **4**, p. 169-210 (2000).

Hoyer, P., "Semiconductor nanotube formation by a two-step template process,"*Adv.Mater.*, **8**, p. 857 (1996).

Huang, S., A. W. H. Maŭ, T. W. Turney, P. A. White and L. Dai, "Patterned growth of well-aligned carbon nanotubes: A soft-lithographic approach," *J.Phys.Chem.B.*, **104**, p. 2193 (2000).

Huang, S. M., L. M. Dai and A. W. H. Mau, "Patterned growth and contact transfer of well-aligned carbon nanotube films," *Journal of Physical Chemistry B.*, **103**, p. 4223-4227 (1999).

Hunter, I. W. and S. Lafontaine, "A comparison of muscle with artificial actuators." *Tech. Digest, IEEE Solid-State Sensor Actuator Workshop.*, p. 178-185 (1992).

Iijima, S., "Helical microtubules of graphitic carbon," *Nature*, **354**, p. 56 (1991).

Iijima, S. and T. Ichihashi, "Single-shell carbon nanotubes of 1-nm diameter," *Nature*, **363**, p. 603-5 (1993).

Ivanov, V., A. Fonesca, J. B. Nagy, A. Lucas, P. Lambin, D. Bernaerts and X. B. Zhang, "Catalytic production and purification of nanotubules having fullerene-scale diameters,"*Carbon*, **33**, p. 1727-38 (1995).

Jang Y.-T., Ahn J.-H., Ju B.-K. and Lee Y.-H., *Solid State Communications* 2003, **126**, p. 305-308.

Joselevich, E. and Lieber, C.M., "Vectorial growth of metallic and semiconducting single-wall carbon nanotubes," *Nano.Lett.*, **2**, 1137-1141 (2002).

Journet, C., W. K. Maser, P. Bernier, A. Loiseau, M. Lamy de la Chapells, S. Lefrant, P. Deniard, R. Lee and J. E. Fischer, "Large-scale production of single-walled carbon nanotubes by the electric-arc technique," *Nature*, **388**, p. 756-758 (1997).

Kavan, L., "XIVth International Winter School on Electronic Properties of Novel Materials," 2000: Unpublished.

Kim, P. and C. M. Lieber, "Nanotube nanotweezers," *Science*, **286**, p. 2148-2150 (1999).

Kong, J., N. R. Franklin, C. W. Zhou, M. G. Chapline, S. Peng, K. J. Cho and H. J. Dai, "Nanotube molecular wires as chemical sensors,"*Science*, **287**, p. 622-625 (2000).

Li, D. C., L. Dai, S. Huang, A. W. H. Mau and Z. L. Wang, "Structure and growth of aligned carbon nanotube films by pyrolysis," *Chem.Phys.Lett.*, **316**, p. 349 (2000).

Li, J., Cassell, A., Delzeit, L., Han, J., Meyyappan, M. "Novel three-dimensional electrodes: Electrochemical properties of carbon nanotube ensembles," *J. Phys. Chem.*, **106**, p. 9299 (2002).

Li, W. Z., S. S. Xie, L. X. Qian, B. H. Chang, B. S. Zou, W. Y. Zhou, R. A. Zhao and G. Wang, "Large-scale synthesis of aligned carbon nanotubes," *Science*, **274**, p. 1701-1703 (1996).

Liu, C.-Y., A. J. Bard, F. Wudl, I. Weitz and J. R. Heath, "Electrochemical characterization of films of single-walled carbon nanotubes and their possible application in supercapacitors," *Electrochem. Solid-State Lett.*, **2**, p. 577-578 (1999).

Liu, M. and J. M. Cowley, "Structures of carbon nanotubes studied by HRTEM and nanodiffraction," *Ultramicroscopy*, **53**, p. 333-42 (1994).

Luo, H., Shi, Z., Li, N., Gu, Z., Zhuang, Q., "Investigation of the electrochemical and electrocatalytic behavior of single-wall carbon nanotube film on a glassy carbon electrode," *Anal., Chem.*, **73**, p. 915 (2001).

Nikolaev, P., Bronikowski, M.J., Bradley, R. K., Rohmund, F., Colbert, D.T., Smith, K.A. and Smalley, R.E. "Gas-phase catalytic growth of single-walled carbon nanotubes from carbon monoxide,"*Chem. Phys. Lett.*, **313**, p. 91 (1999).

Nissan, A. H. and G. L. Batten, Jr. "Unified theory of the mechanical-properties of paper and other H-bond-dominated solids," *Tappi Journal*, **70**, p. 128-31 (1987).

Niu, C.M., E.K., Sichel, R. Hoch, D. Moy, H. Tennet, "High power electrochemical capacitors based on carbon nanotube electrodes," *Appl. Phys. Lett.*, 70, p. 1480-1482 (1997).

Nixon, D. E. and G. S. Parry, "The expansion of the carbon-carbon bond length in potassium graphites," *J.Phys. C: Solid State Phys.*, **2**, p. 1732 (1969).

Oren, Y., I. Glatt, A. Livnat, O. Kafri and A. Soffer, "The electrical double layer charge and associated dimensional changes of high surface area electrodes as detected by moire deflectometry," *J. Electroanal. Chem.*, **187**, p. 59-71 (1985).

Page, D. "A theory for tensile strength of paper," *Tappi Journal*, **52**, p. 671-681 (1969).

Pan, Z. W., S. S. Xie, L. Lu, B. H. Chang, L. F. Sun, W. Y. Zhou, G. Wang and D. L. Zhang, "Tensile tests of ropes of very long aligned multiwall carbon nanotubes," *Appl. Phys. Lett.*, **74**, p. 3152-3154 (1999).

Poulin, P., Vigolo, B., Launois, P., "Films and fibers of oriented single wall nanotubes," *Carbon*, **40**, p. 1741-1749 (2002).

Qidao, C. and L. Dai, "Plasma patterning of carbon nanotubes, " *Appl. Phys. Lett.*, **76**, p. 2719 (2000).

Qin, L. C., "CVD synthesis of carbon nanotubes," *J. Mater. Sci. Lett.*, **16**, p. 457-459 (1997).

Radin, J.-P. and E. Yeager, "Differential capacitance study on the basal plane of stress-annealed pyrolytic graphite," *J. Electroanal. Chem. Interfacial Electrochem.*, **36**, p. 257 (1972).

Rao, A. M., P. C. Eklund, S. Bandow, A. Thess and R. E. Smalley, "Evidence for charge transfer in doped carbon nanotube bundles from Raman scattering," *Nature*, **388**, 257-259 (1997).

Rao, C. N. R., R. Sen, B. C. Satishkumar and A. Govindaraj, "Large aligned-nanotube bundles from ferrocene pyrolysis,"*J. Chem. Soc. Chem. Commun.*, p. 1525-1526 (1998).

Ren, Z.F., Huang, Z.P., Xu, J.W., Wang, J.H., Bush, P., Siegal, M.P. and Provencio, P.N., "Synthesis of large arrays of well-aligned carbon nanotubes on glass," *Science*, **282**, 1105-1107 (1998).

Resasco, D.E., Alvarez, W.E., Pompeo, F., Balzano, L., Herrera, J.E., Kitiyanan, B. and Borgna, A. "A scalable process for production of single-walled carbon nanotubes (SWNTs) by catalytic disproportionation of CO on a solid catalyst," *J. Nanoparticle Res.*, **4**, p. 131 (2002).

Rinzler, A.G., Liu, J., Dai, H., Nikolaev, P., Huffman, C.B., Rodriguez-Macias, F.J., Boul, P.J., Lu, A.H., Heymann, D., Colbert D.T. et al., "Large-scale purification of single-wall carbon nanotubes: process, product, and characterization," *Appl. Phys. A: Mater. Sci. Process.*, A67, p. 29-37 (1998).

Roschier, L., J. Penttila, M. Martin, P. Hakonen, M. Paalanen, U. Tapper, E. I. Kauppinen, C. Journet and P. Bernier, "Single-electron transistor made of multiwalled carbon nanotube using scanning probe manipulation," *Appl. Phys. Lett.*, **75**, p. 728-730 (1999).

Saito, Y., T. Yoshikawa, M. Okuda, N. Fujimoto, K. Sumiyama, K. Suzuki, A. Kasuya and Y. Nishina, "Carbon nanocapsules encaging metals and carbides,"*J.Phys.Chem. Solids*, **54**, p. 1849 (1993).

Salvetat, J.-P., G. A. D. Briggs, J.-M. Bonard, R. R. Bacsa, A. J. Kulik, T. Stockli, N. A. Burnham and L. Forro, "Elastic and shear moduli of single-walled carbon nanotube ropes," *Phys. Rev. Lett.*, **82**, p. 944-947 (1999).

Senthil K.M. et al., "DC electric field assisted alignment of carbon nanotubes on metal electrodes, " *Solid-State Electronics,* **47**, p. 2075-2080 (2003).

Shen, W., B. Jiang, B. S. Han and S.-S. Xie, "Investigation of radial compression of carbon nanotubes with a scanning probe microscope," *Phys. Rev. Lett.*, **84**, p. 3634-3637 (2000).

Soundarrajan, P. *MSc Thesis of The University of Akron* (2003).

Spinks, G. M., Wallace, G. G., Fifield, L. S., Dalton, L. R., Mazzoldi, A., Rossi, D. D., Khayrullin, I. I. and Baughman, R. H. "Pneumatic carbon nanotube actuators," *Adv. Mater.*, **14**, p. 1728-1732 (2002).

Terrones, M., N. Grobert, J. Olivares, J. P. Zhang, H. Terrones, K. Kordatos, W. K. Hsu, J. P. Hare, P. D. Townsend, K. Prassides, et al., "Controlled production of aligned-nanotube bundles," *Nature*, **388**, p. 52-55 (1997).

Terrones, M., N. Grobert, J. P. Zhang, H. Terrones, J. Olivares, W. K. Hsu, J. P. Hare, A. K. Cheetham, H. W. Kroto and D. R. M. Walton, "Preparation of aligned carbon nanotubes catalyzed by laser-etched cobalt thin films,"*Chem. Phys. Lett.*, **285**, p. 299-305 (1998).

Tersoff, J. and R. S. Ruoff, "Structural Properties of a carbon-nanotube crystal,"*Phys. Rev. Lett.*, **73**, p. 676-679 (1994).

Thess, A., R. Lee, P. Nikolaev, H. Dai, P. Petit, J. Robert, C. Xu, Y. H. Lee, S. G. Kim and et al., "Crystalline ropes of metallic carbon nanotubes," *Science*, 273, p. 483-487 (1996).

Treacy, M. M. J., T. W. Ebbesen and J. M. Gibson, "Exceptionally high Young's modulus observed for individual carbon nanotubes," *Nature*, **381**, p. 678-680 (1996).

Tsang, S. C., P. Deoliveira, J. J. Davis, M. L. H. Green and H. A. O. Hill, "The structure of the carbon nanotube and its surface topography probed by transmission electron microscopy and atomic force microscopy," *Chem. Phys. Lett.*, **249**, p. 413-422 (1996).

Vigolo, B., Penicaud, A., Coulon, C., Sauder, C., Pailler, R., Journet, C., Bernier, P., Poulin, P., "Macroscopic fibers and ribbons of oriented carbon nanotubes," *Science*, 290, p. 1331-1334 (2000).

Vigolo, B., Poulin, P., Lucas, M., Launois, P., Bernier, P., "Improved structure and properties of single-wall carbon nanotube spun fibers," *Applied Physics Letters*, **81**, p. 1210-1212 (2002).

Walters, D. A., L. M. Ericson, M. J. Casavant, J. Liu, D. T. Colbert, K. A. Smith and R. E. Smalley, "Elastic strain of freely suspended single-wall carbon nanotube ropes,"*Appl. Phys. Lett.*, **74**, p. 3803-3805 (1999).

Wang, J., Li, M., Shi, Z., Li, N., Gu, Z., "Direct electrochemistry of cytochrome c at a glassy carbon electrode modified with single-wall carbon nanotubes," *Anal. Chem.*, **74**, p. 1993 (2002).

Wood, J. R. and H. D. Wagner, "Single-wall carbon nanotubes as molecular pressure sensors," *Applied Physics Letters*, **76**, p. 2883-2885 (2000).

Yakobson, B. I., C. J. Brabec and J. Bernholc, "Nanomechanics of carbon tubes: instabilities beyond linear response," *Phys. Rev. Lett.*, **76**, p. 2511-2514 (1996).

Yang, Y., S. Huang, W. He, A. W. H. Mau and L. Dai, "Patterned growth of well-aligned carbon nanotubes: A photolithographic approach," *J. Am. Chem. Soc.*, **121**, Art. No. 10832 (1999).

Yang, Z.-H., Wu, H.Q. "The electrochemical impedance measurements of carbon nanotubes," *Chem. Phys. Lett.*, **343**, p. 235 (2001).

Yu, M., M. J. Dyer, G. D. Skidmore, H. W. Rohrs, X. Lu, K. D. Ausman, J. R. Von Ehr and R. S. Ruoff, "Three-dimensional manipulation of carbon nanotubes under a scanning electron microscope," *Nanotechnology*, **10**, p. 244-252 (1999).

Zhang, Q., Z.-Y. Cheng, V. Bharti, T.-B. Xu, H. S. Xu, T. X. Mai and S. J. Gross, "Piezoelectric and electrostrictive polymeric actuator materials, " Proceedings of SPIE, **3987**, p. 34-50 (2000).

Zhang, Q. M., V. Bharti and X. Zhao, *Science* 1998, **280**, p. 2101-2104.

Zhang, Y. and S. Iijima, "Giant electrostriction and relaxor ferroelectric behavior in electron-irradiated poly(vinylidene fluoride-trifluoroethylene) copolymer," *Phys. Rev. Lett.*, **82**, p. 3472-3475 (1999).

Zheng, B., Lu, C., Gu, G., Makarovski, A., Finkelstein, G., Liu, J. "Elastic response of carbon nanotube bundles to visible light," *Nano Lett.*, **2**, 895 (2002).

TOPIC 3.3

Molecular EAP

CHAPTER 9

Molecular Scale Electroactive Polymers

Michael J. Marsella
University of California at Riverside

9.1 Introduction

This chapter reviews the so-called "single-molecule" approach to the design of artificial muscles. More specifically, it will focus on the development of artificial muscles that are capable of function from the bulk (macroscale) to the single-molecule (nanoscale) level. Such materials can only exist if each building block is encoded with all the information necessary to perform an expansion/contraction (or related) cycle (single-molecule approach). Thus, a distinction between intrinsic and bulk properties must first be addressed. Furthermore, the systems reviewed herein are limited to well-defined organic molecules or macromolecules that have some analogy to biological systems. For the sake of clarity, this section uses the term *engine* to include muscles and *all other related forms of biological devices that can transduce one form of energy to mechanical energy*. An excellent review on biological engines has recently appeared in the literature, and describes not only muscles, but also biological springs and ratchets [Mahadevan et al, 2000]. All such engines have synthetic counterparts at the molecular level, and are described herein. It should be noted that carbon nanotubes, another class of well-defined actuators, are discussed elsewhere in this book.

9.1.1 Overview: Intrinsic vs. Bulk Properties

The distinction between an intrinsic engine and a bulk engine can be realized by the following conceptual illustration. Shown in Fig. 1 is an assembly of blocks, with each block defined as the smallest repeat unit of the assembly. The dark shading represents some desired function (such as electromechanical actuation). As the larger assembly is divided into smaller pieces, two possibilities exist. In one case, the desired function ceases to exist (light shading) at the level of the individual unit (bulk property). The second possibility is that the function is maintained down to the individual unit (intrinsic property). In the former case, the function is a bulk property that requires multi-unit interactions to exist in order for the desired function to occur, and is clearly undesirable from the standpoint of molecular devices.

Figure 1: Bulk vs. intrinsic properties, illustrated conceptually.

9.1.2 Bulk Engines: an Example

A concrete example of a "bulk engine" is that of traditional conducting polymer based electromechanical actuators, herein referred to as electroactive polymers (EAPs) [Baughman, 1996]. Bulk EAPs are described elsewhere in this book, and will be addressed only briefly here. The expansion and contraction of bulk conducting polymers arises from the intercalation of requisite counter-ions (and coordinating solvent) during the polymer's oxidation/reduction cycle (Fig. 2). Thus, the intercalating species increases the volume of the bulk polymer, which results in an overall expansion. Conversely, expulsion of the intercalated species during the reverse electrochemical event causes the polymer to contract to its original size. Of key importance is the fact that the intercalating species affects the mean distance between polymer chains. The expansion and contraction are not necessarily related to the *intrinsic* end-to-end change in distance of each polymer chain. This is not to say that conformational changes in individual polymer chains do not occur; it simply points to the fact that the same mechanism cannot be in effect at both the bulk and single-molecule levels. In contrast, the mechanism of an "intrinsic engine" is maintained at all levels. The latter engine is the focus of this section.

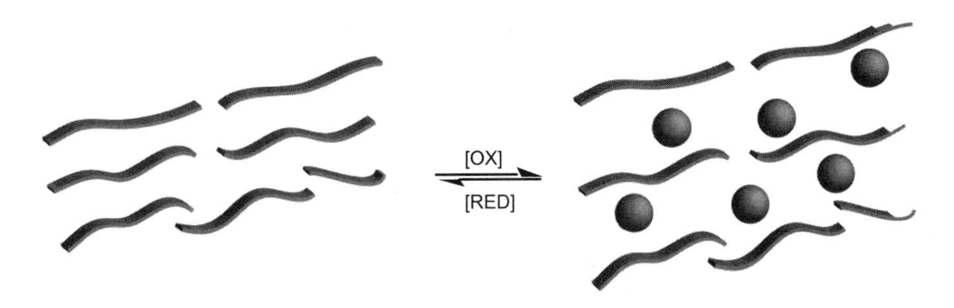

Figure 2: Depicting the bulk expansion/contraction cycle of an electroactive polymer. Ribbons represent polymer chains and spheres represent counter-ions.

9.2 Intrinsic Properties and Macroscale Translation

The unifying design strategy of engines described herein is that all information required for the system to function is encoded within a single molecule. By analogy to biological engines, this design requires that some stimulus (chemical, electrochemical, photochemical, etc.) induce a length-altering conformational change to occur within the molecule. Aggregates of such molecules would work in sync to amplify this stimulus-induced response, thus allowing a continuum of the same function up to the macroscale. Such a concept generates an intriguing question: is there an analogy between molecular engines vs. their macroscale counterparts and quantum vs. classical mechanics? This concept can be addressed directly by reviewing Kelly's pioneering work toward a molecular ratchet.[Kelly et al., 1998; Kelly et al., 1997]

A ratchet is a common mechanical device that allows only unidirectional rotation of a gearwheel. Two essential features of a ratchet are a pawl and wheel with *asymmetric* indentations. The asymmetry of the wheel works in unison with the pawl to allow rotation to occur readily in only one direction [see Fig. 3(a)]. Rotation at the *molecular level* implies rotation about a single bond. For example, rotation about the carbon-carbon bond of ethane is shown in Fig. 3(b), and generates an energy profile reflective of both spatial (steric) and electronic interactions between the two CH_3 units as they pass each other in space. Due to the symmetry of the molecule, each energy barrier is itself symmetrical about the midpoint. *Although technically incorrect to relate the symmetry of this energy barrier to the ease of directional rotation* (this will be explained later), it must be noted that rotation does occur with equal ease in both directions, as would be expected. As such, ethane is not a molecular ratchet.

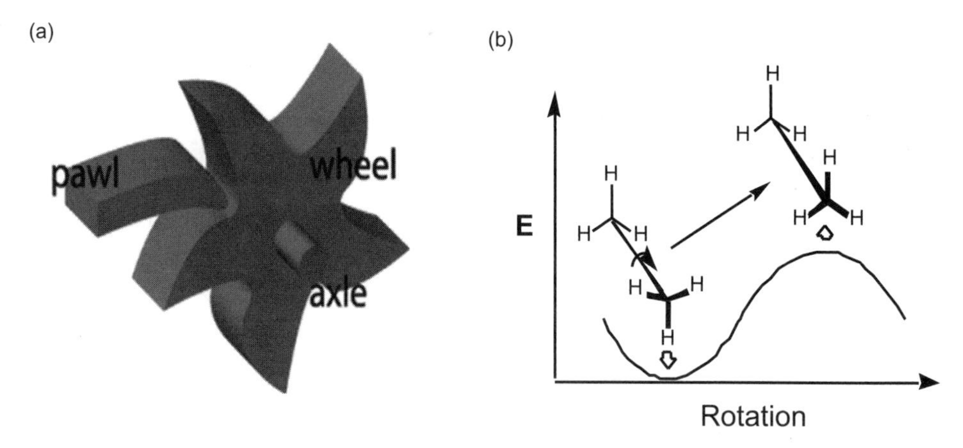

Figure 3: (a) A prototypical ratchet. (b) Energy profile for the rotation about the carbon-carbon bond of ethane.

Molecular gears have been designed such that synchronous rotation of two groups connected by a single bond (axis) is preferred over their independent rotation. Compound **1** exemplifies such a system [Iwamura et al., 1988]. As conceptually illustrated in Fig. 4, compound **1** is analogous to a gear in that the teeth of the macroscopic and molecular units (two tryptycene moieties) are locked into an interpenetrating gearing mechanism that dictates rotation. Once again, the symmetry of compound **1** mandates that the energy profile associated with rotation is symmetrical (similar to ethane, above), and rotation occurs with equal ease in either direction. As with ethane, compound **1** is not a ratchet.

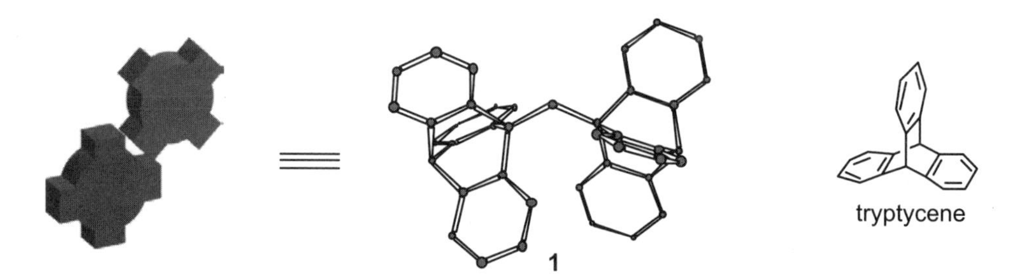

Figure 4: A molecular gear composed of two interdigitated tryptycene units.

In an oversimplification of Kelly's molecular ratchet design, one of the gears of compound **1** is replaced with a nonplanar molecular pawl. The pawl is in fact a helicene, and has both a concave and convex surface. When positioned between two teeth of a gear, rotation in one direction passes over the convex face, while rotation in the opposite direction passes over the concave face. The corresponding energy profile for each direction is asymmetric. It should be noted, however, that the start and end points for the process of passing one tooth completely over the pawl are identical, and are identical in energy. The energy profile is shown conceptually in Fig. 5. While classical mechanics might predict that rotation in the direction of the steeper slope would be disfavored, one cannot overlook the Second Law of Thermodynamics. Thus, at the molecular level, it is not the slope of the energy barrier that is important it is simply the height of the energy barrier. Since the energy barrier to rotation is equal in both directions (in Fig. 5, note symmetry of start and end point; see dashed line in energy profile), the molecular ratchet does not function as planned.

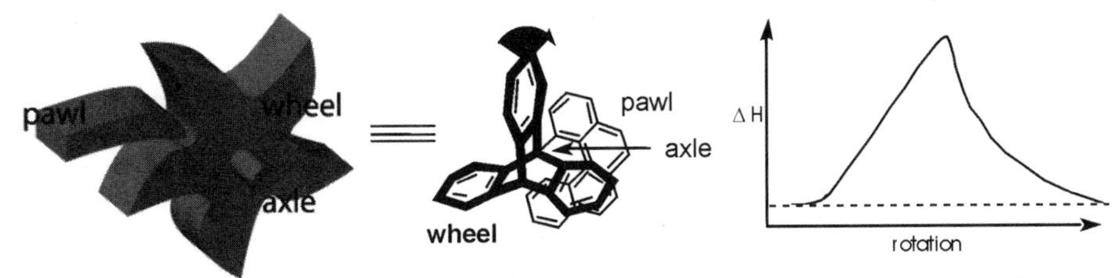

Figure 5: Kelly's "molecular ratchet" and corresponding energy profile for 120° rotation.

The previous example is illustrative of the potential shortcomings of the "macroscale translation" approach to designing molecular engines. When trying to translate a macroscopic device to a molecular device, it is imperative to realize that the molecule's *intrinsic properties* will dominate at the single-molecule level. In contrast, the "single molecule" design strategy first harnesses the *intrinsic* properties of the molecule, and then translates it across all levels. It should be noted, however, that Kelly was ultimately successful in accomplishing unidirectional rotation of a similar ratchet by a very clever mechanochemical process [Kelly et al., 1999]. Specifically, chemically reactive groups were attached to both the wheel and pawl of the original molecular ratchet. Ensuing chemical reactions locked the system in a relatively high-energy state, rendering only unidirectional rotation energetically favorable. This is even more fascinating when realizing that natural muscle is indeed a mechanochemical engine itself!

9.3 Stimulus-Induced Conformational Changes within the Single Molecule

Refocusing on the realm of biological muscles, it is now clear that a stimulus-induced conformational change is the foundation of most biological molecular engines. In order to mimic natural muscle function in a synthetic system, a molecule must be designed such that a macroscopic stimulus initiates a conformational change that in turn alters point-to-point (atom-to-atom) distance. Furthermore, these points must be capable of bonding together such that amplification of the conformational change can be achieved via a concerted response. In other words, what is needed is a polymer composed of monomers each capable of independent actuation. *Many molecules exhibit such a property, but few have been described for use in the design of artificial muscles.* Systems are discussed below based on their similarity to naturally occurring engines, *even for cases where their use as artificial muscles has not been explicitly stated in the original report.* The italics of the last sentence are included to draw attention to the fact that single-molecule muscles are still quite rare in the literature.

The spasmoneme spring is reported to be among the most powerful of biological engines [Mahadevan et al., 2000]. Upon exposure to calcium, this helical organelle undergoes change from an extended rod-like state (via rotation) to a contracted state. As such, this system can be classified as a mechanochemical engine. Moore has reported a random coil-to-helix transformation in synthetic *meta*-connected phenylacetylene oligomers as a function guest molecule [Nelson et al., 1997; Prince et al., 2000; Prince et al., 1999]. For example, when the *X* moiety in Fig. 6 is a cyano group, Ag^{+} ions induce helix formation. Moore has also demonstrated helix formation of *meta*-connected phenylacetylene oligomers induced via small chiral molecules [Prince et al., 2000]. Although the aforementioned systems were not described by the author as potential muscles, the fact remains that the endpoints of the oligomers should alter their distance upon transformation from random coil to helix. Since the conformation of Moore's system is a function of chemical stimuli, analogies to the spasmoneme spring are readily apparent. Possibilities for motility based on such a spring system are intriguing; especially in light of the fact that Moore's system and natural muscle could both be classified as mechanochemical engines.

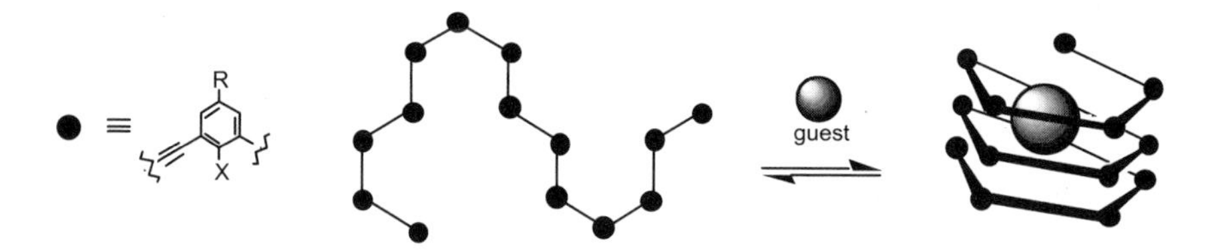

Figure 6: Moore's guest-induced random coil-to-helix interconversion.

Photochromic molecules are another class of molecules that exhibit stimulus-induced conformational changes. Despite this fact, the majority of literature pertaining to photochromes deals with their defining property of photochemical interconversion between two states that differ in optical absorbance. Although molecular switches are the more common photochrome-based devices [Irie, 2000], Stille and Zimmerman have utilized the stilbene photochrome in a pioneering work in the field of single-molecule *photomechanical* actuation [Zimmerman et al., 1985].

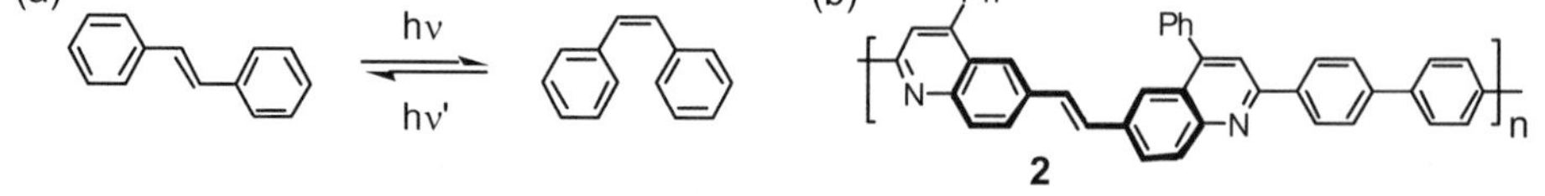

(a) hν hν' (b) Ph Ph N N n **2**

Figure 7: (a) Photoisomerization of stilbene. (b) Stilbene-based polymer exhibiting photomechanical actuation (stilbene moiety emphasized in bold).

Stilbene exists as two photochemically interconvertible isomers: *trans* and *cis* [Fig. 7(a)]. Clearly, the length of each isomer differs as a function of state, making stilbene an intrinsic photomechanical engine. Zimmermann and Stille utilized photoresponsive polyquinolines containing isomerizable stilbene units to demonstrate such a concept [Fig. 7(b)]. Accordingly, polymer **2** is rendered photomechanical due to the concerted action of the individual photoresponsive building blocks. Bulk studies of this polymer were performed. Due to dense packing of polymer chains, the studies were performed above the glass transition temperature of the polymer. In this state, individual polymer chains exhibit fluid-like mobility, thus allowing greater conformational freedom. Irradiation of the sample resulted in 4.8% contraction of the bulk. In contrast to Kelly's original top-down design, the single-molecule approach used here successfully translated intrinsic molecular properties to the macroscale.

EAPs have been briefly discussed herein (*vide supra*) and are reviewed elsewhere in this book. However, the majority of existing reviews focus on traditional EAPs that function via the bulk process of ion-solvent intercalation/expulsion. As described above, such a mechanism is not an intrinsic property, and cannot be extrapolated to the single-molecule level. Driven in part by the recent interest in single-molecule devices, we have pursued an alternate mechanism for EAPs that *is* based on intrinsic molecular properties [Marsella et al., 1999]. The remainder of this section examines this alternate approach. Unless otherwise noted, the following account of the cyclooctatetrathiophene (COTTh) system summarizes our previously reported results, and the reader should be directed to the original literature for a more detailed description of this work [Marsella et al., 1999].

Aromaticity is a property of planar, cyclic organic molecules that possess 4n + 2 number of electrons in the p_z-orbitals. It is well known that aromatic compounds exhibit greater stability than their nonplanar or acyclic counterparts. Thus, given the opportunity to exist in either a nonplanar or planar conformation, an aromatic compound will favor the latter. Conversely, antiaromaticity is achieved when the electron count in similar systems is changed to 4n number of electrons. As suggested by name, antiaromatic compounds contrast aromatic compounds and are consequently less stable than their acyclic or nonplanar counterparts. Given the identical aforementioned conformational option, antiaromatic compounds will avoid planarity. The redox chemistry of the eight-membered ring, cyclooctatetraene (COT), is a classic study of the effect of electron count on conformation, and must be briefly addressed [Fry, 1986].

COT can be envisioned to exist in two distinct conformations: tub and planar (Fig. 8) [Garratt, 1986]. With each of eight carbon atoms contributing one electron into a p_z orbital, COT would be an antiaromatic system if planar. Consequently, in the neutral state, COT exists in the tub conformation. Aromaticity can be achieved by either removal or addition of two electrons (+2 or –2 overall charge, respectively), resulting in the requisite conformational change from tub to planar. This is, of course, an oversimplification as electron count is only one factor in conformational preference. Indeed, addition of only

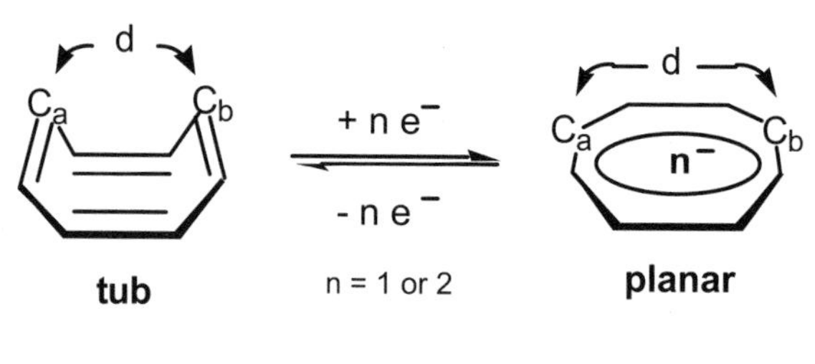

Figure 8: The redox-induced tub-to-planar conformational change of cyclooctatetraene (COT).

one electron to the system has shown to result in a flattening of the ring. This is actually advantageous, since lower oxidation states could provide electromechanical actuation. Excellent reviews regarding the physical organic chemistry of COT exist [Paquette, 1992], and greater detail will not be presented here. Of key importance is the fact that a distance change occurs between carbons as a function of oxidation state. Thus, COT is an intrinsic electromechanical actuator.

In principle, poly(cyclooctatetraene) should be an intrinsic EAP. However, practical issues must be addressed. First, is the synthesis of poly(cyclooctatetraene) a facile venture? Second, if prepared, would the polymer be stable to redox cycles under ambient conditions? In both cases, the answer is probably negative. The synthesis of a poly(cyclooctatetraene) precursor (monomer) would not be straightforward, nor would an all-carbon based conducting polymer be predicted to have high stability. Case in point would be polyacetylene, a highly conductive polymer possessing poor environmental stability [Skotheim, 1986]. Of conducting polymer candidates, polythiophene is a leading candidate [Skotheim, 1986; Roncali, 1992]. The chemistry of both the thiophene ring system and polythiophenes in general is well established in the literature [McCullough, 1998; Roncali, 1992]. Furthermore, depending on substitution, polythiophenes are among the most environmentally robust conducting polymers. Consequently, the design of a single-molecule EAP appears to have newfound requirements.

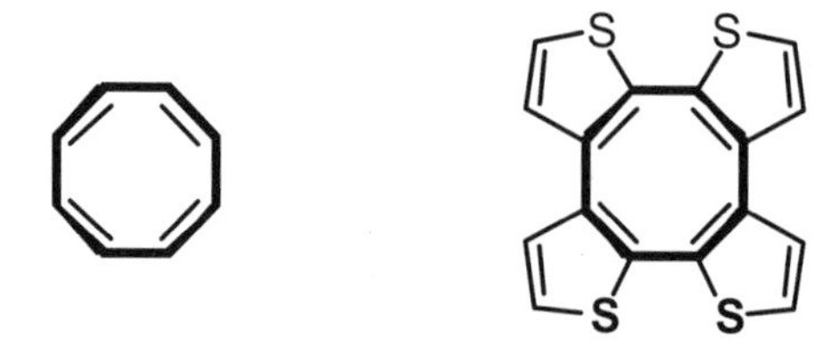

Figure 9: COT and cyclooctatetrathiophene (COTTh) – a "masked" COT.

In 1978, Kaufman prepared a novel ring system: cyclooctatetrathiophene [Kauffmann et al., 1978]. In this molecule, four thiophene rings are joined at their 2- and 3-positions to form a non-planar, cyclic system. At its core, cyclooctatetrathiophene is a "masked" cyclooctatetraene: it possesses eight contiguous sp^2 carbons that establish a non-planar ring of eight p_z electrons (Fig. 9). The tub, or saddle, shape of the molecule is evident from its crystal structure. Viewed as a monomer, the polymerization of any two α-positions on the system would result in a polythiophene derivative. Hence, Kaufmann's compound would make the ideal building block for a single-molecule EAP *as long as it maintains the redox induced conformational changes of its parent cyclooctatetraene.*

Prior to our report, little was known about the cyclooctatetrathiophene system. Indeed, even the crystal structure of the neutral compound had not been reported. Although we were able to solve the structure for the neutral state, confirmation of the oxidized states remained elusive. Because of this fact, the conformational consequences of redox chemistry were investigated by means of molecular modeling.

AM1 semi-empirical calculations were first performed [Foresman et al., 1996] on the neutral species in order to validate the level of modeling by comparing the calculated results with those obtained from x-ray crystallography. For clarity, the calculated redox-induced conformational perturbations are reported herein as a function of the two S-C-C-S dihedral angles, ϕ (Fig. 10). As this angle approaches 0°, the molecule approaches planarity, and the two carbons denoted above move further apart in space. Indeed, AM1 calculations ($\phi = 45.9°$) are in good agreement with x-ray data (two unique values, $\phi^1 = 47.6°$ and $\phi^2 = 46.4°$), with the differences being ascribed to the comparison of solid state (x ray) vs. gas phase (modeling). As described, oxidation of the "masked" cyclooctatetraene moiety should perturb its conformation. By analogy to cyclooctatetraene, removal of one or two electrons should promote a conformational change toward planarity. In the case of one electron oxidation, a decrease of ϕ is driven by better p_z orbital overlap, thus better delocalizing charge. For two-electron oxidation, the same argument is true; however, planarity is further augmented by attaining some degree of aromatic character from the newly formed 4n + 2 ring system. It should be noted that discussion is kept to oxidation only. Furthermore, it is reduction of cyclooctatetraene that abounds in the literature. It is not contradictory, but simply a matter of fact that cyclooctatetrathiophene is more stable when oxidized, and cyclooctatetraene is more stable when reduced. The difference in charge does not inherently prevent a qualitative comparison between the two systems. Most importantly, AM1 calculations do indeed predict the desired trend of a conformational change toward a more planar system ($\phi^1 = 31.7°$ and $\phi^2 = 44.6°$ for $\mathbf{1}^{+1}$ (compound **1**, +1 charge); and $\phi = 29.7°$ for $\mathbf{1}^{+2}$). Given the unique value of ϕ associated with each oxidation state, it follows that poly(COTTh) should exhibit a redox-dependent attenuation of conformation, which should translate into an overall attenuation of electromechanical actuation. The advantages between attenuation of response and an all-or-nothing response should be noted.

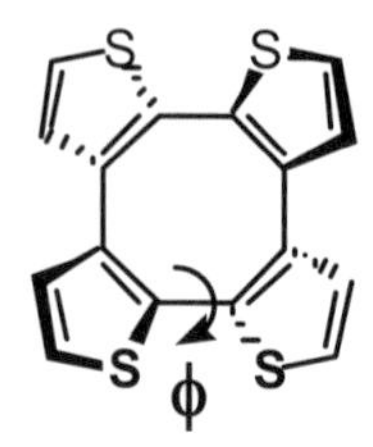

Figure 10: Dihedral angle, ϕ, which implicates the degree of ring planarity.

The connectivity of the corresponding polymer backbone has profound consequences on the behavior of the system. Two scenarios are possible, and are shown below in Fig. 11. In the first case, helical polymer architecture may result. Flattening of the individual monomers would then result in a compression of the helix. In other words, the expansion/contraction cycle would be analogous to a spring. With regard to electrical conductivity, it is this connectivity that is most favored, based on the fact that long-range conjugation is maintained. The second scenario links the monomer at the two carbons that undergo the largest distance change during a redox cycle. In this system, the polymer is expected to expand and contract in a linear manner, similar to the bellows of an accordion. Unfortunately, this connectivity destroys long-range conjugation, and may limit electrical conductivity and/or electrochemical stability. The remainder of this discussion focuses on the putative helical arrangement.

Realizing that cyclooctatetrathiophene exhibits the responses desirable for a single- molecule EAP, its unique nature among benzoannulated cyclooctatetraenes should be addressed. Recently, Rathore and Kochi have described a molecular switch based on tetramethoxytetraphenylene [Rathore et al., 2000]. This compound is structurally similar to COTTh, with the exception that the aromatic thiophene rings are replaced by the classically aromatic benzene ring (Fig. 12). In this substitution, the peripheral aromatic rings are expanded from 5-membered rings to 6-membered rings. The spatial consequences of this ring expansion are evident from the resulting electrochemical processes. In contrast to COTTh, Kochi's system does not move toward planarity upon reduction. In fact, a reversible carbon-carbon bond is formed that further locks the system into a tub shape. An x-ray crystal structure for both the neutral and reduced forms confirms this process. Clearly, the seemingly minor structural change of replacing thiophene rings with those of benzene completely inhibits the COTTh EAP mechanism.

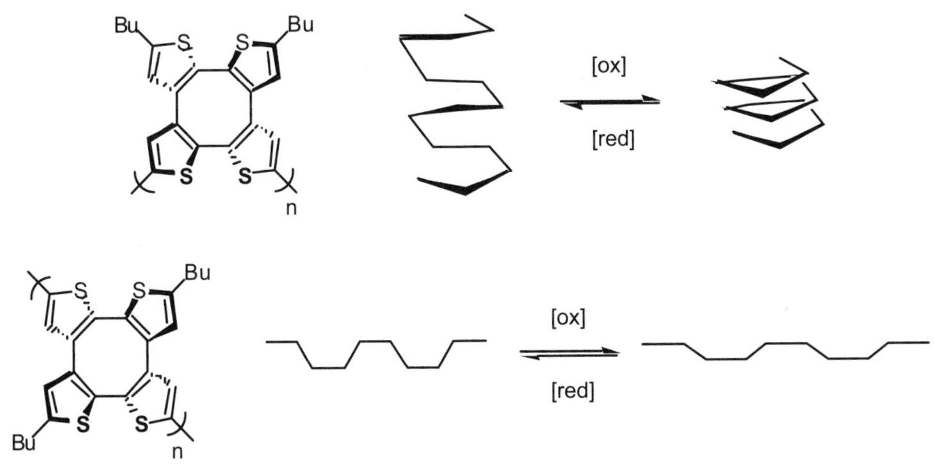

Figure 11: The consequences of alternate backbone connectivity in poly(COTTh).

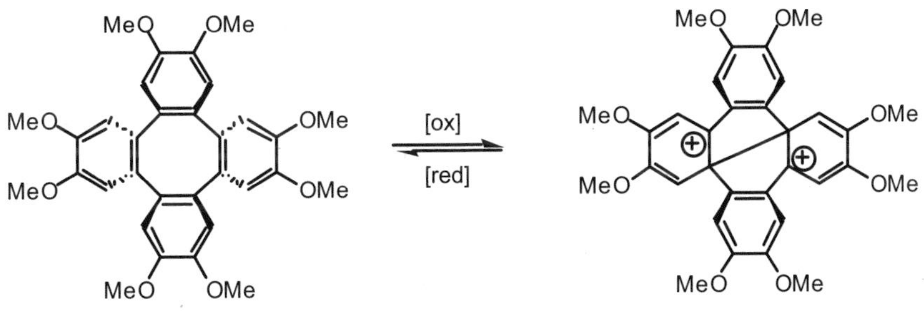

Figure 12: The consequences of redox chemistry on tetramethoxytetraphenylene.

Thin films of poly(COTTh) were prepared, and both their electrical conductivity as a function of oxidation state and their cyclic voltammetry were investigated. In the case of the former, the peak conductivity was in good agreement with other poly(thiophenes) reported to have a similar degree of backbone non-planarity (as determined by ϕ) [Skotheim et al., 1998]. Despite similarities in electrical conductivity, the cyclic voltammetry of poly(COTTh) is in stark contrast to traditional polythiophenes. First, the peak oxidation potential of the polymer is quite high, occurring at 1.6 V vs. a silver wire reference. Secondly, the oxidation is electrochemically irreversible. More specifically, the kinetics of electron transfer seem to be quite slow. Both of the aforementioned observations indicate a conformational change occurring within the polymer. The fact that the electrochemistry is being performed in the solid state mandates that the interchain interactions slow conformational changes (recall the discussion of the glass transition temperature, above). Furthermore, the relatively high oxidation potential corroborates with Paquette's electrochemical studies of substituted COTs; the greater the substitution the more difficult the reduction [Fry, 1986]. Illustrative examples of Paquette's studies are given in Fig. 13.

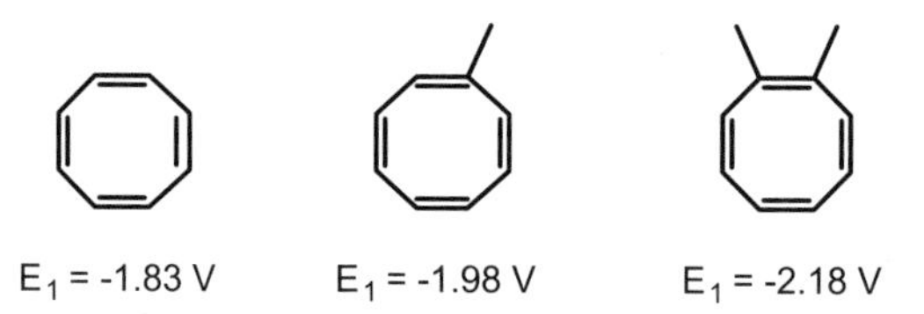

Figure 13: Reduction potentials of COT and its substituted analogs.

Of course, the key question is whether or not poly(COTTh) does indeed behave as a single-molecule EAP. This is not an easy question to answer. Why? Recall the chain of events: a film of poly(COTTh) is placed on an electrode, and electrochemical oxidation is performed in a suitable electrolyte solution. This is a bulk process that results in a charged polymer. Balancing the charge are the requisite counter-ions and solvating species intercalated into the bulk from solution. Consequently, the "traditional" mechanism of counter-ion intercalation/expulsion also takes place! This process overwhelms the more subtle conformational changes, and the two cannot be discerned using the traditional tools of conducting polymer EAP analysis. Indeed, true analysis of single-molecule EAP in poly(COTTh) requires studies on single, isolated polymer chains. While these studies are being pursued, molecular modeling can again be used to qualitatively shed light on single-molecule EAP.

As before, the value of ϕ corresponding to the peak conductivity of poly(COTTh) (1.6 V vs. Ag°) can be estimated by calculating the structure of the repeat unit at the appropriate oxidation state. From the CV, it can be determined that the charge per repeat unit at 1.6 V is +0.6. Approximating this oxidation state by calculating the structure of both COTTh-dimer and COTTh-dimer^{+1} reveals a change in ϕ from $\phi = 44.1°$ to $\phi = 40.5°$ (respectively). Although this approximation underestimates the charge per repeat unit at peak conductivity, it does follow the expected trend predicted by the modeling studies of COTTh (*vide supra*), and confirms that redox-induced conformational changes in poly(COTTh) are expected, and should lead to single-molecule EAP. Studies into this system continue.

Recently, Swager has reported an alternate single-molecule design strategy for single-molecule EAP (Fig. 14) [Yu et al., 2000]. In this design, a flexible calix[4]arene serves as a repeating pivot within a polymer, on an individual basis being analogous to an elbow joint. Continuing with this analogy, the forearm and biceps (engine) are composed of tetrathiophene units and are directly connected to the elbow. In the neutral state, the system is relaxed and the distance between the two distal positions of the tetrathiophenes is 31.5 Å. Upon oxidation, it is envisioned that pi-pi stacking occurs to delocalize the positive charge. In the process, the aforementioned distance shrinks to 3.5 Å. As with poly(COTTh), the analogy here is again reminiscent of the bellows of an accordion. It is important to note that these distances were determined by molecular mechanics calculations, and, as with poly(COTTh), the physical demonstration of this dimensional change is still under investigation. If proven successful, the expected increase in distance should lead to the largest reported volume change of any such electromechanical actuators investigated to date.

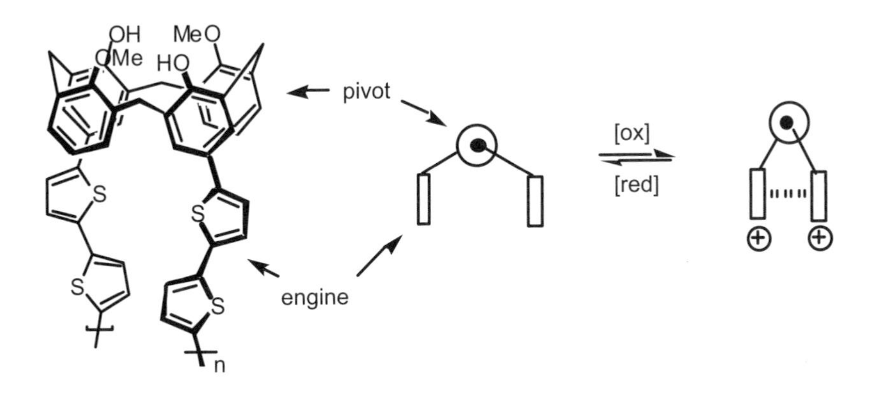

Figure 14: A calixarene-thiophene hybrid polymer anticipated to exhibit very large differences between expanded and contracted states.

9.4 Final Comments

All previous examples were presented as an introduction to the exciting possibilities that exist in the development of single-molecule mechanical devices, and were not intended to constitute a comprehensive review. Interestingly, however, reviews of this area do exist, although they are masked via alternate semantics. Specifically, the fact remains that single-molecule EAP evolves from intrinsic properties of molecules. At the simplest level, all that is required for single-molecule EAP is a molecule that undergoes a stimulus-induced conformational change. Such systems abound, and have been studied for purposes other than those described herein. Since EAP at the nanolevel is an entirely new concept, it is not surprising that the link between new and old is only now becoming apparent. Regardless, the foundation of single-molecule EAP has, in many respects, been previously established by numerous researchers working "ahead of their time." The present focus of research in this area would benefit greatly from harnessing the intrinsic properties of existing and well-characterized molecules. With such a great pool of resources with which to build, it is anticipated that rapid advances in single-molecule actuation (electromechanical, photomechanical, and chemomechanical) should be attained within the near future.

9.5 References

Baughman, R. H., *Synthetic Metals* **78**, 339 - 353 (1996).

Foresman, J. B., Frisch, A. *Exploring Chemistry with Electronic Structure Methods*, Gaussian, Inc., Pittsburgh, 2nd ed., 1996. The corresponding software used in this study was Gaussian 98W (Revision A.7), Frisch, M. J.; Trucks, G. W.; Schlegel, H. B.; Scuseria, G. E.; Robb, M. A.; Cheeseman, J. R.; Zakrzewski, V. G.; Montgomery, J. A.; Stratmann, R. E., Burant, J. C. Dapprich, S.; Millam, J. M.; Daniels, A. D.; Kudin, K. N.; Strain, M. C.; Farkas, O.; Tomasi, J.; Barone, V.; Cossi, M.; Cammi, R.; Mennucci, B.; Pomelli, C.; Adamo, C.; Clifford, S.; Ochtersji, J.; Petersson, G. A.; Ayala, P. Y.; Cui, Q.; Morokuma, K.; Malik, D. K.; Rabuck, A. D.; Raghavachari, K.; Foresman, J. B.; Cioslowski, J.; Ortiz, J. V.; Stefanov, B. B.; Liu, G.; Liashenko, A.; Piskorz, P.; Komaromi, I.; Gomperts, R.; Martin, R. L.; Fox, D. J.; Keith, T.; Al-Laham, M. A.; Peng, C. Y.; Nanayakkara, A.; Gonzolez, C.; Challacombe, M.; Gill, P. M. W.; Johnson, B. G.; Chen, W.; Wong, M. W.; Andres, J. L.; Head-Gordon, M.; Replogle, E. S.; Pople, J. A., Gaussian, Inc., Pittsburgh PA, 1998.

Fry, A. J., in *Topics in Organic Electrochemistry* A. J. Fry, W. E. Britton, Eds. Plenum Press, New York, pp. 1 - 32 (1986).

Garratt, P. J. *Aromaticity,* John Wiley & Sons, New York (1986).

Irie, M., *Chemical Reviews* **100**, 1683 (2000).

Iwamura, H., K. Mislow, *Acc. Chem. Res.* **21**, 175 - 182 (1988).

Kauffmann, T., B. Greving, R. Kriegesmann, A. Mitschker, A. Woltermann, *Chem. Ber.* **111**, 1330-6 (1978).

Kelly, T. R., H. De Silva, R. A. Silva, *Nature* **401**, 150-152 (1999).

Kelly, T. R., J. P. Sestelo, I. Tellitu, *Journal of Organic Chemistry* **63**, 3655-3665 (1998).

Kelly, T. R., I. Tellitu, J. P. Sestelo, *Angewandte Chemie-International Edition in English* **36**, 1866-1868 (1997).

Mahadevan, L., P. Matsudaira, *Science* **288**, 95-99 (2000).

Marsella, M. J., R. J. Reid, *Macromolecules* **32**, 5982-5984 (1999).

McCullough, R. D. *Advanced Materials* **10**, 93 (1998).

Nelson, J. C., J. G. Saven, J. S. Moore, P. G. Wolynes, *Science* **277**, 1793-1796 (1997).

Paquette, L. A., *Advances in Theoretically Interesting Molecules* **2**, 1 - 77 (1992).

Prince, R. B., S. A. Barnes, J. S. Moore, *Journal of the American Chemical Society* **122**, 2758-2762 (2000).

Prince, R. B., T. Okada, J. S. Moore, *Angewandte Chemie-International Edition* **38**, 233-236 (1999).

Rathore, R., P. Le Magueres, S. V. Lindeman, J. K. Kochi, *Angewandte Chemie-International Edition* **39**, 809-812 (2000).

Roncali, J. *Chem. Rev.* **92**, 711-38 (1992).

Skotheim, T. A., R. L. Elsenbaumer, J. R. Reynolds, Eds., *Handbook of Conducting Polymers,* Marcel Dekker, Inc., New York (1998).

Skotheim T. J., Ed., *Handbook of Conducting Polymers*, Dekker, New York (1986).

Yu, H., A. E. Pullen, T. M. Swager, "Toward new actuating devices: synthesis and electrochemical studies of Poly(11,23-bis[2,2'-bithiophen]-5-yl)-26,27-dimethoxycalix[4] arene-25,27-diol)," Charbonneau, Ed., National Meeting of the American Chemical Society, Washington, D.C. American Chemical Society (2000).

Zimmerman, E. K., J. K. Stille, *Macromolecules* **18**, 321-327 (1985).

TOPIC 4

MODELING ELECTROACTIVE POLYMERS

CHAPTER 10

Computational Chemistry

Kristopher E. Wise
National Research Council, NASA Langley Research Center

10.1 Introduction

Enhancement of the capability of electroactive polymer (EAP) materials to induce large strain and stress can be accomplished by conventional enhancement techniques as well as through computational chemistry. Enhancing the load carrying capability of EAP materials can be obtained by increasing their stiffness with a minimum compromise of their displacement actuation. Such an enhancement requires taking into account the properties of the produced material and their compatibility with the operation conditions. Additional enhancement can be obtained by allowing for internal molecular realignment in response to Coulomb forces. Such improvement can be applicable to the electrostrictive polymers however obtaining improvement of other types of materials requires fundamental research at the molecular level as can be obtained by computational chemistry.

Modeling EAPs can be performed on a number of different levels, including:

(1) Atomistic—Using quantum mechanics and molecular mechanics, the molecular sizes, shapes and flexibility as well as the electronic properties (spectroscopy, ionization, etc.), can be predicted;
(2) Molecular—Molecular dynamics techniques can be used to determine the bulk material polarization, diffusion, etc.;
(3) Meso—Monte Carlo techniques are employed to study the thermodynamics, phase behavior, and large-scale motions;
(4) Macro—This is the level of continuum mechanics and finite element modeling (FEM) that determines the mechanical behavior of EAP.

Important insights arise when interdisciplinary research is carried across these levels. Some of the current research efforts are concentrated on developing methods to understand the nature and behavior of charge carriers in polymers [Borodin and Smith, 1998]. The results are helping to facilitate the development and optimization of electrostrictive, piezoelectric, redox-driven and conducting EAPs. Quantum mechanics (primarily semiempirical parameterizations) are used to determine molecular conformations and flexibility of segments of electroactive polymers, and electrical properties (dipole moments, polarizabilities, etc.). To simulate bulk properties, cells on the order of 20 Å on a side are being built with periodic boundary conditions. New methods are being sought to handle long-range Coulomb forces in molecular dynamics (MD) simulations.

Recent work at the NASA LaRC's Computational Materials Program used accurate quantum chemistry calculations to determine force fields for a range of polymers including polyimides. The calculated force field was experimentally verified (through thermophysical and ultrasonic measurements). The method was used to predict response to electric fields, mechanical stresses, and temperature. Planes of large spheres represent the metal electrodes and are used to simulate the poling field. The properties from atomistic simulations are fed into large-scale finite-element models (see Fig. 1). So far, successful models at the atomistic, micromechanical, and continuum levels have been developed.

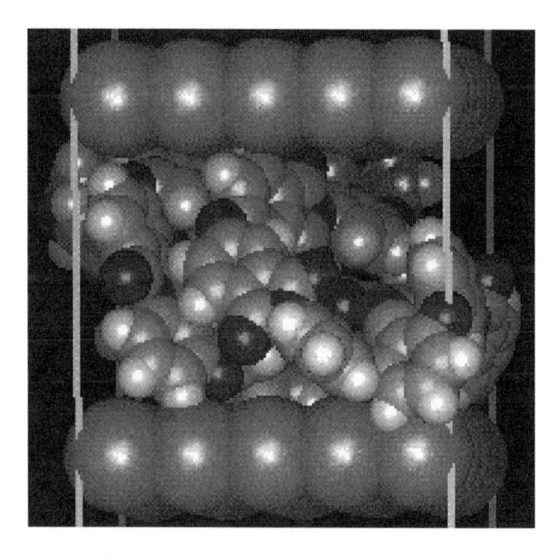

Figure 1: Snapshot from an MD simulation of a piezoelectric polyimide.

10.2 Overview of Computational Methods

The aim of this chapter is to provide a summary of the various approaches most commonly employed in studies of EAP materials. While little has appeared in the literature regarding atomistic modeling of artificial muscles, reports on general polymer modeling could easily fill a large volume [Bicerano, 1992; Binder, 1995; Raabe, 1998]. To maintain a manageable length, this chapter focuses primarily on methods useful in predicting polymer response to external electric fields. Understanding this response is particularly important because piezoelectric polymers, an important class of materials in artificial muscle research, must be poled in a strong electric field before they exhibit their desired properties [Bauer, 1996; Eberle et al., 1996].

Over the past decade, computational chemistry has emerged as an indispensable component in most materials research programs. Computational chemistry, or molecular modeling, encompasses a collection of methods that are selectively applied based on the spatial or temporal size of the problem and on the degree of accuracy that is required. In some instances, one might be willing to accept qualitative accuracy in exchange for the ability to model much larger systems. Another study, however, might demand near quantitative results, albeit on a single molecule scale. Fortunately, well-developed techniques are available to treat the continuum of problems encountered by materials researchers.

This chapter begins by describing the most accurate methods currently available for treating small systems, particularly ab initio molecular orbital methods and density functional theory. This is followed by a description of molecular dynamics and Monte Carlo techniques, classical methods relying on accurate parameterization to either ab initio results or experimental data. These methods are very popular in polymer modeling due to their ability to generate reliable property predictions for systems containing thousands of atoms. Finally, a brief discussion of the rapidly developing field of mesoscale modeling is provided. Positioned between atomistic and continuum mechanical techniques, this relatively new field appears to be a promising avenue for future work.

All of these methods have in common the twin problems of scale and accuracy, and the spatial or temporal scale limits each method so that it may reasonably accommodate and/or be as accurate as it is capable of achieving. Present research is focused on extending the applicability of the individual methods and on finding ways of combining them to take advantage of their respective strengths. The potential payoff of this work is enormous, particularly in terms of design efficiency. Computational chemistry allows researchers to achieve desired materials properties while saving considerable amounts of time and money in the development process. The obvious benefits to device design and development are compounded by advances in physical insight, which lead to improved theoretical methods. This section attempts to highlight both the present challenges and potential rewards of the application of computational chemistry to EAPs.

As mentioned above, piezoelectric polymers are a promising class of materials for use in artificial muscles. A common technique used in their production is poling, which is often done while simultaneously stretching and/or heating the film. In the poling process, the permanent dipoles in the polymer preferentially align in the direction of a strong electric field applied across the thin film. Poly(vinylidene fluoride), PVDF, after stretching and poling, is a semicrystalline polymer exhibiting both crystalline and amorphous regions. Because of its high piezoelectric response and complex morphology, PVDF (shown in Fig. 2) has been studied extensively both experimentally and theoretically [Sessler, 1981; Bauer, 1996; Eberle et al., 1996]. For this reason, the examples given to illustrate the computational methods described below are drawn, wherever possible, from the PVDF literature.

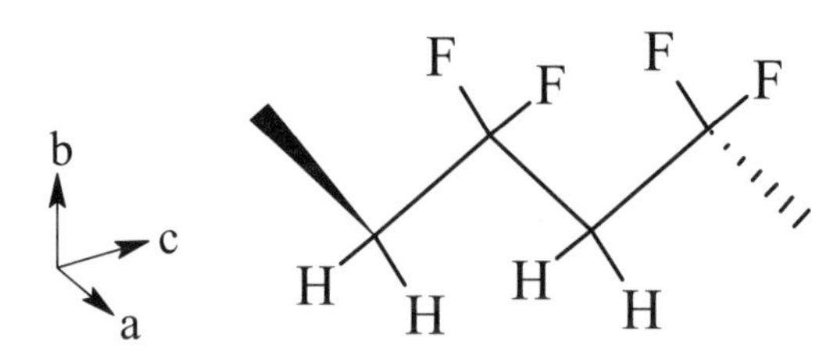

Figure 2: Dimer of poly(vinylidene fluoride) illustrating connectivity and crystal coordinate system referred to in the text and used in Fig. 3.

In preparing this chapter, it was assumed that the reader would not be an expert in the methods described. With this in mind, it seemed appropriate to include a moderate level of technical detail in hopes of exposing some of the mathematical machinery too often hidden from the user of modeling software. A basic knowledge of the physical models and assumptions used in deriving the various computational methods will hopefully instigate a healthy skepticism of the results of these calculations. When deprived of these details, researchers have little choice but to blindly accept or reject computational predictions. An informed user, on the other hand, can make productive use of computational chemistry, within its current limitations. As the artificial muscle research community uses these methods and communicates the results, future reviews of this topic will be much more application oriented.

10.3 Quantum Mechanical Methods

In studies requiring quantitatively accurate structural, energetic, or spectroscopic property predictions, quantum mechanical methods are used almost exclusively. Ab initio molecular orbital (MO) methods [Hehre et al., 1986; Szabo et al., 1996] and density functional theory (DFT) [Parr et al., 1994], in conjunction with appropriate basis sets, are capable of producing results accurate to within experimental error in many cases. While the size of the molecule that may be studied is limited by computer resources, calculations on molecules containing 20–30 heavy atoms are not uncommon.

The starting point for MO methods is, within the Born-Oppenheimer approximation, the electronic Schrodinger equation:

$$\hat{H}\Psi = E\Psi .$$

The description given here is applicable to a system containing n electrons moving in a static external field, $\upsilon(r)$, generated by the group of A nuclei and any external field source. The Hamiltonian for the system may be written as:

$$\hat{H} \equiv \hat{T} + \hat{V} + \hat{U} \,,$$

where [Hehre et al., 1986; Szabo et al., 1996]:

$$\hat{T} \equiv -\frac{1}{2}\sum_{i=1}^{n}\nabla^2(i)$$ is the electronic kinetic energy;

$$\hat{V} \equiv \sum_{i=1}^{n}\upsilon(r_i) = \sum_{i=1}^{n}\sum_{\alpha=1}^{A}\frac{-Z_\alpha}{|R_\alpha - r_i|}$$ is the electron-nuclear Coulombic attraction;

$$\hat{U} \equiv \sum_{i<j}^{n}\frac{1}{|r_i - r_j|}$$ is the electron-electron Coulombic repulsion.

In ab initio MO calculations, the total molecular wave function is represented by a combination of single-electron wave functions, referred to as molecular orbitals [Hehre et al., 1986]. Each of these molecular orbitals has an associated orbital energy. The total energy of a molecule is calculated by summing the products of the molecular orbital energies with their respective occupation numbers. The molecular orbitals used in most calculations are constructed by forming linear combinations of a number of atomic orbitals, usually represented by Gaussian functions with the appropriate radial and angular symmetry (s, p, d, f...). Obtaining an accurate representation of the electron density distribution around even a moderately sized molecule may require hundreds of atomic orbitals, leading to a corresponding increase in the amount of computational effort required.

The results of MO calculations may be systematically improved by including post-Hartree-Fock electron correlation effects and by increasing the size and flexibility of the basis set [Hehre et al., 1986; Szabo et al., 1996]. Inclusion of correlation effects, commonly treated using Moller-Plesset perturbation theory or configuration interaction techniques, can dramatically improve both calculated structures and energetics, particularly in molecules with complex electronic structure and/or low lying excited states. Both of these situations are common in EAPs. Basis set improvement is typically achieved by including basis functions that allow diffuse density distributions (higher principal quantum number) or polarization functions (higher orbital angular momentum quantum number). By variationally optimizing the occupation of these higher energy atomic orbitals, a better overall description of the molecular charge distribution may be obtained. Extensive treatments of basis sets and their relative accuracy may be found in several excellent reviews [Davidson et al., 1986; Hehre et al., 1986; Feller et al., 1991; Shavitt 1993]. While employing these two techniques can significantly improve accuracy, they have the unfortunate side effect of increasing computational cost exponentially.

After being used for many years in the solid state community, density functional theory (DFT) has recently become popular with chemists and molecular physicists [Dreizler et al., 1991; Labanowski et al., 1991; Ziegler, 1991; Parr et al., 1994; Kohn et al., 1996]. The reason for its success lies in its ability to generate results often on par with correlated ab initio MO calculations in about the same time as a normal Hartree-Fock calculation. DFT differs from MO methods primarily in the use of the ground state electron density as the basic variable rather than an approximate wave function. This is possible because of the work of Hohenberg and Kohn [1964], which shows that there exists a direct correlation between the external potential, $\upsilon(r)$, the electron density, $\rho(r)$, and the ground-state wave function, Ψ_0. If the exact ground state electron density is known, solving the Schrodinger equation in this way yields an exact electronic energy of:

$$E[\rho(r)] = \int \upsilon(r)\rho(r)dr_1 + F[\rho(r)],$$

or equivalently,

$$E[\rho(r)] = \sum_{\alpha} \int \frac{Z_\alpha \rho(1)}{r_{1\alpha}} dr_1 + F[\rho(r)],$$

where

$$F[\rho(r)] \equiv \left\langle \Psi[\rho(r)] \middle| \hat{T} + \hat{U} \middle| \Psi[\rho(r)] \right\rangle.$$

These equations can be expanded to give the familiar expression:

$$E_{el} = -\frac{1}{2}\sum_{i=1}^{n} \left\langle \psi_i(r_1) \middle| \nabla^2 \middle| \psi_i(r_1) \right\rangle - \sum_A \int \frac{Z_A \rho(r_1)}{|r_1 - R_A|} dv_1 +$$

$$\frac{1}{2}\iint \frac{\rho(r_1)\rho(r_2)}{|r_1 - r_2|} dr_1 dr_2 + E_{xc}[\rho],$$

where the first term accounts for the kinetic energy of the n electrons, calculated as independent particles while using the same electron density as the actual system of interacting electrons. The second term represents the electron-nuclear attractive forces and the third accounts for the repulsive Coulomb interaction between charges at r_1 and r_2. The final term, the exchange-correlation energy, accounts for all remaining nonclassical effects not included in the other terms. Elimination of correlation from this final term gives a formal expression very similar to that of Hartree-Fock theory.

The orbitals, ψ_i, in the preceding equation are not the familiar ones of molecular orbital theory. Rather, they are the Kohn-Sham single particle orbitals [Kohn et al., 1965], determined self-consistently by solving:

$$\left(-\frac{1}{2}\nabla^2 + \upsilon(r) + \int \frac{\rho(r_2)}{|r_1 - r_2|} dr_2 + \upsilon_{xc}(r)\right) = \varepsilon_j \psi_j(r_1),$$

subject to the constraints:

$$\rho(r) = \sum_{j=1}^{n} |\psi_j(r)|^2,$$

and

$$\upsilon_{xc}(r) = \frac{\delta E_{xc}[\rho(r)]}{\delta \rho(r)}.$$

Assigning physical meaning to these orbitals is ambiguous, and they are generally treated as convenient mathematical tools to simplify evaluating the energy. Similarly, the eigenvalues, ε_j, of the Kohn-Sham orbitals should not be interpreted as molecular orbital energies.

Within the context of the assumptions described above, the only term in these equations that presently cannot be determined exactly is the exchange-correlation functional. As its exact form is unknown, evidently its corresponding energetic contribution to the overall electronic energy of the system can only be approximated. Clearly, finding accurate expressions for the exchange-correlation energy is an important task. Unfortunately, there is presently no known way of systematically improving these functionals. In MO methods, on the other hand, systematic improvement is possible by approaching the limit of a complete basis set and full treatment of configuration interaction. However, even for small systems, this sort of treatment is not possible, thus placing a practical limitation on the accuracy of these methods. As mentioned above, DFT methods have the distinct advantage of requiring only slightly more time for energy evaluation than the Hartree-Fock method. Therefore, if an exact exchange-correlation functional could be found, it would be possible to calculate exact energies in a very expedient manner.

Several classes of functionals have been developed over the years and are referred to by the way in which they utilize the electron density distribution. The simplest method is the local density approximation [Kohn et al., 1965]. In this method, the energy of each particle is evaluated by considering only the influence of the electron density distribution in its immediate vicinity. Another step forward was the realization that considering not just the value of the electron density surrounding a particle, but also its gradient could improve energy evaluation. This approach determines the rate of increase or decrease in the density distribution on moving away from the particle of interest, and factors this into the expression used to evaluate the potential, thus allowing the more distant (or non-local) portions of the density distribution to influence the energy evaluation, resulting in improved accuracy [Labanowski et al., 1991; Parr et al., 1994; Kohn et al., 1996].

It turns out that each of these different approaches tends to be better at calculating some properties than others. It has been found, however, that taking linear combinations of the energies produced by different functionals can give a composite energy that is more accurate than the result of any single method. One of the most successful is the B3LYP method [Becke, 1988; Becke, 1993].

The final class of quantum mechanical approaches commonly employed in polymer studies is the semiempirical molecular orbital methods. By incorporating certain simplifying approximations and parameterizing to experimental data, semiempirical molecular orbital methods expand the size of treatable systems to hundreds of atoms. Caution is required when using semiempirical techniques, however, as their accuracy can vary between near quantitative to qualitatively incorrect, depending on the system. This variance in accuracy is due to both the nature of the simplifying approximations made and the types of molecules considered in parameterizing the methods. Detailed examinations of the strengths and weaknesses of semiempirical methods in calculating various molecular properties can be found in a number of excellent reviews and need not be repeated here [Clark, 1985; Thiel, 1988; Stewart, 1989; Stewart, 1990a; 1990b].

It should now be apparent that MO and DFT methods are very attractive because of their accuracy, but very limited in polymer applications due to their computational expense. There are, however, two areas in which they are very useful: accurate calculations on crystalline polymers with translational symmetry and force field parameterization for classical simulations. Each of these is described briefly below.

While it is currently impossible to perform accurate ab initio calculations on polymer chains of any realistic length, crystalline polymers can be examined by borrowing techniques developed for bulk crystal calculations in solid-state physics [Ashcroft et al., 1976; Hoffmann, 1987]. By considering a unit cell of the crystalline polymer containing one or a few repeat units, the size of the calculation can be reduced from thousands of atoms to 10–20 atoms.

The difference between this approach and a typical molecular orbital calculation lies in the use of crystal orbitals, indexed over the number of bands, j:

$$\Psi_j(\mathbf{k}) = \sum_i c_{ij}(\mathbf{k})\phi_i(\mathbf{k}).$$

In this expression, the usual atomic orbitals are replaced by Bloch sums of the constituent atomic orbitals of each unit cell (χ_j):

$$\phi_i(\mathbf{k}) = N^{-\frac{1}{2}} \sum_R e^{i\mathbf{k}\cdot\mathbf{R}} \chi_i(\mathbf{r}-\mathbf{R}).$$

Here $\mathbf{R} = n \bullet \mathbf{a}$, where **a** is the unit cell vector and the summation is over the N unit cells of the crystal. The $e^{i\mathbf{k}\cdot\mathbf{R}}$ term provides for the phase difference between the orbitals in the reference unit cell and the cell at **R**, and depends on the wave vector, **k**. This formulation allows the calculation to be performed at a limited number of **k** points, the number depending on the nature of the specific problem. Each of these calculations is proportional in cost to the number of orbitals in the unit cell.

The primary advantage of this approach is that it allows for very accurate calculations of polymer properties that would otherwise be inaccessible due to computational demands. The major limitation of this method is that it is restricted to systems with periodic symmetry, i.e., crystalline polymers. Within its realm of applicability, however, this technique is very useful for determining accurate electronic and electromechanical properties.

In one application of this method, the experimental photoelectron spectrum [Delhalle et al., 1977] of the β-form of PVDF (see Fig. 3 for unit cell) was predicted with excellent accuracy [Mintmire et al., 1987]. This work, which actually used a density functional theory based analog of the technique describe above, corroborated an experimental peak assignment which was in conflict with an extended Huckel crystal orbital calculation reported in the original paper [Delhalle et al., 1977].

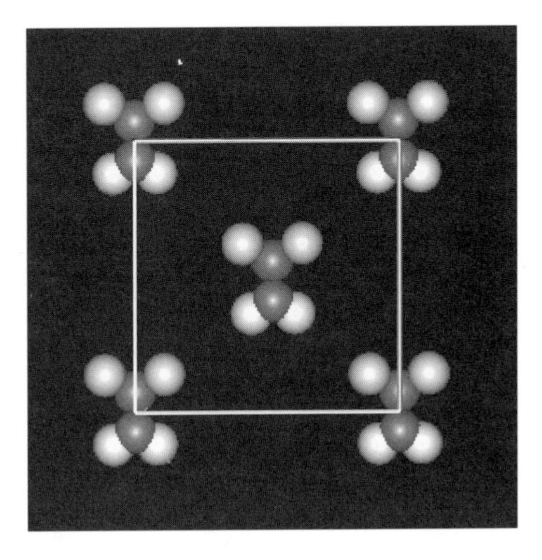

Figure 3: View of PVDF unit cell along *c* axis.

The second area in which quantum mechanical methods are commonly applied is in the parameterization of force fields to be used in classical simulations. These force fields, described in more detail in the next section, specify the potential energy surface of a polymer molecule and are used in the evaluation of forces at each step in the simulation. They typically consist of bonded (stretches, bends, and torsions) and nonbonded (van der Waals and electrostatic) energy parameters. Obtaining accurate parameters for a force field amounts to fitting a set of experimental or calculated data to the assumed analytical form. In situations where experimental data is either unavailable or very difficult to measure, quantum mechanical calculations can be a critical source of information.

Force constants for bond stretching, angle bending, and torsional deformations can be straightforwardly obtained from quantum mechanical calculations. First, the Cartesian force constant matrix is calculated, at the optimized minimum energy conformation of the molecule of interest. This is often done by numerically calculating the second derivatives of energy with respect to nuclear positions, although analytical second derivative methods are available for many MO and DFT methods. Next, a set of 3N – 6 internal coordinates are selected for the molecule [Boatz et al., 1989]. Finally, the Cartesian force constant matrix, from the quantum mechanical calculation, is converted to a set of internal force constants that yield the same vibrational frequencies. It should be noted that these force constants are only accurate within the harmonic approximation and lose validity when coordinate displacements outside this region are encountered in simulations. This is usually not a problem for stretches and bends, but torsions typically require different treatment. Torsional force constants, about the bond in question, are determined by calculating the energy of the molecule at a series of torsion angles within one complete rotation and fitting the potential energy profile to an assumed function, typically a cosine function.

Nonbonded parameters are also frequently fitted to quantum mechanical data. The electrostatic part of most force fields is restricted to the interaction of point charges centered at the nuclei. These charges are easily obtained from either MO or DFT calculations, although the various partitioning methods have generated controversy over the years. The two most commonly employed methods for obtaining charges are Mulliken population analysis and fitting atomic charges to reproduce the molecular electrostatic potential at some predefined distance from the nuclei. Mulliken charges [Mulliken, 1955], while considered intuitively accurate by many chemists, have the disadvantage of being heavily basis-set dependant and somewhat arbitrary in the way interatomic (bonding) charge density is partitioned. Electrostatic potential derived charges [Breneman et al., 1990], on the other hand, yield reliable charges for atoms at the surface of molecules, but can underestimate charges for interior atoms. Finally, van der Waals parameters may be estimated by quantum mechanical calculation, but these are usually the least accurately determined parameters. Due to the large influence of electron correlation and the importance of having a complete basis set, these are very expensive calculations even for small molecules. Fitting to experimental data remains the most reliable way of obtaining these parameters at this time.

Fitting a force field for PVDF is complicated by the existence of two phases in the bulk semicrystalline polymer, crystalline and amorphous. A force field parameterized to reproduce properties of the crystal will not give reliable results for the amorphous phase, and vice versa. To illustrate this, two recent parameterizations are described here [Karasawa et al., 1992; Byutner et al., 1999; Byutner et al., 2000]. In the first, Karasawa et al. were primarily interested in studying crystalline PVDF. They derived the stretch, bend, and torsional terms from Hartree-Fock calculations on 1,1,1,3,3-pentafluorobutane. Atomic charges were derived from the ab initio electrostatic potential and the van der Waals parameters were chosen by analogy from previous publications. The final force field proved capable of accurately reproducing unit cell parameters, although comparison to experiment for other physical properties was complicated by the presence of both crystalline and amorphous regions in the experimental samples. Several new polymorphs of PVDF were also predicted, based on the results of these calculations.

In the second example, Byutner et al. developed a force field for use in simulations of the amorphous phase. As conformational flexibility is much more important in the amorphous phase than in the crystalline, these authors devoted more attention to carefully parameterizing the torsional potential. This was done by extending the ab initio calculations to five and seven carbon backbones, rather than the chain of four used in the previous study. These authors also sought to parameterize the van der Waals terms using ab initio calculations rather than by fitting to experiment. This was done by performing MP2/6-311G** calculations on two dimers, $CH_4 - CF_4$ and $CF_4 - CF_4$. As in the force field described above, electrostatic potential-derived charges were used in this study. When corrected for differences in the degree of polymerization, this force field was found to accurately reproduce the experimental density of amorphous PVDF.

10.4 Classical Force Field Simulations

This section describes three simulation techniques that are related by their use of a parameterized potential energy function to describe the conformational space of a polymer molecule. As described in the previous section, this potential energy function is composed of bonded and nonbonded terms parameterized to reproduce either experimental or quantum-mechanically-calculated data. The bonded part of the potential is typically similar in form to:

$$\text{Ebonded} = \sum_{\text{bonds}} K_r (r - r_0)^2 + \sum_{\text{angles}} K_\theta (\theta - \theta_0)^2$$

$$+ \sum_{\text{torsions}} \frac{V_n}{2} \left[1 + \cos(n\phi - \phi_0)\right].$$

In this expression, K_r and K_θ are the bond stretching and angle bending force constants, respectively, and V_n is the barrier to rotation about a given dihedral angle. Complex rotational barriers occasionally necessitate the use of Fourier terms in this expansion. Many force fields also include terms for improper dihedrals, which enforce planarity about planar-coordinated atoms.

The most commonly encountered expression for the nonbonded energy in polymer simulations is of the form:

$$\text{Enonbonded} = \sum_{i,j} \frac{Aij}{Rij^{12}} - \frac{Bij}{Rij^{6}} + \sum_{i,j} \left[\frac{qiqj}{\varepsilon Rij} \right].$$

The first term is the familiar Lennard-Jones, or 6-12 potential. Short-range repulsion is described by the first term inside the brackets. The second, negative term in the brackets accounts for long range, nonelectrostatic attractive interactions. While the Lennard-Jones potential is very popular due to its accuracy and relatively simple analytical form, several other, usually exponential, forms are also commonly encountered. A number of review articles discuss the relative merits of the various van der Waals potentials [Vitek, 1996; Brenner, 1996, and references therein].

The second sum in the nonbonded potential is the classical Coulombic interaction. As mentioned above, most implementations employ point charges in this term. Both of these summations are carried out over all nonbonded pairs of atoms in the molecule.

Once an appropriate force field has been defined, it may be used in any of the simulation methods mentioned in the introduction: molecular mechanics, molecular dynamics, or Monte Carlo. The first of these, molecular mechanics, is similar in effect to a quantum mechanical geometry optimization. Beginning with an initial guess of the molecular geometry and the force field, the structure is optimized to that having the minimum potential energy. Obviously, the quality of the predicted structure is dependent on the accuracy of the force field. If sufficient experimental data is available for fitting, the results can be better than high-level quantum mechanical calculations. If, on the other hand, the force field is incompletely parameterized or if the system being studied is substantially different from those used in parameterization, the results can be very poor. This constraint can make it quite difficult to obtain force field parameters that produce good results while maintaining transferability among the various bonding situations in which a given element may be found.

The second method, molecular dynamics, is based on Newton's second law. Beginning with the initial positions and velocities of each atom in the system, the equations of motion can be integrated to give a time dependent trajectory containing positions (**r**), velocities (v), and accelerations (a) of all atoms at each time step:

$$\mathbf{r}(t) = \mathbf{r}_0 + v_0 t + a t^2.$$

Where acceleration is the derivative of the potential energy, evaluated from the force field, as a function of position,

$$a = -\frac{1}{m}\frac{dV}{d\mathbf{r}}.$$

The integrations are done numerically, using one of several common algorithms, such as the Verlet, velocity Verlet, or leap-frog [Allen, 1987].

The initial velocities are generally assigned randomly from a Maxwell-Boltzmann or Gaussian distribution at a given temperature subject to the constraint that the overall momentum of the system be zero. The temperature of a system can be maintained at a defined value during a simulation by periodically scaling the atomic velocities. Pressure can also be maintained by scaling the size of the simulation box. The simulation box referred to in the previous paragraph deserves a brief comment. Since even the largest simulations currently possible fall far short of including a realistic number of particles ($\sim 10^{23}$), calculated results will include spurious contributions due to surface effects. One way of circumventing this artificial truncation of the physical system is by using periodic boundary conditions [Born et al., 1912]. In this approach, an infinite lattice is constructed by conceptually replicating the cubic simulation cell in each direction, as depicted in Fig. 4.

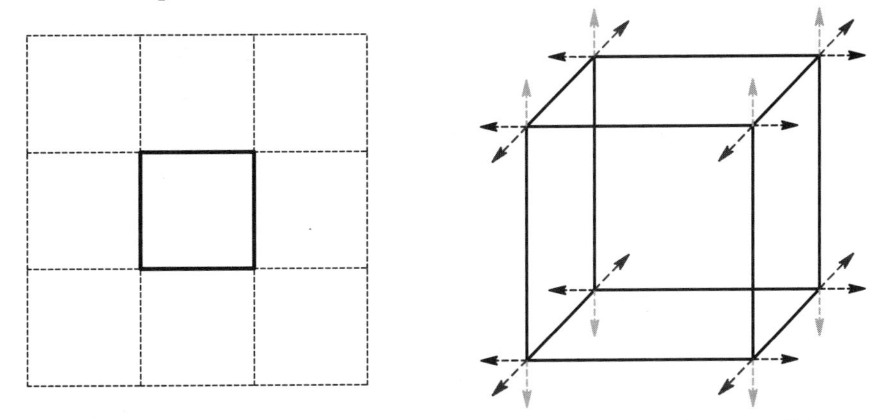

Figure 4: (left) Simplified 2D representation of periodic boundary condition (PBC). Central cell mirrored infinitely in each direction. (right) 3D representation of PBC.

All virtual copies of a particular real particle move in exactly the same way as the original. Therefore, if a particle migrates out of the left side of the central simulation cell, an identical copy will enter on the right side. As every side of the simulation cell adjoins another cell, no surface effects are present. Once the trajectory has been calculated for a system over a sufficiently long time span, average values for various system properties can be calculated.

Molecular dynamics simulations of PVDF in its crystalline [Karasawa et al., 1992] and amorphous [Byutner et al., 1999; Byutner et al., 2000] forms have recently been reported. Using the force fields described above, these studies illustrate the necessity of tailoring the parameterization to the physical situation of interest. Both force fields yield good results in their realm of applicability, but would perform poorly outside of it. Finally, the crystalline force field [Karasawa et al., 1992] was used in a lattice dynamics study, which produced good predictions of several piezoelectric and pyroelectric parameters, such as the stress and strain tensors [Carbeck et al., 1996].

The final classical simulation technique considered here is the Monte Carlo method. As there exist a number of excellent reviews on this topic and its application in polymer studies [Allen, 1987; Bicerano, 1992; Binder, 1995; Raabe, 1998], only a brief overview will be given here. Like a molecular dynamics simulation, a Monte Carlo calculation employs an empirical force field function to evaluate the energy of a given configuration and uses the current configuration to generate the coordinates for the following step. They differ in how the choice of new coordinates is made. Whereas a molecular dynamics calculation uses the equations of motion to propagate structural changes, Monte Carlo methods use statistical sampling of possible configurations. The result, in either case, is a series of system configurations that constitute a trajectory in phase space. Since time is not an independent variable in Monte Carlo simulations, the ensemble-averaged properties are calculated by sampling the statistical ensemble of configurations. Ensemble averages in molecular dynamics calculations, on the other hand, are time averages, over the length of the simulation. Despite these differences between the methods, both are expected to yield the same results for a given system, assuming both are run long enough to achieve statistical validity and that both employ the same force field.

As mentioned above, the objective of a Monte Carlo calculation is the creation of a trajectory through phase space composed of a large number of configurations, each of which is energetically accessible from the immediately preceding configuration of the system. The most common method for generating these configurations is the metropolis method [Metropolis, 1953; Allen, 1987]. In this approach, at each step a random particle is selected and moved a random distance (within a predefined range of acceptable step size) in a random direction. The energy is calculated using the force field function and compared to the energy of the preceding step. If lower, the move is accepted as valid and incorporated into the running average. If the energy is higher, the step may still be accepted with a probability proportional to the energy difference:

$$\text{Probability} \approx \mathbf{e}^{-(\Delta E/kT)}.$$

It is obvious from this expression that moves with the greatest increase in energy relative to the previous step are the least likely to be accepted. Conversely, moves resulting in small energy changes are likely to be accepted and incorporated into the ensemble average.

The metropolis algorithm has a significant computational advantage in that it reduces the size of the integrations that must be carried out to determine ensemble average properties. Average values for a property $M(\mathbf{r}^N)$, where N is the number of particles in the system and $P(\mathbf{r}^N)$ is the ensemble distribution function are calculated as:

$$\langle M(\mathbf{r}^N)\rangle = \int M(\mathbf{r}^N)P(\mathbf{r}^N)d\mathbf{r}^N .$$

Clearly, evaluation of this integral is an imposing task, given both its high dimensionality and the number of points in phase space over which it must be calculated. Limiting the population to only energetically accessible states reduces the problem somewhat.

Temperature enters a Monte Carlo simulation in two places: the acceptance probability function given above and in the estimation of average kinetic energy. In the first case, increasing the temperature of a simulation will clearly result in increased acceptance of higher energy configurations than would be found at a lower temperature. In the second case, kinetic energy is calculated using standard statistical mechanical techniques [McQuarrie, 1976] and added to the potential energy, calculated with the force field function, to yield total system energy. Extensive treatments of the many subtle issues involved in Monte Carlo simulations and in calculating various average system properties may be found in several excellent review articles and books [Allen, 1987; Bicerano, 1992; Binder, 1995; Raabe, 1998].

10.5 Mesoscale Simulations

Mesoscale modeling is a new (relative to any of the methods mentioned above), but emerging subfield of molecular simulation, which is already beginning to make important contributions in polymer science. The methods of primary interest in the context of this book are ones that allow for the treatment of very large systems over much longer time spans than are possible with traditional force field techniques. The spatial scale problem may be attacked by two methods, often used in tandem: discretizing space through the use of a lattice which defines possible particle locations, and course-graining the molecular representation by working with rigid groups of atoms rather than individuals. In the first of these, the spatial range of a simulation cell is filled with a conceptual lattice and particles (either atoms or larger subunits) are only allowed to move from lattice point to lattice point. This has the effect of greatly reducing the available conformational space of a molecule, which significantly accelerates calculations. Since only certain bond distances, angles, and torsional angles are possible, force fields can be quite simple and intramolecular energy evaluations may be done by lookup table rather than by actual calculation. The second approach, course-graining the structural representation, is also quite common. Rather than considering each atom individually in a calculation, larger groups of atoms, such as monomers or groups of monomers (e.g., Kuhn segments), are taken collectively. This obviously requires appropriate modification of the force field. This approach also allows for larger time steps, as will be described below.

Besides spatial limitations, one must also consider the time length that may be simulated. The basic difficulty is that the most rapidly occurring phenomena in a simulation determine the length of time step that may be taken. For example, molecular vibrations that occur on a time scale of $\sim 10^{-14}$ sec determine the upper bound on the time step in a molecular dynamics simulation in which motion between chemically bound atoms is allowed. Comparing this time step to the typical time scales on which many polymer processes occur ($\sim 10^{-2} - 10^{2}$), it is readily apparent that meso- or macroscopic phenomena cannot realistically be treated with atomistic detail. One may significantly increase the allowable step size, however, by freezing molecular vibrations and considering only larger scale molecular motions. It is for this reason that course graining of a molecular structure can greatly increase the accessible time span of a simulation.

One interesting application of this methodology to crystalline PVDF systems recently appeared describing the polarization reversal of a 2D film [Koda et al., 2000]. In this work, all internal degrees of freedom in the PVDF chains were held fixed except the intermonomer rotation angles, which were allowed to flip in increments of 180°. Using the Monte Carlo method, these authors studied the way in which the initially fully poled system reversed its polarization when the sign of the poling field was reversed. While at the small end of the scale of mesoscopic simulations, this study illustrates how making certain simplifying approximations may allow the study of much larger and more complex phenomena than is presently possible with conventional molecular dynamics or Monte Carlo methods.

Many extensions on these basic ideas are active research areas and the representation given above is only a brief overview. Interested readers are referred to several excellent reviews of the current state of the art [Bicerano, 1992; Binder, 1995; Rabbe, 1998].

10.6 References

Allen, M. P., Tildesley, D. J., *Computer Simulation of Liquids*, Oxford University Press: New York, 1987.

Ashcroft, N. W., Mermin, N. D., *Solid State Physics,* Holt, Rinehart, and Winston: Philadelphia, 1976.

Bauer, S., *J. Appl. Phys.* 1996, *80*, 5531-5558.
Becke, A. D., *J. Chem. Phys.* 1993, *98*, 5648-5652.
Becke, A. D., *Phys. Rev. A* 1988, *38*, 3098-3100.
Bicerano, J., *Ed. Computational Modeling of Polymers*, Marcel Dekker: New York, 1992.
Binder, K., *Ed. Monte Carlo and Molecular Dynamics Simulations in Polymer Science*, Oxford University Press: New York, 1995.
Boatz, J. A., Gordon, M. S. *J. Phys. Chem.* 1989, *93*, 1819-1826.
Born, M., Von Karman, T., *Physik. Z.* 1912, *13*, 297-309.
Borodin O., Smith, G. D., "Molecular dynamic simulations of Poly(ethylene oxide)/LiI melts. I - structural and conformational properties," *Macromolecules* 1998, 31, 8396-8406.
Breneman, C. M., Wiberg, K. B., *J. Comp. Chem.* 1990, *11*.
Brenner, D. W., in *Interatomic Potentials for Atomistic Simulations*, Voter, A. F., Ed. Materials Research Society, 1996.
Byutner, O. G., Smith, G. D., *Macromolecules* 1999, *32*, 8376-8382.
Byutner, O. G., Smith, G. D., *Macromolecules* 2000, *33*, 4264-4270.
Carbeck, J. D., Rutledge, G. C., *Polymer* 1996, *37*, 5089-5097.
Clark, T. A., *Handbook of Computational Chemistry*, John Wiley & Sons: New York, 1985.
Davidson, E. R., Feller, D., *Chem. Rev.* 1986, *86*, 681-696.
Delhalle, J., Delhalle, S., Andre, J.-M., Pireaux, J. J., Riga, J., Caudano, R., Verbist, J. J., *J. Electron Spectrosc. Relat. Phenom.* 1977, *12*, 293.
Dreizler, R. M., Gross, E. K. U., *Density Functional Theory*, Springer: Berlin, 1990.
Eberle, G., Schmidt, H., Eisenmenger, W., *IEEE Transactions on Dielectrics and Electrical Insulation* 1996, *3*, 624-646.
Feller, D., Davidson, E. R., in *Review of Computaional Chemistry Volume 1*, Lipkowitz, K. B. and Boyd, D. B., Eds. VCH Publishers: New York, 1990.
Flekkoy, E. G., Coveney, P. V., *Phys. Rev. Lett.* 1999, *83*, 1775-1778.
Hehre, W. J., Radom, L., Schleyer, P. V. R., *Ab Initio Molecular Orbital Theory*, Wiley Interscience: New York, 1986.
Hoffmann, R., *Angew. Chem. Int. Ed. Eng.* 1987, *26*, 846-878.
Hohenberg, P., Kohn, W., *Phys. Rev. B* 1964, *136*, 864-871.
Karasawa, N., Goddard, W. A., III, *Macromolecules* 1992, *25*, 7268-7281.
Koda, T., Shibasaki, K., Ikeda, S., *Comp. Theor. Polym. Sci.* 2000, *10*, 335-343.
Kohn, W., Becke, A. D., Parr, R. G., *J. Phys. Chem.* 1996, *100*, 12974-12980.
Kohn, W., Sham, L., *Phys. Rev. A* 1965, *140*, 1133-1138.
Labanowski, J. K., Andzelm, J., *Density Functional Methods in Chemistry*, Springer-Verlag: New York, 1991.
McQuarrie, D. A., *Statistical Mechanics*, Harper and Row: New York, 1976.
metropolis, N., Rosenbluth, A. W., Rosenbluth, M. N., Teller, A. H., Teller, E., *J. Chem. Phys.* 1953, *21*, 1087-1092.
Mintmire, J. W., Kutzler, F. W., White, C. T., *Phys. Rev. B* 1987, *36*, 3312-3318.
Mulliken, R. S., *J. Chem. Phys.* 1955, *23*, 1833-1840.

Parr, R. G., Yang, W., *Density-Functional Theory of Atoms and Molecules,* Oxford University Press: New York, 1994.
Raabe, D., *Computational Materials Science*, Wiley-VCH: Weinheim, 1998.
Sessler, G. M., *J Acoust. Soc. Am.* 1981, *70*, 1596-1608.
Shavitt, I., *Isr. J. Chem.* 1993, *33*, 357-367.
Stewart, J. J. P., [a] *J. Computer-Aided Mol. Design* 1990, *4*, 1-105.
Stewart, J. J. P., [b] in *Reviews of Computational Chemsitry Volume 1*, Lipkowitz, K., Boyd, D. B., Eds. VCH Publishers: New York, 1990.
Stewart, J. J. P., *J. Comp. Chem.* 1989, *10*, 221.
Szabo, A., Ostlund, N. S., *Modern Quantum Chemistry*, Dover: NY, 1996.
Thiel, W., *Tetrahedron* 1988, *44*, 7393.
Vitek, V., in *Interatomic Potentials for Atomistic Simulations*, Voter, A. F., Ed. Materials Research Society, 1996.
Ziegler, T., *Chem. Rev.* 1991, *91*, 651-667.

CHAPTER 11

Modeling and Analysis of Chemistry and Electromechanics

Thomas Wallmersperger, Bernd Kröplin
University of Stuttgart (Germany)

Rainer W. Gülch
University of Tübingen (Germany)

11.1 Introduction

The growing interest of engineers in gels has arisen from the promising perspectives and possibilities of provoking drastic phase transitions by inducing small changes in their external conditions [Hamlen, 1965; Tanaka, 1982; Yannas, 1973]. For technical applications gels, which respond with considerable swelling or shrinkage, are predestinate. These changes in volume can be obtained by external stimulation, e.g., by applying an external electric field, through temperature changes, light, or a change of the chemical milieu of the solvent in which the gels are immersed. By combining sensoric and actuatoric properties, gels can even be classified as typical adaptive materials. They can fill the gap between the conventional mechanical and solid-state actuators, although at the cost of precision.

Electroactive polymers (EAP) are a particularly attractive class of actuation materials with great similarity to biological contractile tissues [De Rossi et al., 1992]. The backbone of these gels consists of polymers or other long-chain molecules, chemically or physically cross-linked to create a tangled network (Fig. 1).

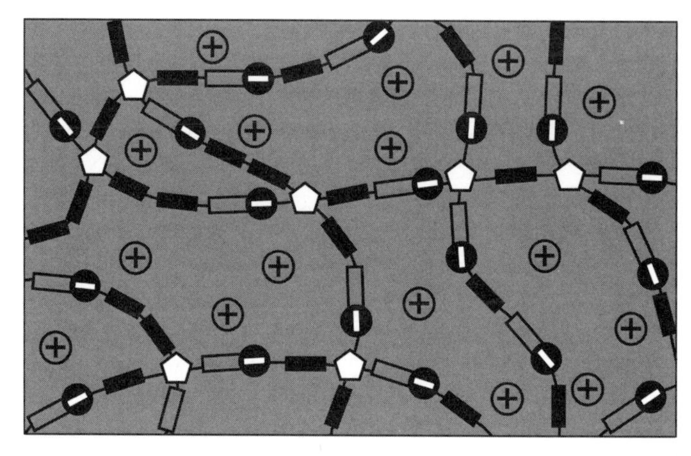

	Chemical Composition	*Example*
	backbone monomer	acrylamide
	electrolyte comonomer	acrylic acid
	mobile counterion	H^+
	cross-linking comonomer	bisacrylamide
	solvent	H_2O

Figure 1: Schematic drawing of a typical polyelectrolyte gel. A polymer network with fixed anionic groups is interconnected by a cross-linker. The polymer network holds the liquid in place in which the gel is immersed. The freely mobile counter-ions as well as ions entering from outside the gel determine the osmotic pressure that is responsible for swelling or deswelling of the gel.

The shape stability of the gel strongly depends on the interaction between its elastic network and the liquid medium in which it is immersed. The liquid prevents the network from collapsing. On the other hand, the network retains the liquid and imparts solidity to the gel. Having pseudomuscular actuators in mind, ionizable polymer gels, so-called polyelectrolyte gels, which can be stimulated by electrical or chemical interventions, must be favored for technical applications (see Chapter 5). The physico-chemical properties of such gels depend on their capability to dissociate the ionizable polymer moiety. When immersed in water or other polar solvent, polyelectrolyte molecules are inclined to dissociate in polyvalent macro-ions and in their freely mobile counter-ions. The polyacidic, polybasic, or polyampholytic moieties of the gels undergo reversible dissociation or association. This depends on the electrolytic composition of the solution surrounding the gel, which may be changed by chemical or electrical interferences. The degree of dissociation modulates the total osmotic pressure in the gel and therefore the degree of swelling. In this manner, polyelectrolyte gels can execute physical work as electro-chemo-mechanical converters under the influence of external electric fields.

To understand the physico-chemical mechanisms underlying these conversion principles, the theoretical and experimental facts behind the processes of swelling have to be investigated. By presenting model calculations and direct measurements in gels, we want to shed more light on the swelling behavior of a typical class of polyelectrolyte gels under chemical and electrical stimulation. Based on these models and direct recordings of the electric potentials in gels, the bending kinetics of gel strips in an electric field can be simulated and predicted mathematically.

11.2 Chemical Stimulation

To learn more about the fundamental transition processes in EAP, a closer look at the effect of changes in the chemical milieu surrounding a polyelectrolyte gel should be taken. Any kind of stimulation that is able to induce such transition processes is supposed to affect the gels by altering the ionic compositions and, therefore, the osmotic pressure inside and outside the gel. In this section, we present a numerical calculation and an experimental determination of the swelling ratio of polyionic gels, based on the ionic composition and concentrations of their surroundings.

11.2.1 Statistical Theory

Based on the theory of "swelling of network structures" by Flory [1953], the change of the ambient conditions can be represented by a change of Gibbs free energy ΔF. The total free energy is the sum of the free energies of mixing ΔF_M, of elastic deformation ΔF_{el}, and of the one depending on the different concentrations inside and outside the gel ΔF_{ion} (see Rička and Tanaka [1984]; Schröder et al. [1996]).

The equilibrium state is characterized by a minimum of the total free energy. In this state, the chemical potentials μ_1 of the solvent in both the solution and the gel phase are identical:

$$\Delta\mu_1 = \left(\frac{\partial \Delta F_M}{\partial n_1}\right)_{p,T} + \left(\frac{\partial \Delta F_{el}}{\partial n_1}\right)_{p,T} + \left(\frac{\partial \Delta F_{\text{ion}}}{\partial n_1}\right)_{p,T} = \underbrace{\Delta\mu_{1,M}}_{\substack{\text{mixture}\\ \text{potential}}} + \underbrace{\Delta\mu_{1,el}}_{\substack{\text{elastic}\\ \text{potential}}} + \underbrace{\Delta\mu_{1,ion}}_{\substack{\text{mobile ion}\\ \text{potenial}}} = 0 \,. \tag{1}$$

In the unswollen state of the gel, the solvent and the polymer matrix are completely separated. In the swollen state, solvent has migrated into the gel; the polymer and the solvent are no longer separated. The mixing rate of the chemical potential is obtained by

$$\Delta\mu_{1,M} = \left(\frac{\partial \Delta F_M}{\partial n_1}\right)_{p,T} = RT\left(\ln(1-\phi_p) + \phi_p + \chi\phi_p^2\right), \tag{2}$$

where ϕ_p represents the molar fraction of polymer in the gel, and χ stands for the Flory-Huggins interaction parameter.

The total swelling ratio q is given by the initial swelling ratio q_v and the relative swelling ratio q_r, such that

$$q = q_v \cdot q_r = \phi_p^{-1} \,. \tag{3}$$

In the swollen state the mixing of solvent and polymer leads to an increase of possible conformations of polymer chains in the total system. This provokes a rise of entropy. The (Gaussian) elastic potential may be given as

$$\Delta\mu_{1,el} = \nu^* RT v_1 s q_v^{-1} q_r^{-\frac{1}{3}}, \tag{4}$$

where ν^* and s are matter constants and v_1 is the molar volume of the solvent.

The elastic deformation consists of two successive processes: first, elongation as a result of free swelling and second, stretch at constant total volume while keeping the enthalpy of the system constant. The expansion of the network leads to a decrease of the chain entropy. Note, the sum of the changes of the entropy based on the elastic, mixing and mobile-ion terms is zero, in the equilibrium state. For large swelling ratios, the non-Gaussian theory [Treloar, 1958] should be more appropriate to compute the elastic potential $\Delta\mu_{1,el}$. The experimental results, however, confirm the validity of the present theory.

The chemical potential, depending on differences in charge density, is given by

$$\Delta\mu_{1,\text{ion}} = -RT v_1 \sum_{\alpha} (c_\alpha^{(g)} - c_\alpha^{(s)}), \tag{5}$$

where $c_\alpha^{(g)}$ are the concentrations of the mobile ions in the gel and $c_\alpha^{(s)}$ in the solution. For simplification, concentrations are used instead of chemical activities. The ion distribution in the gel and in the solution must meet the condition of neutrality

$$\sum_\alpha (z_\alpha c_\alpha) = 0, \tag{6}$$

which denotes neutrality of the charge throughout the domain. For a NaCl solution surrounding the gel, this leads to

$$c_{\mathrm{Na^+}}^{(s)} + c_{\mathrm{H^+}}^{(s)} = c_{\mathrm{Cl^-}}^{(s)} + c_{\mathrm{OH^-}}^{(s)} \tag{7}$$

in the solution, and

$$c_{\mathrm{Na^+}}^{(g)} + c_{\mathrm{H^+}}^{(g)} = c_{\mathrm{OH^-}}^{(g)} + c_{\mathrm{Cl^-}}^{(g)} + c_{\mathrm{A^-}}^{(g)} \tag{8}$$

in anionic gels. For cationic gels, a similar equation can be formulated.

Additionally, for gels with carboxylic groups A^- as fixed charges, the dissociation coefficient

$$K_c = \frac{c_{\mathrm{H^+}}^{(g)} \cdot c_{\mathrm{A^-}}^{(g)}}{c_{\mathrm{HA}}^{(g)}}, \tag{9}$$

and the dissociation equilibrium

$$\frac{c_{M,0}}{q} = c_{\mathrm{HA}}^{(g)} + c_{\mathrm{A^-}}^{(g)}, \tag{10}$$

have to be considered. $c_{M,0}$ denotes the concentration of the polyelectrolyte in xerostate and $c_{\mathrm{HA}}^{(g)}$ the concentration of the acrylic acid compounds of the gel. For completely dissociated gels ($c_{\mathrm{HA}}^{(g)} = 0$), Eq. (10) can be simplified.

Identical electrochemical potential in the gel and in the solution is given by

$$\mu_\alpha^{(g)} + z_\alpha F \Psi^{(g)} = \mu_\alpha^{(s)} + z_\alpha F \Psi^{(s)}, \tag{11}$$

for all species α. This leads to the Donnan equation, where the ratio between the concentrations of ions inside and outside the gel is expressed by

$$\frac{c_\alpha^{(g)}}{c_\alpha^{(s)}} = \exp\left[-z_\alpha \frac{F}{RT}\left(\Psi^{(g)} - \Psi^{(s)}\right)\right], \tag{12}$$

in which Ψ is the potential in the gel or solution, respectively; F represents the Faraday constant, and μ_α is the chemical potential of the α'th component of a mixing phase. For univalent ions, this leads to

$$\frac{c^{(g)}_{\mathrm{Na}^+}}{c^{(s)}_{\mathrm{Na}^+}} = \frac{c^{(g)}_{\mathrm{H}^+}}{c^{(s)}_{\mathrm{H}^+}} = \frac{c^{(s)}_{\mathrm{Cl}^-}}{c^{(g)}_{\mathrm{Cl}^-}} = \frac{c^{(s)}_{\mathrm{OH}^-}}{c^{(g)}_{\mathrm{OH}^-}} \,. \tag{13}$$

Finally, the swelling ratio q is obtained by Eq. (1), using the ion concentrations in the gel and in the solution. The calculated results for PAAm/PNa$^+$A$^-$ and PAAm/PAA gels are plotted in Fig. 2.

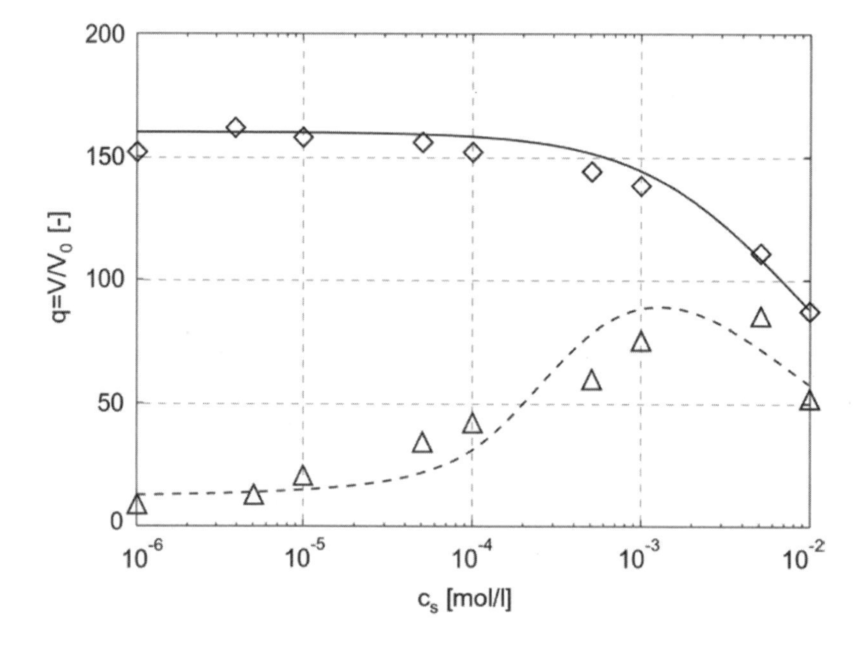

Figure 2: Static swelling ratio $q = V/V_0$ of PAAm/PNa$^+$A$^-$ gel (◊) and of PAAm/PAA gel (Δ) as a function of the salt (NaCl) concentration c_s in the solution. The symbols represent measured values. The two curves show the calculated values based on the following parameters: for PAAm/PNa$^+$A$^-$ gels v_1 = 0.018 M^{-1}; $c_{M,0}$ = 4 M; χ = 0.4; complete dissociation, for PAAm/PAA gels: v_1 = 0.018 M^{-1}; $c_{M,0}$ = 2 M; χ = 0.45; dissociation coefficient $K_c = 10^{-5}$ M.

11.2.2 Comparison between Experimental and Numerical Results

The basic property of polyelectrolyte gels for technical applications is the extraordinary swelling behavior, induced by alterations in the ionic composition of the surrounding liquid phase. This phenomenon, representing the actual drive for deformation and thus, for the execution of mechanical work, is characterized by the swelling ratio $q = V/V_0$, where V stands for the actual volume of the gel immersed in solutions of different ionic concentrations, and V_0 represents the volume of the absolutely dry gel or xerogel. The swelling must be described as a kinetic process, depending on the kind of gel, the dimensions of the gel, and the concentrations inside and outside the gel. The swelling ratios are best related to the equilibrium.

Certain polyacrylamide compounds form a family of polymer gels, being very appropriate for such investigations. The solidity, elasticity, sign, and distribution of their ionic comonomers can be varied over a wide range by the composition of the reaction solution for the polymerization [Buchholz, 1992]. The backbone of these gels is a polymer matrix of cross-linked polyacrylamide (PAAm), in which selected ionic comonomers are incorporated. The polymerization reaction can be described very briefly: adding the initiator ammonium peroxodisulfate and the accelerator tetramethylethylenediamine to an aqueous solution of the monomer acrylamide (AAm), of the cross-linker bisacrylamide (BAAm), and of anionic comonomers [e.g., acrylic acid (AA), acrylate (A^-), acrylamidomethyl-propanesulfonic acid (AMPS)], or of cationic comonomers [e.g., diallyldimethylammonium chloride (DADMAC)], a chain reaction of polymerization is initiated. This leads to a complex network of polyacrylamide-polyelectrolyte chains interconnected by the cross-linker BAAm.

By reducing the portion of the cross-linker, while keeping the other moieties constant, the swelling ratio of these polymer gels can be increased considerably, however, at the cost of stability.

On the other hand, the dissociation of the polyelectrolyte plays another central role in the swelling process. This is demonstrated in Fig. 2. The swelling behavior of the PAAm/PAA gel, characterized by a very low dissociation constant, is compared to the PAAm/PNa^+A^- gel, showing a much higher dissociation constant. The higher the dissociation the more the polyelectrolyte gel will swell. Therefore, the swelling ratio is high under low ionic strength in the surrounding solution because of enhanced dissociation of the polyelectrolyte. The swelling ratio continually decreases with increasing salt concentrations in the bathing solution. For the weak polyacrylic acid, an augmented cationic concentration in the bath solution [Gregor, 1955; Suzuki, 1999] is supposed to increase the dissociation constant, and thus the dissociation for a given ionic composition; the swelling ratio is consequently elevated. An optimum will result when these two opposing effects are superimposed. In the calculations, for the PAAm/PNa^+A^- gel, complete dissociation of the anionic rest groups, and for the PAAm/PAA gel, a constant dissociation coefficient was used. Both model calculations and measurements are in very good agreement. For the stronger electrolytes like PAMPS or PDADMAC, the swelling ratios are altogether higher.

11.3 Electrical Stimulation

Although various chemical or physical interventions are known to induce phase transitions in gels, the electrical stimulation must play a central role in technical applications. As in muscles, considered the biological standard for soft actuators, an external electric field can cause a redistribution of ions inside polyelectrolyte gels and, as a consequence, alterations in internal potentials leading to swelling or deswelling, elongation or contraction. Therefore the knowledge about the interactions between the electric potential and the ion distributions in the two phases is essential for the understanding of the electrical stimulation.

With a mathematical multifield formulation, the ion concentrations and the electric potential in polyelectrolyte gels and in the solution are calculated under the influence of an applied electric field. Direct measurements of the potential inside and outside, with and without an external electric field, are presented and compared with the results of the model calculations.

11.3.1 Chemo-Electro-Mechanical Multifield Formulation

Using a quasi-static formulation, the dynamics of ionic polymer gels in an electric field were investigated by Doi [1992], Shiga [1990], and Shahinpoor [1995]. Segalman et al. [1994] also derived coupled formulations for the kinetics of the ions and theories for large displacements. Nemat-Nasser et al. [2000] have presented an electromechanical model for ionic polymer metal composites, while Neubrand [1999] and Grimshaw et al. [1990] have given electrochemical and electromechanical formulations for ion-exchange membranes and gels.

One possibility for describing the dynamic swelling behavior of polymer gels, as well as the ion concentrations in the gel and in the solution, is to use a coupled chemo-electro-mechanical formulation. The description of the chemical field is based on the accounting equation of the flux J [Yeager, 1983] for the species α:

$$J_i = \underbrace{-D_\alpha c_{\alpha,i}}_{\substack{\textit{diffusive}\\ \textit{flux}}} \underbrace{- \frac{F}{RT} z_\alpha c_\alpha D_\alpha \Psi_{,i}}_{\textit{migrative flux}} + \underbrace{c_\alpha v_i}_{\substack{\textit{convective}\\ \textit{flux}}} \quad , \tag{14}$$

where i denotes the spatial direction x_i, and subscript i the spatial derivative $\partial/\partial x_i$, and also, on the conservation of mass:

$$\dot{c}_\alpha = -J_{i,i} + r_\alpha \, . \tag{15}$$

The change of the amount of the species α contained in the volume with respect to time t is given by the difference between the fluxes entering and leaving the reference volume. The first term of Eq. (14) represents the diffusive flux, which is based on concentration gradients in the domain. The second term originates from the gradient of the electric potential, while the third term stems from an applied convection of the solvent.

If we consider the interaction of the different ions, we obtain the following convection-diffusion equation:

$$\dot{c}_\alpha = \left[\sum_\beta (D_{\alpha\beta} c_{\beta,i} + z_\beta c_\beta \mu_{\alpha\beta} \Psi_{,i}) \right]_{,i} - (c_\alpha v_i)_{,i} + r_\alpha(c_\beta), \tag{16}$$

which describes the migration of the ions of different species α inside the domain. c_α denotes the concentration of the species α; $D_{\alpha\beta}$ is the diffusion constant of the species α induced by the species β; $\mu_{\alpha\beta}$ is the unsigned mobility of the ions and z_α their valence; Ψ represents the applied electric potential; and r_α denotes the source term resulting from the chemical conversion of the molecules.

In dilute solutions and in gel phases containing a small amount of polymer chains, there is little interaction of the ions. For this reason, the secondary diagonal terms in Eq. (16) can be neglected. This yields

$$\dot{c}_\alpha = (D_\alpha c_{\alpha,i} + z_\alpha c_\alpha \mu_\alpha \Psi_{,i})_{,i} - (c_\alpha v_i)_{,i} + r_\alpha(c_\alpha). \tag{17}$$

The electric field is described by the Poisson equation

$$\Psi_{,ii} = -\frac{F}{\varepsilon\varepsilon_0} \sum_\alpha (z_\alpha c_\alpha). \tag{18}$$

The velocity of propagation in the electric field is much higher than that occurring in the convection-diffusion equation. Hence, the quasi-static form of the Poisson equation is adequate. Equation (18) is based on the second and fourth Maxwell equation

$$E_i = -\Psi_{,i}\ , \ D_{i,i} = \rho_{el}\ , \tag{19}$$

where $\boldsymbol{D}$ is the tensor of the dielectric displacement, $\boldsymbol{E}$ the tensor of the electric field strength, ρ_{el} the density of volume charge, ε the dielectric constant, ε_0 the permittivity of free space, and e the electric elementary charge. The condition of neutrality, Eq. (6), is obtained, when $\Psi_{,ii}$ is explicitly set to zero. In this case, no boundary layer is obtained.

In order to describe the swelling behavior and the strain, the equation of motion

$$\rho \ddot{u}_i + f \dot{u}_i = \sigma_{ij,j} + \rho b_i , \tag{20}$$

with the coupling equations

$$\sigma_{ij} = C_{ijkl}(c_\alpha)\gamma_{kl}^{dev} - p\delta_{ij} , \tag{21}$$

$$\gamma_{kl} = \gamma_{kl}^{dev} + \alpha_v \delta_{kl} , \tag{22}$$

$$p = RT \sum_\alpha (c_{\alpha 0} - c_\alpha) \tag{23}$$

is used. $\boldsymbol{u}$ represents the vector of the displacements, $\boldsymbol{\sigma}$ the stress tensor, $\boldsymbol{\gamma}$ the strain tensor, $\boldsymbol{C}$ the elasticity tensor in general form, ρ the density and $\boldsymbol{b}$ the body forces; p is the osmotic pressure, $c_{\alpha 0}$ the stress-free state of the ions, and α_v the volumetric strain of the polymer gel network; f represents the viscous damping coefficient between solvent and polymer-network. Due to a very slow swelling process, the inertia term in Eq. (20) is very small compared to the damping term.

11.3.1.1 Coupling of PDEs of Different Orders in Time

The system of differential equations consists of three coupled partial differential equations (PDEs):

(1) the convection-diffusion equations, Eq. (17), PDEs of first order in time for the concentrations c_α of all species α,

(2) the Poisson equation, Eq. (18), a PDE in space for the electric field Ψ,

(3) the equation of motion, Eq. (20), a PDE of second order in time for the displacement unknowns $\boldsymbol{u}$. Note, for a very slow swelling process, the mechanical field can be described by a PDE of first order in time.

The differential equations for the concentrations, Eq. (17), are solved simultaneously for the concentrations c_α of all the species (e.g., for $c_{A^-}, c_{Cl^-}, c_{Na^+}, c_{H^+}, c_{OH^-}$). The chemo-electric field can be computed depending on the strength of coupling using different strategies. In our case, the convection-diffusion and the electric potential equations are solved simultaneously.

Subsequently, the equation of motion, Eq. (20), a PDE of second order in time, is solved. The osmotic pressure, obtained from the concentration differences of all species, can be regarded as a source term or an external prescribed force.

The resulting displacements lead to a deformation of the domain. In order to solve the chemo-electric field, these deformations have to be considered iteratively, i.e., the solution of the chemo-electric field depends on the actual domain-configuration of the gel (see chemo-electro-mechanical coupling scheme, Fig. 3).

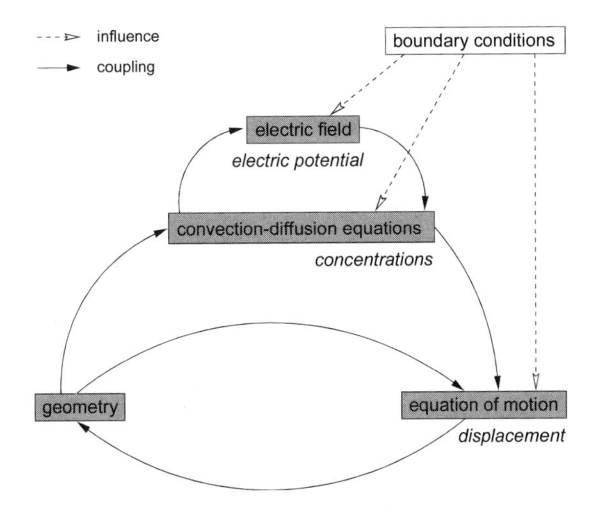

Figure 3: Chemo-electro-mechanical coupling scheme.

11.3.1.2 Discretization

For the discretization of the differential equations, Eq. (17), (18), and (20), stabilized space-time finite elements are used. This discretization method has been investigated for elastodynamics [Hughes, 1988], structural dynamics [Hulbert, 1992], and general second-order hyperbolic problems [Johnson, 1993]. The time-discontinuous Galerkin (TDG) finite elements yield implicit, unconditionally stable and higher-order accurate discretization. Stabilized space-time finite elements have been developed for flow problems [Hughes, 1989]; they have also been applied to elastodynamics [Hughes, 1998], structural dynamics [Hulbert, 1992], and to coupled problems (aeroelasticity) [Grohmann, 2001] employing the deforming-spatial-domain/space-time (DSD/ST) procedure in order to handle deforming domains [Tezduyar, 1992].

For the weighting functions W_h as well as for the unknowns U_h, space-time interpolations are used in space and time. The interpolations are continuous in space in each time-slab $Q_N = I_N \times \Omega(t)$, where $I_N = \{t \,|\, t_n \leq t \leq t_{n+1}\}$, but discontinuous in time between subsequent time-slabs (Fig. 4). For problems using a dynamic mesh for the discretization of domain deformations, space-time interpolations may also be used for the finite element geometry.

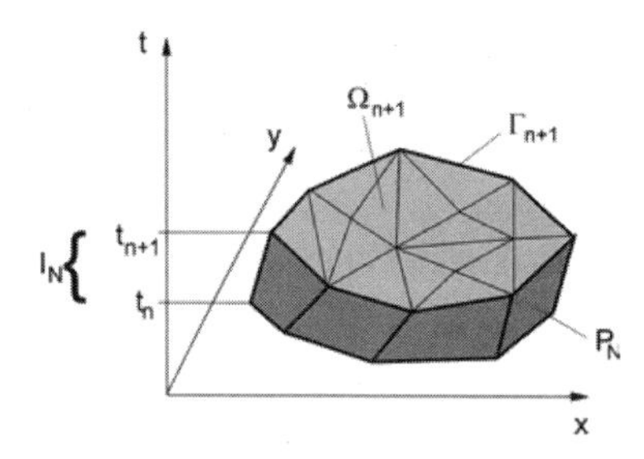

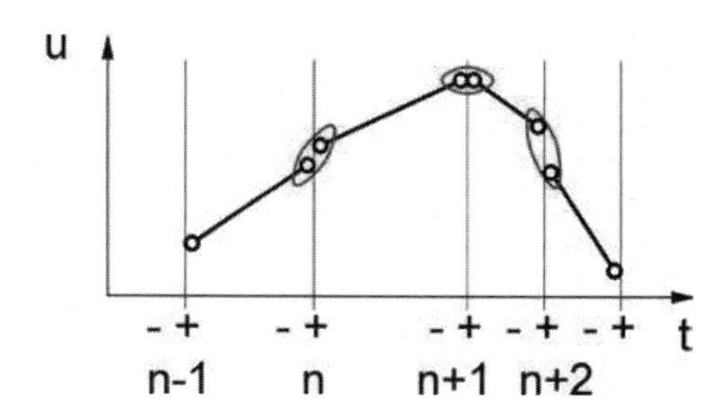

Figure 4: Space-time-domain (left) and time-discontinuous linear interpolation functions (right).

The time-discontinuous Galerkin formulation of Eq. (17), integrated by parts in space and time in order to obtain a conservative formulation, yields

$$\begin{aligned} &-\int_{I_N}\int_{\Gamma(t)} \delta c_\alpha^T (D_\alpha c_{\alpha,i} + z_\alpha c_\alpha \mu_\alpha \Psi_{,i}) \cdot n_i \, d\Gamma\, dt - \int_{I_N}\int_{\Omega(t)} \delta \dot{c}_\alpha^T c_\alpha \, d\Omega\, dt \\ &+\int_{I_N}\int_{\Omega(t)} \delta c_{\alpha,i}^T (D_\alpha c_{\alpha,i} + z_\alpha c_\alpha \mu_\alpha \Psi_{,i}) \, d\Omega\, dt - \int_{I_N}\int_{\Omega(t)} \delta c_\alpha^T r_\alpha \, d\Omega\, dt \\ &+\int_{\Omega(t_{n+1})} \delta c_\alpha^T\Big|^- c_\alpha\Big|^- d\Omega - \int_{\Omega(t_n)} \delta c_\alpha^T\Big|^+ c_\alpha\Big|^- d\Omega = 0\,. \end{aligned} \tag{24}$$

The first term represents the boundary integral, resulting from integration by parts in space. The next three terms report the PDE inside the domain. The last term describes the temporal jump term (integrated by parts in time), enforcing the identity of $c_\alpha|_n^+$ and $c_\alpha|_n^-$ in the weak form, where $c_\alpha|_n^\pm = \lim_{\varepsilon\to 0} c_\alpha(t_n \pm \varepsilon)$.

For more details about the discretization method, different formulations of the space-time finite elements, and the interpolation functions, refer to Wallmersperger et al. [1999].

11.3.1.3 Numerical Results

In this section, a fixed polymer gel strip immersed in solution is considered. In a 2D simulation, the gel (5×5 mm^2) is placed between two electrodes in a solution bath (15×15 mm^2), see Fig. 5 (top, left). In the first test case, no electric potential is applied. The concentration of the bound anionic A^--groups in the gel is $c_{A^-}^{(g)} = 10$ mM. For the Na^+ and Cl^- ions, the boundary conditions in the solution at the electrodes are set to $c_{Na^+} = c_{Cl^-} = 1$ mM.

Equations (17) and (18) are applied to the problem. The convective flux is neglected and no chemical conversion is considered. A stationary solution is obtained after a certain number of time steps. The resulting concentrations c_{Na^+},

c_{Cl^-} and c_{A^-} (where c_{H^+}, $c_{OH^-} << c_{Na^+}$, c_{Cl^-}) as well as the electric potential in the gel and in the solution outside the boundary layer are given in Table 1.

Table 1: Ion concentrations and electric potential in the gel and the solution.

		solution	**gel**
c_{A^-}	[mM]	0	10
c_{Cl^-}	[mM]	1	0.099
c_{Na^+}	[mM]	1	10.099
Ψ	[mV]	0	-58.4

In the second test case, the same boundary conditions as in the first test case are chosen for the mobile ions. The concentration of the bound anionic rest-groups is again set to $c_{A^-}^{(g)} = 10$ mM, constant throughout the domain. An external electric field is applied. In the solution, next to the anode, a time-constant electric potential of 0.1 V and next to the cathode, a time-constant electric potential of –0.1 V is prescribed.

It should be noted that the electric field is quite low, no perfusion of the gel is considered and the gel is not in direct contact with the electrodes. Therefore the effects of electrolysis at the electrodes are also small and are neglected in the numerical simulation.

The stationary solution with an up-scaled boundary layer is plotted in Fig. 5. On the cathode side of the gel, the concentration of Cl^- is larger than without an applied electric field. The same result is obtained for Na^+ due to the electroneutrality. The smaller differences of concentrations between gel and solution on the cathode side result in a smaller difference in the electric potential, while on the anode side, a larger electric potential difference is obtained.

Because these differences are responsible for the alterations of the swelling ratios, this result gives a reasonable explanation for the larger swelling of gels on the anode side and the lower swelling on the cathode side for freely movable gels. A more precise examination of the electric potential is given in Sec. 11.3.2, where the experimental and numerical results are compared.

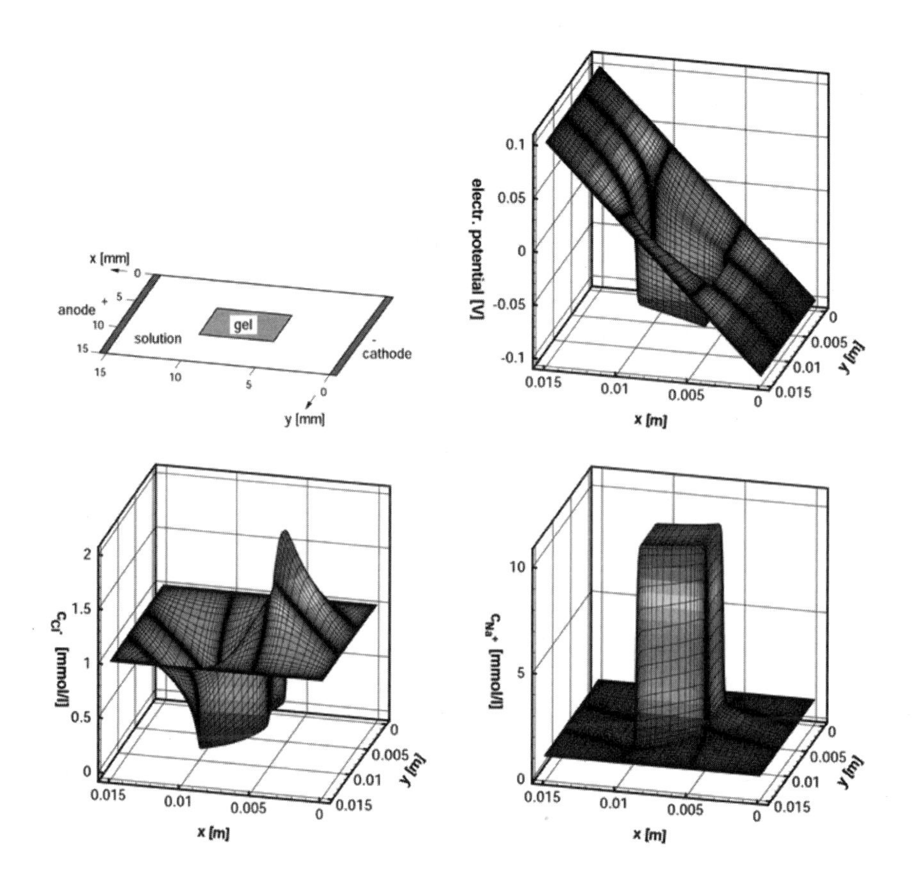

Figure 5: 2D calculation of the ion concentrations [Cl^- (bottom, left), Na^+ (bottom, right)] and electric potential (top, right) in the gel and the solution, resulting from an applied external electric field.

11.3.2 Potential Measurements

The swelling of polyelectrolyte gels is associated with an electrical imbalance between the solution (in which the gel is immersed) and the gel. The Donnan potential expresses this. This potential difference can be detected using 3 M KCl-filled glass microelectrodes, as commonly practiced in electrophysiology, when measuring transmembrane potentials in living cells. In contrast to the conventional technique, the microelectrodes impaled into a gel must be pushed deeper in a continuous and absolutely vibrationless manner, to guarantee stable recordings [Gülch, 2000]. This is achieved by a dc-motor driven micromanipulator. A typical recording of the electric potential in an anionic gel is shown in Fig. 6. For the penetration velocity of 0.1 mm/s chosen in this case,

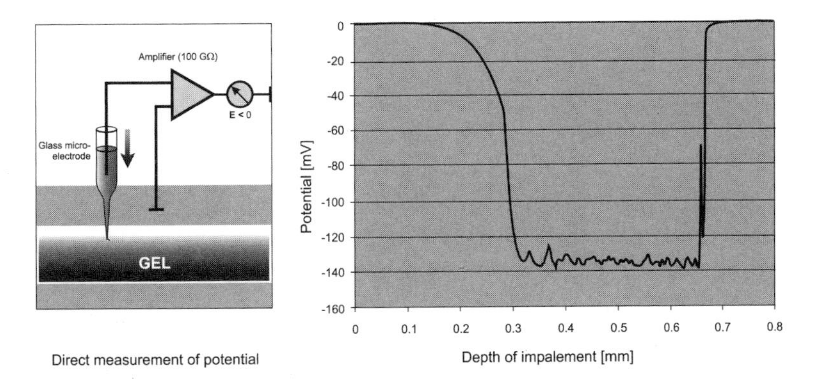

Figure 6: Direct measurement of electric potentials in the gels. A glass microelectrode filled with 3 M KCl is impaled into a gel and moved deeper in a continuous and absolutely vibrationless manner (left). A representative original recording of the electric potentials of an anionic polyelectrolyte gel PAAm/PK$^+$A$^-$ is shown in the right diagram. The velocity of impalement was 0.1 mm/s. The surrounding bathing solution contained 1 mM KCl.

the time scale corresponds with the depth scale perpendicular to the surface of the gel. The signal represents the potential difference between the gel and the surrounding solution. After quickly removing the electrode from the gel, the potential must regain its initial zero level.

Using this technique, series of potential recordings have been performed in the anionic PAAm/PK$^+$A$^-$ [Fig. 7(a)] and the cationic PAAm/PDADMAC gels [Fig. 7(b)], equilibrated for a sufficiently long time in solutions of different KCl concentrations. On first approximation, the experimental values fit the calculated values resulting from the conventional theory of the Donnan potential. For a more precise matching, activities should be used instead of concentrations. The actual dissociation and the spatial distribution of the fixed charges of the dissociated polyelectrolyte, which in principal are not shielded completely by their mobile counter-ions, should be taken into consideration as well.

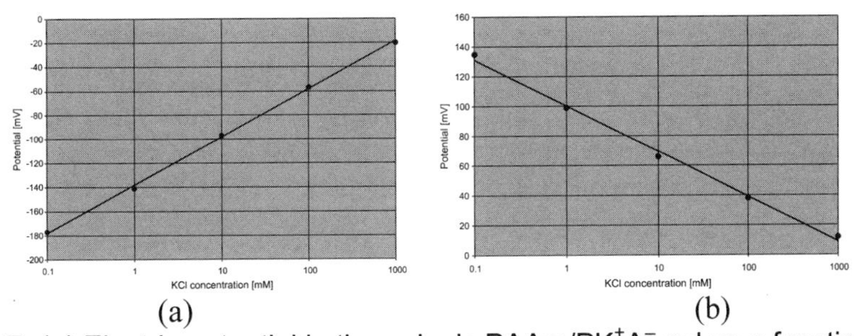

Figure 7: (a) Electric potential in the anionic PAAm/PK$^+$A$^-$ gel as a function of the KCl concentration in the solution. The measured points are fitted convincingly by a logarithmic regression curve. (b) Electric potential in the cationic gel PAAm/PDADMAC as a function of the KCl concentration in the solution. The line represents the best logarithmic fit to the measured values.

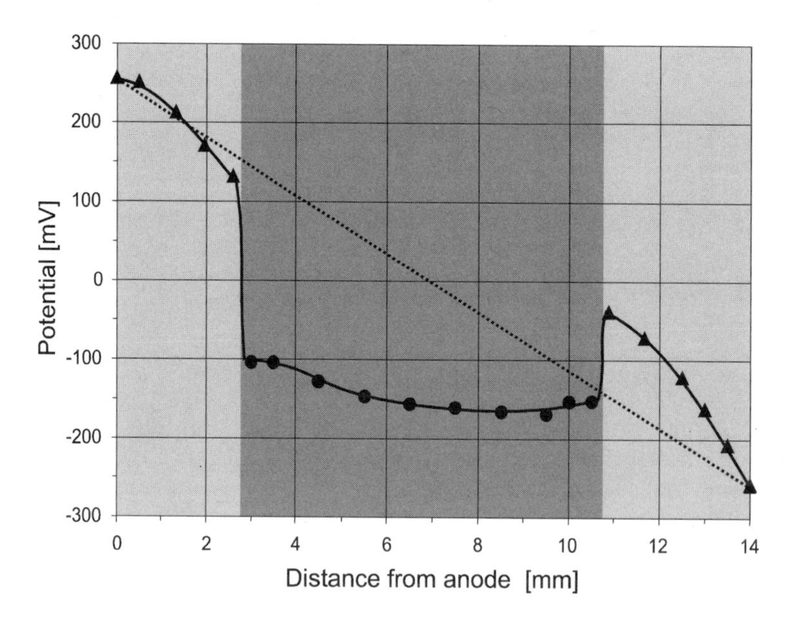

Figure 8: The potential profile measured in the bathing solution containing 0.1 mM KCl and in a PAAm/PK^+A^- gel strip under the influence of an almost homogeneous electric field applied through a pair of parallel platinum electrodes. It is plotted vs. the distance from the anode. The gray rectangle represents dimension and position of the gel strip. The dotted line describes the potential profile in the solution without the gel.

The modified microelectrode technique opens the possibility of recording spatial profiles of the potential in a gel. We have tried to record potentials at different positions within a PAAm/PK^+A^- gel probe exposed to an external homogenous electric field. A gel strip was centered between a pair of platinum electrodes and aligned parallel to them. A dc electric field was applied. In Fig. 8, the potential profile in the solution and in the gel is plotted as a function of the distance from the anode. The measurements were performed under steady-state conditions and under a stimulation dc of 0.15 $\mu A/mm^2$. The potential in the solution declines almost linearly as expected. The potential of the intragel boundary layer is more negative at the anode side, in the sense of a "hyperpolarization," and less negative at the cathode side, in the sense of a "depolarization." This means that the Donnan equilibrium potential is severely disturbed by the external electric field. This leads to a redistribution of the mobile ions within the gels, especially at the boundaries. These differences in the potential are responsible for locally different swelling ratios: an additional slight swelling on the anode side and a slight deswelling on the cathode side, provoking the bending of a gel strip in an external electric field, which is discussed in the next paragraph.

The calculated electric potential in the mid-plane (y=7.5 mm) of the gel probe in Fig. 5 is plotted in Fig. 9. The dashed line (1) represents the linear

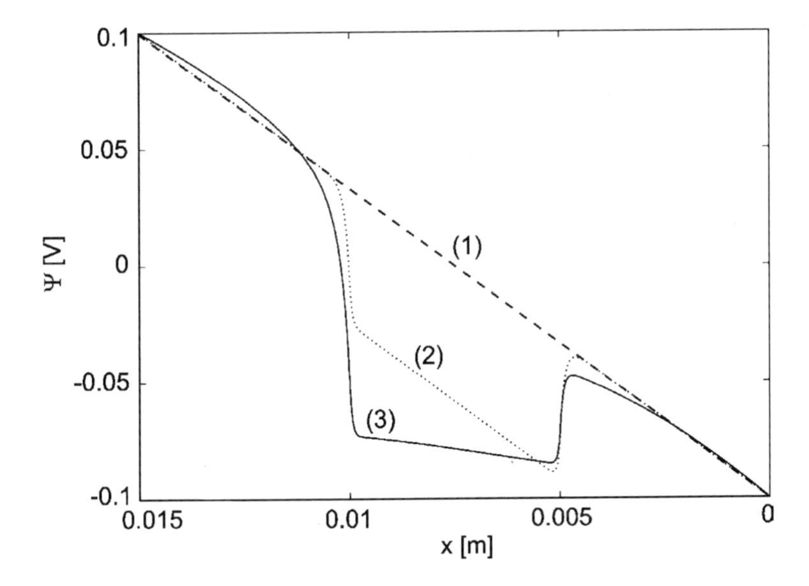

Figure 9: Calculated electric potential Ψ in the gel and the solution at y = 7.5 mm (scaling related to Fig. 5). Line (1), electric potential in the solution; line (2), superposition of the potential in the gel without an external electric field and of line (1); line (3), potential profile in the solution and in the gel under an external electric field.

decrease of the electric potential in the solution (without gel); the dotted line, (2), shows a superposition of the course of the electric potential in the gel without applied electric field and of the potential in the solution, line (1). The solid line, (3), represents the electric potential in the gel and in the adjacent solution with applied external electric field (coupled simulation). Lines (2) and (3) show the same qualitative behavior in the solution outside the gel. Inside the gel, the electric potential given by line (3) shows a smaller decrease than the one represented by line (2); this stems from the higher conductivity due to the larger number of mobile ions in the gel than in the solution.

The smaller increase of the electric potential in the gel is compensated by a larger step at the anode side and a smaller step at the cathode side of the gel. The calculated profiles are in close (qualitative) agreement with the experimental results. Hyperpolarization on the anode side of the gel and depolarization on the cathode side were found under an applied external electric field, in both simulation and experiments. In cationic polyelectrolyte gels like PAAm/PDADMAC, all effects are inverse, an additional swelling on the cathode side and a deswelling on the anode side will bend a cationic gel strip away from the cathode.

11.3.3 Swelling and Bending Behavior of Ionic Gels

While in Sec. 11.3.1, the ion concentrations, the osmotic pressure, and the electric potential for fixed gel probes have been considered, in this section our interest is focused on the swelling and bending behavior of a single gel strip. In order to describe the elongation as well as the bending of the gel under chemical as well as electrical stimulation, at first a one-dimensional model is given. Additionally, in Sec. 11.3.3.3, a two-dimensional model is presented.

11.3.3.1 Isotropic Swelling

In this section, the swelling of a gel under chemical stimulation is investigated. First, the equation of motion, Eq. (20), is solved for a quasi-isotropic gel by a one-dimensional numerical simulation.

The concentration of anionic rest groups is set to $c_{\mathrm{A}^-} = 0.1\,\mathrm{M}$. At t = 0, the concentration of mobile ions in the solution $c^{(s)}_{\mathrm{Na}^+} = c^{(s)}_{\mathrm{Cl}^-} = c_s = 1\,\mathrm{M}$ was reduced to $c_s = 0.001\,\mathrm{M}$. The calculation was done with normalized material parameters; the damping parameter was chosen in such a way that aperiodic damping occurred. Note that an osmotic pressure depending on the position within the gel leads to a space-dependent swelling (elongation) of the gel.

The time history of the elongation of the gel fiber is plotted in Fig. 10. The length of the fiber increases from the initial value $l_0 = 4\,\mathrm{cm}$ to the stationary value $l = l_0 + \Delta l = 12\,\mathrm{cm}$.

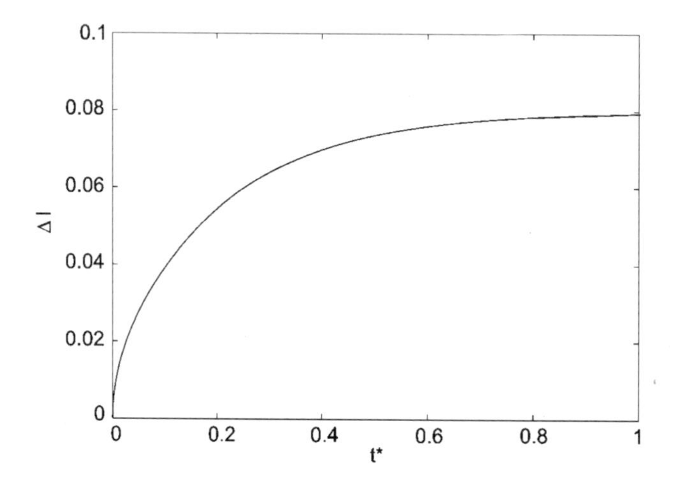

Figure 10: Time history of the elongation of a gel fiber (initial length l = 0.04 m).

11.3.3.2 Bending in a Homogeneous Electric Field

The second model describes the bending behavior of a polyelectrolyte gel in a homogeneous electric field, oriented parallel to the electrodes in the solution (see Fig. 11).

In order to simulate the bending behavior that is the dominant phenomenon of this process, a nonlinear model based on the Timoshenko-beam theory, is formulated. Measurements [Gülch, 2000] have shown that the elastic modulus is nearly constant over a wide range of salt concentrations. Therefore, a constant Young's modulus is employed in the following calculations.

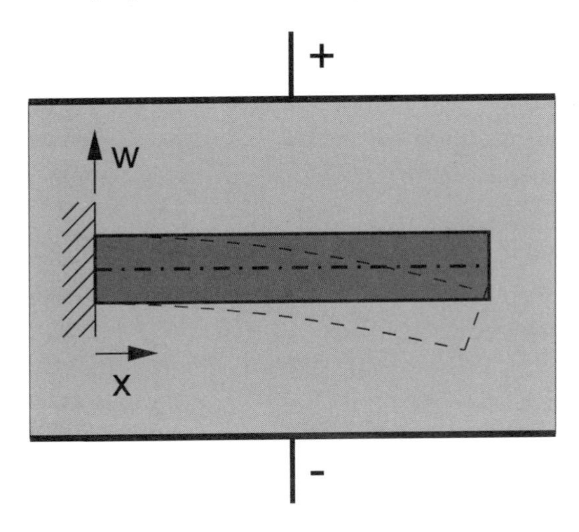

Figure 11: Model for the polymer gel in electric field.

The principle of virtual power yields the following weak form of the equation of motion:

$$\begin{aligned}&\int_{I_N}\int_{\Omega(t)}\Big(\delta\dot{\varepsilon}\,N+\delta\dot{\gamma}\,Q+\delta\dot{\kappa}\,M+\delta\dot{u}\{\rho A\ddot{u}+bA\dot{u}-p_x\}\\&+\delta\dot{w}\{\rho A\ddot{w}+bA\dot{w}-p_z\}+\delta\dot{\varphi}\{\rho I\ddot{\varphi}+bI\dot{\varphi}-m\}\Big)d\Omega\,dt=0,\end{aligned}\tag{25}$$

where the geometric nonlinear strains (according to von Kármán)

$$\varepsilon=u_{,x}+\frac{1}{2}w_{,x}^2,\quad \gamma=w_{,x}+\varphi,\quad \kappa=\varphi_{,x},\tag{26}$$

and the forces and moments, incorporating the actuator loads

$$N=EA(\varepsilon-\overline{\varepsilon}),\quad Q=GA_S\gamma,\quad M=EI(\kappa-\overline{\kappa}),\tag{27}$$

are employed.

The integration is carried out over the time interval $I_N = \{t \mid t_n \le t \le t_{n+1}\}$ and the entire domain $\Omega(t) = \{x \mid x_{\min}(t) \le x \le x_{\max}(t)\}$ of the beam. Appropriate natural (traction) boundary conditions and essential (displacement) boundary conditions are applied on $\Gamma = \partial\Omega$.

Integration by parts of Eq. (25) in time yields:

$$\begin{aligned} &-\int_{I_n}\int_x \delta\ddot{U}_i \mathsf{M}_{ij}\dot{U}_j \, dx\, dt + \int_{I_n}\int_x \delta\dot{U}_i \left(\mathsf{B}_{ij}\dot{U}_j + \mathsf{K}_{ij}U_j - \mathrm{F}_i\right) dx\, dt \\ &+ \int_x \delta\dot{U}_i\Big|_{n+1}^{-} \mathsf{M}_{ij}\dot{U}_j\Big|_{n+1}^{-} dx - \int_x \delta\dot{U}_i\Big|_n^{+} \mathsf{M}_{ij}\dot{U}_j\Big|_n^{-} dx = 0, \end{aligned} \tag{28}$$

where M_{ij}, B_{ij} and K_{ij} are the spatial differential operators corresponding to the mass, damping and stiffness matrix and U represents the vector of unknowns $\boldsymbol{U} = [u, w, \varphi]^T$. Details of the discretization method are given in Grohmann et al. [1998] and Wallmersperger et al. [1999].

For the calculated (see Fig. 13) or measured Donnan potentials at the gel-solution interface, the swelling ratios q_i can be determined. The one-dimensional swelling ratios α_i may be given by the relation $\alpha = q^{1/n}$. For the one-dimensional simulation, $n = 3$ has been chosen, in order to simulate the isotropic swelling.

The conversion of the one-dimensional swelling ratios α_i into the resultant curvature $\bar{\kappa}$ can be formulated according to Fig. 12:

$$\bar{\kappa} = \frac{1}{r} = \frac{2}{d}\frac{\alpha_2 - \alpha_1}{\alpha_1 + \alpha_2}. \tag{29}$$

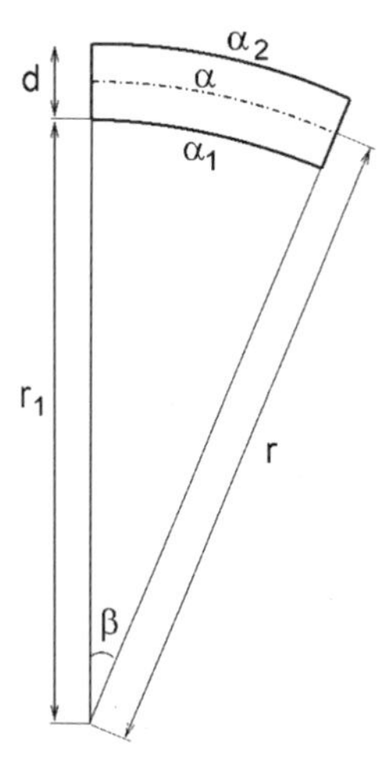

Figure 12: Curvature $\bar{\kappa}$ = $1/r$ in dependence of the one-dimensional swelling ratios α_i.

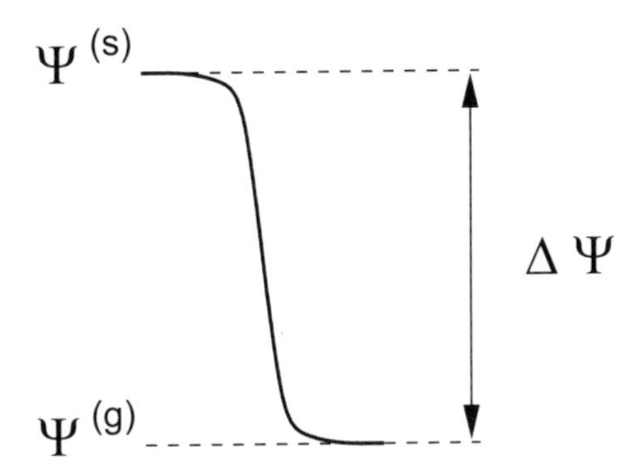

Figure 13: Determination of Donnan potential $\Delta\Psi$.

In the following test case, starting from the measured Donnan potentials $\Delta\Psi_{\text{anode}}$ and $\Delta\Psi_{\text{cathode}}$, the two calculated swelling ratios $\alpha_{\text{anode}} = 5.24$ and $\alpha_{\text{cathode}} = 5.10$ are determined.

The properties and parameters for the numerical simulation are

$$l = 0.04\,m,\ \ d = 0.002\,m,\ \ A = 3.14\cdot 10^{-6}\,m^2,\ \ A_s = 2.516\cdot 10^{-6}\,m^2,\ \bar{\kappa} = 13\,m^{-1},$$

$$I = 7.854\cdot 10^{-13}\,m^4,\quad E = 120000\,\frac{N}{m^2},\quad G = 40000\,\frac{N}{m^2},\quad \rho = 1000\,\frac{kg}{m^3},\ \text{ and}$$

$$b = 50000\,\frac{kg}{m^3\,s}.$$

These parameters have been measured on the gel cylinder demonstrated in Fig. 18.

The gel beam is discretized with nine elements using quadratic interpolations for the displacements in space. Figure 14 (a) shows the time history of the vertical displacement $w(x,t)$ of the gel. The steady-state solution, obtained after approximately 250 time steps of $\Delta t = 0.02\,s$, is plotted in Fig. 14 (b).

It should be noted that normally the mechanical properties (e.g., the elastic modulus) may depend on the hydration—which is dependent on the pH or the salt concentration in the solution—of the polymer gel sample. This dependency is quite small, see for example, Fig. 6 in Gülch et al. [2000] and therefore this effect has been neglected.

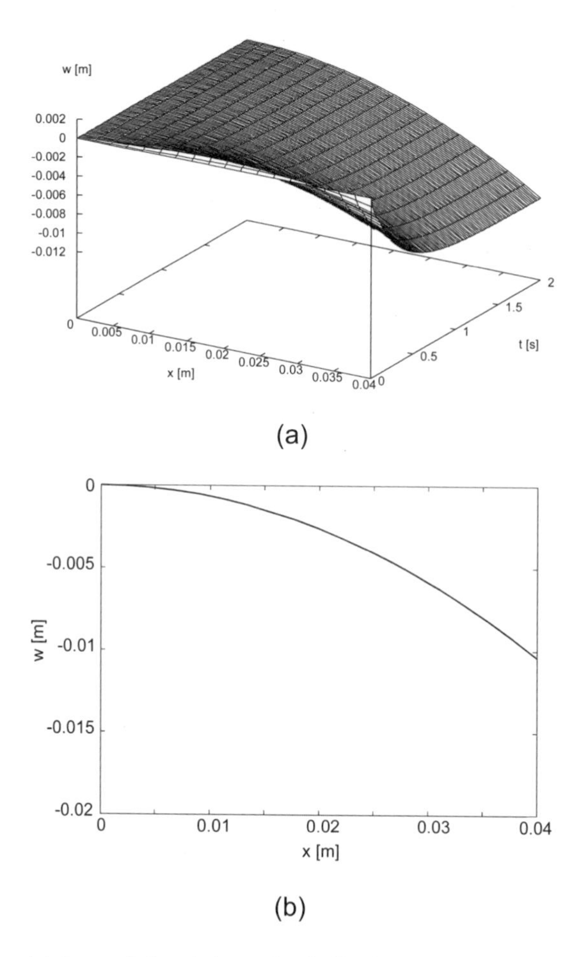

Figure 14: Time history of the (a) vertical displacement $w(x,t)$, and (b) the time history of the steady-state solution.

11.3.3.3 Two-dimensional Simulation of Swelling and Bending

As a consequence of the concentration differences (of the mobile ions) between gel and solution, the osmotic pressure difference—according to Eq. (23)—can be given as

$$\Delta p = RT\sum_{\alpha}(c_{\alpha}^{(s)} - c_{\alpha}^{(g)}) \tag{30}$$

From Eq. (21) follows

$$\sigma_{ij} = C_{ijkl}(\varepsilon_{kl} - \overline{\varepsilon}_{kl}),$$

where

$$\overline{\varepsilon}_{kl} = -\mathrm{const}_{kl}\frac{\Delta p}{RT}. \tag{31}$$

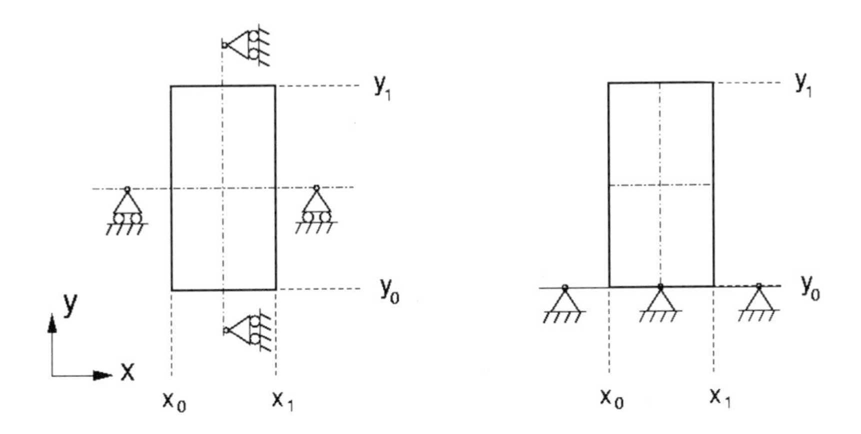

Figure 15: Mechanical boundary conditions of test case A (left) and test case B (right).

In the following test case, the cathode is placed at the left boundary of the solution ($x = 0$), the anode is placed at the right one ($x = 0.015m$); see Fig. 15.

To give the qualitative solution of the deformation of the gel film, the prescribed mechanical strains $\boldsymbol{\varepsilon}$ are chosen as (for more details refer to Wallmersperger et al. 2002])

$$\bar{\varepsilon}_{11} = \bar{\varepsilon}_{22} = -0.06 + 0.13\frac{x - x_0}{l_x},$$

where $x_0 = 0.005m$ and $l_x = x_1 - x_0 = 0.005m$.

For this two-dimensional test case, two different bearings have been investigated. First, the gel is fixed in the x-direction at $x = 0.5(x_0 + x_1) = .0075m$ and fixed in y-direction at $y = 0.5(y_0 + y_1) = 0.0075m$; see Fig. 15 (left), test case A. The simulation is conducted on a three-level multigrid using 24×24 shell elements.

Figure 16 (left) depicts the undeformed gel and Fig. 16 (right) the stationary solution of the deformed gel fiber. A deswelling on the cathode side (left) and an additional swelling on the anode side, leading to a bending toward the cathode, can be seen.

Second, the gel is fixed in x- and y-direction at $y = 0.0025m$; see Fig. 15 (right), test case B. In Fig. 17 (right), the stationary solution of the deformed gel fiber is plotted. Due to this kind of bearing, the bending deflection of the gel toward the cathode is more distinct.

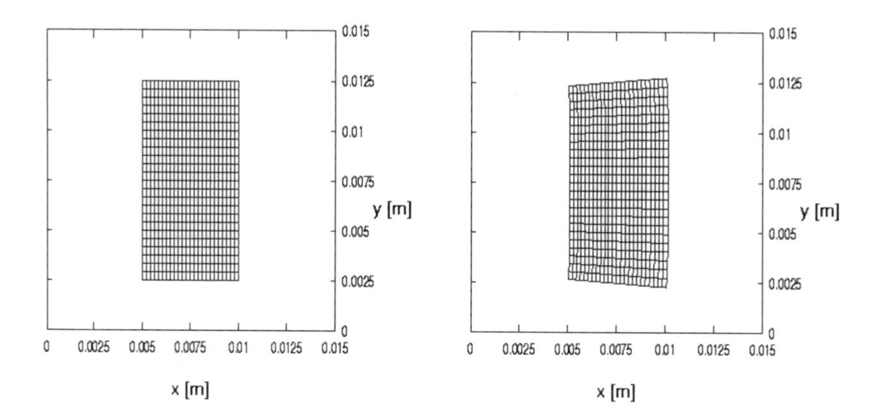

Figure 16: Test case A: undeformed (left) and deformed gel fiber (right) discretized by 24×24 shell elements; symmetric bearing.

Comparing both test cases, we can see the same qualitative behavior. The first example (A) represents an actuator, for which a different extension in the vertical direction is desired; a total decrease of the gel area is obtained when applying the mechnical boundary conditions at the right boundary of the gel ($x = 0.01m$), whereas fixing the gel on the opposite side ($x = 0.005m$) leads to a total increase of the area. The second case (B) can be considered as a typical example of a bending actuator. Comparing this behavior with the example in Sec. 13.3.3.2, we are able to describe the deformation of each finite element in both directions while in the one-dimensional example only a constant curvature over the whole gel beam can be modeled.

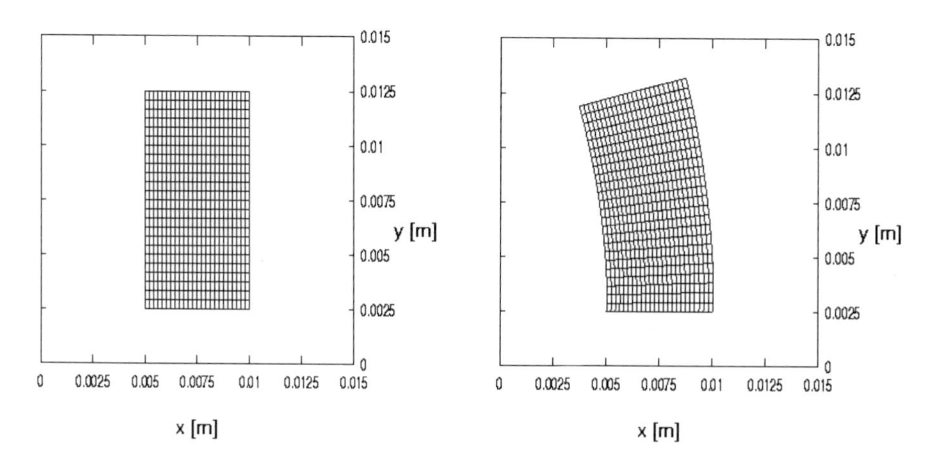

Figure 17: Test case B: undeformed (left) and deformed gel fiber (right); bearing at the bottom of the gel ($y = 0.0025m$).

11.3.4 Experiments

As described in the preceding paragraph, an external electric field induces a bending of a gel strip, provoked by small differences in the swelling ratios. The bending of an anionic PAAm/PK^+A^- gel cylinder is shown in Fig. 18 for different KCl concentrations in the solution. Instead of the bending velocity, the time elapsed for a certain bending angle is plotted as a function of the salt concentration in the solution. The bending process is speeded up with higher electric current. The bending of the anionic gel is directed away from the anode and characterized by its longitudinal axis, being the neutral line. In a cationic gel, opposite movements are obtained. Projecting the properly scaled bending curves of Fig. 14 upon the photographed gel cylinders deflected by the electric field, the numerical curves fit with the measured contours.

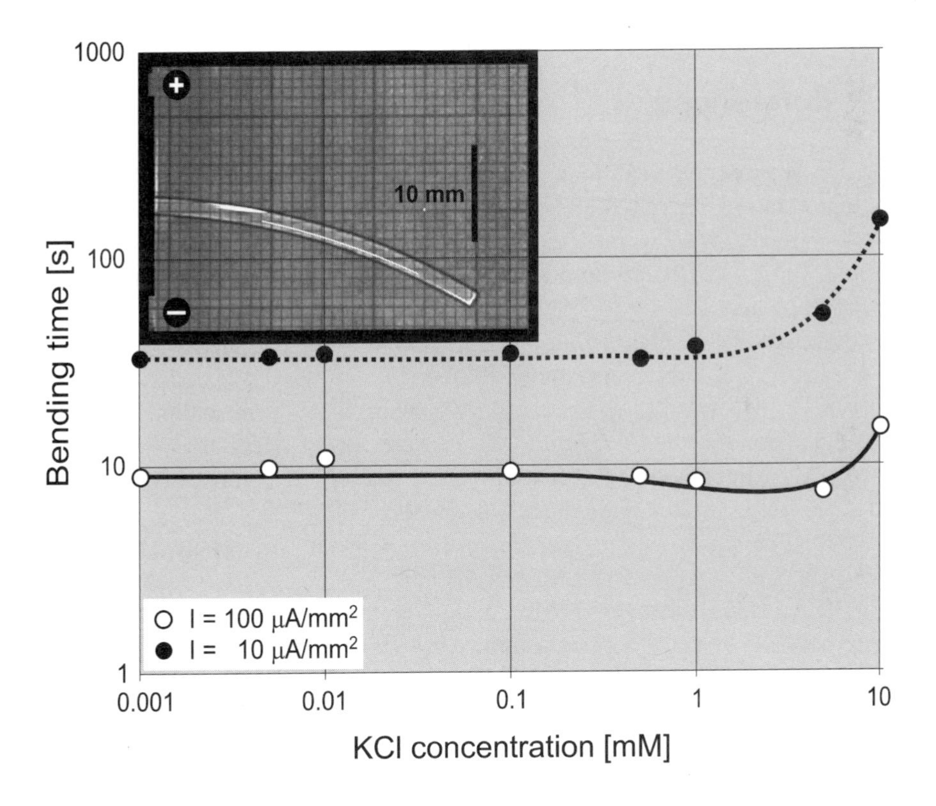

Figure 18: Bending of a PAAm/PK^+A^- gel cylinder (length = 30 mm, diam = 2 mm) in an electric field. The gel cylinder was fixed at one end, the other end could freely move. The bending time is plotted as a function of the KCl concentration in the bathing solution for two different stimulation dc currents. A snapshot of a typical bending experiment is shown in the inset.

11.4 Conclusion

Bending responses of polyelectrolyte gels can be easily induced by direct electrical stimulation. But it is generally almost impossible to actuate appreciable in-toto elongation or contraction of ionic gels by external electric fields, except for the hydrolytic effects of the electric current. In order to better understand the swelling behavior of polyelectrolyte gels in electric fields, one should know more about the direct effects of electric currents on the distribution of ions and about the resulting alterations in the electric potentials involved. This could be achieved by the synopsis of model calculations on the basis of a chemoelectromechanical multifield formulation and of direct measurements of the potential in the gel. The important findings of these investigations are that the bending movements of polyelectrolyte gels exposed to electric fields are supposed to originate from the redistribution of ions and consequently from local differences in the swelling behavior of the gels.

11.5 References

Brock, D., W. Lee, D. Segalman, W. Witkowski, "A dynamic model of a linear actuator based on polymer hydrogel," *J. Intell. Mater. Syst. and Struct.* 5: 764–771, 1994.

Buchholz, F.L. "Polyacrylamides and poly(acrylic acids)," in "Plastics, Properties and Testing to Polyvinyl Compounds," B. Elvers, S. Hawkins, G. Schulz, eds., Vol. 21A, *Ullmann's Encyclopedia of Industrial Chemistry*, pp. 143–156. VCH Publishers, Inc., 1992.

De Rossi, D., M. Suzuki, Y. Osada, P. Morasso, "Pseudomuscular gel actuators for advanced robotics," *J. Intell. Mater. Syst. Struct.* 3: 75–95, 1992.

Doi, M., M. Matsumoto, Y. Hirose, "Deformation of ionic polymer gels by electric fields," *Macromolecules*, 25: 5504–5511, 1992.

Flory, P.J., *Principles of Polymer Chemistry*, Cornell University Press, Ithaca, NY, 1953.

Gregor, H.P., L.B. Luttinger, E.M. Loebl, "Metal-polyelectrolyte complexes. I. The polyacrylic acid- copper complex," *J. Phys. Chem* 59: 34–39, 1955.

Grimshaw, P.E., J.H. Nussbaum, A.J. Grodzinsky, M.L. Yarmush, "Kinetics of electrically and chemically induced swelling in polyelectrolyte gels," *J. Chem. Phys*. 93(6): 4462–4472, 1990.

Grohmann, B., D. Dinkler, "Nonlinear panel flutter phenomena in supersonic and transonic flow," *Nonlinear Dynamics*, 2001, (in press).

Grohmann, B., T. Wallmersperger, R. Dornberger, B.-H. Kröplin, "Time-discontinuous stabilized space-time finite elements for Timoshenko beams," Structures, Structural Dynamics and Materials Conference, AIAA/ASME/ASCE/AHS/ASC, Long Beach, AIAA-98–2017, 1998.

Gülch, R.W., J. Holdenried, A. Weible, T. Wallmersperger, B. Kröplin, "Polyelectrolyte gels in electric fields. A theoretical and experimental approach," in *Smart Structures and Materials 2000: Electroactive Polymer*

Actuators and Devices (EAPAD), Y. Bar-Cohen, ed., Proc. SPIE Vol. 3987, 193–202, 2000.
Hamlen, R.P., C.E. Kent, S.N. Shafer, "Electrolytically activated contractile polymer," *Nature* 206: 1149–1150, 1965.
Hughes, T.J.R., G.M. Hulbert, "Space-time finite element methods for elastodynamics: formulations and error estimate," *Comput. Methods Appl. Mech. Eng.* 66: 339–363, 1988.
Hughes, T.J.R., L.P. Franca, G.M. Hulbert, "A new finite element formulation for computational fluid dynamics: VIII. The Galerkin/least-squares method for advective-diffusive equations," *Comput. Methods Appl. Mech. Eng.* 75: 173–189, 1989.
Hulbert, G.M., "Time finite element methods for structural dynamics," *Int. J. Numer. Meth. Eng.* 33: 307–331, 1992.
Johnson, C., "Discontinuous Galerkin finite element methods for second order hyperbolic problems," *Comput. Methods Appl. Mech. Eng.* 107: 117–129, 1993.
Nemat-Nasser, S., J.Y. Li, "Electromechanical response of ionic polymer-metal composites," in *Smart Structures and Materials 2000, Electroactive Polymer Actuators and Devices, (EAPAD)*, Y. Bar-Cohen, ed., Proc. SPIE Vol. 3987, 82–91, 2000.
Neubrand, W., "Modellbildung und Simulation von Elektronenmembranverfahren," PhD thesis, Universität Stuttgart, 1999.
Rička, J., T. Tanaka, "Swelling of ionic gels: Quantitative performance of the Donnan theory," *Macromolecules* 17: 2917–2921, 1984.
Schröder, U.P., W. Oppermann, "Properties of polyelectrolyte gels," in *Physical Properties of Polymeric Gels*, J.P. Cohen Addad, ed., pp. 19–38. John Wiley and Sons, 1996.
Shahinpoor, M., "Micro-electro-mechanics of ionic polymeric gels as electrically controllable artificial muscles," *J. Intell. Mater. Syst. and Struct.* 6: 307–314, 1995.
Shiga, T., T. Kurauchi, "Deformation of polyelectrolyte gels under the influence of electric field," *J. Appl. Polymer Science* 39: 2305–2320, 1990.
Suzuki, H., B. Wang, R. Yoshida, E. Kokufuta, "Potentiometric titration behaviors of a polymer and gel consisting of n-isopropylacrylamide and acrylic acid," *Langmuir* 15: 4283–4288, 1999.
Tanaka, T., I. Nishio, S.-T. Sun, S. Ueno-Nishio, "Collapse of gels in an electric field," *Science* 218: 467–469, 1982.
Tezduyar, T.E., M. Behr, "A new strategy for finite element computations involving moving boundaries and interfaces – The deforming-spatial-domain/space-time procedure: I. The concept and the preliminary numerical tests," *Comput. Methods Appl. Mech. Eng.* 94: 339–351, 1992.
Treloar, L.R.G., *The Physics of Rubber Elasticity*, Oxford University Press, 1958.
Wallmersperger, T., B.A. Grohmann, B. Kröplin, "Time-discontinuous stabilized space-time finite elements for PDEs of first- and second-order in time,"

European Conference on Computational Mechanics (ECCM '99), München, Germany, GACM, 1999.

Wallmersperger, T., M. D'Ottavio, B. Kröplin, R.W. Gülch "Coupled Multi-Field Formulation for Ionic Polymer Gels.[Invited Lecture]," WCCM V - Fifth World Congress on Computational Mechanics, Vienna, Austria, July 7th - 12th, 2002.

Yannas, I.V., A.J. Grodzinsky, "Electromechanical energy conversion with collagen fibers in an aqueous medium," *J. Mechanochem. Cell. Motility* 2: 113–125, 1973.

Yeager, E., J. O'M. Bockris, B. E. Conway, S. Sarangapani, eds., "Electrodics," Volume 6 of *Comprehensive Treatise of Electrochemistry*, Plenum Press, 1983.

CHAPTER 12

Electromechanical Models for Optimal Design and Effective Behavior of Electroactive Polymers

Kaushik Bhattacharya, Jiangyu Li, and Yu Xiao
California Institute of Technology

12.1 Introduction

Electroactive polymers (EAPs) offer the promise of creating the flexible, low-mass actuators that can form the basic building blocks of artificial muscles. Their main attractive features are relatively small density, compatibility with large deformations, and ease of manufacture and modification. These features have been highlighted by the current generation of actuators [Bar-Cohen, 2000]. Unfortunately, these materials typically produce low force, and the current state of the art is insufficient for most applications. However, the current generation of actuators has been fabricated to demonstrate attractive large deformation features, and little effort has been placed on optimizing the configurations to generate large forces. This chapter discusses a theoretical framework to model the electromechanical behavior of these materials with the goals of

- Optimizing the design of these actuators to produce large force. Such optimal design is essential to produce practically viable artificial muscle.
- Obtaining reduced constitutive models of these actuators. Such reduced models are essential for designing controllers for these actuators and also for the broader mechanical design of the devices.

This chapter focuses on mesoscopic and macroscopic electromechanical behavior, and thus the electrochemistry is often ignored or treated in a very simplistic manner. The reader is referred to the other chapters where these issues are discussed at length. A challenge for the future is integrating these various aspects of EAPs.

For the purposes of this chapter, we will divide EAPs into two broad categories:

(1) *Electrostatic actuators*
Here an elastomer is embedded with a configuration of electrodes. Typically, the electrodes are metallic in such actuators. The elastomers are largely insulating so the electric field creates a mechanical force on the electrodes, which is transmitted to the elastomer causing it to deform. In such actuators, two crucial aspects have to be taken into account in choosing the configuration of the electrodes:

- maximizing the capacitance to generate favorable electrical fields,
- maximizing the mechanical force transfer between the electrodes and the elastomer.

(2) *Ionomeric actuators*
Here a conductive EAP is embedded with a configuration of electrodes. The electrodes may be either metallic or polymeric. These polymers conduct electricity through ion transport, and change shape depending on the local ionic composition. Thus, actuation is generated through the transport of ions, and the crucial aspects in choosing the configuration of the electrodes are:

- designing ionic transport pathways for maximum force,
- configuring the time-dependant electric field to produce necessary ionic transport.

12.2 Introduction to Finite Elasticity

EAPs undergo large deformations, and it is necessary to use a finite deformation theory to account for them. We provide a brief introduction to such a theory for elastomers and refer the reader to Spencer [1980], Billington and Tate [1981], Ogden [1984], and others for a detailed description.

Consider a body, and let it occupy the region Ω in the reference configuration. We label the particles of this body using the position x of the particle in the reference configuration as shown in Fig. 1. We describe the deformation of this body as a mapping $y = y(x)$, where $y(x)$ describes the position of the particle x after deformation. Note that this is a vector-valued function of a vector-valued variable:

$$x = \begin{pmatrix} x_1 \\ x_2 \\ x_3 \end{pmatrix}, \qquad y(x) = \begin{pmatrix} y_1(x_1, x_2, x_3) \\ y_2(x_1, x_2, x_3) \\ y_3(x_1, x_2, x_3) \end{pmatrix}, \tag{1}$$

with respect to a rectangular Cartesian coordinate system. The displacement of the particle x is defined as $u(x) = y(x) - x$.

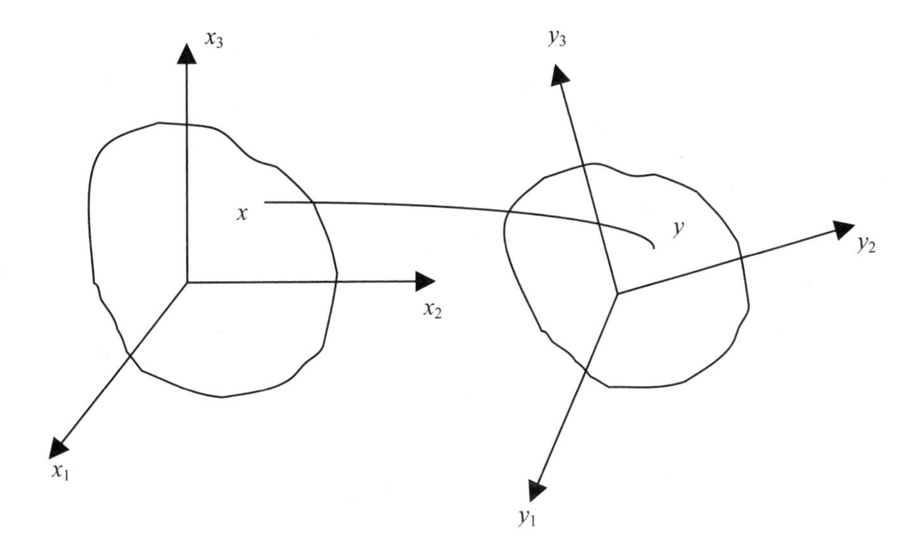

Figure 1: The reference and deformed configuration.

A simple example, which we will find useful later, is the tri-axial extension shown in Fig. 2. Here the reference configuration is a block $(0,L) \times (0,W) \times (0,H)$, and deformation is

$$y_1 = \lambda_1 x_1; \quad y_2 = \lambda_2 x_2; \quad y_3 = \lambda_3 x_3, \tag{2}$$

so that the deformed configuration is the block $(0,l) \times (0,w) \times (0,h)$ with

$$l = \lambda_1 L, \quad w = \lambda_2 W, \quad h = \lambda_3 H. \tag{3}$$

The relative deformation of the body in a small neighborhood around a particle is described by the deformation gradient $F = \nabla_x y$ which is the 3×3 matrix of partial derivatives of y with respect to x. The components of F are given by

$$F_{ij} = \frac{\partial y_i}{\partial x_j}. \tag{4}$$

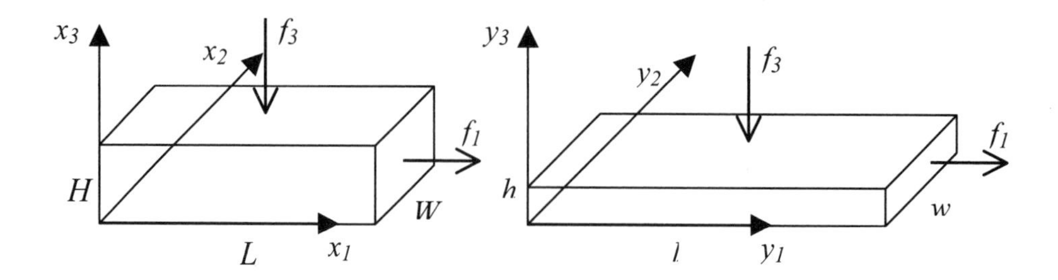

Figure 2: Tri-axial extension.

Typical elastomers are incompressible and have to satisfy the constraint det $F = 1$. The matrix F describes both the rotation and the distortion (stretch) of the body. Only distortion can contribute to the stress or stored energy of the body and it is therefore necessary to "factor out" the rotation. This is accomplished by considering either the Cauchy-Green matrix, which is defined as $F^T F$, or the Lagrangian strain matrix

$$E = \frac{1}{2}\left(F^T F - I\right) , \tag{5}$$

where I is the identity matrix, and superscript T is used to denote matrix transpose.

We note in passing that infinitesimal or linear elasticity approximates the Lagrangian strain matrix E with the (infinitesimal) strain matrix e. It is conventional here to use the displacement $u(x) = y(x) - x$ instead of the deformation y. It is easy to see that the displacement gradient $\nabla_x u = F - I$ and that

$$E = \frac{1}{2}\left(\nabla_x u + (\nabla_x u)^T + (\nabla_x u)^T \nabla_x u\right) \approx \frac{1}{2}\left(\nabla_x u + (\nabla_x u)^T\right) = e \ , \tag{6}$$

when the displacement and displacement gradient are small.

In the setting of finite deformations there are two notions of stress. The first is the nominal or Piola-Kirchhoff stress S: this describes the force per unit reference area. The second is the Cauchy stress T: this describes the force per unit current area. They are related through the relationship

$$T = \frac{1}{\det(F)} S F^T . \tag{7}$$

The stress is related to the deformation gradient through a constitutive relation. For incompressible materials, the constitutive relation has the form

$$T = -pI + \tilde{T}(F) \, , \tag{8}$$

where p is an arbitrary pressure field that is obtained from equilibrium and $\tilde{T}(F)$ is a given function that satisfies the conditions of frame indifference and materials symmetry. Note that this is a matrix-valued function of a matrix-valued variable.

Deformation and stress also cause a stored or elastic energy in the body (this is the Helmholtz free energy in the language of thermodynamics). This can again be described as a constitutive relation $W(F)$, which must also satisfy the conditions of frame indifference and materials symmetry. These constitutive functions are not independent; instead one can obtain T from $W(F)$ through the formula

$$\tilde{T}(F) = \frac{1}{\det F} \frac{\partial W}{\partial F} F^T \, . \tag{9}$$

A simple example of the constitutive relation is the so-called neo-Hookean relation

$$W(F) = \frac{\alpha}{2}\left(tr(FF^T) - 3 \right) \, , \tag{10}$$

or equivalently

$$T = -pI + \alpha FF^T \, , \tag{11}$$

where α is a material constant. This one constant model has been found to be remarkably accurate for many incompressible elastomers describing accurately complex behavior including cavitation, bifurcation, and other instabilities.

The frequency of operation of the actuators and the other processes in EAPs is very slow compared to the speed of sound in these materials. Therefore we may regard the polymer to be at equilibrium at each time. We can obtain the deformation that the body undergoes under a given system of loads by solving the equations of equilibrium, which may be written in reference configuration as

$$\nabla_x \cdot S + b = 0 \, , \tag{12}$$

or equivalently in the deformed configuration as

$$\nabla_y \cdot T + \tilde{b} = 0 \ , \tag{13}$$

where b and $\tilde{b}$ are body force per unit reference and deformed volume, respectively. One has to solve these highly nonlinear equations subject to suitable force or displacement boundary conditions. For example, we may assume that part of the boundary $\partial_1\Omega$ is subjected to a fixed deformation y^0 while another part $\partial_2\Omega$ is subjected to a given traction $t^0(y)$:

$$y(x) = y^0, \quad x \in \partial_1\Omega, \ T\hat{n} = t^0, \quad y \in \partial_2 y(\Omega), \tag{14}$$

where $\hat{n}$ is the unit vector normal to the surface. Note that the deformed configuration is unknown a priori, so it is easier to solve the former equations. However, often the boundary conditions depend on the deformation [as in Eq. $(14)_2$] and this adds to the complexity of this problem.

It is often advantageous to formulate this as a variational problem:

$$\min_{y(\partial_1\Omega)=y^0} \left(\int_\Omega W(\nabla_x y)\, dV - \int_{\partial_2 y(\Omega)} t^0 \cdot y\, da \right). \tag{15}$$

Mathematically, the equilibrium Eqs. (12) and (13) are the Euler-Langrange equations associated with this variational problem.

This is clearly a very difficult nonlinear problem, and, in general, one has to resort to numerical methods to solve it. There are very well-developed finite element methods and even commercial packages like ABAQUS which work very well in most problems.

There is also a well-developed "semi-inverse" method pioneered by R. S. Rivlin, which gives analytic solutions to many problems of interest. We close this section with a simple problem illustrating this method. Suppose our block of elastomer shown in Fig. 2 is constrained so that it cannot deform in the x_2-direction, but subjected to total forces f_1 and f_3 in the x_1- and x_3-directions respectively. We look for a solution of the form (2) described above. The constraint in the x_2-direction and incompressibility gives us

$$\lambda_2 = 1, \quad \lambda_1\lambda_2\lambda_3 = 1; \quad \text{so } \lambda_1 = \frac{1}{\lambda_3} = \lambda . \tag{16a}$$

The force conditions give us

$$\frac{f_1}{hw} = -p + \alpha\lambda^2, \quad \frac{f_3}{lw} = -p + \frac{\alpha}{\lambda^2} . \tag{16b}$$

We can then find the following relation between f_1, f_3, λ, and the dimensions of the block:

$$\frac{f_1\lambda}{HW} - \frac{f_3}{\lambda LW} = \alpha\left(\lambda^2 - \frac{1}{\lambda^2}\right). \tag{17}$$

We see that the response is inherently nonlinear and it is not easy to solve explicitly for λ.

12.3 Optimal Design of Electrostatic Actuators

Let us now consider the electromechanical problem relevant to electrostatic actuators. Consider an elastomer occupying the region Ω in the reference configuration embedded with electrodes $C = C^+ \cup C^-$ as shown schematically in Fig. 3. We shall model the elastomer as before but consider the electrodes to be two-dimensional elastic (material) surfaces that are perfectly conducting. Suppose the electrodes are subjected to some voltage ϕ^+ on C^+ and ϕ^- on C^-. Then we have to solve the electrostatic problem to find the forces that the electrodes apply to the elastomer, but we need to solve this in the deformed configuration, which is unknown a priori. In other words, the electromechanical problem is nontrivially coupled.

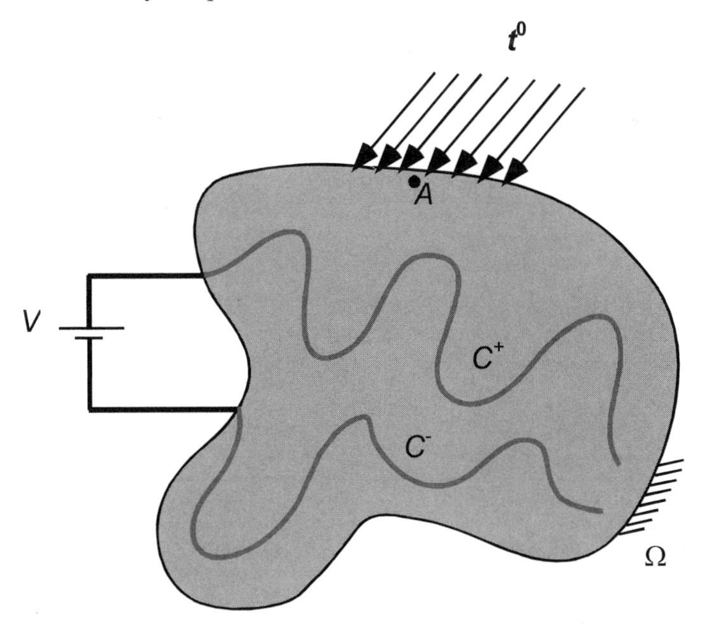

Figure 3: An EAP actuator.

Let us now formulate the electrostatic problem assuming that the actuator has undergone a deformation y so that the elastomer occupies the region $y(\Omega)$ and the electrodes occupy the surfaces $y(C)$. We have to solve Poisson's equation

$$-\nabla_y \cdot \varepsilon \nabla_y \phi = \rho \tag{18}$$

on all of space subject to the boundary conditions

$$\phi(y) = \phi^+, \quad y \in y(C^+);\ \phi(y) = \phi^-, \quad y \in y(C^-); \tag{19}$$

for the electric potential ϕ. Above, ρ is the charge density on the electrodes, and ε is the dielectric constant, which takes the value ε_1 in the elastomer and the value of free space ε_0 outside it. We can now obtain the forces applied by the electrodes on the elastomer as follows. First, we find the forces normal to the conductor using the Maxwell stress tensor

$$M = \varepsilon \nabla_y \phi \otimes \nabla_y \phi - \frac{\varepsilon}{2} |\nabla_y \phi|^2 I. \tag{20a}$$

Second, we find a mixed boundary condition in the tangential direction based on c, the elastic constant of the electrodes. Therefore, we have

$$\hat{n} \cdot T\hat{n} = \hat{n} \cdot M\hat{n}, \quad \hat{t} \cdot T\hat{n} = c\,|F\hat{t}|, \tag{20b}$$

where $\hat{n}$ is the normal to the electrode, and $\hat{t}$ is any vector tangential to it.

To solve the coupled electromechanical problem, we have to solve the coupled problems [Eqs. (12) and (18)] subject to the boundary conditions [Eqs. (14) and (19)]. We can do so iteratively:

- start with the reference configuration,
- solve the electrostatic problem [Eqs. (18) and (19)] and find the forces according to Eqs. (20a) and (20b),
- solve the finite elasticity problem [Eqs. (12) and (14)] for these forces to find a new configuration,
- iterate.

We have implemented such a procedure using the commercial package ABAQUS for some common geometries.

Alternately, we can formulate this variationally, and this is useful from the point of view of theoretical considerations of optimal design: minimize the potential energy

$$\int_{\Omega} W(\nabla_x y)\, dV - \int_{\partial_2 y(\Omega)} t^0 \cdot y\, da + \tfrac{1}{2} \int_{R^3} \varepsilon |\nabla_y \phi|^2\, dV \ , \tag{21}$$

among all deformations satisfying Eq. $(14)_1$, where t^0 is the traction applied by both the electrodes and external loads, and ϕ is the electric potential obtained by solving Eqs. (18) and (19) as before for any given y. Note that this is a variational problem with an auxiliary partial differential equation. Such variational problems with auxiliary equations were developed in micromagnetics [James and Kinderlehrer, 1993] and their potential was recently demonstrated in the electromechanical context when Shu and Bhattacharya [2000] used this framework to model ferroelectrics.

We have thus far discussed the electromechanical problem for a given conductor configuration. We now turn to the optimal design problem of finding the conductor configuration that optimizes a given criterion. For now, let us assume that the criterion is maximizing the elongation in some direction $\hat{i}$. So we choose as our objective function

$$\int_{\Omega} |(\nabla_x y)\hat{i}\,|\, dV \ . \tag{22}$$

The optimal design problem then is the following. Maximize the objective [Eq. (22)] among all possible conductor configurations C that are allowed ($C \subset C_0$), where for each configuration C, the deformation y is obtained by solving the coupled electromechanical problem [Eqs. (12) and (18)] subject to boundary conditions [Eqs. (14), (19), and (20b)]. Here, C_0 gives us the overall geometric constraints on the allowable electrode configuration—it may say, for example, that the electrodes should be confined to some preassigned surfaces.

We can implement this numerically as follows:

- start with a test configuration of the electrodes,
- solve the coupled electromechanical problem,
- calculate the configurational forces [variational derivative of Eq. (22) with respect to C] on the electrodes,
- evolve the electrode configuration and iterate.

This turns out to be not only very expensive but also wrong! Optimal design problems like these are notorious for their ill-posedness [Kohn and Strang, 1986; Bendsøe and Kikuchi, 1988; Allaire et al., 1997; Sigmund, 1997; Cherkaev, 2000]. In this context this means that the optimal configuration C may not be a nice surface, but one that is highly broken up. This is easily illustrated in the flat-sheet actuator shown in Fig. 4. Here we start with a sheet of elastomer, coat it with electrodes on the two faces and apply a potential difference between the two electrodes. The charge induced by the electric field causes an attractive force between the electrodes, which in turn squeezes the elastomer and thereby elongates it. However, the electrodes resist the elongation, and the extension is biaxial with unpatterned electrodes while applications demand uniaxial elongation. A way to achieve large uniaxial elongation is by patterning the electrodes as shown in Fig. 4. The natural question is: what is the optimal pattern, width, and spacing? Consider now a thought experiment. Suppose we cut the electrode along the dashed line in Fig. 4. It would relieve all the constraint for deformation in one direction, but would not significantly change the constraint in the other direction or the electric field. Therefore it is easy to see that the more cuts the better, and the optimal solution is an infinite number of parallel strips which have infinitesimal width and infinitesimal separation — this is the emergence of microstructure, common in such an ill-posed optimal design problem. This configuration gives the largest possible electric field — no constraint in one direction and full constraint in the other, so that all the thickness compression is "channeled" in one direction.

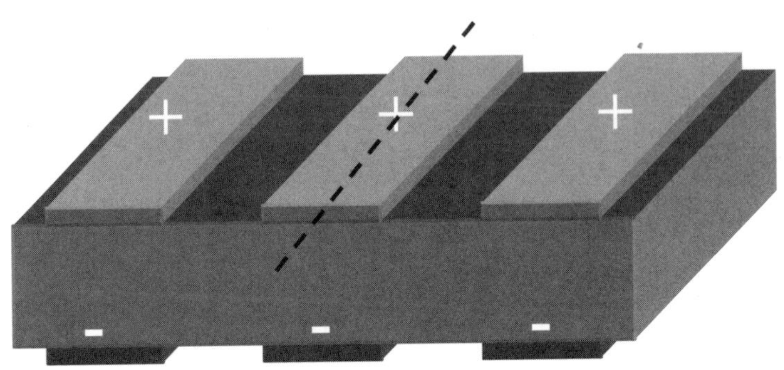

Figure 4: A flat sheet actuator.

In this optimal design, the flat sheet of elastomer is subjected to a constant electric field (neglecting the end loss which is very reasonable if the thickness to length ratio is small) and thus a uniform compression on the top and bottom surfaces, and it is constrained against lateral elongation. It then resembles the tri-axial problem that we have studied earlier (Fig. 2) with force $f_3 = -\varepsilon \frac{V^2 \lambda^3 WL}{H^2}$, and we can obtain the relationship between the applied voltage and elongation from Eq. (17):

$$\left(1 - \varepsilon \frac{V^2}{\alpha H^2}\right) \lambda^4 - \frac{f_1}{\alpha HW} \lambda^3 - 1 = 0, \tag{23}$$

where V is the applied voltage. Note that this is inherently nonlinear and it is difficult to solve explicitly. Figure 5 shows the maximum possible elongation (longitudinal strain λ–1) as a function of elastomer properties (Young's modulus 3α and relative dielectric constant ε) assuming a dielectric breakdown of 150

V/μm. These results are in general agreement with the recent observations by Pelrine et al. [1998, 2000], where such a design was implemented using carbon grease and parallel conducting fibers. An alternative may be using photolithography to deposit patterned electrodes. Once we know the elongation λ_V with applied voltage V and the elongation λ_0 without applied voltage, we can find the work done by the actuator in lifting the load f_1:

$$\text{Work per cycle} = f_1 L(\lambda_V - \lambda_0)\,. \tag{24}$$

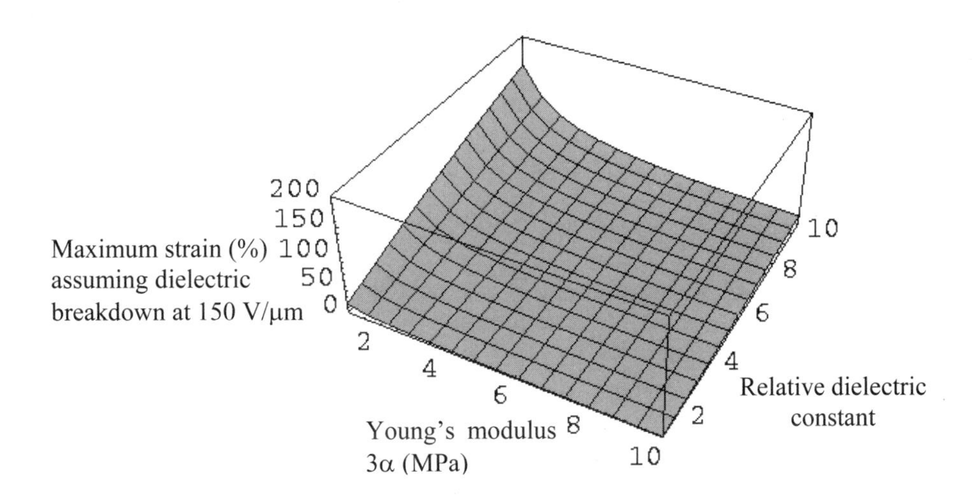

Figure 5: The elongation of an optimal flat-sheet actuator as a function of material properties.

Figure 6 shows the work per cycle for various thicknesses and Young's moduli for a fixed applied voltage (5000 V and force (1N).

Let us now return to the general problem. The ill-posedness arises as the conductor tries to maintain the electric field, but relieves suitable mechanical constraint by developing microstructure or by becoming highly tattered. This can be cured by "relaxation" [Kohn and Strang, 1986; Bendsøe and Kikuchi, 1988; Allaire et al., 1997; Sigmund, 1997; Cherkaev, 2000]. The basic idea in this context is to say that the elastic constant of the conductor is not fixed at c, but it can take a range of values, and treat this as an optimization variable. The range of values is all effective constants that are possible by the development of microstructure. This range of values is known; in particular, it can become anisotropic and vary in a given direction from 0 to c. This relaxed problem has no numerical difficulties, though in general it is computationally expensive. The detailed analysis is beyond the scope of this chapter: interested readers are referred to Xiao and Bhattacharya [2000].

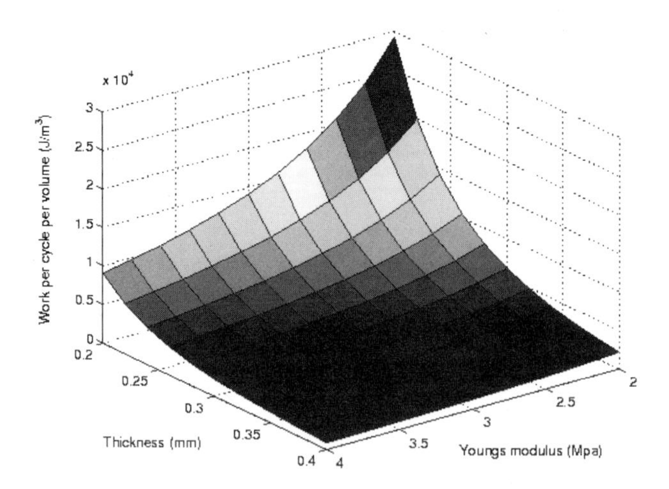

Figure 6: The work per volume per cycle of an optimal flat-sheet actuator.

We conclude this section by discussing a few examples of cylindrical geometry, shown in Fig. 7. We have found through computations that these are significantly less efficient than the flat sheet, with decreasing efficiency as we go from tubular to spiral penetrating to spiral wrapped. In the spiral wrapped, the electrodes are wrapped in a spiral around a cylindrical elastomer ligament. The electric field is very inhomogeneous, so the configuration has low capacitance and the electrodes develop smaller forces. Further, the elastomer has inactive regions, which along with the electrodes constrain against actuation. The spiral-penetrating electrodes generate larger forces, but not to the extent of the flat sheet, and the inactive regions and electrodes continue to constrain against actuation. The tubular, which is similar to the spiral penetrating except for a hollow core, removes some of these constraints, but the overall behavior does not reach the levels of the flat sheet. This configuration is also difficult to manufacture. Therefore we conclude that if one wants a cylindrical actuator, the optimal strategy is to make a sheet and then roll it up to make a rope.

12.4 Models of Ionomer Actuators

We now turn to ionomer actuators. Here ionic fluxes lead to the deformation of the polymer in the stress-free state. We thus have an ionic diffusion problem in a finitely deformable body. Let us try building each of these elements one by one. First let us study ionic diffusion in a rigid body. Assume for now that the positive ions are mobile, but the negative ions are fixed or immobile. The transport of positive ions described by the equations [Lakshminarayanaiah, 1969]

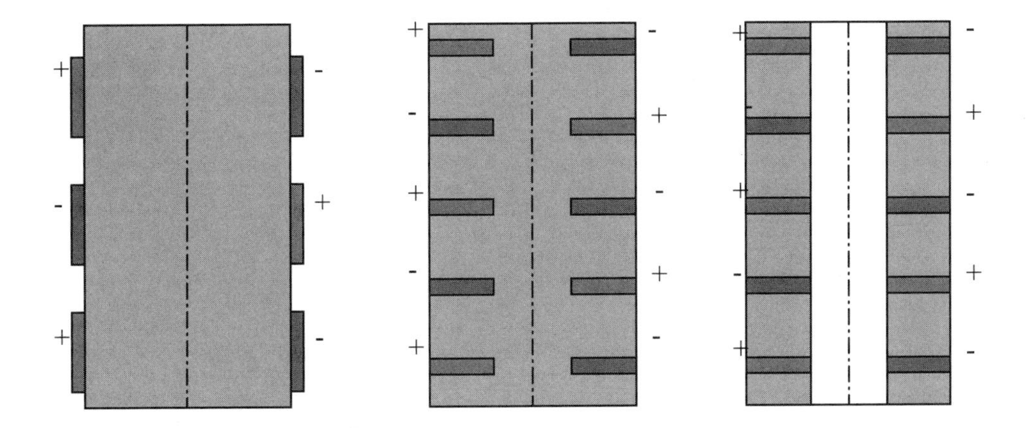

Figure 7: A spiral-wrapped electrode, a spiral-penetrating electrode, and a tubular actuator.

$$J = -d(\nabla_x c^+ + \frac{zc^+ F_{Far}}{RT} \nabla_x \phi), \ \frac{\partial c^+}{\partial t} + \nabla_x \cdot J = 0, \tag{25}$$

subject to the boundary conditions

$$J = 0. \tag{26}$$

Here, c^+ is the concentration of positive ions, z is valency of ion, J is the ion flux, d is the diffusion coefficient, F_{Far} is Faraday constant, R is universal gas constant, and T is absolute temperature. ϕ is the electrostatic potential obtained by solving Poisson's equation [Eq. (18)] with $\rho = zF_{Far}(c^+ - c^-)$, where c^- is the concentration of negative ions.

These materials are useful because as they accumulate ions, the polymer deforms. For example, polyelectrolyte gels swell as the positive ion concentration goes up [Grimshaw, et al., 1990]. Thus, choosing our reference configuration to be the neutral state, we can describe this swelling using the deformation y such that the deformation gradient F satisfies

$$\det F = \delta(c^+), \tag{27}$$

where $\delta(c^+)$ is the relative volume of a polymer with ionic concentration c^+. We will regard this $\delta(c^+)$ as a given function (it can be determined by the microscopic modeling described in the other chapters, or see Nemat-Nasser and Li [2000]).

Suppose we take this swollen polymer and apply some force to it. It deforms further, but now the stress is zero, not when det F = 1, but when $\det F = \delta(c^+)$. So we use as our constitutive relation

$$T = -pI + \tilde{T}\left(\tilde{F}\right), \qquad \tilde{F} = \frac{1}{\left(\delta\left(c^+\right)\right)^{1/3}} F, \tag{28}$$

where $\tilde{T}$ is as defined earlier. In particular, if the polymer is neo-Hookean we can say

$$\boldsymbol{T} = -p\boldsymbol{I} + \alpha\tilde{\boldsymbol{F}}\tilde{\boldsymbol{F}}^T. \tag{29}$$

We can now solve the finite deformation problem if we are given the ionic concentration c^+. We solve the equilibrium Eq. (13) subject to appropriate boundary condition where the stress T is given by Eq. (28).

We now combine the ionic transport with the finite deformation. This is a little tricky because the electrostatic problem [Eq. (18)] has to be solved in the current configuration. In other words, the Laplacian in Eq. (18) is with respect to y, not x. So we can try the following:

- Start with the reference configuration and solve the ionic transport [Eq. $(25)_2$] and electrostatic Eq. (18) in the reference configuration,
- obtain $\delta(c^+)$ and solve the finite elasticity problem [Eqs. (13), (14), and (28)],
- solve the electrostatic problem [Eq. (18)] in the current configuration and obtain the electrostatic potential in the current configuration,
- transform the electrostatic potential to reference variables and solve the ionic transport Eq. $(25)_2$ to obtain c^+,
- iterate.

Note that this is a very difficult coupled time-dependent problem. In other words, it tells us how a polymer would deform at different times. Often actuators are tested at a constant applied voltage, and the polymer evolves until it reaches an equilibrium configuration. This equilibrium configuration describes the maximal deformation and generates the maximal force for a given voltage, and that is the quantity we would like to optimize. So let us try to find this configuration directly without solving the time-dependent problem. The equations that describe it are the same as before, except we drop the term dc^+/dt in Eq. $(25)_2$. It is advantageous to write this as a variational problem:

$$\int_{\Omega} W(c^+, \nabla_x y) dV_0 - \int_{\partial_2 y(\Omega)} t^0 \cdot y da + \int_{\Omega(t)} [\varphi - \frac{\varepsilon}{2} |\nabla_y \phi|^2] dV_t, \tag{30}$$

where φ is the electrochemical potential of mobile ions given by $\varphi = RTc^{+}(y)\ln\frac{c^{+}(y)}{c^{-}(y)} + zF_{Far}[c^{+}(y) - c^{-}(y)]\phi$. The variational procedure for a given arrangement of electrodes is to minimize Eq. (30) over all y and c^{+}.

We are interested in optimizing the actuator configuration to maximize elongation as we did in the case of elastomers. We can do so in two steps, since we note that the mechanical problem knows about the ionic concentration and electrostatic potential only through δ. So, in step one we optimize over all possible δ, and in step two we solve an inverse problem to find the arrangement of electrodes to induce δ.

Note that in step one, the optimization over δ is a purely mechanical problem. To gain some insight into the nature of this optimization problem let us assume that the strains are infinitesimal and linearize these equations. Recall that above we used the relative deformation gradient $\tilde{F} = \left(\delta(c^{+})\right)^{-1/3} F$ to calculate the stresses. Setting $F^{*} = \left(\delta(c^{+})\right)^{-1/3} I$, we obtain

$$F = \tilde{F}F^{*}, \tag{31}$$

which leads to

$$E(F) = F^{*T}E(\tilde{F})F^{*} + E(F^{*}) .(32)$$

Ignoring the higher order terms gives us

$$e = \tilde{e} + e^{*}, \tag{33}$$

where e^{*} is the *eigenstrain*, i.e., the value of the strain at which the stress is zero. This e^{*} depends on the ionic composition c^{+} and represents the strain created by the swelling of the polymer under zero stress. The linearized strain energy and constitutive equations are

$$W(\tilde{e}) = \frac{1}{2}(e - e^{*})c(e - e^{*}), \tag{34}$$

and

$$T = c(e - e^{*}), \tag{35}$$

where c is the elastic modulus of the material. The use of eigenstrain has proven to be a very powerful idea in various problems of micromechanics including composite materials, phase-transforming materials, and plasticity [Mura, 1987; Nemat-Nasser and Hori, 1999].

We now turn to the optimal design problem. Let us assume that our actuator is fixed at some point O and we wish to move a load at a different point x_0 in the direction $\hat{i}$. Therefore we choose as our objective function

$$\hat{i} \cdot \left(u(x_0) - u(O)\right) = \int_L \hat{i} \cdot e\hat{t} dl , \tag{36}$$

where L is a line through the body joining O and x_0 and $\hat{t}$ is the unit tangent to this line. Our optimal design problem is then to find a function e^* that maximizes Eq. (36) where e is obtained for any given e^* by minimizing the functional

$$\int_\Omega \left(e - e^*\right) \cdot c\left(e - e^*\right) dV , \tag{37}$$

subject to suitable boundary conditions. We can write this expression using Lagrangian multipliers as

$$\max_{e^*, e} \left\{ \int_L \hat{i} \cdot e\hat{t} dl - \lambda \int_\Omega \left(e - e^*\right) \cdot c\left(e - e^*\right) dV \right\}, \tag{38}$$

where $\lambda > 0$. We have to solve this numerically as we did for the electrostatic actuators. However, to understand the nature of the problem, we notice that the elastic modulus c is positive-definite and the second term is always nonnegative. This motivates us to try and set it to be exactly equal to zero. This is possible if and only if the eigenstrain e^* is a compatible strain field, i.e., if it can be obtained as the symmetrized gradient of a displacement, or equivalently if it satisfies the strain compatibility condition

$$\nabla \times \nabla \times e^* = 0 . \tag{39}$$

Recall that the eigenstrain in most EAPs is dilitational (liquid crystal elastomers are an exception), and so let us assume for now

$$e^* = e_v I , \tag{40}$$

where e_v is a scalar and denotes the dilitational eigenstrain created by the swelling of the polymer under zero stress. Putting Eq. (40) into Eq. (39) we obtain

$$\frac{\partial^2 e_v}{\partial x_1^2} = \frac{\partial^2 e_v}{\partial x_2^2} = \frac{\partial^2 e_v}{\partial x_3^2} = \frac{\partial^2 e_v}{\partial x_1 \partial x_2} = \frac{\partial^2 e_v}{\partial x_2 \partial x_3} = \frac{\partial^2 e_v}{\partial x_3 \partial x_1} = 0\,, \tag{41}$$

from which we conclude that

$$e_v = Ax_1 + Bx_2 + Cx_3\,, \tag{42}$$

for some constants A, B, C. It is important to recall that this is not the solution to the optimal design problem [Eq. (38)], but only an approximation that provides good insight into it. This distribution of dilitational eigenstrain provides us with the objective that we seek to achieve in designing the electrodes. It is also a benchmark against which one can test a particular device. The most commonly used configuration of EAPs is in the shape of a flat sheet where the electric field is applied perpendicular to the sheet. This will give us a dilitational eigenstrain distribution which is constant across the lateral extent of the sheet, but varies across the thickness, or

$$e_v = e_v(x_3)\,, \tag{43}$$

where x_3 is the thickness direction. Comparing this with our optimal design criterion [Eq. (36)], we conclude that the best we can do is to make this distribution linear in the thickness direction:

$$e_v = e_v(x_3) = Cx_3\,. \tag{44}$$

In ionomeric polymer-metal composites (IPMC), preliminary detailed models show that the distribution is not quite linear, instead there is a tendency to accumulate charges and consequently dilatation near the electrodes [Nemat-Nasser and Li, 2000; Xiao and Bhattacharya, 2000]. We are in the process of exploring methods of overcoming this. We are also in the process of benchmarking various other polymers and configurations, and working on the general nonlinear problem.

12.5 Reduced Models

So far we have seen detailed models, which are useful in understanding and designing polymer and conductor configurations as actuators. However, these models are very complicated and completely useless to an engineer who wants to use the actuator as a component, and design a device using it or controller for it. This engineer does not need to know all the details of electrode design, but needs a simple model that describes the overall behavior of the actuator. Such a model is known as a reduced model, and it may be derived from a detailed model or obtained semi-empirically. Such a model would have macroscopic inputs and outputs, and a few easily characterizable constants.

Consider, for example, the flat-sheet elastomer actuator described in Sec. 12.3. The overall elongation as a function of applied voltage is given by Eq. (23). We may regard this as a simple example of a reduced model, which was derived from a more detailed model.

We close with an example of a semi-empirical model appropriate for a narrow strip of ionomer. This model will also highlight the concept of eigenstrain. Consider a narrow strip (for example, 0.17 mm × 2 mm × 20 mm) of ionomer with electrodes on the two faces (for example, IPMC). This strip bends as a voltage is applied, and since the strip is narrow the bending is largely in plane. Therefore, we can use a nonlinear Euler-Bernoulli beam model, augmented with an eigenstrain.

Consider the strip shown in Fig. 8, and assume for now that the applied force f_3 is zero. The strip bends in response to an applied electric field, and Fig. 9 shows that the curvature is uniform [Bar-Cohen and Leary, 2000]. We call this curvature the load-free curvature, material curvature, or eigen curvature κ. If we apply a step voltage to the beam, this curvature changes according to the formula

$$\kappa(t) = \kappa_v V - (\kappa_v V - \kappa_0)\exp\left(-\frac{t}{\tau}\right), \tag{45}$$

where κ_v (the saturation curvature per unit applied voltage) and τ (the time constant) are material constants, and κ_0 is the initial curvature. For a more general time-dependent applied voltage, the response may be described by the first order ordinary differential equation

$$\frac{d\kappa}{dt} = \frac{1}{\tau}(\kappa_v V - \kappa). \tag{46}$$

Note here that we have assumed that the IPMC has only one time constant. Some specimens are known to have two [Grimshaw et al., 1990] and we can incorporate them easily by modifying Eqs. (45) and (46).

Let us now consider more general deformations of the beam. Assuming that it is inextensible, we can describe the deformation of the beam using a function $\theta(s)$ where θ is the angle shown in the figure and s is the arc-length along the beam. For given eigen curvature κ, the (excess) strain energy of the beam is

$$W = \int_0^l \tfrac{1}{2} EI(\theta' - \kappa)^2 \, ds, \tag{47}$$

where prime denotes differentiation with respect to s and EI is the flexural rigidity. For an applied force f_3 as shown in Fig. 8, the total potential energy of the beam is

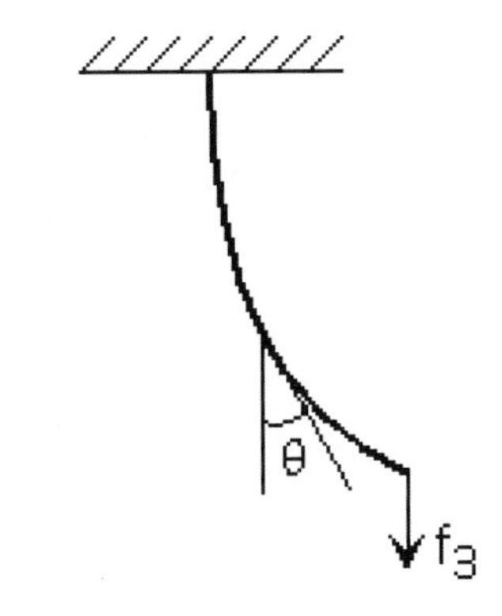

Figure 8: An IPMC strip.

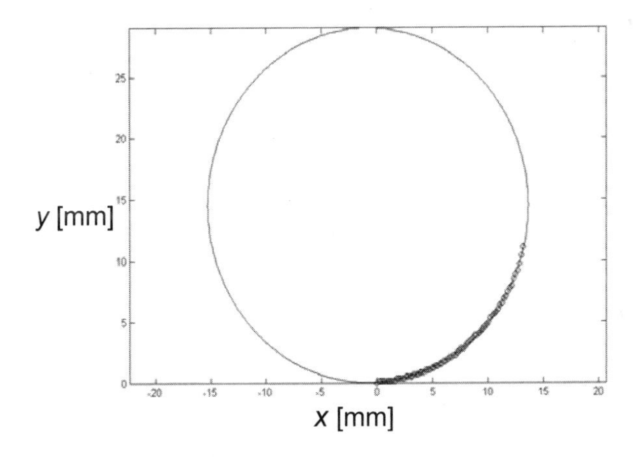

Figure 9: The deformed shape of an IPPC strip with no forces but constant voltage is quite perfectly circular [Bar-Cohen and Leary, 2000].

$$\begin{aligned} L &= W - \textit{work done by external force} \\ &= \int_0^l \left(\tfrac{1}{2} EI \left(\theta' - \kappa \right)^2 + f_3 \left(1 - \cos\theta \right) \right) ds. \end{aligned} \tag{48}$$

From the principle of virtual work $\delta L = 0$, we obtain the equilibrium equations and boundary conditions

$$EI\theta'' - f_3 \sin\theta = 0, \qquad \theta(0) = 0, \quad \theta'(l) = \kappa. \tag{49}$$

This is a simple second order semilinear ODE, which we can solve semi-analytically for any given κ.

This completes the simple reduced model, which has three material constants: E, κ_v, and τ (respectively the elastic modulus, the saturation curvature at unit applied voltage, and the time constant). The solution procedure for a fixed applied load f_3 and given time-dependant voltage $V(t)$ is as follows. At each time t, we integrate Eq. (46) to obtain the eigen curvature κ. For this eigen curvature we solve Eq. (49) to obtain the deformation $\theta(x)$. We repeat for each time t.

By applying different time-dependant voltages and different forces and fitting them to the model above, we can evaluate all the material constants E, κ_v, and τ. Using experimental data from Bar-Cohen and Leary [2000] for zero and 25-mg tip mass and applying the inverse problem we determined the following values

$$\kappa_v = 53.4\ (\text{mV})^{-1},\ \ \tau = 53.4\ \text{sec}\quad E = 72\ \text{Mpa}\,. \tag{50}$$

The comparison to experiments is shown in Fig. 10.

We can easily adapt this reduced model to various other loading situations (for example when the force is applied laterally) and extend this methodology to membranes.

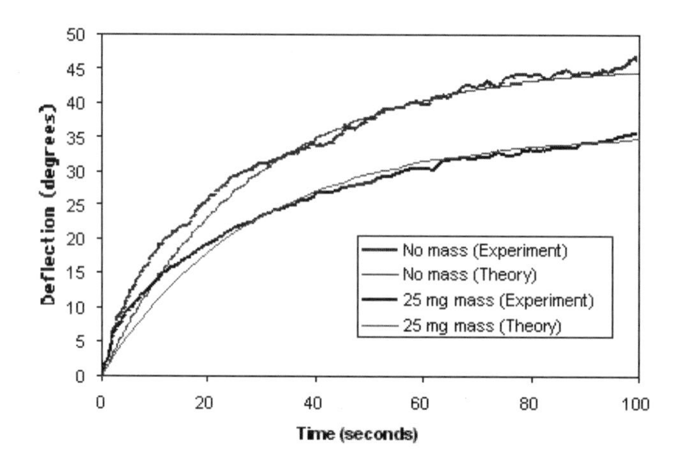

Figure 10: Comparison of the reduced model with experiment.

12.6 Conclusion

This chapter has described the modeling of EAPs from a mesoscopic and macroscopic electro-mechanical point of view. The following is a list of the salient features.

- Elastomer-based actuators can be modeled by combining finite elasticity with classical electrostatics, and there are well-developed numerical techniques for this problem.
- The optimal design of elastomer-based actuators is well understood.
- There is an emerging understanding of microscopic models of ionomers.
- The optimal design of ionomer-based actuators remains a challenge.
- There are faithful reduced models of elastomer-based actuators.
- Faithful reduced models of ionomer-based actuators are available for strips and are being developed for sheets.

12.7 Acknowledgment

We gratefully acknowledge the partial financial support of the Defense Advanced Research Projects Agency and the Caltech President's Fund.

12.8 References

Allaire, G., Bonnetier, E., Francfort, G., and Jouve, F., "Shape optimization by the homogenization method," *Numer. Math.*, v. 76, 1997, p. 27-68.

Bar-Cohen, Y., Editor, *Smart Structures and Materials 2000*: *Electroactive Polymer Actuators and Devices (EAPAD)*, Proceedings of SPIE's 7th Annual International Symposium on Smart, Structures and Materials, SPIE, 2000.

Bar-Cohen, Y. and Leary, S., "Electroactive polymers (EAP) characterization methods," *Smart Structures and Materials 2000*: *Electroactive Polymer Actuators and Devices (EAPAD)*, SPIE Proc. Vol. 3987, p. 12-16, 2000.

Bendsøe, M. P and Kikuchi, N., "Generating optimal topologies in structural design using a homogenization method," *Comput. Methods Appl. Mech. Engg.*, v. 71, 1988, p. 197-224.

Billington, E. W. and Tate, A., *The Physics of Deformation and Flow*, McGraw Hill, 1981.

Cherkaev, A., *Variational Methods for Structural Optimization*, New York: Springer-Verlag, 2000.

Grimshaw, P. E., Nussbaum, J. H, Grodzinsky, A. J., Yarmush, M.L., "Kinetics of electrically and chemically-induced swelling in polyelectrolyte gels," *Journal of Chemical Physics*, v. 93, no. 6, 1990, p. 4462-4472.

James, R. D. and Kinderlehrer, D., "Theory of magnetostriction with application to $Tb_xDy_{1-x}Fe_2$," *Philosophical Magazine B.*, v. 68, no. 2, 1993, p. 237-274.

Kohn, R. V. and Strang, G., "Optimal-design and relaxation of variational problems 1, 2, and 3," *Comm Pure Appl. Math.*, v.39, 1986, p. 113-137, 139-182 and 353-377.

Lakshminarayanaiah, N., *Transport Phenomena in Membranes*, New York: Academic Press, 1969.

Mura, T., *Micromechanics of Defects in Solids*, Dordrecht: Nijhoff Publishers, 1987.

Nemat-Nasser, S. and Hori, M., *Micromechanics: Overall Properties of Heterogeneous Materials*, Amsterdam: Elsevier, 1999.

Nemat-Nasser, S. and Li, J. Y., "Electromechanical response of ionic polymer-metal composites," *Journal of Applied Physics*, v. 87, no. 7, 2000, p. 3321-3331.

Ogden, R. W., *Non-linear Elastic Deformations*, New York: Halsted Press, 1984.

Pelrine, R., Kornbluh, R., and Joseph, J., "Electrostriction of polymer dielectrics with compliant electrodes as a means of actuation," *Sensors and Actuators*, v. A64, 1998, p. 77-85.

Pelrine, R., Kornbluh, R., Pei, Q. B., and Joseph, J., "High-speed electrically actuated elastomers with strain greater than 100%," *Science* v. 287, 2000, p. 836-839.

Shu, Y. C. and Bhattacharya, K, "Domain patterns and macroscopic behavior of ferroelectric materials," submitted, 2000.

Sigmund, O., "On the design of compliant mechanisms using topology optimization," *Mech. Struct. Mach.*, v. 24, 1997, p. 493-524.

Spencer, A. J. M., *Continuum Mechanics*, New York : Longman, 1980.

Xiao, Y. and Bhattacharya, K., "The optimal design of electrodes in an electro-static actuator," in preparation, 2000.

CHAPTER 13

Modeling IPMC for Design of Actuation Mechanisms

Satoshi Tadokoro, Masashi Konyo
Kobe University (Japan)

Keisuke Oguro
Osaka National Research Institute, AIST (Japan)

13.1 Models and CAE Tools for Design of IPMC Mechanisms

Ionic polymer metallic composite (IPMC, which is also known as ICPF[1]) [Oguro et al., 1992; Shahinpoor et al., 1999a] is one of the electroactive polymers that have shown potential for practical applications. This material was described in detail in Chapter 6, and the focus of the present chapter is on the modeling aspects of electroactivation, providing a foundation for the design of actuation mechanisms. IPMC is an electroless plated electroactive polymer (EAP) material that bends when subjected to a voltage across its thickness (see Fig. 1). IPMC has several attractive EAP characteristics that include:

(1) Low drive voltage is (1.0 – 5.0 V).
(2) Relatively high response (up to several hundreds of Hertz).
(3) Soft material ($E = 2.2 \times 10^8$ Pa).
(4) Possible to miniaturize (< 1 mm).
(5) Durability to many bending cycles (> 1×10^6 bending cycles).
(6) Can be activated in water or in a wet condition.
(7) Exhibits distributed actuation, allowing production of mechanisms with multiple degrees of freedom.

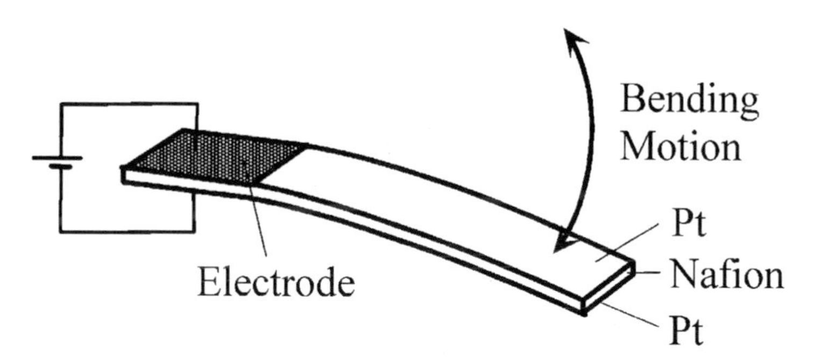

Figure 1: Ionic polymer metal composite (IPMC) actuator shown for Pt/Nafion composite EAP.

IPMC generates a relatively small force where a cantilever-shaped actuator (2 × 10 × 0.18 mm) can generate about 0.6 mN, and therefore its applications need to be scoped accordingly. Some of the applications that were investigated for IPMC include an active catheter system [Guo et al., 1995], a distributed actuation device [Tadokoro et al., 1997; Tadokoro et al., 1998; Tadokoro et al., 1999b], an underwater robot [Guo et al., 1998], micromanipulators [Tadokoro et al., 1999a and 2000a], a micropump [Guo et al., 1999], a face-type actuator [Tadokoro et al.,

[1] Kanno and Tadokoro named the Nafion-Pt composite ICPF (Ionic Conducting Polymer gel Film) in 1992. In the field of robotics, most researchers use the name ICPF, and it is well recognized. However, in this book, IPMC will be used for name consistency.

1999b], a wiper of an asteroid rover [Bar-Cohen et al., 2000; and Fukuhara et al., 2000], and a tactile haptic display for virtual reality [Konyo et al., 2000]. The actual number of applications that were considered is still small, but the list is expected to grow in the coming years with the emergence of requirements that account for the limitations while taking advantage of the unique capabilities. The challenges to the practical application of IPMC include:

(1) *Material development and improvement*: It is essential that EAP materials have performance characteristics that cannot be obtained by existing actuators. To become an alternative actuator of choice, EAP researchers would be expected to overcome constraints of conventional actuators and develop capabilities that are not possible with existing materials. However, it is also essential that a thorough understanding of the material behavior is established and that its output power, response speed, and environmental restriction are improved and optimized.
(2) *Stabilization and optimization*: The processes and characteristics of the material should be optimized and made robust and repeatable.
(3) *Effective design methodologies*: In order for designers to consider application of such materials, they need to be able to use established design methodologies: e.g., analytical models, CAE tools, etc. Such methodologies can support the development of new applications for EAP benefiting from the unique characteristics of these materials as actuators that are different from a conventional one.
(4) *Market development of applications*: The unique characteristics of EAP offer the potential for new technologies and materials, leading to effective actuators that comply with market demands.

Generally, it is difficult to develop mechanisms that use new materials without effective models even if these materials have many characteristic advantages. Effective analytical modeling of EAP is one of the most important foundations for establishing design methodologies and can contribute as follows:

(1) Allow analysis and evaluation of the designed mechanism performance characteristics before a prototype is made and physically tested.
(2) Minimize the number of repetitive trial- and error- sequences.
(3) Reduce design cost and the period of new product development.
(4) Rapid prototyping.
(5) Accelerate design cycle.

As a model for such development, piezoelectric ceramics have become popular in many products for which, in addition to the robust material characteristics, a relatively well-established analytical model and design methodologies are available. Existing models can predict the behavior of piezoelectric material in response to electrical excitation, and there are commercially available software packages that support finite element analysis and interactive designs [ABAQUS, 1999]. Further, CAD methods are used to study shape-memory alloys [Ikuta and Shimizu, 1993] allowing developers to address design parameters that affect the performance of such actuators in response to mechanical and electrical parameters.

As described in Chapter 6, many investigators have studied models for IPMC, with the largest number addressing Nafion-Pt composite EAP. However, there are still issues that need to be addressed with regards to understanding the material behavior. Kanno et al. [1994] made a time-series black-box model. Guo and Fukuda [1995] approximated the motion by a circular shape. Kanno and his co-investigators [1996a and 1997] proposed the Kanno-Tadokoro model that consists of distributed current in the membrane and internal stress induced with a time delay. Shahinpoor and his co-investigators [Shahinpoor, 1998 and 1999; and Shahinpoor et al., 1999] proposed a model consisting of four forces produced by ionic motion. Firoozbakhsh et al. [1998] modeled ionic interactions, and Shahinpoor [1999] considered electrostatic force caused by the surface microstructure. Tadokoro and his co-investigators [Tadokoro, et al., 2000b and 2000c] proposed the Yamagami-Tadokoro model, which considers the volume change by water content and electrostatic force generated by ionic migration in the membrane. Nemat-Nasser and Li [2000] modeled conformation change of IPMC assuming a sphere shape of ionic clusters of Nafion. More modeling efforts are still expected to evolve that address the fundamental mechanism of actuation and the design of devices using this material.

13.2 A Physicochemical Model Considering Six Phenomena

In this chapter, the focus of the analysis is on Nafion/platinum-type IPMC material for which the authors have developed an extensive body of knowledge. The adaptation of the mythology to other types of IPMC can be made by addressing the related differences. The Yamagami-Tadokoro model is introduced and is used herein as a basis for the design of IPMC-type actuators. This model uses six physicochemical phenomena:

(1) ionic motion induced by electric field
(2) water motion induced by ion drag
(3) swelling and contraction of the membrane
(4) momentum effect
(5) electrostatic force
(6) conformation change.

13.2.1 Hypotheses on Motion Principles

In developing the model for the actuation of IPMC, the following hypotheses are assumed regarding the mechanisms of motion:

(1) The Nafion has a chemical structure as shown in Fig. 2.
(2) An electrostatic force causes sodium ions to migrate from the anode to the cathode.

(3) Hydration causes migration of water molecules with the sodium ions.

The following forces act on the gel membrane due to the travel of the sodium ions and the water molecules.

(1) Swell and contraction caused by the water content change.
(2) Electrostatic force generated by the deviation of the fixed charge of sulfonic acid groups.
(3) Momentum conservation effect related to the ionic migration and the water travel.
(4) Conformation change of the polymer structure resulting from ionic migration.

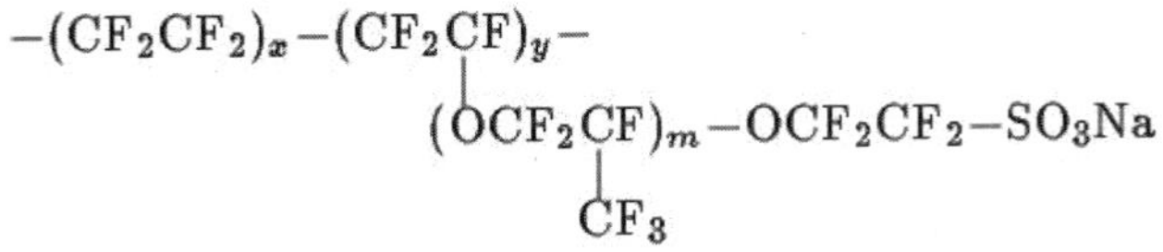

$$-(CF_2CF_2)_x-(CF_2CF)_y- \quad (OCF_2CF)_m-OCF_2CF_2-SO_3Na \quad CF_3$$

Figure 2: Chemical structure of Nafion.

13.2.2 Model of Sodium Ion Travel

The travel of the sodium ions (shown in Fig. 3) can be model using the Gaussian law,

$$E(x,t) = \frac{1}{\varepsilon}\left\{\frac{1}{S_x}\int_0^t i(\tau)d\tau + \int_0^x \rho(\xi,t)d\xi\right\}, \tag{1}$$

where ε is a dielectric constant of the membrane containing water, ρ is the volume density of electric charges in the membrane, and $i(t)$ is the current for the surface area S_x that is supplied from the outside. It is assumed for simplicity that the surface charge is zero at the initial condition.

Let the concentration of sodium ions and sulfonic acid groups be denoted $c(x,t)$ and c_0, respectively. The volume density $\rho(x, t)$ is

$$\rho(x,t) = Ne(c(x,t) - c_0), \tag{2}$$

where N is Avogadro's constant, and e is the elementary charge.

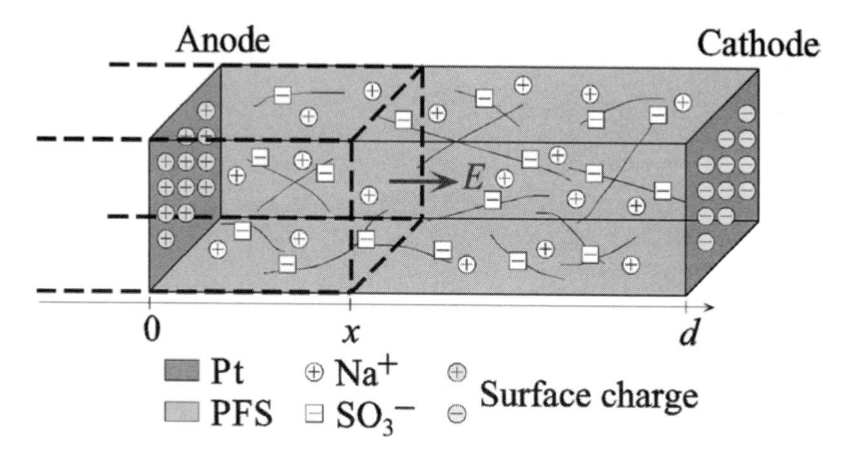

Figure 3: Ionic migration through the membrane.

The equilibrium of forces that is applied to a water molecule is expressed as

$$eE(x,t) = \eta v(x,t) + kT\frac{\partial \ln c(x,t)}{\partial x} + nkT\frac{\partial \ln w(x,t)}{\partial x}, \tag{3}$$

as shown in Fig. 4 [Hanai, 1978]. The left side is electrostatic force applied on a sodium ion having electric charge e. The first, second, and third terms on the right side indicate viscous resistance force applied on a sodium ion traveling with velocity $v(x,t)$ (η: a viscous resistance coefficient), diffusion force caused by the deviation of sodium ions (k: Boltzmann constant, T: the absolute temperature), and diffusion force caused by the deviation of water ($w(x,t)$: water concentration, n: average number of water molecules of a hydrated sodium ion). Transient characteristics can be ignored because the inertial force is negligible.

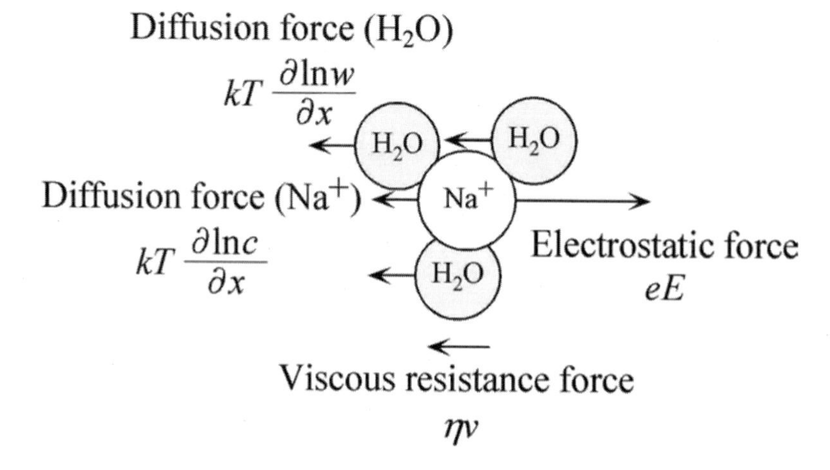

Figure 4: Forces acting on a sodium ion.

A flux $J(x, t) = c(x, t)\ v(x, t)$ satisfies an equation of continuity,

$$\frac{\partial c(x,t)}{\partial t} = \frac{\partial J(x,t)}{\partial x}. \tag{4}$$

The total electric charge of sodium ions existing in the half space R (–inf, x) in Fig. 3 is

$$Q(x,t) = NeS_x \int_0^x c(\xi,t)d\xi. \tag{5}$$

By the above relations, the model of sodium ion migration is expressed by a partial differential equation of $Q(x, t)$,

$$\begin{aligned}\eta\frac{\partial Q(x,t)}{\partial t} &= kT\frac{\partial^2 Q(x,t)}{\partial x^2}\\ &+\left\{kT\frac{\partial \ln w(x,t)}{\partial x} - \frac{e}{\varepsilon S_x}\left(\int_0^t i(\tau)d\tau + Q(x,t) - Q(x,0)\right)\right\}\frac{\partial Q(x,t)}{\partial x}.\end{aligned} \tag{6}$$

Initial condition:

$$Q(x,0) = NeS_x c_0 x \ \ (0 \le x \le d,\ t = 0). \tag{7}$$

Boundary conditions:

$$Q(0,t) = 0 \ \ (x = 0,\ t > 0),\ Q(d,t) = NeS_x c_0 d \ \ (x = d,\ t > 0). \tag{8}$$

When the current $i(\tau)$ is given as an input, $Q(x,t)$ is obtained by solving this partial differential equation, simulating the change of electric charge distribution of sodium ions.

Using the finite difference method,

$$\begin{aligned}\eta\frac{Q_m^{p+1} - Q_m^p}{\tau} &= kT\frac{Q_{m+1}^{p+1} - 2Q_m^{p+1} + Q_{m-1}^{p+1}}{h^2}\\ &-\left\{\left(\frac{e}{\varepsilon S_x}\sum_{k=0}^{p}\frac{i^{k+1}+i^k}{2}\tau + Q_m^{p+1} - Q_m^0\right) - nkT\frac{\ln w_{m+1}^{p+1} - \ln w_m^{p+1}}{h}\right\}\frac{Q_{m+1}^{p+1} - Q_m^{p+1}}{h}\end{aligned} \tag{9}$$

satisfies for a minute width h and a short sampling time τ. This equation can be simplified into

$$(-A + B_m^{p+1})Q_{m+1}^{p+1} + (1 + 2A - B_m^{p+1})Q_m^{p+1} - AQ_{m-1}^{p+1} = Q_m^p. \tag{10}$$

The boundary conditions are

$$Q_1^{p+1} = 0,\ Q_{N+1}^{p+1} = NeS_x c_0 Nh. \tag{11}$$

Hence, the following linear equation is derived.

$$\begin{bmatrix} 1+2A-B_2^{p+1} & -A+B_3^{p+1} & & & 0 \\ -A & 1+2A-B_3^{p+1} & -A+B_4^{p+1} & & \\ & \ddots & \ddots & \ddots & \\ & & -A & 1+2A-B_{N-1}^{p+1} & -A+B_N^{p+1} \\ 0 & & & -A & 1+2A-B_N^{p+1} \end{bmatrix} \begin{bmatrix} Q_2^{p+1} \\ Q_3^{p+1} \\ \vdots \\ Q_{N-1}^{p+1} \\ Q_N^{p+1} \end{bmatrix}$$
$$= \begin{bmatrix} Q_2^p \\ Q_3^p \\ \vdots \\ Q_{N-1}^p \\ Q_N^p + (A - B_{N+1}^{p+1}) Q_{N+1}^{p+1} \end{bmatrix}. \tag{12}$$

The Gauss-Seidel method is applied to solve the above equation.

13.2.3 Model of Water Travel

The travel of water through the membrane is modeled using the equilibrium of the diffusion force of water and viscous resistance as follows:

$$kT \frac{\partial \ln w(x,t)}{\partial x} = -\eta' v'(x,t). \tag{13}$$

Applying an equation of continuity in the same way as in the ionic migration model, the water travel is modeled by a partial differential equation of the number of water molecules $W(x,t)$ in the half space R (–inf, x) in Fig. 3,

$$\eta' \frac{\partial W(x,t)}{\partial t} = kT \frac{\partial^2 W(x,t)}{\partial x^2}. \tag{14}$$

Initial condition:

$$W(x,0) = NS_x w_0 x \ \ (0 \le x \le d,\ t = 0). \tag{15}$$

Boundary conditions:

$$\frac{\partial W(0,t)}{\partial x} = NS_x w_0 \ \ (x=0,\ t>0), \ \frac{\partial W(d,t)}{\partial x} = NS_x w_0 \ \ (x=d,\ t>0). \quad (16)$$

The migration of sodium ions and water molecules occurs at the same time. The two partial differential equations [Eqs. (6) and (14)] have to be solved simultaneously. However, the travel of sodium ions is dominated by the electrostatic force because it is much larger than the diffusion force. Therefore, motion of sodium ions [Eq. (6)] is obtained approximately at first, and then water diffusion [Eq. (14)] is computed so that the computation has practical speed.

13.2.4 Stress Generation by Swelling and Contraction Caused by Water Content Change

A change in water content causes stress generation that results in swelling or contraction depending on the electric field polarity. Figure 5 shows the experimental relation between water content C and strain ε by swell and contraction of the gel membrane. This is expressed by a linear equation

$$\varepsilon = 6.46 \times 10^{-3} C. \quad (17)$$

Internal stress is computed by ε when C is known.

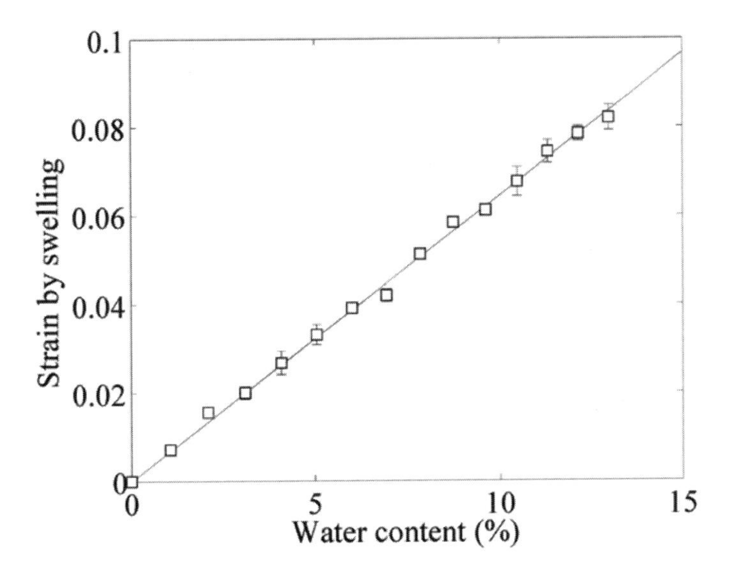

Figure 5: Water content C vs. strain of Nafion membrane ε.

13.2.5 Stress Generation by Electrostatic Force by Fixed Electric Charge

The sodium cations travel to the cathode platinum surface to balance against a negative charge. In contrast, the sulfonic acid groups are fixed to the high polymer chains and cannot travel freely through the membrane. Therefore, negative charge deviates in the membrane and causes electrostatic force, generating internal stress that expands the gel near the anode.

Gaussian law derives an electric field downward in Fig. 6:

$$E(y,t) = \frac{1}{\varepsilon S_y} \int_0^y \rho(\eta,t) S_y d\eta. \tag{18}$$

The total electrostatic force $F_b(y, t)$ applied on the fixed electric charge existing in the half space R_2 is

$$F_b(y,t) = E(y,t) \int_y^l \rho(\eta,t) S_y d\eta. \tag{19}$$

The internal stress caused by electrostatic force is

$$\sigma(y,t) = \frac{F_b(y,t)}{S_y}. \tag{20}$$

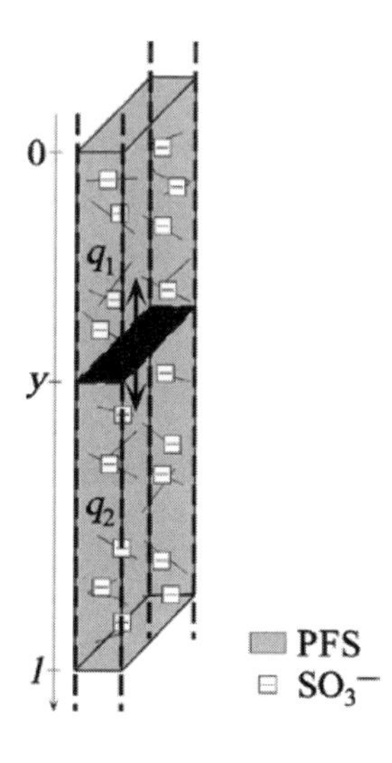

Figure 6: Electrostatic force caused by fixed negative charge of sulfonic acid groups.

13.2.6 Stress Generation by Momentum Conservation

The momentum conservation law of hydrated sodium ions and gel membrane as shown in Fig. 7 gives

$$NmS_x \int_0^d c(x,t)u(x,t)dx - MU(t) = 0, \tag{21}$$

where m is the mass of a hydrated sodium ion, M is the mass of the gel membrane containing water of S_x in cross section and d in thickness, $u(x, t)$ is the velocity of a hydrated sodium ion, and $U(t)$ is the velocity of the gel membrane with respect to the inertial coordinate. Therefore,

$$U(t) = \frac{NmS_x \int_0^d c(x,t)u(x,t)dx}{NmS_x c_0 d + M}. \tag{22}$$

The reaction force $F_a(t)$ applied on the gel membrane in the direction of thickness from the cathode to the anode is

$$F_a(t) = M\frac{dU(t)}{dt}. \tag{23}$$

Quantitative analyses revealed that the effect of $F_a(t)$ can be neglected in analyses for most applications.

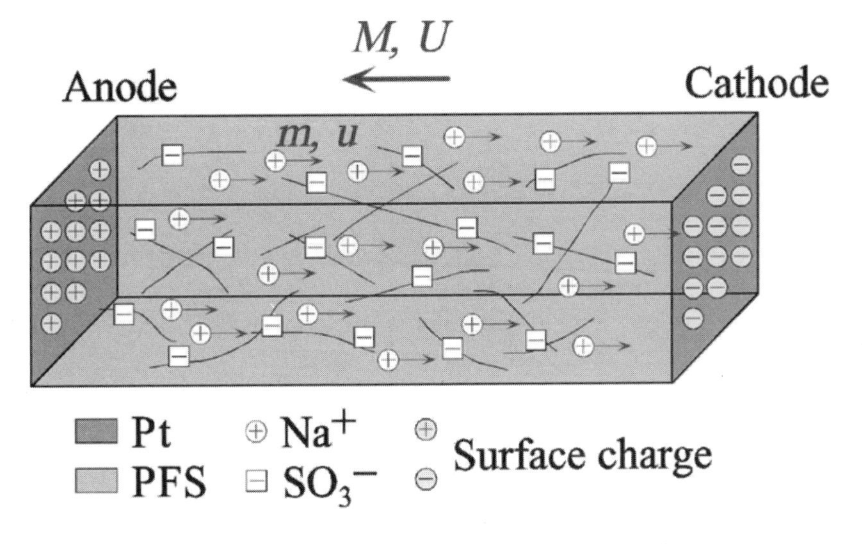

Figure 7: Momentum conservation in the membrane.

13.2.7 Stress Generation by Conformation Change

Conformation change of the polymer also causes internal stress. However, it is difficult to formulate it because the related phenomena are still unknown. The conformation change is ignored in this Yamagami-Tadokoro model, which may cause modeling errors.

The Yamagami-Tadokoro model can predict motion of IPMC including nonlinear characteristics and can be used to create computer simulation. The results can give an insight into material itself. However, this model is not as useful for mechanical and control design and it is too complex for designers to have an intuitive idea of how to create effective mechanisms. A lower-order model is necessary to design controllers. For this purpose, a gray-box macroscopic model can be used to address the mechanical and control design issues.

13.3 Gray-Box Macroscopic Model for Mechanical and Control Design

Modeling strategies depend on the application. In mechanical and control design, detailed models reflecting physicochemical phenomena, such as the Yamagami-Tadokoro model, are not always effective. A superficial approximation model of motion, maintaining sufficient accuracy for application, is an important issue. Simple models sometimes aid intuitive consideration to create new applications since complexity tends to conceal the substantial points. In the case of piezoelectric materials, a large number of applications depend on the piezoelectric equations and CAE systems that are used. Even if principles of piezoelectric phenomena are not precisely taken into account, they provide good abstraction of the macroscopic phenomena and they are useful for application design. It is desirable for EAP materials to have such design models in addition to detailed models. The Kanno-Tadokoro model is one such example for Nafion-platinum composite (IPMC or ICPF). It assumes that the phenomena consist of the following three stages:

(1) *Electrical stage*: Applied voltage to the electrodes generates distributed current all over the membrane.
(2) *Stress generation stage*: Distributed current is converted into distributed internal stress in the membrane.
(3) *Mechanical stage*: Distributed internal stress causes strain according to viscoelastic properties.

Figure 8 shows the step response curve by a voltage of 1.5 V. All the experiments in this chapter were performed in water in order not to change the condition. The actuator motion is as follows: after the voltage is applied, the actuator immediately bends toward the anode side, reaches the maximum displacement after 0.2 sec, bends back toward the cathode side, passes the initial position, reaches the maximum displacement toward the cathode side, and then gradually returns to the initial position. The current through the actuator sharply increases as soon as the voltage is applied and then decreases exponentially [Kanno et al., 1994].

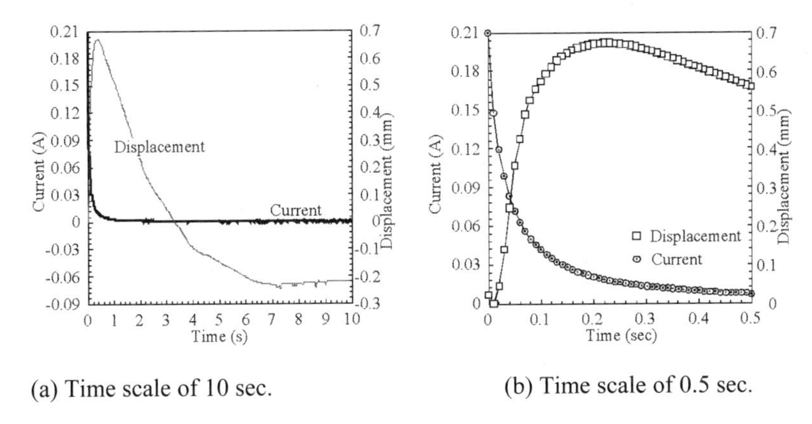

(a) Time scale of 10 sec. (b) Time scale of 0.5 sec.

Figure 8: Step response under voltage input.

13.3.1 Electrical Stage

The discrete model of the relation of the input voltage and the output current is shown in Fig. 9. The authors call this relation an electrical stage, as part of the whole actuator model. R_a and R_b are measurable surface resistances of the actuator. R_x is the resistance of the polymer gel layer as an electric conductor. The connection of R_c and C expresses the remaining characteristics of the exponential step response curve of the current. This combination forms a distributed parameter system. Simulation results, as can be seen in Fig. 10, show the agreement of response of the electrical model with experimental results [Kanno et al., 1996a and 1996c]. Nonlinearity ignored in this model causes small discrepancies of the results.

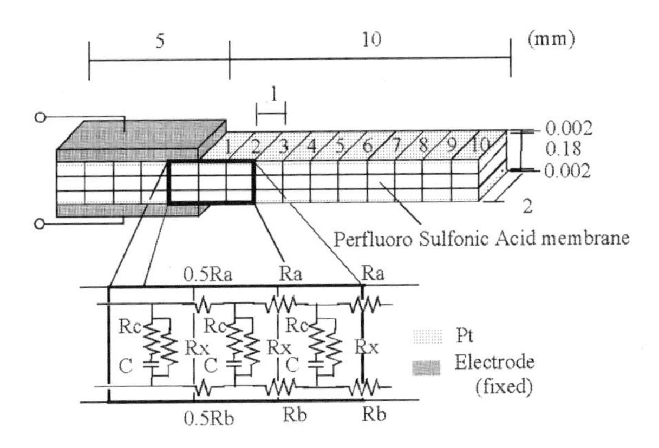

Figure 9: Electrical model.

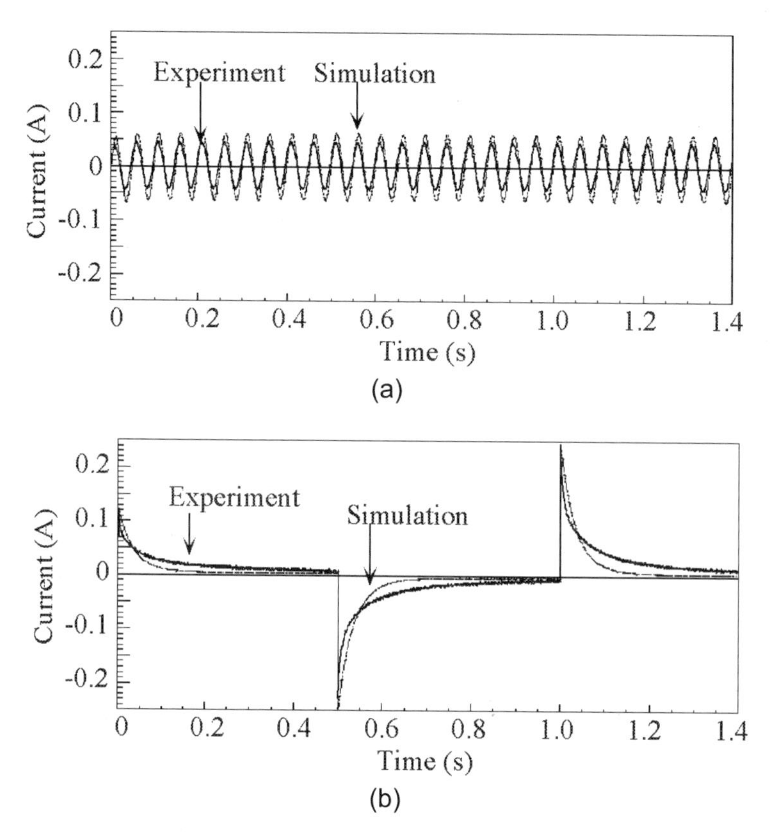

Figure 10: Electrical response (voltage-current relation). (a) Sinusoidal voltage input (0.5 V, 20 Hz), and (b) square voltage input (1.0 V, 0.1 Hz)

13.3.2 Stress Generation Stage

This model assumes that the distributed current causes distributed internal stress. Compressive stress is induced on a surface, and tensile stress is generated on the other surface because of the current. Both the maximum current and the maximum displacement are in proportion to the magnitude of the step voltage. Figure 11 shows the experimental relation of the magnitude of the step voltage to the maximum displacement of the anode side. Figure 12 shows the experimental maximum current through the actuator. Both of them are approximately proportional to the voltage.

Frequency characteristics from current to displacement as shown in Fig. 13 are obtained by subtracting the electrical frequency characteristics from the whole frequency characteristics from voltage to displacement. They do not depend on the amplitudes of input voltage. Therefore, the current-stress relation of the stress generation stage is assumed as linear, and the stress-strain relation of the mechanical stage is modeled as elastic.

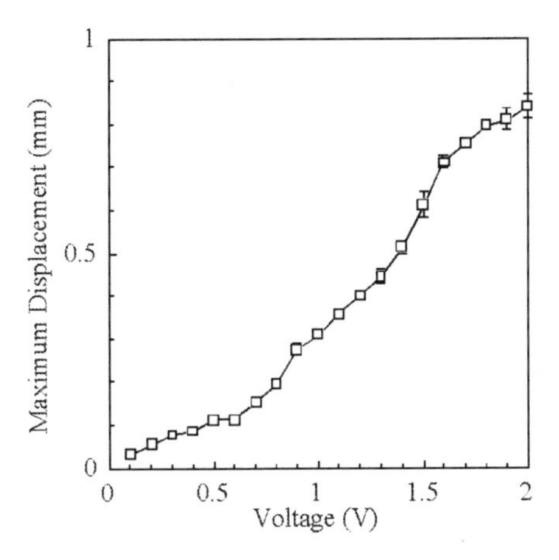

Figure 11: Maximum displacement to anode side under step voltage input.

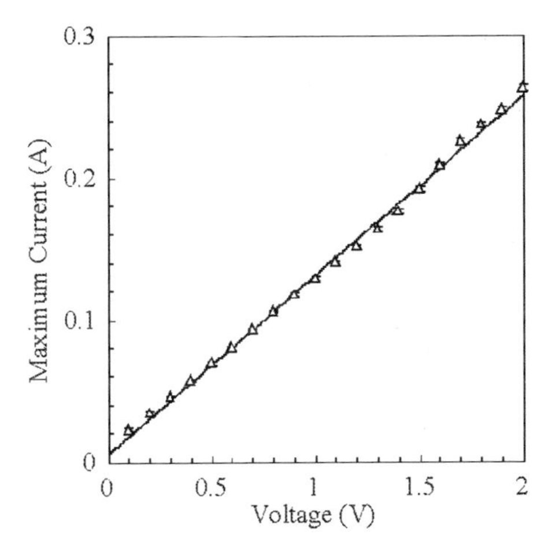

Figure 12: Maximum current under step voltage input.

Observation of experimental step displacement curves in time domain shows that the current and the stress are not simply proportional. If the current itself were proportional to the internal stress, internal stress for the swing-back motion would not be generated. A number of curve fitting trials would fail if this was assumed.

Therefore, the following model is proposed:

(1) Distributed current is differentiated by time and is transformed into internal stress for the linear bending motion.
(2) When internal stress is generated, there exists a delay because the energy transformation occurs in polymer gel with permeated water.
(3) The internal stress generates volumetric strain of expansion and contraction. Then the actuator bends.

This is also supported by the fact that the actuator did not continue to bend while constant current was applied.

The following mechanical equation expresses the above model in a way similar to the piezoelectric equation:

$$\sigma = D(s)\varepsilon - ei\frac{\omega_n^2 s}{s^2 + 2\zeta\omega_n s + \omega_n^2}, \tag{24}$$

where σ is a stress vector, D is a mechanical characteristic matrix, ε is a strain vector, e is a stress generation tensor, and $\boldsymbol{i}$ is the current through the actuator. ζ and ω_n are parameters of the second-degree delay of the internal stress generation.

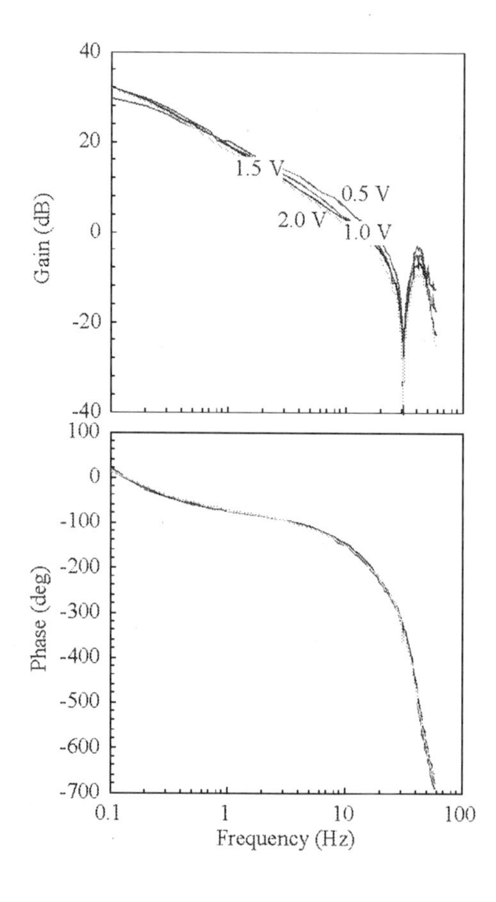

Figure 13: Frequency characteristics of the stress generation stage.

13.3.3 Mechanical Stage

For dynamic simulation, a dynamic finite element model using mass matrices and Rayleigh dumping was adopted. For example, a finite element model for a cantilever actuator is constructed as follows:

(1) The 10 elements in the electrical model are divided into three layers.
(2) The strain is produced at the boundary between the gel and the platinum because the membrane never bends by applying voltage without the platinum plating. The outside two layers in the FEM model are assumed to generate the strain by current input.
(3) The middle layer does not generate strain but has the same mechanical characteristics as the outside layers.

Stress generation tensors by which one side expands and the other side contracts in the x and the z directions by electron movement in the y direction are given to the outsides layers. (The coordinate system is defined in Fig. 16.) Tensor elements are all 0 except e_{yxx} and e_{yzz}. Because expansion and contraction are isotropic in the actuator surfaces, e_{yxx} and e_{yzz} are equal. The signs of e_{yxx} and e_{yzz} in one layer and those in the other layer are reverse. Therefore, applying a voltage, a layer expands and the other layer contracts. Then, the actuator bends. Assuming Rayleigh damping, a damping matrix C is given by

$$C = \alpha M + \beta K, \tag{25}$$

where M is a mass matrix, K is a stiffness matrix, and α and β are Rayleigh constants.

The Young modulus of the material is measured by experiments. Parameters e, ζ, ω_n, α, and β are determined so that simulation displacement curves at a point J agree with experiments. Table 1 shows the identified parameters.

Table 1: Model parameters of Nafion-Pt composite.

Parameter	Value
Actuator size	$10.0 \times 2.0 \times 0.184$ mm
R_a	5.4Ω (Resistivity: $2.2 \times 10^{-2}\Omega$ mm)
R_b	5.0Ω (Resistivity: $2.0 \times 10^{-2}\Omega$ mm)
R_x	$3.8 \times 10^{3}\Omega$
R_c	$9.5 \times 10\Omega$
C	2.9×10^{-4} F
Density	2.7×10^{3} kg/m^3
Young's modulus	2.2×10^{8} N/m^2
Poison's ratio	0.3
Rayleigh damping constant	$\alpha = 0.5$ s^{-1}, $\beta = 0.1$ s
Stress generation coefficients	$e_{yxx} = e_{yzz} = \pm 1.8$ Ns/m^{2}A
Parameters of 2nd-degree delay	$\zeta = 0.6, \omega_n = 1.2$ rad/s

13.4 Simulation Demonstration by Models

In this section, the Kanno-Tadokoro model is verified by experimental data. Figure 14 shows the experimental curve and the simulation curve at the point *J* under step voltage input of 1.5 V. The model response expresses the experimental response very well.

Figure 15 shows simulation results of displacement responses of multiple points. Dynamic bending shape change of simulation is shown in Fig. 16. These simulation results agree with the experimental results and the curl of edges. The shape of whole membrane and dynamic response are expressed well.

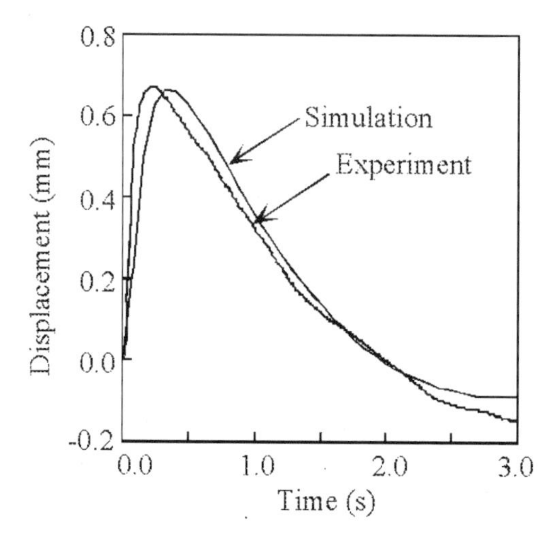

Figure 14: Step response of motion of the point 1 mm from the tip.

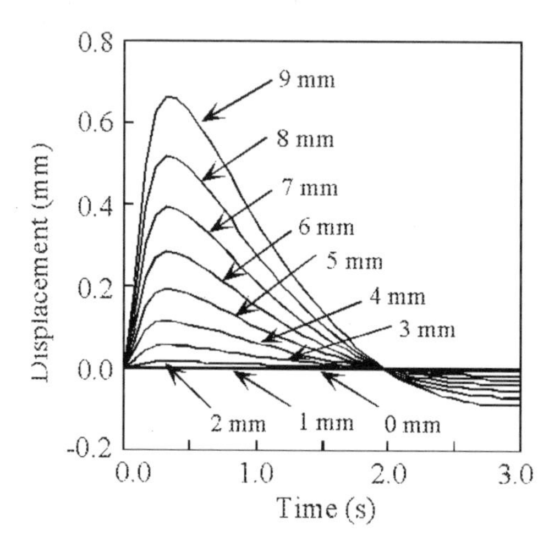

Figure 15: Simulation of displacement of multiple points under step voltage input.

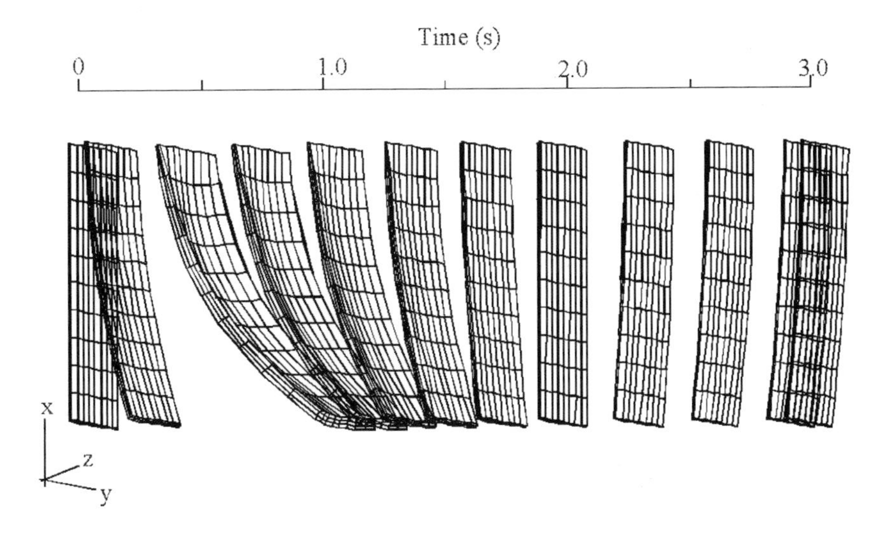

Figure 16: Simulation of bending shape change by step voltage input.

The frequency of the current response at each part of the actuator in the Kanno-Tadokoro model agrees with the distribution of the experimental displacement dynamics. Figure 17 shows simulation results of the model current responses at each part to a step voltage input. Figure 18 shows the experimental displacement curve of each part of the actuator under the same step voltage. Response frequencies near the base are high and those near the tip are low in both of the results. Distribution of the maximum current in the actuator estimated by the model corresponds to distribution of experimental curvature data. Figure 19 shows the simulation result of the maximum current under step voltages. Figure 20 shows the experimental curvature distribution of actuator shapes at the maximum displacement to the anode side under the same step voltages. Both the curvature and the maximum current are large near the base and small near the tip. These experimental results support the validity of the proposed model.

The approaching speed to the initial position in the simulation result is faster than that in the experimental result. Hysteresis and internal friction exist in the actuator material because it is a polymer gel. The Kanno-Tadokoro model does not include these factors because it is a linear model. Modeling the nonlinear factors should improve the model accuracy. Figure 21 shows the result of the same simulation using the Yamagami-Tadokoro model considering nonlinear physicochemical phenomena. The result shows more accuracy, demonstrating the advantage of considering physicochemical phenomena. More simulation results are shown in Tadokoro et al. [2000b and 2000c].

However, this result does not mean that the Yamagami-Tadokoro model is better than the Kanno-Tadokoro model. It is difficult in a practical design stage to use the physicochemical model because of its complexity. A simple nonlinear model is desirable for a mechanical design.

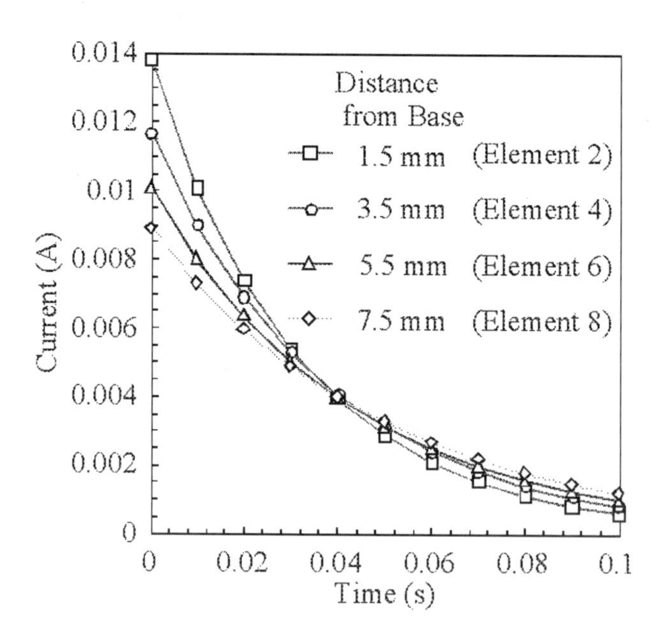

Figure 17: Simulation of current through each part under step voltage input (1.5 V).

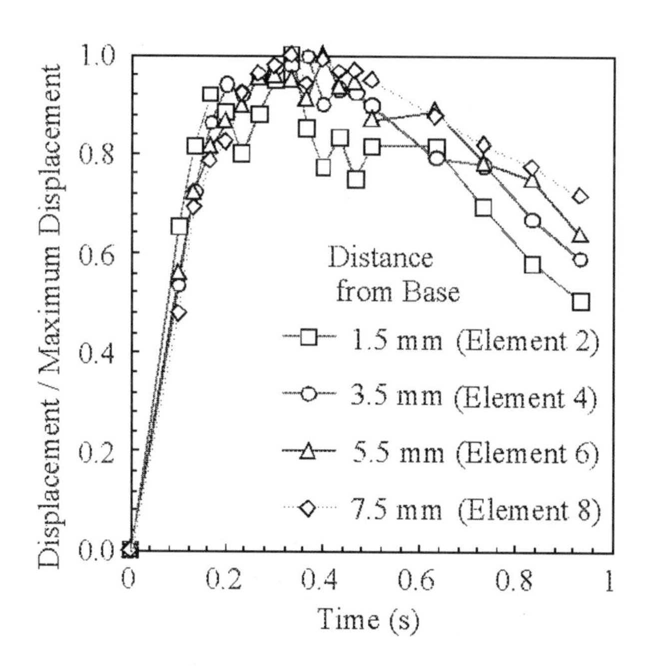

Figure 18: Dynamics of displacement response in each part under step voltage input (1.5 V).

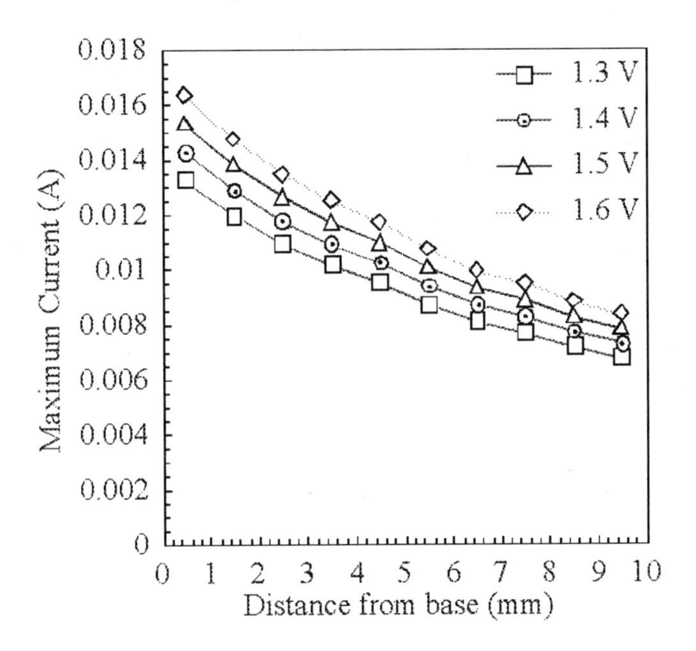

Figure 19: Simulation of maximum current in each element.

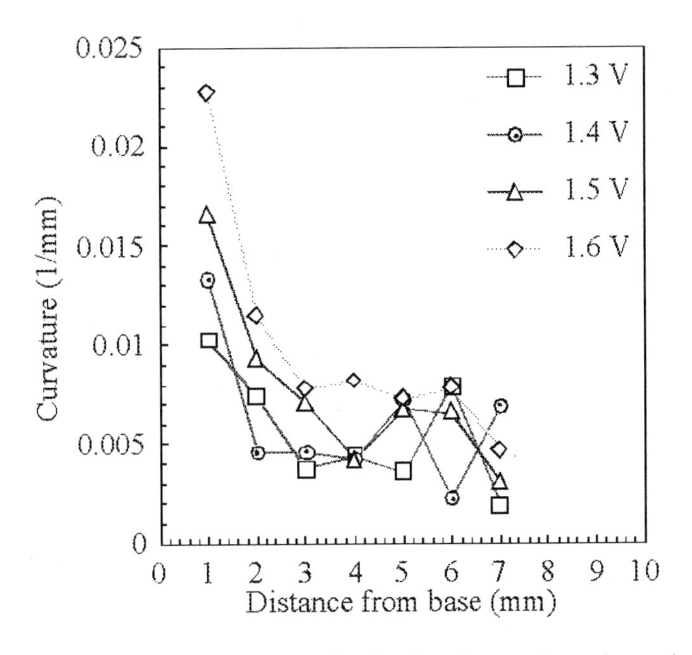

Figure 20: Experimental curvature of actuator shape at maximum bending.

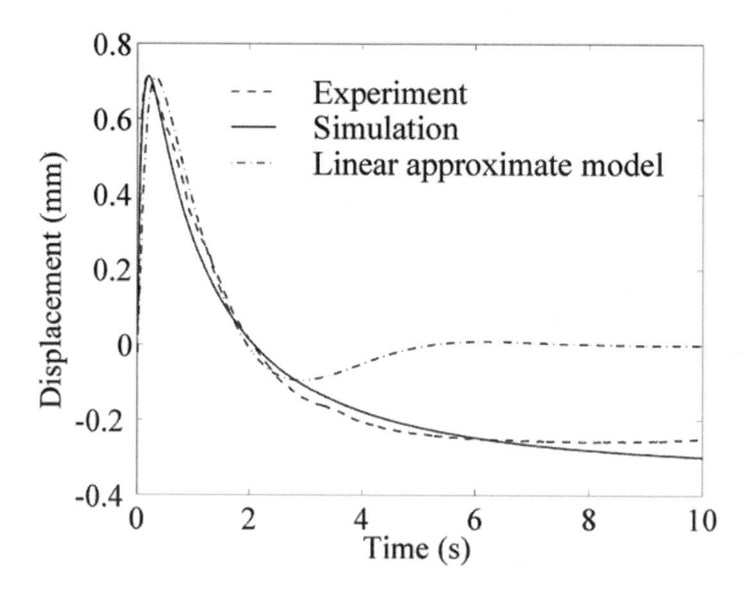

Figure 21: Simulation result by Yamagami-Tadokoro model.

13.5 Applications of the Model

The Kanno-Tadokoro model was used for the following application developments.

(1) Distributed-actuation device [Tadokoro et al., 1998 and 1999b].
(2) Soft micromanipulation devices with 3 and 6 degrees of freedom [Tadokoro et al., 1999a and 2000a].
(3) Dust wiper prototype of MUSES-CN nanorover [Bar-Cohen et al., 2000; Fukuhara et al., 2000].
(4) Haptic interface for virtual tactile display [Konyo et al., 2000, 2002 and 2003].
(5) Face-actuation device [Tadokoro et al., 1999b].
(6) EFD element [Tadokoro et al., 1997].
(7) MultiDOF fin for fish robots.

The first four applications are introduced briefly here.

13.5.1 Distributed Actuation Device

Softness of end-effectors is important in manipulation of soft objects like organs, food materials, micro objects, etc. This softness can be actualized using two approaches: (1) drive by hard actuators with soft attachments, and (2) direct drive by soft actuators by themselves. The former looks to be a sure method because of present technological development. However, to create micromachines or compact machines like miniature robot hands, the former is limited so it is difficult to find a breakthrough. The problem with the latter is that a readily available soft actuator material does not exist. However, the material revolution currently underway will surely result in the discovery of an appropriate material in the near future. For these reasons, it is meaningful to study methodologies for the effective use of such materials for manipulation with an eye to future applications.

A promising candidate for such a soft actuator material is gel. Many gel materials for actuators have been studied up to the present. The Nafion-platinum composite (IPMC or ICPF) is a new material that is closest to satisfying the requirements for our applications. Because such materials are soft, it is impossible to apply large forces/moments at only a few points on an object, contrary to the case with conventional robot manipulation. At the same time, however, it is an advantage that large pressures cannot be applied actively or passively. So as not to detract from this feature, a number of actuator elements should distribute for applying the driving force.

The distributed drive is also desirable from the viewpoint of robust manipulation. Even if there are elements that cannot generate appropriate force, in principle it is possible for the other elements to compensate for them. This signifies insensitivity to environmental fluctuation. In human bodies, for example, numerous cilia perform the excretion of alien substances by a whipping motion. Paramecia move by paddling their cilia. Centipedes crawl by the cooperative wavy motion of a number of legs. Any of these can robustly accomplish their objectives irrespective of environmental change.

An elliptic friction drive (EFD) element is an actuator element that generates driving force by friction using bending actuators. Figure 22 shows an experimental development using the Nafion-Pt composite. It has two actuator parts with platinum plating for actuation and one Nafion part without plating for elastic connection. The whole structure is fixed to form the shape of an arch.

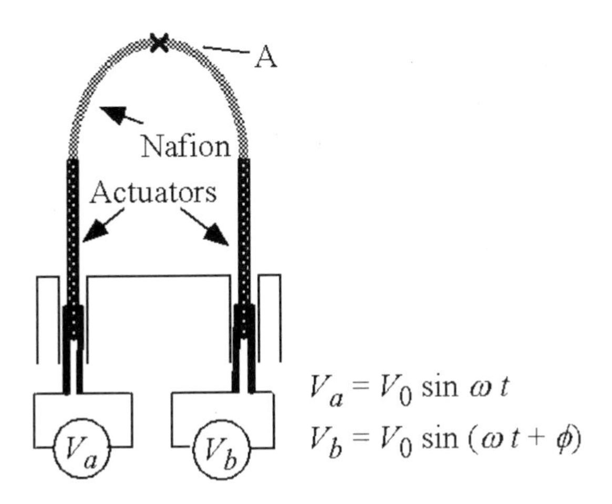

Figure 22: Structure of EFD actuator element.

When sinusoidal voltages with a phase difference are applied to the two actuators, the excited sinusoidal bending motions also have a phase difference. This results in an elliptical motion in the top point (A) of the connecting part. Figure 23 shows a developed distributed EFD device. It has 5 × 8 EFD elements on a plate. They cooperatively apply a driving force to an object.

The driving principle is shown in Fig. 24. Adjacent elements make elliptic motions with a phase difference of π (a two-phase drive). On the planar contact face, a frictional force in the x direction is generated alternately by adjacent elements, and then the object is driven.

This element could be applied to a robot hand, for example, as shown in Fig. 25. The Nafion-Pt composite is produced by a process consisting of surface roughening, adsorption of platinum, reduction and growth on a Nafion membrane. A masking technique using crepe paper tape with a polyethylene coating can be used to form any arbitrary shape of actuator on the Nafion. This technique is called the pattern plating method. It is an essential technique for creating the various shapes in the gel material required for the actuator. It is also important for supplying electricity efficiently.

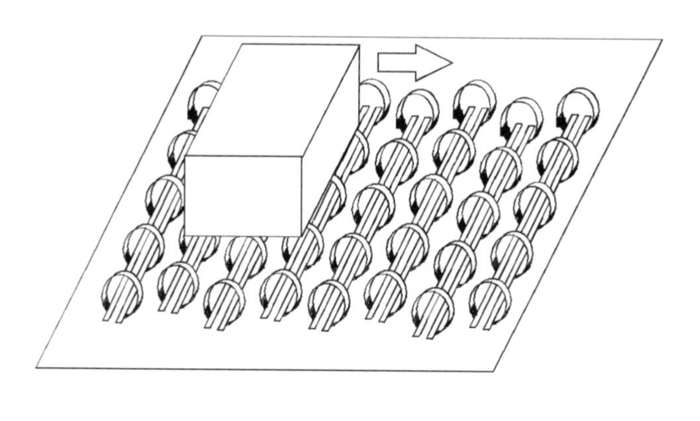

Figure 23: Distributed actuation device consisting of multiple EFD elements.

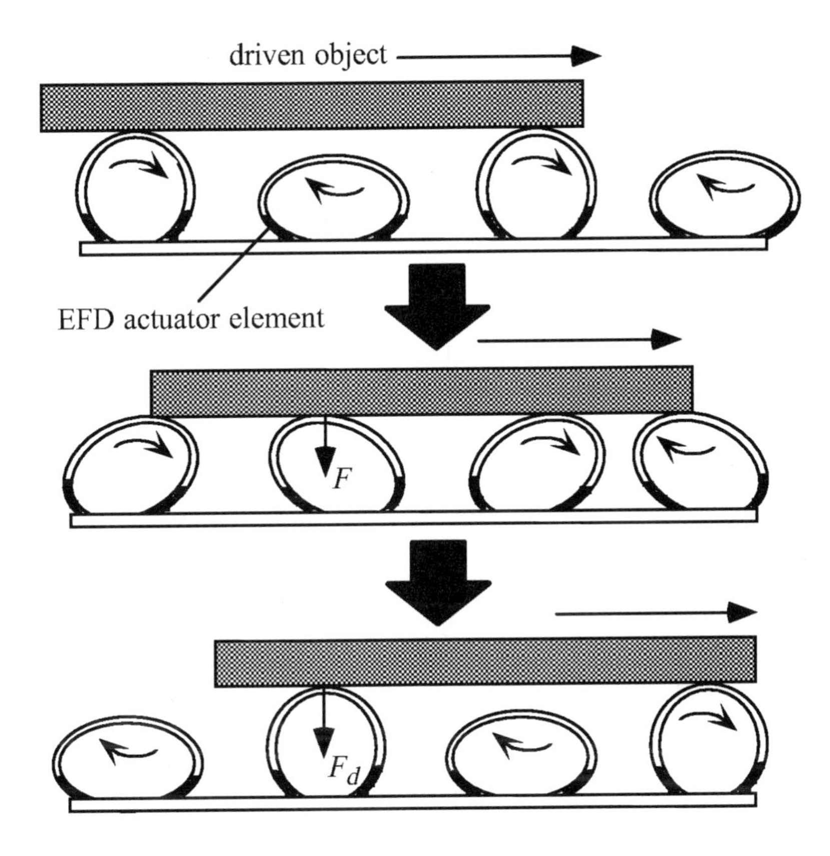

Figure 24: Principle of distributed drive.

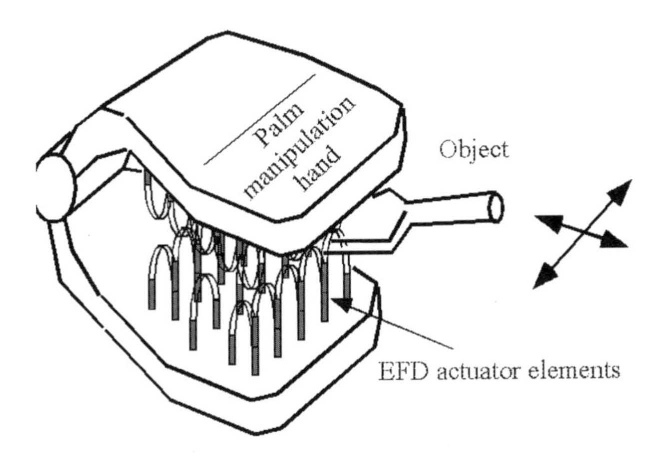

Figure 25: Application of developed device to robotic hand (palm manipulation hand).

The manufacturing process is as follows. First, composite actuators are produced on the supports of a Nafion membrane cut in a ladder shape. The ladder is formed into rings by rolling the membrane a round a column in a hot water bath. The resultant multiple EFDs are fixed together on a plate. Then, they are wired electrically and the shape of each element is adjusted.

An EFD element has many design parameters of mechanism and control, as shown in Fig. 26. These parameters have effects on the performance of the element. It is difficult for analytical methods to give an optimal design of these parameters, because

(1) the actuator part is not a point,
(2) motion of each part of actuator is not uniform (output internal stress and time constant), and
(3) function of each part interferes each other.

Investigating changing parameters by trial and error such as in Fig. 27 is necessary. Models and simulation tools minimize the number of experiments.

Design parameters of the mechanism and control are determined by simulation and analysis using the Kanno-Tadokoro model and an assumption of a viscous friction driving mechanism as shown in Fig. 28. The mechanical parameters are listed in Table 2.

Figure 29 shows the experimental result obtained by varying the phase difference, ϕ, between the sinusoidal input voltages for each EFD element. The shape of the elliptical motion changes according to this phase difference. At the same time, the velocity of the plate changes. Consideration of the analytical and experimental results shows that the elliptical motion in the x direction becomes large when the two phases are similar, and that the y displacement increases when the difference between them is close to π. The speed of the object is determined by these two displacements. The optimal value in this experiment is found to be $\phi = \pi/2$.

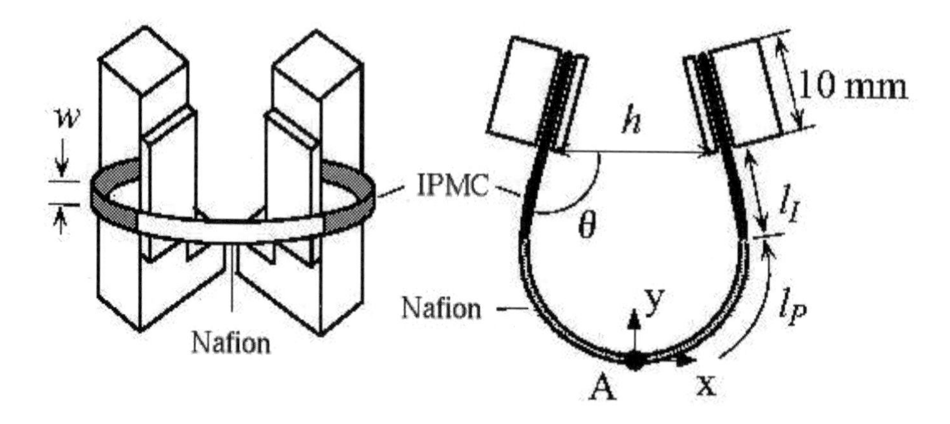

Figure 26: Design parameters of EFD element.

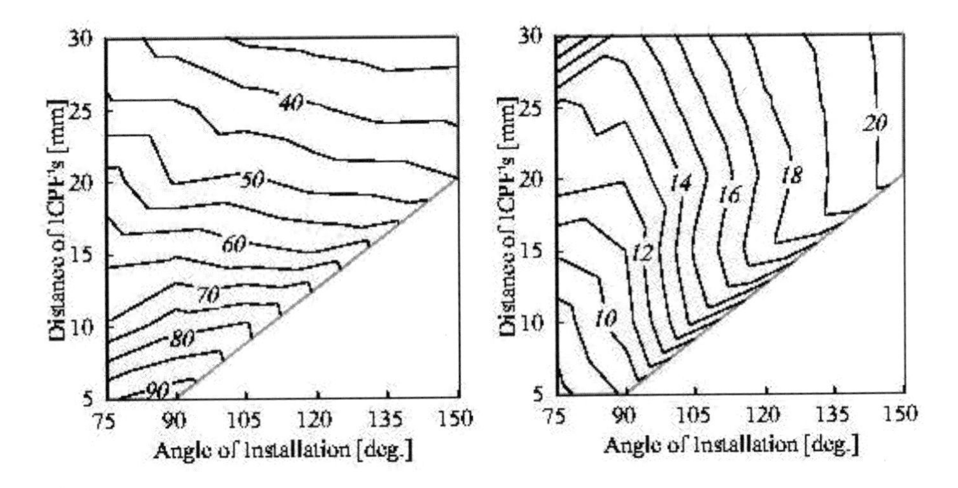

Figure 27: Motion of EFD tip vs. design parameters (1 = 62 μm).

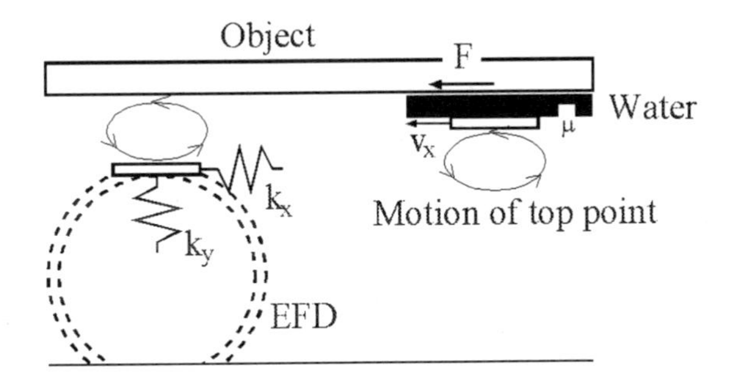

Figure 28: Mechanism of distributed drive.

Table 2: Mechanical design parameters of distributed device.

Parameter	Value
Number of elements (N)	40
Thickness (t)	180 μm
Width (w)	1 mm
Length of ICPF part (l_I)	2 mm
Length of PFS part (l_P)	11 mm
Distance between supports (h)	7 mm
Angle of supports (θ)	180°
Element interval in the x direction (d_x)	14 mm
Element interval in the y direction (d_y)	4 mm

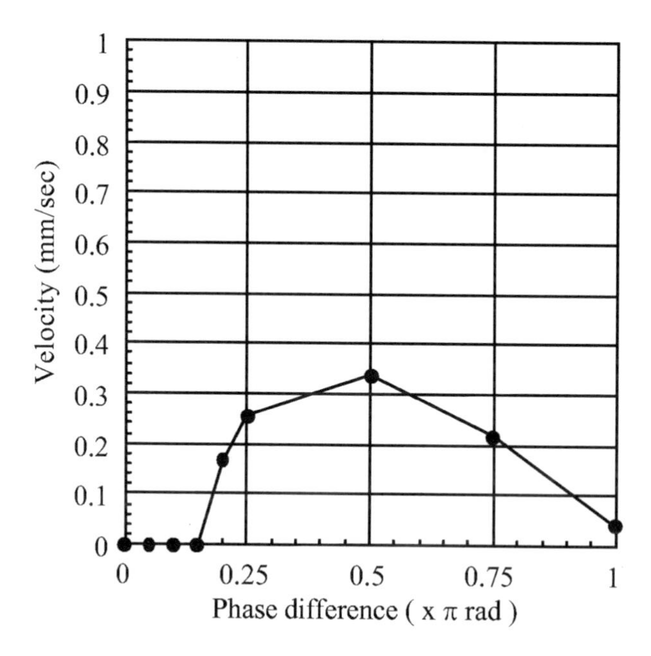

Figure 29: Effect of phase difference between the input voltage on velocity of plate transfer (f = 2 Hz, V_0 = 1.5 V).

The results from varying the frequency of the sinusoidal input are shown in Fig. 30. Analytical results, assuming elasticity of the material, indicate that the resonant frequencies of the element are approximately 5 Hz, 10 Hz, and 16 Hz. The experimental curve shows a good correspondence because the speed is high at these frequencies. Viscous resistance of water is so minor that complex flow near the device can be ignored. The major analysis error results from a modeling error of the contact and the element shape. Under 3 Hz, the experimental result indicates that the velocity increases linearly with the frequency. In this range, the speed of the elements is proportional to the frequency because the effect of distributed flexibility is small.

The resonant frequencies vary by 20% depending on the shape of the elements. These frequencies depend on the initial shape and the pressure exerted by the object being manipulated.

13.5.2 Soft Micromanipulation Device with Three Degrees of Freedom

Micromanipulators require the following elements: (1) compact micro mechanisms, (2) passive softness, (3) many DOF of motion, and (4) multimodal human feedback. The microactuator is one of the key issues for actualization of such advanced micromanipulators. They are difficult to construct using conventional actuators [Carrozza et al., 1998; Ono and Esashi, 1998; Zhou et al., 1998].

The important features of the actuators are (1) soft material, (2) force output, (3) ease of miniaturization and machining, and (4) multi-DOF motion ability. A 3-DOF-manipulation device is developed by crossing a pair of EFD elements perpendicularly at the end point as shown in Fig. 31.

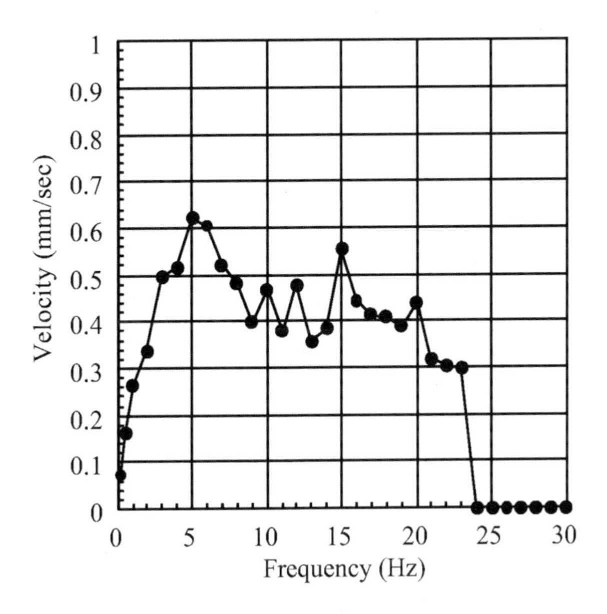

Figure 30: Effect of input voltage frequency on velocity of plate transfer (V_0 = 1.5 V, ϕ = 0.5 π rad).

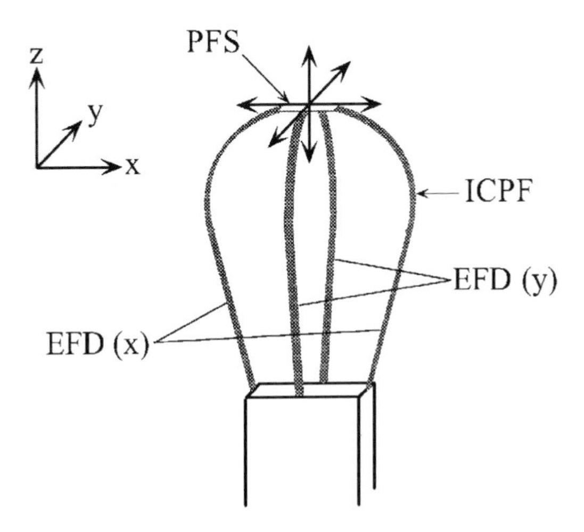

Figure 31: Structure of the 3-DOF micromotion device developed.

The most important factors in its design are: (1) high flexibility and softness, (2) minimum internal force, and (3) large displacement (especially in 2 DOF).

According to a characteristic synthesis using the Kanno-Tadokoro model, the width of the actuator is designed to be as thin as possible (w = 0.4 mm), and the pair of actuators is installed to a fixture in parallel. When external force and moment $\boldsymbol{f}_i = [f_x, f_y, f_z, n_x, n_y, n_z]^T$ are exerted at the end of the i'th actuator segment, it makes a minute translation and rotation $\boldsymbol{x}_i = [x, y, z, a_x, a_y, a_z]^T$. The relation is approximated by

$$x_i = C_i f_i, \tag{26}$$

where $\boldsymbol{C}_i$ is a compliance matrix. It is a constant matrix if the actuator can be modeled as an elastic body as in Kanno et al., [1996a]. Considering the viscoelastic property, it becomes a time-varying matrix depending on the deformation history.

Because minute translation and rotation of each actuator are expressed by

$$x_i = D_i x_0 \tag{27}$$

using minute motion of the end-effector $\boldsymbol{x}_o$, force and moment exerted on the end $\boldsymbol{f}_o$ have a relation of a compliance $\boldsymbol{C}_o$ of the end-effector

$$x_0 = C_0 f_0 \tag{28}$$

$$C_0 = \left(\sum_i D_i^T C_i^{-1} D_i \right)^{-1}. \tag{29}$$

Using analysis based on the Kanno-Tadokoro model, it was revealed that the developed device was much softer than conventional micromanipulators, because the compliance matrix under a static equilibrium condition was

$$C_0 = \begin{bmatrix} 1.9 & 0 & 0 & 0 & 9.3\times10 & 0 \\ 0 & 1.9 & 0 & -9.3\times10 & 0 & 0 \\ 0 & 0 & 3.5\times10^{-1} & 0 & 0 & 0 \\ 0 & -9.3\times10 & 0 & 2.2\times10^{4} & 0 & 0 \\ 9.3\times10 & 0 & 0 & 0 & 2.2\times10^{4} & 0 \\ 0 & 0 & 0 & 0 & 0 & 1.6\times10^{4} \end{bmatrix} \text{[m/N, rad/Nm]} \tag{30}$$

Therefore, this device is safer than a conventional device in manipulation of fragile microstructures. Only the dynamics of the micromanipulator itself were considered here. The actual compliance is effected by surface tension, Van der Waals force, electrostatic force, etc. Addressing this issue is beyond the scope of this chapter because these parameters depend on the working environment and cannot be estimated for the manipulator.

Figure 32 shows a relation between the magnitude of step input voltage and the resultant maximum displacement. Frequency characteristics under sinusoidal input are shown in Fig. 33. It is observed that the displacement increases by a quadratic curve according to the voltage. Displacement in the z direction is small because the IPMC (or ICPF) actuators are installed in parallel, and this characteristic has been predicted at the stage of design. Displacement in the xy direction (diagonal motion in the direction 45° to the x axis) is larger than the others because of the effect of the section shape and reduction of internal friction caused by water molecule movement in the material. The maximum displacement observed is 2 mm, and the available frequency range is up to 13 Hz. These performances are sufficient for the micromanipulation application. These characteristics can be predicted by computer simulation using models. In this application, rough estimation was performed using the Kanno-Tadokoro model. However, the accuracy was insufficient to determine the final design.

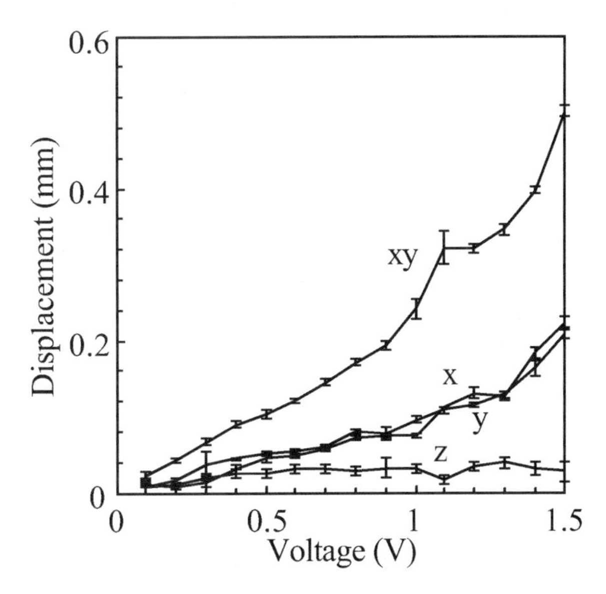

Figure 32: Effect of the step input voltage on the displacement.

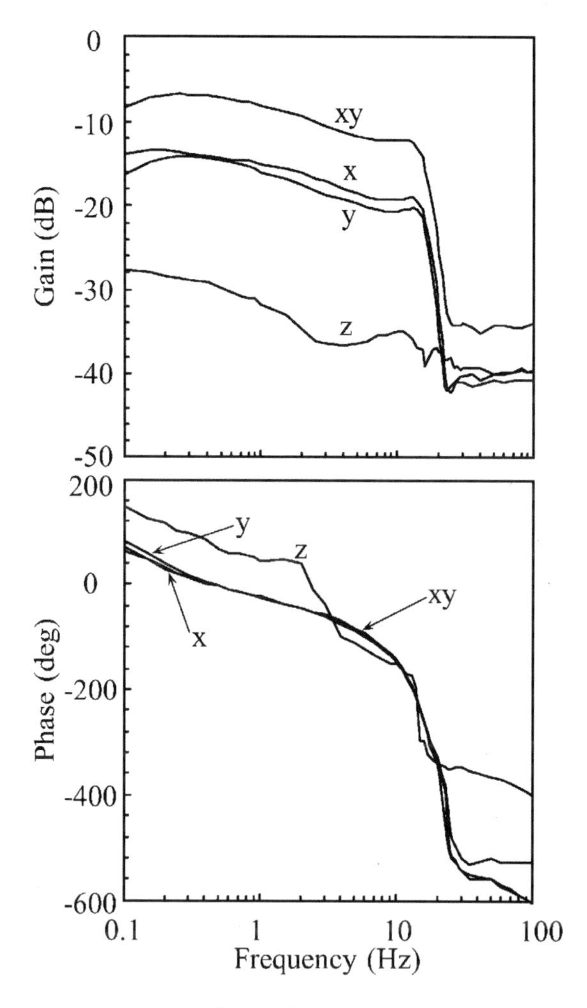

Figure 33: Frequency characteristics [input: 1.5 V sinusoidal waves, output: displacement (mm)].

Figure 34 illustrates experimental Lissajou figures in the *xy* plane. This demonstrates that the developed device has sufficient ability of motion for micromanipulation.

The feasibility of telemanipulation by the 3-DOF device was proven using the experimental setup shown in Fig. 35. Motion commands from an operator are given by a joystick and transmitted to the device via a PC. Actual motion is fed back to the operator by a microscopic image on a monitor. Figure 36 is a photographic view of the manipulation setup. The following are revealed as a result of various motion tests:

(1) If the joystick is moved without rest, the device responds to high-speed motion commands completely, and the operator can control the motion very easily.

(2) When the joystick stops, the device does not stop and return to the initial position. The period for which the device could stop at arbitrary points was 3 seconds. The latter characteristic is because the Nafion-Pt composite material is not a position-type actuator but a force-type actuator.

Experiments have shown that the new device is capable of supporting dynamic micromanipulation strategies where dynamic conditions, such as adhesion and pushing, can be major. Gripping or grasping is possible using this manipulator, particularly for short-duration applications that require soft handling. While effective operation was observed, issues of control still require improvement.

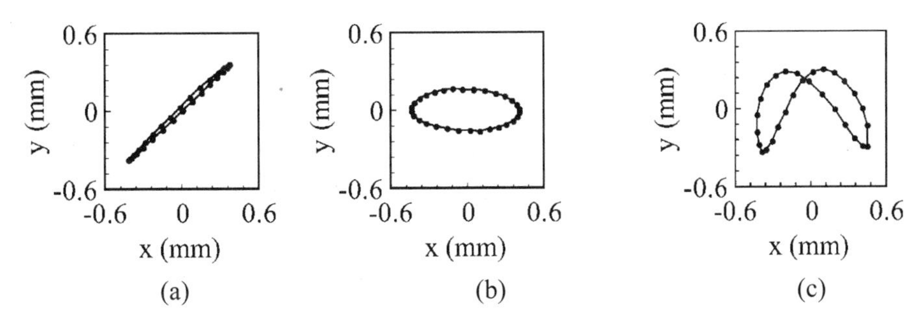

Figure 34: Lissajou motion of the 3-DOF device in *xy* plane. (a) V = 1.5 V, (b) $\phi = 0.5\pi$, V = 0.9, 1.5 V, (c) f = 1 Hz and 2 Hz

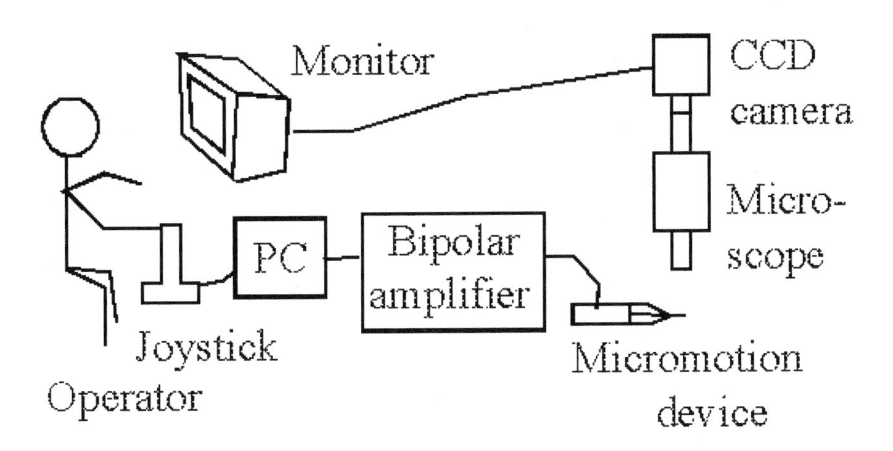

Figure 35: Experimental setup for telemanipulation.

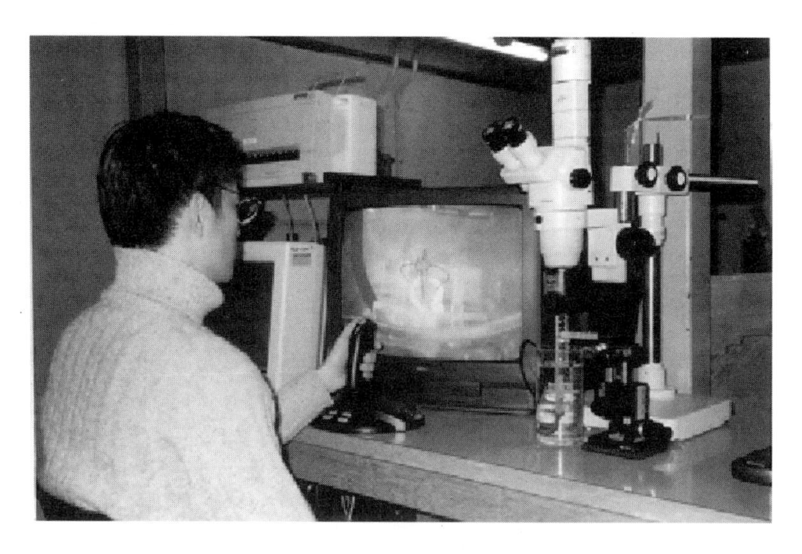

Figure 36: Experiments of direct operator control.

13.5.3 Haptic Interface for Virtual Tactile Display

Haptic interfaces for presenting the human sense of touch were developed using IPMC actuators [Konyo et al., 2000, 2002, and 2003]. To express delicate tactile sensitivity, including qualitative information such as tactile pressure or material feeling, there is a need to control the sensory fusion of elementary sensations generated by different sensory receptors. This basic concept was proposed as a selective-stimulation method, which makes stimulation on several different sensory receptors selectively and at the same time, successfully generating simple integrated sensations using magnetic oscillators [Shinoda et al., 1998]. EAP actuators are effective at expressing more realistic tactile sensations according to this unique concept.

Conventional tactile displays could hardly express such a delicate sensitivity, because it was difficult to make finely distributed stimuli on human skin under the limitation of actuators such as magnetic oscillators, piezoelectric actuators, shape-memory alloy actuators, pneumatic devices, and so on. In contrast, EAP materials have many effective characteristics as a 'soft' and 'light' actuator for such a stimulation device. IPMC is suitable for the following requirements:

(1) *High spatial resolution.* The spatial resolution of sensory receptors, especially Meissner's corpuscle in a fingertip, is less than 2 mm. IPMC films are easy to shape and their simple operation mechanism allows miniaturization of stimulators to make high-density distributed structures. Conventional actuators can hardly control such minute forces because of their heavy identical mass and high mechanical impedance. By using the passive material property, IPMC is soft enough that no special control algorithm is required.

(2) *Wide frequency range.* Each sensory receptor has a different response characteristic for vibratory stimulation (Fig. 37). The response of IPMC is fast enough to make a vibratory stimulation on skin higher than 200 Hz. This means that IPMC can stimulate several receptors selectively by using the difference of the range of response characteristics on each receptor.

(3) *Stimuli in multiple directions.* Each sensory receptor has selectivity for the direction of mechanical stimuli. Especially Meissner's corpuscle, which detects only the shearing stress toward the surface of a skin. Figure 38 shows that the bending motions of IPMC, which contacts with a surface of skin in a tilted position, causes stress in both the normal direction and shearing direction.

(4) *Wearability.* In human tactual perception, an active perceptual process based on hand contact motion is very important. For the generation of virtual tactile feeling, a subject must move freely and actively his hand and receive appropriate stimuli in response. For conventional mechanical stimulation displays, it is difficult to attach the device to a finger so the subjects cannot perform contact motion freely in a 3D space. An IPMC-based wearable display was successfully developed, which was made so small in size and weight that there was no interference with hand movements.

(5) *Safety.* The low-driving voltage (less than 3 V) is safe enough to touch directly with a human finger.

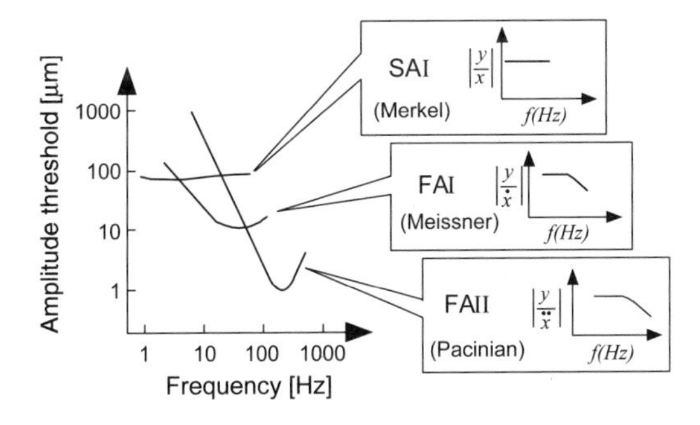

Figure 37: Thresholds of tactile receptors for vibratory stimulus [Maeno, 2000]. (Courtesy of the Robotics Society of Japan.)

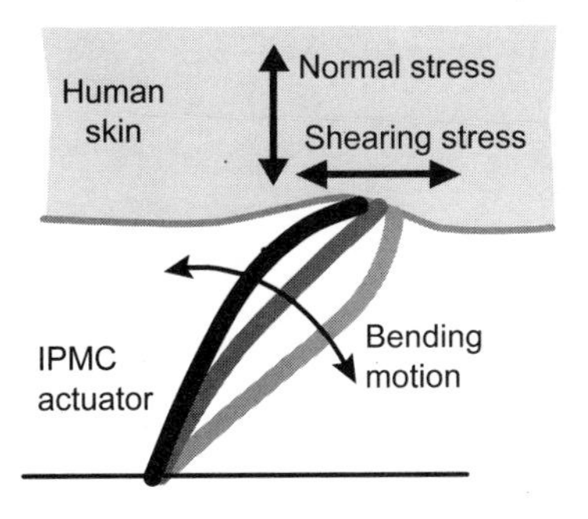

Figure 38: Multidirectional stimulation to a human skin using IPMC actuator.

The structure of the wearable stimulation device is shown in Fig. 39. The ciliary part is provided with IPMC actuators, where each cilium is 3-mm long and 2-mm wide, in 12 rows leaving at 1-mm gaps horizontally and 1.5-mm gaps vertically. All cilia are tilted 45 deg to transmit mechanical stimuli efficiently in both the normal and the tangential directions to the surface of the skin, as shown in Fig. 38. The use of silicon rubber of 25 × 25 × 8 mm applied to the base of the ciliary part has made it possible to lighten the device to approximately 8 g, including the flexible wiring board.

In an early prototype, tactile sensation has been presented as shown in Fig. 40(a). In this case, subjects obtained only passive tactual perception because contact motion could not be performed.

As shown in Fig. 40(b), the new wearable device can be attached to the tip of a finger. The power-supply line of IPMC is provided with a flexible wiring board in order to minimize restrictions laid on the hand, so the fingertip can bend easily. IPMC need to be kept moistened since its actuators are operated by ionic migration. However, the device can provide stimuli sufficiently even in the air for several minutes, even in humid conditions.

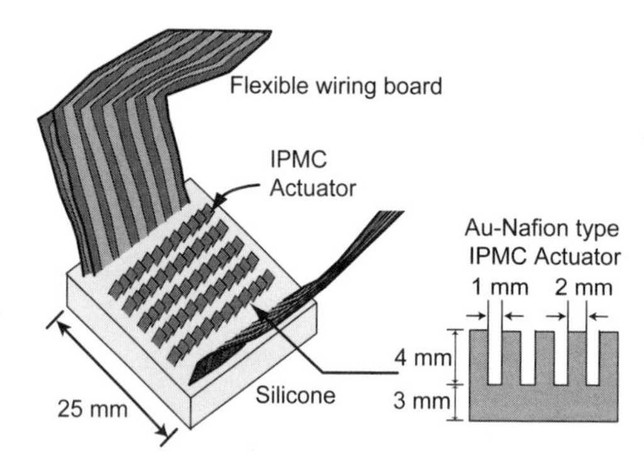

Figure 39: Structure of ciliary device using IPMC actuators.

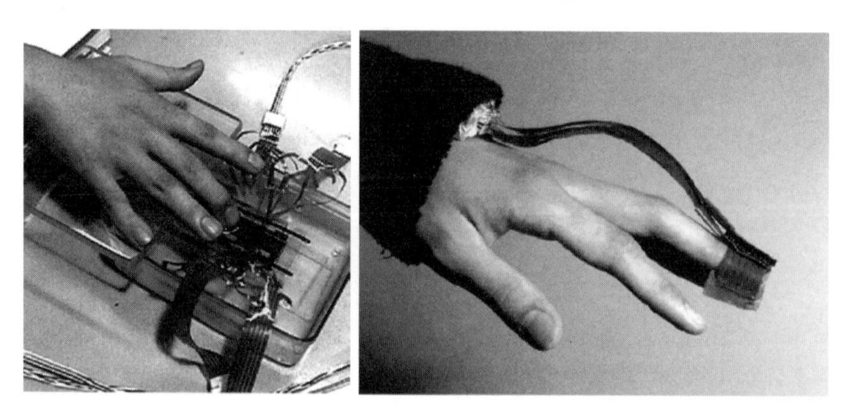

Figure 40: Overview of the tactile displays. (a) Fixed-type device, and (b) wearable device

To create arbitrary tactile sensation, the following two concepts were applied:

(1) Selective stimulation to tactile receptors.
(2) Simulation in response to active hand movements.

13.5.3.1 Selective Stimulation

In human skin, sensory receptors generates elementary sensations such as touch, pressure, vibration, pain, temperature sensing, and so on. Tactile impression is an integrated sensation of these elementary sensations. The four elementary sensations given by mechanoreceptors are classified into FA I, FA II, SA I, and SA II, which are respectively related to Meissner's corpuscle, Pacinian corpuscle, Merkel's disk, and Ruffini endings. To present arbitrary tactile sensation, stimuli applied to these receptors should be controlled selectively and quantitatively.

For the IPMC tactile display, a selective stimulation is realized by changing drive frequencies, using the receptors' response characteristics (Fig. 37). It is confirmed by the subject's introspection that the contents of sensation vary with the change of drive frequency as follows (anticipated receptors related to the sensation are shown in the parenthesis):

(1) Less than 10 Hz: pressure sensation, but very small. (SA I)
(2) 10-100 Hz: pressure and fluttering sensation, as if the surface of a finger is wiped with something rough material. (SA I and FA I)
(3) More than 100 Hz: simple vibratory sensation. (FA II)

Figure 41 shows the experimental results of the perceptual range of vibratory sensations for (a) fixed-type display and (b) wearable display. Sinusoidal waves are applied to determine the change in tactile sensation caused by changes in frequency, set to 0 Hz at first and increased with steps of 5 Hz every 3 seconds. The following three items are examined: (1) the minimum frequency at which the

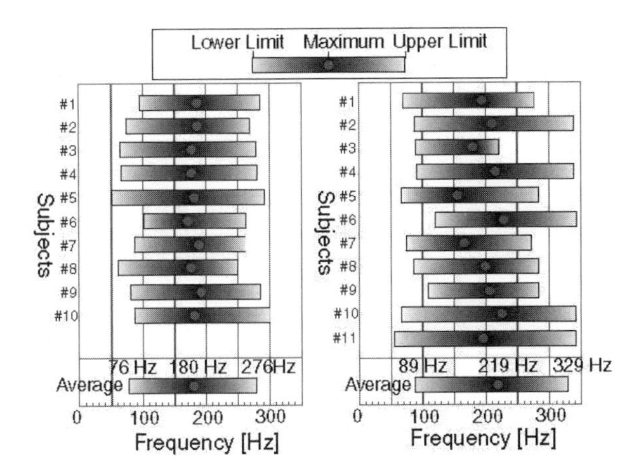

Figure 41: Perceptual range of simple vibratory sensation.

subject begins to feel a simple vibratory sensation, (2) the frequency at which the vibratory sensation reaches the maximum, and (3) the maximum frequency at which the subject can feel the vibratory sensation. The subjects began to feel a simple vibratory sensation when the information from FA II exceeds that of FA I. Figure 37 shows that the detection threshold of FA II exceeds that of FA I in the vicinity of the frequency range from 50 to 100 Hz. This agrees with the results of the perceptual range of vibratory sensation.

For creating integrated sensations, stimulation by composite waves of several frequencies was proposed. Composite waves can stimulate the different kind of tactile receptors at the same time based on the selective stimulation method.

In the earlier experiment using the fixed-type IPMC display, composite waves of high and low frequencies that present both pressure sensation and vibratory sensation at the same time was applied. Eight of the ten subjects sensed some special tactile sensation, which is clearly different from simple vibratory sensation against the composite waves consist of the 180 Hz high-frequency component and one low-frequency component from 30 Hz to 80 Hz. It suggests that the high-frequency component generates FA II, and the lower component generates FA I successfully.

The artificial tactility, which was generated by the above composite waves, was also evaluated by comparison with e11 sample materials. The subjects chose the most similar material compared with each artificial sensation by using both hands at the same time, the right hand touching the IPMC display and the left hand touching the real materials. Figure 42 shows the experimental results for the 6 artificial tactile feeling generated by the composite waves that have different low-frequency components. The graph illustrates the 6 materials selected by more than 4 subjects out of 17. The most frequently selected materials of each artificial sensation changed according to the low-frequency components. Notably, about half of the subjects responded to a towel and denim with the same feeling. This result demonstrates that changing the component of frequencies produces variation of tactile response, expressing various sensations similar to those of cloth materials.

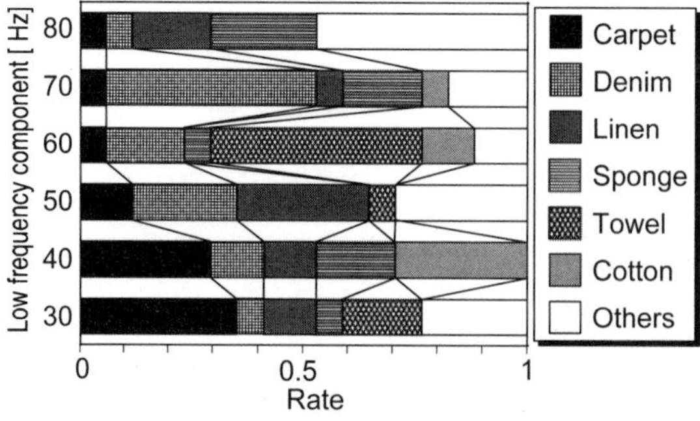

Figure 42: Low-frequency component vs. tactile sensation by comparison with real cloth materials.

13.5.3.2 Stimulation in Response to Active Hand Movements

In human-tactual perception, an active perceptual process based on contact motion dominates. An active touch in connection with contact motion excels passive sensory perception both qualitatively and quantitatively [Gibson, 1962]. In the nervous system, tactual sensation is sent to sensory perception analysis when afferent sensory information and the movement control signal are copied [Loomis, 1986]. This means that hand movement essentially takes part in the generation of tactile feeling. For the generation of virtual reality of tactile feeling, a subject must move his hand freely and receive appropriate stimuli in response to movements.

A tactile display system in response to hand movements was developed using the wearable IPMC device (Fig. 43). The stimulation device is attached to the tip of the middle finger. The system is designed to read information about hand position using Polhemus' FASTRAK, which can read information according to a magnetic field.

The efficiency of the wearable display was evaluated by measuring the changes in sensitivity of both passive touch and active touch using detection thresholds of vibratory stimuli. As for the presentation of passive touch, subjects are asked to put their hand in a water tank and to keep it still while vibratory stimulation is applied. For the presentation of active touch, subjects can perform contact motion in the water tank, and vibratory stimulation is applied only when the velocity of hand movement exceeds a specified value (200 mm/s). The vibratory stimulations are generated by sinusoidal waves of frequencies that varied from 10 to 60, 130, and 200 Hz. The amplitude threshold is determined by a method of limits. Seven males in their twenties were the subjects.

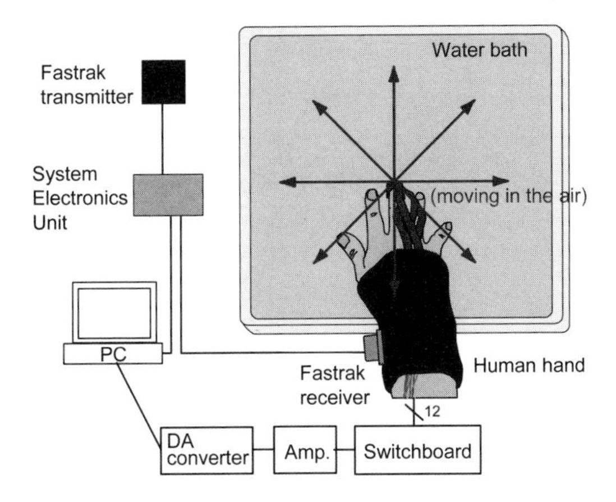

Figure 43: Wearable display in response to virtual contact motion.

Figure 44 shows the average of the thresholds obtained from passive and active stimulation. At each frequency, the detection threshold decreases more by active touch than by passive touch. This improvement of sensitivity seems to have been brought about both (1) by active perception and (2) by detection of variable components in relation to hand movements. The results showed that the efficiency of stimulation is improved clearly in comparison with our conventional passive display.

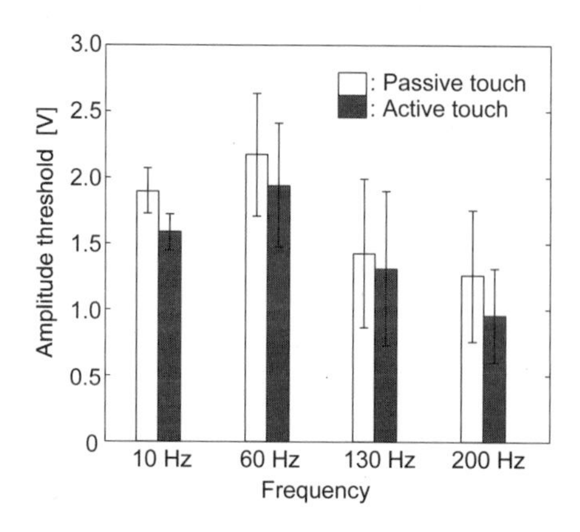

Figure 44: Changes in detection threshold due to active touch.

As to a stimulation method in response to hand movements, perceptive characteristics of touch were used. It is known that against mechanical stimuli applied to the skin, the FA I unit responded to components of velocity, and the FA II unit responded to components of acceleration. A method of generating stimulation with composite waves was proposed by which low-frequency vibratory stimulation is presented in response to the velocity of hand movements and high-frequency vibratory stimulation is presented in response to acceleration for stimulating corresponding receptors. If the periodic function of a low-frequency component and a high-frequency component are represented by *L* and *H*, respectively, the composite wave can be expressed by

$$V = \alpha|v|L(2\pi f_L) + \beta|a|H(2\pi f_H), \tag{31}$$

where V is the output voltage to the device, v and a are the velocity and the acceleration of hand movements, f_L and f_H are the frequency of the periodic function L and H, and α and β are the proportional constants when the velocity and the acceleration are converted to the amplitude and frequency.

It is confirmed by the subject's observation that several kinds of composite waves express the natural feeling of contact such as the starting, stopping, and sliding feel of hand movements. In the future, we plan to clarify such relationships between stimulation to receptors and sensation of contact motions.

The haptic interface developed through our study makes it possible to realize a wearable device applicable to the generation of stimulation in multiple directions, which was not accomplished by studies conducted in the past. This device is indispensable for reproducing interactive tactile perception in 3D space. It will greatly expand the range of application to virtual reality.

13.6 References

ABAQUS/Standard User's Manual, Ver. 5.4, 1999.

Bar-Cohen Y., S.P. Leary, K. Oguro, S. Tadokoro, J.S. Harrison, J.G.Smith, and J. Su, "Challenges to the application of IPMC as actuators of planetary mechanisms," *Proc. SPIE 7th International Symposium on Smart Structures, Conference on Electro-Active Polymer Actuators and Devices*, pp. 140-146, 2000.

Carrozza M. C., P. Dario, A. Menciassi, and A. Fenu, "Manipulating biological and mechanical micro-objects using LIGA-microfabricated end-effectors," *Proc. IEEE International Conference on Robotics and Automation*, pp. 1811-1816, 1998.

Firoozbakhsh K., M. Shahinpoor, and M. Shavandi, "Mathematical modeling of ionic-interactions and deformation in ionic polymer-metal composite artificial muscles," *Proc. SPIE* ,3323, pp. 577-587, 1998.

Fukuhara M., S. Tadokoro, Y. Bar-Cohen, K. Oguro, and T. Takamori, "A CAE approach in application of Nafion-Pt composite (ICPF) actuators: Analysis for surface wipers of NASA MUSES-CN nanorovers," *Proc. SPIE 7th International Symposium on Smart Structures, Conference on Electro-Active Polymer Actuators and Devices*, pp. 262-272, 2000.

Gibson J. J., "Observation on active touch," *Psychological Rev.*, Vol.69-6, pp.477-491, 1962.

Guo S., T. Fukuda, K. Kosuge, F. Arai, K. Oguro, and M. Negoro, "Micro catheter system with active guide wire," *Proc. IEEE International Conference on Robotics and Automation*, pp. 79-84, 1995.

Guo S., T. Fukuda, N. Kato, and K. Oguro, "Development of underwater microrobot using ICPF actuator," *Proc. IEEE International Conference on Robotics and Automation*, pp. 1829-1835, 1998.

Guo S., S. Hata, K. Sugimoto, T. Fukuda, and K. Oguro, "Development of a new type of capsule micropump," *Proc. IEEE International Conference on Robotics and Automation*, pp. 2171-2176, 1999.

Hanai T., "Membrane and ions, Theory and computation of material migration," *Kagaku-Dojin* Publ., 1978 (in Japanese).

Ikuta K. and H. Shimizu, "Two dimensional mathematical model of shape memory alloy and intelligent SMA-CAD," *Proc. IEEE Micro Electro Mechanical Systems Conference*, pp. 87-92, 1993.

Kanno R., A. Kurata, M. Hattori, S. Tadokoro, and T. Takamori, "Characteristics and modeling of ICPF actuator," *Proc. Japan-USA Symposium on Flexible Automation*, pp. 692-698, 1994.

Kanno R., S. Tadokoro, T. Takamori, M. Hattori, and K. Oguro, "Linear approximate dynamic model of an ICPF (ionic conducting polymer gel film) actuator," *Proc. IEEE International Conference on Robotics and Automation*, pp. 219-225, 1996a.

Kanno R., S. Tadokoro, M. Hattori, T. Takamori, and K. Oguro, "Modeling of ICPF (ionic conducting polymer gel film) actuator, Part 1: Fundamental characteristics and black-box modeling," *Trans. of the Japan Society of Mechanical Engineers*, Vol. C-62, No. 598, pp. 213-219, 1996b (in Japanese).

Kanno R., S. Tadokoro, M. Hattori, T. Takamori, and K. Oguro, "Modeling of ICPF (ionic conducting polymer gel film) actuator, Part 2: Electrical characteristics and linear approximate model," *Trans. of the Japan Society of Mechanical Engineers*, Vol. C-62, No. 601, pp. 3529-3535, 1996c (in Japanese).

Kanno R., S. Tadokoro, T. Takamori, and K. Oguro, "Modeling of ICPF actuator, Part 3: Considerations of a stress generation function and an approximately linear actuator model," *Trans. Japan Society of Mechanical Engineers*, Vol. C-63, No. 611, pp. 2345-2350, 1997 (in Japanese).

Kita H. and K. Uozaki, "Fundamental of electrochemistry," *Gijutsudo Publ.*, 1983 (in Japanese).

Konyo M., S. Tadokoro, T. Takamori, and K. Oguro, "Artificial tactile display using soft gel actuators," *Proc. IEEE International Conference on Robotics and Automation*, pp. 3416-3421, 2000.

Konyo M., S. Tadokoro, M. Hira, and T. Takamori, "Quantitative evaluation of artificial tactile display integrated with visual information," *Proc. IEEE International Conference on Intelligent Robotics and Systems*, pp. 3060-3065, 2002.

Konyo M., K. Akazawa, S. Tadokoro, T. Takamori, "Wearable haptic interface using ICPF actuators for tactile display in response to hand movements," *Journal of Robotics and Mechatronics*, Vol. 15, No. 2, pp. 219-226, 2003.

Loomis J. M. and S. J. Lederman, "Tactual perception," *Handbook of Perception*, Vol. II, 31-3, John Wiley & Sons, New York, 1986.

Maeno T., "Structure and function of finger pad and tactile receptors," *J. of Robot Society of Japan*, Vol. 18, No.6, pp.772-775, 2000 (in Japanese).

Nemat-Nasser S. and J. Y. Li, "Electromechanical response of ionic polymer metal composites," *Proc. SPIE Smart Structures and Materials 2000, Conference on Electro-Active Polymer Actuators and Devices*, Vol. 3987, pp. 82-91, 2000.

Oguro K., Y Kawami, and H. Takenaka, "Bending of an ion-conducting polymer film-electrode composite by an electric stimulus at low voltage," *J. of Micromachine Society*, Vol. 5, pp. 27-30, 1992.

Ono T., and M. Esashi, "Evanescent-field-controlled nano-pattern transfer and micro-manipulation," *Proc. IEEE International Workshop on Micro Electro Mechanical Systems*, pp. 488-493, 1998.

Shahinpoor M., "Active polyelectrolyte gels as electrically controllable artificial muscles and intelligent network structures, structronic systems: smart structures, devices and systems, part II: systems and control," *World Scientific*, pp. 31-85, 1998.

Shahinpoor M., "Electro-mechanics of iono-elastic beams as electrically-controllable artificial muscles," *Acta Mechanica*, 1999.

Shahinpoor M., Y. Bar-Cohen, J.O. Simpson, and J. Smith, "Ionic polymer-metal composites (IPMC) as biomimetic sensors, actuators and artificial muscles -- a review," *Field Responsive Polymers*, American Chemical Society, 1999.

Shinoda H, N. Asamura and N. Tomori, "A tactile display based on selective stimulation to skin receptors, *Proc. IEEE*, pp. 435-441, 1998.

Tadokoro S., T. Murakami, S. Fuji, R. Kanno, M. Hattori, and T. Takamori, "An elliptic friction drive element using an ICPF (ionic conducting polymer gel film) actuator," *IEEE Control Systems*, Vol. 17, No. 3, pp. 60-68, 1997.

Tadokoro S., S. Fuji, M. Fushimi, R. Kanno, T. Kimura, T. Takamori, and K. Oguro, "Development of a distributed actuation device consisting of soft gel actuator elements," *Proc. IEEE International Conference on Robotics and Automation*, pp. 2155-2160, 1998.

Tadokoro T., S. Yamagami, M. Ozawa, T. Kimura, T. Takamori, and K. Oguro, "Multi-DOF device for soft micromanipulation consisting of soft gel actuator elements," *Proc. IEEE International Conference on Robotics and Automation*, pp. 2177-2182, 1999a.

Tadokoro S., S. Fuji, T. Takamori, and K. Oguro, "Distributed actuation devices using soft gel actuators," *Distributed Manipulation*, Kluwer Academic Press, pp. 217-235, 1999b.

Tadokoro S., S. Yamagami, T. Kimura, T. Takamori, and K. Oguro, "Development of a multi-degree-of-freedom micro motion device consisting of soft gel actuators," *J. of Robotics and Mechatronics*, 2000a.

Tadokoro S., S. Yamagami, T. Takamori, and K. Oguro, "Modeling of Nafion-Pt composite actuators (ICPF) by ionic motion," *Proc. SPIE 7th International Symposium on Smart Structures, Conference on Electro-Active Polymer Actuators and Devices*, pp. 92-102, 2000b.

Tadokoro S., S. Yamagami, T. Takamori, and K. Oguro, "An actuator model of ICPF for robotic applications on the basis of physicochemical hypotheses," *Proc. IEEE International Conference on Robotics and Automation*, pp. 1340-1346, 2000c.

Zhou Y., B. J. Nelson, and B. Vikramaditya, "Fusing force and vision feedback for micromanipulation," *Proc. IEEE International Conference on Robotics and Automation*, pp. 1220-1225, 1998.

TOPIC 5

PROCESSING AND FABRICATION OF EAPs

CHAPTER 14

Processing and Fabrication Techniques

Yoseph Bar-Cohen, Virginia Olazábal, José-María Sansiñena
Jet Propulsion Lab./California Institute of Technology

Jeffrey Hinkley
NASA Langley Research Center

14.1 Introduction

Transition of EAP materials to practical applications requires effective processing techniques to allow the fabrication of electroded actuators that are shaped to the desired configuration. These methods need to produce consistent material, maximize the actuation capability and assure the performance and durability of the material. Once EAP actuators are made they may need to be integrated with sensors and packaged with electronic and mechanical components to form a functional system. As described in Chapter 1, the availability of effective processing methods is a critical part of the EAP technology and its infrastructure. Optimization of the response of existing materials or the development of new EAPs requires the use of systematic approaches rather than a hit-or-miss evolutionary process that is subject to false starts and wasted efforts. Computational chemistry [Chapter 10] offers the hope to enhance the understanding of the phenomena responsible for the response of the various EAP materials and may provide tools for the design of effective materials. Once a new EAP material or structure is analytically predicted to have an efficient response, a synthesis method is required to produce the predicted material. Polymer modeling tools in the form of commercial software packages are available and have been in use for a number of years by manufacturers of plastic materials and drugs. The available software packages are capable of predicting thermal expansion coefficients and bulk moduli of crystallike arrays or amorphous assemblies of simple molecules at a specified density. Studies are currently under way to apply similar techniques to model high-performance polymeric materials.

As an emerging field, there are no established processing techniques for fabricating or mass-producing EAP materials. Producing such materials can rely on existing techniques of making polymer-based materials and structures. Some modifications may be needed to account for the unique requirements of the specific EAP materials. In preparing this chapter, efforts were made to identify reported processes of producing EAP and methods that seemed to be applicable for making such materials at various scales. Some of the methods that are described include ink-jet printing, ionic self-assembled monolayering (ISAM), spincoating, and lithography.

14.2 Synthesis and Material Processing

In general, polymers are divided into thermoplastics and thermosets depending on the reversibility of their properties in relation to the glass transition point. Processing each of these two categories of polymers may require different techniques as described below.

14.2.1 Processing Thermoset and Thermoplastic EAPs

Thermoplastic materials can be synthesized in solution, suspension, emulsion, or bulk and then treated by different polymer processing methods such as injection molding, solvent casting, or extrusion. An example of an automated injection molding system is shown in Fig. 1. During this process, molten polymer is injected into a mold cavity at high speeds and then the material is allowed to cool and solidify, followed by ejection of the part from the mold. Examples of thermoplastic materials include polyamide (PA) and polyvinylidene fluoride (PVDF) ferroelectric polymers and some conductive polymers (CP).

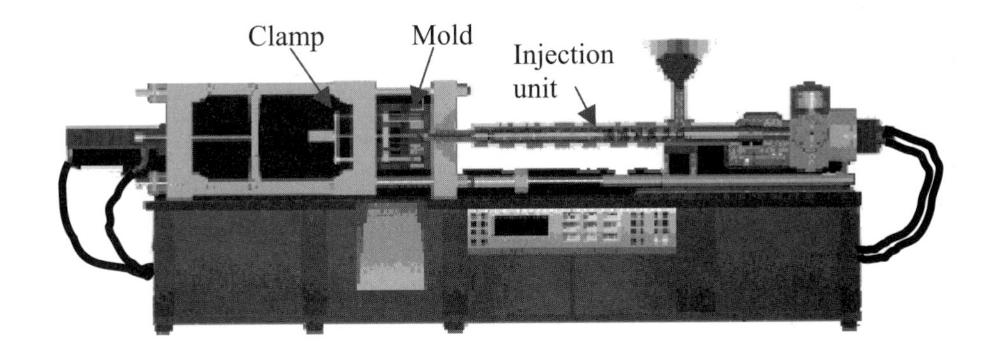

Figure 1: A schematic view of an injection molding process. (Courtesy of David Waters, Polymer Training Resources, LLC.)

In contrast to thermoplastic materials, the thermosets are fabricated at net shape since chemical cross-linking renders them insoluble and infusible. Often, the starting material is a mixture of monomers, one of which is multifunctional, and the processing leads to polymer chain branching and cross-linking. In addition to materials that are thermally cross-linked, we include in this class insoluble materials that are made using photoinitiators or even ionizing radiation. An example of EAP from the thermoset class, gel actuators (e.g., polyacrylamide and polyacrylic acid cross-linked gels), are sometimes synthesized in a test tube or similar mold and formed as cylindrical rods. Hybrid approaches are also used as in the case of thermoplastic polyacrylonitrile that is spun into fibers (commercial acrylic yarn is made this way), then oxidatively cross-linked and finally hydrolyzed into a responsive gel.

14.2.2 Composite EAP

Making a composite EAP can enhance the performance of dielectric EAPs to exhibit strong electrostrictive behavior [Chapters 4 and 16]. Since there is a limit to the capability of existing materials, the use of mixtures of various materials offers an important design and fabrication capability. In general, dielectric polymers require a large activation electric field in the range of 150 V/μm, which significantly affects their potential applications. This limitation hampers the application of dielectric EAPs in spite of their relatively large stored mechanical energy that can exceed the level of 0.2 MPa. In order to obtain large strains under low voltages it is necessary to use thin films and increase the elastic modulus and the dielectric constant. Increasing these material properties may be achieved by making an electroceramic/elastomer composite (see Fig. 2) [Shiga et al., 1993], where using stiff ceramic filler can increase the elastic modulus. Further, a higher dielectric constant is obtained, leading to a stronger electroactive response resulting from high Coulomb forces. One may need to take into account the possibility that the breakdown field may become lower as a result of producing such composite material. Controlling the polarization direction of the filler material during processing enhances the response of the composite EAP produced, providing larger response along the direction of the activation field. For this purpose, the electrorheological phenomenon [Randall et al., 1993] can be used to control the fabrication process and to tailor the dielectric and mechanical properties. For example, one can use PANi particles suspended in a fluid with electrorheological properties [Bohon and Krause, 1998]. The variety of useful composite structures that can be made by blending two materials is almost limitless.

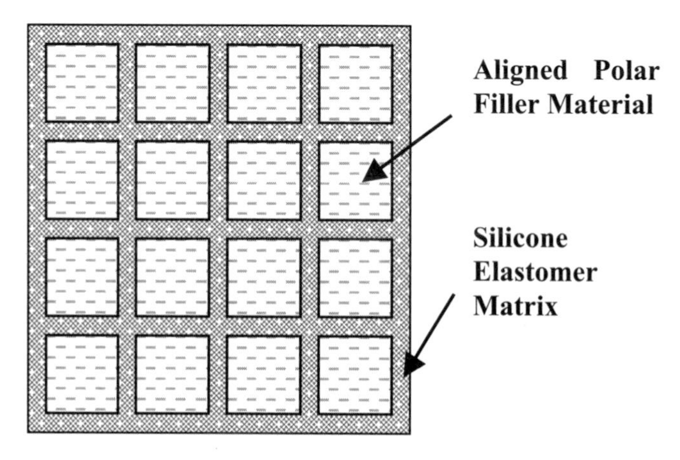

Figure 2: Composite elastomer with polarized filler.

Another form of producing a composite EAP can be enabled by using biaxial stretching to form lightweight films with microvoids that have a large internal surface area. Such a film was used by Peltonen et al., [1999a] to produce to produce an electrostrictive actuator and one may be able to fill the voids with a second polymer phase to create a composite membrane. Causing thermodynamic phase separation during solidification or cross-linking can also create porosity and voids. Using this process, one can produce electroactive gels with vastly improved transport properties [Suzuki and Hirasa, 1993].

14.3 Fabrication and Shaping Techniques

It is important to realize that one of the key benefits of polymers as materials is the relative ease with which their molecular structure can be manipulated. Stretching to form fibers and films leads to anisotropy or crystallinity while improving the tensile strengths of the material. As an example, fibers made of polyaniline are over 40-fold stronger than in the bulk form [Pomfret et al., 2000]. Generally, stretched semicrystalline polymers can exhibit interesting morphologies. Most EAP materials that are currently used as actuators are shaped as a film or a layered structure. There is a wide variety of methods that can be used to produce polymer films including vacuum evaporation [Touihri et al.,

1997; and Misra et al., 1992], plasma polymerization [Shi, 1996; and Chowdhury and Bhuiyan, 2000], synchrotron radiation photodecomposition [Katoh and Zhang, 1999], spincoating [Frank, 1998; & Wang and Gan, 2000], vapor-phase deposition [Forrest, 1997], plasma sputtering [Wydeven et al., 1998], vacuum deposition [Burrows et al., 1998] and pulsed laser deposition [Heitz et al., 1998; and Piqué et al., 1999].

Thermoplastic embossing and forming are currently practiced on an industrial scale and can be used to produce films that are as thin as 0.6-μm. This method was used by Dreuth [1999] to prepare intricate electrostatic actuators. Hot embossing has been established as a flexible process for the fabrication of polymer microcomponents, e.g., microfluidic devices for total analysis systems (TAS) applications [Becker et al., 1998]. A hot embossing process [Becker and Heim, 2000] was also used to fabricate large-area polycarbonate (PC) and polymethylmetacrylate (PMMA) polymeric structures with high aspect ratios (Fig. 3).

To create complex shape polymers directly from computer models one can use solid free-form fabrication methods. This technique can be applied directly to extrude a molten thermoplastic polymer layer-wise [Calvert and Liu, 1997]. This technique is also referred to as "solid free-form fabrication" or "computer automated manufacturing" or "layered manufacturing." A related 3DP method utilizes ink-jet printing technology to create a solid object by printing a binder into selected areas of sequentially deposited layers of powder. This technology is applied, for example, to fabricate polymeric drug delivery systems [Wu et al., 1996; and Sastry et al., 2000]. Recently, 3DP was used to fabricate injection-molding tooling with cooling channels [Sachs et al., 2000]. Stereolithography (STL) is another type of free-form fabrication method that can be used to produce complex shape polymers. This process allows rapid prototyping, where a laser is used to stimulate photopolymerization of a liquid resin and to convert it into a solid model [Bedal and Nguyen, 1996].

14.3.1 Forming Membranes and Films

Most of the EAP materials that have been reported so far have a film shape. Therefore, forming uniform films in a mass-production process can be an important processing method for the transfer of this technology to commercial use. The simplest methods of producing a film are to dissolve the material in a solvent or to melt it and press it between two plates. Membranes based on cellulose used in renal replacement therapy are manufactured either by wet solution or melt spinning and may be considered as a hydrogel. In this case, the membranes were produced in the form of hollow fibers incorporated into haemodialysers. The fibers were made by a wet solution spinning of an aqueous cuprammonium solution of cellulose derived from cotton. For this purpose, the material was formed from this solution using a spinneret in which the cellulose was coextruded with a core liquid (isopropyl myristrate) [Hoenich and Stamp, 2000].

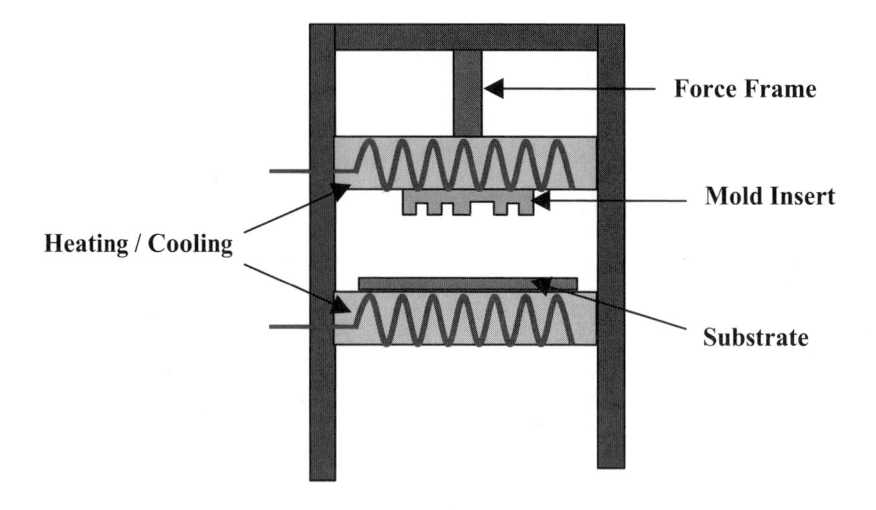

Figure 3: Hot embossing equipment.

14.3.2 Producing Fibers

Producing EAPs in a fiber shape and bundling the fibers to form an artificial muscle allows emulating biological muscles and provides the benefits of redundancy (i.e., robustness) and resilience (resulting from the maintained flexibility). Fibers can be made in a variety of processes. In spinning, the material is first converted to a liquid or semiliquid state by dissolving it in a solvent or heating it until molten. The resulting liquid is extruded through small holes in a device known as a spinnerette. The fine jets of liquid that emerge harden to form a long fiber or filament. Several spinning techniques are used widely to produce fibers, including solution spinning (wet or dry), melt spinning (Fig. 4), and emulsion spinning.

In solution spinning a viscous solution of polymer is pumped through a filter and then passed through the fine holes of a spinnerette. The solvent is subsequently removed leaving a fiber that can be stretched to orient it as desired. Wet spinning is studied to produce fibers of conductive polyaniline and its derivatives in order to improve the processability. In this way, wet spinning of polyaniline/poly(p-phenylene terephthalamide) fibers from homogeneous solutions in concentrated sulphuric acid is reported by Andreatta et al., [1990]. Poly(o-toluidine) was spun from sulphuric acid, m-cresol and N-methyl-2-pyrrolidinone (NMP) [Hsu et al., 1997]; high molecular weight emeraldine base (EB) was spun from NMP [Mattesa et al., 1997] and N,N'-dimethyl propylene urea (DMPU) [Tzoua et al., 1995]; and leucoemeraldine base (LB) was spun, post-oxidized, then doped to give the conductive emeraldine salt (ES) [Chacko et al., 1997].

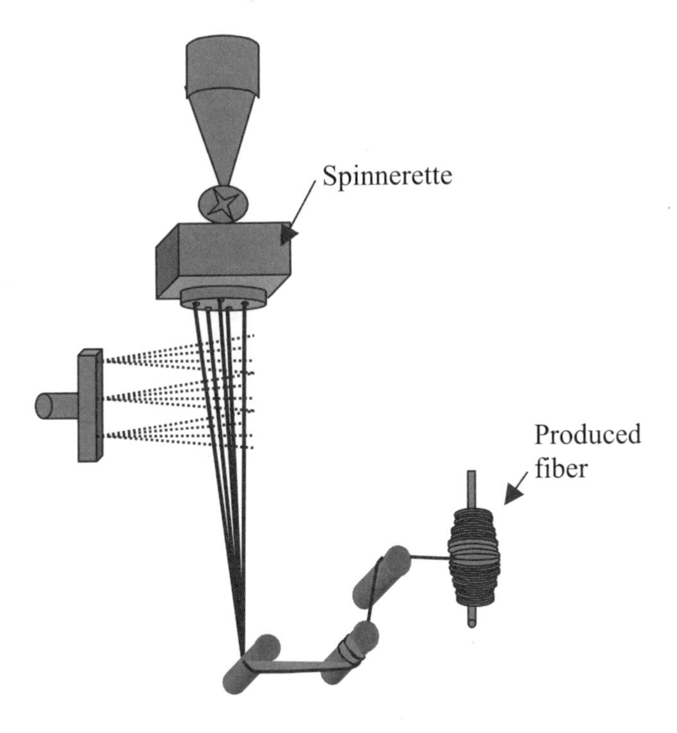

Figure 4: Melt spinning process.

A more economical method is melt spinning, since it does not require recovery of the solvent, as does solution spinning. This method is used, for example, in the production of fibers that are made of thermotropic cellulose derivatives [Gilbert et al., 2000]. Also, high-speed bicomponent spinning of poly(ethylene terephthalate) (PET) (core) and polypropylene (PP) (sheath) [Kikutani et al., 1996], and polyethylene sheath/poly(ethylene terephthalate) core fibers was performed. In these cases, when two polymers are coextruded in bicomponent spinning, the stress and thermal histories of each component are caused by the mutual interaction of two components, contrary to those in single-component spinning. Thus, by selecting a suitable combination of polymers, it is possible to improve the structure of high-speed spun fibers [Cho et al., 2000].

To process nonmelting and insoluble polymers one can use the method of emulsion spinning, where such polymers are ground to a finely divided powder, mixed into a solution of another polymer, and solution-spun to fibers. Such a process is used to produce fibers from fluorocarbons such as Teflon, which have extremely high melting points. This technique can also be used to spin fibers that are made of nonpolymer-based materials, including inorganic materials (e.g., ceramics), where these materials can be suspended in a solution of an inexpensive polymer (e.g., cellulose esters).

To produce nanofiber polymers with diameters that are approximately 50 nm, one can use the electrospinning process, which involves high-voltage charging of suspended droplets produced from a polymer solution. When the droplet reaches several thousand volts, a fine jet of the polymer solution shoots out toward a grounded target and forms a continuous multifilament fiber of the polymer. The fiber is splayed out as it reaches the target, dried, and collected as an interconnected web of small fibers. Electrospinning from liquid crystal or other disentangled systems can be used to produce fibers containing few molecules, most of which are on the surface. Moreover, this process can be used to produce fibers that are as thick as 5 μm, which is the size of textile fibers. This process of electrospinning was used at the University of Akron to produce fibers from over 30 types of polymer materials including polyethylene oxide, DNA, polyaramids, and polyaniline [Doshi and Reneker, 1995; Reneker and Chun, 1996; and Fong et al., 1999].

Using electrospinning, biologically inspired temperature-responsive gels can be used to produce filaments having diameters as small as 200 nm [Huang et al., 2000]. At present, the fibers are usually collected as a random 2D mat, but fabrication of more organized structures may become feasible. Genetic engineering is now inspiring a variety of biocompatible structures with exciting possibilities for highly specific, molecularly designed materials. Recently, using such an approach, a biomimetic polymer gel was synthesized by *self-assembly* under conditions of controlled pH and concentration [Qu et al., 2000].

Scientists at the University of Pisa, Italy, are currently studying methods of producing fibers or layers of polyelectrolyte gels and combining them together with conducting polymer elements [Chiarelli and De Rossi, 1996]. The emphasis of these researchers' study is on the development of conductive polymers that operate as dry actuators [Mazzoldi et al., 1998]. Figure 5 shows such a dry actuator consisting of a polyaniline (PANi) fiber covered by a solid polymer electrolyte and using a Cu wire as a counterelectrode. The actuator was examined at several different stimulation conditions—including cyclic voltammetry (CV), square wave potential (SWP), square wave current (SWC)—and showed interesting performance. The tensile isometric stress exerted by the contractile fibers is expected to be about 10 times higher than typical human skeletal muscles. Under low-voltage drive (<2 V), a linear isotonic strain of 0.3% is obtained during CV and of 0.2% during SWP. The use of the polyelectrolyte gels and CP is important because they are naturally predisposed to reciprocal interaction. The ionic exchange process of these components can be used to realize actuators with a strain of several percent and stress of >100 MPa. Using polyelectrolyte gels, a strain of $< 50\%$ and stress of 20 MPa can be obtained. Tests on polypyrrole and PAN showed work density of 7×10^5 J/m^3 for polypyrrole [Chiarelli et al., 1994] and 4×10^5 J/m^3 for PAN [Casalino et al., 1991].

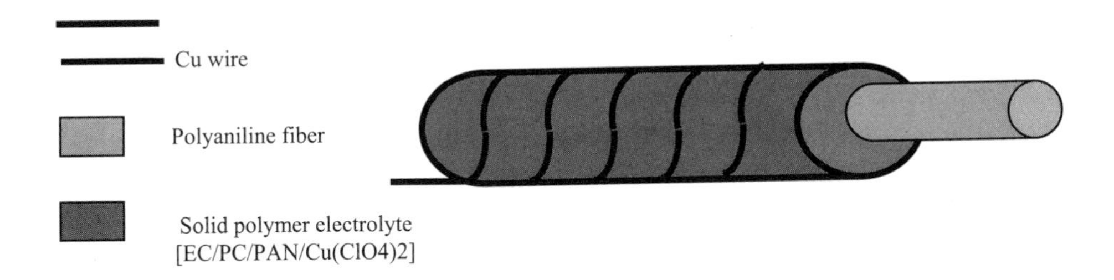

Figure 5: PANi dry polymer actuator (based on De Rossi et al. [1999]).

14.3.3 Ink-Jet Printing of Complex Configurations

Material printing offers a powerful means of producing, configuring, and electroding EAP on the microscale level. The phenomenon of forming uniform drops from a stream of liquid that is emitted from an orifice was first observed in the nineteenth century [Savart, 1833]. Generally, pressing fluid through a 50- to 80-μm-diam orifice causes it to break up into uniform drops. Applying an elastic wave onto a capillary tube through which the fluid jet travels enhances this drop-producing process significantly. The drops are charged as they break off from the jet and an electrostatic field directs them to the desired location (onto a catcher or substrate). This "continuous" type system (shown in Fig. 6) produces drops with a trajectory that is controlled by their applied charge.

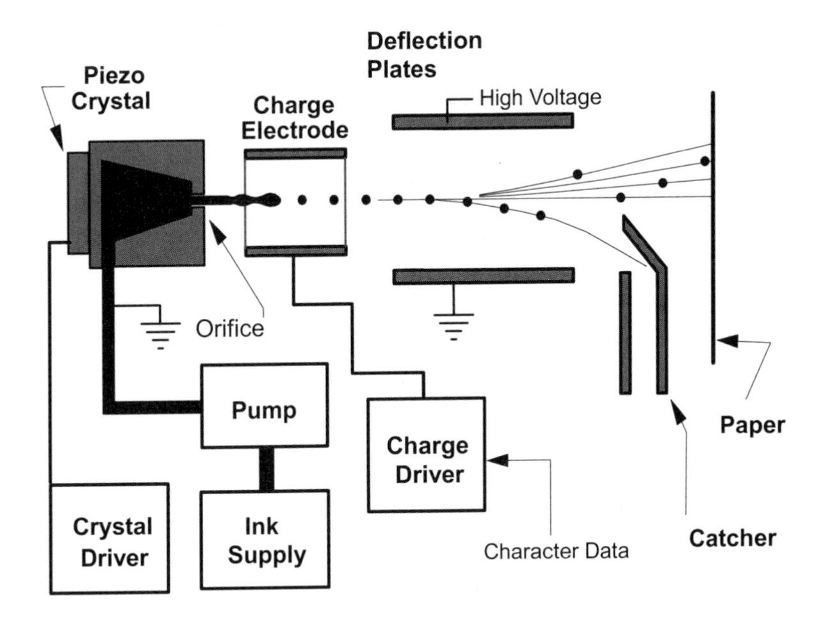

Figure 6: Continuous type ink-jet printing. (Courtesy of D. Wallace, MicroFab Technologies, Inc.)

Figure 7 shows a photo-micrograph of a 50-μm-diam jet of water breaking up into 100-μm-diam drops at a rate of 20 Kdrops/sec. Microdrop ejection can also be induced by a pulsed piezoelectric actuator that causes pressure/velocity transients to produce drops on-demand (DOD) [Wallace, 1989]. Figure 8 shows type 5-μm-diameter drops of ethylene glycol generated at a rate of 2-Kdrops/sec. DOD systems are used primarily in office printers, whereas continuous systems are widely used for product labeling.

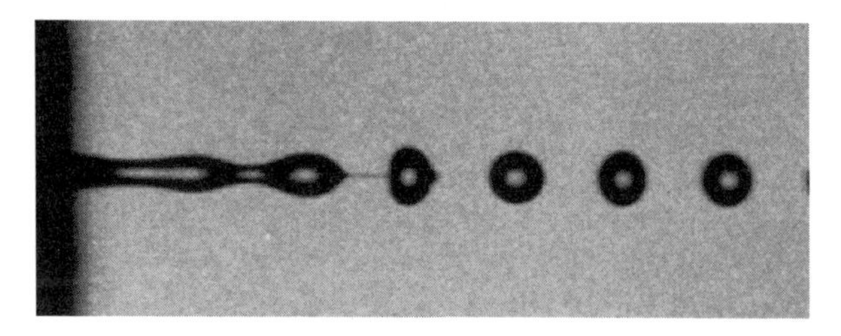

Figure 7: A 5-μm-diam water jet breaking up in continuous mode into 100-μm-diam droplets. (Courtesy of D. Wallace, MicroFab Technologies, Inc.)

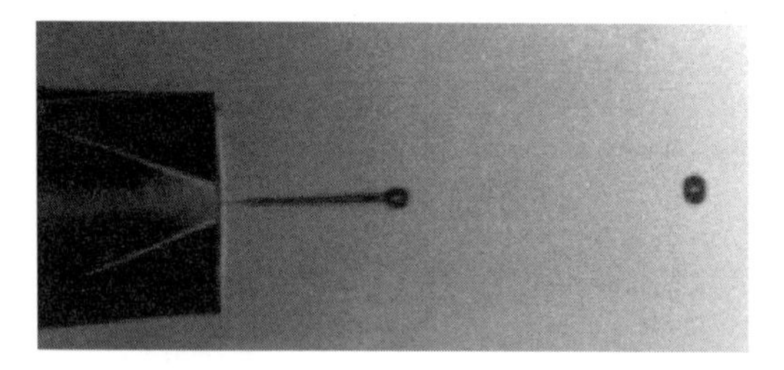

Figure 8: DOD generating 50-μm-diameter drops of ethylene glycol from a 50-μm orifice. (Courtesy of D. Wallace, MicroFab Technologies, Inc.)

Forming microscale structures using ink-printing enables, production of unique electroded EAP materials with complex shapes. Electroded actuators can be "printed" in the form of films, arrays, and fibers, and special jet-heads can be designed to address specific fabrication requirements. Often the jet can consist of a single-channel drop-on-demand ink-jet [Wallace, 1989]. Individual spots of each material can be printed and overprinted onto substrates, and the spot characteristics can be monitored as a function of the jet parameters. Heated systems can be employed to lower the viscosity of the polymer that is being dispensed.

Ink-jet printing has been used to directly deposit patterned luminescent-doped polymers and has been used to successfully fabricate organic light-emitting diodes (OLED) from ink-jet-deposited organic films [Hebner et al., 1998]. Ink-jet printing has also been applied to fabricate devices where conductive polymers were patterned in 2D planar configurations [De Rossi,

1996; and McDiarmid, 1996]. Recently, Pede et al. [1998] reported an ink-jet stereolithographic technique that was used to construct conjugated polymer devices. Using this technique, the polymer dissolved in a volatile solvent (chloroform and trichloroethylene) and ejected by the printing head drop-by-drop onto a substrate that is made of alumina. As the drops reach the surface, the solvent evaporates very quickly, leaving a solidified polymer. In the case of polymers that are not soluble, an alternative deposition method can be employed in which the ejected liquid is a monomer solution and the liquid is exposed to oxidant vapors to induce both polymerization and doping of the monomer pattern. This simple technique allows the deposition of polymeric patterns onto a wide variety of different surfaces while providing the resolution that can be obtained by the print head. By simply depositing different layers of polymer, one on top of the other, this technique also enables the fabrication of 3D structures.

14.4 Electroding Techniques

Electrodes that are applied onto EAP materials need to be capable of being easily deformed under both compression and tension while sustaining large strains over numerous cycles. Such electrodes need to have high conductivity and preferably need to be ultrathin to avoid constraining the activated EAP. Electroding techniques that are used for such materials as piezoelectric ceramics and PVF_2 have also been adapted to EAP. These techniques include lithography, deposition etching, ink-jet printing, and self-assembled monolayer, and they can be used to electrode both planar and cylindrical configurations. Methods of 2D control of EAP surfaces can be applied using synthesized conductive polymer electrodes. A dipping machine can be used for rapid fabrication of multiple elements simultaneously with nonconductive layers to prevent electrical crosscoupling. Further, patterns can be applied to surfaces of materials using microstamping that contains 2D patterns. In lithography, a layer of photoresist can be applied to the polymer wafer and processed to mask the desired electrode patterns.

Complex electroding can be applied using photolithographic techniques. A schematic description of a basic process is shown in Fig. 9, where a substrate (a) is coated with the layer of material (b) that is to be patterned. This coating is often an evaporated metal film that is used for interconnects, but with slight variations to the process it can be a polymer or inorganic oxide dielectric layer for insulation or actuation. Photoresist, a photo- or e-beam sensitive polymer, is deposited next, (c), and it is exposed through a patterned mask, (d). Then it is "developed" to wash away the area that has been solubilized by exposure to light, (e). At the final stage, wet chemicals or plasma are used to etch the patterns of the desired material, (f), and then the photoresist (g) material is removed. Numerous ingenious variations of this process have been introduced to allow preparation of suspended membranes detached from the substrate as well as other specialized structures.

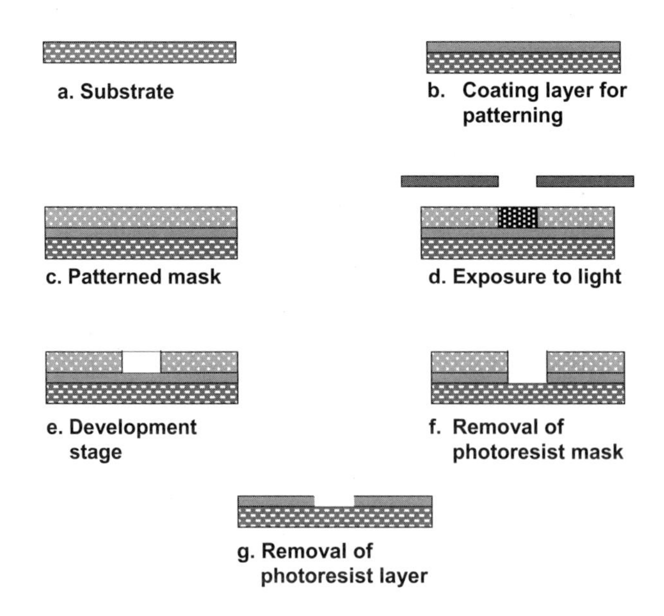

Figure 9: Method of constructing multilayers and electroding.

Generally, it is preferable to use flexible electrodes to avoid the fatigue difficulties associated with the use of metallic electrodes, to allow easy interconnection to electronic drive circuits, and to simplify the application of the excitation field. For this purpose, conductive polymers offer attractive characteristics of flexibility and high conductivity. In recent years, the electrostatic layer-by-layer self-assembly technique has been reported as a good method to achieve this aim. The control of the surface architecture of electrodes at the molecular scale can be performed by self-assembly techniques and Langmuir-Blodgett (LB) technique [Petty, 1996; and Ulman, 1991]. Conductive polymers can be electrochemically polymerized, electroded, and passivated in successive steps to produce complete microscale devices [Smela et al., 1993]. Gel microactuators have been directly polymerized using similar techniques [Beebe et al., 2000], where the polymer itself was made photosensitive and exposed through a mask. For this purpose, polypyrrole (PPy) can be readily electropolymerized in the shape of any desired planar electrode.

In recent years, several patterning methods have been reported for conjugated polymers [Smela, 1999]. Earlier techniques included conventional optical lithography [Abdou et al., 1991] and photoelectrochemical polymerization [Yoshino et al., 1990]. Bard et al. [1989] developed a direct-patterning method based on the use of selective aniline polymerization with a scanning electrochemical microscope (SECM) was. However, the pattern size was limited and the deposited polymer patterns could not be aligned with other microstructures. Using electron beam exposure (at 50 keV in a scanning electron microscope) of a water-soluble polyaniline resist [Angelopoulos et al., 1993] were able to obtain 1-mm-wide features. By utilizing the conductivity loss in a conjugated polymer following argon ion implantation Schiestel and his coresearchers [Schiestel et al., 1994] fabricated microstructures of copper on polypyrrole using an electroplating technique.

Ultraviolet (UV) lasers have been used widely for the patterning and processing of nonconducting polymers [Brannon, 1997]. However, their application to conjugated polymers has been more limited [Taguchi, 1988]. To overcome this limitation, Van Dyke and coresearchers [1992] used 308-nm radiation from a xenon chloride (XeCl) excimer laser and a mask to pattern 50-μm-diam holes in polyaniline. Further, using 442-nm radiation from a helium-cadmium (He-Cd) laser, Abdou and coresearchers patterned thin films of poly(3-hexylthiophene) by direct writing [1992].

Applying complex pattern electrodes to the surface of fibers can produce an element of artificial muscle. A potential fiber with helical electrodes is shown in Fig. 10. A pair of electrodes is helically applied to the fiber with their ends terminating separately at the two ends of the fiber to allow activation of the EAP fiber from its two poling ends. Such EAP fibers take advantage of polymers' resilience, large strain actuation, low mass, distributability, packageability, and high inherent vibration damping. Constructing an array of such fibers and interrogating them individually allows control of a membrane surface shape. An example is shown schematically in Fig. 11 where an array of fibers is shown to provide a method of controlling the focus of a deformable mirror. Under electrical excitation, such a helically electroded composite EAP fiber (HECEF) actuator [Bar-Cohen et al., 1999] would contract similar to a biological muscle. Coulomb forces induce the contraction of the fiber between the two helically wound electrodes [Kornbluh et al., 1998]. In contrast to the piezoelectric effect, where the displacement changes linearly with the electric field, this electroactive phenomenon (i.e., electrostriction and/or electrostatics) is related to the square of the field. The use of fiber bundles provides redundancy, increased reliability, and a large induced actuation force. Making helical shape electrodes is a challenge that may be accomplished by "macro-MEMS" electroding as shown in Fig. 12 [Jackman, 1997]. Other methods of applying helical electrodes may include the application of parallel strips and twisting of the fiber after it is electroded.

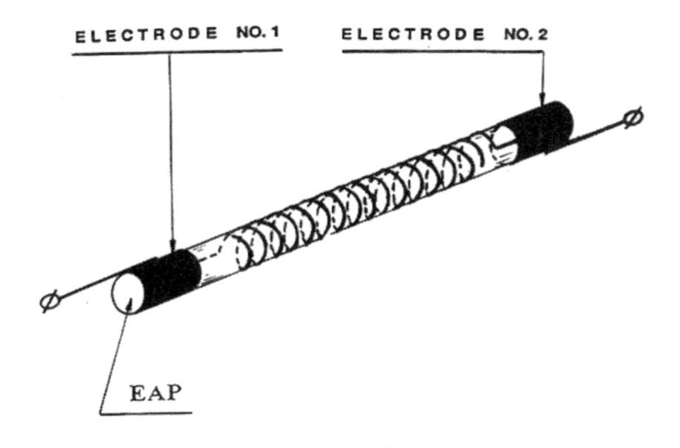

Figure 10: Helically electroded EAP fiber offers an electrically contractile actuator.

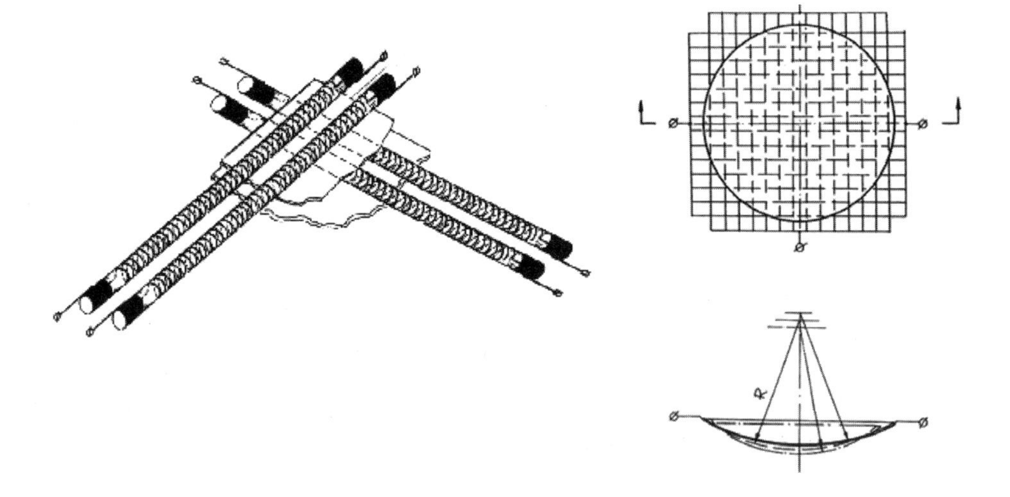

Figure 11: Schematic view of the concept of surface shape control using helically electroded EAP fibers as actuators.

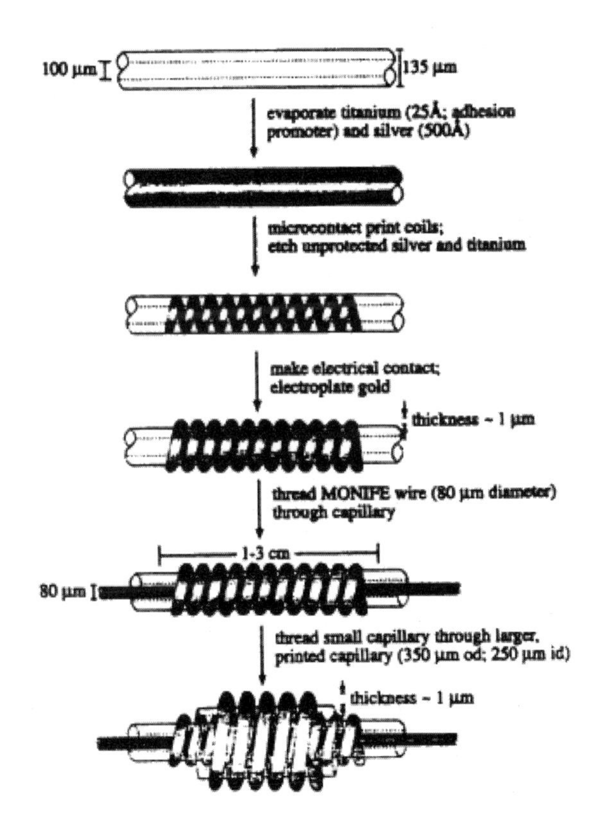

Figure 12: Potential macro-MEMS electroding configurations [Jackman, 1997]. (Copyright 1997, IEEE.)

14.4.1 Ionic Self-Assembled Monolayering (ISAM) and Thin Films

The technique of ionic self-assembly was originally developed by Decher and Hong [1991] for polyelectrolytes, and later extended to doped conjugated polymers [Fou et al., 1996; & Vargo et al., 1995], fullerenes [Mirkin and Caldwell, 1996] and other materials. This method involves processing multifunctional nanoparticle/polymer multilayer thin films and it can be used to produce combined active piezoelectric and conducting thin films. It allows the construction of multilayer assemblies by consecutively alternating adsorption of multiply charged cationic and anionic species. This technique has been reported as a simple, inexpensive, and effective process of controlling nanoscale molecular structure and macroscopic material properties that allows the application of various anionic and cationic molecular species. Generally, self-assembled thin films exhibit useful properties such as quantum-size effects, controllable thickness of films, desirable surface roughness, well-defined patterning surfaces, and tunable layer architectures [Kleinfield and Ferguson, 1994]. An important aspect of this processing technique is that it produces layers with properties that are independent of the nature, size, and topology of the substrate. Further, it allows the creation of highly tailored thin film polymers with a nearly unlimited range of functional groups incorporated within the structure of the film. In recent years, the process of self-assembling was used to apply various multilayers including polyelectrolytes, metal colloids, biological molecules, dyes, conducting polymers, and light-emitting polymers [Ferreti et al., 2000; Conticello and Qu, 1999; and Zhai et al., 2000;]. The use of self-assembled mono- and multi-layers (SAMs) is increasing rapidly in various fields of research, and this applies especially to the construction of biosensors [Wink et al., 1997].

The basic concept of ISAM is shown schematically in Fig. 13 based on Liu et al. [1999]. On the left, a substrate surface is thoroughly cleaned, and functionalized to effectively charge (negative) the outermost surface layer. This net surface-charged substrate is dipped into a solution containing water-soluble "cation" polymer molecules that have net positively charged functional groups fixed to the polymer backbone. Because the polymer chain is flexible, it is free to orient its geometry with respect to the substrate so a relatively low energy configuration is achieved. According to Claus and his researchers' simple model [Liu et al., 1999; and Rosidian et al., 1998], some of the positively charged functional groups along the polymer chain experience attractive ionic forces toward the negative substrate, and the polymer chain is bent as shown in response to those forces. The net negative charge on the substrate is thus masked from other positive groups along the polymer chain. Due to the fixed positive functional groups bonded to the substrate surface, these groups feel a net force and therefore they move away from that surface. In the process, they form a net positive charge distribution on the outermost surface of the coated substrate. Since the total polymer layer is neutral, negative charges with relatively loose binding to the polymer network pair up with positive ions. Subsequent polyanion and polycation monolayers are added to produce the multilayer structure shown.

The properties of the multilayer thin films fabricated using this method may be determined by both the chemistry of the individual monolayers and the physical ordering of the multiple monolayers. The electrostatic interactions between the polyion in solution and the surface are the key to the final structure of the polyion layered thin film. However, secondary, shorter range forces also play a role in determining the film thickness, the final morphology of the film, the surface properties, and in some cases they can determine whether or not stable multilayers form at all. These secondary interactions can also play a role in the selective deposition of polymers on surfaces, the formation of acentric polar structures, and the nature of permeation and ion transport within the film.

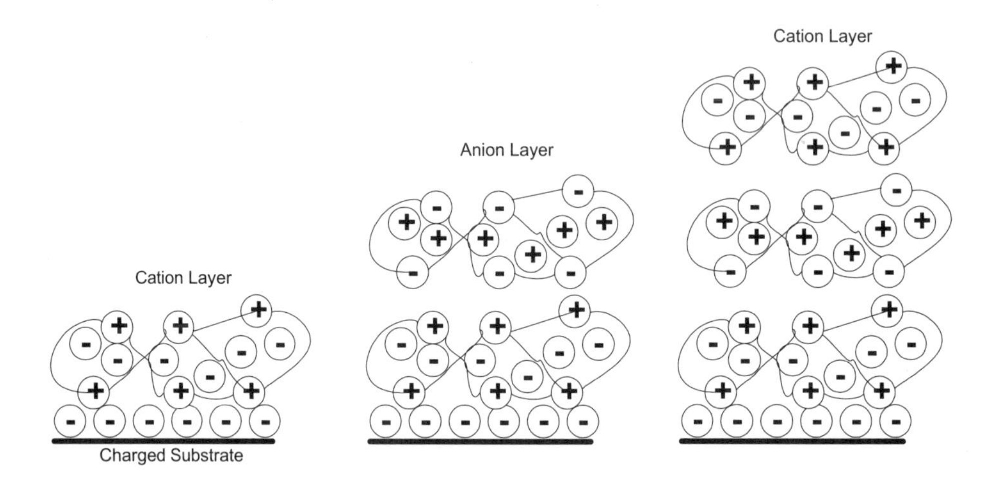

Figure 13: ISAM buildup of multilayer assemblies by consecutive adsorption of anionic and cationic molecule-based polyelectrolytes.

In the creation of two- and three-dimensional microstructures, layer-by-layer systems become the basis for patterned devices, encapsulated particles, and dimensional objects. The ability to direct the deposition of layer-by-layer films onto specific regions of a surface was introduced by Hammond and Whitesides [1995] and Clark et al. [1997] as a means of creating patterned films. To achieve a chemically patterned substrate, self-assembled monolayers may be microcontact printed [Kumar et al., 1994], or applied to surfaces using writing or other techniques to form micron scale patterns. An ionizable surface group such as an acid terminated SAM (COOH) acts as a site for polyion deposition, and an oligoethylene oxide (EG) or polyethyleneoxide surface acts as a resist to deposition of charged groups in an aqueous solution. The result is the buildup of uniform, regular microstructures on the surface.

The creation of 3D microstructures with controlled, ultrathin thickness has been investigated by Ghosh and Crooks, [1999] in the patterned assembly of alternating polyanhydrides and polyamidoamine dendrimers utilizing chemically patterned surfaces as templates. In this layer-by-layer process, the ionic interactions are replaced with covalent bonds and cross-links are formed at a high temperature between the multiple functional groups of the multiply branched dendrimer and the anhydride groups, to form polyamide networks. On the other hand, Huck et al. [1999] have recently illustrated that these types of structures can be used as effective etch resists, and may actually be removed from the surface by etching away an underlying gold layer, obtaining dimensional micron-sized objects with excellent mechanical integrity. Chemical sensors and other molecular devices can be fabricated by this technique. Electroactive nanocomposite ultrathin films of polyaniline (PAN) and isopolymolybdic acid (PMA) were fabricated by a self-assembling method to produce a sensor for humidity, NO_2 and NH_3 [Li et al., 2000].

Heterostructures of multilayer thin films comprised of electroactive polymers such as polypyrrole, poly(thiophene acetic acid), sulfonated polyaniline, poly(pyridinium acetylene), and a poly(p-phenylene vinylene) precursor have been successfully fabricated via spontaneous self-assembly of conjugated polyions onto a substrate [Collard and Sayre, 1997; Sayre and Collard, 1997; and Gorman et al., 1995]. This technique was also used to form electroluminiscent polymers with a layer-by-layer film structure [Ferreira et al., 1995; & Gao et al., 1997] and to improve the transport properties at the electrode-polymer interface in LEDs [Onitsuka et al., 1996]. Recently, an ultrathin self-assembled layer of PAN has been used to control the charge injection and electroluminescence efficiency in polymer light-emitting diodes [Ho et al., 1998]. It was also reported that this method improves the stability of organic pigments in the films leading to a photoluminescence enhancement compared with spin-cast films [Sun et al., 1999].

14.4.2 Langmuir-Blodgett (LB) Technique

In addition to the self-assembly method for producing nanostructures and molecular organization of functional polymers, one can also use the Langmuir-Blodgett (LB) technique [Petty, 1996; and Ulman, 1991]. The LB technique was developed in the 1930s as a popular method for depositing organic thin films on a solid substrate and is widely used for the fabrication of biosensors. It essentially exploits the amphiphilic properties of certain molecules, such as phospholipids and fatty acids. A monomolecular layer of phospholipid can be spread at the air-water interface in a trough (Langmuir trough) filled with water or an electrolyte solution. When a film is compressed using a barrier, the phospholipid molecules are compacted and the hydrophilic polar head groups are immersed in water (the subphase), while the hydrophobic nonpolar hydrocarbon chains are pointing upward into the air. The film can be transferred to a solid support, such as a glass substrate coated with a thin and transparent layer of metal, which becomes one of the two electrodes. The second electrode can also be evaporated onto the LB film. Multi-layered LB films can be engineered with various orientations (Fig. 14).

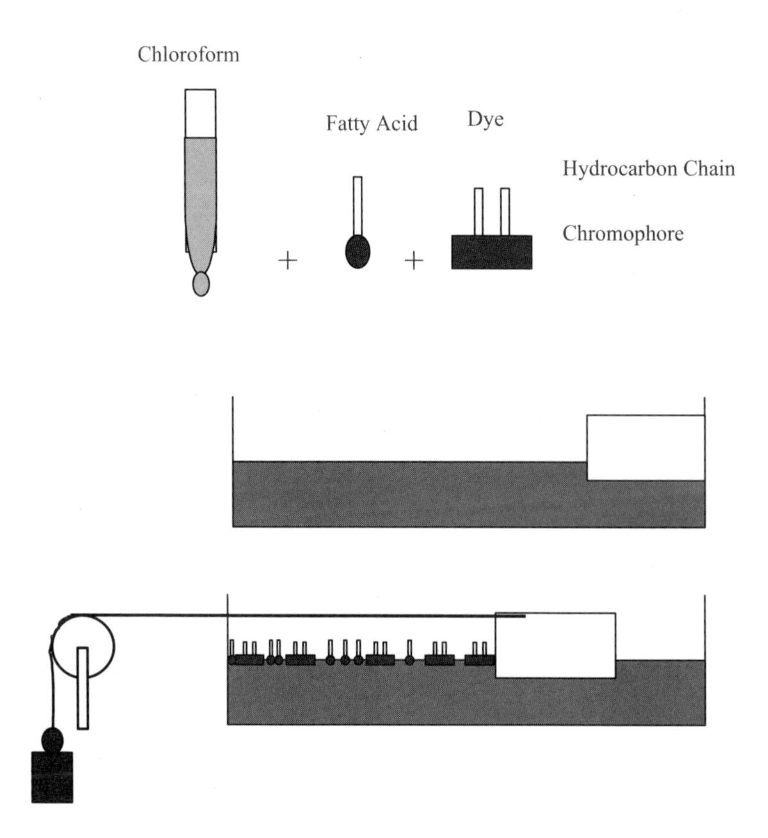

Figure 14: LB technique of forming multiple orientated layers of organic thin films on glass substrate. Dyestuff dissolved in chloroform and mixed with fatty acid, is spread on air-water interface on Langmuir trough. Mechanical compression forms compact monolayer. (Reproduced from Hong [1999] by permission of Elsevier.)

The Langmuir-Blodgett (LB) technique offers the possibility to fabricate highly ordered films with monolayer-by-monolayer control of the thickness and with low surface roughness. This technique can be used to develop composite materials with the advantage of the intimate nature of mixing that is possible at the molecular level. This can be exploited either in the form of a mixed Langmuir film containing both species or as an ordered molecular laminate formed by controlled sequential deposition of the two species. Such intimate interaction between the active groups and the matrix may give rise to more robust materials and also to other useful properties as far as composites are concerned. Examples of constructing composites of conducting polymers were reported by [Srinivasan and Jing, 1998].

To apply this technique it is necessary to have the appropriate materials with the proper balance between hydrophilic and hydrophobic properties and also between rigid (shape-persistent) and flexible moieties to facilitate the formation of a stable, liquid, crystallinelike monolayer phase at the air-water interface. Examples of such materials include fatty acids, phospholipids, hairy rod polymers, and some polymers with long hydrophobic side chains. A more diverse set of materials such as nanoparticles, porphyrins, phthalocyanines, oligothiophenes, proteins, and many others form stable monolayers at the air-water interface. Some of these materials can be transferred to solid substrates as LB monolayers, sometimes even as multilayers, and many can be co-transferred with transferable molecules to produce composite LB multilayers. Langmuir monolayers of alkyl-substituted monomers such as pyrrole, aniline, and thiophene have been extensively studied in order to understand the packing structures and electronic properties after they are transferred to solid substrates [Iyoda et al., 1987; Granholm et al., 1997; Schmelzer et al., 1993; Dabke et al., 1998; Kim et al., 1999]. Also, this technique has been investigated to control the molecular architecture and superstructure of suitably modified conducting polymers [Tabuchi et al., 1999; Ochiai et al., 1998; Rikukawa and Rubner, 1998].

14.5 System Integration Methods

EAPs are typically fabricated as freestanding sheets. For many existing macroscopic applications these free-formed materials can be used to assemble functional devices using mechanical clamping to integrate fixtures, actuators, and sensors. To fully realize the potential benefit of EAPs, efficient fabrication and integration techniques on both the micro- and macro- scales are needed to produce functional systems. This process may require introducing patterns of electrodes to enable specific deformations. LIGA techniques offer potential capabilities for the integration of miniature EAP structures, as well as producing monolithic components and systems for smart structures. Integrated 3D EAP functional structures could realize complex actuation modes for such items as active legs, wings, and fins.

14.5.1 LIGA

The acronym LIGA comes from the abbreviation of the German words Lithographie, Galvanoformung, and Abformung, which mean lithography, electroplating and molding. Ehrfeld introduced the LIGA technique in the early 1980s at the FZK center in Germany [Becker et al., 1986 and Megtert et al., 1996]. This process is a powerful tool for the fabrication of pseudo-3D microstructures and microsystems with high aspect ratios (up to 100). The combination of LIGA and x-ray lithography is used increasingly in microtechnologies [Menz, 1996] particularly to: fabricate high-aspect-ratio structures, and process materials other than silicon. X-ray LIGA operates like modern deep-etch technology [Bustillo et al., 1998] to produce high-aspect-ratio structures with perpendicular sidewalls and low surface roughness that are superior to any other known techniques. This method needs synchrotron radiation, however, and it is not readily available. With this technique it is possible to produce microstructures that have aspect ratios that are as high as 100—very small structures in the range of submicrons and that have smooth walls of optical quality (having surface roughness that is smaller than 50 nm).

In the process as originally developed, a special kind of photolithography using x rays (x-ray lithography) is used to produce patterns in very thick layers of photoresist. The lithography steps involve subjecting a thick photoresist layer that is sensitive to x rays (usually PMMA or epoxy [Lorenz et al., 1998]). This photoresist covers a conductive substrate [Fig. 15(a)]. The irradiated resist structure is then developed by a wet organic chemical rinse, dissolving the resist regions of lower molecular weight [Fig. 15(b)]. Then, in the electroforming step, the formed pattern is electroplated and filled with metal [Fig. 15(c)]. The produced metal structures can be the final product; however, it is common to produce a metal mold [Fig. 15(d)] that can be used as a tool in a molding step. This mold is then filled with a suitable material, such as a plastic [Fig. 15(e)], to produce the finished product [Fig. 15(f)].

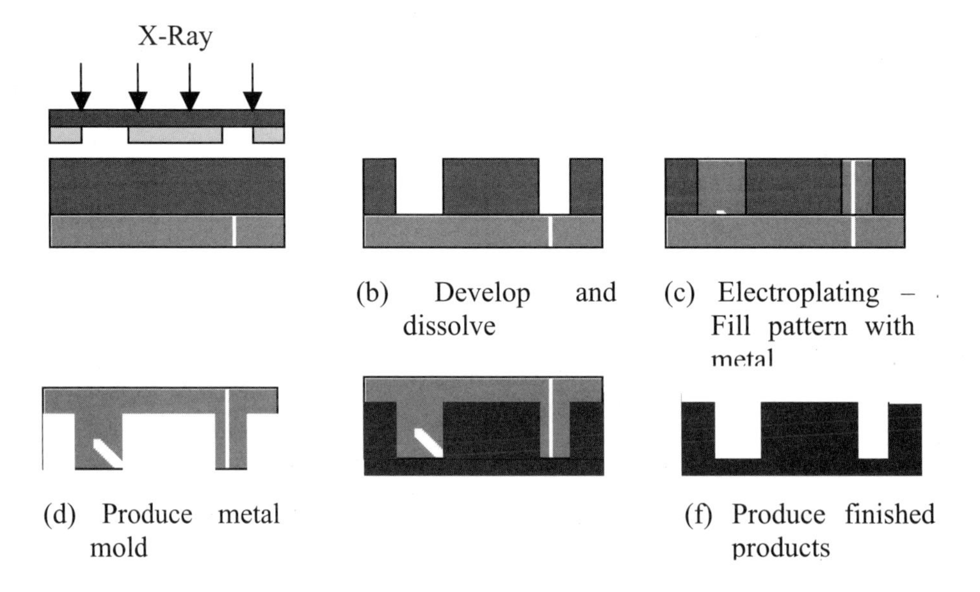

Figure 15: X-LIGA photolithographic process.

As the synchrotron source makes LIGA expensive, alternatives are being developed. These include high-voltage electron beam lithography, which can be used to produce structures of the order of 100-μm high, excimer lasers capable of producing structures of up to several hundred microns high, and deep UV lithography. Deep UV is a new technique that is still under development. It needs special resists allowing partial transparency to UV light but it also sensitive enough so that high-resolution structures can be produced. The sensitivity of a resist can be improved many-fold using chemical amplification, a process where a single photon initiates a cascade of chemical reactions to increase scissioning (for positive resists), or cross-linking (for negative resists). Negative resists such as SU-8 are proving to be very successful for deep-UV lithography. Currently, using deep UV lithography aspect ratios of 10 can be achieved.

Those techniques have been applied in the development of microfluidic devices such as microcolumns, micropumps [Schomburg et al., 1994], microvalves [Ruzzu et al., 1998a], tapered microvalves [Tsuei et al., 1998], bearings [Samper et al., 1998], microfluidic calorimeters [Köhler et al., 1998] and also in MEMS technology [Koch et al., 1999]. X-ray/UV LIGA may provide access in the future to more biocompatible devices [Ford et al., 1998]. Examples of such biological applications include microtechnologies like DNA chip sequencing [Woolley and Mathies, 1995] and microtechnical interfaces to neurons. Medical applications of x-ray LIGA include [Wallrabe et al., 1998] like microturbines [Wallrabe et al., 1996] and cutters for minimally invasive therapy [Ruzzu et al., 1998b], fluid-driven micromotors [Gebhard et al., 1996], or reconstruction of tissues [Weibezahn et al., 1995].

14.5.2 Microstereolithography

Several techniques are being developed for the production of 3D microstructures that can be integrated to produce miniature systems [Becker et al., 1986 and Beluze et al., 1999]. Even though the LIGA process has been used for some years, due to its limited industrial accessibility and high operational cost, it has not found a large number of industrial applications. In recent years, an alternative process to produce high-aspect-ratio and complex 3D parts has been introduced, and this processing method is called microstereolithography, (μSL). This method (Fig. 16) can be used to produce sophisticated 3D parts by scanning a UV beam on the liquid monomer resin and curing the resin layer-by-layer into solid polymer. μSL is an additive process, which enables one to fabricate high-aspect-ratio microstructures with novel smart materials [Ikuta et al., 1993] however the limitation of this technique is its low accuracy [Bertsch et al., 1999].

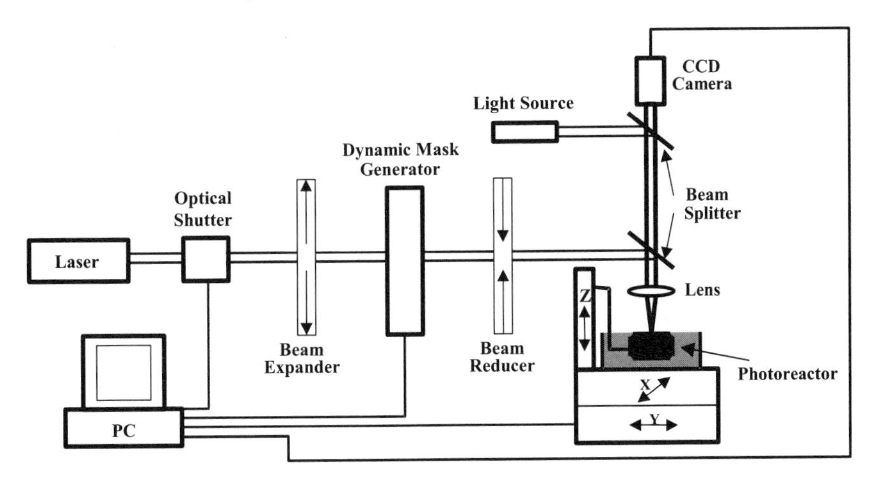

Figure 16: A schematic description of a microstereolithography system.

Improving the resolution of the stereolithography process has led to the design of different microstereolithography processes [Zissi et al., 1996]. The developed microstereolithography process is based on a vector-by-vector tracing of each layer, and it is obtained by moving a focused light beam onto the surface of a photopolymerizable liquid [Ikuta et al., 1994; and Nakamoto et al., 1996]. Recently, an "integral" process was conceived in which a complete layer was built in a single irradiation process [Bertsch et al., 1997]. Combining microstereolithography and conventional microstructuring techniques opens the way to integrating new functions on planar structures. Moreover, the direct processing of 3D, complex shaped polymer microstructures by microstereolithography on top of microparts patterned by silicon-based technologies eliminates the manipulations that are associated with microassembling separate components.

The fabrication of micropolymeric parts and subsequent electroplating of micrometallic parts have been explored by several investigators including Ikuta et al. [1994] and Maruo et al., [1997]. Using these methods microceramic structures have been fabricated successfully and adaptation to EAP will require future studies.

14.6 EAP Actuators

While demonstrating the responsiveness of polymer systems can be straightforward, in many cases making a useful actuator poses a number of technological challenges. Some of the issues are associated with the packaging of the material to assure its operation in a useful environment. A concomitant challenge is to make sure the package is compliant enough to minimize hysteresis and other losses in efficiency. The change in volume that results from the activation of ionic gels can be changed to linear motion through systems of rods and pistons [Tatara, 1987], or if a volume-responsive gel is packaged in an elastomeric envelope, a musclelike actuator can be achieved [Ricard et al., 1998]. An electrically actuated sphincterlike device makes direct use of the same type of volume transition [de Rossi et al., 1985]. One example of a linear actuator uses an assembly of PANi fibers [Mazzoldi et al., 1998], whereas Pelrine et al. [1997] demonstrate films, stacks, and tubes [Chapter 16].

Small-scale and even micro-scale devices are a natural application for EAPs for which the electric fields can be quite high, the absolute forces small, and response times quick. Individually addressable microactuators have been used to manipulate micrometer-sized objects as reported by Jager et al. [2000] and Tadokoro et al. [1999]. One such example is a six-degree-of-freedom micromanipulator that was made from IPMC based actuator strips [Chapter 13]. Rashidian and Allen [1991] used on-chip poling to fabricate piezoelectric microsensors and actuators. However, they observed that conventional piezopolymers are not compatible with silicon-based processing. Higher-temperature materials hold some promise to overcoming this incompatibility [Ounaies et al., 1999].

To obtain large strains, planar bilayer or sandwich structures are frequently employed to turn volume expansion of, for example, conductive polymers into bending deformation [Otero and Sansiñena, 1998]. Multiple benders (see Fig. 17) can be assembled into bellowslike structures to amplify the deformation [Bohannan et al., 1999]. Polypyrrole sandwich structures were constructed to take advantage of the fact that the polymer can be driven with either polarity [Sansiñena et al., 1997; Lewis et al., 1999].

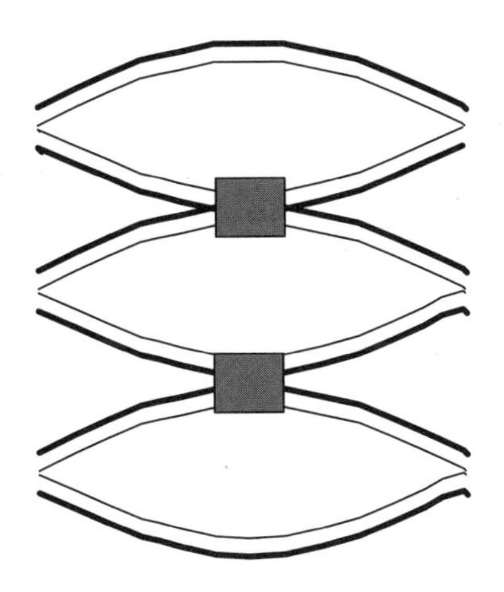

Figure 17: A bellow-shaped actuator using multiple benders can provide accumulative actuation.

14.7 Concluding Remarks

The technologies of synthesizing, producing, and shaping EAP materials are expected to continue to evolve as improved or new materials are developed. In this chapter, the authors took the challenge of sorting and documenting the reported as well as the potential techniques. This chapter covered the topic by following the sequence of fabricating an EAP actuator from the stage of synthesizing and processing the material, fabricating various shapes (emphasizing films and fibers), electroding, system integration, and last, providing a brief description of the methods for fabricating actuators. Efforts were focused on methods that were reported as useful, specifically to making EAP materials. Computation chemistry may play a greater role in future development of such materials toward optimization of their performance and efficiency. The predicted materials will require effective synthesis and processing techniques. Methods that are commonly used to fabricate polymers will continue to serve as a resource for the fabrication of EAP materials. Potentially, such methods as ink-jet printing may play a role in rapid prototyping and production of miniature devices in 3D details. Issues of fabrication scaling and mass production are expected to be addressed in future years as the field evolves.

14.8 References

Abdou, M. S. A., Diaz-Guijada, G. A., Aroyo, M. I., and Holdcroft, S. (1991), Chemistry of Materials 3, p.1003.

Abdou, M.S.A., Xie, Z., Leung A., and Holdcroft, S. (1992), "Laser, direct-write microlithography of soluble polythiophenes," *Synthetic Metals* 52, pp. 159-170.

Andreatta A., Heeger A.J., Smith P., (1990), "Electrically-conductive polyblend fibres of polyaniline and poly(p-phenylene terephthalamide)," *Polymer Communications*, 31, pp. 275-278.

Angelopoulos, M., Patel, N., Shaw, J.M., Labianca, N.C., and Rishton, S.A. (1993), "Water soluble conducting polyanilines: Applications in lithography," *Journal of Vacuum Science & Technology B: Microelectronics and Nanometer Structures* 11(6), pp. 2794-2797.

Bar-Cohen Y., Leary, S., and Knowles, T. (1999), "Helical electroded composite EAP fiber (HECEF) actuators emulating biological muscles," (1999), *New Technology Report*, Docket No. 20909, Item No. 0505b, November 23.

Bard, A.J., Fan, F.R., Wuu, Y.M. (1989), "High resolution deposition of polyaniline on pt with the scanning electrochemical microscope," *Journal of Electrochemical Society*, 136(3), pp. L885-886.

Becker, E.W., Ehrfeld, W., Hagmann, P., Maner, A., and Münchmeier, D. (1986), "Fabrication of microstructures with high aspect ratios and great structural heights by synchrotron radiation lithography, galvanoforming, and plastic moulding (LIGA process)," *Microelectronic Engineering* 4(1) pp. 35-56.

Becker, H., and Heim, U. (2000), "Hot embossing as a method for the fabrication of polymer high aspect ratio structures," *Sensors and Actuators A: Physical*, 83 (1-3), pp. 130-135.

Becker, H., Dietz W., and Dannberg, P. (1998), "Microfluidic manifolds by polymer hot embossing for -TAS applications," in *Proceeding of Micro-TAS '98*, Banff, Canada, pp. 253-256.

Bedal, B., and Nguyen, H. (1996), in *Stereolithography and Other RP&M Technologies* (Jacobs, Paul F., Ed.). Society of Manufacturing Engineers, Dearborn, Michigan, p.156.

Beebe D.J., Moore, J.S., Bauer, J.M., Yu, Q., Liu, R.H., Devadoss, C., and Jo, B-H. (2000), "Functional hydrogel structures for autonomous flow control inside microfluidic channels," *Nature*, 404 (6778), pp. 588-590.

Beluze, L., Bertsch, A., Renaud, P. (1999), "Microstereolithography: a new process to build complex 3D objects," *Symposium on Design, Test and Microfabrication of MEMs/MOEMs*, Proc. SPIE Vol. 3680, pp.808-817.

Bertsch, A., Jézéquel J.Y., André, J.C. (1997), "Study of the spatial resolution of a new 3D microfabrication process: the microstereo-photolithography using a dynamic mask-generator technique". *J. Photochem. Photobiol. A: Chem.* 107, pp. 275-281.

Bertsch, A., Lorenz, H., and Renaud, P. (1999), "3D microfabrication by combining microstereolithography and thick resist UV lithography," *Sensors and Actuators: A,* 73, pp.14-23.

Bohannan, G., Schmidt, H., Brandt, D., Mooibroek, M., (1999), "Piezoelectric polymer actuators for active vibration isolation in space applications," *Ferroelectrics*, 224(1-4), pp. 639-645.

Bohon, K., Krause, S. (1998), "An electrorheological fluid and siloxane gel based electromechanical actuator: working toward an artificial muscle," *Journal of Polymer Science part B: Polymer Physics*, 36(6), 1091-1094.

Brannon, J.H. (1997), "Excimer-laser ablation and etching," *IEEE Circuits and Devices Magazine*, 6(5), pp. 18-24.

Burrows, P. E., Bulovi, V., Gu, G., Kozlov, V., Forrest, S. R., Thompson, M. E. (1998), "Light emitting devices using vacuum deposited organic thin films," *Thin Solid Films*, 331, (1-2), pp. 101-105.

Bustillo, J.M., Howe, R.T., Muller, R. S. (1998), "Surface micromachining for microelectromechanical systems," *Proceedings of the IEEE*, 86(8), pp. 1552-1574.

Calvert, P., Liu, Z. (1997), "Extrusion freeform fabrication of bonelike mineralized hydrogels and musclelike actuators," *Solid Freeform Fabrication Proceedings* (D.L. Bourell, J.J. Beaman, R.H. Crawford, H.L. Marcus and J.W. Barlow, eds), U. Texas, Austin pp. 11-15.

Casalino G., Chiarelli. P., De Rossi, D., Genuini, G., Morasso, P., Solari, M. (1991), *Progress in the design and controll of pseudomuscular linear actuator*, (Dario et. al, Eds.), Nato ASI serie, Springer Verlag, Berlin, p. 495.

Chacko, A.P., Hardaker, S.S., Gregory, R.V., Samuels, R.J. (1997), "Viscoelastic characterization of concentrated polyaniline solutions: New insights into conductive polymer processing," *Synthetic Metals,* 84(1-3), pp. 41-44.

Chiarelli P., De Rossi, D. (1996), "Bi-phasic Elastodynamics of Electron Conducting Polymers," *Polymer Gels and Networks*, 4(5-6), pp. 499-508.

Chiarelli, P., De Rossi, D., Della Santa, A., Mazzoldi, A. (1994), "Doping induced volume change in a -conjugated conducting polymer," *Polymer Gels and Networks*, 2(3-4), pp. 289-297.

Cho, H.H., Kim, K.H., Kang, Y.A., Ito, H., Kikutani, T. (2000), "Fine structure and physical properties of polyethylene/poly(ethylene terephthalate) bicomponent fibers in high-speed spinning. I. Polyethylene sheath/poly(ethylene terephthalate) core fibers," *Journal of Applied Polymer Science*, 77 (10), pp. 2254-2266.

Chowdhury, F. -U. -Z., Bhuiyan, A. H. (2000), "An investigation of the optical properties of plasma-polymerized diphenyl thin films," *Thin Solid Films*, 360(1-2), pp. 69-74.

Clark, S.L., Montague, M., Hammond, P.T. (1997), "Selective deposition in multilayer assembly: SAMs as molecular templates," *Supramolecular Science*, 4(1-2), pp. 141-146.

Collard, D. M., Sayre, C. N., (1997), "Micron-scale patterning of conjugated polymers on microcontact printed patterns of self-assembled monolayers, *Synthetic Metals*, 84(1-3), pp. 329-332.

Conticello, V.P., Qu, Y. (1999), "Synthesis and characterization of a self-assembling polypeptide block copolymer," American Chemical Society Polymer Preprints, Division of Polymer Chemistry, 40(2), pp. 1047-1048.

Dabke, R.B., Dhanabalan, A., Major, S., Talwar, S.S., Lal, R., Contractor, A.Q. (1998), "Electrochemistry of polyaniline Langmuir-Blodgett films," *Thin Solid Films*, 335(1-2), pp. 203-208.

De Rossi, D. (1996), "Polymeric Smart Sensors and Actuators," Workshop on Multifunctional Polymers and Smart Polymer Systems, Wollongong, Australia.

De Rossi, D., Mazzoldi, A. (1999), "Linear fully dry polymer actuators," *Proceedings of the Electroactive Polymer Actuators and Devices*, Conf., *6th SPIE Smart Structures Symposium*, 3669, pp. 35-44.

De Rossi, D., Parrini, P., Chiarelli, P., Buzzigoli, G. (1985), "Electrically-Induced Contractile Phenomena In Charged Polymer Networks: Preliminary Study on the Feasibility of Musclelike Structures," *Transaction of American Society of Artificial Internal Organs*, vol. XXXI, pp. 60-65.

Decher, G., Hong, J-D. (1991), "Buildup of ultrathin multilayer films by a self-assembly process, Consecutive adsorption of anionic and cationic bipolar amphiphiles on charged surfaces," *Makromolecular Chemistry Macromolecular Symposium*, 46, pp. 321-327.

Doshi, J., Reneker, D. H. (1995), "Electrospinning Process and Applications of Electrospun Fibers," *Journal of Electrostatics*, 35 (2-3), pp. 151-293.

Dreuth, H., Heiden, C. (1999), "Thermoplastic structuring of thin polymer films", *Sensors and Actuators, A* 78(2-3), pp. 198-204.

Ferreira, M., Rubner, M.F., Hsieh, B.R. (1995), "Luminescence behavior of self-assembled multilayer heterostructures of poly(phenylene-vinylene)," 328, pp. 119-124.

Ferretti, S; Paynter, S; Russell, D.A., Sapsford, K.E., Richardson, D.J. (2000) "Self-assembled monolayers: a versatile tool for the formulation of bio-surfaces," *TrAC Trends in Analytical Chemistry*, 19 (9), pp. 530-540.

Fong, H; Chun, I., Reneker, D. H. (1999), "Beaded nanofibers formed during electrospinning," *Polymer*, 40(16), pp. 5-2.

Ford, S.M., Karr, B., McWorther, S., Davies, J., Soper, S.A., Klopf, M., Calderon, G., Saile, V. (1998), "Microcapillary electrophoresis devices fabricated using polymeric substrates and X-ray lithography," *Journal of Microcolumn Separations* 10(5), pp. 413-422.

Forrest, S. R. (1997), "Ultrathin organic films grown by organic molecular beam deposition and related techniques," *Chemical Reviews*, 97(6), pp. 1793-1896.

Fou, A.C., Onitsuka, O., Ferreira, M., Rubner, M.F. (1996), "Fabrication and properties of light-emitting diodes based on self-assembled multilayers of poly(phenylene vinylene)," *Journal of Applied Physics*, 79(10), pp. 7501-7509

Frank, C.W. (1998), *In Organic Thin Films ACS Symposium Series 695*, (Frank, C.W. ed.), American Chemical Society, Washington, DC.

Gao, M., Richter, B., Kirstein, S. (1997), "White light electroluminescence from self-assembled Q-CdSe/PPV multilayer structures," *Advanced Materials* 9(10), pp. 802-805.

Gebhard, U., Günter, R., Just, E., Ruther, P. (1996), In *Proceedings of Microsystem Technologies'96*, VDE Verlag, Berlin, pp. 609-614.

Ghosh P, Crooks RM (1999), "Covalent grafting of a patterned, hyperbranched polymer onto a plastic substrate using microcontact printing," *Journal of the American Chemical Society* 121, pp. 8395-8396.

Gilbert, R.D., Venditti, R.A., Zhang, Ch., Koelling, K.W. (2000), "Melt spinning of thermotropic cellulose derivatives," *Journal of Applied Polymer Science*, 77(2), pp. 418-423.

Granholm, P., Paloheimo, J. and Stubb, H., 1997. "Conducting Langmuir-Blodgett films of polyaniline: fabrication and charge transport properties," *Synthetic Metals*, 84, pp. 783-784.

Hammond, P.T., Whitesides, G.M. (1995), "Formation of polymer microstructures by selective deposition of polyion multilayers using patterned self-assembled monolayers as a template," *Macromolecules* 28, pp. 7569-7571.

Hebner T. R., Wu C. C., Marcy D., Lu M. H., Sturm J. C. (1998), "Ink-jet printing of doped polymers for organic light emitting devices," *Applied Physics Letters*, 72(5), pp. 519-521.

Heitz, J., Li, S. T., Arenholz, E., Bäuerle, D. (1998), "Pulsed-laser deposition of crystalline Teflon (PTFE) films," *Applied Surface Science*, 125(1), pp. 17-22.

Ho, P.K.H., Granström, M., Friend, R.H. Greenham, N. C. (1998), "Ultrathin self-assembled layers at the ITO interface to control charge injection and electroluminescence efficiency in polymer light-emitting diodes," *Advanced Materials* 10(10), pp. 769-774.

Hoenich, N. A., Stamp S., (2000), "Clinical investigation of the role of membrane structure on blood contact and solute transport characteristics of a cellulose membrane," *Biomaterials*, 21(3), pp. 317-324.

Hong, F.T. (1999), "Interfacial photochemistry of retinal proteins," *Progress in Surface Science*, 62 (1-6), pp. 1-237.

Hsu, C.-H., Epstein, A.J. (1997), "Processing of poly(o-toluidine) into fibers and their properties," *Synthetic Metals*, 84(1-3), pp. 51-54.

Huang, L., McMillan, R.A., Apkarian, R.P., Pourdeyhimi, B., Conticello, V.P., Chaikof E.L. (2000), "Generation of synthetic elastin-mimetic small diameter fibers and fiber networks," *Macromolecules,* 33(8), pp. 2989-2997.

Huck, W.T.S., Yan, L., Stroock, A., Haag, R., Whitesides, G.M. (1999), "Patterned polymer multilayers as etch resists," *Langmuir,* 15(20), pp. 6862-6867.

Ikuta, K., Hirowatari, K. (1993), "Real three dimensional micro fabrication using stereo lithography and metal molding," *Micro Electro Mechanical Systems, 1993, MEMS'93 Proceedings*, An Investigation of Micro Structures, Sensors, Actuators, Machines and Systems. IEEE, pp. 42-47.

Ikuta, K., Hirowatari, K., Ogata, T. (1994), "Three dimensional micro integrated fluid systems (MIFS) fabricated by stereo lithography," *Micro Electro Mechanical Systems, MEMS'94 Proceedings*, IEEE Workshop on, pp. 1-6.

Iyoda, T., Ando, M., Kaneko, T., Othani,A., Shimidzu, T., Honda, K. (1986), "Electrochemical polymerization in Langmuir-Blodgett film of new amphiphilic pyrrole derivatives," *Tetrahedron Letters*, 27(46), pp.5633-5636.

Jackman R.J. (1997), "Fabrication and characterization of concentric cylindrical microtransformer," *IEEE Trasactions on Magnetics,* 33(4) 97, pp. 2501-2503.

Jager, E.W.H., Smela, E., Ingana, O. (1999), "On-chip microelectrodes for electro-chemistry with moveable PPy bilayer actuators as working electrodes," *Sensors and Actuators B,* B56(1-2), pp. 73-78.

Katoh T., Zhang, Y. (1999), "Deposition of teflon-polymer thin films by synchrotron radiation photodecomposition," *Applied Surface Science*, 138-139, pp. 165-168.

Kikutani, T., Radhakrishnan, J., Arikawa, S., Takaku, A. Okui, N., Jin, X., Niwa, F., Yosuke, Y. (1996), "High-speed melt spinning of bicomponent fibers: mechanism of fiber structure development in poly(ethylene terephthalate)/polypropylene system," *Journal of Applied Polymer Science*, 62 (11), pp. 1913-1924.

Kim, Y-H., Wurm, D.B., Kim, M.W., Kim, Y-T. (1999), "Electrochemistry of Langmuir-Blodgett and self-assembled monolayers of pyrrole derivatives," *Thin Solid Films*, 352(1-2), pp. 138-144.

Kleinfeld, E.R.; Ferguson, G.S. (1994), "Stepwise formation of multilayered nanostructural films from macromolecular precursors," *Science,* 265(5170), pp. 370-373.

Koch, M., Schabmueller, C.G.J., Evans, A.G.R., Brunnschweiler, A. (1999), "Micromachined chemical reaction system," *Sensors and Actuators A*, 74(1-3), pp. 207-210.

Köhler, J.M., Zieren, M. (1998), "Chip reactor for microfluid calorimetry," *Thermochimica Acta,* 310(1-2), pp. 25-35.

Kornbluh R., Pelrine, R., Joseph, J. (1998), "Electrostriction of polymer dielectrics with compliant electrodes as a means of actuation," *Sensors and Actuators A,* 64(1), pp. 77-85.

Kumar, A., Biebuyck, H.A., Whitesides, G.M. (1994), "Patterning self-assembled monolayers: applications in materials science," *Langmuir* 10, pp. 1498-1511.

Lewis, T.W., Kane-Maguire, L.A.P., Hutchison, A.S., Spinks, G.M., Wallace, G.G. (1999), "Development of an all-polymer, axial force electrochemical actuator," *Synthetic Metals*, 102 (1-3), pp. 1317-1318.

Li, D., Jiang, Y., Wu, Z., Chen, X., Li, Y. (2000), "Self-assembly of polyaniline ultrathin films based on doping-induced deposition effect and applications for chemical sensors," *Sensors and Actuators B: Chemical*, 66(1-3), pp. 125-127.

Liu Y., Rosidian, A., Lenahan, K., Wang, Y-X., Zeng, T., Claus, R.O. (1999), "Characterization of electrostatically self-assembled nanocomposite thin films," *Smart Materials and Structures* 8(1), pp. 100-105.

Lorenz, H., Despont, M., Fahrni, N., Brugger, J., Vettiger, P., Renaud, P. (1998), "High-aspect-ratio, ultrathick, negative-tone near UV photoresist and its applications for MEMS," *Sensors and Actuators, A: Physical,* 64 (1), pp. 33-39.

Maruo, S., Kawata, S. (1997), "Two-photon-absorbed photopolymerization for 3D microfabrication," *Proceedings of IEEE MEMS'97*, pp. 169-174.

Mattesa, B.R., Wanga, H.L., Yanga, D., Zhub, Y.T., Blumenthalb W. R., Hundley, M.F. (1997), "Formation of conductive polyaniline fibers derived from highly concentrated emeraldine base solutions," *Synthetic Metals,* 84(1-3), pp. 45-49.

Mazzoldi A., Degl'Innocenti, C., Michelucci, M., De Rossi, D. (1998), "Actuative properties of polyaniline fibers under electrochemical stimulation," *Materials Science and Engineering. C,* 6(1), pp. 65-72.

McDiarmid, A. (1996), International Conference on Synthetic Metals.

Megtert, S., Liu, Z.W., Kupka, R., Labeque, A., Casses, V., Basrour, S., Bernede, P. (1996), In Proceedings of AIP Conference 392, pp. 745-748.

Menz, W. (1996), "LIGA and related technologies for industrial application," *Sensors and Actuators A: Physical,* 54(1-3), pp. 785-789.

Mirkin, Ch.A., Caldwell, W.B. (1996), "Thin film, fullerene-based materials," *Tetrahedron*, 52(14), pp. 5113-5130.

Misra, S.C.K., Ram, M.K., Pandey, S.S., Malhotra, B.D., Chandra, S. (1992), "Vacuum-deposited metal/polyaniline Schottky device," *Applied Physics Letters*, 61(10), pp. 1219-1221.

Nakamoto, T., Yamaguchi, K., Abraha, P.A., Mishima, K. (1996), "Manufacturing of 3D micro-parts by UV laser induced polymerization," *Journal of Micromechanics and Microengineering*, 6(2), pp. 240-253.

Ochiai, K., Tabuchi, Y., Rikukawa, M., Sanui, K; Ogata, N. (1998), "Fabrication of chiral poly(thiophene) Langmuir-Blodgett films," *Thin Solid Films*, 327-329, pp. -.

Onitsuka, O., Fou, A.C., Ferreira, M., Hsieh, B.R., Rubner, M.F. (1996), "Enhancement of light emitting diodes based on self-assembled heterostructures of poly(p-phenylene vinylene)," *Journal of Applied Physics*, 80(7), pp. 4067-4071.

Otero T.F., Sansiñena, J.M. (1998), "Soft and wet conducting polymers for artificial muscles," *Advanced Materials,* 10(6), pp. 491-494.

Ounaies Z., Young, J.A., Harrison, J.S. (1999), In Chapter 6 in field responsive polymers, ACS Symposium Ser., (I. M. Khan and J. S. Harrison, eds), p. 726.

Pede D., Serra G., De Rossi, D. (1998), "Microfabrication of conducting polymer devices by ink-jet stereolithography," *Materials Science Engineering: C*, 5(3-4), pp. 289-291.

Pelrine, R., Kornbluh, R., Pei, Q., Joseph, J. (2000), "High-speed electrically actuated elastomers with strain greater than 100%," *Science*, 287 (5) pp. 836-839.

Peltonen, J.P.K., Paajanen, M., Lekkala, J. (1999), "Nanomechanical properties of electromechanical polypropylene films as determined by SPM," *Book of Abstracts, 218th ACS National Meeting*, New Orleans, August, pp. 22-26.

Petty, M.C. (1996), "Langmuir-Blodgett films: an introduction," *Cambridge University Press*, Cambridge.

Piqué, A., McGill, R.A., Chrisey, D.B., Leonhardt, D., Mslna, T.E., Spargo, B.J., Callahan, J. H., Vachet, R.W., Chung, R., Bucaro, M.A. (1999), "Growth of organic thin films by the matrix assisted pulsed laser evaporation (MAPLE) technique," *Thin Solid Films*, 355-356, pp. 536-541.

Pomfret, S.J., Adams, P.N., Comfort, N.P., Monkman, A.P. (2000), "Electrical and mechanical properties of polyaniline fibers produced by a one-step wet spinning process," *Polymer* 41(6), pp. 2265-2269.

Qu, Y., Payne, S.C., Apkarian, R.P., Conticello, V.P. JACS 2000.

Randall, C.A., Miller, D.V., Adair, J.H., Bhalla, A.S. (1993), "Processing of electroceramic - polymer composites using electrorheological effect," *Journal of Materials Research*, 8(4), pp. 899-904.

Rashidian, B., Allen, M. (1991), "Integrated piezoelectric polymers for microsensing and microactuation applications," *American Society of Mechanical Engineers, Dynamic Systems and Control Division* DSC, 1-6, pp. 171-179.

Reneker, D.H., Chun, I. (1996), "Nanometre diameter fibres of polymer, produced by electrospinning," *Nanotechnology*, 7(3), pp. 216-223.

Ricard, A., Tondu, B., Lopez, P., Vial, D., Conscience, M. (1998), "Artificial muscle based on polymeric contractile gels," *Mater. Tech.* (Paris), 86, pp. 43-48.

Rikukawa, M., Tabuchi, Y., Ochiai, K., Sanui, K., Ogata, N. (1998), "Langmuir-Blodgett films manipulation of regioregular poly(3-alkylthiophene)s," *Thin Solid Films*, 327-329, 469-472.

Rosidian A., Wang, Y-X., Liu, Y., Claus, R. (1998), "Ionic self-assembly of ultrahard zirconia/polymer nanocomposite thin flms," *Advanced Materials,* 10(14), 1087-1090.

Ruzzu, A., Fahrenberg, J., Heckele, M., Schaller, Th. (1998a), Microsystem Technologies 4, 128-131.

Ruzzu, A., Fahrenberg, J., Müller, M., Rembe, C., Wallrabe, U. (1998b), in Proceedings of MEMS'98, pp. 499-503.

Sachs, E., Wylonis, E., Allen, S., Cima, M., Guo, H. (2000), "Production of injection molding tooling with conformal cooling channels using the three dimensional printing process," *Polymer Engineering and Science*, 40(5), pp. 1232-1247.

Samper, V.D., Sangster, A.J., Wallrabe, U., Reuben, R.L., Grund, J.K. (1998), "Advanced LIGA technology for the integration of an electrostatically controlled bearing in a wobble micromotor," *Journal of Microelectromechanical Systems,* 7(4), pp. 423-427.

Sansiñena, J.M., Olazábal, V., Otero, T.F., Polo da Fonseca, C.N., De Paoli, M-A. (1997), "A solid state artificial muscle based on polypyrrole and a solid polymeric electrolyte working in air," *Journal Chemical Society Chemical Communication*, pp. 2217-2218.

Sastry, S.V., Nyshadham, J.R., Fix, J.A. (2000), "Recent technological advances in oral drug delivery -a review," *Pharmaceutical Science & Technology Today*, 3(4), pp. 138-145.

Savart, F. (1833), *Annales de Chimie et de Physique*, 53, p. 337-386. (in Italian)

Sayre, C.N., Collard, D.M. (1997), "Deposition of polyaniline on micro-contact printed self-assembled monolayers of ω–functionalized alkanethiols," *Journal of Materials Chemistry*, 7(6), pp. 909-912.

Schiestel, S., Ensinger, W., Wolf, G.K. (1994), "Ion bombardment effects in conducting polymers," *Nuclear Instruments and Methods in Physics Research, Section B: Beam Interactions with Materials and Atoms,* 91, pp. 473-477.

Schmelzer, M., Roth, S., Bauerle, P., Li, R. (1993), "2D arrangement of thiophene: highly ordered structures prepared by the Langmuir-Blodgett technique," *Thin Solid Films*, 229(2), pp. 255-259.

Schomburg, W.K., Vollmer, J., Büstgens, B., Fahrenberg, J., Hein, H., Menz, W. (1994), "Microfluidic components in LIGA technique," *Journal of Micromechanics and Microengineering* 4(4), pp. 186-191.

Shi, F.F. (1996), "Recent advances in polymer thin films prepared by plasma polymerization Synthesis, structural characterization, properties and applications," *Surface and Coatings Technology*, 82(1-2), pp. 1-15.

Shiga, T., Okada, A., Kurauchi, T. (1993), "Electroviscoelastic effect of polymer blend consisting of silicone elastomer and semiconducting polymer particles," *Macromolecules*, 26, pp. 6958-6963.

Smela, E. (1999), "Microfabrication of PPy microactuators and other conjugated polymer devices," *Journal of Micromechanics and Microengineering,* 9(1), pp. 1-18.

Smela, E., Inganäs, O., Lundström, I. (1993), "Conducting polymers as artificial muscles: challenges and possibilities," *Journal of Micromechanics and Microengineering*, 3(4), pp. 203-205.

Srinivasan, M.P., Jing, F.J. (1998), "Composite Langmuir-Blodgett films containing polypyrrole and polyimide," *Thin Solid Films*, 327-329, pp. 127-130.

Sun, J., Zou, S., Wang, Z., Zhang, X., Shen, J. (1999), " Layer-by-layer self-assembled multilayer films containing the organic pigment, 3,4,9,10-perylenetetracarboxylic acid, and their photo- and electroluminescence properties," *Materials Science and Engineering: C,* 10 (1-2), pp. 123-126.

Suzuki, M., Hirasa, O. (1993), "An approach to artificial muscle using polymer gels formed by microphase separation," *Advanced Polymer Science*, 110 (Responsive Gels: Volume Transitions II), pp. 241-61.

Tabuchi, Y., Rikukawa, M., Sanui, K., Ogata, N. (1999), "Self-Organized Properties of Conductive Langmuir-Blodgett Films," *Synthetic Metals*, 102(1-3), pp. 1441-1442.

Tadokoro, S., Yamagami, S., Ozawa, M., Kimura, T., Takamori, T., Oguro, K. (1999), "Soft micromanipulation device with multiple degrees of freedom consisting of high polymer gel actuators," 12th *IEEE International Conferences of the Micro-Electro-Mechanical System*, Tech. Dig., pp. 37-42.

Taguchi S., Tanaka T. (1988), "Deep UV photolithography for forming fine patterns with conjugated polymers," *European Patent Application EP 261991.*

Touihri, S., Safoula, G., Bernède, J. C., Leny, R., Alimi, K. (1997), "Comparison of the properties of iodine-doped poly(N-vinylcarbazole) (PVK) thin films obtained by evaporation of pure powder followed by iodine post-deposition doping and iodine pre-doped powder," *Thin Solid Films*, 304(1-2), pp. 16-23.

Tsuei, T.W., Wood, R.L., Khan M.C., Donnelly, M.M., Fair, R.B., (1998), "Tapered microvalves fabricated by off-axis X-ray exposures," *Microsystem Technologies,* 4(4), pp. 201-204.

Tzoua, K.T., Gregorya, R.V. (1995), "Improved solution stability and spinnability of concentrated polyaniline solutions using N,N'-dimethyl propylene urea as the spin bath solvent," *Synthetic Metals,* 69(1-3), pp.109-112

Ulman, A. (1991), "An Introduction to Ultrathin Organic Films," Academic Press, New York.

VanDyke, L.S., Brumlik, C.J., Martin, C.R., Yu, Z.G., Collins, G.J. (1992), "UV Laser ablation of electronically conductive polymers," *Synthetic Metals,* 52(3), pp. 299-304.

Vargo, T. G., Calvert, J. M., Wynne, K. J., Avlyanov, J. K., MacDiarmid, A. G., Rubner, M. F., (1995), "Patterned polymer multilayer fabrication by controlled adhesion of polyelectrolytes to plasma-modified fluoropolymer surfaces," *Supramolecular Science*, 2(3-4), pp. 169-174.

Wallace, D.B. (1989), "A Method of Characteristics Model of a DOD Ink-Jet Device Using an Integral Drop Formation Method," *ASME Pub. 89*-WA/FE-4.

Wallrabe, U., Mohr, J., Tesari, I., Wulff, W. (1996), In Proceedings of the IEEE MEMS'96, pp. -466.

Wallrabe, U., Ruther, P., Schaller, T., Schomburg, W.K. (1998), "Microsystems in medicine," *International Journal of Artificial Organs,* 21(3), pp. 137-146.

Wang G., Gan, F. (2000), "Optical parameters and absorption studies of azo dye-doped polymer thin films on silicon," *Materials Letters*, 43(1-2), pp. 6-10.

Weibezahn, K.F., Knedlitschek, G., Dertinger, H., Bier, W., Schaller, T., Schubert, K. (1995), "Reconstruction of tissue layers in mechanically processed microstructures," *Journal of Experimental & Clinical Cancer Research*, 14(1), pp. 41-42.

Wink, T., Van Zuilen, S.J., Bult, A., Van Bermekom, W.P. (1997), "Self-assembled monolayers for biosensors," *Analyst* 122, pp. R43-50.

Woolley, A.T., Mathies, R.A. (1995), *Analytical Chemistry,* 67(20), pp. 3676-3680.

Wu, B.M., Borland, S.W., Giordano, R.A., Cima, L.G., Sachs, E.M., Cima, M.J. (1996), "Solid free-form fabrication of drug delivery devices," *Journal of Controlled Release*, 40(1-2) pp. 77-87.

Wydeven, T., Golub, M.A., Lerner, N.R. (1998), "Etching of plasma-polymerized tetrafluoroethylene, polytetrafluoroethylene, and sputtered polytetrafluoroethylene induced by atomic oxygen [O(3P)]," *Journal of Applied Polymer Science*, 37(12), pp. 3343-3355.

Yoshino K., Kuwabara T., Manda Y., Nakajima S., Kawai T. (1990), "Photoinduced solubilization of conducting polymer and its application to etching and recording," *Japanese Journal of Applied Physics*, 29, pp. L1716-L1719.

Zhai, J., Wei T., Huang, C-H. (2000), "Fabrication of a new photoelectric conversion system by using a self-assembling technique," *Thin Solid Films,* 370(1-2), pp. 248-252.

Zissi, S., Bertsch, A., Jézéquel, J.Y., Corbel, S., André, J.C., Lougnot, D.J. (1996), "Stereolithography and microtechniques," *MicrosystemTechnolgies,* 2, pp. 97-102.

TOPIC 6

Testing and Characterization

CHAPTER 15

Methods of Testing and Characterization

Stewart Sherrit , Xiaoqi Bao and Yoseph Bar-Cohen
Jet Propulsion Lab./California Institute of Technology

15.1 Introduction

Implementing EAP materials as actuators requires the availability of a properties database and scaling laws to allow the actuator or transducer designer to determine their response at the operational conditions. A metric for the comparison of these materials' properties with other electroactive materials and devices is needed to support users in the implementation of these materials as actuators of choice. In selecting characterization techniques it is instructive to look at the various classes of electroactive Polymers and the source of their strain-field response. Generally, two main classes can be identified [Chapter 1 and Topic 3]:

(1) *Electronic EAP materials* – These are mostly materials that are dry and are driven by the electric field or Coulomb forces. This category includes piezoelectrics, and electrostrictive and ferroelectric materials. Generally these materials are polarizable with the strain being coupled to the electric displacement. The strain of electrostrictive and ferroelectric materials is proportional to the square of the polarization or electric displacement. In piezoelectric materials the strain couples linearly to the applied field or electric displacement. Charge transfer in these materials is in general electronic and at dc fields they behave as insulators. These properties have been studied for over a century in single crystals and for over three decades in polymers.

(2) *Ionic EAP materials* – These materials contain electrolytes and they involve transport of ions/molecules in response to an external electric field. Examples of such materials include conductive polymers, IPMCs, and ionic gels. The field controlled migration or diffusion of the various ions/molecules results in an internal stress distribution. These internal stress distributions can induce a wide variety of strains, from volume expansion or contraction, to bending. In some conductive polymers the materials exhibit both ionic and electronic conductivities. These materials are relatively new as actuator materials and have received much less attention in the literature than the piezoelectric and electrostrictive materials. At present, due to a wide variety of possible materials and conducting species, no generally accepted phenomenological model exists and much effort is underway to determine the commonalities of the various materials systems. A clearer understanding of the characterization techniques would help immensely in determining underlying theories and scaling laws for these actuator materials.

15.2 Characterization of EAP with Polarization-Dependent Strains

A significant body of knowledge is available for the characterization of polar polymer electromechanical materials. This includes general information garnered from other electromechanical materials as well as a significant body of work dealing with polymer transduction materials in the last thirty years. The characterization of the material properties of polar materials involves both the acquisition of the data (e.g., strain, stress, charge, and field) and the interpretation of the data using the appropriate constitutive equations.

15.2.1 Piezoelectricity

A phenomenological model derived from thermodynamic potentials mathematically describes the property of piezoelectricity. The derivations are not unique and the set of equations describing the piezoelectric effect depends on the choice of potential and the independent variables used. An excellent discussion of these derivations is found in Mason [1958]. In the case of a sample under isothermal and adiabatic conditions and ignoring higher order effects, the elastic Gibbs function may be described by

$$G_1 = -\tfrac{1}{2}\left(s_{ijkl}^D T_{ij} T_{kl} + 2g_{nij} D_n T_{ij}\right) + \tfrac{1}{2}\beta_{mn}^T D_m D_n , \qquad (1)$$

where g is the piezoelectric voltage coefficient, s is the elastic compliance, and β is the inverse permittivity. The independent variables in this equation are the stress T and the electric displacement D. The superscripts of the constants designate the independent variable that is held constant when defining the constant, and the subscripts define tensors that take into account the anisotropic nature of the material. The linear equations of piezoelectricity for this potential are determined from the derivative of G_1 and are

$$\begin{aligned} S_{ij} &= -\frac{\partial G_1}{\partial T_{ij}} = s_{ijkl}^D T_{kl} + g_{nij} D_n \\ E_m &= \frac{\partial G_1}{\partial D_m} = \beta_{mn}^T D_n - g_{nij} T_{ij} , \end{aligned} \qquad (2)$$

where S is the strain and E is the electric field. The above equations are usually simplified to a reduced form by noting that there is a redundancy in the strain and stress variables (see Nye [1957] or Cady [1942] for a discussion detailing tensor properties of materials). The elements of the tensor are contracted to a 6×6 matrix with 1, 2, 3 designating the normal stress and strain and 4, 5, and 6 designating the shear stress and strain elements. Other representations of the linear equations of piezoelectricity derived from the other possible thermodynamic potentials are shown below (Ikeda 1990, Mason 1958). This set of equations includes Eq. (2) in contracted notation:

$$\begin{aligned} S_p &= s_{pq}^D T_q + g_{pm} D_m \\ E_m &= \beta_{mn}^T D_n - g_{pm} T_p \end{aligned} \qquad (3)$$

$$
\begin{aligned}
S_p &= s^E_{pq} T_q + d_{pm} E_m \\
D_m &= \varepsilon^T_{mn} E_n + d_{pm} T_p
\end{aligned}
\tag{4}
$$

$$
\begin{aligned}
T_p &= c^E_{pq} S_q - e_{pm} E_m \\
D_m &= \varepsilon^S_{mn} E_n + e_{pm} S_p
\end{aligned}
\tag{5}
$$

$$
\begin{aligned}
T_p &= c^D_{pq} S_q - h_{pm} D_m \\
E_m &= \beta^S_{mn} D_n - h_{pm} S_p ,
\end{aligned}
\tag{6}
$$

where d, e, g, and h are piezoelectric constants, s and c are the elastic compliance and stiffness, and ε and β are the permittivity and the inverse permittivity. The relationship described by each of these equations can be represented in matrix form as shown below for Eq. (4):

$$
\begin{bmatrix} S_1 \\ S_2 \\ S_3 \\ S_4 \\ S_5 \\ S_6 \\ D_1 \\ D_2 \\ D_3 \end{bmatrix}
=
\begin{bmatrix}
s^E_{11} & s^E_{12} & s^E_{13} & s^E_{14} & s^E_{15} & s^E_{16} & d_{11} & d_{12} & d_{13} \\
s^E_{12} & s^E_{22} & s^E_{23} & s^E_{24} & s^E_{25} & s^E_{26} & d_{21} & d_{22} & d_{23} \\
s^E_{13} & s^E_{23} & s^E_{33} & s^E_{34} & s^E_{35} & s^E_{36} & d_{31} & d_{32} & d_{33} \\
s^E_{14} & s^E_{24} & s^E_{34} & s^E_{44} & s^E_{45} & s^E_{46} & d_{41} & d_{42} & d_{43} \\
s^E_{15} & s^E_{25} & s^E_{35} & s^E_{45} & s^E_{55} & s^E_{56} & d_{51} & d_{52} & d_{53} \\
s^E_{16} & s^E_{26} & s^E_{36} & s^E_{46} & s^E_{56} & s^E_{66} & d_{61} & d_{62} & d_{63} \\
d_{11} & d_{12} & d_{13} & d_{14} & d_{15} & d_{16} & \varepsilon^T_{11} & \varepsilon^T_{12} & \varepsilon^T_{13} \\
d_{21} & d_{22} & d_{23} & d_{24} & d_{25} & d_{26} & \varepsilon^T_{12} & \varepsilon^T_{22} & \varepsilon^T_{23} \\
d_{31} & d_{32} & d_{33} & d_{34} & d_{35} & d_{36} & \varepsilon^T_{13} & \varepsilon^T_{23} & \varepsilon^T_{33}
\end{bmatrix}
\begin{bmatrix} T_1 \\ T_2 \\ T_3 \\ T_4 \\ T_5 \\ T_6 \\ E_1 \\ E_2 \\ E_3 \end{bmatrix} .
\tag{7}
$$

The matrix shown in Eq. (7) is a generalized representation of the equations shown in Eq. (4). Fortunately due to symmetry, many of the elements shown in the matrix in Eq. (7) are either zero or not independent. The matrix above expresses the relationship between the material constants and the variables S, T, E, and D. Ideally, under small fields and stresses and for materials with low losses within a limited frequency range these constants contain all the information that is required to predict the behavior of the material under the application of stress, strain, electric field, or the application of charge to the sample surface. In practice, however, most materials display dispersion, nonlinearity, and have measurable losses. For the material constants of the matrix representing Eq. (7), a more accurate representation for these constants would be a functional relationship of the form

$$s_{kl}^{E} = s_{kl}^{E'}\left(\omega, E_i, T_{ij}, T, t\right) + i s_{kl}^{E''}\left(\omega, E_i, T_{ij}, T, t\right), \tag{8}$$

$$d_{ij} = d_{ij}''\left(\omega, E_i, T_{ij}, T, t\right) + i d_{ij}''\left(\omega, E_i, T_{ij}, T, t\right), \tag{9}$$

$$\varepsilon_{ij}^{\mathrm{T}} = \varepsilon_{ij}^{T'}\left(\omega, E_i, T_{ij}, T, t\right) + i \varepsilon_{ij}^{T''}\left(\omega, E_i, T_{ij}, T, t\right), \tag{10}$$

where the constants are written in terms of the real and imaginary (losses) components as a function of the frequency, electric field, stress, temperature, and time. This is a generalized representation, which accounts for the field E_i, stress T_{ij} and frequency dependence ω. Equations 8–10 include the dependence on temperature T and time t. Many of practical piezoelectric polymer materials used for transduction are also ferroelectric and have an associated Curie point that marks a phase change (e.g., ferroelectric to paraelectric). In these materials the elastic, piezoelectric, and dielectric properties depend on the proximity of the measurement temperature to the Curie temperature. These materials are poled by the application of a field greater than the coercive field E_C (field at which orientation noticeably begins to switch). The poling requires a finite time, and after the application of the poling field a relaxation occurs which is described by an aging curve, which is generally logarithmic in time t. In some polymer materials, a uniaxial stress is applied to aid in the reorientation of the dipoles. Operation of the sample at high field and high stress may therefore accelerate the relaxation and the sample may be partially depoled. These aging/poling/depoling processes require a measurable time that depends on the present and prior conditions to which the material is/was subjected. This time and temperature dependence is closely associated with the ferroelectric nature of the material. In general, ferroelectric polymers have a large coercive field and are mechanically compliant, which means that these materials show little change under dc bias fields and larger stress dependencies.

Berlincourt, Curran, and Jaffe [1964] noted that a piezoelectric material which can be poled could be considered to be mathematically identical to a biased electrostrictive material (strain is quadratic with field) with the internal field of the poled piezoelectric supplying the bias field. As can be seen from these equations, the material constants of piezoelectric materials may have a variety of dependencies. These may introduce significant errors in device design and operation if it is assumed that the materials are lossless, linear, and frequency independent.

15.2.1.1 Resonance Analysis

The most widely used technique for measuring the material constants of piezoelectric materials is the resonance method, which is outlined in the *IEEE Standard on Piezoelectricity* [1987]. A piezoelectric sample of specific geometry is excited with an ac voltage. The phase and the magnitude of the current with respect to the excitation voltage is monitored and the ac impedance of the sample as a function of the frequency of the ac voltage is found. The impedance spectrum is complex with both a resistance R and a reactance X. A typical impedance spectrum for a polyvinylidene difluoride/trifluoroethylene P(VDF/TrFE) sample is shown in Fig. 1. The impedance spectra contain resonances, which are the result of ultrasonic standing waves in the piezoelectric material. The parallel resonance frequency f_p is defined to be the frequency at which a maximum in the resistance occurs. The sideband frequencies $f_{+1/2}$ and $f_{-1/2}$ occur at the maximum and minimum of the reactance, respectively. The admittance spectrum for the P(VDF/TrFE) sample is also shown in Fig. 1. The resonance frequency of this spectrum is the series resonance frequency f_S (maximum in the conductance G). The sideband frequencies $f_{+1/2s}$ and $f_{-1/2s}$ occur at the maximum and minimum of the susceptance B.

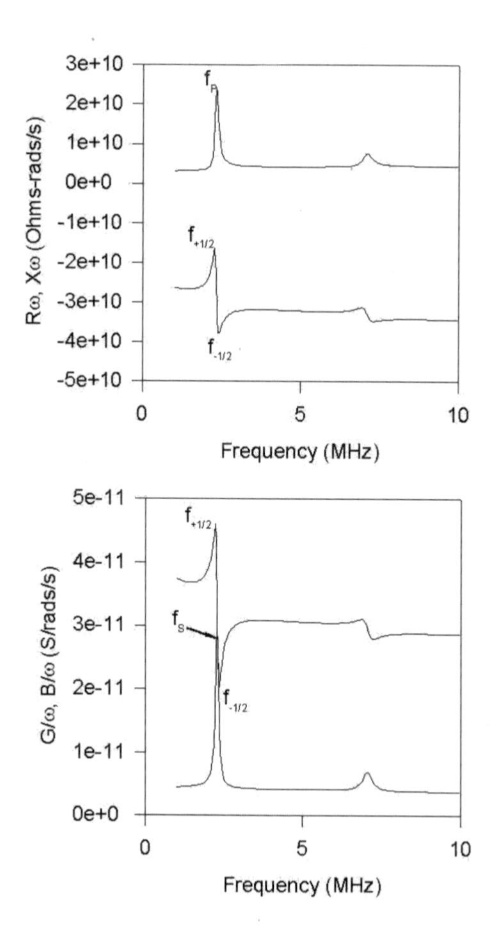

Figure 1: The impedance spectra of a P(VDF/TrFE) plate sample in the thickness mode. (Sample thickness t = 0.000462 m) The parallel/series resonance frequency and the sideband frequencies are marked.

Common modes of excitation used for piezoelectric material analysis are shown in Fig. 2 along with the recommended geometrical aspect ratio for samples used to determine each mode. The arrow marked on each sample indicates the poling direction. For example, for a thickness mode resonator, the thickness should be at least 1/10 the smallest lateral dimension ($t \leq l / 10 \leq w / 10 \leq a / 10$) where t is the thickness, l and w are the length and width of the plate, or if the sample is a disk, a is the radius of a disk. These aspect ratios ensure that the sample is excited in a mode where the one-dimensional approximation is valid and that coupling between modes is limited.

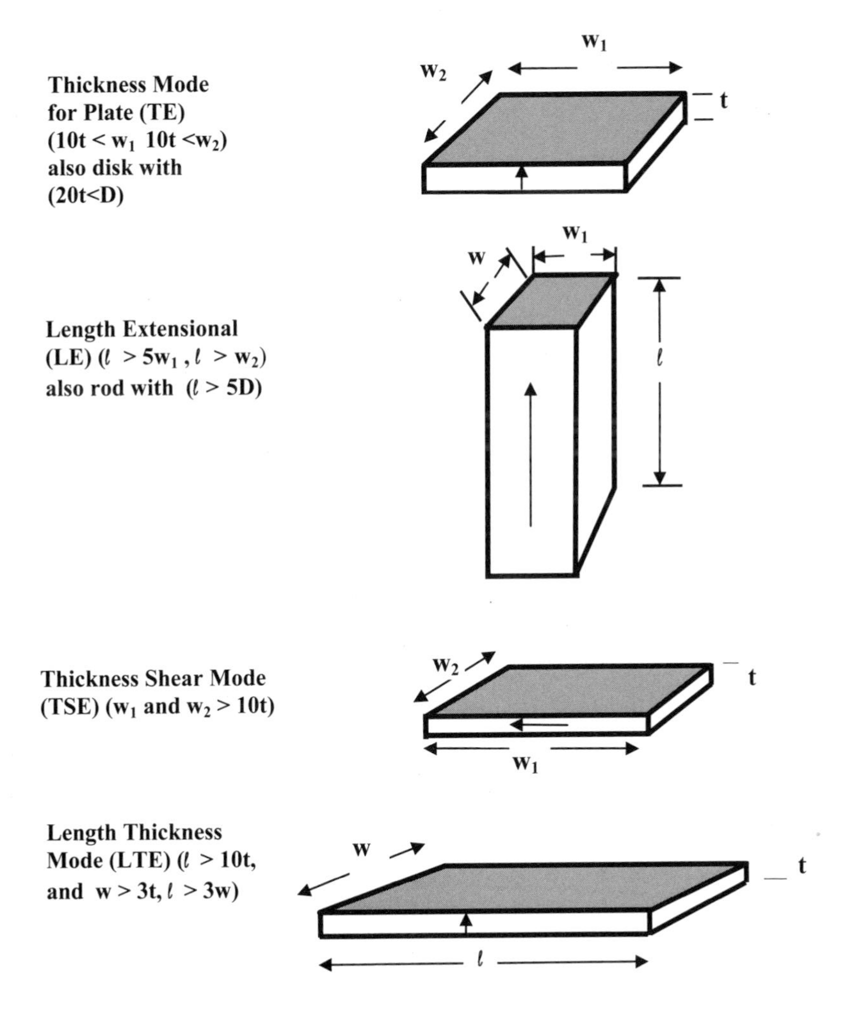

Figure 2: The geometry and poling direction for common modes of piezoelectric resonators for materials characterization. Note: The thickness resonance can be excited in a flat disk and the length extensional resonator may also be of the form of a long rod.

The impedance equation governing resonance spectra similar to the spectra shown in Fig. 1 is derived from phenomenological theory based on the linear equations of piezoelectricity and the wave equation. The previous derivation of the thickness, thickness shear, length, and length thickness impedance equations by Berlincourt, Curran, and Jaffe [1964] was for materials with real material constants (lossless materials). Holland [1967] showed that the losses of a piezoelectric material could be taken into account by representing the material constant as a complex quantity in the frequency domain. In the following sections, losses are accounted for by the addition of an imaginary component to the various material constants.

15.2.1.1.1 Thickness extensional mode (TE)

The thickness equations governing the thickness resonance are derived from the linear piezoelectric equations and the thickness mode material constants

$$T_3 = c_{33}^D S_3 - h_{33} D_3 \tag{11}$$

$$E_3 = -h_{33} S_3 + \beta_{33}^s D_3 \tag{12}$$

where T, S, E, and D are the stress, strain, electric field, and the electric displacement and c_{33}^D, h_{33} and β_{33}^S are the elastic stiffness (constant D), the piezoelectric constant, and the inverse permittivity at constant strain. During an impedance measurement, a sinusoidal electric voltage is placed across the sample. The above equations represent the spatial and time component of the T, S, E, and D variable governing the piezoelectric material. The time dependence of the fields T, S, E, and D can be taken into account by multiplying by an $\exp(i\omega t)$ term

$$T_3(t) \Rightarrow T_3 e^{-i\omega t} \tag{13}$$

$$S_3(t) \Rightarrow S_3 e^{-i\omega t} \tag{14}$$

$$D_3(t) \Rightarrow D_3 e^{-i\omega t} \tag{15}$$

$$E_3(t) \Rightarrow E_3 e^{-i\omega t}\,. \tag{16}$$

The linear piezoelectric equations above and the wave equation

$$\frac{\partial^2 u_3}{\partial t^2} = \frac{c_{33}^D}{\rho} \frac{\partial^2 u_3}{\partial x_3^2}\,, \tag{17}$$

are coupled since the strain is defined as $S_3 = \frac{\partial u_3}{\partial x_3}$, where u_3 is the displacement along the 3 direction (x_3). The solution of the wave equation is of the form

$$u_3 = \left[A \sin\left(\frac{\omega x_3}{v_D}\right) + B \cos\left(\frac{\omega x_3}{v_D}\right)\right] \exp^{-i\omega t}, \tag{18}$$

where

$$v_D = \sqrt{\frac{c_{33}^D}{\rho}}. \tag{19}$$

Taking the derivative of u_3 with respect to x_3 to get S_3 and substituting this value into the first linear equation for a linear piezoelectric material [Eq. (11)], and using the boundary conditions $T_3 = 0$ at $x_3 = 0$ and $T_3 = 0$ at $x_3 = l$, an equation for S_3 in terms of D_3 can be written. Using Eq. (12) we find the voltage on the sample is

$$V = -\int_0^l E_3 dx_3 = -\int_0^l \left(-h_{33} S_3 + \beta_{33}^s D_3\right) dx_3, \tag{20}$$

which simplifies to

$$V = -\beta_{33}^s D_3 l + \frac{h_{33}^2 D_3 v_D}{c_{33}^D \omega}\left[2 \tan\left(\frac{\omega l}{2 v_D}\right)\right]. \tag{21}$$

The current may be found by taking the time derivative of the electric displacement

$$I = A\frac{dD_3}{dt} = -i\omega A D_3, \tag{22}$$

where A is the electrode area. The impedance of the resonator is given by $Z = V/I$

$$Z = \frac{\beta_{33}^s D_3 l - \frac{h_{33}^2 D_3 v_D}{c_{33}^D \omega}\left[2 \tan\left(\frac{\omega l}{2 v_D}\right)\right]}{i\omega A D_3}, \tag{23}$$

or

$$Z = \frac{\beta_{33}^{s} l - \frac{h_{33}^{2} v_D}{c_{33}^{D} \omega}\left[2\tan\left(\frac{\omega l}{2 v_D}\right)\right]}{i\omega A}. \tag{24}$$

Equation (24) can be simplified by using

$$k_t^2 = \frac{h_{33}^2}{\beta_{33}^S c_{33}^D} \text{ and } 4f_p = \frac{2v_D}{l} \text{ and } \beta_{33}^S = \frac{1}{\varepsilon_{33}^S}, \tag{25}$$

where k_t is the thickness electromechanical coupling constant and f_p is the complex parallel resonance frequency. The square of the coupling constant k_t defined in Eq. (25) is a measure of the conversion efficiency in the material. In general, the square of the coupling constant in the quasi-static limit is the ratio of the available elastic energy in the material divided by the input electrical energy, or, conversely, the available electric energy divided by the input mechanical energy. For a more extensive treatment of the electromechanical coupling see Berlincourt et al. [1964] or refer to the *IEEE Standard on Piezoelectricity* [1987].

Using the relationships in Eq. (25), the impedance equation can be rewritten with complex material constants in a form identical to the form presented in the *IEEE Standard on Piezoelectricity* [1987], however, the parallel resonance-frequency constant-coupling and permittivity are now all complex:

$$Z = \frac{l}{i\omega A \varepsilon_{33}^S}\left[1 - \frac{k_t^2 \tan\left(\frac{\omega}{4f_p}\right)}{\frac{\omega}{4f_p}}\right]. \tag{26}$$

15.2.1.1.2 Thickness shear extensional (TSE) mode

The thickness shear extensional mode derivation is mathematically identical to the thickness extensional mode derived above; the only difference is the set of material constants used. The resonance is described by an equation of the form

$$Z = \frac{t}{i\omega A \varepsilon_{11}^S}\left[1 - \frac{k_{15}^2 \tan\left(\frac{\omega}{4f_p}\right)}{\frac{\omega}{4f_p}}\right], \tag{27}$$

where t is the sample thickness, A is the electrode area, ε_{11}^{S} is the clamped permittivity and k_{15} is the complex thickness shear extensional coupling constant. The complex parallel frequency constant is the physically significant root (real part positive) and defined by

$$f_p = \left[4\left(1-k_{15}^2\right)\rho s_{55}^{\mathbf{E}} t^2\right]^{-1/2} = \left[4\rho s_{55}^{D} t^2\right]^{-1/2} = \left[\frac{c_{55}^{\mathbf{E}}}{4\left(1-k_{15}^2\right)\rho t^2}\right]^{1/2} = \left(\frac{c_{55}^{D}}{4\rho t^2}\right)^{1/2}, \quad (28)$$

and the complex thickness shear electromechanical coupling constant is defined by

$$k_{15}^2 = \frac{e_{15}^2}{\varepsilon_{11}^{S} c_{55}^{D}}. \quad (29)$$

In many cases, due to the thickness limitation of poled polymer samples, laminated resonators are produced to create the appropriate boundary conditions to allow for the investigation of a shear resonator mode as described by Wang et al. [1993] and Ohigashi [2000].

15.2.1.1.3 Length extensional mode (LE)

The length extensional mode derivation is mathematically identical to the thickness extensional mode derived above. The only difference is the set of material constants used. The resonance is described by an equation of the form

$$Z = \frac{l}{i\omega A \varepsilon_{33}^{S}}\left[1 - \frac{k_{33}^2 \tan\left(\dfrac{\omega}{4f_p}\right)}{\dfrac{\omega}{4f_p}}\right], \quad (30)$$

where l is the sample length, A is the electrode area, ε_{33}^{S} is the clamped permittivity, and k_{33} is the complex length extensional coupling constant. The complex parallel frequency constant is the physically significant root (real part positive) and defined by

$$f_p = \left[4\left(1-k_{33}^2\right)\rho s_{33}^{E} l^2\right]^{-1/2} = \left[4\rho s_{33}^{D} l^2\right]^{-1/2}, \quad (31)$$

and the complex length electromechanical coupling constant is defined by

$$k_{33}^2 = \frac{d_{33}^2}{\varepsilon_{33}^T s_{33}^E}. \tag{32}$$

In general, due to the limitation of the thickness of most piezoelectric polymer samples LE resonators would require sheets of the polymer to be oriented and laminated to create a stack resonator with the appropriate mechanical boundary conditions [Sherrit et al., 2000].

15.2.1.1.4 Length thickness extensional (LTE) mode

The admittance equation governing the resonance of the length thickness extensional mode is

$$Y(\omega) = \left(\frac{i\omega\varepsilon_{33}^T w_1 w_2}{t}\right)\left\{1 - k_{13}^2\left[1 - \frac{\tan\left(\frac{\omega}{4f_s}\right)}{\frac{\omega}{4f_s}}\right]\right\}, \tag{33}$$

where t is the sample thickness, $A = w_1 w_2$ is the electrode area where w_1 is the longest lateral dimension of the electrode and w_2 is the smaller lateral dimension, ε_{33}^T is the complex free permittivity, and k_{13} is the complex length thickness extensional coupling constant. The complex series frequency constant is the physically significant root (real part positive) and defined by

$$f_s = \left(\frac{1}{4\rho s_{11}^E w_1^2}\right)^{1/2} \tag{34}$$

is a function of the width w_1, the density ρ and the elastic compliance $s_{11}^E = s_{11}^D/(1-k_{13}^2)$.

The *IEEE Standard on Piezoelectricity* [1987] uses the impedance equations discussed in the previous section and critical frequencies derived from these equations to determine the real parts of the material constants. The losses associated with the dielectric constant are determined away from resonance, and the losses associated with the elastic constants are determined using equivalent circuits recommended by the standard. In the case of high-loss polymer materials the *IEEE Standard on Piezoelectricity* [1987] may not be sufficient and other techniques based on inverting specific data points [Smits, 1976] or nonlinear regression [Lukacs et al., 1999; Kwok et al., 1997] may be required. As was shown above, the impedance equation for the thickness resonator has the form shown in Eq. (35) where t is the sample thickness, A is the electrode area and

$\mathbf{c}_{33}^{D}$, k_t, and ε_{33}^{S} are the elastic stiffness, thickness electromechanical coupling constant and clamped dielectric permittivity.

$$Z = \frac{t}{i\omega A\varepsilon_{33}^{S}}\left[1 - \frac{k_t^2 \tan\left(\frac{\omega t}{2}\sqrt{\frac{\rho}{c_{33}^{D}}}\right)}{\frac{\omega t}{2}\sqrt{\frac{\rho}{c_{33}^{D}}}}\right]. \tag{35}$$

The fundamental resonance in the impedance spectrum (first maximum in the resistance spectra) occurs at $f = f_p$ when the argument of the tangent function in Eq. (35) is equal to $\pi/2$, or when

$$f = \frac{1}{2t}\sqrt{\frac{c_{33}^{D}}{\rho}} = f_p. \tag{36}$$

Rearranging Eq. (36), the elastic stiffness c_{33}^{D} is given in terms of the parallel resonance frequency by

$$c_{33}^{D} = 4\rho\, t^2 f_p^2. \tag{37}$$

The electromechanical coupling factor can be determined in a similar manner. The series resonance frequency f_s in the admittance spectrum is the frequency where the maximum in the conductance versus frequency occurs, or when

$$\left[1 - \frac{k_t^2 \tan\left(\frac{\omega t}{2}\sqrt{\frac{\rho}{c_{33}^{D}}}\right)}{\frac{\omega t}{2}\sqrt{\frac{\rho}{c_{33}^{D}}}}\right] = \left[1 - \frac{k_t^2 \tan\left(\frac{\omega}{4f_p}\right)}{\frac{\omega}{4f_p}}\right] = 0. \tag{38}$$

At the series resonance $\omega = \omega_s$ and Eq. 38 can be written as

$$\frac{\omega_s}{4f_p} = k_t^2 \tan\left(\frac{\omega_s}{4f_p}\right), \tag{39}$$

which can be arranged to determine an equation for the electromechanical coupling constant in terms of the series and parallel resonance frequencies of the sample. k_t is then

$$k_t^2 = \frac{\pi}{2}\frac{f_s}{f_p}\tan\left[\frac{\pi}{2}\left(1-\frac{f_s}{f_p}\right)\right]. \quad (40)$$

The value of the clamped permittivity is then found by noting that it is equal to the high frequency permittivity above resonance

$$\varepsilon_{HF} = \varepsilon_{33}^S, \quad (41)$$

or by noting that the low-frequency permittivity measured below the thickness resonance is of the form

$$\varepsilon_{LF} = \frac{\varepsilon_{33}^S}{1-k_t^2}. \quad (42)$$

The complex part of the dielectric permittivity is determined by the loss tangent $D = \tan\delta$. It is apparent from Eqs. (41) and (42) that the dielectric loss can only be determined from the high frequency permittivity shown in Eq. (41) since the low-frequency loss has an electromechanical coupling term. Only the real part of the electromechanical coupling constant is determined from Eq. (40) and the loss component is assumed to be small. Ignoring the loss component of k_t may introduce a significant error in the imaginary component of the clamped permittivity ε_{33}^S and a smaller error in the real part of the clamped permittivity. The piezoelectric constant governing the thickness mode may be found using the definition of the thickness electromechanical coupling constant

$$h_{33} = k_t\sqrt{\frac{c_{33}^D}{\varepsilon_{33}^S}}. \quad (43)$$

The other material constants determined from the thickness resonator are

$$e_{33} = h_{33}\varepsilon_{33}^S, \quad (44)$$

and

$$c_{33}^E = c_{33}^D(1-k_t^2), \quad (45)$$

where e_{33} is the piezoelectric coefficient, and c_{33}^E is the elastic stiffness at constant electric field. The *IEEE Standard on Piezoelectricity* [1987] equations for analyzing the other modes are summarized in Table 1.

The development of the *IEEE Standard on Piezoelectricity* [1987] from the preceding standards [1949, 1957, 1958, and 1976] began before the discovery of new piezoelectric polymer materials, which has taken place over the last 30 years. These developments included low Q material such as PVDF [Kawai, 1969], P(VDF/TrFE) copolymers [Murata and Koizumi, 1989], low Q piezoelectric composites [Newnham et al., 1978 and Banno 1983], piezoelectric polymides [Ounaies et al., 1999], liquid crystalline elastomers [Biggs et al., 2000] and the odd nylons [Litt et al., 1977; Newman et al., 1980]. Prior to these developments most materials had high mechanical Q and low dielectric dissipation. The techniques developed in the *Standard* were therefore adequate to determine the materials properties to a reasonable degree of accuracy. With the introduction of piezoelectric polymer materials having high loss in the elastic, dielectric, and piezoelectric material constants many of the techniques of the current standard fall short in determining accurately these material constants and their loss. As has been mentioned previously, Holland [1967] showed that the material constants could be represented as complex quantities with the real part designating the in phase component of the constant and the imaginary part designating the out of phase component of the linear relationship (phase shift). With this extension, the impedance equations of the *IEEE Standard on Piezoelectricity* [1987] can be written in terms of complex material constants. In complex notation the material constants governing the thickness mode can be written as

$$
\begin{aligned}
c_{33}^{D} &\Rightarrow \mathbf{c}_{33}^{D} = c_{33r}^{D} + i c_{33i}^{D} \\
h_{33} &\Rightarrow \mathbf{h}_{33} = h_{33r} + i h_{33i} \ , \\
\varepsilon_{33}^{S} &\Rightarrow \boldsymbol{\varepsilon}_{33}^{S} = \varepsilon_{33r}^{S} + i \varepsilon_{33i}^{S}
\end{aligned}
\tag{46}
$$

where the extra subscripts r and i refer to the real and imaginary components. Two other common forms of reporting the loss component of the material constant stem from historical usage. The imaginary component of the elastic constant is sometimes reported as the mechanical Q_M where the mechanical Q is the ratio of the real to the imaginary component of the elastic constant. In the case of the elastic stiffness at constant D governing the thickness extensional resonance the mechanical Q is defined by

$$
Q_M = c_{33r}^{D} / c_{33i}^{D} \, . \tag{47}
$$

The piezoelectric and dielectric permittivity losses are sometimes reported in this manner and labeled the piezoelectric Q_P and the dielectric Q_D, and each represents the ratio of the real to imaginary component.

Table 1: The *IEEE Standard on Piezoelectricity* [1987] equations for analyzing impedance resonance of the thickness, thickness shear, length, and length thickness modes.

Thickness Extensional Mode				
$c_{33}^D = 4\rho t^2 f_p^2$	$k_t^2 = \frac{\pi}{2}\frac{f_s}{f_p}\tan\left[\frac{\pi}{2}\left(1-\frac{f_s}{f_p}\right)\right]$	$\varepsilon_{HF} = \varepsilon_{33}^S$	$\varepsilon_{LF} = \frac{\varepsilon_{33}^S}{1-k_t^2}$	$h_{33} = k_t\sqrt{\frac{c_{33}^D}{\varepsilon_{33}^S}}$
Thickness Shear Extensional Mode				
$c_{55}^D = 4\rho t^2 f_p^2$	$k_{15}^2 = \frac{\pi}{2}\frac{f_s}{f_p}\tan\left[\frac{\pi}{2}\left(1-\frac{f_s}{f_p}\right)\right]$	$\varepsilon_{HF} = \varepsilon_{11}^S$	$\varepsilon_{LF} = \frac{\varepsilon_{11}^S}{1-k_{15}^2}$	$h_{15} = k_{15}\sqrt{\frac{c_{55}^D}{\varepsilon_{11}^S}}$
Length Extensional Mode				
$s_{33}^D = \frac{1}{4\rho l^2 f_p^2}$	$k_{33}^2 = \frac{\pi}{2}\frac{f_s}{f_p}\tan\left[\frac{\pi}{2}\left(1-\frac{f_s}{f_p}\right)\right]$	$\varepsilon_{HF} = \varepsilon_{33}^S$	$\varepsilon_{LF} = \frac{\varepsilon_{33}^S}{1-k_{33}^2} = \varepsilon_{33}^T$	$d_{33} = k_{33}\sqrt{s_{33}^E \varepsilon_{33}^T}$
Length Thickness Extensional Mode				
$s_{11}^E = \frac{1}{4\rho l^2 f_s^2}$	$\frac{k_{13}^2}{1-k_{13}^2} = \frac{\pi}{2}\frac{f_p}{f_s}\tan\left[\frac{\pi}{2}\left(\frac{f_p}{f_s}-1\right)\right]$	$\varepsilon_{HF} = \varepsilon_{33}^T(1-k_{13}^2)$	$\varepsilon_{LF} = \varepsilon_{33}^T$	$d_{13} = k_{13}\sqrt{s_{11}^E \varepsilon_{33}^T}$

Another representation that stems from designating the dielectric loss is the use of the dissipation or tanδ notation. In the field of dielectrics it is common to represent the dielectric loss as the dissipation or tanδ. In the case of the clamped permittivity at constant strain the dissipation or tanδ is defined as

$$\tan\delta = \varepsilon_{33i}^S / \varepsilon_{33r}^S \, . \tag{48}$$

Similarly, the piezoelectric and elastic losses are sometimes reported in this manner and labeled the piezoelectric $\tan\delta_P$ and the elastic $\tan\delta_M$ dissipation, and each represents the ratio of the imaginary to real component.

15.2.1.1.5 Resonance Analysis of Unclamped Stacks

Piezoelectric stacks are used in a variety of applications that require relatively high force and larger displacement than single element piezoelectric transducers can produce for a limited voltage. The zero-bond-length stack solution, where internal electrodes are very thin and have little or no effect on the elastic properties of the stack, was derived by Martin [1964a and 1964b]. The model was derived from a network of Mason's equivalent circuits of n layers connected mechanically in series and electrically in parallel.

The general solution for the admittance of a piezoelectric stack of area A, n layers and total length nL after simplification of the network is

$$Y(\omega) = i\omega n C_0 + \frac{2N^2}{Z_{ST}} \tanh\left(\frac{n\gamma}{2}\right), \tag{49}$$

where

$$C_0 = \frac{\varepsilon_{33}^T A}{L}(1 - k_{33}^2), \tag{50}$$

$$N = \frac{A}{L}\frac{d_{33}}{s_{33}^E}, \tag{51}$$

$$Z_{ST} = \left[Z_1 Z_2\left(2 + \frac{Z_1}{Z_2}\right)\right]^{1/2}, \tag{52}$$

$$\gamma = 2 \operatorname{arcsin} h\left[\left(\frac{Z_1}{2Z_2}\right)^{1/2}\right], \tag{53}$$

$$Z_1 = i\rho v^D A \tan\left(\frac{\omega L}{2v^D}\right), \tag{54}$$

$$Z_2 = \frac{\rho v^D A}{i \sin\left(\frac{\omega L}{v^D}\right)} + \frac{iN^2}{\omega C_0}, \tag{55}$$

and $Y(\omega) = i\omega n C_0 + \frac{2N^2}{Z_{ST}} \tanh\left(\frac{n\gamma}{2}\right)$ is the acoustic velocity at constant electric displacement. The constants ε_{33}^T, s_{33}^D, d_{33} are the free permittivity, the elastic compliance at constant electric displacement and the piezoelectric charge coefficient, respectively. Using Eqs. (49) to (55), Martin [1964b] demonstrated that in the limit of large n ($n > 8$), the acoustic wave speed in the material was determined by the constant-field elasticity constant s_{33}^E ($v^E = 1/\sqrt{\rho s_{33}^E}$). In the limit of $n > 8$, an analytical equation for admittance was presented, which allowed for direct determination of material constants from the admittance data. In the work by Sherrit et al. [2000], nonlinear regression techniques were presented and used to fit the data for a stack with an arbitrary number of layers n. In addition the constants ε_{33}^T, s_{33}^D, d_{33} were defined as complex to account for losses in the stack material. From the fitted data one can determine the "average" or effective material properties ε_{33}^T, s_{33}^D, d_{33} and use these to calculate the off resonance displacement $d_{33}^{stack} = nd_{33}$ and low-frequency unclamped capacitance [$C = Y/i\omega$; see Eq. (49)] as well as the overall elastic compliance of the stack and mechanical Q.

15.2.1.1.6 Resonance Analysis of Unclamped Bimorphs

Under the application of an ac field a piezoelectric bimorph will resonate at its natural resonance frequency, which is seen as resonance in the admittance or capacitance of the sample. The expression for the capacitance of an unloaded piezoelectric bimorph as a function of frequency was recently given by Smits, Choi, and Ballato [1997]. For a bimorph of length L, width w, and thickness $2h$, the equation for the capacitance is

$$C = \frac{Lw}{2h}\left(\varepsilon_{33}^{T} - \frac{d_{31}^{2}}{s_{11}^{E}}\right) + \frac{3w}{8h}\frac{d_{31}^{2}}{s_{11}^{E}\Omega}\frac{\cosh(\Omega L)\sin(\Omega L) + \cos(\Omega L)\sinh(\Omega L)}{1 + \cos(\Omega L)\cosh(\Omega L)}, \tag{56}$$

where ε_{33}^{T}, d_{31}, and s_{11}^{E} are the free permittivity, transverse piezoelectric constant and the elastic stiffness at the constant electric field, respectively. The constant Ω was defined as

$$\Omega = \left(\frac{\omega^{2}\rho A_{cs}}{EI}\right)^{1/4} = \sqrt{\frac{\omega}{a}}, \tag{57}$$

where ω, ρ, A_{cs}, E, and I are the angular frequency, effective density, cross-sectional area ($A_{cs}=2hw$), Young's modulus, and moment of inertia of the cross section (in units of m^4), respectively. The square of the constant a is the flexural rigidity $EI/\rho A_{cs}$. If the moment of inertia I, cross-sectional area A_{cs} and the density are known, then the Young's modulus can be determined from the flexural rigidity. The admittance of a resonator of capacitance C is $Y = i\omega C$ and was shown to be:

$$Y = \frac{i\omega L w \varepsilon_{33}^{T}}{2h}\left[1 + k_{31}^{2}\left(\frac{3}{4}\frac{F(\Omega L)}{\Omega L} - 1\right)\right], \tag{58}$$

where the function $F(\Omega L)$ is

$$F(\Omega L) = \frac{\cosh(\Omega L)\sin(\Omega L) + \cos(\Omega L)\sinh(\Omega L)}{1 + \cos(\Omega L)\cosh(\Omega L)}. \tag{59}$$

The analytical solution for the flexural rigidity and coupling k_{31} is then determined using methods similar to those previously used (Sherrit et al. [1999]). The constant that controls the frequency of the flexural resonance is the flexural rigidity given above. At the first series, resonance frequency ω_s, the admittance of the bimorph tends to infinity. This rapid increase in Y is due to the denominator of Eq. (59) approaching zero as the argument approaches the first root of

$$1 + \cos(\Omega L)\cosh(\Omega L) = 0, \tag{60}$$

as was pointed out by Smits and Choi [1993]. They found the first root of this equation to be $R_1 = 1.8751$ and presented a relationship for the flexural rigidity as a function of length, the series resonance frequency, and R_1:

$$a^2 = \left(\frac{EI}{\rho A_{cs}} \right) = \left(\frac{L^2 \omega_s}{R_1^2} \right)^2 . \tag{61}$$

Smits, Choi, and Ballato [1997] presented a table relating the normalized resonance frequencies to the electromechanical coupling constant k_{13}. We have found that an analytical solution can be found by using Eq. (61) to determine the normalized frequency argument of the impedance equation in terms of the series resonance frequency. Upon substitution of this relationship into Eq. (58), the impedance of the bimorph resonator becomes

$$Y = \frac{i\omega L w \varepsilon_{33}^T}{2h} \left[1 + k_{31}^2 \left(\frac{3}{4} \frac{F\left(\frac{\Omega R_1}{\Omega_s} \right)}{\frac{\Omega R_1}{\Omega_s}} - 1 \right) \right] . \tag{62}$$

At the normalized parallel resonance frequency Ω_p, the admittance Y goes to zero as the terms in the outer bracket of Eq. (62) go to zero. Upon rearranging these terms we find

$$k_{13} = \left(1 - \frac{3}{4} \frac{F(x)}{x} \right)^{-1/2} , \tag{63}$$

where x is

$$x = \frac{\Omega_p}{\Omega_s} R_1 = \left(\frac{\omega_p}{\omega_s} \right)^{1/2} R_1 , \tag{64}$$

and the function $F(x)$ was defined in Eq. (59). Equation (63) as reported earlier by Sherrit, et al. [1999] gives the same values for k_{13} as the graphically determined values of Table I (Smits, Choi, and Ballato [1997]), to the accuracy of the values reported in the table.

Numerous authors have proposed techniques to determine the real and imaginary components of the elastic and dielectric constants. Holland and Eernisse [1969] developed a gain bandwidth technique that used a result of Land et al. [1964], relating the bandwidth of the resonance to the mechanical Q_M. More recent papers by Ohigashi et al. [1988] and Tsurumi et al. [1990] have studied alternative methods to determine these losses.

The first technique to determine the complex material constants including the piezoelectric loss component involved applying the results of the gain-bandwidth technique to a Taylor's series expansion of the resonance equation [Holland, 1970]. The data had to be chosen judiciously, and the method was cumbersome if a large number of samples had to be analyzed.

Smits [1976] was the first to determine a general technique applicable to the thickness, thickness shear, length, and length-thickness resonators. The analysis of the material constants for the thickness extensional resonator uses three points around resonance: $Z_0(R_0, X_0, \omega_0)$, $Z_1(R_1, X_1, \omega_1)$, $Z_2(R_2, X_2, \omega_2)$. The points are usually chosen so that ω_0 is near the frequency of maximum resistance of the spectra; ω_1 and ω_2 are chosen at frequencies above and below the resonance. Two of the points—plus an initial guess for the elastic constant using a relationship between the mechanical Q and the resonance bandwidth described by Land et al., [1964]—are used to calculate the electromechanical coupling constant and permittivity. Using the coupling constant, the permittivity, and a third point a new elastic constant is calculated and the process is repeated until all constants converge. The method helps to evaluate the material constants in a limited region about the resonance, and the values determined are only valid within this region. One of the disadvantages of Smits' technique is that it relies on the values of the impedance or admittance around resonance [Smits, 1976 and 1985]. Because these values can increase by several orders of magnitude around resonance, care must be taken to ensure that the measuring instrument is not overburdened. The values of the material constants can also depend on the points chosen to analyze the spectrum, although the values may only vary slightly, depending on the degree of dispersion and measurement error. This variance can usually be corrected by repeated analysis using different points and averaging the results. A more recent investigation of Smits' technique has been published [Alemany et al., 1994], which looks at procedures to obtain convergence during iterations.

An alternative technique by Sherrit et al. [1992] is noniterative and determines the material constants from critical frequencies in the spectrum. The technique is similar to the *IEEE Standard on Piezoelectricity* [1987] method for determining the real elastic, dielectric, and piezoelectric constants, but all values are treated as complex. Land, Smith, and Westgate [1964] showed that the mechanical Q_M of the resonator was inversely related to the bandwidth of the resonance

$$Q_M = \frac{f_S}{\Delta f_S} = \frac{\mathrm{Re}(c)}{\mathrm{Im}(c)}, \tag{65}$$

where c is the elastic stiffness and $\Delta f_S = (f_{+1/2} - f_{-1/2})$ is the frequency separation between the high and low half-power frequencies about the series resonance. This equation was developed for the length extensional vibrator and utilized a relationship between the mechanical Q_M of the elastic compliance and the series resonance frequency. The relationship is only approximate since the elastic constant is controlled by the parallel resonance frequency for this mode, and it assumes that the length-extensional electromechanical-coupling constant loss component is insignificant. The mechanical Q_M is more accurately determined from the parallel resonance frequency since for the length extensional mode the

Q_M value is independent of the loss component of the electromechanical coupling:

$$Q_M = \frac{f_P}{\Delta f_P} = \frac{\mathrm{Re}(c)}{\mathrm{Im}(c)}. \tag{66}$$

This approach has been extended to determine a complex frequency constant in terms of the critical frequencies in the impedance and admittance spectra. The complex frequency constants are defined as

$$f_P = f_P\left(1 - i\frac{\Delta f_P}{f_P}\right)^{-1/2} \tag{67}$$

$$f_S = f_S\left(1 - i\frac{\Delta f_S}{f_S}\right)^{-1/2}, \tag{68}$$

where f_S and f_P are the series resonance frequency (maximum in G/ω) and the parallel resonance frequency (maximum in $R\omega$). The bandwidths $\Delta f_P = (f_{+1/2P} - f_{-1/2P})$ and $\Delta f_S = (f_{+1/2S} - f_{-1/2S})$ are shown in Fig. 1. They correspond to the frequency difference between the maximum and the minimum in the product of the reactance and angular frequency $X\omega$ spectra, or to the difference in frequency of the maximum and the minimum of the quotient of the susceptance and the angular frequency spectra B/ω. It should be noted that the current *IEEE Standard on Piezoelectricity* [1987] defines these frequencies using the maximum and minimum in the R, X, G, and B spectra rather than the $R\omega$, $X\omega$, G/ω, and B/ω spectra. It is argued that the latter are more accurate. Consider a typical impedance or admittance equation for a piezoelectric resonator. The equation has the form

$$Z \propto \frac{D}{\omega}\left(1 - \frac{k^2 \tan b\omega}{b\omega}\right), \tag{69}$$

where D is a function of the dielectric permittivity and the geometry and b is a function of the elastic constant and the density and length of the resonator. The technique proposed by the *IEEE Standard on Piezoelectricity* [1987] for calculating the elastic and electromechanical-coupling constant was derived using the fact that the part of the equation in brackets goes to infinity or to zero at the parallel or series resonance frequency. For a high mechanical Q_M material the resonance is quite sharp, and the $1/\omega$ term outside the bracket has little effect on the position of f_P and f_S determined from Z or Y. If, on the other hand, the resonance has a low mechanical Q_M, then the resonance is quite broad. The maximum and the minimum in the impedance determined from Z or Y can be shifted substantially, and an error in the determination of the elastic and electromechanical coupling constant can occur. If the critical frequencies are defined in terms of the $R\omega$, $X\omega$, G/ω, and B/ω spectra, the error caused by the shift in the frequencies is removed and a more accurate determination of the elastic and electromechanical coupling constant is made.

It is generally assumed that the analysis of the fundamental resonance is all that is required to determine the "small-signal coefficients" of a material. The frequency dependence of material constants governing a resonance is usually ignored even though it is well known that piezoelectric polymers display large frequency dispersion in the dielectric permittivity. However, dispersion is not limited to the permittivity and the piezoelectric and the elastic constants can have a measurable dispersion. This can readily be seen in spectra of samples (e.g., PVDF) that have been fit to the resonance impedance equations using the fundamental resonance. The fit is generally very good around the fundamental resonance; however, at higher-order resonance the fit may become poor and the peaks in the fit and the data may not overlap.

The technique proposed by Smits [1976] to analyze the material constants can be extended to fit the impedance equation to higher-order resonances. To do this one must analyze higher-order resonances one has to ensure that the correct quadrant is used to evaluate the function for the correct resonance order. This translates to adding a δ term to the real component of the arctan function, where δ is a function of the resonance order n of the form

$$\delta = \frac{(2n-1)\pi}{2}, \tag{70}$$

where the resonance order is $n = 1$ for the fundamental resonance ($n = 2$ for the second resonance, etc.). The material constants can then be analyzed at the fundamental and higher-order resonances, and the material constants at each frequency over the frequency range of the spectrum can be determined. The result is a set of three complex constants (elastic, dielectric, piezoelectric) determined at discrete frequencies. If the dispersion mechanism is known, the results can then be fitted to the dispersion model (e.g., viscoelastic, debye, etc.). If the specific mechanism is as yet unknown, these constants may be fit to a general polynomial to determine a functional form for the dispersion. The piezoelectric, dielectric, and elastic constant then can be written as polynomials of the form

$$\operatorname{Re} x(f) = a_0 + a_1 f + a_2 f^2 \ldots a_n f^n \tag{71}$$

$$\operatorname{Im} x(f) = b_0 + b_1 f + b_2 f^2 \ldots b_n f^n, \tag{72}$$

where x may be the piezoelectric coefficient (d, g, e, h), the elastic constant (s or c) or the dielectric constant (ε or β). The maximum value of n in the polynomial equals the number of resonance peaks that have been analyzed so that $n_{\max} = n$ is the largest order of the resonance analyzed. The fit of the material constants to the polynomial in f is arbitrary, and a polynomial in $\log f$ or other functional forms of the dispersion can be used.

15.2.1.2 Quasi-Static Measurements

The linear model of piezoelectricity cannot explain some of the behavior of piezoelectric materials and nonlinear effects have been reported by a variety of authors. Berlincourt and Krueger [1959] and Woolett and Leblanc [1973] looked at the general aspects of nonlinearity as a function of stress and field, while Krueger [1954, 1967, 1968a, and 1968b], Brown and McMahon [1962 and 1965], Cao and Evans [1993] and Woolett and Leblanc [1973] studied the stress dependence of the material properties of piezoelectric ceramics. Fukada et al. [1988] investigated nonlinearities in certain directions in PVDF, while Land, Smith, and Westgate [1964], Gdula [1968], and Viera [1986] investigated different aspects of the electric field dependence of piezoelectric ceramics. Recent work by Vinogradov and Holloway [1998a, 1998b] and Vinogradov [1999] investigated the mechanical and viscoelastic properties of PVDF as a function of stress, time, and temperature.

The majority of the studies quoted above were done under quasi-static conditions where a stress or electric field excitation is applied over some time and the properties are monitored as a function of the stress or electric field. One of the general representations of the linear equations of piezoelectricity is shown in Eq. (57):

$$\begin{aligned} S_p &= s^E_{pq} T_q + d_{pm} E_m \\ D_m &= \varepsilon^T_{mn} E_n + d_{pm} T_p \end{aligned} . \tag{73}$$

These are coupled linear equations and due to the coupling it is apparent that typical nonlinear curves such as the ferroelectric hysteresis curve ignore an important variable. Consider the case where a rod sample has electrodes on the major faces and is free to expand ($T = 0$) under the application of an electric field. For a typical ferroelectric hysteresis measurement, both the charge Q and the field E applied to the sample are measured using the method of Sawyer and Tower [1930]. Under an electric field the sample both expands and forms charges on the face of the electrodes. In the case of an ideal piezoelectric material one would see a linear strain and a linear charge build up as a function of the electric field. However, measurement of the strain S and electric displacement D in real materials show that the curves are nonlinear and display hysteresis due to the ferroelectric nature of the material (domain switching). The converse effect can be monitored as a function of stress by measurement of the current and strain. Some of the possible different quasi-static measurements are listed in Table 2.

Table 2: Some of the various quasi-static measurements that can be made on a piezoelectric material with one of the variables set to zero.

Linear Equations	Boundary Condition	Simultaneous Equations	
$S_p = s^E_{pq}T_q + d_{pm}E_m$ $D_m = \varepsilon^T_{mn}E_n + d_{pm}T_p$	$T = 0$ (unclamped) Apply E-Measure S and D	$S = dE$	$D = \varepsilon^T E$
$S_p = s^E_{pq}T_q + d_{pm}E_m$ $D_m = \varepsilon^T_{mn}E_n + d_{pm}T_p$	$E = 0$ (short circuit) Apply T-Measure S and D	$S = s^E T$	$D = dT$
$S_p = s^D_{pq}T_q + g_{pm}D_m$ $E_m = \beta^T_{mn}D_m - g_{pm}T_p$	$T = 0$ (unclamped) Apply D (Charge electrode) -Measure S and E	$S = gD$	$E = \beta^T D$
$S_p = s^D_{pq}T_q + g_{pm}D_m$ $E_m = \beta^T_{mn}D_m - g_{pm}T_p$	$D = 0$ (open circuit) Apply T-Measure S and E	$S = s^D T$	$E = -gT$

One variable is experimentally set to zero by appropriate choice of boundary conditions in order to simplify the measurement. A mixed measurement is possible but technically the problem of measuring two variables while adjusting two other variables is quite difficult and problems arise in the interpretation of the results. There is no sound reasoning to suggest that these effects are independent, and in the case of ferroelectric materials it is highly likely that they are not. If the electric field is set to zero and the strain and electric displacement are monitored as a function of the stress, different results would be expected than if both a stress and electric field were applied, since both stress T and field E can affect the polarization in the material.

It is obvious from the preceding discussion that care should be taken when choosing the sample geometry to match the boundary conditions shown in Table 2. In many cases in the literature these restrictions are ignored due to practical concerns. An engineer may need to know the high field strain of a thin plate, or in the case of thin film, evidence of ferroelectricity may be sought by measuring the D-E hysteresis curve. The majority of quasi-static measurements use the boundary conditions shown in the first two rows of Table 2. An experimental setup is shown in Fig. 3, which allows for the simultaneous measurement of the strain S and electric displacement D as a function of the electric field E. A variety of instruments can be used to measure the strain including interferometers, capacitance dilatometers, linear variable displacement transducers, optical levers, fiber optic sensors, and direct optical methods. If the sample area is A and the thickness is t, the field experienced by the sample is $E = (V - V_c)/t$ and the electric displacement is $D = CV_c/A$ where C is the standard capacitance in series with the sample. The strain S may be monitored by measuring an analog output of the strain sensor $S = kV_{\Delta X}$ where k is a linear constant that relates the strain to the output voltage of the strain sensor. Alternately the controller of the strain sensor may have a RS232 or GPIB for digital output directly to the computer.

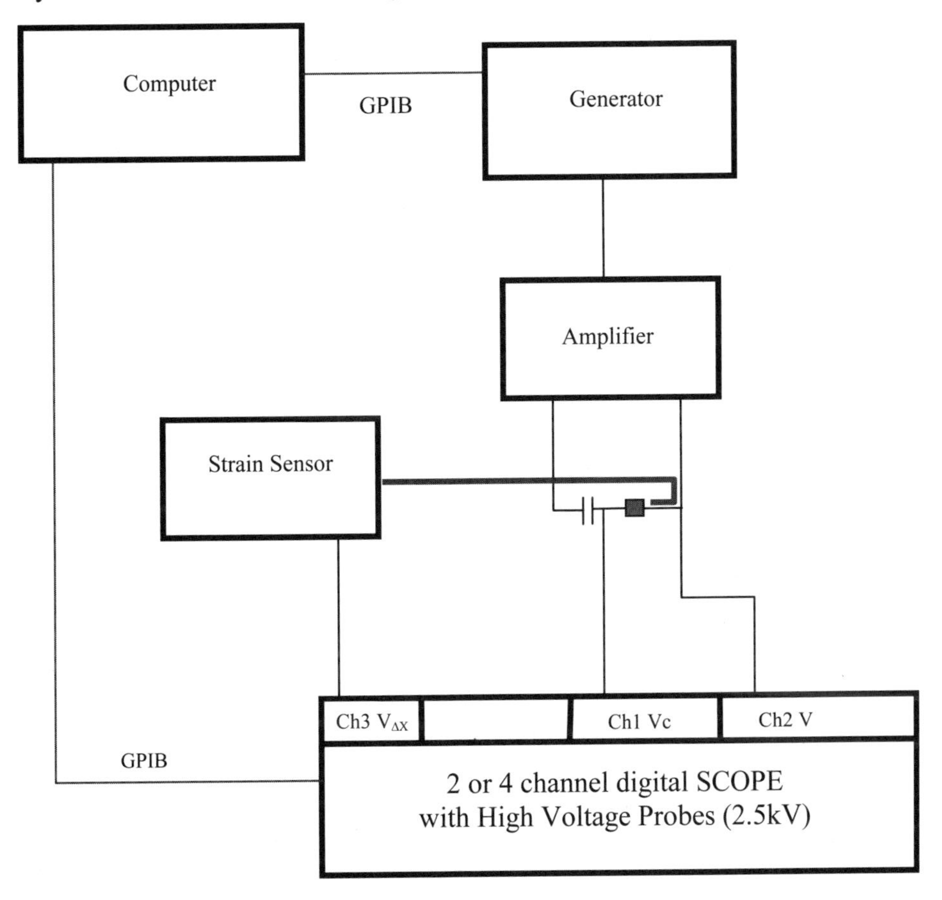

Figure 3: Schematic of a system to measure electric displacement as a function of the electric field (*D* vs. *E*) (Sawyer Tower Circuit) and simultaneously the strain as a function of the electric field (*S* vs. *E*).

A schematic of a measurement system for the strain S and electric displacement D as a function of the transverse stress is shown in Fig. 4. A sheet of sample material is fastened between grippers. A dc force is applied to take up slack in the sample and a ac force is applied. The charge generated in the sample is collected on a large capacitor connected in parallel with the sample. The strain is measured by monitoring the displacement of the free gripper. If the sample has a thickness t, length l, and width w and the electrode length is l_e, then the stress T is $T = F/wt$. The electric displacement D is $D = CV_C/wl_e$, and the strain S is $S = \Delta x/l$. It should be noted that due to the mass of the gripper, an upper frequency limit exists on the frequency of the ac forces.

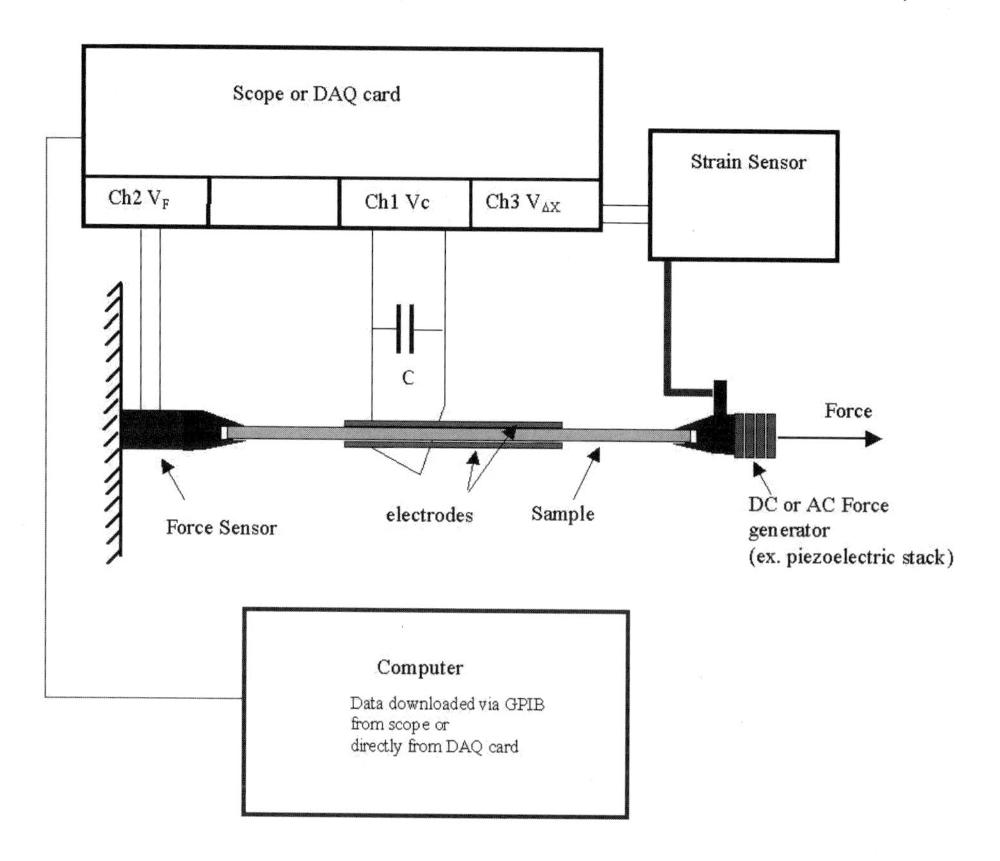

Figure 4: A schematic of an apparatus to measure the strain *S* and electric displacement *D* as a function of the transverse stress *T*. Similar arrangements can be used to measure compressive longitudinal stress dependencies [Sherrit et al., 1992c]. (Copyright 1992, IEEE.)

Another technique discussed by Tiersten [1978, 1981], Hruska and Kucero [1986], and Aleksandrov et al., [1982] is the study of the piezoelectric impedance resonance under a high electric bias field. These studies looked at single-crystal piezoelectric materials and the effect of the bias field on the resonance frequencies. In all cases the materials were assumed to be lossless and in some of the studies the field dependence of the piezoelectric and dielectric coefficients was assumed to be small. The use of a dc bias field with a small ac field is very sensitive to nonlinearities because the measurement is differential. It should be noted that only reversible contributions to the piezoelectric, elastic, and dielectric response were measured. The dc impedance of the samples is usually large and heating effects due to the dc field are much smaller than for an ac field of the same order, therefore the isothermal conditions used to derive the equations describing the experiment were maintained. The technique requires additional protection circuitry to isolate and protect the impedance analyzer from the dc field [Sherrit and Mukherjee, 1998]. Since this technique is more suited for studying nonlinear behavior, we discuss the technical details of the measurement in more detail in the next section.

15.2.2 Quadratic Response—Maxwell Stress, Electrostriction

The equations for the electric field E_i and strain S_{ij}, proposed for a nonlinear dielectric material described by Mason [1958], are:

$$\begin{aligned} E_i &= -2Q_{klij}T_{kl}D_j + \left[\beta_{ij}^T + R_{ijmnkl}T_{mn}T_{kl}\right]D_j + \mathrm{K}_{ijkl}D_jD_kD_l + \mathrm{K}_{ijklmn}D_jD_kD_lD_mD_n \\ S_{ij} &= \left[s_{ijkl}^D + R_{ijmnkl}D_mD_n\right]T_{kl} + Q_{ijmn}D_mD_n \end{aligned} \quad (74)$$

In this description, E, S, T, and D are the electric field, strain, stress, and electric displacement. The coefficients Q, β, R, s and K are the electrostrictive, inverse permittivity, electroelastic, compliance, and morphic correction terms. The superscripts define the boundary conditions of the specific constant and the subscripts designate directional dependencies of the constants. Variations of Mason's model have been published by Hom et al., [1994] who use a set of equations relating the stress T_{ij} and polarization P_i to the strain S_{ij} and electric field E_i. In their model they use the tanh saturation model proposed by Zhang and Rogers [1993] to describe the polarization at high fields. Piquette and Forsythe [1997] have put forth an alternative model based on saturation in the electric displacement, D, which varies as E divided by the square root of a quadratic in E. Under mechanical free conditions, the Mason's model shown above reduces to

$$\begin{aligned} E_i &= \varepsilon_{ij}^T D_j + \mathrm{K}_{ijkl}D_jD_kD_l + \mathrm{K}_{ijklmn}D_jD_kD_lD_mD_n \\ S_{ij} &= Q_{ijmn}D_mD_n \end{aligned} \quad (75)$$

Kloos [1995 and 1995b] published an alternate derivation for linear dielectric materials based on the Gibb's function with the electric field E and stress T as independent variables. For a linear dielectric, nonpiezoelectric material under isothermal conditions, the model is

$$\begin{aligned} D_i &= \varepsilon_{ij}^T E_j + 2\gamma_{klij}T_{kl}E_j + \frac{1}{2}\varsigma_{klmni}T_{kl}T_{mn} \\ S_{ij} &= s_{ijkl}^E T_{kl} + \frac{1}{2}\eta_{ijklmn}T_{kl}T_{mn} + \gamma_{ijmn}E_mE_n + \varsigma_{ijklm}T_{kl}E_m \end{aligned} \quad (76)$$

This model is valid for linear dielectrics and this representation is used to a large extent for polymer characterization. Under stress-free conditions this model reduces to

$$
\begin{aligned}
D_i &= \varepsilon_{ij}^T E_j \\
S_{ij} &= \gamma_{ijmn} E_m E_n
\end{aligned} \quad . \tag{77}
$$

Typical plots of the response to excitation by an electric field of an electrostrictive material for a linear and nonlinear dielectric are shown in Fig. 5.

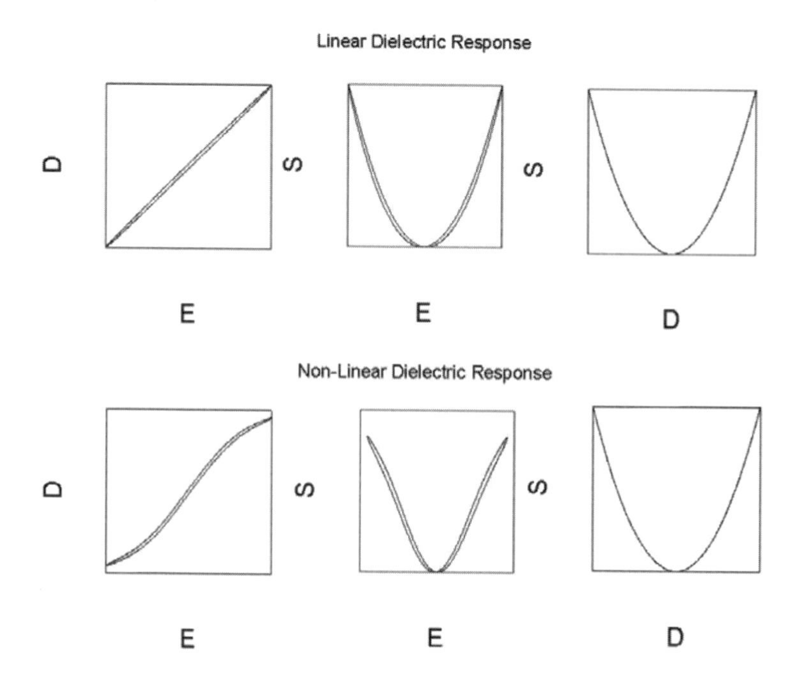

Figure 5: The electrostrictive response to a linear and nonlinear dielectric material. The strain is quadratic in *D*, however in the nonlinear dielectric material saturation is seen in both the electric displacement and the strain as a function of the electric field.

In both cases, the strain vs. electric displacement are quadratic. The strain vs. electric field for the nonlinear dielectric displays saturation, and higher-order even terms in field are required to model the strain-field behavior. It should also be noted that any hysteresis (loss) in the linear dielectric term would produce a hysteresis in the strain field data independent of any loss in the electrostrictive coefficient Q. The response shown assumes no hysteresis in the electrostrictive Q coefficient. Losses in the electrostrictive Q coefficient may be represented by a complex electrostrictive constant $Q = Q_r + iQ_i = |Q|e^{i\theta}$. The addition of this phase can be shown to introduce an additional hysteresis in the strain-electric displacement plots.

The electrostrictive coefficients determined from these techniques contain contributions that behave like electrostriction but are caused by other effects such as Maxwell stresses, viscoelastic effects, and thermal stress. In order to determine the true electrostriction, correction terms must be applied, as outlined by Kloos [1995 and 1996]. The electrostrictive coefficient as a function of the electric field measured by any instrument is therefore

$$\gamma_{ijkl}^{\text{MEAS}} = \gamma_{ijkl} + \gamma_{ijkl}^{\text{MS}} + \gamma_{ijkl}^{\text{TS}}, \tag{78}$$

the sum of electrostriction, Maxwell, and thermal stress terms. The correction for Maxwell stresses was recently reported by Kloos [1995] and Krakovsky et al., [1999] for an isotropic polymer with compliant electrodes. In the low-frequency limit the Maxwell stress correction in the direction of the applied field for an isotropic polymer is reported as

$$\gamma^{\text{MS}} = -\frac{\varepsilon(1+2\nu)}{2Y}, \tag{79}$$

where Y is the Young's modulus, ν is Poisson's ratio, and ε is the free(unclamped) permittivity of the polymer. In the transverse direction Krakovsky, Romijn, and Posthuma deBoer [1999] report that the Maxwell stress correction is

$$\gamma^{\text{MS}} = +\frac{\varepsilon}{2Y}. \tag{80}$$

The thermal stress correction for an isotropic dielectric is dependent on the thermal boundary conditions (isothermal/adiabatic), coefficient of thermal expansion and heat capacity [Kloos, 1995]. Typically this correction is quite small, however, if the coefficient of thermal expansion is large or the heat capacity (constant stress) is small the order of this correction should be calculated to determine its significance.

In some elastomers the quadratic response is due primarily to the presence of the Maxwell stress in the material [Zhang et al., 1997; Pelrine et al., 1998, 2000; Su et al., 2000; Kornbluh et al., 2000]. These materials can exhibit very large lateral strains of the order of 10 to 215%. It should be noted that when dealing with strain levels of this order that the engineering strain approximation ($S = \Delta l/l$) is no longer a valid approximation to the Lagrangian/Eulerian strain [Saada, 1974], and one should use $\Delta l/(l + \Delta l)$ as was noted by Pelrine et al., [1998]. It should be kept in mind that due to the coupling between the longitudinal and transverse strains, an area correction is required to determine the proper dielectric response since the capacitance is dependent on the area of the film.

A variety of technical issues arise when trying to characterize these materials. The primary problem is the large dispersion that is present [Wang et al., 1994; and Zhang et al., 1997] in the elastic, dielectric, and electrostriction constants. The properties change as a function of frequency, and unless the specific relaxation mechanism is known one cannot, in general, extrapolate results measured at one frequency to other frequency ranges. This requires characterizing the material over the frequency range to be used by the transducer/actuator in order to correlate measured material properties to performance of the transducer or actuator. Another complication is that at higher field levels, higher-order terms in the thermodynamic potentials will affect the overall response of the material. In quasi-static measurements these would be seen as saturation in the response of a strain field plot or frequency components in the strain time curve that are greater than 2ω assuming the applied field is $E_0\cos(\omega t)$. One approach that has been used to characterize electrostrictive ceramics for high-frequency transducer materials is the biased resonance measurement. A schematic of an experimental system for measuring the elastic, dielectric, and induced piezoelectric properties is shown in Fig. 6. A large dc bias voltage is applied to the sample. The impedance analyzer is shielded from the high voltage by a pair of high-voltage, high-capacity blocking capacitors. Since the sample capacitance is a small fraction of the capacitance of the blocking capacitors, the impedance of the blocking capacitors is generally negligible. By analyzing the impedance resonance curves and plotting the results as a function of frequency, one can separate the dependencies of the elastic, dielectric, and induced piezoelectric constant on the applied field [Sherrit and Mukherjee, 1998]. It is doubtful that the length extensional resonator can be tested due to the boundary conditions and the extremely large voltages required. However, the thickness (TE), length thickness (LT), and radial mode (RAD) resonators can be measured. It should be noted that under a large bias field the material would no longer be isotropic but rather isotropic in the plane perpendicular to the bias field. As in the case of the quasi-static measurements, the geometry (area and thickness) must be known as a function of the field to evaluate the fields and material constants when the quasi-static strains exceed 1%.

15.2.3 Higher-Order Effects—Ferroelectricity

Although the phenomenon of ferroelectricity is not generally used to couple electric to mechanical energy, a switching strain is associated with a ferroelectric material driven to field levels above its coercive field. In general these data are useful for the transducer/actuator designer since it puts limits on the size of the ac drive field or alternatively determines the size of the bias field required to inhibit switching. Typical ferroelectric polarization and strain response curves are shown in Fig. 7. The electric displacement as a function of the electric field is a typical hysteresis figure and is characterized by the coercive field E_C, saturation D_S, and remnant D_r displacement. The strain response is characterized by the switching strain ΔS and the coercive field. A variety of analytical models have been presented for the *D-E*, *P-E* hysteresis loops. An example is the model proposed

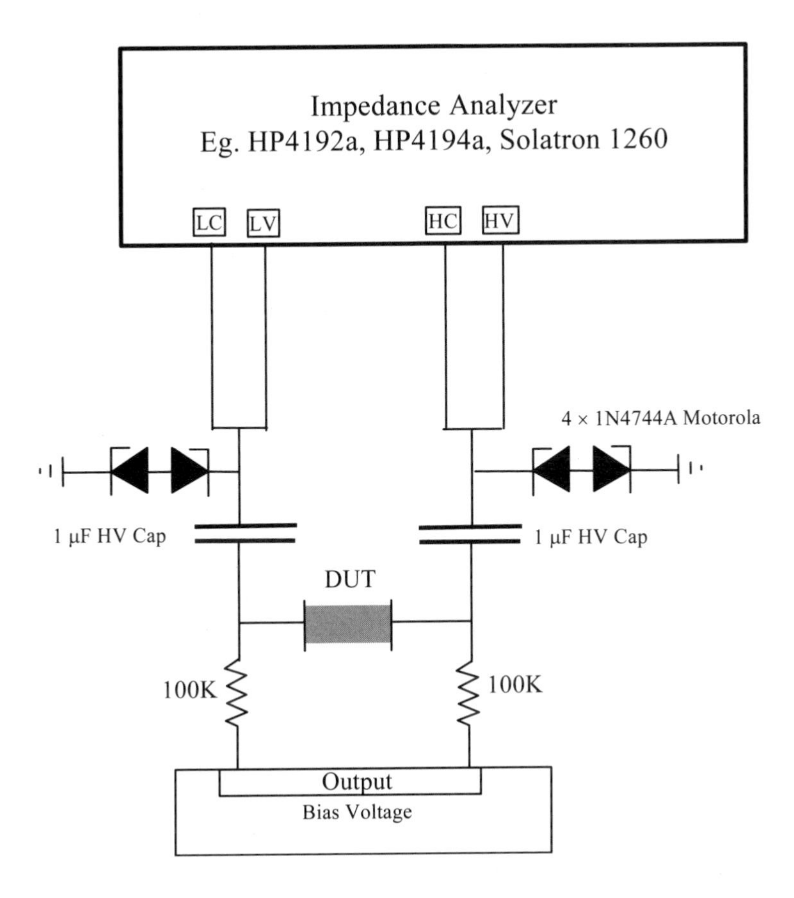

Figure 6: Schematic diagram of the isolation circuitry for the dc-biased resonance measurement. The protection diodes are optional but recommended.

piecewise for increasing and decreasing the field. It assumes switching between two orientations. Although this assumption is generally an oversimplification, the model typically will fit the data well enough by Chen et al., [1994]. The model uses the tanh(x) saturation function and is to determine critical features such as E_C, D_S and D_r. A more compact form can be written if the equation is made parametric in field E and the time derivative of the field $\dot{E}$. ε is the high field permittivity and is used to account for hysteresis curves, which do not completely saturate.

$$D(E) = D_S \tanh\left[k(E - \frac{\dot{E}}{|\dot{E}|} E_C) \right] + \varepsilon E . \tag{81}$$

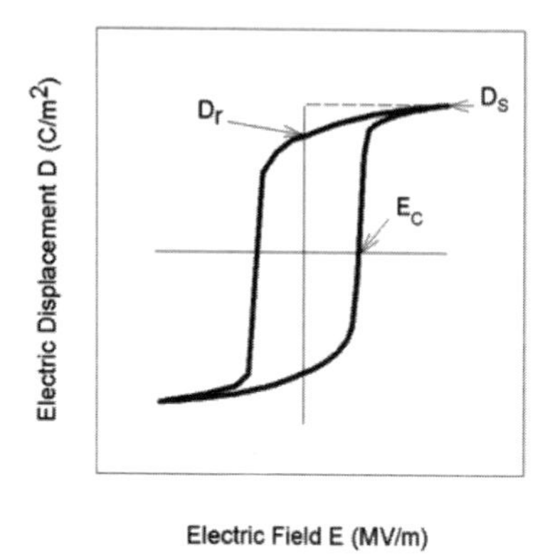

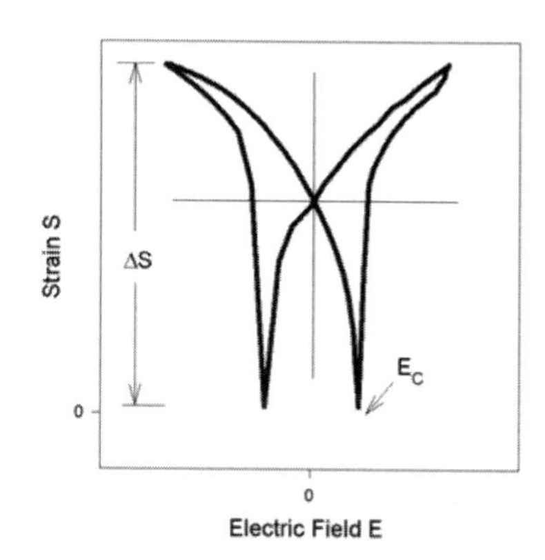

Figure 7: Typical dielectric and strain response of a ferroelectric material with a positive electrostriction coefficient *Q* driven at field levels above E_C.

The strain $S(E)$ shown in Fig. 7 can be modeled using $S(E) = Q[D(E)]^2$. The hysteresis plots shown in Fig. 7 can be measured with the apparatus shown in Fig. 3. For a discussion on the various models used to describe ferroelectric hysteresis see the recent articles by Smith and Ounaies [1999]. Two recent books [Herbert et al., 1988; Nalwa, 1998] give a good general overview of the ferroelectric polymer field.

15.3 Characterization of Ionic EAP with Diffusion-Dependent Strain

Characterization of the properties of the ionic EAP materials, which involve diffusion-dependent strain, poses unique challenges to the development of test methods [Bar-Cohen, 2000]. This section describes the available methods with emphasis on IPMC [see Chapter 6]. IPMC consists of Nafion [Tant et al., 1997] or Flemion [Oguro et al., 1999] as membranes made of fluorocarbon backbones and mobile cations (counter-ions). The exact mechanism that is responsible for the electroactivation is still the subject of a series of studies. However, recently significant progress has been made toward understanding the related phenomena [Nemat-Nasser and Li, 2000; Chapter 6]. When a voltage (<5 V) is applied to a hydrated IPMC sample, the large ionic conductivity may promote electro-osmosis [Helfferich, 1995] and/or hydrolysis [Holze and Ahn, 1992]. The former response manifests itself as a bending of the film toward the positive electrode (anode) and can be exploited in actuation applications [Sewa et al., 1998]. The induction of electrolysis is an undesired electrochemical reaction that consumes power and may damage the electrode by producing gas. Kanno et al., [1994] have shown that the bending response of Pt-electroded Nafion (Na+ counter-ion) is complicated by relaxation processes. If a dc voltage is applied for sufficient time duration, the primary deflection will gradually return to its initial position. This phenomenon is thought to be due to the excess concentration of water near the cathode and its subsequent back-flux [Okada et al., 1998]. It is interesting to note that this behavior is not evident in Au–electroded Flemion (tetra-n-butylammonium counter-ion) [Oguro et al., 1999]. The large size of the cation and its sluggish mobility may provide an explanation.

The large bending deflections, the required hydration, and the relaxation processes that are involved with IPMC electroactivation make the task of electromechanically characterizing such materials difficult. This section focuses on techniques that were used for testing the response of gold-electroded Flemion (tetra-n-butylammonium counter-ion). Similar tests can be applied to other ionic EAP materials such as polypyrrole [Otero and Sasinena, 1997] and also electronic EAP materials. Due to a lack of standard test procedures no reliable data are currently available and therefore this section focuses on characterization tools.

15.3.1 Micrography

Micrography is a well-established field. A large number of test methods are available for the examination of microscopic details of test objects. Various methods are used including optical, such as visual microscopy, as well as enhancement imaging allowing one to see details beyond the capability of visual techniques. Scanning electron microscopy (SEM) and similar techniques are examples of such techniques. These techniques are useful when examining ionic EAP materials since insight can be gained into the microstructure supporting the efforts to understand the mechanism of actuation, determine the quality of the material, and assure the conformance to a standard once it is established. An operator can perform the evaluation of images, particularly at the stage of research when there is a need to understand the characteristic structure when there is no sufficient database. IPMC is one of the materials that was studied with the aid of such techniques. An example is given in Fig. 8. A gold electroded IPMC (made of 0.18-mm Nafion117 with perfluoro-carboxylate) was examined micrographically after the gold layer was applied in seven cycles and a dendritic structure was observed. Generally, gold electroding exhibited superior electromechanical response compared to platinum electroding [Abe et al., 1998]. The electrode microscopic view of its cross section shows a difference from the uniform diffusion appearance that is obtained by such researchers as Shahinpoor et al. [1998]. To understand the effect of the electrode structure other methods need to be used to determine its effect on the material behavior. Generally, micrographic techniques allow measurement of the thickness of the polymer layer(s) and electrodes, determination of the internal structure, identification of defects, determination of the uniformity of the material, and possibly identification of anisotropic structures.

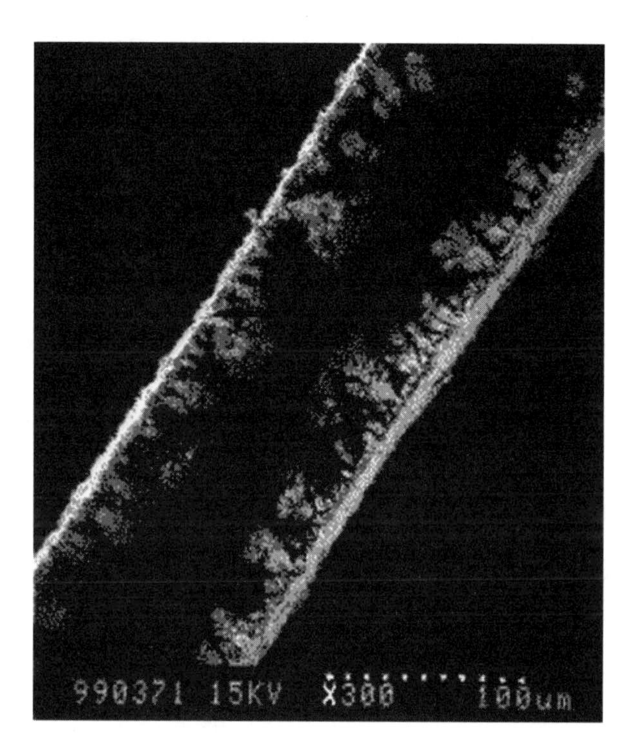

Figure 8: An SEM view of a cross section of an IPMC made of perfluorocarboxylate membrane with tetra-n-butylammonium cation and seven cycles of ion exchange and reduction (resulting dendritic growth) of the gold electrodes [Courtesy of Keisuke Oguro, Osaka National Research Institute, Japan].

15.3.2 Voltammetry

Voltammetry refers to techniques in which the relationship between voltage and current is observed during an electrochemical process. The voltage is applied and the current is measured. A plot of the current vs. the voltage is called a voltammogram. Peaks in these curves represent electrochemical reactions. Voltammograms are used to access reaction/deposition rates, reversibility, and reaction potentials. The voltage sweep rate can be adjusted to determine relaxation processes of deposition. In addition to the electrochemical reaction, the electronic conductivity and capacitance can also contribute to the shape of the voltammogram and these parameters need to be taken into account.

15.3.3 Sheet Resistance

Sheet resistance is an indicator of the quality of the electrodes, and it is commonly measured using a four-probe system. A schematic view is shown in

Fig. 9. This test system is ideal for measuring the sheet resistance or conductivity of metal films. A current-supply forces current through the sheet electrodes. This arrangement allows for an accurate determination of the impedance of a conductive material by eliminating contact resistance from the measurement. The outer current electrodes are used to force a current through the sample. The inner electrodes measure the voltage drop between two fixed points on the sample. Since the input impedance of the voltage probes is very large compared to the voltage drop due to contact resistance, an accurate measurement of the sample impedance can be made while excluding the contact resistance. If the voltage probes are separated by a distance l and the width, or w, and the thickness of the film is t, respectively, the resistance of the film is

$$R = \frac{\rho \; l}{wt}\,, \tag{82}$$

where ρ is the resistivity of the film.

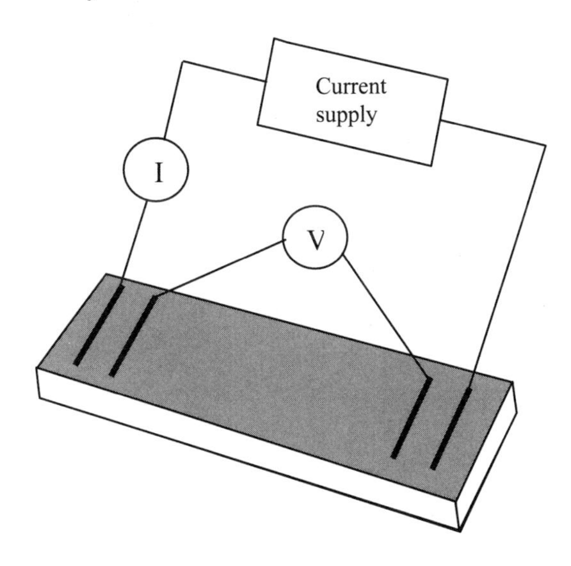

Figure 9: Schematic diagram of a four-probe sheet resistance measurement.

It is common practice when discussing sheet resistance to report data in units of ohms per square (area). In this case, the length of the conductor is equal to the width, the resistance is

$$R_{\square} = \frac{\rho \; w}{wt} = \frac{\rho}{t}\,. \tag{83}$$

To convert sheet resistances in ohms per square to a resistivity value multiply the sheet resistance by the thickness of the film

$$\rho = R_{\square} t . \tag{84}$$

In determining the sheet resistance of an electrode material on an ionic conducting substrate, care should be taken to assure that only electronic contributions are being measured. This can be accomplished by keeping the voltage below any reaction potentials and measuring the voltages and current after a steady state has been reached. Since the bending moment is proportional to the local current density, in an ideal ionic bender the voltage drop along the electrode should be a small fraction of the voltage drop across the ion conduction membrane (in order to ensure a constant bending along the membrane). In terms of the total electrode resistance R and in plane resistance of the ion membrane R_i this occurs when

$$R_i >> R. \tag{85}$$

If this condition is not met then the current density along the membrane will be larger at the base of the membrane due to the electrode resistance.

Large electrode resistance can be caused by poor conductivity, insufficient electrode thickness, microcracks in the electrode, and inhomogeneous deposition. Large cyclic tensile and compressive stresses on the electrode during bending may cause fatigue and further increase the sheet resistance. In Fig. 10, a schematic view shows the effect of the sheet resistance on the curvature. The density of arrows indicates local current density. In the case where the electrode resistance is much smaller than the resistance of the ion membrane the current density and curvature are constant. In the case where $R >> R_i$, the curvature and current density are found to be a maximum near the electrical connections at the base of the bender. Kanno et al [1995a, 1995b, 1996a, 1996b] discuss this effect in some detail along with additional effects related to the conduction length.

15.3.4 Mechanical Testing

Mechanical testing of polymers involves measuring the stress-strain behavior as a function of frequency f, temperature ϑ, stress T, time t, and relative humidity for ionic EAP. A variety of standard tests are available for the mechanical testing of polymers for various properties. These include:

- Stress Analysis
- Ultimate Strength
- Energy Dissipation and Damping
- Impact Testing
- Fatigue Behavior
- Elasticity
- Glass Transition and Thermal Behavior
- Creep

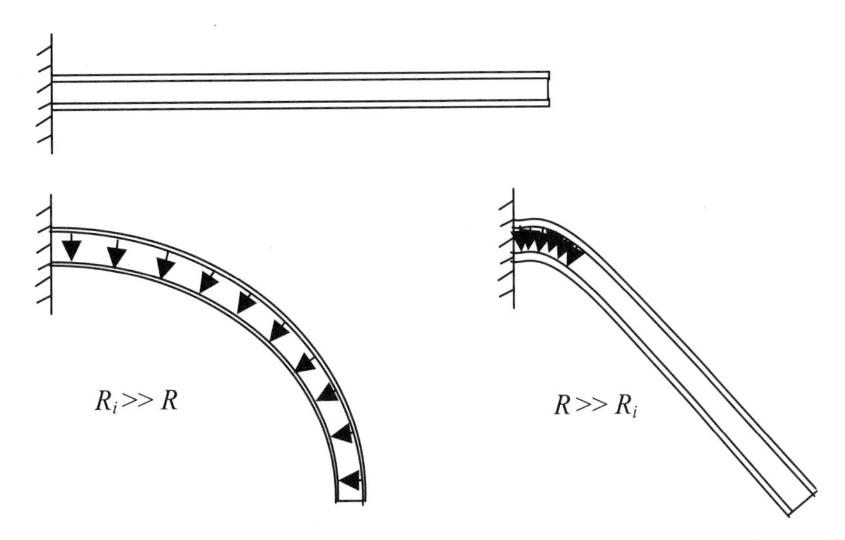

Figure 10: The effect of sheet resistance on the current density and local bending moment in ionic bender material.

The American Society for Testing and Measurement (ASTM) has published variety of standards for testing of materials, including ASTM Standard E1640-99, ASTM Standard D4065-95, and ASTM Standard D6049-96. Also, a number of books on mechanical characterization of polymers has been published by Swallowe [1999, Ward et al., [1993], and Lakes [1998]. The mechanical properties of EAP are tested in a similar manner to other polymer materials, however, in the case of IPMC and other ionic EAP materials, the relative humidity has to be controlled or assessed during the experiment [Yeo and Eisenberg, 1977; Nakano and MacKnight, 1984].

15.3.5 Displacement Measurement Techniques

Although IPMC show longitudinal and transverse strains under the application of the applied voltage, the effect is found to be much smaller than the developed bending strain. Measurement of the small longitudinal and transverse strains can be accomplished using the same apparatus that is used for piezoelectric materials, however the measurement procedure for IPMC is complicated by the need for a "wet" system. A schematic diagram of a setup using visual tracking of strained EAP is shown in Fig. 11. In this figure, a Flemion membrane is shown submerged in deionized water. Counting the number of pixels along the dimensions of the rectangular part of the sample holder and comparing the results with the known size in millimeters determines the scale. This scaling factor is determined to be 0.15 mm/pixel. The outline of the area traversed by the bending sample under the influence of applied 3.0 V, 0.05-Hz cosine signal is marked on the image. The plot in Fig. 12 shows the swept region of Fig. 11 with scaled units. The circles respresent pixel positions determined by an edge detection

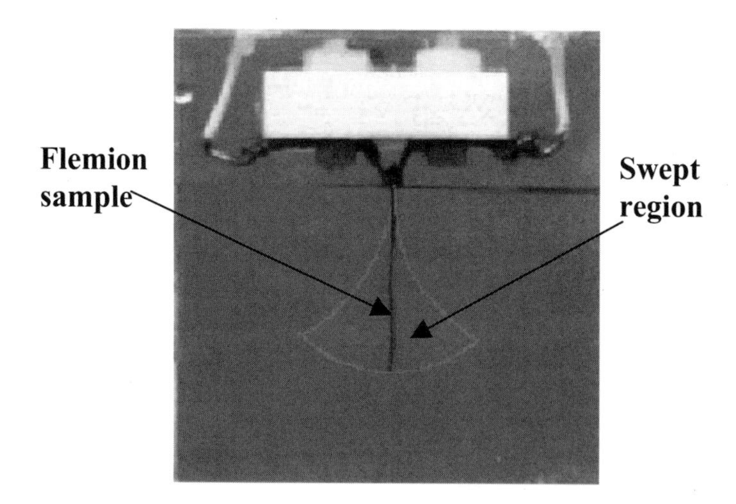

Figure 11: Image displaying electrically induced bending deformation of Flemion sample.

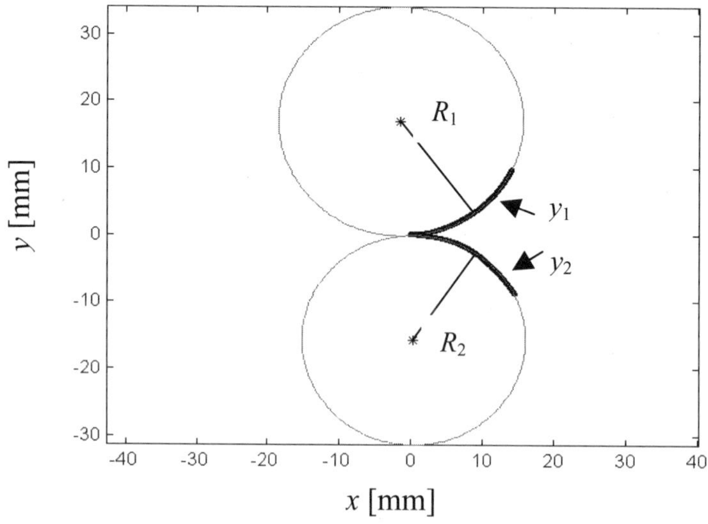

Figure 12: The bending response can be approximated as constant curvature, which simplifies modeling.

algorithm employing a radon transform [Lim, 1990]. The sampling frequency (number of pictures taken per second) was 2 Hz. Using least squares curve fitting resulted in analytic expression, y_1 and y_2, for the extreme bending deflections. The shaded area was calculated to be 125 mm^2. An asymmetry in the bending response is evident and a circular curvature appears when a pristine IPMC is tested. However, experiments have shown that subjecting the sample to pulsed activations and dehydration would lead to significant and inconsistent permanent deformation.

When a cantilevered beam is subjected to pure bending, its edge can be described by arc of a circle with a curvature, $1/R$, of the neutral surface. This can be expressed as

$$\frac{1}{R} = \frac{PL}{EI}, \tag{86}$$

where P is a concentrated load acting perpendicular to the free end of the cantilever of length L. The "stiffness" of the beam is proportional to the product of the elastic modulus E and moment of inertia I, and is commonly referred to as the *flexural rigidity*. To a first approximation the bending curvature of the Flemion samples subjected to a voltage of 3.0 V, 0.05-Hz cosine wave, can be assumed constant along its length. Figure 13 shows the two deflections, y_1 and y_2, fitted with circles of radii $R_1 = 17.1$ mm and $R_2 = -15.6$ mm. With appropriate values of elastic modulus and neglecting viscous forces of the surrounding water, it is possible to estimate the induced tip force generated by a given voltage using Eq. (86). There are a number of different ways to express the bending motion. Figure 14 shows tip deflection as a function of time as measured by (b) linear displacment from the y-axis, and (c) angular displacment. It is also possible to monitor the change in curvature [Fig. 14(b)].

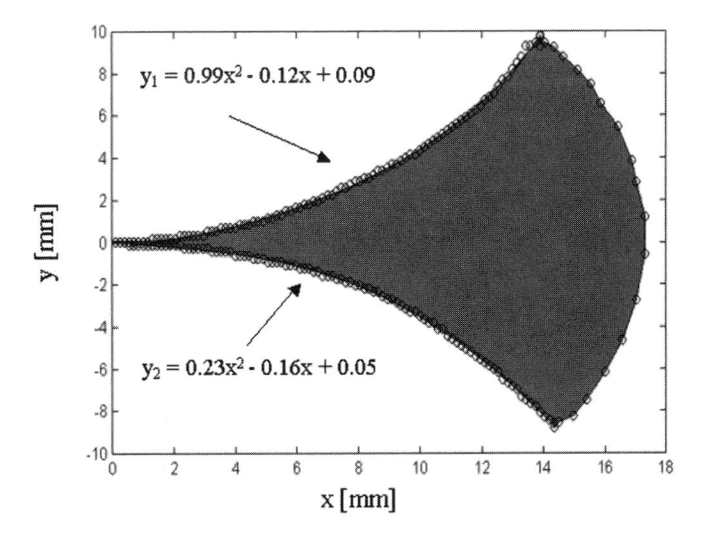

Figure 13: Bending response of Nafion to applied voltage of 3.0 V, 0.05 Hz. The shaded region represents the swept area equal to 125 mm^2.

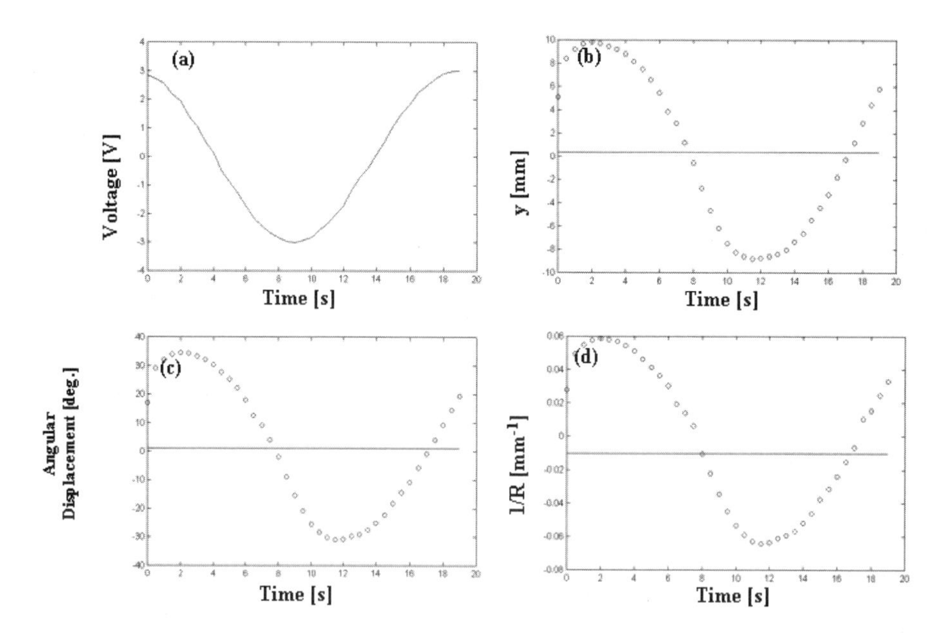

Figure 14: Bending response to (a) applied voltage can be quantified in a number of ways including: (b) displacement of tip from *y*-axis, (c) angular displacement of tip, (d) curvature.

The hysteresis between angular tip deflection and applied voltage is shown in Fig. 15 for 1.0 V, 2.0 V, and 3.0 V at 0.05 Hz. Figure 16 shows the response to a dc voltage applied at t = 0 s. Both of these figures imply that the mechanism for deformation is sluggish, probably due to the large size and low mobility of the tetra-n-butylammonium counter-ion. The relaxation process is quite different from that observed for Pt-Nafion (Na^+ counter-ion) [Kanno et al., 1994]. Tip force was also measured. The sample was removed for water and 3 mm of free end was in contact with the load cell. A 5-V, 0.05-Hz cosine wave was applied and the sampling frequency was 1.0 Hz. From Fig. 17 it is evident that the nominal value of force is quite small (~0.6 mN peak). The horizontal line in the plots represents the initial displacement of the sample before voltage was applied.

Using a simple macroscopic model described in detail in Chapter 12 of this book [Bhattacharya, 2001], three material constants can be determined: the elastic modulus, *E*, the saturation curvature at unit-applied voltage, *c*, and the time-constant, τ, using different time-dependant voltages and different forces. Using experimental data for zero and 25-mg tip mass and applying the inverse problem, the following values were determined: $c = 53.4\ (\text{mV})^{-1}$, $\tau = 53.4$ sec and $E = 72$ MPa. This methodology of determining the relation between the force, displacement, and the material elastic constants is under development, and it requires addressing the nonlinearity of the behavior, as well as the fact that the problem is a two-dimensional one.

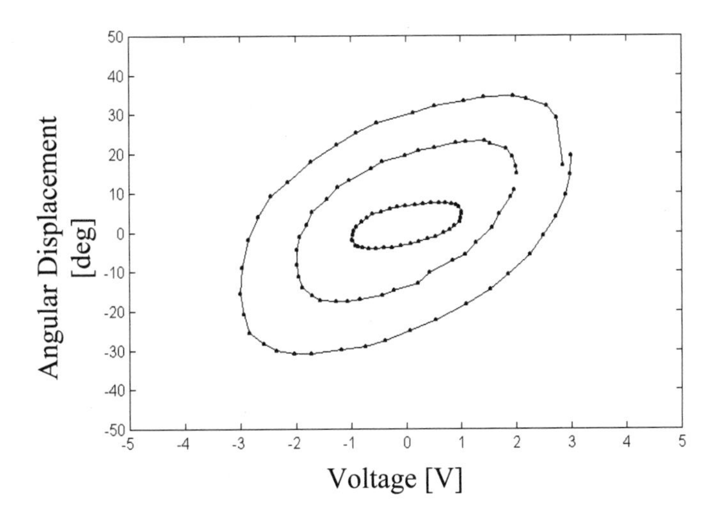

Figure 15: Hysteresis of tip displacement for 1.0 V, 2.0 V, and 3.0 V at 0.05 Hz.

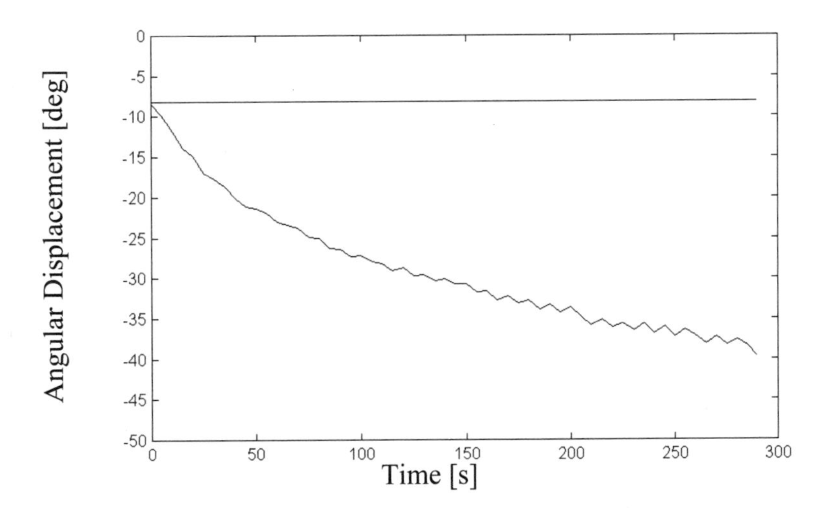

Figure 16: Tip displacement response to applied dc voltage of 2.0 V. The horizontal line represents the initial deformation present before voltage is applied.

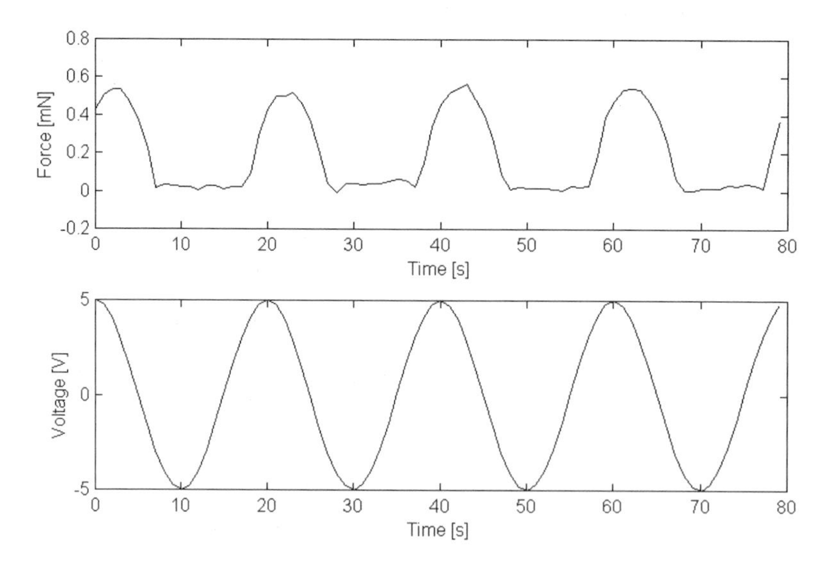

Figure 17: Tip force and applied voltage.

15.3.6 Time Response, Macro Model, and Parameter Extraction

The IPMC as an ionic EAP material type has a much slower response than electronic EAP materials. The actuation and input time responses can be measured by monitoring the displacement and the input current while applying a proper voltage. Using a step voltage, time domain responses can be measured directly. To obtain the parameters of the material properties, macro models are needed to define the relative parameters. Then, the parameter values can be determined by finding the best fit to the experimental data.

15.3.6.1 Electromechanical response and relaxation

15.3.6.1.1 Introduction

Studies indicate that the response of IPMC strongly depends on its backbone polymer and ionic content [Nemat-Nasser et al., 2001; and Onishi et al., 2001], especially the counter ion. Basically, they can be divided into two categories based on their cations size: (a) small cations such as Li^+, Na^+ and K^+, and (b) large cations such as alkyl ammonium ions. The typical actuation responses of IPMC with these two types of cations are shown in Figs. 18 and 19. It is interesting to note that the curve in Fig. 18 is very similar to the mechanical creep under static loading conditions in linearly viscoelastic materials and rheological fluids. The diagrams in Fig. 19(b) are identical to creep and recovery responses to the application and removal of mechanical loads. However, the process behind the phenomenon is different. The IPMC with small cation of Li^+ has a quick response to the applied voltage and a slow back relaxation. The IPMC with large cation of tetra-n-butylammonium$^+$ (TBA) responds slowly to applied voltage but with no relaxation. It is believed that it is easier for the small cations to move over the polymer backbone. The fast movement of the cations toward the cathode together with associated water molecules results in a quick bending toward to the anode. This response is followed with a relaxation that may be caused by water leakage near the cathode that flows toward the anode and through channels in the polymer backbone, resulting from a high-pressure layer. The process stops when water equilibrium is reestablished. On the contrary, large cations migrate significantly slower and present slow reaction to the electric field. Thus, no relaxation is observed, which may be the result of ions blocking the channels, or because the equilibrium with concentrated cations require more water. Although the ion and water movements described above may not include all physical mechanisms of the time-dependent behavior of IPMC, this simple explanation is still a good guide for macro-model construction.

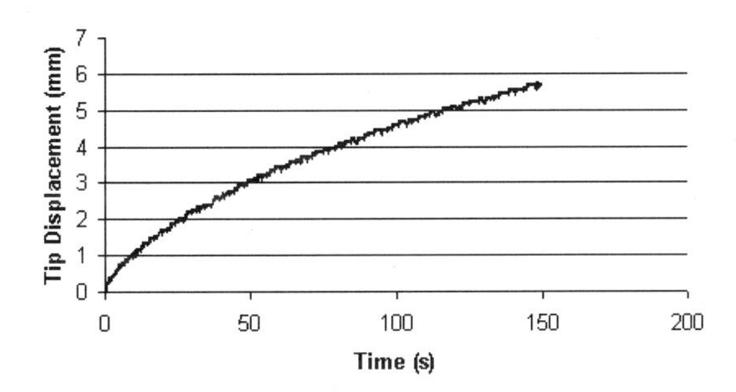

Figure 18: Response of tetra-n-butylammonium/Flemion.

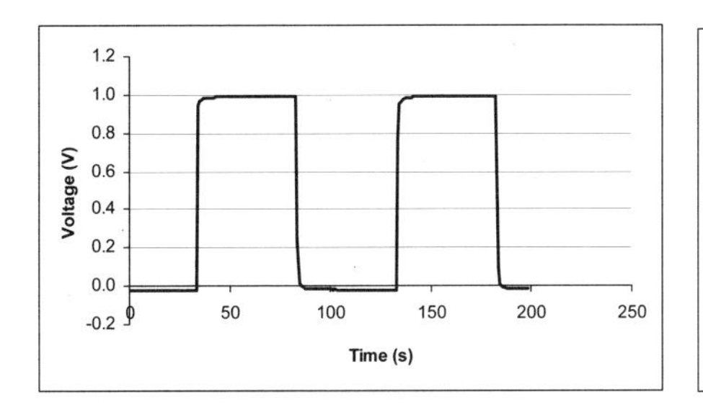

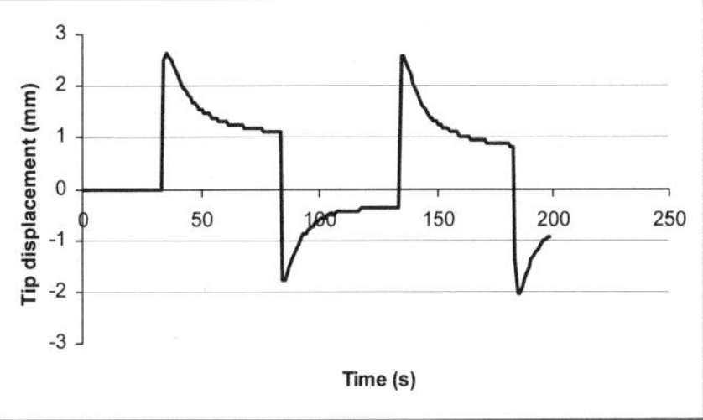

Figure 19: Response of a Li^+/Nafion strip sample of 25x3x0.19 mm^3 to 1V square wave voltage. (a) Applied voltage, (b) tip displacement.

15.3.6.1.2 Model for IPMC with relaxation

Bhattacharya et al. (2001) developed a phenomenological model for electromechanical response of the IPMC without relaxation. Bao et al. (2002) presented a macro model for an IPMC strip with relaxation, and considered the IPMC reaction without relaxation as a special case of the model. This is described in detail below.

Under a step voltage, after a very quick bending toward the anode, the strip shows a slow relaxation toward the cathode, indicating the existence of two time constants. The model was constructed based on the understanding of the possible underlying mechanism of the phenomenon. It is assumed that ions bring water from the anode to the cathode when moving under electric field. Then, there is a diffusion of the water back to the anode after initial moving toward the cathode. The time dependence of the material strip curvature can be expressed as follows:

$$\frac{dk}{dt} = K_1 \frac{dq}{dt} - \frac{1}{\tau_2}(k - K_2 q) , \tag{87}$$

where k is the curvature of the strip, q the electric charge, K_1 is the coefficient for the initial bending effect of the charge moving to the electrode, K_2 is the coefficient for bending effect of the charge in equilibrium state, and τ_2 is the relaxation time constant.

When a simple clumped RC circuit can represent the electric input response of the sample, the electric charge on the electrode can be calculated as

$$R\frac{dq}{dt} = V - \frac{q}{C} . \tag{88}$$

The solution for the step voltage is

$$q = VC(1 - e^{-t/RC}) = VC(1 - e^{-t/\tau_1}) , \tag{89}$$

where τ_1 is time constant of the RC circuit which is equal to RC. Substituting Eq. (89) with Eq. (87) and assuming the initial condition $k(0) = 0$, we have

$$k = V\left[K_{V2} - \frac{K_{V1}\tau_2 - K_{V2}\tau_1}{\tau_2 - \tau_1} e^{-t/\tau_1} + \frac{\tau_2(K_{V1} - K_{V2})}{\tau_2 - \tau_1} e^{-t/\tau_2} \right] , \tag{90}$$

where $K_{V1} = CK_1$ and $K_{V2} = CK_2$.

This model defines four parameters of IPMC materials, K_{V1}, K_{V2} , τ_1 and τ_2. The parameters can be determined by curve fitting the experiment data. Figure 20 shows the fitting results for a Li^+/Nafion strip. It is interesting to point out that the theoretical curve and the experimental data are showing an excellent agreement. The parameters obtained are $\tau_1 = 0.3$ s, $\tau_2 = 10.7$ s, $K^t_{V1} = 2.87$ mm/V, $K^t_{V2} = 1.07$ mm/V, where K^t_{V1} and K^t_{V2} are the coefficients of the tip displacement over voltage corresponding to the K_1 and K_2 in Eq. (90).

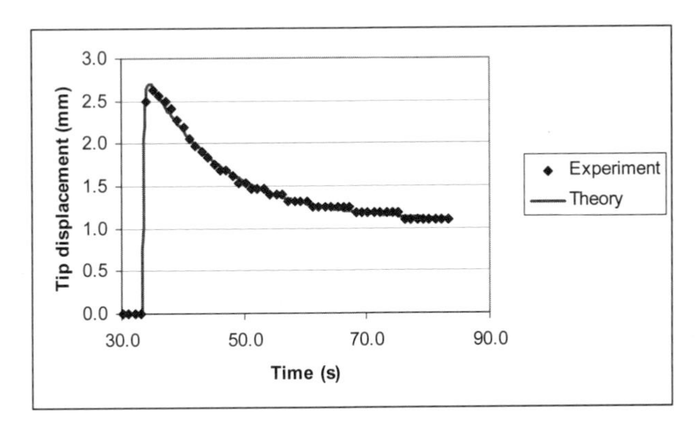

Figure 20: Comparison of the theoretical model with the experimental data of a Li^+/Nafion strip from ERI.

When $K_1 = K_2$, this four-parameter model is reduced to a two-parameter model representing actuation response without relaxation. The ordinary differential equation becomes

$$\frac{dk}{dt} = \frac{1}{\tau}(K_V - k). \tag{91}$$

The response for a step voltage is written as

$$k = VK_V(1 - e^{-t/\tau}). \tag{92}$$

The two-parameter case shows the same result as obtained by the model developed by Bhattacharya et al. [2001].

15.3.6.2 Input characteristics of IPMC

15.3.6.2.1 Clumped RC model for input current/voltage response of IPMC

An IPMC actuator consists of two parallel electrodes and an electrolyte between the electrodes. This condition can be described as two double-layer capacitors that are formed on the interfaces between the two electrodes and the electrolyte. The electrolyte between the electrodes may introduce an internal resistance. An RC circuit can simplify this series circuit of CRC. Once the possibility of leakage is taken into account, a (RC)||R circuit can be constructed as shown in Fig. 21, where r_0 is the internal resistance of the voltage source. In this model, there are three parameters to describe the input characteristics.

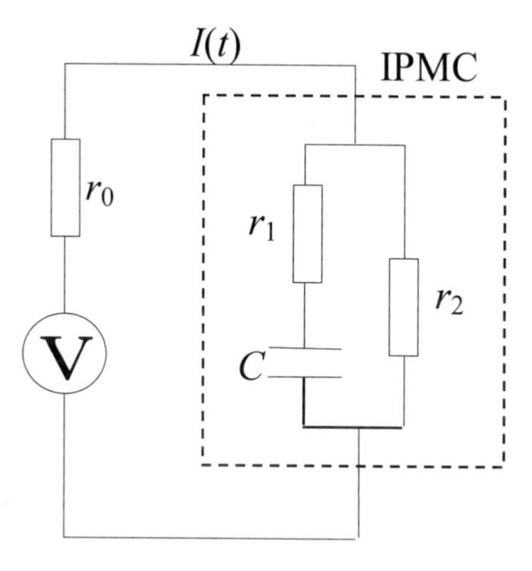

Figure 21: Clumped RC model for input response.

Under step voltage V, the response of input current is derived as

$$I(t) = \frac{V}{r_0 + r_2}\left[1 + \left(\frac{r_1 + r_2}{R} - 1\right)e^{-\alpha t}\right], \tag{93}$$

where

$$R = r_1 + \frac{r_0 r_2}{r_0 + r_2}, \tag{94}$$

and

$$\alpha = \frac{1}{\mathrm{RC}}. \tag{95}$$

This model was used to fit the experimental data with three adjustable parameters of r_1, r_2 and C. The model fits the data that was obtained for some types of IPMC samples well and an example is presented in Fig. 22. In this figure the sample that is shown consists of a 30 × 3 × 1 mm Li^+/ Nafion IPMC strip (made by ERI). The parameters that were determined by best fitting are r_1 = 232 Ω, r_2 = 700 Ω, and C = 10,000 μF.

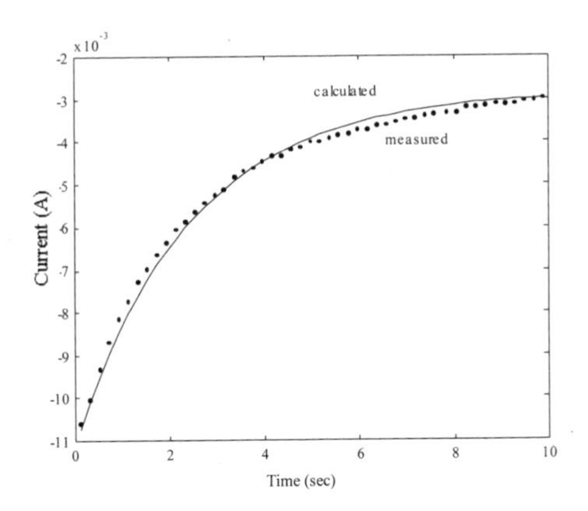

Figure 22: Time response of input current of a Li^+/Nafion sample. The dotted line is the measured data, and the solid line is the curve fitted by the clumped RC model.

15.3.6.2.2 Distributed RC model for input current/voltage response of IPMC

Different types of IPMC have different input properties. The clumped RC model fits the characteristics of specific types of IPMC while it would show poor fit for other types. An example is shown in Fig. 23. In this case, the sample is a 32 × 3.4 × 0.17 mm TBA^+ /Flemion IPMC strip (made by AIST, Japan). When using the above model, the best fitting curve is obtained with r_1 = 340 Ω, r_2 = 8280 Ω, and C = 7580 μF. However, there is a significant difference between the model and the experimental curves as can be seen Fig. 23, suggesting that there is a need to develop a different model to present this type of IPMC material.

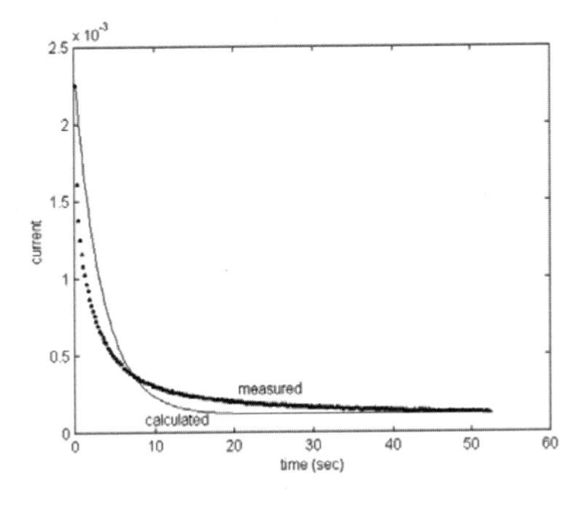

Figure 23: Time response of input current of a TBA^+/Flemion sample. The dotted line is the measured data, and the curved line fits the clumped RC model.

The experimental curve shows that the actual current drops faster than the model prediction at the beginning and slower after few seconds. This behavior may indicate that the capacitance increases with the time. Onishi et al. [2000] reported a similar phenomenon i.e., the capacitance of IPMC sample is "varying" with the rising time of the applied voltage, from 400 μF/cm2 at 1 V/s to 1300 μF/cm^2 at 0.1 V/s for a Li^+/ Flemion sample. Actually, a formula in the form of $I(t) = a + b(t - t_0)^{-n}$ fits the experimental data much better than the exponential curve that is predicted by the clumped RC model (see Fig. 24). For the data shown, the best fit is obtained for n equal to 0.45.

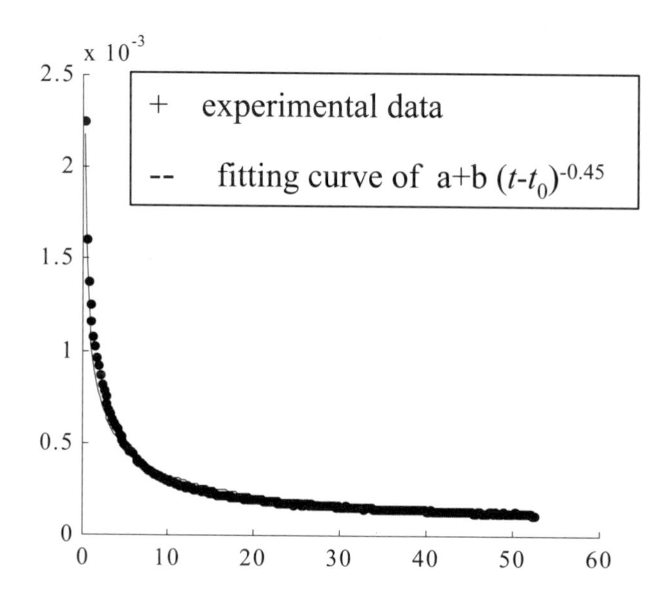

Figure 24: Time response of input current of the TBA^+/Flemion sample fitted by formula of $I(t) = a + b(t - t_0)^{-0.45}$.

The cause for the "varying" capacitance may be the fractallike microstructure of the electrodes (see Fig. 25), which are diffused into the polymer. The microstructure of the electrode strongly depends on the fabrication process. The interfaces of electrode and electrolyte have unequal paths to the electric source and to the electrolyte body. Parts of the double-layer capacitors are located at the main branches of the electrode, whereas other parts are located at its thin or even tiny branches. As a result, there is the formation of double-layer capacitances associated with different resistances behaving as a "varying" capacitance.

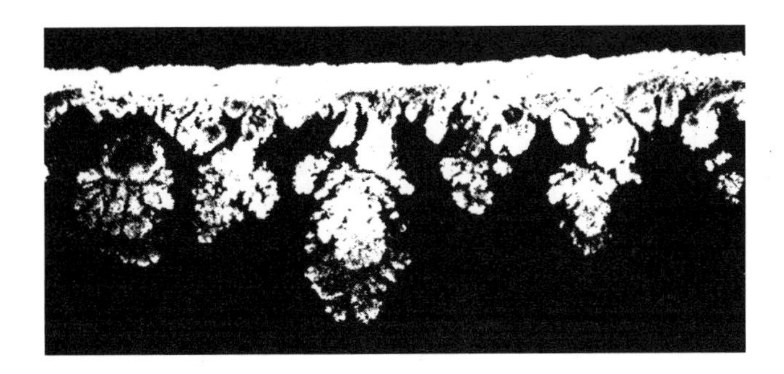

Figure 25: SEM micrograph showing fractallike structure of a gold electrode. (Courtesy of AIST.)

Bao et al. [2002] proposed a distributed RC line to present the double-layer capacitance of fractallike electrodes. This model is shown in Fig. 25 where the resistances in the circuit represent resistances both in the metal of the electrode and, maybe more significantly, in the ionomeric polymer.

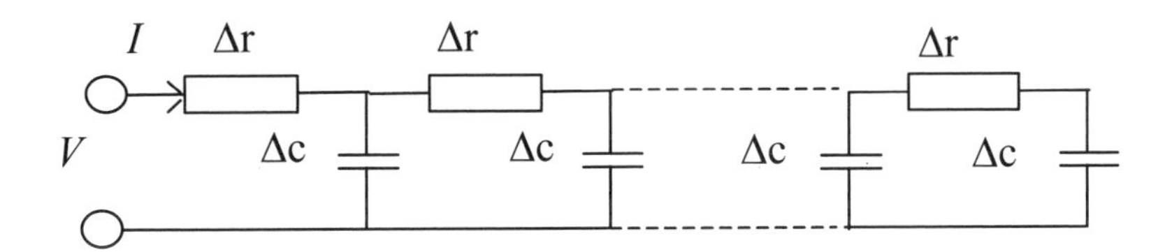

Figure 26: Distributed RC line model for fractallike electrode in IPMC.

The partial differential equation of this distributed RC line is derived as

$$\frac{\partial V}{\partial t} = Z_c^{\,2} \frac{\partial^2 V}{\partial r^2}, \tag{96}$$

where

$$Z_c^{\,2}(r) = 1 / \frac{dC}{dr}. \tag{97}$$

In the simplest case of $Z_c(r)$ being constant, the equation has an explicit general solution as

$$V(r,t) = c_1 e^{st - \sqrt{s}r/Z_c} + c_2 e^{st + \sqrt{s}r/Z_c} + c_3, \tag{98}$$

and a corresponding current as

$$I(r,t) = c_1 \frac{\sqrt{s}}{Z_c} e^{st-\sqrt{s}r/Z_c} - c_2 \frac{\sqrt{s}}{Z_c} e^{st+\sqrt{s}r/Z_c}. \tag{99}$$

If the line is infinite, the response of input current to a step voltage input is

$$I(0,t) = \frac{V_0}{2Z_c\pi j} \int_{\sigma-j\infty}^{\sigma+j\infty} \frac{e^{st}}{\sqrt{s}} ds = \frac{V_0}{Z_c\sqrt{\pi t}}, \tag{100}$$

while assuming the voltage on all capacitors is zero at the beginning. Equation (100) shows that the current response decreases with time as $t^{-0.5}$, which is close to the best fit of the experimental data that was found to be $t^{-0.45}$. Using a varying $Z_c(r)$ should reduce the discrepancy between the model results and the fit to the experimental data. This result implies that the distributed RC line may be a more accurate model for IPMC with diffused electrodes than the clumped RC model.

15.4 Summary of Test Methods

Establishing a database for EAP materials requires standard test procedures that provide data about the active and passive material properties. The range of properties is extensive; an example is shown in Table 3. The type and the properties that are measured as well as the metric for the measurement are shown.

15.5 Conclusion

Accurate information about the properties of EAP materials is critical to designers who are considering the construction of mechanisms or devices using these materials. In order to assess the competitiveness of EAPs for specific applications, there is a need for a properties matrix. This matrix needs to provide performance data that are presented in such a way that designers can scale the properties for incorporation into their models of the device under design. In addition, such a matrix needs to show the EAP material properties in such a form that allows the users to assess the usefulness of the material for the specific application. This data needs to include properties and information that can be compared with the properties of other classes of actuators, including piezoelectric ceramic, shape-memory alloys, hydraulic actuators, and conventional motors. The range of actuation and stress generation of the various types of EAPs is quite large, and the excitation field that is required for these materials can vary by five orders of magnitude. Some of the macroscopic properties that can be included in the matrix are maximum strain, maximum blocking stress, response time, maximum electric and mechanical energy density as well as maximum energy efficiency. In addition, due to the mechanical interaction that is associated with the electroreaction, there is a need to characterize both the passive and electroactive properties. The properties that may be of significance when characterizing EAPs were described in Table 3. While some of the properties (particularly those that are driven by polarization mechanisms) have relatively well-established methods of characterization, the ionic materials (particularly IPMC) still require new techniques. These materials pose the greatest challenge to developers of characterization methods due to their complex behavior. This complex response is associated with the mobility of the cations on the microscopic level, the strong dependence on the moisture content, as well as the nonlinear and the hysteresis behavior of the material. The technology related to the characterization of EAPs is expected to evolve as the field is advanced, and methods for determining standards will need to be established in the coming years.

Table 3: The properties that need to be characterized for EAP materials and the assumed metric.

Measurement		Properties	Metric
Mechanical		Tensile strength [Pa]	Mechanical strength of the actuator material
		Stiffness [Pa]	Required to calculate blocking stress, mechanical energy density, and mechanical loss factor/bandwidth
		Coefficient of thermal expansion [ppm/C]	Affects the thermal compatibility and residual stress
Electrical		Dielectric breakdown strength [V]	Necessary to determine limits of safe operation
		Impedance spectra [ohms and phase angle]	Provides both resistance and capacitance data. Used to calculate the electrical energy density; electrical relaxation/dissipation and equivalent circuit.
		Nonlinear current [A]	Used in the calculation of electrical energy density; quantify nonlinear responses/driving limitations
		Sheet resistance [ohms per square]	Used for quality assurance
Microstructure Analysis		Thickness (electrode and EAP), internal structure, uniformity and anisotropy as well as identify defects	These are features that will require establishment of standards to assure the quality of the material
Electro-active Properties	Strain	Electrically induced strain [%] or displacement [cm]	Used in calculation of "blocking stress" and mechanical energy density
	Stress	Electrically induced force [*g*], or charge (*C*)	Electrically induced force/torque or stress-induced current density
	Stiffness	Stress/strain curve	Voltage controlled stiffness
Environmental Behavior		Operation at various temperatures, humidity and pressure conditions	Determine material limitations at various conditions

15.6 Acknowledgments

The research at Jet Propulsion Laboratory (JPL), California Institute of Technology, was carried out under a Defense Advanced Research Projects Agency (DARPA) contract with the National Aeronautics and Space Agency (NASA). The authors would like to thank Dr. Sean Leary and Dr. Shih-Shiuh Lih for providing some of the experimental data that were used in this chapter, and to acknowledge the contribution of samples from Dr. Keisuke Oguro that were used to acquire the data. Also, the authors would like to express their appreciation to Prof. Kaushik Bhattacharya, Dr. Jiangyu Li, and Ms. Xiao Yu, Caltech, for their IPMC analytical data.

15.7 References

Aleksandrov, K.S, Zaitseva, M.P., Sysoev, A.M. and Kokorin, Yu.I. (1982), "The piezoelectric resonator in a dc electric field," *Ferroelectrics*, **41**, pp. 3-8.

Alemany, C., Pardo, L., Jimenez, B., Carmona, F., Mendiola, J., and Gonsalez, A.M. (1994), "Automatic iterative evaluation of complex material constants in piezoelectric ceramics," *J. Phys. D* (Appl. Phys.), **27**, pp.148-155.

Alemany, C., Gonsalez, A.M., Pardo, L., Jimenez, B., Carmona, F., and Mendiola, J. (1995), "Automatic determination of complex constants of piezoelectric lossy materials in the radial mode," *J. Phys. D* (Appl. Phys.), **28**, pp.945-956.

Bhattacharya K. (2001), Li J., X. Yu X., "Electro-mechanical models for optimal design and effective behavior," *Electroactive Polymer (EAP) Actuators as Artificial Muscles - Reality, Potential and Challenges*, Y. Bar-Cohen (Ed.), SPIE Press, Vol. PM98, pp. 309-330.

Bao X. (2002), Bar-Cohen Y., Lih S., "Measurements and macro models of ionomeric polymer-metal composites (IPMC)," Proceedings of SPIE's 9th Annual International Symposium on Smart Structures and Materials, 6-9 March, 2002, San Diego, CA. Paper No. 4695-27.

Banno, H., (1983), "Piezoelectric and dielectric properties of composites of synthetic rubber and $PbTiO_3$ and PZT," *Japanese J. of Appl. Phys.*, **22,** Suppl. 22-2, pp. 67-69.

Bar-Cohen Y., "Electroactive polymers as artificial muscles - capabilities, potentials and challenges," Keynote Presentation, Robotics 2000 and Space 2000, Albuquerque, NM, USA, February 28 - March 2, (2000), pp. 188-196.

Beige, H., and Schmidt, G., (1982), "Electromechanical resonances for investigating linear and nonlinear properties of dielectrics," *Ferroelectrics*, **41**, pp. 39-49.

Berlincourt, D. and Krueger, H. H.A. (1959), "Domain processes in lead titanate zirconate and barium titanate ceramics," *J. Appl. Phys.*, **30**, (11), pp. 1804-1810.

Berlincourt, D.A., Curran, D.R., and Jaffe, H. (1964), "Piezoelectric and piezomagnetic materials and their function in transducers," in *Physical*

Acoustics I Part A, Chapter 3, W.P. Mason,- editor, Academic Press pp. 169-270.

Biggs, A.G., Blackwood, K.M., Bowles, A., Dailey, S., and May, A., (2000), "A study of liquid crystalline elastomers as piezoelectric devices," Materials Research Society Symposium Proceedings, Vol. 600, *Electroactive Polymers (EAP),* pp. 159-164.

Bowen, L.J., French, K.W., (1992), "Fabrication of piezoelectric ceramic/polymer composites by injection moulding," Proceedings of the IEEE International Symposium on the Application of Ferroelectrics, Greenville, SC, pp. 160-163.

Brown, Lewis F. and Carlson, David L. (1989), "Ultrasound transducer models for piezoelectric polymer films," *IEEE Trans. on Ultrasonics, Ferroelectrics and Frequency Control,* **36**, (3), (May, 1989) pp. 313-318.

Brown, R.F., and McMahon, G.W (1962), "Material constants of ferroelectric ceramics at high pressure," *Can. J. Physics*, **40**, pp.672-674.

Brown, R.F. and McMahon, G.W (1965), "Properties of transducer ceramics under maintained planar stress," *J. Acoustical Society of America*, **38**, pp.570-575.

Brown, W.T., Chen, P.T., (1978), "On the nature of the electric field and the resulting voltage in axially loaded ferroelectric ceramics," *J. Appl. Phys*., **49**, pp 3446-3450.

Cady, W.G., (1964), *Piezoelectricity; An Introduction to the Theory and Applications of Electromechanical Phenomena in Crystals*, New York, Dover Publications.

Cao H., and Evans, A. G. (1993), "Nonlinear deformation of ferroelectric ceramics," *J. of the American Ceramic Society*, **76**, (4), pp. 890-896.

Chan, H.L.W., Ramelan, A.H., Guy, I.L., and Price, D.C., (1991), "Piezoelectric copolymer hydrophones for ultrasonic field characterization," *Rev. Sci. Instrum*., **62**, pp.203-207.

Chen, D-Y, Azuma, M., McMillanm L.D., Paz de Araujo, C.A.(1994)," A simple unified analytical model for ferroelectric thin film capacitor and its applications for nonvolatile memory operation," Proceedings of the IEEE International Symposium on the Application of Ferroelectrics, State College, PA, pp. 25-28.

D4065 (1995), ASTM Standard–Standard Practice for Determining and Reporting Dynamic Mechanical Properties of Plastics.

D6049 (1996), ASTM Standard - Standard Test Method for Rubber Property--Measurement of the Viscous and Elastic Behavior of Unvulcanized Raw Rubbers and Rubber Compounds by Compression between Parallel Plates.

E1640 (1999), ASTM Standard–Standard Test Method for Assignment of the Glass Transition Temperature By Dynamic Mechanical Analysis.

Fukada, E., Date, M., Neumann, H.E., and Wendorff, J.H. (1988), "Nonlinear piezoelectricity in poly(vinylidene flouride)," *J. Appl. Phys*., **63**, (5), pp. 1701-1704.

Gdula, R.A. (1968), "High field losses of adulterated lead zirconate-titanate piezoelectric ceramics," *J. of the American Ceramic Society*, **51**, (12), pp.683-687.

Goodman, G., (1952), *Am. Ceram. Soc. Bull.*, **31**, pp. 113.

Helfferich, F., (1995), *Ion Exchange*, Dover: New York, pp. 323-336.

Herbert, J.M., Wand, T.T., Glass, A.M., eds. (1988), *The Applications Of Ferroelectric Polymers*, Published By Routledge Chapman and Hall, New York

Holland, R., (1967), "Representation of the material constants of a piezoelectric ceramic by complex coefficients," *IEEE Transactions on Sonics and Ultrasonics*, **SU-14**, pp.18-20.

Holland R. and Eernisse E. P. (1969), "Accurate measurements of coefficients in ferroelectric ceramics," *IEEE Trans on Sonics and Ultrasonics*, **SU-16**, (4), pp. 173-181.

Holze, R. and Ahn, J.C. (1992), "Advances in the use of perfluorinated cation-exchange membranes in integrated water electrolysis and hydrogen – oxygen fuel systems," *J. Membrane Sci.*, vol. 73, no.1, pp.87-97.

Hom, C.L., Pilgrim, S.M., Shankar, N., Bridger, K., Massuda, M., Winzer, S.R., (1994), "Calculation of the quasi-static electromechanical coupling coefficient for electrostrictive ceramic materials," *IEEE Trans. on Ultrasonics, Ferroelectrics and Frequency Control*, **41** (4), pp. 542-550.

Hruska C. K.. and Kucera, M. (1986), "The dependence of the polarizing effect on the frequency of quartz resonators," *J. of the Canadian Ceramic Society*, (**55**) pp. 38-41.

IEEE Standards on Piezoelectric Crystals (1949), [49 IRE14. S1] Proceedings of the I.R.E. **37**, pp. 1378-1395, December.

IEEE Standards on Piezoelectric Crystals (1957), [57 IRE14.S1] Proceedings of the I.R.E. **45**, pp. 353-358, March.

IEEE Standards on Piezoelectric Crystals (1958), [58 IRE14.S1 Proceedings of the I.R.E. **46**, pp. 764-778, April.

IEEE Standard on Piezoelectricity (1987), [ANSI/IEEE Standard 176-1987].

Ikeda T., (1990), *Fundamentals of Piezoelectricity*, Oxford University Press, Oxford.

Ikeda T. (1985), "Mechanical loss and depolarizing field in a piezoelectric medium," Proceedings of the 5th Symposium on Ultrasonic Electronics, *Japanese J. of Appl. Phys.*, **24,** Suppl. 24-1, pp. 3-6.

Jaffe, B., Roth, R.S., and Marzullo, S., (1954), *J. Appl. Phys.*, **25**, (10), pp. 809-810.

Kahn, M., Dalzell, A., Kovel, B., (1987), "PZT ceramic-air composites for hydrostatic sensing," *Adv. Ceram. Mat.*, **2**, pp. 836-840.

Kanno, R., Kurata, A., Oguro, K., (1994),"Characteristics and modeling of ICPF actuator," *Proc. Japan-USA Symposium on Flexible Automation*, pp. 692-698.

Kanno R., Tadokoro S., Takamori T., (1995a), Modeling of ICPF (ionic conducting polymer gel film) actuator: modeling of electric characteristics,"

Twenty-First Annual Conference of the IEEE Industrial Electronics Society, Orlando, pp. 913-918.
Kanno R., Tadokoro S., Hattori M., Takamori T., Cotsaftis M. and Oguro K., (1995b), "Dynamic model of ICPF (ionic conducting polymer gel film) actuator," 1995 IEEE International Conference on Systems, Man and Cybernetics, pp.177-182.
Kanno R., Tadokoro, Takamori T., Hattori M., Oguro K., (1996a), "Linear approximate dynamic model of an ICPF (ionic conducting polymer gel film) actuator," Proc. 1996, IEEE International Conference on Robotics and Automation, pp. 219-225.
Kanno R., Tadokoro S., Hattori M., Oguro K. and Takamori T., (1996b), "3-dimensional dynamic modeling of ICPF (ionic conducting polymer gel film) actuator," Proc. 1996 IEEE International Conference on Systems, Man and Cybernetics.
Kato T. and Imai A., (1987), "Measurements of equivalent circuit parameters of ceramic resonators," Proceedings of the 6th Meeting on Ferroelectric Materials and their Applications, Kyoto, *Japanese J. of Appl. Phys.*, **26,** Suppl. 26-2, pp. 191-194.
Kawai, H., (1969), *Japanese J. of Applied Phys*, **8**, pp.975- 976.
Kloos G., (1995), "The corrections of interferometric measurements of quadratic electrostriction for cross effects," *J. Phys. D.* (Appl. Phys.), **28**, pp. 939-944.
Kloos G., (1995b), "The dependence of the electrostatic stresses at the surface of a dielectric on its orientation in a electric field," *J. Phys. D.* (*Appl. Phys.*), **28**, pp. 2424-2429
Kloos G., (1996), "A thermodynamic approach to studying the influence of viscoelasticity on the apparent coefficient of quadratic electrostriction," *J. Polymer Science Part B.* (Polymer Phys.), **34**, pp. 683-689.
Kornbluh, R., Pelrine, R., Joseph, J., Pei, Q., Chiba, S., (2000), "Ultra-high strain response of elastomeric polymer dielectrics," Materials Research Society Symposium Proceedings, Vol. 600, *Electroactive Polymers (EAP)*, pp 119-130.
Krakovsky, I., Romijn, T., Posthuma Deboer, A., (1999), "A few remarks on the electrostriction of elastomers," *J. Appl. Phys.*,**85**, (1), pp. 628-629.
Krueger, H. H. A. (1954), "Mechanical properties of ceramic barium titanate," *Phys. Rev.*, **93**, pp. 362.
Krueger, H. H. A (1967), "Stress sensitivity of piezoelectric ceramics: part 1., sensitivity to compressive stress parallel to the polar axis," *J. Acoustical Society of America*, **42**, (3), pp. 636-645.
Krueger, H. H. A (1968a), "Stress sensitivity of piezoelectric ceramics: part 2. heat treatment," *J. Acoustical Society of America*, **43**, (3), pp. 576-582.
Krueger, H. H. A. (1968b), "Stress sensitivity of piezoelectric ceramics: part 3., sensitivity to compressive stress perpendicular to the polar axis," *J. Acoustical Society of America*, **43**, (3), pp. 583-591.

Kwok K.W., Chan H.L., Choy C.L. (1997), "Evaluation of the material parameters of piezoelectric materials by various methods," *IEEE Trans. on Ultrasonics, Ferroelectrics and Frequency Control*, **44**, (4), pp. 733-742.

Lakes, R., (1998), *Viscoelastic Solids*, CRC Press, Boca Raton.

Land, C.E., Smith, G.W., and Westgate, C.R. (1964), "The dependence of the small-signal parameters of ferroelectric ceramic resonators upon the state of polarization," *IEEE Trans on Sonics and Ultrasonics,* (**SU-11**), pp. 8-19.

Lim, S.J., (1990), *Two Dimensional Signal and Image Processing*, Prentice Hall: Englewood Cliffs, NJ.

Lines, M.E., Glass, A.M., (1977), *Principles and Applications of Ferroelectrics and Related Materials*, Oxford University Press, Oxford, U.K.

Litt, M.H. Hsu, C.H., Basu, P., (1977), "Pyroelectricity and piezoelectricity in Nylon 11," *J. Appl. Phys*., **48**, pp. 2208-2212.

Lukacs M., Olding T., Sayer M., Tasker R., Sherrit S. (1999), "Thickness mode material constants of a supported piezoelectric film," *J. Appl. Phys*. **85**, pp. 2835-2843.

Mason W.P., (1958), *Physical Acoustics and the Properties of Solids*, D. Van Nostrand Co. Inc., Princeton, New Jersey.

Martin G. E. (1954), "Determination of equivalent circuit constants of piezoelectric resonators of moderately low Q by absolute-admittance measurements," *J. Acoustical Society of America*, **26**, (3), pp. 413-420.

Martin G.E. (1964a), "On the theory of segmented electromechanical systems" *JASA*, **36**, pp 1366-1370.

Martin G.E. (1964b), "Vibrations of coaxially segmented longitudinally polarized ferroelectric tubes," *JASA* **36**, pp. 1496-1506,

Meeker T.R. (1972), "Thickness mode piezoelectric transducers," *Ultrasonics*, **10**, pp. 26-36.

Murata, J., Koizumi, N., (1989), "Ferroelectric behavior of vinylidene fluoride-tetrafluoroethylene copolymers," *Ferroelectrics*, **92**, pp. 47-54.

Nakano, Y., Macknight, W.J. (1984) "Dynamical mechanical properties of perfluorocarboxylate ionomers," *Macromolecules* **17**, pp. 1585-1591.

Nalwa H.S., ed. (1998), *Ferroelectric Polymers*, Marcel Dekker.

Nemat-Nasser, S., and Li, J. Y. (2000), "Electromechanical response of ionic polymer-metal composites," *Journal of Applied Physics*, Vol. 87, No. 7, pp. 3321-3331.

Nemat-Nasser S. (2001), and C. Thomas "Ionic polymer-metal composite (IPMC)," Topic 3.2, Chapter 12, "*Electroactive Polymer (EAP) Actuators as Artificial Muscles - Reality, Potential and Challenges*," Y. Bar-Cohen (Ed.), SPIE Press, Vol. PM98, pp. 405-453

Newman, B.A., Chen, P., Pae, K.D., Scheinbeim, J.A., (1980), "Piezoelectricity in Nylon 11," *J. Appl. Phys*., **51**, pp. 5161-5164.

Newnham, R.E., Skinner, D.P.,Cross, L.E., (1978), "Connectivity and piezoelectric-pyroelectric composites," *Mat. Res. Bull*., **13**, pp. 525-536.

Newnham, R.E., Skinner, D.P., Klicker, K.A., Bhalla, A.S., Hardiman, B. Gururaja, T.R., (1980), "Ferroelectric ceramic-plastic composites for piezoelectric and pyroelectric applications," *Ferroelectrics*, **27**, pp. 49-55.

Nix E.L. (1986), "A direct method for measurement of the film-thickness piezoelectric coefficient of polyvinylidene flouride," *Ferroelectrics*, **67**, pp.125-130.

Nix E.L. and Ward I.M. (1986), "The measurement of the shear piezoelectric coefficients of polyvinylidene flouride," *Ferroelectrics*, **67**, pp.137-141.

Nye, J.F., (1957), *Physical Properties of Crystals*, Oxford Science Publications, Claredon Press, Oxford, Reprinted 1993.

Oguro, K., Fujiwara, N., Asaka, K., Onishi, K. and Sewa, S., (1999), "Polymer electrolyte actuator with gold electrodes," *Smart Materials and Structures,* Proc. SPIE Vol. 3669, p. 64-71.

Ohigashi H. (1976), "Electromechanical properties of polarized polyvinylidene flouride films as studied by the piezoelectric resonance method," *J. Appl. Phys.*, **47**, (3), pp. 949-955.

Ohigashi, H., Itoh, T., Kimura, K., Nakanishi, T., and Suzuki, M. (1988), "Analysis of frequency responce characteristics of polymer ultrasonic transducers," *Japanese J. of Appl. Phys.*, **27**, 3, pp. 354-360.

Ohigashi H. (2000), "Structure properties, and applications of single crystalline films of vinylidene fluoride and trifluoroethylene copolymers," Materials Research Society Symposium Proceedings, Vol. 600, *Electroactive Polymers (EAP),* pp 23-34.

Okada, T., Xie, G., Gorseth, O., Kjelstrup, S., Nakamura, N., Arimura, T. (1998), "Ion and water transport of nafion membranes as electrolytes," *Electroch. Acta*, vol. 43, no. 24, pp. 3741-3747.

Onishi K. (2000), Sewa S., Asaka K., Fujiwara N., Oguro K., "Morphology of electrodes and bending response of the polymer electrolyte actuator," *Electrochimica Acta* **46**, pp. 737-743.

Onishi K. (2001), Sewa S., Asaka K., Fujiwara N., Oguro K., "The effects of counter ions on characterization and performance of a solid polymer electrolyte actuator," *Electrchimica Acta* **46**, pp. 1233-1241.

Otero, T.F., Sasinena, J.M. (1997), "Bilayer dimensions and movement in artificial muscles," *Bioelectrochemistry and Bioenergetics*, vol. 42, pp.117-122.

Ounaies Z., Cheol P., Harrison J.S., Smith, J.G., and Hinkley, J., (1999), "Structure-property study of piezoelectricity in polyimides," Proc. SPIE Vol. 3669, pp. 171-178.

Pelrine, R.E, Kornbluh, R.D., and Joseph J.P.,(1998), "Electrostriction of polymer dielectrics with compliant electrodes as a means of actuation," *Sensors and Actuators* A, **64**, pp. 77-85.

Pelrine R, Kornbluh R, Pei Q, Joseph J, (2000), "High-speed electrically actuated elastomers with strain greater than 100%," *Science*, **287**, 5454, pp.836-839.

Piquette, J.C., Forsythe, S.E. (1997), "A nonlinear material model of lead magnesium niobate (PMN)," *J. Acoust. Soc. Am.*, **101**, pp. 289-296.

Richard, C., Eyraud, P., Eyraud, L., Richard, M., Grange, G., (1992), "1-3-1 PZT polymer composites for high pressure hydrophones," *Ferroelectrics*, **134**, pp. 59-64.

Saada, A.S., (1974), *Elasticity: Theory and Applications*, Pergamon Press Inc., New York.

Safari, A., Young, H.L., Halliyal, A., Newnham, R.E., (1987), "0-3 piezoelectric composites prepared by co-precipitated $PbTiO_3$ powder," *Am. Ceram. Soc. Bull.*, **66**, pp. 668-670.

Sakaguchi, K., Sato, T., Koyama, K., Ikedda, S. Yamamizu, S., Wada, Y., (1986), "Wide band multilayer ultrasonic transducers made of piezoelectric films of vinylidene fluoride - trifluoroethylene copolymer," *Japanese J. of Appl. Phys.*, **25**, Supp.25-1, pp. 91-93.

Sawyer, C.B., Tower, C.H., (1930), *Phys. Rev.*, **35**, pp. 269-273.

Scheinbeim, J.I., (1981), "Piezoelectricity in γ form Nylon 11," *J. Appl. Phys.*, **52**, pp. 5939-5942.

Seo, I., Sasaki, M., Kosaka, T., Tamura, Y., Saito, K., and Miyata, S. (1985), "Piezoelectricity in vinylidene cyanide/vinyl acetate copolymer and its application to ultrasonic transducers," Proceedings of the 5th International Symposium on Electrets, Heidelberg, pp. 808-812.

Sewa, S., Onishi, K., Asaka K., Fujiwara, N., and Oguro, K., (1998), "Polymer actuator driven by ion current at low voltage applied to catheter system," *Proc. IEEE 11th Workshop on Microelectronic Mechanical Systems (MEMS 98),* pp.148-153.

Shahinpoor M., Y. Bar-Cohen, J. O. Simpson and J. Smith, "Ionic polymer-metal composites (IPMC) as biomimetic sensors, actuators & artificial muscles- a review," *Smart Materials & Structures Journal*, Vol. 7, No. 6, (December, 1998) pp. R15-R30.

Sherrit, S., Gauthier, N., Wiederick, H.D., and Mukherjee, B.K. (1991), "Accurate evaluation of the real and imaginary material constants for a piezoelectric resonator in the radial mode," Ferroelectrics, **119**, pp.17-32.

Sherrit, S., Wiederick, H.D., Mukherjee, B.K. (1992a), "Non iterative evaluation of the real and imaginary material constants of piezoelectric resonators," *Ferroelectrics*, **134**, pp.111-119.

Sherrit, S., Wiederick, H.D., Mukherjee, B.K. (1992b), "A polynomial fit for calculating the electromechanical coupling constants of piezoelectric materials using the method of Onoe et. al. [J. Acoust. Soc. Am., **35**, pp. 36-42, 1963]," *J. Acoustical Society of America*, **91**, (3), pp.1770-1771.

Sherrit, S., Van Nice, D.B., Graham, J.T., Wiederick, H.D., Mukherjee, B.K., (1992c), "Domain wall motion in piezoelectric materials under high stress," Proceedings of the 8th International Symposium on the Application of Ferroelectrics, Greenville, South Carolina, pp. 167-170.

Sherrit, S., Wiederick, H.D., Mukherjee, B.K., Sayer, M. (1997), "An accurate equivalent circuit for the unloaded piezoelectric vibrator in the thickness mode," *J. Phys. D* (Appl. Phys) **30**, pp. 2354-2363.

Sherrit, S., Mukherjee, B.K., (1998), "Electrostrictive materials: characterization and applications for ultrasound," Proc. SPIE Vol. 3341, pp. 196-207.

Sherrit S, Mukherjee B.K., Tasker R., (1999):"An Analytical Solution for the Electromechanical Coupling constant of an Unloaded Piezoelectric Bimorph" IEEE Trans. Ultrason., Ferroelectrics and Freq. Control, 46 (3), pp. 756-757

Sherrit, S., Leary, S.P., Bar-Cohen, Y., Dolgin, B.P., Tasker, R. (2000), "Analysis of the impedance resonance of piezoelectric stacks," Proceedings of the IEEE Ultrasonics Symposium, pp. 1037-1040, San Juan, Puerto Rico.

Shimizu H., Saito, S. (1985), "An improved equivalent circuit of piezoelectric transducers including the effect of dielectric loss," *J. Acoustical Society of Japan* (E), **6**, (3), pp. 225-239.

Smith R.C., Z. Ounaies, Z. (1999), "A domain wall model for hysteresis in piezoelectric materials," NASA/CR-1999-209832 ICASE Report No. 99-52, Institute for Computer Applications in Science and Engineering, NASA Langley Research Center, December 1999, pp. 31.

Smith, W.A., Shaulov, A.A., Singer, B.M., (1984), "Properties of composite piezoelectric materials for ultrasonic transducers," Proceedings of the IEEE Ultrasonics Symposium, Dallas, pp. 539-544.

Smits, J.G., (1976), "Iterative method for accurate determination of the real and imaginary parts of materials coefficients of piezoelectric ceramics," *IEEE Trans. on Sonics and Ultrasonics*, (**SU-23**),(6), pp. 393-402.

Smits, J.G., (1985), "High accuracy determination of real and imaginary parts of elastic, piezoelectric, and dielectric constants of ferroelectric PLZT (11/55/45) ceramics with iterative method," *Ferroelectrics*, **64**, pp. 275-291.

Smits, J. G., Choi, W. S. (1993), "Dynamic behavior and shifting of resonance frequencies of ZnO ON Si_3N_4 bimorphs", Ferroelectrics, 145, pp.73-82.

Smits, J. G., Choi, W. S. , Ballato, A. (1997): "Resonance and Antiresonance of Symmetric and Asymmetric Cantilevered Piezoelectric Flexors", IEEE Transactions on Ultrasonics, Ferroelectrics, and Frequency Control, 44, 2, pp.250-258.

Su, J., Harrison, J.S. St. Clair, T.L. Bar-Cohen, Y., and Leary, S.,(2000), "Electrostrictive graft elastomers and applications," Materials Research Society Symposium Proceedings, Vol. 600, *Electroactive Polymers (EAP)*, pp. 131-136

Swallowe, G.M. (1999) *Mechanical Properties and Testing of Polymers - An A-Z Reference*, Polymer Science And Technology Series, Volume 3, Kluwer Academic Publishers.

Tant, M.R., Mauritz, K.A., and Wilkes, G.L., Eds. (1997), *Ionomers: Synthesis, Structure, Properties, and Applications*, Blackie Academic and Professional Press.

Tiersten, H.F. (1978), "Perturbation theory for linear electroelastic equations for small fields superimposed on a bias," *J. Acoustical Society of America*, **64**, (6), p. 832.

Tiersten, H.F (1981), "Electroelastic interactions and the piezoelectric equations," *J. Acoustical Society of America*, **70**, pp.1567-1576.

Vieira, S. (1986), "The behavior and calibration of some piezoelectric ceramics used in the STM," *IBM J. Res. Develop.*, **30**, (5), pp.553-556.

Vinogradov, A., Holloway, F., (1999), "Electro-mechanical properties of the piezoelectric polymer PVDF," *Ferroelectrics*, **226**, pp. 169-181.

Vinogradov, A., Holloway, F., (2000), "Dynamic mechanical testing of the creep and relaxation properties of polyvinylidene fluoride," *Polymer Testing*, **19**, pp. 131-142.

Vinogradov, A., (1999), "A constitutive model of piezoelectric polymer PVDF," Proceedings of the SPIE Smart Structures Meeting, Vol. 3667, pp.711-718.

Von Hippel H., (1967), *Handbook of Physics, 2nd Ed.*, Chapter 7-Dielectrics, E.U. Condon and H. Odishaw, eds., McGraw-Hill, New York, pp. 4-102-4-106.

Waller, D.J., Safari, A., (1992), "Piezoelectric lead zirconate titanate ceramic fiber/polymer composites," *J. Am. Ceram. Soc.*, **75**, pp.1648-1655.

Wang, H., Zhang, Q., Cross, L.E., (1993), "A high sensitivity, phase sensitive d_{33} meter for complex piezoelectric constant measurement," *Japanese Journal of Applied Physics*, **32**, pp. L1281-L1283.

Wang, H., Zhang, Q. M., Cross, L.E., Ting, R., Coughline, C., Rittenmyer, K., (1994), "The origins of electromechanical response in polyurethane elastomers," Proc. Int. Symp. Appl. Ferro. 1984, pp. 182-185.

Ward, I.M., Hadley, D.W., and Ward, A.M. (1993), *An Introduction To The Mechanical Properties Of Solid Polymers*, John Wiley & Son Ltd.

Wiederick, H.D., Sherrit, S., Stimpson, R.B., Mukherjee, B.K., (1996), "An optical lever measurement of the piezoelectric charge coefficient," Presented at 8th European Meeting on Ferroelectricity, Nijmegen, 1995, *Ferroelectrics*, **186**, pp. 25-31.

Woollett, R.S., and Leblanc, C.L., (1973), "Ferrroelectric nonlinearities in transducer ceramics," *IEEE Trans on Sonics and Ultrasonics*, (**SU-20**), pp. 24-31.

Yeo, S.C., Eisenberg, A. (1977), "Physical properties and supermolecular structure of perfluorinated ion-conducting (Nafion) polymers," *J. of Appl. Polymer Science*, **21**, pp. 875-898.

Zhang, Q.M., Pan, W.Y., Jang, S.J. And Cross, L.E., (1988), "Domain wall excitations and their contributions to the weak signal response of doped lead zirconate titanate ceramics," *J. Appl. Phys.* **64**, pp. 6445-6451.

Zhang, Q.M., Su, J., Kim, C.H., Ting, R., and Capps, R., (1997), "An experimental investigation of electromechanical responses in a polyurethane elastomer," *J. Appl. Phys.* **81**, pp. 2770-2776.

Zhang, X.D., and Rogers, C.A., (1993), *J. Intell. Syst. and Struct.*, **4**, 307.

TOPIC 7

EAP Actuators, Devices, and Mechanisms

CHAPTER 16

Application of Dielectric Elastomer EAP Actuators

Roy Kornbluh, Ron Pelrine, Qibing Pei, Marcus Rosenthal, Scott Stanford, Neville Bonwit, Richard Heydt, Harsha Prahlad, Subramanian V. Shastri

SRI International

16.1 Introduction

Electroactive polymers (EAPs) that are suitable for actuators undergo changes in size, shape, or stress state upon the application of an electrical stimulus. Much research in the field of EAPs tends to focus on the development and understanding of the polymer materials themselves. However, practical devices require that changes in dimension and stress state be effectively exploited to produce the desired functionalities (e.g., driving the motion of a robot limb or simply changing appearance or surface texture). This chapter focuses on those issues that must be considered in implementing EAP materials in practical devices.

For purposes of discussion we will focus on one particular type of electroactive polymer: *dielectric elastomers*. In the literature [e.g., Liu, Bar-Cohen, and Leary, 1999] and elsewhere in this book, dielectric elastomers are also known as *electrostatically stricted polymers*. Dielectric elastomers are a type of electronic EAPs as defined in Chapter 1 of this book—in that their operation is based on the electromechanical response of polymer materials to the application of an electric field. They have demonstrated good performance over a range of performance parameters and thus show potential for a wide range of applications. Dielectric elastomers were pioneered by SRI International, but several research groups around the world are actively investigating applications of this technology [e.g., Wingert et al., 2002; Sommer-Larsen et al., 2001; Jeon et al., 2001]. While the principle of operation of dielectric elastomer EAPs is not used with all EAPs, many of the issues we will discuss are common to all. These issues include the high compliance and large strains that EAPs can produce, as well as the necessity of simultaneous consideration of both the electrical properties and mechanical properties of materials.

This chapter is organized as follows. First, we consider the specifications used to match actuation technologies with applications, and when it makes sense to consider EAPs. Next, we discuss the basic principles of dielectric elastomer technology. We then consider design issues that may affect the actuation performance of dielectric elastomer EAPs, as well as the operational characteristics of EAPs and how they may affect an application. We present several examples of dielectric elastomer actuators for a wide range of applications, highlighting both the potential advantages of EAPs and the challenges associated with their use. Finally, we conclude with a brief summary of the subchapter and a discussion of the future of EAP application.

16.1.1 Motivation for a New Actuator Technology

To understand the issues involved with the application of dielectric elastomer actuators, one must first understand the motivation for developing a new actuation technology. Table 1 summarizes parameters that are commonly used to select actuators. EAPs in general and dielectric elastomers in particular do not offer the best performance all performance metrics. However, considering the full parameter space of actuator specifications, EAPs can offer performance and characteristics that cannot be reproduced by other technologies. For example, their combination of large strains, high energy density, fast speed of response, and good efficiency is unique. Table 2 compares dielectric elastomer EAPs with other actuator technologies. It should be noted that the values in Table 2 have been collected from several different sources and are not based on uniform measurement standards. In some cases, the reported values are projections of performance, rather than actual measurements. Efficiency, in particular, is seldom measured directly. Thus, Table 2 should be considered a preliminary comparison. Another comparison of EAPs with other actuator technologies is given by Wax and Sands [1999]. Topic 5 discusses performance measurements of EAPs in more detail. In the case of dielectric elastomers, the values in Table 2 are the maximum performance achieved in dielectric elastomer materials. Practical dielectric elastomer actuators will likely have lower performance.

Other desirable features of actuators are not easily expressed by metrics such as those in Table 1. These features include ruggedness, the ability to be mass-produced on a single substrate, and compatibility with electronic circuit manufacturing techniques and materials. Again, EAPs may excel in these areas.

Although usually disregarded by many researchers, cost can often drive the selection of actuators for a given application. Fortunately, the materials and manufacturing cost of EAP materials and devices is expected to be low in contrast to that of more exotic materials such as magnetostrictive ceramics, single-crystal piezoelectrics, or even shape memory alloys.

Figure 1 graphically illustrates application areas where EAPs can address the shortcomings of traditional actuators and open up a range of new applications. Many new application opportunities may involve the mere replacement of traditional actuators with EAP actuators so as to improve the performance or lower the cost of a device. Table 3 shows several applications where EAP actuators can effectively replace traditional electric actuator technologies. Several of these applications, as well as more exotic actuators with applications related to other application areas shown in Fig. 1, are discussed in Sec. 16.5 below.

Table 1: Common actuator specifications.

Parameter	Scale-Invariant Version	Comments
Energy and Power		
Energy output	Specific energy	Over full cycle
Power output	Specific power	Average or instantaneous
Energy conversion efficiency	Same	Energy out over a full cycle / energy in over a full cycle (excluding recovered energy)
Response time	Same (response time is not generally scale invariant)	For one direction or full cycle
State Variables		
Displacement	Strain	Instantaneous or maximum over a cycle
Force	Stress, pressure	Instantaneous or maximum during a cycle
Velocity	Strain rate	Instantaneous or maximum
Impedance and Controllability		
Stiffness	Elastic modulus	Usually nonlinear (not a constant)
Damping	Specific damping, loss factor, loss tangent	Usually nonlinear (not a constant)
Accuracy (displacement or force) (%)	Percentage of strain or stress	Usually percentage of maximums
Repeatability (%)	Same	Usually percentage of maximums
Linearity or sensitivity (%)	Same	Deviation from linear input-output relationship
Operational Characteristics		
Environmental tolerance	Same	Recommended ranges of temperature, humidity, etc., or effect of variations on temperature, humidity, etc., on the above parameters
Durability, reliability	Same	Number of cycles before degradation threshold or total failure, degradation per cycle or time
Input impedance (power supply requirements)	Specific impedance	Voltage and current requirements or pressure and flow, depending on power mode

Table 2: Comparison of candidate artificial muscle actuator technologies with natural muscle.

	Actuator Type (specific example)	Max. Strain (%)	Max. Pressure (MPa)	Specific Elastic Energy Density (J/g)	Elastic Energy Density (J/cm^3)	Max. Efficiency (%)	Relative Speed (full cycle)
ELECTROACTIVE POLYMER	Dielectric elastomer [1]						
	Acrylic	380	8.2	3.4	3.4	60–80	Medium
	Silicone	63	3.0	0.75	0.75	90	Fast
	Electrostrictive Polymer						
	(P(VDF-TrFE-CFE)[2]	4.5	45	>0.6	1.0	–	Fast
	Graft Elastomer[3]	4	24	0.26	0.48	–	Fast
	Electrochemo-mechanical Conducting Polymer (Polyaniline)[4]	10	450	23	23	< 1%	Slow
	Mechano-chemical Polymer/Gels (Polyelectrolyte)[5]	> 40	0.3	0.06	0.06	30	Slow
	Piezoelectric Polymer (PVDF)[6]	0.1	4.8	0.0013	0.0024	n/a	Fast
NON-EAP TRANSDUCERS	Electrostatic Devices (Integrated Force Array)[7]	50	0.03	0.0015	0.0015	> 90	Fast
	Electromagnetic (Voice Coil)[8]	50	0.10	0.003	0.025	> 90	Fast
	Piezoelectric						
	Ceramic (PZT)[9]	0.2	110	0.013	0.10	> 90	Fast
	Single Crystal (PZN-PT)[10]	1.7	131	0.13	1.0	> 90	Fast
	Shape Memory Alloy (TiNi)[11]	> 5	> 200	> 15	> 100	< 10	Slow
	Shape Memory Polymer[12]	100	4	2	2	< 10	Slow
	Thermal (Expansion)[13]	1	78	0.15	0.4	< 10	Slow
	Magnetostrictive (Terfenol-D, Etrema Products)[14]	0.2	70	0.0027	0.025	60	Fast
	Natural Muscle (Human Skeletal)[15]	> 40	0.35	0.07	0.07	> 35	Medium

[1] Pelrine, Eckerle, and Chiba, 1992.
[2] Xia et al., 2003.
[3] Su, J., personal communication
[4] Bobbio et al., 1993.
[5] These values are based on an array of 0.01-m thick voice coils, 50% conductor, 50% permanent magnet, 1 T magnetic field, 2 ohm-cm resistivity, and 40,000 W/m^2 power dissipation.
[6] PZT B, at a maximum electric field of 4 V/μm.
[7] Park and Shrout, 1997.
[8] PVDF, at a maximum electric field of 30 Vμm.
[9] Hunter et al., 1991.
[10] Tobushi, Hayashi, and Kojima, 1992.
[11] Aluminum, with a temperature change of 500 °C.
[12] Baughman et al., 1990.
[13] Shahinpoor, 1995.
[14] Terfenol-D Etrema Products.
[15] Hunter and Lafontaine, 1992.

Table 3: Representative applications for dielectric elastomer EAPs.

Application	Common Existing Electrical Actuator Technology	Potential Advantages of EAPs
Motors	Electromagnetics	Higher power density, lower-speed operation (eliminates gearing); low-cost materials
Linear Actuators	Electromagnetics, ceramic piezoelectrics (small stroke)	Higher energy density, higher power density, greater efficiency at low speeds; lower-cost materials and production; greater variety of shapes and sizes
Loudspeakers	Electromagnetic, ceramic, electrostatic (air gap)	Lighter weight; more flexibility in design shape (e.g., conformal, flat), higher power output; low-cost materials
Pumps	Electromagnetics	Higher power density, lower speed operation, novel designs; lower cost materials
Robotic Actuators	Electromagnetic motors	Higher energy and power densities; ability to operate in direct drive; greater simplicity; ability to be configured in novel shapes more easily (e.g., snakelike robots); lower cost; ability to combine actuation, sensing, and structure in a single material; ability to match the passive behavior of natural muscle for biomimetics
MEMS	Electrostatics	Higher energy and power density; simpler design; imperviousness to dust and other contaminants; ability to combine actuation and structure

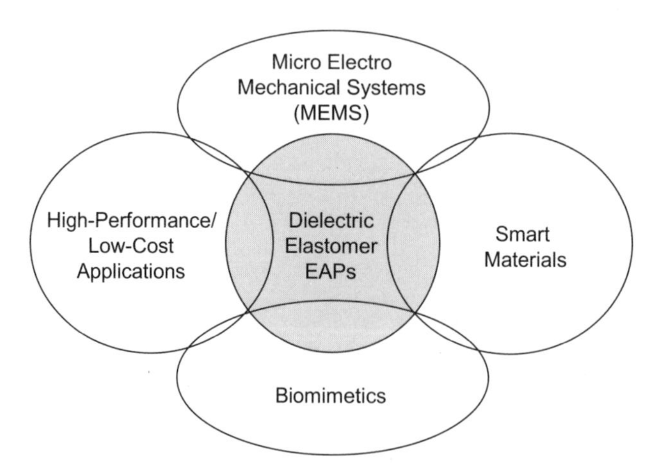

Figure 1: Application opportunities well suited to dielectric elastomer EAPs.

16.2 Dielectric Elastomer EAP—Background and Basics

An understanding of the operation of dielectric elastomers can help elucidate issues surrounding the application of this technology to actuators. All actuators based on dielectric elastomer technology operate on the simple principle shown in Fig. 2. When a voltage is applied across the compliant electrodes, the polymer shrinks in thickness and expands in area.

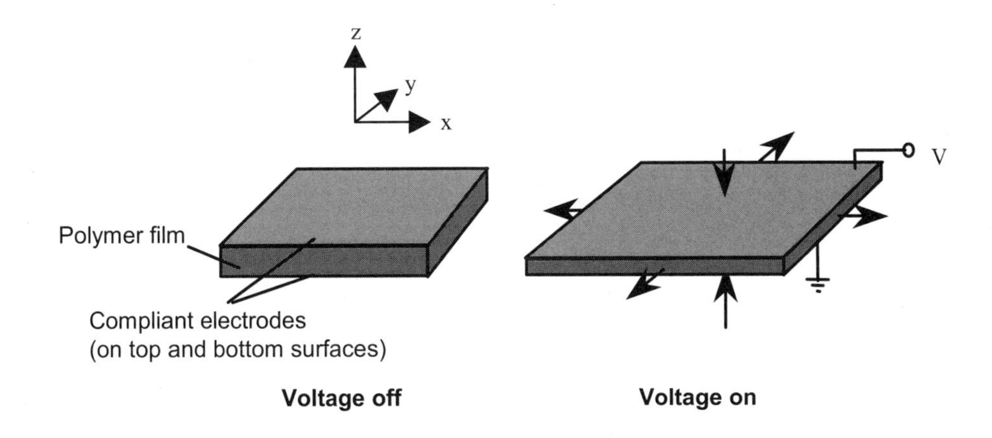

Figure 2: Principle of operation of dielectric elastomer actuators.

The net volume change of the polymer materials that we investigated is small [Pelrine, Kornbluh, and Kofod, 2000] because of their high bulk modulus. Therefore, the electrodes must be compliant, to allow the film to strain.

The observed response of the film is caused primarily by the interaction between the electrostatic charges on the electrodes [Pelrine, Kornbluh, and Kofod, 1999]. Simply put, the opposite charges on the two electrodes attract each other, while the like charges on each electrode repel each other. This domination of the observed response by these electrostatic effects differentiates this technology from others based on the electromechanical response of certain semicrystalline elastomers [e.g., Zhenyi et al., 1994].

Using this simple electrostatic model, we can derive the effective pressure produced by the electrodes on the film as a function of the applied voltage [Pelrine et al., 1995]. This pressure, p, is

$$p = \varepsilon_r \, \varepsilon_o E^2 = \varepsilon_r \, \varepsilon_o \, (V/t)^2 , \quad (1)$$

where ε_r and ε_o are the relative permittivity of the polymer (dielectric constant) and the permittivity of free space , respectively; E is the applied electric field; V is the applied voltage; and t is the film thickness. Strictly speaking, the effective pressure in Eq. (1) is the result of compression in thickness and tension in the planar directions of the film, but because the two modes are coupled in a thin film, it is more convenient to consider single effective pressure acting in thickness compression. The response of the polymer is functionally similar to that of electrostrictive polymers in that the response is directly related to the square of the applied electric field.

The resultant strains produced in the polymer are dependent on the boundary conditions and loads on the polymer. Further, the strains depend on the elastic modulus of the polymer, which may be nonlinear for elastomers at large strains. A further complication is that the polymers are often actuated with large initial prestrains. Such prestrains may be anistropic and thus may cause the effective elastic modulus to also be anistropic. For these reasons it is not possible to write a simple general equation for the resultant strain.

Fortunately, large strains are easy to measure by optical means. Since we typically operate with large prestrains, it is more convenient to use the relative strain, which is the strain difference from the prestrain conditions. Figure 3 shows the extremely large strain response that some actuator designs make possible. Strains over 100% have been observed in both acrylic and silicone elastomers. Table 4 summarizes representative high-strain responses of the materials that have been found to produce the largest strain response. Kornbluh et al. [1999] showed that the strain vs. field behavior of silicone elastomers is consistent with the electrostatic model given by Eq. (1). The deformation of the polymer film can be used in many ways to produce musclelike linear actuation. For example, the film and electrodes can be formed into a tube, rolled into a scroll, stretched over a frame, or laminated to a flexible substrate to produce bending. The best configuration depends on the application and properties of the film. Several configurations made by the authors are shown in Fig. 4.

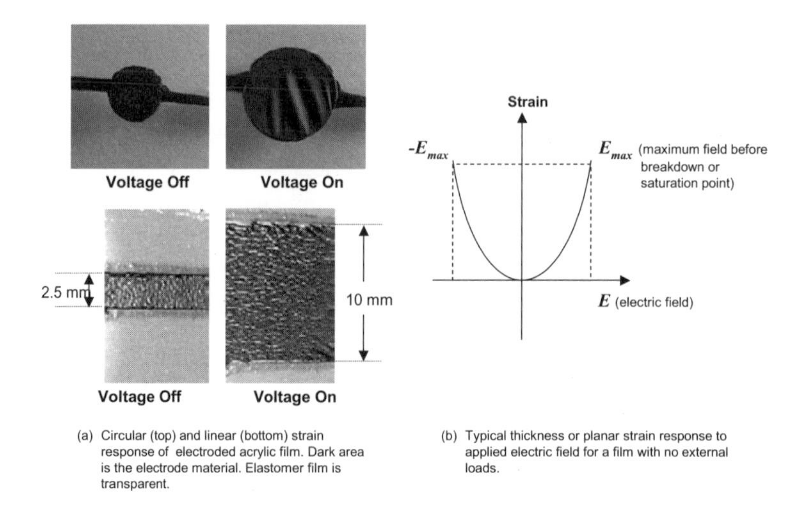

(a) Circular (top) and linear (bottom) strain response of electroded acrylic film. Dark area is the electrode material. Elastomer film is transparent.

(b) Typical thickness or planar strain response to applied electric field for a film with no external loads.

Figure 3: Strain response of acrylic film.

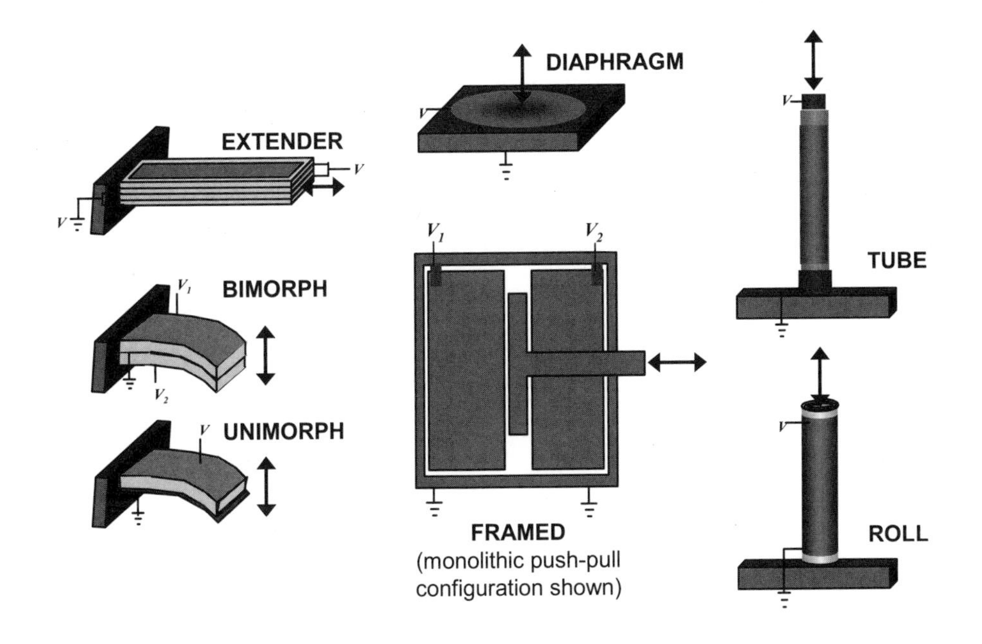

Figure 4: Dielectric-elastomer actuator configurations.

Table 4: Strain response (circular and linear) of electroelastomers.

Material	Relative Thickness Strain (%)	Relative Area Strain (%)	Test Field Strength MV/m	Pressure (MPa)	Estimated Elastic Energy, *e* (MJ/m^3)
Circular Response					
Silicone A [1]	48	93	110	0.3	0.098
Silicone B [2]	39	64	350	3.0	0.75
Acrylic [3] Elastomer	61	158	412	7. 2	3.4
Linear Response					
Silicone A[1]	54	117	128	0.4	0.16
Silicone B [2]	39	63	181	0.8	0.2
Acrylic[3] Elastomer	68	215	240	2.4	1.4

[1] NuSil CF19-2186, NuSil Corporation Carpinteria, California
[2] Dow Corning HS III, Dow Corning Corporation, Midland, Michigan
[3] 3M VHB 4910, 3M Corporation, St. Paul, MN

Many of these configurations are similar to those used for piezoelectric materials. This is not surprising, since piezoelectric materials are also typically actuated as dielectric layers between electrodes. However, elastomeric polymer material is much softer and the amount of deformation is much greater than can be achieved with piezoelectric ceramics, allowing for a greater variety of actuator configurations.

The nonlinear mechanical material behavior, large strain, and anisotropic boundary conditions of these actuators make it difficult to develop simple analytical relationships for their performance. Nonetheless, we can make several assumptions about the actuators that, while simplifications, at least qualitatively show the relationship between the applied electric field, material properties, and resulting stresses and strains in the presence of an external load.

For purposes of illustration, we will approximate the dielectric elastomer materials as linearly elastic, and we will ignore second-order nonlinearities resulting from the extremely large strains that may be produced. Ignoring mechanical constraints on the deformation of the film the strain in thickness, s_z, planar strain, s_x, of a single layer of film is

$$s_z = (^{-}p - p_{z,\mathrm{load}})/Y - 0.5p_{x,\mathrm{load}}/Y - 0.5p_{y,\mathrm{load}}/Y$$
$$s_x = 0.5(p - p_{z,\mathrm{load}})/Y + p_{x,\mathrm{load}}/\mathrm{Y} - 0.5p_{y,\mathrm{load}}/Y, \tag{2}$$

where p_{load} is the pressure from the load of the actuator on the EAP material and Y is the Young's modulus of the material. We have assumed that the polymer is incompressible and thus has a Poisson's ratio of 0.5. One can write similar equations for the strain in the other planar direction of the film by using a generalized Hooke's law. A different set of equations could be developed for partially constrained deformation of the film, such as the deformation that occurs in unimorph and bimorph configurations. Other actuator configurations where the film is constrained in one direction will be discussed in the following section.

By combining Eqs. (1) and (2), we can calculate the electric field that must be applied across the EAP material in order to produce a given load at a given stroke:

$$\Delta l = l\,(0.5p - F_{\mathrm{load}}/\,\mathrm{wt})/Y = 1\,(0.5\varepsilon_r\,\varepsilon_o\,\mathrm{E}^2 - F_{\mathrm{load}}/\,\mathrm{wt})/Y, \tag{3}$$

where l and w are the length and width of the film, t is the film thickness, F_{load} is the axial force acting on the actuator, and Δl is the stroke of the actuator. This force vs. stroke performance of an EAP actuator is shown graphically in Fig. 5. This equation applies to actuator types such as an extender, a tube, or a roll. Also shown in Fig. 5 is the load line for a spring or opposing dielectric actuator, and a constant load, which would be experienced, for example, when lifting a weight. The implications of operating against different types of loads are discussed in the next section.

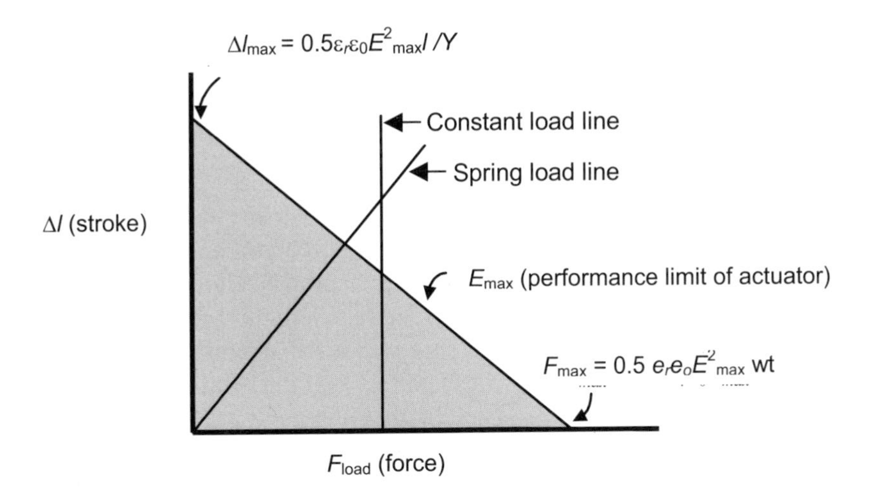

Figure 5: Simplified actuator force vs. stroke curve.

16.3 Actuator Design Issues

In contrast to many other electroactive materials such as piezoelectric ceramics and magnetostrictive ceramics, electroactive polymers are relatively compliant and capable of extremely large strains. For example, one type of dielectric elastomer, acrylic, is capable of actuating with more than 300% strain and has an effective elastic modulus of roughly 2–5 MPa when prestrained. PZT, a type of piezoelectric ceramic, can produce only 0.1% strain and has an elastic modulus of roughly 65 GPa. Even smaller-strain (but higher-force) field-activated electroactive polymers such as P(VDF-TrFE) have strains up to the 5% range, much higher than ceramic piezoelectrics, so many of the issues discussed here are not unique to dielectric elastomers. This high compliance and large strain capability present both opportunities and design challenges.

These design challenges include

1. The selection of actuator configurations and geometries that effectively load the polymer material so that stresses and strains induced in directions orthogonal to the output direction do not adversely affect the desired stresses and strains in the output direction,
2. Impedance matching between the highly compliant elastomer materials and the load,
3. Bandwidth limitations due to the lower natural frequency of the actuator and load.

Each of these challenges is discussed below.

16.3.1 Loading Condition

The importance of proper loading of the polymer material can be illustrated with a simple example. Imagine trying to push a heavy weight across a table by pumping water into a balloon that is in contact with the weight. The balloon would tend to bulge out at the sides rather than push the weight. To a lesser extent, the same thing happens when the electrostatic compression of a dielectric elastomer is poorly coupled to an actuator output. In our balloon analogy, proper loading would restrain the bulging in the undesirable lateral directions.

This restraint could be achieved several ways. Rings could be built into the balloon film that would restrain the lateral bulging while permitting the expansion in the desired direction. The balloon could be of a shape that is already flattened against the weight. Alternatively, an external mechanism could allow the lateral expansion to push the weight in parallel with the expansion of the balloon in the desired direction of the push. Such a mechanism could consist of a loosely woven tube that is narrower toward the center than at the ends.

Like the water in the balloon, the dielectric elastomer material is essentially incompressible. In the most basic actuation configuration, such as that depicted in Fig. 2, the electrodes uniformly impart an effective compressive stress in the z direction, which tends to decrease the thickness of the polymer film. The incompressible polymer film expands in area so that the total volume of polymer between the electrodes is conserved. Depending on the loading, the area expansion can occur equally in both planar directions (such as the circular expansion shown in Fig. 3, top) or in a single planar direction (such as the linear expansion shown in Fig. 3, bottom).

In many applications, motion is desired only in a single planar direction. Thus, as in the balloon example, we can restrain expansion in the undesirable direction or couple both directions of expansion into the desired output direction.

Undesirable deformation can be restrained by adding anisotropy into the material, or by changing the geometry, such as making the active area much wider than it is long. This latter approach is shown in Fig. 6.

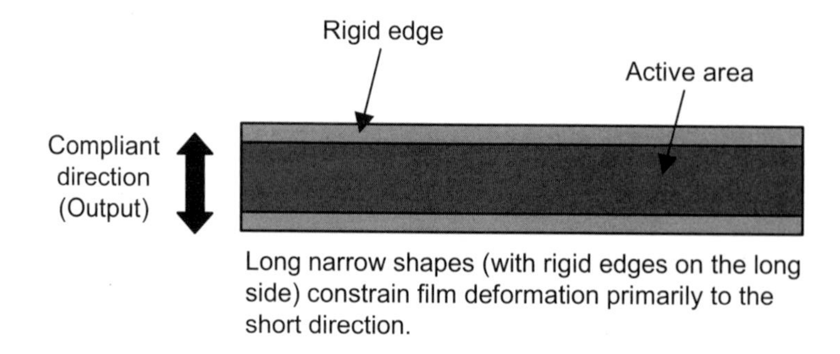

Figure 6: Constraining the deformation of polymer film to a single output direction.

Both planar directions of deformation can also be coupled into a single desirable output direction. Two methods are shown in Fig. 7. Some of these configurations are related to those developed for piezoelectric devices, such as the moonie or cymbal actuators [Uchino, 1997].

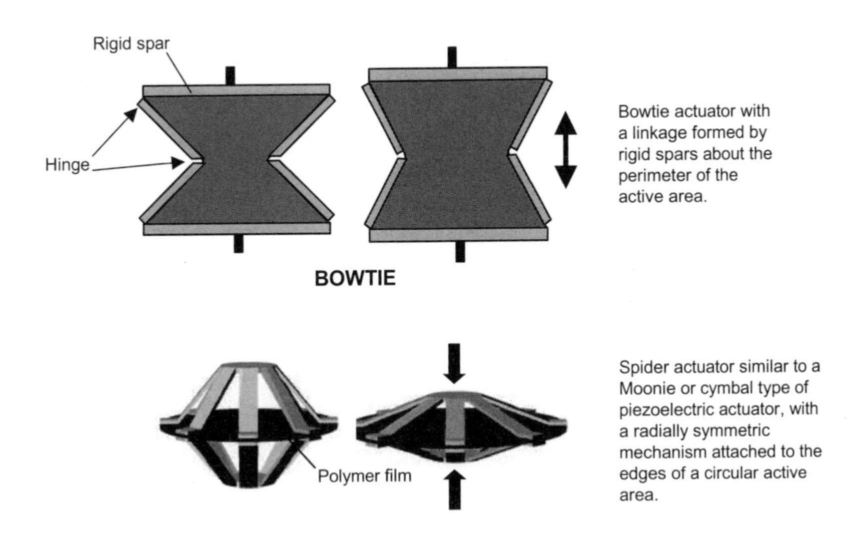

Figure 7: Actuator configurations that couple both directions of planar deformation into the output.

Note that it is not always necessary to couple two directions of planar expansion into a single direction of output. For example, in diaphragm actuators that operate across a pressure gradient such as might be used for a pump, the two directions of planar deformation are implicitly coupled to the output. In this case, the planar deformations displace the diaphragm and thereby impart momentum to the fluid being pumped. In some instances, the output of the actuator is produced by only a change in the geometry of a material or coating, with little or no force applied. In some cases it is desirable to expand the dielectric elastomer uniformly in both planar directions, such as in a solid-state aperture.

Note that the pressure that is produced in a dielectric elastomer, as given by Eq. (1), does not depend on the elasticity of the polymer. Thus, while the dielectric elastomer materials themselves are relatively soft, the pressure or force need not be small. If the elastomer is properly coupled to the load, then the pressure or forces that can be produced are determined by the strength of the electric field that can be produced in the material. Returning to our balloon analogy, we note that even a balloon could lift a car if properly constrained (e.g., by placing it inside a hydraulic or pneumatic cylinder or bellows). In fact, automobile tires are basically well-constrained balloons.

16.3.2 Impedance Matching

It is well known that the maximum energy output of an actuator can be achieved when the impedance of the actuator matches that of the load that it is driving (including all intervening transmission components). One must therefore consider the high compliance of EAP materials such as dielectric elastomers when selecting the actuator configuration and size.

The dynamic behavior and corresponding impedance of dielectric elastomer actuators can be difficult to model accurately. Dielectric elastomers, like many elastomers, have nonlinear elastic and viscous properties. Additionally, the large strains produced by dielectric elastomers further complicate the production of simple analytical expressions relating the force and displacement of the actuator. The dynamics and impedance of the driven load can also be difficult to model.

While an accurate model of the actuator and load may not usually be feasible, we can simplify the model so that it illustrates design principles and can be used in many applications. For this purpose we can assume that the actuator and driven load can be represented by a lumped-parameter model. We further assume that the force vs. displacement behavior can be linearized about an operating point. Finally, we consider a quasi-static analysis: that is, we ignore inertial and viscous forces. With these simplifications, the actuator is modeled as a stiffness k_{act}, while the environment is modeled by a stiffness k_{env}. We now perform a simple analysis that shows the importance of the stiffness of the actuator and the stiffness of the environment in determining the actuator output. Our analysis is based on that of Giurgiutiu, Rogers, and Choudhry [1996]. Figure 8 shows the basic setup.

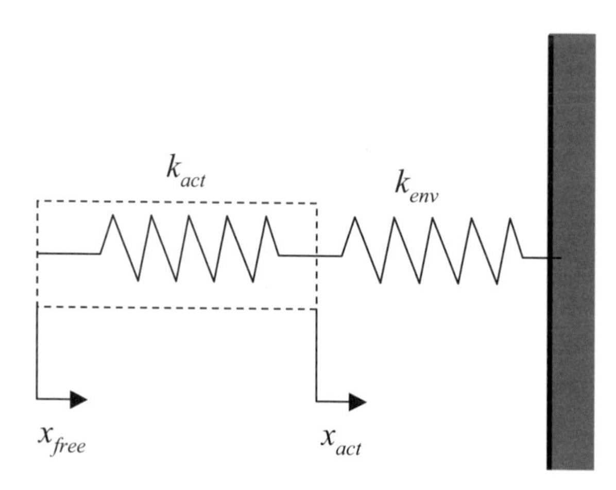

Figure 8: Simple quasi-static model of a compliant actuator operating on a compliant environment.

The free stroke of the actuator is given by x_{free}. This is the amount that the actuator would move if there were no contact with the environment. However, because of the stiffness of the environment, the actuator actually moves x_{act}:

$$x_{act} = x_{free} - F/k_{act}, \tag{4}$$

where F is the force between the actuator and the environment. Since $F = k_{env}x_{act}$, we can express x_{act} as

$$x_{act} = [1/(1 + k_{env}/k_{act})]\, x_{free}\,. \tag{5}$$

It can now be seen that the amount the actuator can move is greater in a less stiff environment. This result is no great surprise. However, the point at which the energy output is greatest is less apparent. The energy output (for half of a full cycle) is given by

$$E_{act} = \tfrac{1}{2}\, k_{env}x_{act}^{\,2} = [k_{env}/k_{act}/(1 + k_{env}/k_{act})^2](\, \tfrac{1}{2}\, k_{act}x_{free}^{\,2})\,. \tag{6}$$

This energy is maximized when $k_{env} = k_{act}$. In other words, as expected, the energy output is maximum when the impedance of the actuator matches the impedance of the environment. The maximum energy output is

$$E_{actmax} = \tfrac{1}{4}\,(\, \tfrac{1}{2}\, k_{act}x_{free}^{\,2}). \tag{7}$$

This quasi-static analysis suggests that only 25% of the energy that can be generated internally in the actuator (the energy of deformation with no external load) is available for driving a load. Dynamic factors and other design features, however, can increase the energy output per cycle under some conditions. For example, at resonance, the effective free strain of the actuator can be increased over that in the quasi-static analysis. Nonetheless, the basic principle of the importance of impedance matching has been illustrated.

Even if we do not wish to optimize the energy output, it should be noted that both the displacement and the energy output of the actuator can be greatly diminished in the presence of a large impedance mismatch. If we wish to extend this analysis to maximize other parameters, such as power output or speed of response, we must include other factors. Frequency limitations and the impact on power output will be discussed in the following section.

16.3.3 Frequency and Response Speed Limitations

In many applications, not only is the energy per cycle important but the response time (speed of response) or bandwidth can also be critical. It may, for example, be necessary to complete a cycle in a given amount of time, track an input, or counteract high-frequency vibrations. In order to assess or maximize the power output capabilities of EAP actuators, we must also determine their speed of response and frequency limitations (which are based on the selection of the actuator design, geometry, and polymer materials). The instantaneous power is the rate at which energy is being delivered. For cyclic operation, the power output is simply the energy output multiplied by the frequency of operation. Clearly, for maximum power output we want to operate at the highest frequency. In this section we look at the factors that limit this maximum frequency and speed of response.

The electromechanical pressure response of field-activated elastomers themselves is typically very fast [Kornbluh et al., 2000]. To explain why the response is fast, we first note that the electromechanical response is produced by the electrostatic forces on the electrodes, so that no significant delay should be associated with an intrinsic response mechanism of the polymer. This assumption is supported by the fact that the dielectric constant of the elastomers does not change appreciably over the range of 100 Hz to 100 kHz. The fundamental limitation in response speed would then be due to the speed at which the pressure wave can propagate across the thickness of material (the speed of sound). Since the individual polymer layers are thin, this propagation time can be less than a millionth of a second.

Another factor that might limit the speed of response is the time required to get the electrical charge on and off of the electrodes. This response time is closely related to the so-called RC time constant of the actuator. The time required to put charge on or off an electrode would be equal to four or five times the value of the resistance of the electrodes, multiplied by the capacitance of the actuator. Typical surface resistivities of the electrodes are on the order of tens to hundreds to thousands of ohms, so the RC time constant need not be a limiting factor for most small devices. For example, a 10-cm^2 area of silicone film that is 50 μm thick could achieve a 1 kHz response bandwidth with electrodes whose resistivities exceed 10^5 Ω. The speed of response, based on the RC time constant, is size dependent (the capacitance scales with the film area). However, the individual electrodes could be composed of many individual film areas that are electrically connected in parallel, so as to effectively reduce the surface resistivity of the individual electrodes. For example, electrically connecting in parallel the electrodes of many individual film layers of a multilayer actuator would have this effect. However, this approach could also be used on individual electrodes: thin, higher-conductivity traces could be used to distribute charge over the surface of an electrode and thereby reduce the effective resistance. This "structured electrode" approach is described by Kornbluh et al. [1999]. Since the RC time constant can be manipulated by the actuator design, it need not be a fundamental limitation on the actuator performance.

Most commonly, the maximum frequency or speed of response will be limited by the ability of the compliant polymer material to drive a load (the natural frequency of the electroactive polymer elements and the inertial load). Since the stiffness of EAP actuators such as dielectric elastomers is relatively low, these limitations can be significant. Again, for comparison purposes we can develop some simple equations that will enable us to assess the power output capabilities of our polymer materials.

A simple EAP actuator is shown in Fig. 9. This element is attached to an inertia, m. This inertia includes portions of the transmission and possibly a fraction of the load (depending on the effective transmission ratio). Now we can write the natural frequency of the actuator element as

$$f_n = [Y/(0.5 + q)l\rho]^{0.5}/2\pi\,, \tag{8}$$

where l is the actuator length, Y is the effective elastic modulus at the loading condition of the element in the motor, ρ is the mass density of the EAP material, and q is a ratio of the driven mass to the actuator mass. If the driven mass is zero, we assume that half of the total actuator material mass can be represented as a lumped mass that is driven by the actuator.

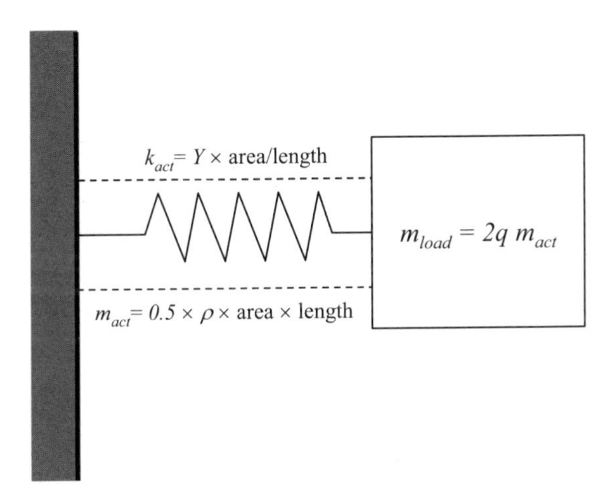

Figure 9: Simple dynamic model of a compliant actuator driving an inertial load.

While we can increase the frequency of Eq. (8) by choosing an actuator of shorter length, we note that the maximum frequency is limited by the speed at which a pressure wave can propagate along the actuator length:

$$f_{n\max} = v_s/2L, \tag{9}$$

where v_s is the speed of sound though the polymer material (v_s = 2000 m/s is a close approximation for most polymers). Strictly speaking, the resonant frequencies are somewhat lower than those given by the equations above, due to damping in the materials, transmission and load. In general, however, it is desirable to keep this damping low so that the deviation from the undamped resonant frequency will be small. Table 5 shows the measured viscoelastic properties of some common dielectric elastomer materials. According to our simple model, the mechanical loss factor, $\tan\delta_m$, is equal to the ratio of the half-power bandwidth divided by the natural frequency. More generally, the effect of the damping on the frequency response is given by

$$\text{strain / strain at low frequency} = 1 / [\, 1 + \tan\delta_m{}^2]^{0.5} \,. \qquad (10)$$

Experiments have shown that the speed of response of silicone polymers can be explained by this simple natural frequency analysis [Kornbluh et al., 2000].

Table 5: Speed of response and efficiency-related characteristics of dielectric elastomer films.

Material	Effective Modulus (MPa)	Relative Dielectric Constant	Dielectric Loss Factor $\tan\delta_e$	Mechanical Loss Factor $\tan\delta_m$	Maximum Electro-mechanical Coupling Factor, k^2 [1] (%)	Overall Maximum Efficiency [2] η_t (%)
Silicone A[3]	0.1	2.8 at 1 kHz	0.005 at 1 kHz	0.05 at 80 Hz	79	82 at 80 Hz
Silicone B[4]	1.0	2.8 at 1 kHz	0.005 at 1 kHz	0.05 at 80 Hz	63	79 at 80 Hz
Acrylic[5]	3.0	4.8 at 1 kHz	< 0.005 at 1 kHz	0.18 at 20 Hz	90	80 at 80 Hz

Source: Kornbluh et al. [2000]
[1] Based on maximum quasi-static strain conditions as shown in Table 4.
[2] Based on estimated mechanical and dielectric losses at about 100-Hz maximum electromechanical coupling conditions, driving circuits with 90% charge recovery are used.
[3] NuSil CF19-2186, NuSil Corporation Carpinteria, California
[4] Dow Corning HS III, Dow Corning Corporation, Midland, Michigan
[5] 3M VHB 4910, 3M Corporation, St. Paul, MN

The acrylic material seems to limit the speed of response in other ways that are not yet fully understood.

From the above analysis of natural frequency limitations we can see that it is desirable to have a short actuator element in order to maximize the resonant frequency. Stroke or energy output requirements may dictate a different geometry, so it may be necessary to compromise on design parameters for particular applications. The selection of the EAP material will also greatly affect the frequency and amplitude of motion. Different dielectric elastomer materials have different elastic properties, so what works best at low frequencies may not be best for higher frequencies.

This simple analysis ignores other dynamic effects including the effect of the electrical driver circuit on the dynamic properties of the actuator and the effect of enhanced strains at resonance. A detailed discussion of these factors is beyond the scope of this chapter; however, we note that the principle of impedance matching described above can be extended to a dynamic analysis and can also include the dynamics of the electrical driver circuit.

This discussion focuses on the speed of response and power output. A similar analysis could determine the factors that affect the electrical-to-mechanical energy conversion efficiency, if that is an important parameter. Generally, actuator designs and operating conditions that maximize energy or power output by impedance matching will also come close to maximizing efficiency.

16.4 Operational Considerations

Aside from issues directly related to the output or performance of an EAP actuator, such as those discussed above, many others related to the conditions under which the actuator must operate can, in the end, determine the suitability of an actuator technology to a particular application.

16.4.1 Input Impedance—Power Supplies and Driver Circuits

A full consideration of implementation issues must include the input requirements of an actuator. The importance of the electrical driver circuit to the dynamic performance of the actuator has already been noted. One way to characterize the electrical input requirements of an actuator is to measure its electrical impedance, which relates the voltage and current. As with output impedance, the electrical impedance may be a complex nonlinear and time-varying expression dependent on the loading conditions on the actuator. Nonetheless, some general observations may be made.

Field-activated EAPs such as dielectric elastomers are highly capacitive loads. Further, the maximum applied voltages are typically quite high. This combination presents challenges for the electronic driver circuits. In theory, a high electric field can be achieved at fairly low voltages by the use of thin films. In practice, such thin films are difficult to produce reliably in large areas, so most applications will employ thicker films that require applied voltages of at least several hundred to several thousand volts.

High voltage is not in and of itself difficult to generate. Many types of dc-dc voltage converters or transformers can be used to generate high-voltage from batteries or line voltage. Numerous successful high-voltage devices can be cited in other fields, such as cathode ray tubes, fluorescent lights, and spark plug and ignition systems in automobiles. Exotic approaches to generating high voltage directly may also be useful. These include generation from series-connected photovoltaic or electrochemical cells, or from capacitive rotary generator equipment.

Some applications where the speed of response, efficiency, or total size and mass of the electronic drivers are not critical can control the voltage applied to the muscle on the low-voltage end of a dc-dc converter. Higher-performance circuits will require means of controlling the high voltage directly. This requirement can present challenges if the cost, efficiency, or size of the electronic driver circuits is an issue. At present, small electronics that operate at voltages of several thousand volts, such as transistors and integrated circuits, are not available. However, there are no fundamental obstacles to the development of high-voltage devices. The currently available selection reflects the existing market demand rather than technology limitations. If field-activated EAPs are to become more widespread, then the development of such circuits will have to parallel the development of the actuators themselves.

Dielectric elastomers and other field-activated EAPs convert only a portion of the applied electrical energy into mechanical work. This conversion proportion, k^2, is often called the coupling efficiency. The dielectric elastomers can, in principle, have coupling efficiencies of up to 90% [Kornbluh et al. 2000]. Other field-activated EAP devices have coupling efficiencies of up to 50% [Zhang, Bharti, and Zhao, 1998]. If high efficiency is a requirement, the driver circuits must include charge recovery in order to avoid wasting the energy that is not converted to mechanical work.

Many electronic driver circuit designs include charge recovery. Most of these have been developed for piezoelectric devices, which, as field-effect materials, also require high-voltage drivers capable of handling a capacitive load.

Aside from the electronic driver circuits, high-voltage operation introduces both advantages and disadvantages for certain applications.

High-voltage operation allows lower current to be used to produce a desired amount of power. This lower current in turn allows the use of smaller connecting wires and reduces concern with connector contact resistance. The resistance of the electrodes themselves can be greater.

By way of example, commercially available wire with a diameter of just 0.6 mm [equivalent to #29 AWG (American Wire Gage)] wire can transmit up to 2500 W at 5000 V since the current is just 0.5 A. If we were to try to drive a 12 V actuator with 2500 W then we would need to use #3 AWG wire. This #3 wire is roughly 20 times the diameter of the #29 wire.

The potential dangers of high voltage must be addressed in many applications. If the actuator is small, so that the total power is low, or if the actuator can be subdivided into separately powered regions of lower power, the danger can be mitigated. High-voltage electrodes must also be coated or isolated. In many instances, a high-voltage electrode can be sandwiched between electrodes at ground potential, isolating it and preventing stray electric fields from producing electromagnetic interface. External packaging for safety, such as that used for other high-voltage applications, is an important consideration.

16.4.2 Environmental Tolerance

An actuator must maintain acceptable performance over the range of environmental conditions it might encounter. In most cases, the environmental conditions of interest are temperature and humidity. In some other applications, such as usage in space, factors such as radiation (e.g., solar), outgassing, and chemical compatibility may be important.

Dielectric elastomers and many other types of EAPs have not yet been rigorously tested over a range of temperatures and humidities, although some studies have been performed. Further, projections can be based on knowledge of the basic physical phenomena. The physical basis of the electromechanical response of dielectric elastomers is electrostatics. Electrostatic forces do not themselves vary with temperature or other environmental factors, so that one can focus on the mechanical and electrical properties of dielectric elastomers in estimating their environmental tolerance. If these properties are not themselves adversely affected by environmental conditions, then the actuator itself will not be affected.

One of the best-performing dielectric elastomers, silicone rubber is known to exhibit stable mechanical properties over a fairly wide range of temperatures. For example, one silicone (CF19-2186, NuSil Corporation, Carpinteria, California) has been shown to have similar performance over a temperature range of –65°C to 240°C [Kornbluh et al., 2003]. While we do not expect the material to sustain its maximum performance over such a temperature range, we do expect that acceptable performance can be sustained over the range of temperatures seen in many applications. Another silicone rubber was shown to be capable of operation at temperatures below –100°C [Kornbluh et al., 2003].

Silicones are also known to have low rates of moisture absorption, so changes in humidity should not affect performance significantly.

Acrylic dielectric elastomer does not have as large a temperature range as the silicones. They have been shown to operate over a range of –10°C to 80°C, enough for many applications.

To some extent, it is possible to select a dielectric elastomer material with a desired temperature range. For example, if high-temperature operation is important, one would likely select a silicone rather than an acrylic.

The compliant electrode materials, like the dielectric elastomers themselves, can be selected to match the needs of the application. The electrode materials typically include silicones or other polymers combined with carbon or other conductive particles. Since the main requirement for electrodes is to remain compliant, there is even greater freedom in selecting electrode materials than dielectric materials and the electrode performance is not typically a limiting factor for most applications.

While normal atmospheric variations in the humidity are not expected to greatly affect the performance of dielectric elastomer materials, long-term immersion might require coatings of moisture-proof materials. It is of course critical that the actuator be designed so that moisture cannot create a short circuit across the electrodes. Since the electrodes are exposed to high voltage, even a small amount of moisture could create a short.

16.4.3 Reliability and Durability

An actuator must also be capable of performing without failure or significant degradation during an acceptable period of time.

Like environmental tolerance, basic physical phenomena and the mechanical and electrical material properties of electrode materials can be used to estimate reliability and durability. The electrostatic mechanism of actuation will not inherently fatigue with time or number of cycles. Therefore, if the mechanical or electrical properties of the electrodes can sustain a large number of cycles or long duration of both mechanical strain and electric field, then we can expect that the actuator will survive.

Some of the dielectric elastomer materials, such as certain silicones, are routinely used as electrical insulators and so are expected to be electrically robust. It should be noted, however, that the film thickness commonly used in dielectric elastomer actuators is less than that of the materials for which long-term tests of the electrical and mechanical properties are made. Further, the electric fields that are applied to these thinner films often greatly exceed the long-term dielectric strength ratings for these materials. Fortunately, it appears that the dielectric strength of thin films is greater than that of the thicker counterparts, a phenomenon well known in the insulation field and often exploited in integrated circuits.

Many of the dielectric elastomer materials are often used commercially in applications where cyclic strains occur. These applications include shock and vibration absorption. Thus, we would expect that the materials are mechanically robust if properly loaded.

While we expect that dielectric elastomers will have good reliability and durability, the reliability and durability of EAP materials, including dielectric elastomers, have not been rigorously tested over the full range of operating conditions that could be expected. More testing under application-oriented conditions is needed. Some preliminary testing has, however, been performed. Silicone-based stretched-film actuators have been tested at strains of 5% to more than 10,000,000 cycles, with no observable degradation in performance. These tests were conducted at a frequency of about 10 Hz. Loudspeakers made from acrylic films have survived many hours of operation with no detectable decrease in performance. The loudspeakers maintain a dc bias voltage on the films and so give some indication of the durability of the acrylic film to long-term applications of dc voltages.

16.4.4 Controllability

Some actuator applications are simply "bang-bang"—full on or full off. Other applications require more precise control of intermediate positions or forces. The latter applications are most frequently used where a constant relationship exists between an applied stimulus and the electromechanical response. As with environmental tolerance, reliability, and durability parameters, the physical phenomena of the electromechanical response can be used to estimate an application's controllability performance. The relationship between the applied electric field and the resulting electrostatic forces, such as given by Eqs. (1), (2), and (3), indicates that position or force can be controlled by the applied voltage. The electrostatic effect expressed in these equations does not vary with time for a constant electric field. Further, there is no inherent electrical hysteresis in this mechanism. The low dielectric loss factors of Table 5, as well as observed strain vs. field behavior [e.g., Kornbluh et al., 1999; Kornbluh et al., 2000], support the notion that electrical hysteresis is not significant.

In general, there is a nonlinear relationship between the displacement or force output and the applied electric field. To a first order, the relationship will be quadratic, as given by Eq. (1). Lookup tables or curve-fitting techniques can be used to determine the correct voltage level for a given output. Alternatively, a closed-loop position of force feedback could be employed if greater control is desired. Occasionally it is desirable to have a response that is more or less linear with the applied voltage. In such cases, it is possible to effectively linearize the quadratic response of the actuator by using a "push-pull" configuration. If the maximum dynamic range can be reduced, the output of the actuator can be linearized to some extent by operation about a bias voltage.

The electrical properties of polymer materials, such as permittivity, can fluctuate slightly with variations in frequency or other factors. In most lower-frequency applications, the greatest source of fluctuations in the electromechanical response is in the mechanical properties of the polymer materials. More specifically, we may focus on the viscoelastic hysteresis (dynamic hysteresis) or creep (static hysteresis) of the materials. This hysteresis or creep can be quite small in some elastomers such as the silicones. Dynamic hysteresis is closely related to the mechanical loss factor described in Sec. 16.3.3. Hysteresis and creep are greater in acrylics, but are still at an acceptable level for many applications.

As with piezoelectrics and other field-activated materials, methods based on closed-loop position feedback and open-loop compensation can help minimize the effects of hysteresis and creep. Often such techniques are needed even in the absence of hysteresis and creep because of load variations, so in these cases compensating for hysteresis and creep may incur little or no additional cost or complexity. Additionally, the actuator design itself affects the amount of hysteresis or creep. For example, loading conditions that maintain a large amount of stress on the polymer material are more subject to creep than designs where the polymer is not highly loaded for long periods of time.

For a closed-loop system, we need only show that the elastomers are capable of achieving the needed resolution. Fortunately, the best-performing silicone and acrylic materials have been shown to have high resolution. In experiments aimed at evaluating the use of dielectric elastomer materials for adaptive optics, the open-loop resolution of both silicone and acrylic was evaluated. Both materials were shown to be capable of operating with a resolution of less than 50 nm. When these materials were laminated to create more rigid materials, the resolution of the resulting structure was less than 1 nm [Kornbluh et al., 2003; see also Sec. 16.8.4 of this chapter].

16.5 Overview of Dielectric Elastomer EAP Actuators and Their Applications

The features of the dielectric elastomer EAPs that present challenges in the design of actuators and devices also offer unique performance and application capabilities. These advantages include:

1. The ability to produce large motions with no bearings or sliding parts.
2. The ability to produce devices with multiple degrees of freedom (DOFs) produced by patterning different electrode regions on a single substrate.
3. The ability to couple output to one or both planar directions of polymer deformation.
4. The ability to match some of the mechanical properties and characteristics of natural muscle.
5. Inherent safety and ruggedness due to high compliance.
6. The ability of an actuator to "run backwards" as a generator or sensor.
7. The ability to act simultaneously as an actuator and a sensor or structure (multifunctionality).
8. The ability to be fabricated mostly by 2D fabrication techniques and integrated with electronic circuits.

In the following sections we present examples of actuators and devices that illustrate the range of these possibilities and exploit many of these unique capabilities. Many common actuator applications, and the potential advantages of replacing existing actuator technologies with dielectric elastomer actuators, have already been highlighted in Table 3. We discuss several of these applications as well as others. Some of the applications we discuss may appear to be exotic in that they are based on devices or systems that do not yet commonly exist outside laboratories or are simply not feasible with traditional actuator technologies. However, exploiting the unique capabilities of EAPs may make such devices and systems, as well as a multitude of others, a reality.

16.6 Artificial Muscles and Applications to Biologically Inspired Devices

16.6.1 "Artificial Muscle" Actuators

16.6.1.1 Is Dielectric Elastomer Really Artifical Muscle?

Since such actuators have a stress-strain behavior similar to that of natural muscle, we often call these actuators *artificial muscle*. Of course, these artificial muscles do not duplicate the performance or characteristics of natural muscle in all respects, so the researcher must be cautious in the usage of the term "muscle." Chapter 3 discusses the similarities between these dielectric elastomer actuators and natural muscle in more detail. This topic is also discussed in more detail by Pelrine et al. [2002]; Full and Meijer [2000]; Meijer, Rosenthal, and Full [2001];

and Kornbluh et al. [2002a]. Here we provide a brief overview of the similarities and differences.

The data presented in Chapter 3 show that dielectric elastomer actuators are indeed capable of reproducing the stress and strain (force and stroke) and the frequency response behavior of many natural muscles. Accordingly, the work and power output of these artificial muscles are similar to those of natural muscle. Unlike the data in Table 2 of this chapter, these data were obtained from actual freestanding actuators by means of apparatus and techniques identical to those used for characterizing natural muscle. None of the other solid-state or conventional actuator technologies of Table 2 have reproduced this behavior (not surprisingly, other EAPs such as electrostrictive polymers, which are similar to dielectric elastomers in many respects, have come closest).

Natural muscle has a number of functions besides conventional actuation. We have already seen how the compliance of dielectric elastomer EAPs must be considered when we design an actuator for a particular application. In the case of locomotion and other biological motions, the compliance and viscoelastic damping of natural muscle helps organisms achieve robust and energy efficient locomotion [see, for example, Alexander, 1988; Dickinson et al., 2000]. By implication, artificial muscle should have similar viscoelastic properties. Figure 10 illustrates a comparison of the viscoelastic properties of silicone and acrylic dielectric elastomer materials with the properties of a cockroach leg muscle. The cockroach is an example of a legged creature with great mobility and is thus an inspiration for robot designers. The compliance of silicone and acrylic is similar to that of natural muscle. The damping of these artificial muscle materials is somewhat less than that of the natural muscle, but it is always easy to add more damping (e.g., by adding inactive viscoelastic materials or transferring energy in the driving electronics, i.e., regenerative braking).

Other functions of natural muscle are known, but fewer quantitative comparisons to dielectric elastomer materials exist. For example, natural muscle can include integrated sensing and structural functions. Dielectric elastomer actuators have also shown similar capabilities as sensors and structural members [Pei et al., 2002]. In many organisms muscle makes up a significant portion of the total mass and provides a significant portion of the rigidity of the structure itself. Examples of actuators and devices that use both the sensory and structural capabilities of EAPs are given in this section.

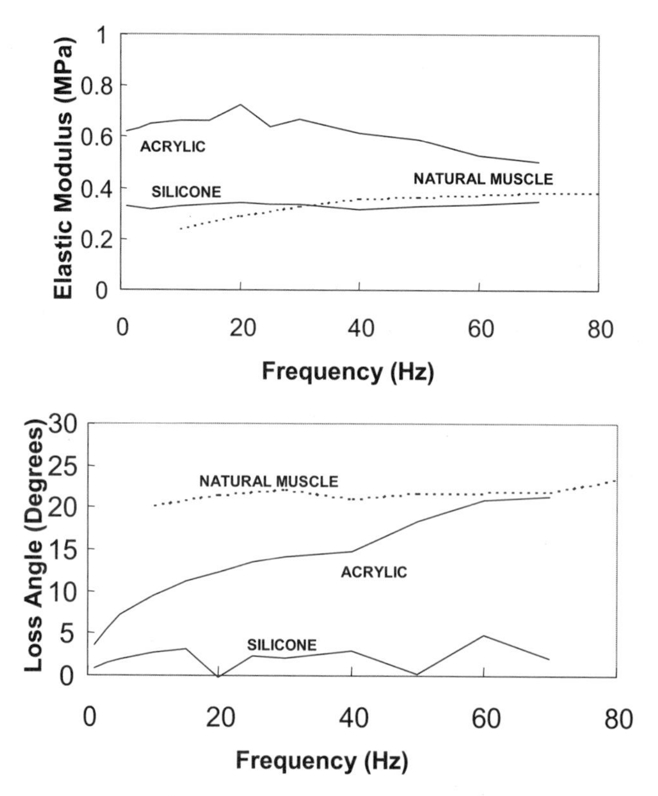

Figure 10: Comparison of viscoelastic properties of natural muscle (cockroach leg muscle) and two dielectric elastomer materials (elastic modulus, top, and mechanical loss angle, bottom) [adapted from Pelrine et al., 2002].

"Artificial muscle" actuators are well suited for application to biologically inspired robots or biomedical applications where the duplication of human muscle is desired. Such biomedical applications offer the promise of lifelike prosthetic limbs, exoskeletons for the mobility impaired, and even artificial hearts. Figure 11 shows a "spring roll" artificial muscle actuator (see Sec. 16.6.1.2) attached to a full size model of the arm of a human skeleton. While the artificial muscle biceps shown in this photo cannot yet match the performance of a natural biceps muscle, it does illustrate the great promise that electroactive polymer artificial muscles offer. Perhaps, in the not too distant future an arm powered by artificial muscles will defeat a human arm in an arm wrestling match.

Although only the first steps have been taken toward such biomedical applications, the actuators are being applied to small robots, as will be described in the next section. In addition to biologically inspired robots, many types of automation robots can benefit from high-energy, long-stroke actuator technology because of the limitations of existing electromagnetic drives.

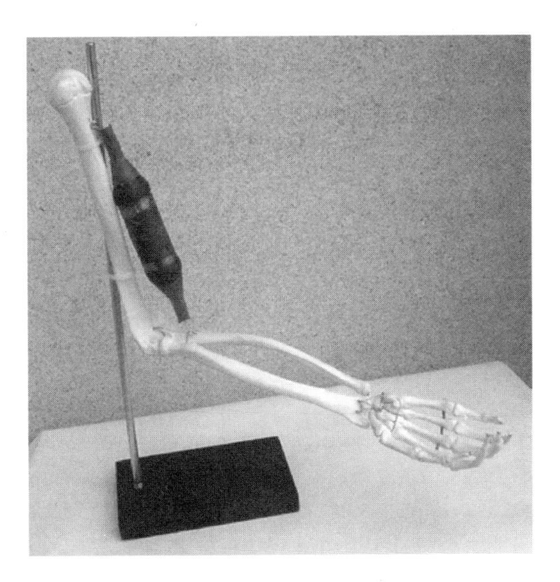

Figure 11: Model of a human skeletal arm powered by an artificial muscle. Although this arm does not yet perform as well as a natural arm, it illustrates the promise that artificial muscles offer for new generations of prosthetics as well as other biologically inspired devices.

16.6.1.2 Examples of Linear Artifical Muscle Actuators

The bow-tie and linear actuators shown in Figs. 6 and 7, and variations on these designs, have been considered for artificial muscles. These actuators are based on simple 2D fabrication procedures. The basic actuator configuration can be stacked in parallel to increase the force produced, or chained in series to increase the stroke. In this sense each actuator is like the active element of a single muscle fiber.

The linear actuator in Fig. 12 is composed of two layers of acrylic film sandwiched together. The basic actuator configuration is that of a "double bow-tie." That is, it is composed of two bow-tie actuators connected in series. This approach allows for longer and thinner actuators that are closer in shape to natural muscles yet do not sacrifice energy coupling. An advantage of the two-layer construction is that the high-voltage electrode can be sealed into the center of the actuator. It is then relatively simple to stack several actuators in parallel in order to produce the required force.

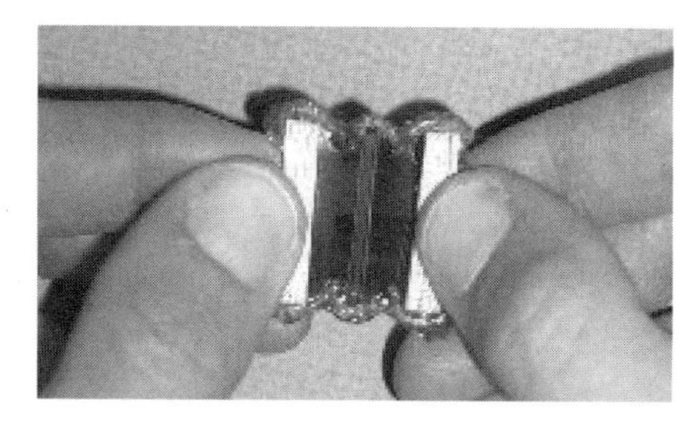

Figure 12: Linear artificial muscle based on an acrylic-film double bow-tie actuator.

Actuators like the ones shown in Fig. 12 have a force output of about 2 N, with a stroke of up to 10 mm. The actuators measure about 3 cm on a side, including the mechanical connections. The mass of the actuators is about 1.5 g; however, the mass of the active area is just 0.12 g. The specific energy density ofthese actuators is more than ten times of those piezoelectric actuators available commercially. Similar actuators have been produced from silicone polymers, although their maximum strain response is not as large.

Figure 13 shows a bow-tie actuator made from a single-layer silicone film about 40-μm thick. This actuator is 7-cm wide and 3-cm in length (including connectors). It is capable of a free stroke of 5 mm and a blocked force of roughly 1 N. Its mass is 1 g, of which the active film mass is less than 0.1 g.

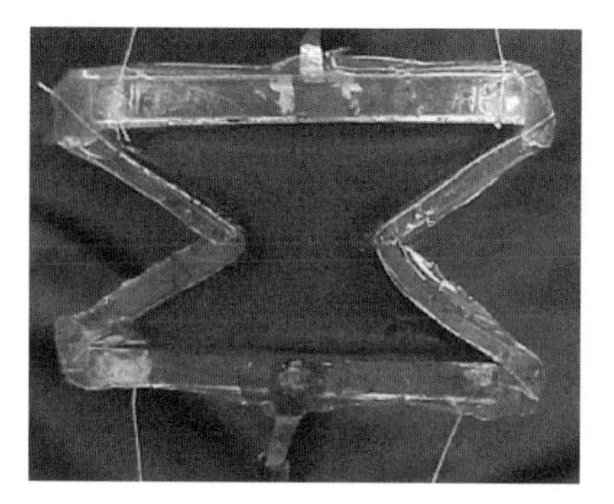

Figure 13: Silicone film bow-tie actuator.

Rolled actuators have high force and stroke in a relatively compact package. Because, like muscle, this package is cylindrical, rolled actuators make good artificial muscles. Rolled actuators incorporate a large cross-sectional area of film and thus can produce a relatively large force. One of the more successful types of rolled actuators is the "spring roll" (Fig. 14) This actuator is formed from an acrylic film that is stretched in tension and rolled around an internal spring.

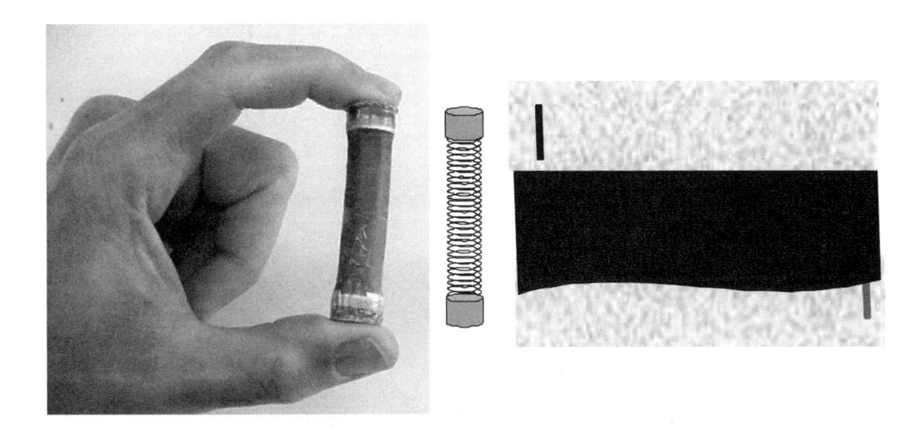

Figure 14: Spring roll linear actuator. The structure of the actuator is shown at right [adapted from Pei et al., 2002 and Kornbluh et al., 2002b].

Rolled actuators have been made in a variety of lengths and diameters with maximum stokes of up to 2 cm and forces of up to 33 N. Figure 14 shows a rolled dielectric elastomer actuator that weighs approximately 15 g and can produce peak forces of more than 5 N with maximum displacements of 5 mm (a strain of about 25%). The maximum performance of dielectric elastomer devices is expected to increase greatly as designs are improved, since these devices are still well below the measured peak performance of the polymers.

The main disadvantage of the rolled actuator is that it is somewhat more difficult to fabricate, since it is no longer a flat structure. Additionally, the energy coupling of a long roll is not as good as that of the bow-tie or trench type of actuator, since only one direction of deformation is coupled to the load.

We have noted that dielectric elastomer actuators can be multifunctional. That is, they can perform functions such as sensing or structure in addition to actuation. The spring roll has been shown to be an effective multifunctional device in that it incorporates sensing. Figure 15 displays the capacitance change of a spring roll as it is subjected to axial tension and compression. Each data set is a continuous loop of either tension and relaxation or compression and relaxation. As can be seen from the graph, the capacitance increases with the length of the spring roll. This multifunctionality can be used to make simple linear servo systems. Since muscle incorporates stretch and other proprioceptive sensors, such capabilities are another aspect of musclelike behavior.

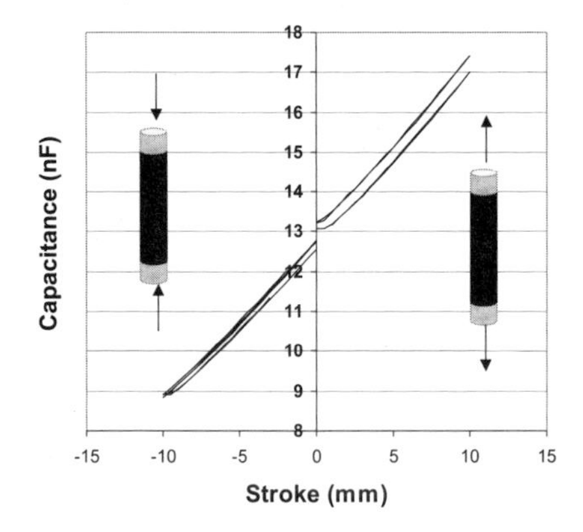

Figure 15: Capacitance change of a spring roll subjected to extension and compression [adapted from Pei et al., 2002].

16.6.1.3 Bending Roll Actuators

Spring roll actuators that bend as well as extend can also be fabricated. This ability is accomplished by patterning the electrodes so that they result in individually addressable segments around the circumference of the actuator. Figure 16 shows the structure and behavior of a 2-degree-of-freedom roll. Figure 17 shows a 3-DOF roll, undergoing bending in different directions according to the electrode areas activated.

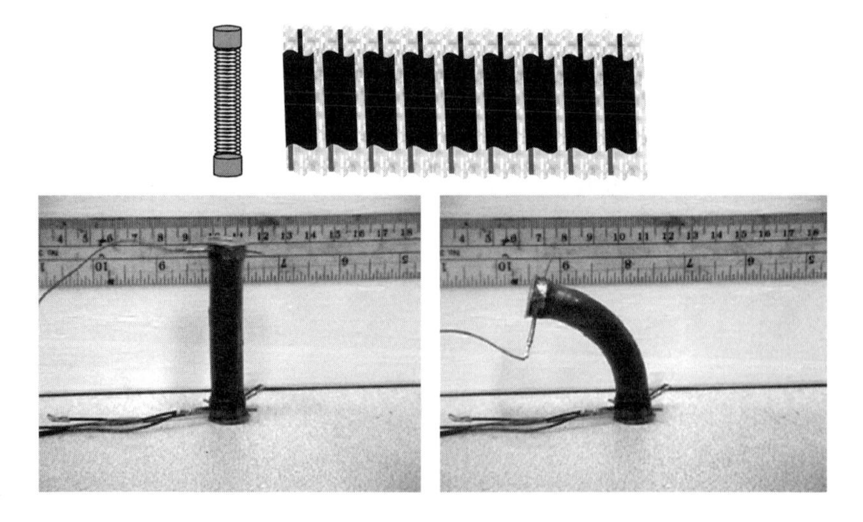

Figure 16: Two-DOF bending spring roll, Schematic of the structure (top) and video stills of the roll bending in response to the actuation of one set of electrodes (left and right) [adapted from Pei et al., 2003].

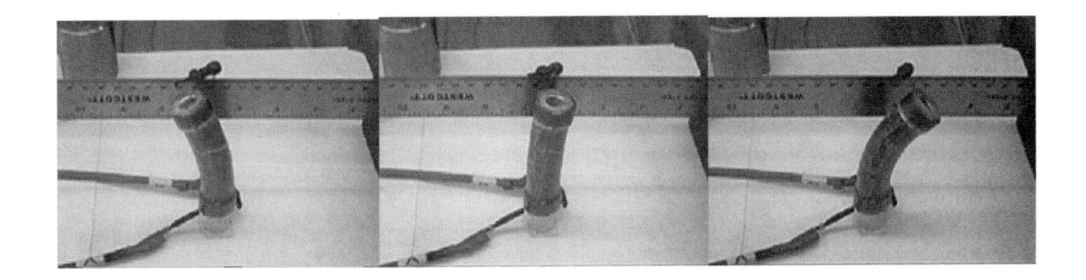

Figure 17: Video stills of a 3-DOF spring-roll segment undergoing bending.

These bending rolls are musclelike in the sense that they not only can produce musclelike motion but also are similar in structure to many appendages found in nature (e.g., the bodies of worms and larvae, elephant trunks, and octopus tentacles). Table 6 shows details of the performance of various multi-DOF rolls we have fabricated. We expect that improvements in fabrication procedures and designs will dramatically improve the performance of such actuators.

Table 6: Performance of bending spring rolls.

Attribute	2-DOF Roll Version 1	2-DOF Roll Version 2	3–DOF Roll Version 1	3-DOF Roll Version 2
Size (length, outside diameter)	9 cm, 2.3 cm	6.8 cm, 1.4 cm	9 cm, 2.3 cm	6.8 cm, 1.4 cm
Weight	29 g	11 g	29 g	11 g
Maximum bending angle	60°	90°	35°	20°
Maximum lateral force	1.68 N	0.7 N	1 N	0.2 N
Lateral stiffness	0.46 N/cm	0.28 N/cm	0.46 N/cm	0.28 N/cm

16.6.2 Biologically Inspired Robotics

Biologically inspired robots exploit the musclelike characteristics of dielectric elastomer EAPs.

One feature common to virtually all biomimetic robot designs for locomotion is the use of direct-drive approaches. Gears and other complex transmissions do not exist in nature, and macroscopic organisms generally have a 1:1 correspondence between their natural muscle motion and the motion of an equivalent locomotion appendage (e.g., a leg or a wing). This does not mean that creatures do not use transmissions, in the sense of mechanical leverage such as bones, to produce a different stroke-force combination than is supplied directly by the muscle, but that appendages generally operate at the same frequency as their corresponding muscles. By contrast, most electromagnetic devices cannot supply sufficient energy or peak power in a single stroke to be used in direct-drive mobile robots. Various means such as gear boxes are employed that enable the electromagnetic device to actuate (e.g., in a motor, rotate) many times per single cycle of a leg or other appendage. However, such use of transmissions cannot increase the low peak power of electromagnetics.

Consistent with the observation that biologically inspired designs are generally direct drive, is that they are back-drivable: for example, pushing on a leg causes the dielectric elastomer to stretch or contract, and the compliance and damping of the leg are directly related to those of the actuator. By contrast, nonbiological designs often use motors with high-gear-ratio transmissions, and these are typically not reversable.

16.6.2.1 Insect-Inspired Legged Robots

Given that most of our comparisons between natural muscle and the dielectric elastomer artificial muscle are based on the leg muscles of the cockroach, it follows that the application of dielectric elastomer artificial muscle to insectlike walking robots is of interest. Figure 18 shows a completely self-contained, battery-powered six-legged robot that was loosely modeled on a cockroach. Each leg has two DOF driven by a single actuator bundle that constitutes an artificial muscle. This robot, known as FLEX 1 [Eckerle et al., 2001] weighs 650 g (including the battery) and is roughly 30 cm in length. In order to minimize the number of muscles required (until manufacturing procedures can be improved), we have opposed the muscles with springs located at the pivot points of the legs. Each muscle is a bundle of three acrylic double-bow-tie actuators (see Figs. 7 and 12 above).

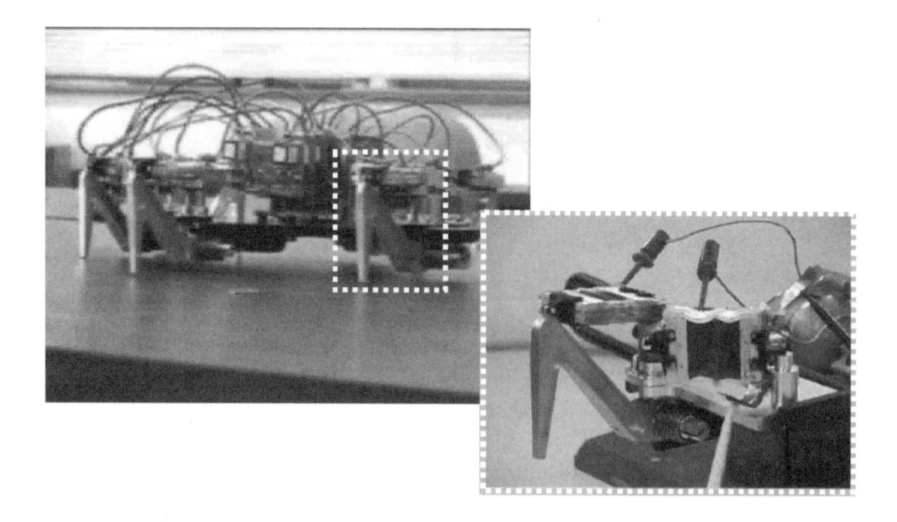

Figure 18: FLEX 1, a self-contained hexapod robot powered by dielectric elastomer EAP artificial muscles.

Each muscle is driven by a small (4 g) dc-dc converter with a maximum output of 5 kV and 500 mW of power. The joint motion is controlled by a peripheral interface controller (PIC) microprocessor. At this point the FLEX 1 robot can only walk slowly over even terrain. Improvements in the driving electronics and in the strength and durability of the muscles themselves are needed for greater mobility. Nonetheless, this robot is significant because it is believed to be the first self-contained walking robot that is powered by EAP actuators.

To address the limitations of FLEX 1, we recently built a second robot, FLEX 2, using the same basic kinematic design (see Fig. 19). To separate the power and integration issues from the actuator and biomimetic aspects of the robot, off-board power was used on FLEX 2. More important is the use of more powerful rolled acrylic actuators to replace the previous bow-tie actuators. The roll actuators proved better in virtually every respect. Speed was increased from an unimpressive few millimeters per second to a respectable 13.5 cm/s. Lifetime and shelf life were also dramatically improved.

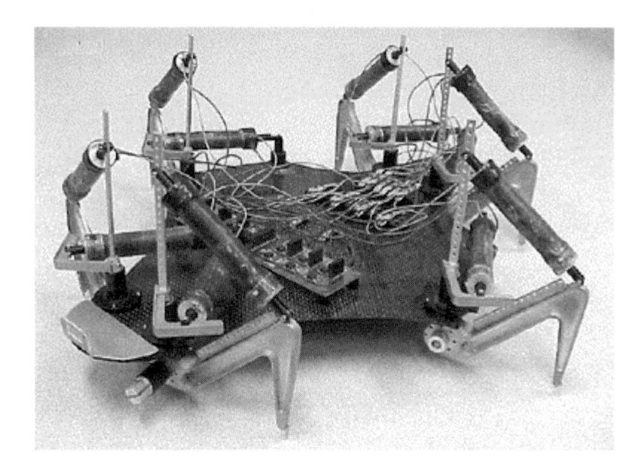

Figure 19: FLEX 2 Robot [adapted from Pelrine et al., 2002].

Figure 20 shows a different legged robot, dubbed Skitter [Pei et al., 2002]. Skitter is based on an earlier pneumatically driven robot dubbed Sprawlita [Clark et al., 2001]. Sprawlita's design was based on the results of research into cockroach locomotion: and in a sense, Sprawlita can be described as a "first-order cockroach." In Skitter, rolled acrylic dielectric elastomer actuators were substituted for the pneumatic cylinders in Sprawlita, primarily to demonstrate the use of this new rolled actuator technology. Skitter uses six rolled actuators to provide six single-degree-of-freedom legs. This robot was successfully demonstrated at a peak speed of approximately 7 cm/s.

Figure 20: Skitter robot using six rolled actuators [adapted from Pelrine et al., 2002].

Unlike those of the FLEX series robots, Skitter's legs can naturally rotate backward about a horizontal axis when it encounters an obstacle in the forward direction. On the other hand, FLEX has two DOF per leg, which makes it more controllable than the simpler robots (e.g., it can go backward). Thus, planning for the next-generation legged robot focuses on achieving the best capabilities of both types of biomimetic robots. We also note that the legs of both types have limited motion for obstacle clearance—a current limitation that should be possible to overcome in future designs.

Robots like FLEX 1, FLEX 2, and Skitter may also serve as testbeds for assessments of the advantages of musclelike actuation. In particular, we hope to exploit the viscoelastic behavior of the muscles to enable robots to reject disturbances due to obstacles or uneven terrain, similar to the dynamics of a cockroach's musculoskeletal system that help it to run over uneven terrain in a stable manner and without large disturbances of its torso [Full et al., 1998]. Eventually, we hope the muscles will operate fast enough that the viscoelastic behavior of the muscles can provide the correct amount of energy storage and absorption while the robot is walking or running.

16.6.2.2 A New Type of Legged Robot

The ability of a multi-DOF roll to act as both a muscle and a structure can enable radically new types of legged robots. The simple robotic system shown in Fig. 21, dubbed MERbot (Multifunctional Electroelastomer Roll robot), is a simple robot that contains six 2-DOF spring rolls, a hexagonal frame (chassis), and wires. The dimensions of the robot are 18 cm × 18 cm × 10 cm. Its weight is 292 g. The power and controls are tethered to the robot. The robot moves in a

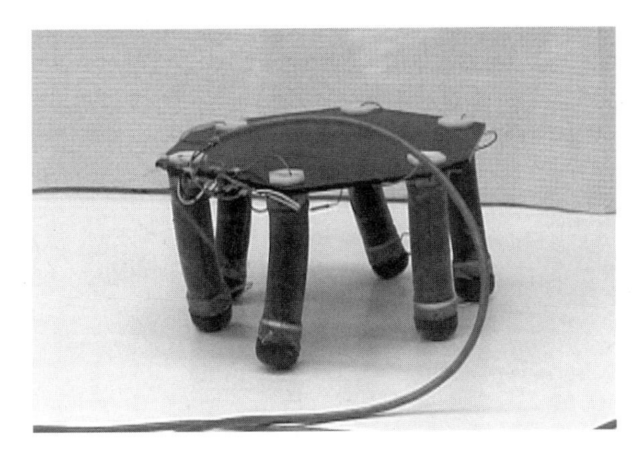

Figure 21: MERbot, a robot using a 2-DOF spring roll as each of its six legs [adapted from Pei et al., 2003].

dual tripod gait. For this gait only four input controls are needed: two controls for bending each tripod forward and two controls for bending each tripod backward. The current maximum speed of MERbot is 13.6 cm/s at 7 Hz, with 5.5 kV of driving voltage.

As has been the case with the other robots, MERbot has not yet demonstrated exceptional performance but highlights the potential of EAPs. In MERbot, the multifunctionality of the dielectric elastomer EAP actuators can enable a very simple robot. In the future, such robots might perform similar to that of spiders or starfish.

16.6.2.3 Serpentine Robot

Bending spring rolls can effectively replicate the motion of body segments of creatures such as snakes and worms. Hence, the application of this type of actuator to serpentine manipulators is not surprising. Dielectric elastomer EAP may be an even better enabling technology for such robots than it is for the other types of biologically inspired robots. Many other researchers have attempted to make serpentine manipulators with electric motors operating through gearing. However, the cost and complexity of these robots limits the number of their degrees of freedom. Further, electric motors do not possess the necessary compliance and peak power density that direct-drive artificial muscles can produce. Although many serpentine manipulators have been developed using conventional actuator technologies such as electric motors, these devices cannot reproduce the dynamics of snake locomotion and therefore cannot reproduce the ability of snakes to effectively traverse a great variety of terrain and overcome a wide range of obstacles.

Figure 22 shows how bending rolls can be concatenated to produce a segment of a serpentine manipulator. This segment is shown undergoing actuation to form an S curve as would occur in lateral undulation, a common mode of snake locomotion.

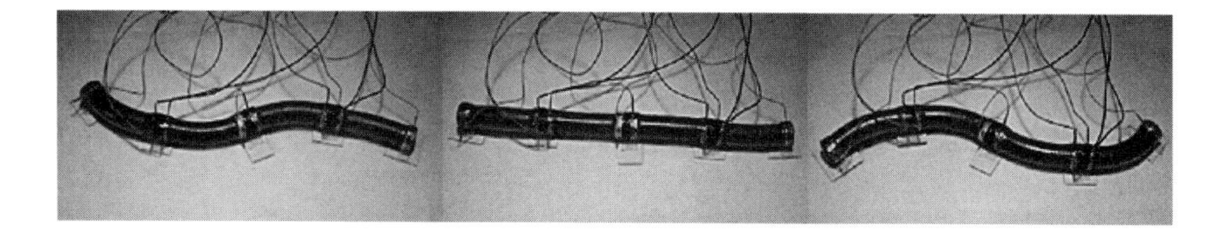

Figure 22: Video stills of a proof-of-concept segment of a snake robot based on concatenated 3-DOF MER segments.

16.6.2.4 Inchworm Robot

Figure 23 shows a small (16-mm long) robotic platform that crawls like an inchworm. The robot's "body" is a rolled silicone actuator with an electrostatic clamp at each end. The clamps enable the inchworm to travel on vertical as well as horizontal surfaces. In tests, the inchworm was able to travel at a maximum speed of about 10 cm/s.

Robotic platforms like the inchworm could eventually be used to make small robots for tasks such as the inspection of narrow pipes. The inchworm takes advantage of the large strain capability of the dielectric elastomer EAP roll. It also shows the multifunctionality of its musclelike actuator. In this instance, the roll functions both as the actuator and the body. Like worms and other creatures, it has no separate, rigid supporting skeleton.

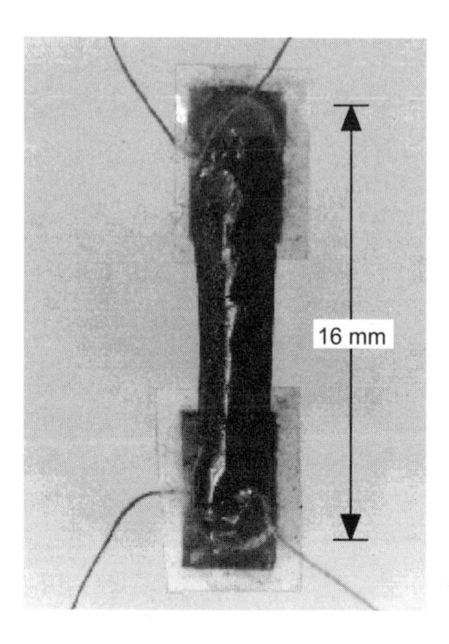

Figure 23: Inchworm-inspired propulsion system based on a silicone dielectric elastomer rolled actuator with electrostatic clamps.

16.6.2.5 Insect-Inspired Flapping-Wing Robot

Figure 24 shows a flapping-wing mechanism whose design was inspired by the mechanics of the many flying insects whose wings are driven indirectly by the muscles located in the thorax. These muscles flex the exoskeleton and move the wings, which are attached to the exoskeleton. In the same way, a bundle of artificial muscles can flex a plastic exoskeleton with wings attached.

Four silicone bow-tie actuators (see Fig. 13) drive the mechanism, which is designed so that its optimum flapping frequency coincides with the resonance of its muscles. Thus, the flapping amplitude can be increased and the power required for flapping the wings minimized, since no energy is needed to counteract the inertia of the wings and mechanism. It has been shown that insect muscle does indeed operate in this manner [Dickinson and Lighton, 1995; Alexander, 1988]. This principle is a biological example of the importance of impedance matching.

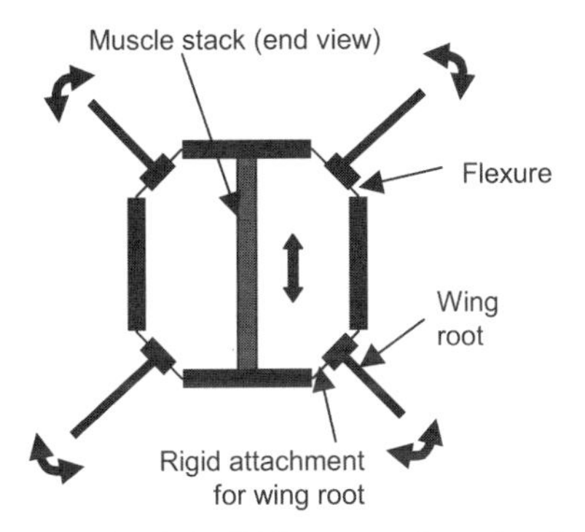

Schematic of the EAP-powered mechanism

Demonstration mechanism with a 6-in wingspan

Conceptual representation of muscle-powered flapping-wing flyer
(Source: Dave Loewen, University of Toronto Institute for Aerospace Studies)

Figure 24: Insect-inspired flapping-wing mechanism powered by artificial muscle.

The eventual goal of this effort is to make simple and robust flying platforms that can be used for reconnaissance in cluttered environments. Such missions will require efficient slow-speed and hovering-flight capabilities. Inspired by nature, flapping-wing flight offers potential advantages in thrust-to-power ratio and stability over conventional flyers based on rotors. If the flyers are to be electrically powered, which is both quieter and logistically advantageous, then lightweight and powerful electric actuation is needed. Dielectric elastomer EAPs, incorporated into a biomimetic flapping mechanism, can form the basis of such an electric flapping-wing propulsion system.

The flapping-wing mechanism has proven to be quite robust because it flaps wings at the spanwise bending resonant frequency of 18 Hz. However, it currently has only four muscles operating in parallel. It will need as many as 25 muscles operating in parallel to produce the power output needed for hovering a 50-g vehicle, where the muscle bundle will resonate with the wings at 40 Hz (the optimal thrust-to-power point for the wings).

16.7 General-Purpose Linear Actuators

Many of the same features that make dielectric elastomer EAPs attractive for artificial muscle applications also make them attractive for general-purpose linear actuators. Such actuators might be used to drive small industrial valves or other industrial automation or consumer products. The roll and bow-tie actuators of the previous section are well suited to these applications.

The dielectric elastomer actuator can be highly competitive with other actuators. It is significantly lighter than existing commercially available actuators based on electromagnetics or other "smart materials." Figure 25 shows graphically the equivalent masses of commercially available actuators that would be required to produce the same amount of work as the roll actuator of Fig. 14. In many applications, the dielectric elastomer roll could replace an electromagnetic solenoid.

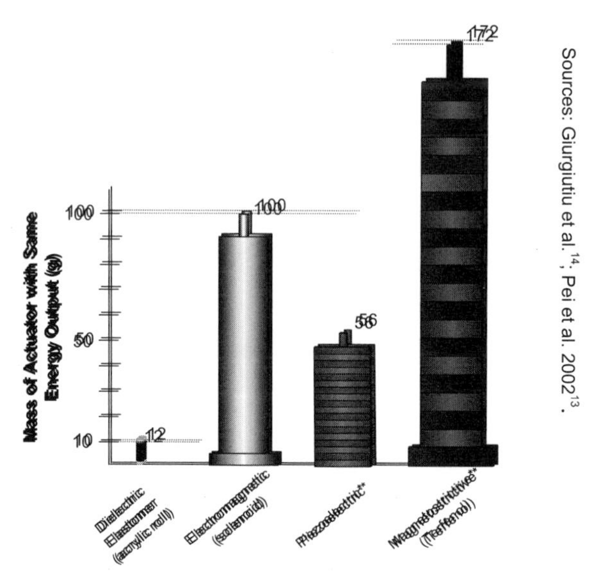

Figure 25: Comparison of actual performance of dielectric elastomer actuators with those of commercially available actuators, in terms of energy density (adapted from Kornbluh et al., 2003).

16.8 Planar and Other Actuator Configurations

Most of the above examples rely on linear actuators with a single direction of output. As we have noted, many other actuator designs exist where two directions of output can be coupled to the load. Still other actuator designs do not rely on an energetic output at all but take advantage of the large dimensional change possible with dielectric elastomer EAP. Both these classes of actuators exploit the unique high-strain capabilities of dielectric elastomer EAPs perhaps even more successfully than linear artificial-muscle actuators. In other cases, the dielectric elastomer material is laminated directly to another material in order to form a bending beam or plate actuator. Examples of these configurations are described below.

16.8.1 Diaphragm Actuators

Dielectric elastomer actuators produce their greatest energy output when they can realize large strains. Diaphragm actuators can take advantage of this large strain. Figure 26 illustrates the structure of a diaphragm actuator.

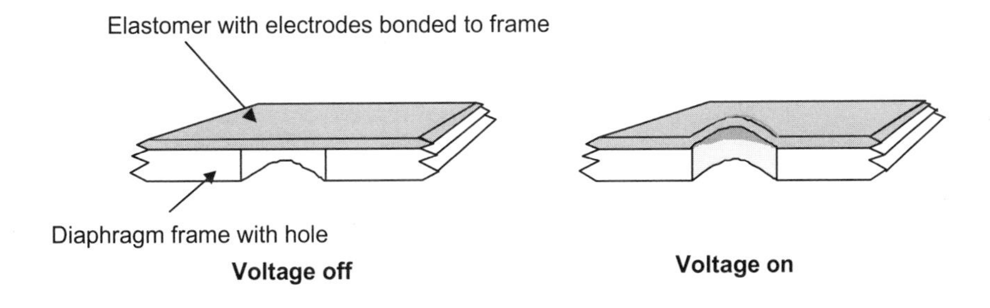

Figure 26: Diaphragm actuator structure (sectional view).

Diaphragms are particularly well suited to pumps with direct energetic coupling to an external load. Diaphragms could also be used for adaptive optics or for controllable surface roughness (for example, on an aerodynamic surface), where the actuator need only create a change in appearance or texture.

For pumps, single-layer, 3-mm-diam diaphragm actuators with silicone films have been demonstrated producing up to 20 kPa (3 psi) pressure. Single-layer acrylic diaphragms with diameters of up to 17 mm have produced pressures of 10 kPa. We have also demonstrated small proof-of-principle pumps using single-layer diaphragms and one-way valves. These pumps produced flow rates of roughly 30–40 ml/min and pressures up to 2500 Pa. Multiple cascaded pumps or thicker diaphragms could be used to increase pressure. An attractive feature of dielectric elastomer diaphragms, as opposed to piezoelectric diaphragms, is that the displacement can be relatively large without the sacrifice of other performance parameters.

Our highest-performing films allow for out-of-plane deflection equal to 50% or more of the diaphragm diameter. Figure 27 shows an acrylic diaphragm actuator undergoing large out-of-plane deformation in which the diaphragm changes shape from flat to hemispherical. In principle, piezoelectrics can achieve large diaphragm strokes, but in practice only very thin piezoelectric diaphragms can do so, because the intrinsic strain of piezoelectrics is so much smaller than that of electrostrictive polymers. The use of very thin piezoelectric diaphragms, however, sacrifices other parameters such as pumping pressure or packaging density, and in most cases significantly reduces the size of piezoelectric diaphragm strokes.

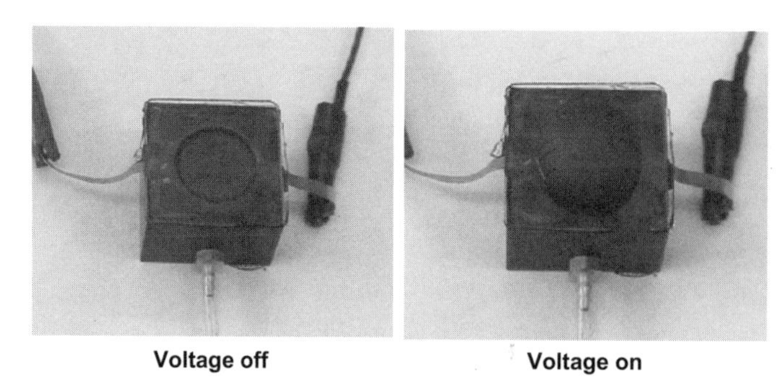

Figure 27: Acrylic diaphragm undergoing actuation.

Diaphragm actuators are well suited to microdevices because they can be fabricated easily in small sizes. Arrays of diaphragms with 56-μm diameters have been demonstrated, as well as in-situ fabrication of diaphragms on silicon wafers. Linear actuators can also be made in this configuration by adding a flexure that presses down on the diaphragm. We used linear diaphragm actuators to demonstrate microlight scanners. Driving voltages of 190–300 V were used to demonstrate scanning, typically at 60–200 Hz, although much higher frequencies can probably be used.

16.8.2 Framed Actuator

A framed actuator consists of a polymer film stretched over a rigid frame. These actuators closely resemble the devices used to measure the strain response described above. Since their design is similar to such devices, their performance would also be similar. In other words, we can expect the generation of extremely large strains.

A framed actuator can make a good linear artificial muscle actuator. Here we focus on an example where no energetic coupling to the environment is necessary. Such actuators can be used for many applications, such as an optical switch, in which an opaque electrode area interrupts a light beam when actuated. The advantage of this approach is the simplicity and therefore low cost of the structure. The apparatus is basically solid state and has just one moving part—the electroded polymer film.

A binary switch of this type can be useful, but in some cases it may be desirable to continuously modulate the amount of light transmitted. This modulation can also be done with a framed actuator. The electrode gradually becomes less opaque as the electroded area increases. Figure 28 shows an example of such an actuator.

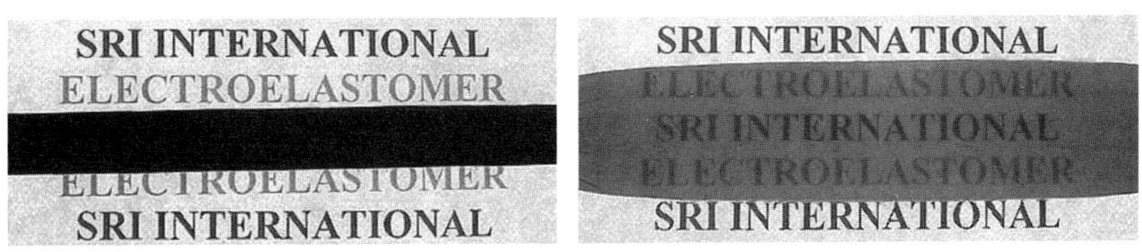

Voltage off **Voltage on**

Figure 28: Solid-state optical aperture based on a framed actuator. The letters behind the electrode are visible only when the voltage is applied (adapted from Kornbluh et al., 2002b).

16.8.3 Acoustic Actuator (Loudspeaker)

The same diaphragm configuration that could be used to pump fluids could also, if actuated at a higher frequency, pump air. In other words, it can create an acoustic output and serve as a loudspeaker. Several variations on this basic diaphragm configuration have been explored. In some of these variations the speaker consists of an array of small, circular, bubblelike diaphragms; in others, a single large diaphragm is used. Both silicone- and acrylic-based materials have served as the dielectric elastomer. Figure 29 shows examples of dielectric elastomer speakers based on a single polymer film.

The speakers built thus far should be considered proof-of-concept devices, in that they have not been subjected to rigorous testing and redesign. Nonetheless, some good performance has been achieved. Qualitatively, these speakers can play music and speech with little obvious distortion. Figure 30 shows the output of a silicone bubble array speaker. The output shows good fidelity and power sensitivity in the upper middle to tweeter ranges. At low frequencies, the fidelity and power output is poor. However, the dielectric elastomer materials are capable of good power output at relatively low frequencies, so that this poor low-frequency performance may not be fundamental and should improve with better speaker design. Fidelity may also be improved by including compensation in the electronic driving circuitry. Heydt et al. [2000] describe the principles of operation and evaluation of this speaker in more detail.

Three acrylic diaphragm speakers.

The larger speaker is 12 in × 12 in.

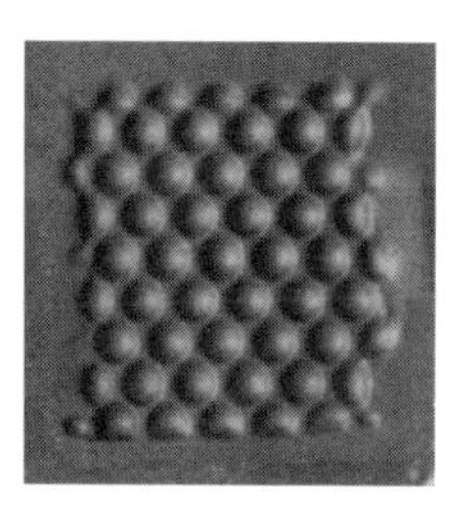

Surface of a speaker based on an array of 5.5 mm diameter bubble-like diaphragm elements formed from an active silicone film stretched over a grid of holes. A slight positive pressure causes the diaphragm elements to form a bubble shape.

Figure 29: Examples of loudspeakers based on dielectric elastomers.

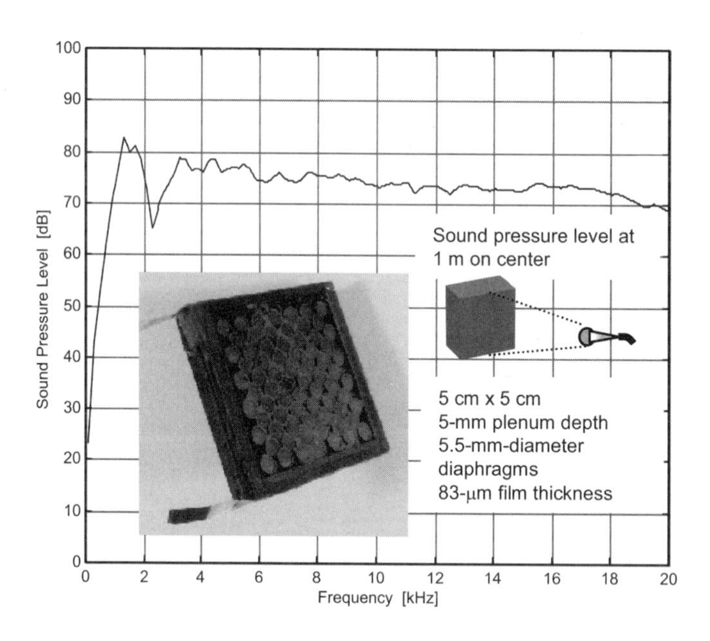

Figure 30: Output of a silicone dielectric elastomer loudspeaker.

16.8.4 Shape Control of Flexible Mirrors

Space-based astronomy and remote-sensing systems would benefit from extremely large aperture mirrors that could permit higher-resolution images. To be cost effective and practical, such optical systems must be lightweight and capable of deployment from highly compact, stowed configurations. Such gossamer mirror structures are likely to be very flexible and therefore present challenges in achieving and maintaining the required optically precise shape. Dielectric elastomers were investigated for use as active materials that could be easily integrated to control the shape of such structures [Kornbluh et al., 2003].

Dielectric elastomers have several properties that make them attractive for such applications: they can operate over a wide range of temperatures (as noted, silicones, in particular, have been demonstrated operating at –100°C to 260°C); they are power efficient and do not require much power to hold a position; they are open-loop stable and precisely controllable; and they can be made in large-area sheets that can be integrated with the mirror structure. Many of these characteristics are ideal for terrestrial optical systems as well.

Figure 31 shows several different types of lightweight mirrors whose shape was controlled by dielectric elastomers. Analytical (finite element) and

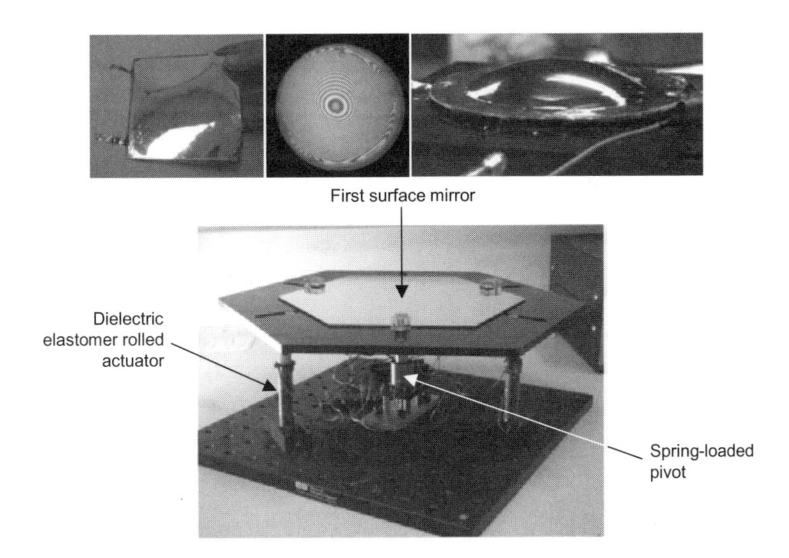

Figure 31: Proof-of-principle devices showing shape control of mirrors with dielectric elastomer actuation. Clockwise from top left: Acrylic elastomer laminated to 4-m-thick polymer mirror; interferogram of surface of 0.5-mm-thick gold-plated silicon wafer with dielectric elastomer layer on reverse side; inflated elastic mirror composed of gold on silicone with powdered graphite electrodes on reverse side (mirror would increase curvature upon the application of a voltage); three rolled acrylic actuators in tripod arrangement to control position and orientation of hexagonal mirror segment (adapted from Kornbluh et al., 2003).

experimental methods suggested that dielectric elastomers could produce the necessary shape change when laminated to the back of a flexible mirror (a unimorph bending plate with multiple active areas on a single substrate, or multiple actuator regions on a single mirror). Interferometric measurements verified the ability of the elastomers to effect controllable shape changes of lightweight flexible mirrors with a resolution of less than the wavelength of light (as would be required for optical applications).

Dielectric elastomers were also incorporated into an inflatable mirror. This design is basically a diaphragm actuator with multiple active areas on a single substrate.

In an alternative design, discrete linear roll actuators were shown to be able to control the position of a rigid mirror segment with a sensitivity of 1800 nm/V, suggesting that subwavelength position control is feasible and that dielectric elastomer actuators can be used as direct replacements for much heavier piezoelectric or magnetostrictive linear actuators.

While initial results are promising, numerous technical challenges remain to be addressed, including the development of shape-control algorithms, the fabrication of optically smooth reflective coatings, dynamic effects such as vibration, methods of addressing large numbers of active areas, and stowability and deployment schemes.

16.8.5 Actuator Arrays for Haptic Displays

The diaphragm array and the large speaker of Fig. 29 suggest that dielectric elastomers can be used to make smart skins or other structures consisting of relatively large area arrays of individually addressable actuator (or sensor) elements.

Haptic displays such as Braille could benefit from such arrays of dielectric elastomer actuators. Present-day refreshable Braille displays typically use piezoelectric actuators to raise the Braille dots, although many actuation methods have been tried. Space limitations associated with piezoelectric actuation, and the cost of manufacturing large actuator arrays, limit most displays to a single line of characters. The cost of such refreshable displays is too high for many visually disabled people.

SRI is developing a refreshable Braille display based on arrays of dielectric elastomer diaphragm actuators [Heydt and Chhokar, 2003]. SRI has demonstrated individually addressable diaphragm actuators at the small scale of Braille dots: 1.5 mm diameter and 2.3 mm center-to-center spacing. Such 2-mm-diameter diaphragm actuators built with acrylic films have produced pressures of up to 25 kPa (3.7 psi). These pressures can result in 10 to 25 g of actuation force on the Braille dot that is needed for easy reading. Figure 32 shows a laboratory prototype actuator for a single cell (character) eight-dot Braille cell as well as a three-cell device. The approach is scalable to large numbers of cells, and is expected to enable the building of refreshable displays with many lines of characters at an affordable price. Dielectric elastomer actuators can also be used for other haptic display devices.

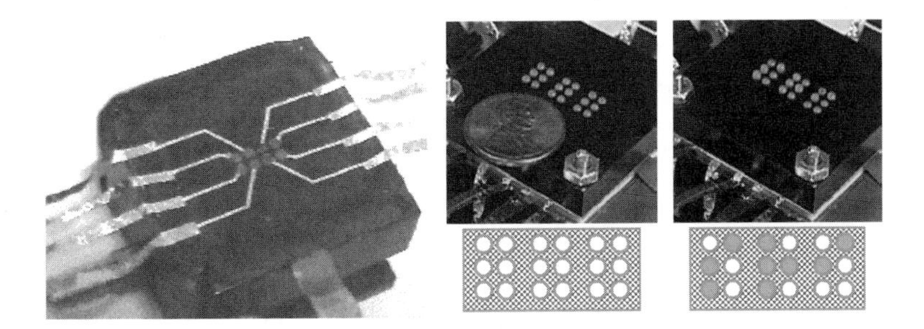

Figure 32: Refreshable Braille cells based on dielectric elastomer: single cell device (left); 3-cell device (right) (adapted from Kornbluh et al., 2002b and Heydt and Chhokar, 2003). (Permission for reprint, courtesy of the Society for Information Display.)

16.8.6 A More General Perspective on Artificial Muscle

It is easy to think of muscle as only a linear actuator that does mechanical work to move or modulate the impedance of an appendage or body segment. In many instances in nature, muscle has other functions. It is interesting to note that in a more general perspective the diaphragm, framed, and acoustic actuators described above can be considered analogs of natural muscle. In our bodies, muscles act as diaphragm actuators to inflate and deflate our lungs. Similarly, muscle changes the shape of our eyes to adjust their focus. In a sense, muscles control some of our surface properties by controlling the erection of hairs on the body and causing goose pimples. The stretched-film actuator with the expanding aperture is functionally similar to the dilating pupils in our eyes. The acoustic actuator is functionally similar to our vocal chords or other means by which animals produce sound.

16.9 Motors

The large energy output per cycle and the fast speed of response of dielectric elastomer actuators can be exploited to make motors that offer very high power density and high specific torque. Resonance also can be used to improve the output. Based on the measured actuator energy density and speed of response, it should be possible to greatly exceed the specific power of electromagnetic motors.

Figure 33 shows a simple proof-of-concept motor as an example. A pair of bow-tie actuators operates 180 deg out of phase to oscillate a shaft. The oscillatory motion is rectified by a one-way clutch to produce rotary motion. This

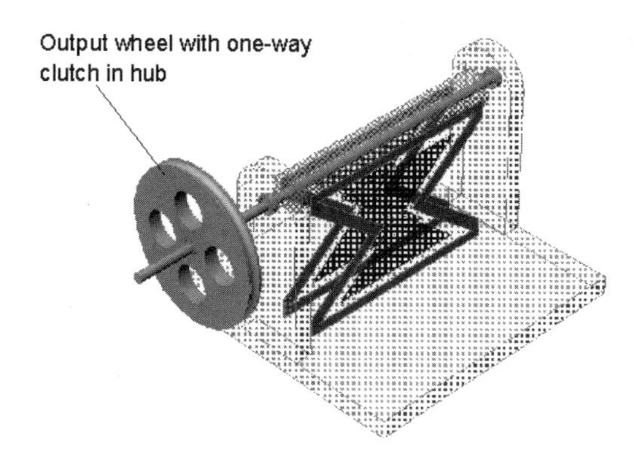

Figure 33: A simple rotary motor based on dielectric elastomers (adapted from Kornbluh et al., 2002b).

motor has produced rotation at speeds up to 650 rpm. Many other more efficient motor designs are being explored.

In addition to high power density, dielectric elastomer motors have good low-speed and stall torque. Unlike electromagnetic motors, which are inductive in nature, dielectric elastomers are capacitive and do not consume much power when stalled. Recently, a motor with an output of 4 W at only 100 rpm has been demonstrated. While not yet equal to the power capabilities of electromagnetic motors, this low-speed performance is encouraging. Dielectric motors are configurable in a wide range of motor sizes and shapes.

16.10 Generators

Thus far, we have described actuator applications of dielectric elastomers. However, it should be noted that the material also functions quite well in generator mode. In generator mode, a voltage is applied across the elastomer and the elastomer is deformed with external work [Pelrine et al., 2001]. As the shape of the elastomer changes, the effective capacitance also changes and, with the appropriate electronics, electrical energy can be generated. Just as the maximum energy density of the dielectric elastomer material as an actuator exceeds that of other field-activated materials, the energy density in generator mode is high. An energy density of 0.4 J/g has been demonstrated with the acrylic elastomer. Performance projections based on material properties are even greater.

Many of the design principles and potential advantages of dielectric elastomers that are true in actuator mode also apply to generator-mode operation. In particular, this type of generation is well suited to applications where electrical power must be produced from relatively large motions. The stress-strain match of the dielectric material to natural muscle further suggests a good impedance match to motions produced by natural muscle, such as human activity.

Figure 34 shows a device that can be used to capture the energy produced in walking or running when the heel strikes the ground. This generator, located in the heel of a shoe, effectively couples the compression of the heel to the deformation of an array of multilayer diaphragms. Other configurations are also being explored. The use of dielectric elastomers is appropriate for a heel-strike generator because of the large deflections produced in the heel with low to moderate pressures. The heel-strike generator has thus far produced a maximum of 0.8 J per cycle. With further development we expect that such devices will be able to generate about 1 W per foot during normal walking. The generator is being developed for the military to help supply the power needs of future soldiers, but also has many commercial applications, including cell phone chargers, and PDAs or power-on-shoe devices such as lights, health, and performance monitors, or navigation devices.

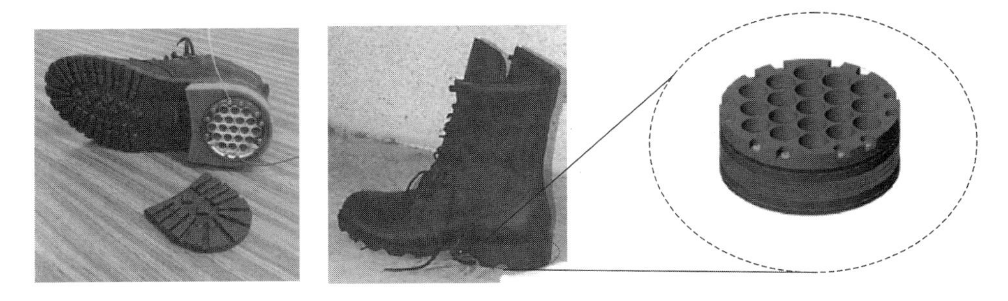

Figure 34: Heel-strike generator located in a boot heel and based on the peformation of dielectric elastomers during walking or running.

16.11 Sensors

We have already noted that dielectric elastomer actuators are also intrinsically position or strain sensors. Dielectric elastomer sensors have certain advantages in their own right, even if the actuation function is not included. The large strain capabilities and environmental tolerance of dielectric elastomer materials allow for sensors that are very simple and robust. In sensor mode, it is not often important to maximize the energy density of the sensor materials since relatively small amounts of energy are converted. Thus, the selection of dielectric materials can be based on criteria such as maximum strain, environmental survivability, and even cost. For example, many polymer materials are virtually unaffected by fluctuations in temperature or humidity. This ability to be tailored to applications is unique to dielectric elastomers. In addition, we are not maximizing energy, so sensors can be operated at low voltage and can interface with existing electronic components and circuits.

As sensors, dielectric elastomers can be used in all of the same configurations as actuators, as well as others. Figure 35 shows examples of several dielectric elastomer sensors, which can be thin tubes, flat strips, arrays of diaphragms, or large-area sheets. In many applications these sensors can replace bulkier and more costly devices such as potentiometers and encoders. In fiber or ribbon form, a dielectric elastomer sensor can be woven into textiles and provide position feedback from human motion. Dielectric elastomer sensors can also be laminated to structures or skins to provide position information for multifunctional smart materials. Sensors such as the diaphragm array can be used to measure force or pressure as well as motion.

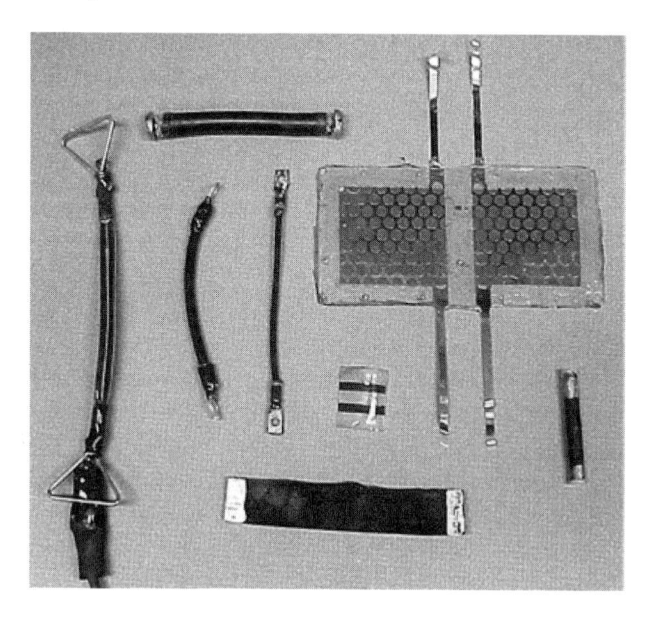

Figure 35: Representative dielectric elastomer sensors (adapted from Kornbluh et al., 2002b).

16.12 Summary and Future Developments

We have discussed how the unique characteristics of EAPs, specifically dielectric elastomers, offer unprecedented opportunities for a wide range of applications. In the future, a product designer might simply assume that EAPs are the actuator of choice from a cost and performance perspective. To get to this point, we must profoundly understand the design issues, performance, and operational characteristics of EAPs. In some cases the high-voltage requirements for dielectric elastomer EAPs will be an obstacle to their application. In other cases, the high voltage will be an advantage. In all cases, the use of these dielectric elastomers will require the parallel development of efficient or low-cost electronic driver circuits.

To this point in our discussion we have not focused on further improvements needed in the dielectric elastomer materials themselves. While many of their performance and other characteristics are already quite good compared to those of other actuator technologies, many questions are still unanswered about their long-term reliability and durability, among other issues. The relationship between their material, electrical, and mechanical properties and the resulting actuation output is still not completely understood. Rigorous studies have not been performed to test the effects of many key parameters. Thus, there are still many opportunities for significant improvements in the basic dielectric elastomer technology itself that can further affect the suitability of this technology to many applications.

16.13 Acknowledgments

The authors would like to thank the many individuals at SRI International who have contributed to the development of the dielectric elastomer actuation technology described above and to the preparation of this manuscript. These individuals include Seajin Oh, Don Czyzyk, Surjit Chhokar, Shari Shepherd, Lynn Mortensen, Bob Wilson, Jon Heim, John Marlow, Prasanna Mulgaonkar, Philip Jeuck, David Flamm, Karen Nashold, David Huestis, Jeff Simons, Tom Cooper, David Watters, Jose Joseph, Seiki Chiba, Peter Marcotullio, and Philip von Guggenberg. Joe Eckerle, in particular, provided many useful comments. Jean Stockett, Peggy Nutsch, and Naomi Campbell contributed greatly to the preparation of the manuscript.

Much of the basic dielectric elastomer technology was developed under the management of the Micromachine Center of Japan under the Industrial Science and Technology Frontier Program, Research and Development of Micromachine Technology of MITI (now known as METI), Japan, supported by the New Energy and Industrial Technology Development Organization. The development of dielectric elastomer artificial-muscle actuators and the application of the technology to large biomimetic robots were supported by the Defense Advanced Research Projects Agency (DARPA), the Office of Naval Research (ONR), and the Naval Explosive Ordinance Technology Division (NAVEODTECHDIV). DARPA also supported the development of the heel-strike generator and some aspects of the loudspeaker. The Department of Education supported the development of the refreshable Braille display. The National Reconnaissance Office supported the investigation of dielectric elastomers for large space optics. SRI visiting scientist Guggi Kofod of the Risø National Laboratory, Denmark, performed work that contributed to this chapter. Biologists Prof. Robert Full and Dr. Kenneth Meijer of the University of California at Berkeley also provided valuable insight and testing of dielectric elastomer artificial muscles. Dr. James DeLaurier, David Loewen, Derek Bilyk, and others at the University of Toronto Institute for Aerospace Studies helped inspire the flapping-wing mechanism.

16.14 References

Alexander, R., *Elastic Mechanisms in Animal Movement*, Cambridge University Press, Cambridge, UK (1988), pp. 56–69.

Baughman, R., L. Shacklette, R. Elsenbaumer, E. Pichta, and C. Becht, "Conducting polymer electromechanical actuators," *Conjugated Polymeric Materials: Opportunities in Electronics, Optoelectronics and Molecular Electronics*, eds. J.L. Bredas and R.R. Chance, Kluwer Academic Publishers, The Netherlands (1990), pp. 559–582.

Bobbio, S., M. Kellam, B. Dudley, S. Goodwin Johansson, S. Jones, J. Jacobson, F. Tranjan, and T. DuBois, "Integrated force arrays," in *Proc. IEEE Micro Electro Mechanical Systems Workshop*, Fort Lauderdale, Florida (February 1993).

Clark, J.E., J.G. Cham, S.A. Bailey, E.M. Froehlich, P.K. Nahata, R.J. Full, and M.R. Cutkosky. "Biomimetic design and fabrication of a hexapedal running robot." *Proc. IEEE Intl. Conf. on Robotics and Automation* (2001),

Dickinson, M., and J. Lighton, "Muscle efficiency and elastic storage in the flight motor of drosophila," *Science*, Vol. 268 (1995), pp. 87–90.

Dickinson, M., C. Farley, R. Full, M. Koehl, R. Kram, and S. Lehman, "How animals move: an integrative view," *Science* 288 (2000), pp. 100–106.

Eckerle, J., J.S. Stanford, J. Marlow, Roger Schmidt, S. Oh, T. Low, and V. Shastri, "A biologically inspired hexapedal robot using field-effect electroactive elastomer artificial muscles," *Proc. of SPIE, Smart Structures and Materials 2001: Industrial and Commercial Applications of Smart Structures Technologies*, Vol. 4332 (2001).

Full, R., D. Stokes, A. Ahn and R. Josephson, "Energy absorption during running by leg muscles in a cockroach," *J. Experimental Biology*, Vol. 201, pp. 997-1012 (1998).

Full, R., and K. Meijer, "Artificial muscle versus natural actuators from frogs to flies," *Proc. of SPIE*, Smart Structures and Materials 2000: Electroactive Polymer Actuators and Devices, ed. Y. Bar-Cohen, Vol. 3987, pp. 2–9 (2000).

Giurgiutiu, V., C. Rogers, and Z. Choudhry, "Energy-based comparison of solid-state induced-strain actuators," *J. Intelligent Materials Systems and Structures*, Vol. 7 (1996), pp. 4–14.

Heydt, R., R. Pelrine, J. Joseph, J. Eckerle, and R. Kornbluh, "Acoustical performance of an electrostrictive polymer film loudspeaker," *J. Acoustical Society of America*, Vol 107, No. 2 (2000), pp. 833–839.

Heydt, R. and Chhokar, S., "Refreshable Braille display based on electroactive polymers," *Proc. 23rd Intl. Display Research Conf.*, Phoenix, Arizona (15–18 September 2003).

Hunter, I., S. Lafontaine, J. Hollerbach, and P. Hunter, "Fast reversible NiTi fibers for use in microrobotics," *Proc. 1991 IEEE Micro Electro Mechanical Systems—MEMS '91*, Nara, Japan, (1991), pp. 166–170.

Hunter, I., and S. Lafontaine, "A comparison of muscle with artificial actuators," *Technical Digest of the IEEE Solid-State Sensor and Actuator Workshop*, Hilton Head, South Carolina, (1992), pp. 178–185.

Jeon, J., K. Park, S. An, J. Nam, H. Choi, H. Kim, S. Bae, and Y. Tak, "Electrostrictive polymer actuators and their control systems," *Proc. of SPIE Smart Structures and Materials 2001: Electroactive Polymer Actuators and Devices*, ed. Yoseph Bar-Cohen, 4329, (2001) pp. 380–388.

Kornbluh, R., R. Pelrine, J. Joseph, R. Heydt, Q. Pei, and S. Chiba, "High-field electrostriction of elastomeric polymer dielectrics for actuation," *Proc. of SPIE, Smart Structures and Materials 1999: ElectroActive Polymer*

Actuators and Devices,, ed. Yoseph Bar-Cohen, Vol. 3669 (1999), pp. 149–161.

Kornbluh, R., R. Pelrine, Q. Pei, S. Oh, and J. Joseph, "Ultrahigh strain response of field-actuated elastomeric polymers," *Proc. of SPIE Smart Structures and Materials 2000: Electroactive Polymer Actuators and Devices*, ed. Y. Bar-Cohen in, Vol. 3987 (2000), pp. 51–64.

Kornbluh, R., R. Full, K. Meijer, R. Pelrine, and S. Shastri, "Engineering a muscle: an approach to artificial muscle based on field-activated electroactive polymers," *Neurotechnology for Biomimetic Robots*, eds. J. Ayers, J. Davis, and A. Rudolph, MIT Press (2002a), pp. 137–172.

Kornbluh, R., R. Pelrine, Q. Pei, R. Heydt, S. Stanford, S. Oh, and J. Eckerle, "Electroelastomers: applications of dielectric elastomer transducers for actuation, generation and smart structures," *Proc. of SPIE Smart Structures and Materials 2002: Industrial and Commercial Applications of Smart Structures Technologies*, ed. A. McGowan, 4698 (2002b), pp. 254–270.

Kornbluh, R.D., D.S. Flamm, H. Prahlad, K.M. Nashold, S. Chhokar, R. Pelrine, D.L. Huestis, J.Simons, T. Cooper, and D.G. Watters, "Shape control of large lightweight mirrors with dielectric elastomer actuation," *Proc. of SPIE Smart Structures and Materials 2003: Electroactive Polymer Actuators and Devices*, ed. Y. Bar-Cohen, 5051 (2003), pp. 143–158.

Liu, C., Y. Bar-Cohen, and S. Leary, "Electro-statically stricted polymers (ESSP)," *Proc. of SPIE Smart Structures and Materials 1999: Electroactive Polymer Actuators and Devices*, ed. Y. Bar-Cohen, 3669 (1999), pp. 186–190.

Meijer, K., M. Rosenthal, and R. Full, "Musclelike actuators? A comparison between three electroactive polymers," *Proc. of SPIE Smart Structures and Materials 2001: Electroactive Polymer Actuators and Devices*, ed. Yoseph Bar-Cohen (2001), pp. 7–15.

Park, S., and T. Shrout, "Ultrahigh strain and piezoelectric behavior in relaxor based ferroelectric single crystals," *J. Applied Physics*, 82 (1997), pp. 1804–1811.

Pei, Q., R. Pelrine, S. Stanford, R. Kornbluh, M. Rosenthal, K. Meijer, and R. Full, "Multifunctional electroelastomer rolls and their application for biomimetic walking robots," *Smart Structures and Materials 2002: Industrial and Commercial Applications of Smart Structures Technologies*, ed. Anna-Maria McGowan, pp. 246–253.

Pei, Q., M. Rosenthal, R. Pelrine, S. Stanford, and R. Kornbluh, "Multifunctional electroelastomer actuators and their application for biomimetic walking robots," *Proc. of SPIE Smart Structures and Materials 2003: Electroactive Polymer Actuators and Devices*, ed. Y. Bar-Cohen, 5051, pp. 281–290 (2003).

Pelrine, R., J. Eckerle, and S. Chiba, "Review of artificial muscle approaches," invited paper, *Proc. of the Third Intl. Symposium on Micro Machine and Human Science*, Nagoya, Japan (1992).

Pelrine, R., R. Kornbluh, J. Joseph, and J. Marlow, "Analysis of the electrostriction of polymer dielectrics with compliant electrodes as a means of actuation," *Sensors and Actuators A: Physical* 64 (1995), pp. 77–85.

Pelrine, R., R. Kornbluh, and G. Kofod, "High-strain actuator materials based on dielectric elastomers," *Advanced Materials* 2000, 12, 16 (2000), pp. 1223–1225.

Pelrine, R., R. Kornbluh, Q. Pei, and J. Joseph, "High-speed electrically actuated elastomers with over 100% strain," *Science*, 287, 5454 (2000), pp. 836–839.

Pelrine, R., R. Kornbluh, J. Eckerle, P. Jeuck, S. Oh, Q. Pei, and S. Stanford, "Dielectric elastomers: generator mode fundamentals and applications," *Proc. of SPIE Smart Structures and Materials 2001: Electroactive Polymer Actuators and Devices*, ed. Y. Bar-Cohen, 4329 (2001), pp. 148–156

Pelrine, R., R. Kornbluh, Q. Pei, S. Stanford, S. Oh, J. Eckerle, R. Full, M. Rosenthal, and K. Meijer, "Dielectric elastomer artificial muscle actuators: toward biomimetic motion," *Proc. of SPIE Smart Structures and Materials 2002: Electroactive Polymer Actuators and Devices*, 4695 (2002), ed. Y. Bar-Cohen.

Shahinpoor, M., "Micro-electro-mechanics of ionic polymer gels as electrically controllable artificial muscles," *J. Intelligent Material Systems and Structures*, 6 (1995), pp. 307–314.

Sommer-Larsen, P., J. Hooker, G. Kofod, K. West, M. Benslimane, and P. Gravesen, "Response of dielectric elastomer actuators," *Proc. of SPIE Smart Structures and Materials 2001: Electroactive Polymer Actuators and Devices*, ed. Yoseph Bar-Cohen, 4329, pp. 157–163 (2001).

Tobushi, H., S. Hayashi, and S. Kojima, "Mechanical properties of shape memory polymer of polyurethane series," *JSME International J., Series I*, 35, 3 (1992).

Uchino, K., *Piezoelectric Actuators and Ultrasonic Motors*, Kluwer Academic Publishers, Norwell, Massachusetts (1997), pp. 142–143.

Wax, S., and R. Sands, "Electroactive polymer actuators and devices," *Proc. of SPIE Smart Structures and Materials 1999: Electroactive Polymer Actuators and Devices*, ed. Y. Bar-Cohen, 3669 (1999), pp. 2–10.

Wingert, A., M. Lichter, S. Dubowsky, and M. Hafez, "Hyper-redundant robot manipulators actuated by optimized binary dielectric polymers," *Proc. of SPIE Smart Structures and Materials 2002: Electroactive Polymer Actuators and Devices*, ed. Y. Bar-Cohen, 4695 (2002).

Xia, F., H. Li, C. Huang, M. Huang, H. Xu, F. Bauer, Z. Cheng, Q. Zhang, "Poly(vinylidene fluoride-trifluoroethylene) based high performance electroactive polymers," *Proc. of SPIE Smart Structures and Materials 2003: Electroactive Polymer Actuators and Devices*, ed. Y. Bar-Cohen, 5051 (2003), pp. 133–142.

Zhang, Q., V. Bharti, and X. Zhao, "Giant electrostriction and relaxor ferroelectric behavior in electron-irradiated poly(vinylidene fluoride-trifluoroethylene) copolymer," *Science*, 280, (1998), pp. 2101–2104.

Zhenyi, M., J.I. Scheinbeim, J.W. Lee, and B.A. Newman, "High field electrostrictive response of polymers," *J. Polymer Sciences, Part B—Polymer Physics*, 32 (1994), pp. 2721–2731.

CHAPTER 17

Biologically Inspired Robots

Brett Kennedy
Jet Propulsion Lab. and California Institute of Technology

Chris Melhuish and Andrew Adamatzky
University of the West of England

17.1 Introduction

In a very real way, EAP and its related active polymer technologies such as McKibben actuators (and SMA to some extent) represent a sea change in humanity's technology. Since the beginning of the industrial age, our technology has been, by and large, a hard technology. The label "hard" is literally true from the standpoint of materials and systems. Structures tend to break rather than bend and systems tend to fail rather than degrade gracefully. Systems and structures tend to be divided physically and conceptually into discrete elements. EAP and its brethren, however, are undeniably "soft." Moreover, they are soft in terms of system control as much as in material properties.

As applied to robotics, the promise of EAP and "musclelike" actuators lies on two major fronts. First, as the moniker "musclelike" suggests, these actuators will make possible robotic systems that more closely resemble biological systems. This resemblance extends not only to the physical structure of the system, but also to the dynamics, control, programming, and, ultimately, use of these systems. As humans have much more varied goals than does nature itself, the biological systems should also inspire designs that expand the use of muscle actuators beyond the applications to be found in nature. Second, muscle actuator technology joins MEMS as a tool for the realization of so-called smart structures. One can easily conceive of robots for which structure, actuation, and sensing are so tightly interwoven as to be one system. With the progress in computing power, such smart structures could be the basis for robots that are flexible, both in form and function.

17.2 Biologically Inspired Mechanisms and Robots

When we first contemplate the application of muscle actuators to robotics, our instinctive reaction is to visualize the creation of direct analogs of the creatures around us; home security is handled by a robotic dog, deep sea exploration is performed by robotic dolphins, and so on. Indeed, a humaniform robot is probably the most easily conceived form of robotics in popular thought. Such biomimicry (biomimetics) may have a place in some situations, notably entertainment robotics, but good scientists and engineers must be somewhat more careful in the application of nature's designs. Three important points must be remembered when looking at nature for instruction. One, Nature's technology is not yet our technology, nor its materials our materials, though muscle actuators have brought us closer. Two, nature has a very different set of success criteria for its designs than we do for ours. Its creations have different jobs, different environments, and different cost functions to optimize. Three, *nature is not perfect.* Though Darwin's paraphrased principle of "survival of the fittest" has borne itself out over the last century and a half, that "fittest" design is derived directly from an existing design and is only measured against other designs that have already been built. Human invention is only limited by our own vision and can be optimized relative to all potential designs. Those caveats stated, nature remains an admirable teacher.

17.3 Aspects of Robotic Design

The design of robotic systems is particularly challenging due to the breadth of engineering expertise required. In general, a robot may be represented schematically as shown in Fig. 1. In brief, the "Outside Control" box refers to

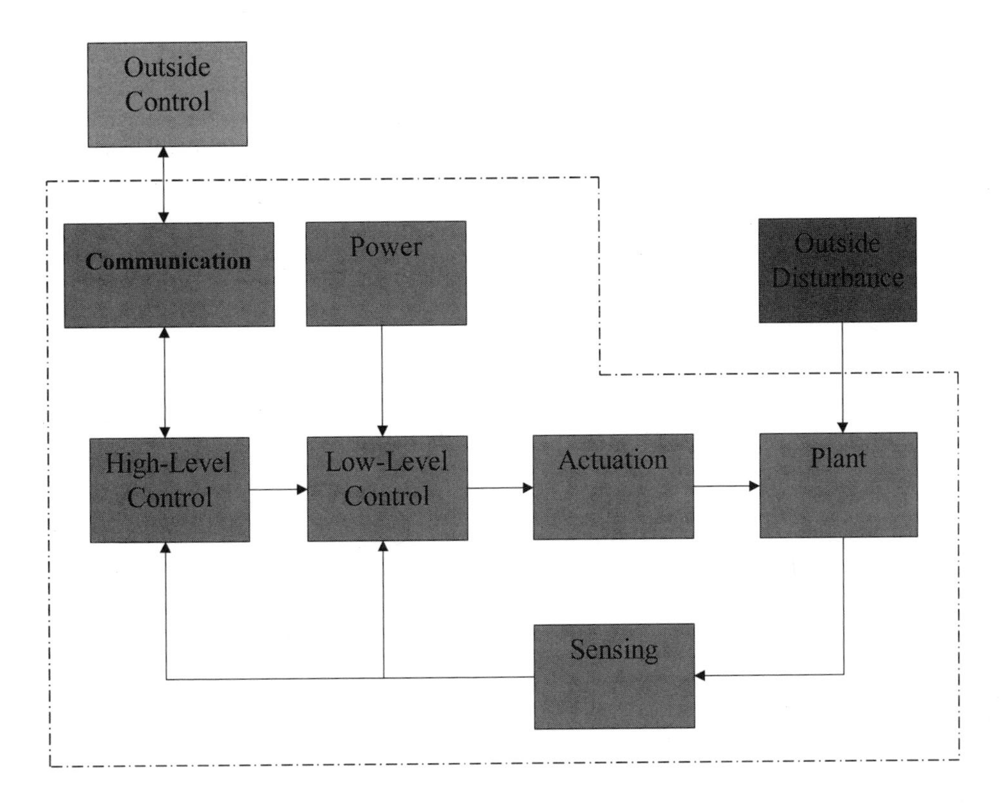

Figure 1: Schematic diagram of a generic control system showing information flow.

any off-board teleoperator, whether it is a computer or a human. Robots can be created without this component or the following "Communications" component, which represents whatever means the robot has to pass and receive information. Communications can be as simple as a tether cable, or as complex as NASA's Deep Space Network of large dish radio transmitters. Presiding over the control of the robot is "High-Level Control," which takes in objectives from the outside (if any) as well as internal objectives, compares those to the information provided by the sensors ("Sensors"), and then issues objectives to "Low-Level Control." Examples of high-level objectives are navigational imperatives, data collection routines, and manipulation directives. The needs of these objectives are translated by high-level control into specific needs for individual actuators, primarily related to motion (position, speed, etc.). Low-level control represents the hardware and software directly responsible for producing the excitation signal (usually some form of energy from the "Power" source) to the system actuators ("Actuators"). Again, information from the sensors is compared to the desired goal, and the appropriate stimulus is fed to the actuators. The actuators, in turn, act on the "Plant," which is a catchall name for the system whose dynamics and statics the controllers seek to modify. In general, the "Plant" refers to the mechanical components of a robot. However, it could just as easily refer to a chemical solution, a magnetic field, or any other of a myriad of physical systems. These components will be discussed in greater detail in the following sections.

17.3.1 Actuation: Designing Robots with Muscles

While the concept of EAP as muscle has been discussed in the introduction, it is useful to look at what a "muscle" might mean to a design engineer. In essence, a muscle is a linear displacement actuator. The closest analogs in general use would be pneumatic or hydraulic cylinders. These actuators are the indisputable champions of high-force applications (think Caterpillar equipment). The only other common solutions involve translating rotational displacement from a motor into linear motion through the use of cables and pulleys. However, if high force is not needed and simplicity is a requirement, muscle actuators may be an appropriate substitute. In fact, they offer several advantages over traditional actuation methods.

As implied, the low moving-part count plays a significant role. Outdoing even the simplicity of a two-piece hydraulic actuator (piston and cylinder), one-piece muscle actuators have no surfaces that slide relative to one another. This attribute should correspond to an increase in reliability by reducing the potential for wear and binding. When measured against rotary-to-linear technologies, this aspect achieves a greater significance. Not only is wear reduced, but force loss due to friction (mechanical efficiency) is also reduced. Depending on actual design, there need not be any mechanical loss in a muscle actuator system, while mechanical loss is inherent in any actuator that relies on some type of transmission, as a rotary-to-linear actuator must.

More interestingly, and with a greater impact on the overall shape of the product, the displacement of the actuator does not have to occur along the same line of action as the part being actuated due to the flexible nature of the material. In other words, the main body of the actuator can be located around a corner from the part on which it acts (Fig. 2). To jump ahead in the discussion, this characteristic can be seen in the muscles of the human forearm acting to control the flexion and extension of the fingers. This example also shows that when it becomes necessary to translate linear motion to rotary (muscle contraction to joint rotation), the mechanism need only be the attachment of the muscle to a lever arm about the joint. And, again, the actuator need not form a straight line from anchor to lever arm pivot.

One major difference between most common actuators and musclelike actuators is that muscles have a preferred direction of actuation (i.e., contraction). Therefore, like in animals, joints in muscle-powered robots must have an antagonistic arrangement of actuators. This arrangement is shown in Fig. 3. Simply, if a muscle (the agonist) moves a part, something (the antagonist), usually another muscle or a spring, must move that part back. In special cases, the force on the part imparted by gravity can also be used. Using a spring as the antagonist provides the simplest solution because only one actuator is needed. However, a spring-antagonist can be undesirable for power and mass considerations. The added demands are due to the fact that the actuator must work against the spring in addition to exerting the necessary force on the environment. That extra force generation may require a larger actuator than otherwise necessary, with the attendant increase in power use. Of course, any restorative force inherent to the actuator can mitigate the design impact of the spring.

The muscle-antagonist requires the added complexity and mass of an entire actuator. However, the power requirements are lower due to the fact that only one of the actuators is activated at a time, providing only the force necessary to impart on the environment. Other advantages also come out of a muscle-muscle arrangement, and they will be discussed in the following section.

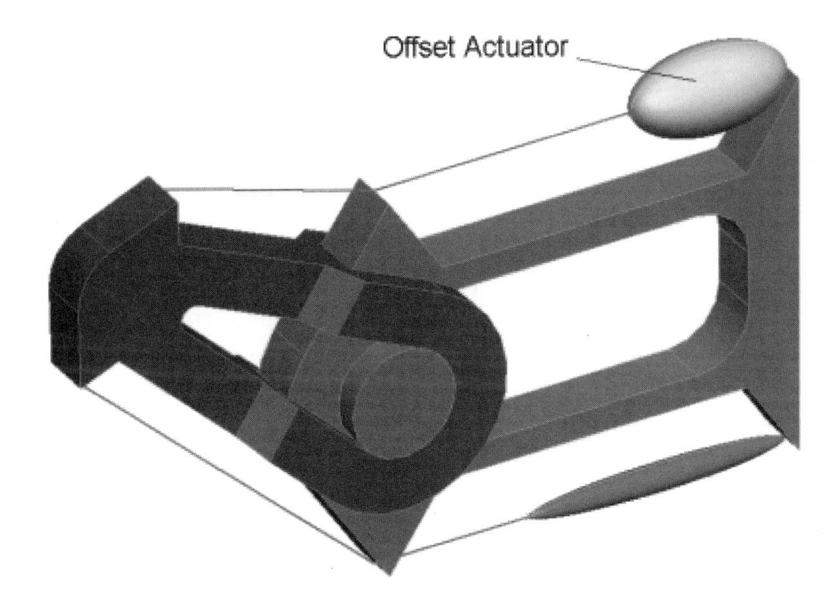

Figure 2: A schematic view of an arm driven by a pair of actuators emulating the operation of muscles.

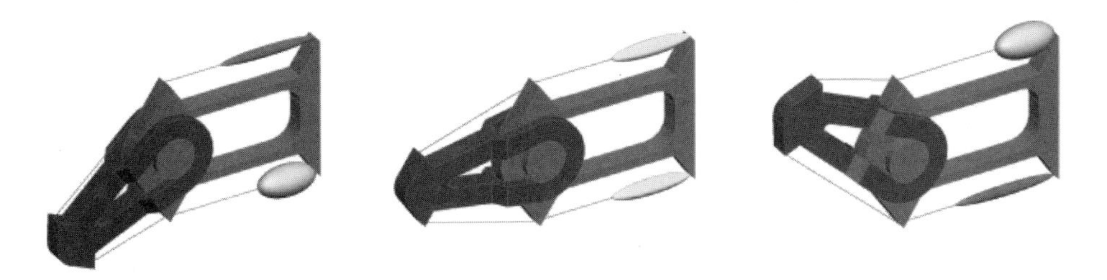

Figure 3: Schematic view of the articulation of an arm using a pair of muscles in an antagonistic arrangement.

17.3.2 Plant: The Effects of EAP on Dynamic System Response

Perhaps the least obvious influence of muscle actuators on systems design is how they affect the nature of dynamic control. In the introduction to this chapter, the system controls for active polymers were referred to as being potentially "soft." This assertion is based both on the actuator's mechanical material properties as well as the response of the materials to their particular actuation stimulus. Unlike the manner in which dynamically "stiff" actuators are treated, muscle actuators' properties encourage that they be treated as part of the structure.

Active polymers present an interesting material due to their combination of low spring rate relative to metals and their passive viscous (speed-dependent) damping. In general, a controllable system must have some level of damping. Due to the low inherent damping of metal structures, classic systems will normally include a shock absorber to damp out unwanted motion or vibration imparted by the environment (think of your car). In the case of polymers, the shock absorber is built directly into the actuator (another reduction in part count!). In addition, the low spring rate allows the actuator to stretch in response to outside influences, storing energy that would have been otherwise imparted to the structure. In essence, active polymers can be thought of as a method of driving the wheels of your hypothetical car and providing the suspension as well, all in one package. If a computer is driving this car along a defined direction, it will have to react less often and less dramatically to the effect of potholes to keep the car going straight. This benefit in turn means that the computer doesn't have to think as fast (processor speed) or expend as much energy. If the example were instead a walking robot, its body would tend to move along a straight line despite the fact that its feet periodically stumbled on obstacles.

The springiness of polymers may provide another benefit. Animals have been shown to store energy in their actuation systems during certain phases of their motion, only to release it constructively in a later phase. In this way, energy that would have to be otherwise dissipated can be stored and then used to the benefit of the system, resulting in a greater overall efficiency. This concept is particularly well illustrated by the energy states of any running animal.

The materials from which polymer-based muscle actuators are made have another major mechanical difference from those of traditional technologies. They are inherently less dense than the generally metallic structure that is commonly used. This aspect gives promise of radically lighter designs leading to less inertia, which plays a significant role in the action of the control system. If the overall system is lighter, less energy is needed to change speed and direction, not to mention the decrease of internal forces and stresses due to the relatively smaller actuation forces.

Potentially one of the most innovative areas for system dynamics is the tailoring of system stiffness with the antagonistic muscle arrangement. The previous section indicated that activating only one actuator at a time could minimize power consumption. However, there are some significant advantages to having both muscles active at the same time. Some polymers have demonstrated a correlation between excitation signal and actuator stiffness. This would imply that the muscle actuators could be used as variable springs.

The schematic in Fig. 4 provides an idealization of the antagonistic muscle arrangement.

The upper view in Fig. 5 represents the system at equilibrium without any outside force. Each actuator is symbolized by a tension spring element with stiffness $\boldsymbol{K}$. The lower view represents the same system that has been displaced $\boldsymbol{\Delta x}$ due to an outside force $\boldsymbol{F}$.

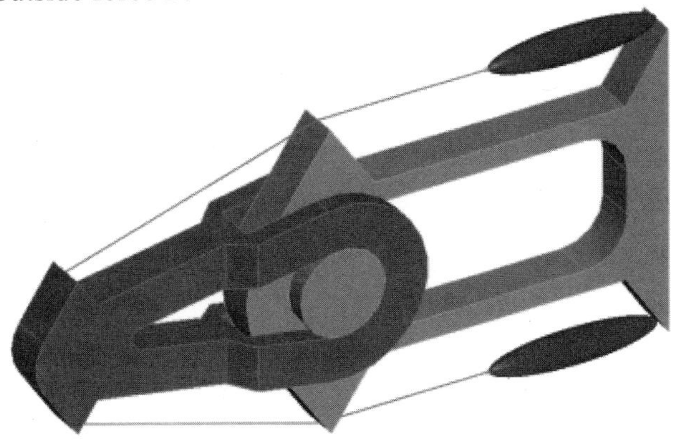

Figure 4: Idealized view of an arm in a rest position.

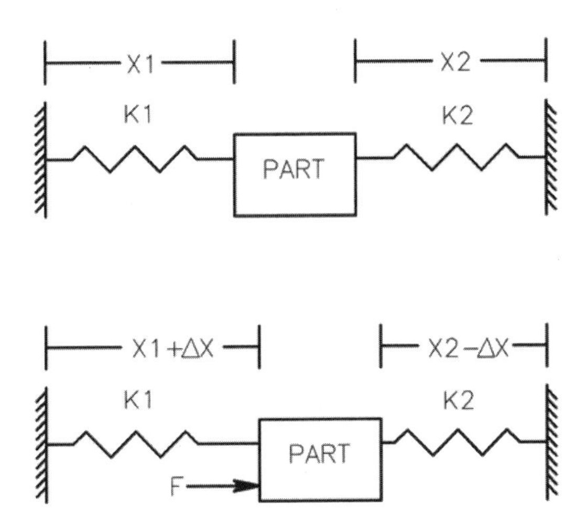

Figure 5: A mass with opposing springs representing the case of antagonistic muscle arrangement.

A force balance for the first case yields:

$$0 = K_1 \cdot x_1 - K_2 \cdot x_2.$$

A force balance for the second case yields:

$$0 = K_1 \cdot (x_1 + \Delta x) - K_2 \cdot (x_2 - \Delta x) - F.$$

Rearranging:

$$0 = (k_1 \cdot x_1 - K_2 \cdot x_2) + (K_1 + K_2) \cdot \Delta x - F.$$

Given that the system was at equilibrium before force was applied,

$$\Delta x = \frac{F}{K_1 + K_2}.$$

This equation supports the common-sense notion that the stiffer the actuators, the more difficult it is for an outside force to disturb the system for a given displacement. To illustrate the importance of this concept, think of the human arm. In its stiffened state, it provides an effective battering ram for a football player's "cold-arm shiver." However, with antagonistic pairs of muscles relaxed, the arm can clean delicate crystal glasses with a cloth. The concept of variable system stiffness can also be cast as a type of force control, which will be explored in the section on low-level control.

17.3.3 Sensors: The Information for Feedback Control

As is implied by the term "feedback control," sensing is extremely important to controlled systems. However, effective direct sensing may be more difficult for musclelike actuators due to their more flexible morphologies. Most methods in common use are not applicable. Traditional motors and pistons are very convenient to monitor with encoders or potentiometers (to name two of the most common) because their areas and modes of motion are distinctly defined. For instance, rotary motors will have a shaft of some type to which a rotary encoder can be mounted. The amount of bend or extent of contraction/expansion in an active polymer is difficult to determine except for special cases. An example of such a case would be a contracting actuator that had a single line of action. It could be treated in the same way as a piston—with the displacement determined with a linear encoder or potentiometer.

One possible direct measurement method would be the incorporation of strain gauges within the actuator. The effectiveness of such a system would be highly dependent on the material properties and the fixturing of the actuator because the strain in the vicinity of gauge must be a good indication of the strain throughout the actuator. Any large discontinuities in strain would create unacceptable errors in measurement.

The sensing for feedback control of active polymers may have to be taken from secondary sources that are monitored accurately more easily. This technique would require examining the entire actuated system for points that are amenable to current sensing methods. A case in point would be our hypothetical rotational joint from Fig. 4. Rather than looking at the contraction of the actuators, the rotation of the joint could be measured with an encoder or potentiometer. Such a system of indirect measurement may prove a problem the less mechanically coupled the point of measurement is to the actuator. Lags in the response of the measured output from the actuator input may result in an unstable system.

When the desired feedback is force rather than displacement or its derivatives, the problem of output monitoring becomes somewhat less problematic. If normal methods of implementing force control are employed, the system need only strain gauges in the structure or load cells at the anchor points of the actuators. However, the antagonistic muscle arrangement mentioned above offers another path. Assuming that actuator stiffness is well correlated to excitation signal, a particular contact force profile could be maintained without any feedback. In fact, if polymer technology allows tailoring of stiffness response, it may be possible to create non-Hookian spring force responses, including constant force regardless of displacement.

17.3.4 Low-Level Control: Making Active Polymers Do What You Want

As the preliminary characterization experiments have shown (see Chapter 10), electroactive polymers tend to demonstrate a nonlinear correlation between input and output. This condition causes complication in the techniques that can be used to accurately control these actuators. While no system that exists beyond the most primitive of experiments can be said to be strictly linear, those mechanisms that use common actuators can usually be approximated as a linear system. As a system moves away from linearity, classic control methods start to break down, and the system becomes inaccurate or unstable.

A short primer on classic control may be useful at this point. Considering a system composed of only the actuator and using simple negative feedback control, the block diagram will look like Fig. 6. Assuming that the system is being controlled for displacement, the set point is the desired position, the excitation might be the voltage level applied to the actuator, the output is the true displacement, and the sensory feedback is the measured displacement. The error, on which the control law operates, is, not surprisingly, the difference between the set-point and the measured output. The plant and sensing have already been discussed, though it should be noted that this diagram differs from Fig. 1 in that

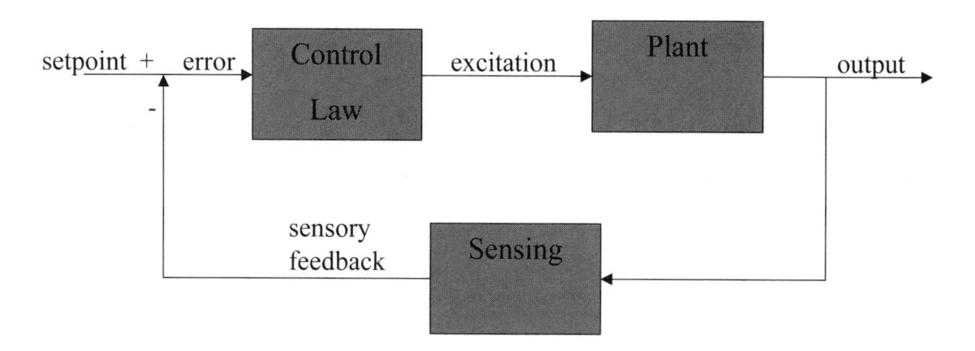

Figure 6: Schematic diagram of control system.

the plant and actuator blocks have been combined because the example system is limited to control of the actuator itself. The control law block represents the mathematical equation that relates the error to the excitation signal. Most commonly (over 90% of installed industrial control systems), the equation used is the Proportional Integral Derivative (PID) equation. As the name suggests, the excitation signal is the sum of the multiple of the error, a multiple of the derivative of the error, and a multiple of the integral of the error.

$$u(t) = K_{\mathrm{p}} e(t) + K_{\mathrm{i}} \int e(t)dt + K_{\mathrm{d}} \frac{de(t)}{dt}$$

The "tuning" of this controller consists of determining values for the "gains" K_P, K_I, and K_D that satisfy stability and performance requirements. Although the PID control law is usually implemented as a software routine, it can actually be built as an analog circuit. For a much more complete and useful description of the PID controller, as well as most aspects of linear control, see Dutton et al [1997].

As mentioned, however, active polymer actuators cannot be expected to be close to linear, and therefore the system cannot be expected to respond in a linear fashion even if a linear control law is implemented. Gains that work through one part of the system input envelope will cause inaccuracy or instability in another part. The next logical step, then, is to break the system response into regions that are themselves somewhat close to linear, and to linearize those regions. In practice, points in the input are chosen that correspond to expected operating points, and then the equation is linearized about those points. Formally speaking, a Taylor expansion is applied to the system response equation, and the linear terms of the expansion are used as the approximated system response for the region around the expansion point. In general, as the input moves further from the point of the expansion, the approximation will introduce larger errors into the system. The size of those errors versus the system design requirements will dictate the number of linearized regions necessary. For each of the regions, a distinct set of gains is applied. The software monitoring the control of the system then switches between sets of gains depending on what region the system is currently operating within. This process is sometimes referred to as "gain scheduling." An example of a graph of a linearized response can be seen in Fig. 7. It shows the equation $y(u)=u^3$ linearized about $u = 5$ and $u = 10$.

The method of linearization described assumes that a sufficient mathematical model exists to create a function. However, the general idea behind the creation of a piecewise linear model applies to statistical models as well. Instead of linearizing an equation, a linear fit can be created for regions of data, creating the same sort of approximated system model.

The type of nonlinearity already discussed includes only continuously differentiable models. However, many of the most troublesome nonlinearities fall under the heading of discontinuous. These complications include saturation, deadzone, polarity (absolute value), quantization, switching, hysteresis, backlash, and friction. As there is no general closed-form solution to the dynamic equations governing systems with discontinuous nonlinearities, there is no general approach to creating a control law for such systems. However, there are a few tools in the controls toolbox. One of the most common is "sliding control." In essence, sliding control uses a bang-bang (on-off) input, which is governed by a law that seeks to stabilize a function of the tracking error. This function is chosen to be linear in respect to the tracking error. A complete description of this method is beyond the scope of this book. However, a treatment of the subject can be found in Slotine and Li [1991].

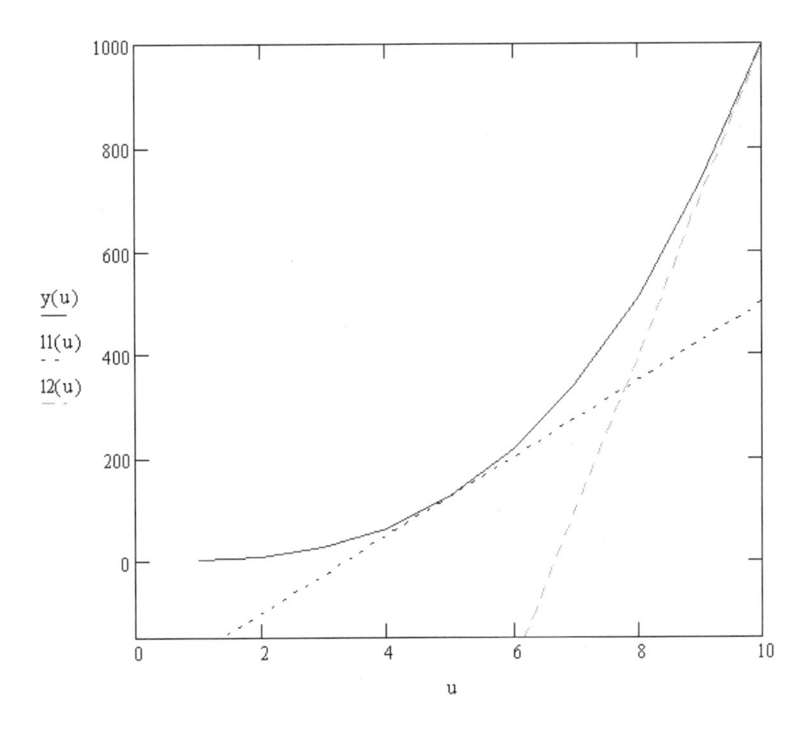

Figure 7: Example of linearization of the response of a system.

There is another interesting point to the control of active polymers. Just as they tend to damp disturbance forces, they should also damp input signals. Inherent input signal damping will protect a system from itself to some degree. Simple controllers tend to increase actuation input to saturation over a very short period of time, producing an internally driven shock to the system. Creating a controller that is more sophisticated as to how it applies actuation is more difficult and requires more powerful processors to implement. If the system automatically limits the rate in change of the input, system components will never see a self-produced shock load condition.

17.4 Active Polymer Actuators in a Traditional Robotic System

A biologically inspired hexapod named LEMUR (Legged Excursion Mechanical Utility Robot) has been developed at the Jet Propulsion Laboratory as a step toward an on-orbit or extraterrestrial maintenance robot. Despite its insectile appearance, it was conceived more as a six-legged primate (as the name suggests). Chief among its attributes is the ability to use its legs and feet as arms and hands. The current configuration, shown in Fig. 8, is capable of walking on six legs or manipulating with two while stabilizing with the remaining four. The sections of the limbs below the knee/elbow have been designed to convert from feet to tools, and a quick-release mechanism has been incorporated allowing easy swapping of tool types.

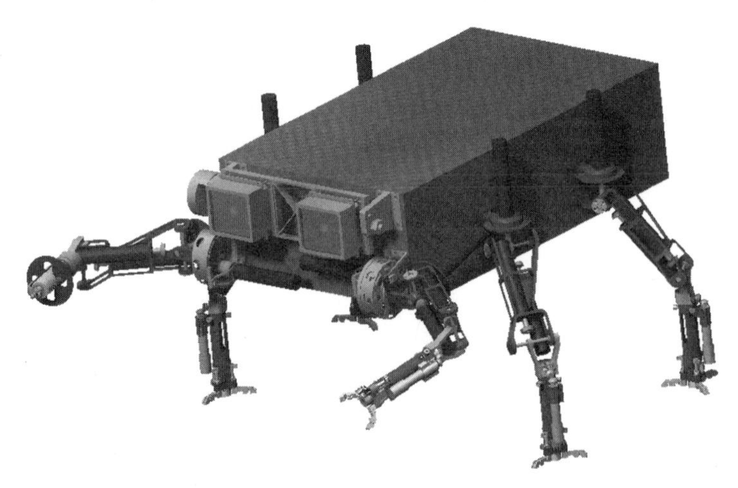

Figure 8: LEMUR can use its front limbs for mobility and manipulation.

Designing LEMUR using only standard electromechanical elements was quite tricky. As LEMUR is small (its body is only the size of a shoe box) for a robot with its level of articulation (20 independent joints with two actuated tools), incorporating a sufficient number of motors and drivetrains proved a challenge. In addition, the desire for a kinematically spherical shoulder joint for the front legs required a relatively complicated mechanism. Moreover, system demands required that the mobility system and structure comprise only about 50% of the overall 5 kg mass budget. While the design problems were eventually solved, a robot in the same class as LEMUR could significantly benefit from the use of capable active polymer actuators in its design for many of the reasons discussed in the preceding sections.

Immediately obvious are the benefits of designing biologically analogous limbs. Two advantages are particularly applicable to LEMUR. First is the potential for a decreased mobility system mass. Given lighter actuators with less drivetrain structure, the overall system should be lighter as well. The second advantage is the simplification of the joint designs. As LEMUR's design doesn't use cable drives with remote actuators, the motors must be housed within the legs. This arrangement causes two problems. One is the placement of relatively massive components (the actuators) out on the end of links, increasing the rotational inertia of the system. Second is the difficulty of simply packaging the actuators. Mechanisms and parts become complicated and therefore less robust, heavier, and more expensive (in general). In contrast, we can look at the design of the spherical shoulder with muscle actuators. Rather than three distinct axes and actuators, one ball joint could be actuated using several artificial muscles. The exact number of muscles would depend on the particular agonist/antagonist scheme implemented. An example of such an arrangement can be seen in Fig. 9, which shows a human arm built by Hannaford's lab at the University of Washington. It should be noted that the Washington lab is concerned with replicating a human arm. A limb designed for a LEMUR-class vehicle would probably be much simpler.

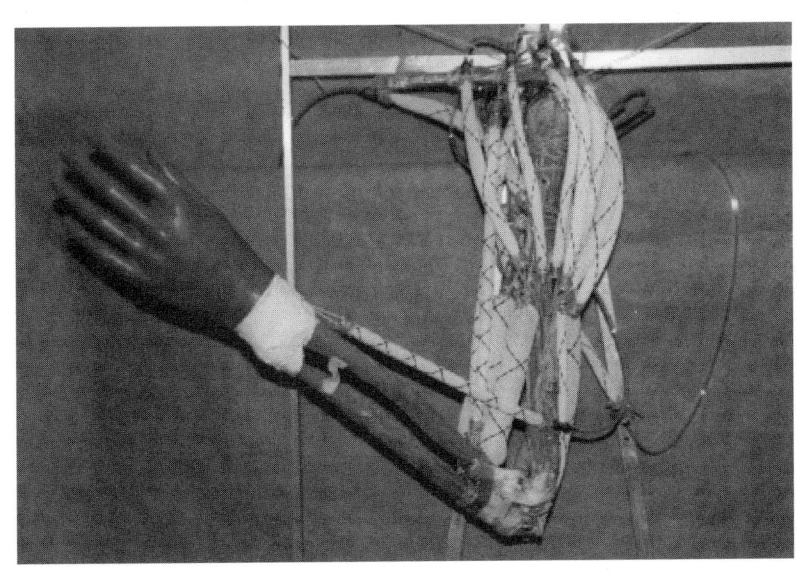

Figure 9: A shoulder with a high degree of freedom created using artificial muscles. (Courtesy of B. Hannaford, University of Washington.)

The modified LEMUR just described could provide a template for a first step in active polymer robots because it only substitutes muscles into the overall controlled system, leaving the computing and software architecture largely unaffected. This substitution is not trivial, of course. Other than the mechanical design aspects, allowances must be made for a different feedback system than that currently implemented (rotational encoders on the motor shafts), the actuator drive electronics must be radically modified, and a control method that takes the dynamics of the muscles into account must be devised. That said, an actuator substitution approach would provide a jumping-off point that would build on known techniques in software and computing. There are, of course, more radical approaches to design and control, which will be covered in the following sections.

17.5 Using Rapid Prototyping Methods for Integrated Design

As any roboticist will tell you—building robust robots is difficult. However, in keeping with the design and manufacture of most artifacts, most robot development follows a development cycle that includes looping around the trial-and-error phase. A great deal of time is often spent in cycling around this particular loop. Two techniques may be useful in the future to reduce the development time: rapid prototyping and artificial evolution. The embryonic technology of 3D printing may, in the future, allow us to 'print' out robot forms in EAP "ink" that have been developed by artificial evolution in a simulated world. The following section takes a closer look at these ideas.

The steps involved in a rapid design and prototyping design loop have already been touched on, but it may be useful to explore the tools that make fabrication possible in more depth. Probably the most important tool is a concept known as shape-deposition manufacturing (SDM), which is being explored as it pertains to robotics by the Dexterous Manipulation Laboratory (DML) at Stanford (directed by Cutkosky). To borrow from a paper from the lab, SDM "is a technology in which mechanisms are simultaneously fabricated and assembled… [T]he basic SDM cycle consists of alternate deposition and shaping…of layers of part material and sacrificial support material" (Bailey et al., 2000). The exact method of the deposition and shaping depends on the scale and application. In the case of the work being done by the DML, robots are fabricated by alternately molding polymeric parts and then machining them with a computer-controlled milling machine. Interspersed with these procedures is the incorporation of sensors and actuators. A finished part and its elements are shown in Fig. 10. To date, SDM sequences are often ad hoc. However, a formalized methodology is under development by the Stanford Rapid Prototyping Lab in conjunction with the DML. This approach is explained in Binnard, [1999].

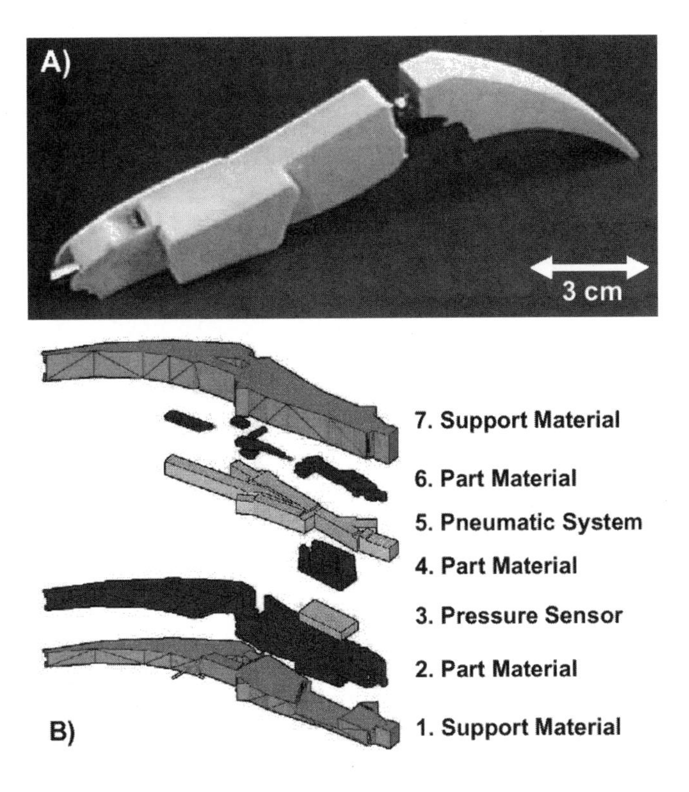

Figure 10: (a) a completed leg fabricated using SDM techniques, (b) sequence of material layers and components (Bailey et al., 2000 with permission).

However, several other part-forming technologies can be used in SDM processes as well. Some of these are stereolithography (SLA), selective laser sintering (SLS), photolithography, and the aforementioned 3D printing. Of these, SLA and SLS have already been proven to be capable of building articulated structures down to a sub-inch scale (Mavroidis, 2000). In the future, 3D printing may be useful for EAP robotics if it can be made to create heterogeneous material layers. With that advance, we could imagine laying down EAP or other active polymers at the same time as the polymeric structure. While current 3D printing lays down relatively thick beads of material, it may be possible to adapt ink-jet printing to perform the same processes down to the micro-inch scale. In addition to the SDM techniques for mechanical elements, related techniques are applied in a concept called molded interconnect devices (MID) for the manufacture of electrical circuits embedded in structures. In particular molding coupled with photolithography (or photoimaging) is used to create traces within polymer materials. Of course, at small enough scales, the fabrication looks much like that for typical IC chips. If already established practices from SDM and MID and/or IC manufacture are combined with the promise of active polymers, a truly integrated robotic system could be created.

17.6 Evolutionary Design Algorithms (Genetic Algorithm Design)

Once a robot is seen as a whole and each subsection as a variable that affects the response of the system, the impulse is to tweak those variables and simulate the response. One can imagine a design loop in which components are specified in the genotype of artificial creatures, the robot form and function is encoded, and the population of robots is then evolved over time through the implementation of a pseudo-Darwinian selection for fitness. This selection process is sometimes referred to as genetic algorithm design. The best-designed robots could be implemented in hardware and continue their evolution in the real world where they would receive stringent testing and assessment. Some of them may return back to the simulated world and join their virtual fellows. Wonderful examples of such evolution supervised by humans can be found in the work of Hasslacher and Tilden (1995) concerning their physical populations of experimental machines. Here the intrinsic mechanics of robot evolution is based on the employment of minimalist electronics, reusability of components, and utilization of solar energy (BEAM, 1999). However, this process takes an inordinate amount of time due to the fact that fitness is measured by performance over time, and one is compelled to ask if there are techniques available that might speed the process. Is there any theoretical, or rather computer-based, method that could offer us reliable techniques for fast evolution and easy prototyping of minimalist robotic devices? If rapid prototyping techniques such as stereolithography were used to produce real-world versions of the robots, the overall scheme of robot evolution would look like: (1) seed a primordial culture of robot components, (2) evolve population of robots, (3) select the best performers, (4) fabricate the robots using rapid prototyping techniques, (5) verify robots' performance in real world, (6) select best designs and inject their virtual copies back to the virtual robots' population, and (7) go back to step (2).

Parts of such a design cycle have been explored by several researchers. Those working purely in the hardware world are represented by Hasegawa and the gibbonlike robot Brachiator. It is built of several links, joints and pneumatic pistons. Brachiator moves from one branch of a tree to another by swinging its body as a pendulum (Hasegawa et al., 1999), as most monkeys do. The actuators of this "monkey" are controlled by artificial neural networks that learn and evolve themselves. In this case, no simulation is performed. The controller attempts to make a particular move using particular actuation profiles. A series of profiles are compared for effectiveness, and the best is retained and built on. Also in this vein of research is the reconstruction of an extinct swimming animal *Anomalocaris*, as described in Usami et al.. (1998). The reconstructed creatures swim by the waving motions of "fins." Each creature is created by simple rules, where every rule gives rise to the action of a particular unit of the creature's body. Some genetic-inspired recombinations of rule strings (the artificial "genes") are employed to increase the morphological diversity that is directly linked to locomotion activity. By employing artificial evolution the better swimmers are selected.

Rather than emulating the design and function of existing animals, some researchers have started from the basic building blocks that nature itself uses. In the context of a simplistic design strategy a novel robotic creature must include endo- or exo-skeleton, muscles, sensors and primitive neuronlike control decision elements. It is quite convenient to consider the skeleton built from stiff cylinders or 'sticks' connected by joints. The muscles can be attached to the sticks as well as sensors and neurons. The most primitive virtual creatures with this minimalist design form a "clone" of Swimbots, swimmers that inhabit Ventrella's artificial ponds (Ventrella, 1998, 1999). In this section, we will discuss in depth the creation of another construct, the virtual stick-based creatures, *Framsticks*, or Ulatowski-Komocinski machines. These are artificial creatures built from three basic components: rigid sticks, flexible muscles and primitive neurons (Komocinski, 2000).

Let us look briefly at the Framsticks design. The body of a *Framstick* is built from sticks. A stick is subdivided into a finite number of control points, which are affected by several forces: gravity, friction, elastic reaction, reaction with ground, and many more. If Framsticks collide with each other then some sticks may be destroyed. In addition to sticks the creatures also have muscles, neurons and receptors. They may also exhibit some kind of metabolism. Using specialized endings of the sticks, Framsticks can assimilate energy from their environment and even ingest their dead fellows! A Framstick starts its life with a certain amount of energy. It dies when energy level is zero. The creatures dissipate their energy when idle; they also spend energy on static and dynamic muscular activity.

There are two types of muscles that join the sticks: bending muscles and rotating muscles. The muscles consume energy and are controlled by neurons. A Framstick gets information about the environment via three types of receptors that can be attached to the sticks: G-receptor, T-receptor and S-receptor.

The G-receptor is an analogue of an otolith, which can be found, in one form or another, in all real creatures from *Protozoa* to humans. The receptor signals if the position of the stick it is attached to is not perfectly horizontal. The T-receptor is a receptor of pressure. The receptor generates constant negative value when it is free. Its value becomes close to null when it touches an object; it can go to a positive scale when pressure increases. The S-receptor is a receptor of smell. It smells equally well presence of food and other Framsticks; however the receptor does not discriminate between different stimuli. The receptors of smell and pressure are sufficient to guide a creature in its world. Theoretically we can avoid using the G-receptors; however, these receptors are quite useful where swimming creatures are concerned.

Framstick neurons are quite primitive. They are attached to sticks and take inputs from other neurons or receptors. They can bear self-inhibiting and self-exciting terminals as well. The neurons send their efferent terminals to the muscles. Being connected to rotation or bending muscles the neurons control rotation or bending of a stick relative to its neighbouring sticks. An excitation function of neurons is based on a weighted sum of input signals. Neuron reactivity can be tuned (via modification of genotype entries) by indicating how fast the neuron updates its state toward weighted sum of its inputs and how long the current state persists. It is possible to achieve oscillatory mode combining the values of neuron reactivity.

The Framsticks have simple hence powerful genetic system, which is instantiated in a separate program module. We could also mentioned that by modifying Framstick genotype one can change each stick's properties (e.g., rotation, twist, curvedness, length, weight, friction, muscle strength), topology of neural network, and functions of receptors.

Genetic evolution is highly controllable. One can set up global intensities of mutations, parameters of genes reparation (which is a bit technical yet useful), features of crossover, mutation probabilities for genes coupled to all parts of Framstick body, e.g., detailed mutation probabilities for neurons, muscles, and receptors. Mating preferences between the creatures are implicitly expressed via similarities of genotypes. The user can also specify capacity of Framstick world, or the maximal size of the population, or rules for deleting genotypes (fitness-based selection, random deleting, or elimination of only worst types).

A selection of creatures is based on four basic characteristics: duration of life, velocity of movement, spanned distance, and size of a creature. A fitness of any particular genotype (read individual) is calculated as a sum of weighted values of four selection characteristics. Sometimes energetic efficiency can also be employed to evaluate the fitness.

Two types of evolution process are implemented in the Framsticks software package: directed evolution (selection) and "spontaneous" evolution. During straightforward selection one can specify parameters, which are used to evolve a best creature. That is, we explicitly define optimization criteria. Thus, for example, if we wish to produce the fastest lightweight crawlers we weight velocity parameter positively and the structure size parameter negatively. Eventually we will get small and fast creatures. In a course of "spontaneous" evolution we may not define any explicit parameters of selection but just general rules of the evolution, such as outcomes of the collisions between the creatures, utilization of dead creatures, aging rate, specifics of muscular work, and so on. Neither of these parameters guides evolution: only the creatures with longest life benefit. A few examples of spontaneously evolved creatures are shown in Fig. 11.

Let us look at a couple of examples of experiments made with the help of the Framsticks software package. At first, assume we decided to breed some nontrivial swimming creature. In the artificial world we set up the water level above zero and put a one-stick creature with no neurons, no receptors, and no muscles in this bath. We adjusted simulation parameters to select creatures by

Figure 11: Examples of creatures evolved in the populations of Framsticks in the dry and flat world. (Courtesy of M. Komocinski, Poznan University of Technology, Poland.)

both velocity and structure size. After several thousand generations of the creatures, during which we intervened in evolution and subjectively chose "the best" examples, the final result was a creature consisting of 20 sticks, 6 neurons, a couple of muscles and one G-receptor, responsible for a "sense of equilibrium." We call him *back-swimmer* because he swims backwards: two branches of joint sticks form a type of "legs" that move and "grab" water allowing the creature to move. As you can see in Fig. 11, an artificial aquarium with *back-swimmers*, several sticks of a creature float on the surface of the water thus keeping the rest of the creature's body in medium layers of the water.

One lab that is attempting to use all the steps of the suggested design cycle is the DEMO lab at Brandeis University. The GOLEM project reported by Lipson and Pollack (2000) has resulted in physical models of robots designed through the use of genetic algorithms in simulation. The structure dictated by the algorithm was directly fabricated from a thermoplastic using a rapid prototyping technique known as 3D printing, and then the actuators were incorporated by hand. In essence, these robots represent a real-world instantiation of the Framsticks, substituting electromagnetic actuators for theoretical muscles. The exciting next step for active polymer researchers would be to lay down the plastic muscles (and perhaps conductive plastic wires) at the same time as the plastic structure, allowing one-stop robot design and fabrication. This concept will be explored more in Sec. 17.7.3.

17.7 EAP Actuators in Highly Integrated Microrobot Design

The traditional model of a robot with discrete components may be joined by a new model in which the lines between controller, plant, and feedback are less rigidly drawn. This new paradigm results in a much more tightly integrated system, which is enabled by up-and-coming technologies such as active polymers, rapid prototyping, and new simulation and design algorithms. The combination of these technologies promises important advances toward the engineering ideals of faster, cheaper, and more effective design cycles.

Although these techniques can be used to create systems of any physical size, they are particularly important—possibly indispensable—tools for the design and manufacture of tiny robots. These microrobots will have the advantages of being small (by definition), lightweight, and easily produced, therefore making them candidates for mass manufacture. As a somewhat whimsical example, consider a system of sub-inch robots equipped with "sticky" cilia composed of EAP that would enable them to move across small "ceilings" and "walls" inside a structure that is difficult for humans to access—perhaps the escalators in a subway. They may be able to power themselves using light, heat, ambient chemistry, etc., or perhaps be built with a small on-board, gel-based battery system. With some judicious tinkering they are given two basic behaviors: a phototactic behavior in the presence of light and a default random walk in its absence. If these robots are "injected" into the machinery, which is then optically sealed, the robots would then carry out a random walk and spread out through the machine. In doing so they may come across small items of grit and dirt, which could become attached to their bodies. After an appropriate time an intense light is shone into the machine from an 'exit' portal. The robots would then switch to their phototactic behavior and move toward the light source. In doing so the robots would bring out the small items of grit and dirt attached to themselves. Perhaps the robots would then be thrown away or even washed and reused! This approach also has the advantage that broken robots themselves could be removed by their peers—but designers would also have to ensure that the robots didn't simply gum up the works.

17.7.1 Control of Microrobot Groups

Of course, the problem of controlling and coordinating the behavior of very small robots must be considered. For the purpose of discussion let us focus on robots with volumes in the range of a cubic millimeter to a cubic centimeter. All we can expect to be able to build in this domain over the next few years are really quite dumb and simple robots, with rudimentary sensing, communication, locomotion, and computation abilities. These robots will experience considerable limitations. As with all autonomous systems, not only will the power provision be problematic, the very capacity of a single tiny robot to achieve anything or even to survive for any length of time will be in doubt; an individual is limited and vulnerable and so it is likely that only collective actions will succeed. On the face of it, the prospects look bleak for an engineer required to build such a system. However, studies of evolved natural systems, particularly the social insects, provide us with an existence proof that collections of relatively "simple," mostly reactive, creatures can achieve remarkable feats that are beyond the capacity of an individual; such organization is underpinned by decentralized mechanisms for decision making as well as for the control and coordination of task-achieving behavior. Although such studies engender optimism we should be cautious about the use of the word "simple" sometimes applied to animals such as ants. The "lowly" ant is a wonderful biological machine, which has been evolving for over 100 million years, machine which is capable, for example, of impressive feats of locomotion, power efficiency, navigation, and strength. This form of decentralized system, which employs simple units, and which appears to collectively solve problems traditionally tackled by a single smart individual, has sometimes been referred to as swarm intelligence. The task of control and coordination of a group or microrobots might be encapsulated thus: "how do we get a lot of dumb robots to collectively do something smart?"

A number of researchers have contributed toward the definitions of swarm intelligence in the context of a distributed system with a large number of autonomous robots. Beni and Wang [1991] expressed the idea as "the essence of the .... problem is to design a system that, while composed of unintelligent units, is capable as a group, to perform tasks requiring intelligence — the so-called Swarm Intelligence." Theraulaz et al., [1990] define a swarm as "a set of (mobile) agents that are liable to communicate directly or indirectly (by acting on their local environment) with each other and which collectively carry out a distributed problem solving." Such a swarm will exhibit functional self-organization [Aron et al., 1990] as a consequence of the collective set of internal dynamics and interaction with the environment. Deneubourg and Goss, [1989] neatly sum up the problem: "The key ... lies in remembering that at each moment the members of an animal group decide, act and interact, both amongst each other and with the environment, permanently changing the state of the group. Just as sociobiology, with its population genetics and games theory, shows the importance of dynamics and individual interactions in the evolution of social behavior we propose the analysis of these interactions as the straightest path to understanding the short term collective behavior of animal (read robot) groups."

The autonomous robots we currently build are mostly rudimentary, often unreliable, incapable of self-repair, and problematic with respect to power budgets. Against this background we speculate that the construction of small robots in the future will necessarily have to incorporate advances in material science including the incorporation of artificial "biological material" such as muscle actuators, sensors, and artificial metabolism, as well as a firm understanding of the principles underpinning the control and coordination strategies of social insects.

The minimalist approach has also been undertaken in other domains and the interested reader is directed to the following papers: kinesis, taxis, and target following [Holland and Melhuish, 1996a, 1996b], secondary swarming [Holland and Melhuish, 1996a, 1997a, Melhuish 1999b], formation of "work gangs," [Holland and Melhuish, 1997a, Melhuish et al., 1998a, Melhuish et al., 1999], collective behavior transition [Holland and Melhuish, 1997a, 1997b]. A brief overview is given in Melhuish, [1999].

17.7.2 Unconventional Locomotion Controllers for Microrobots

When talking about sub-inch-scale robots one would obviously turn to biological analogs. These are living prototypes of microrobots who do not have centralized control, but who sense, decide, and act distributively. The *Protists* offer us examples of ciliates and amoeboids.

The phylum *Ciliophora*, or ciliates, includes unicellular organisms, the bodies of which are covered with cilia. One could refer to *Paramecium caudatum* as a typical example, see for example Melkonian, Anderson and Schnepf, [1991]. Each cilium, attached to a membrane of a protist, sweeps with a power stroke in the direction opposite to the intended direction of organism movement [Fig. 12(a)]. The cilia are usually arranged in rows; this arrangement forms a longitudinal axis of beating [Fig. 12(b)]. Cilia beat coherently in waves and propel the individual forward. A physical integration of strokes of many cilia propel the protist in the direction opposite to the direction of beating.

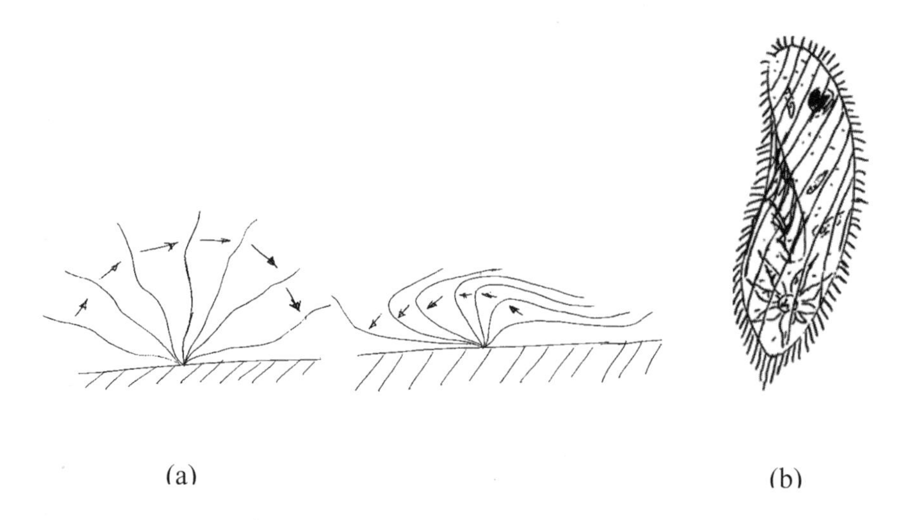

Figure 12: Ciliar mechanics in *Paramecium caudatum*. (a) Cilium beating, (b) arrangements of cilia.

Perhaps the control may be implemented by either information transfer via the sub-membrane network of microtubules or by coordinated contractions of the membrane travelling along the protist's body. In any case, we can speculate about direct analogies between waves of cilia beating caused by excitation waves of the lattice, and waves of cobeating of cilia in protists, that move from front to rear parts of the organism. A working prototype of an artificial ciliate would be ideal. We are building it in the near future. However, in this chapter we will try to explore some idea of employing an excitable medium in the form of a molecular array of sensors and actuators to provide the controller for a microrobot by exploiting decentralized computation.

Whether we look at the conventional models such as sense-model-process-act or the behavior-based architecture [Brooks, 1986], which stresses the tight coupling between sensors and actuators, three key design areas need to be addressed: sensing, actuation, and decision-making. Implementation of sensory input devices is a nontrivial task. However, future research may show that such implementation can either be solved using conventional techniques and devices or using future smart materials, which conflate elements of sensing, actuation, and "processing." Here we make a reference to "unconventional" controllers. Conventional controllers might be considered to employ explicit control algorithms such as in digital computers and are often associated with an explicit symbol processing approach. In contrast, unconventional controllers might employ implicit forms of computation, as in the case of wave computation in which the "result" of local microprocesses is a macroscopic phenomenon that can be used, for example, as an indicator of a quality or quantity of state. A controller for robot navigation can be designed using several techniques, which are discussed briefly below.

We can, of course, couple sensory elements with motor units. This gives quite a wide range of complex robot behavior. These ideas are demonstrated by Braitenberg [1984]. By expeditious coupling of input with output, Braitenberg shows how interesting and seemingly directed behaviors (which he playfully refers to as love, fear, aggression, etc.) such as attraction and repulsion, along with nonlinear dependencies, can result from simple input-output mappings.

An example of such an approach is the "solarbot" shown in Fig. 13. The solarbot consists of two wheels (made from small mobile phone vibrator motors), each coupled to a separate capacitor. Each of the capacitors is connected to a photocell "wing" on the opposite side of the body from the wheel. The robot can accomplish phototaxis by first storing the energy generated by each photocell on a "wing" in its associated capacitor and then releasing the stored energy to each wheel when sufficient power accumulates. In this way, the side of the robot that is nearer to the light source gives its cross-couple motor more energy than the "darker" side, thus making the opposite wheel turn more. By judicious use of time constants to control the charge and release times the robot then makes its way toward a light source in a series of arcs.

Figure 13: Solarbot made at IAS Laboratory at the University of the West of England.

One might also be able to apply this simple "Braitenberg Vehicle" style controller by judicious employment of linking input to output. One approach might be based on the dynamics of excitation in nonlinear media or a so-called wave-based computing [Adamatzky, 2000]. Two types of nonlinear media, either discrete or continuous, are of particular interest: reaction-diffusion media and excitable media. Both of these types of media support waves—either phase waves or diffusion waves—in their evolution. The waves are generated by external stimuli and travel through the medium. The waves interact with each other and form distinctive concentration profiles or dissipative structures as a consequence of their interactions. To implement computation in active wave media we could represent data by the distribution of elementary excitations or concentration profiles. The waves collide with one another, i.e., they implement "computation" since the consequent dissipative structure or distributions of a precipitate can represent the "results" of the computation [Adamatzky and Melhuish, 1999].

Let us discuss the wave-based computing in an array of ciliumlike actuators. Consider the idea of a microrobot constructed as a mobile array, which is required to demonstrate phototaxis. Let us first fabricate a 2D molecular array in such a manner that every molecule of the array is associated with its eight closest neighboring molecules. After that we couple every element of the array with its own propulsive actuator, as shown in Fig. 14. We could perhaps choose molecules that are light sensitive, i.e., excited by photons. Let us assume, for simplicity, that the only molecules at the edges of the array are light sensitive. When such edge molecules are excited they transmit excitation (energy) to the internal molecules of the array. The excitation can be passed between one internal molecule to another internal molecule. Let us restrict the position of the actuators such that they can take up one of eight orientations. If the actuator positions itself away from the direction from which excitation arrives, then the actuator could produce a local propulsive force. The combined propulsive force of the array of actuators provides an overall propulsive vector toward the external light source.

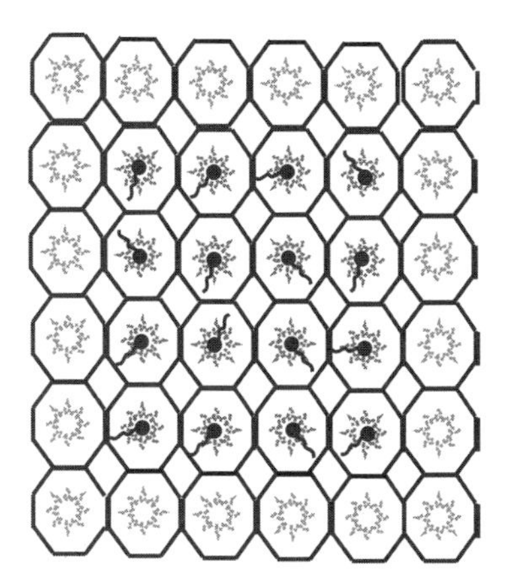

Figure 14: A sketch of an excitable molecular array with incorporated actuators. Molecules are hexagonal. Cilia are shown explicitly.

All molecules of the array update their states in parallel; therefore, we have a parallel array of processing elements coupled with parallel array of actuators. In our model every molecule of the array connects with eight neighboring molecules. Every molecule is excited if a number of its excited neighbors, among these eight neighbors, lies in the interval $[\theta_1, \theta_2]$, where θ_1 lies in the range from 1 to 8 and θ_2 lies in the range from θ_1 to 8. This interval-based parameterization potentially gives us 36 excitation rules. These rules determine various regimes of excitation dynamics on the molecular lattice: from chaotic dynamics to spiral waves to self-localized excitations.

Simulations have been conducted that employed a controller composed of an array of coupled sensors and actuators, as described in the previous section. In computer experiments we place a robot at random in a virtual 2D space with a light source. It was found that the robot demonstrated photactic behavior; typically the robot starts to wander around, performs weird motions, and then begins to move toward a light source along a non-linear trajectory. This behavior of the robot moving through the 2D space toward a light source can be explained by the following chain of events in the excitable medium of the controller. The edge molecules of the controller array are excited by the photons, and patterns of excitation move inward on the array. The movement of the excitation patterns modifies the local orientations of the actuators, attached to the internal molecules of the array. Local propulsive forces are generated by each actuator. The interaction between the local forces generated by each actuator and the environment implicitly create a form of "vector integration" which then causes the rotation and translation motion in the robot. For each of 36 types of molecule sensitivity (due to the parameterization intervals of θ_1 and θ_2) we have recorded the robot trajectories and the space-time dynamic of excitation patterns. From observation of the form of trajectory we partitioned the parameter space into three main groups: graceful, pirouette and cycloidal (Figs. 15 and 16).

To check whether these ideas work in real-world experiments we installed a model of the excitable lattice controller in a mobile robot. The robot is about 23-cm diam and is shown in Fig. 17. The excitable controller (programmed in C) is simulated on the board processors. Ideally, every molecule of the controller must have its own actuator, and every edge molecule should be able to react to light. Unfortunately, the engineering realization of such a setup would be very complicated and costly. Therefore, we employed a realistic and pragmatic approach of employing a "large" robot with two driving wheels and three light sensors (left front, right front, and rear sensor). The left and right sensors are coupled with left and right parts of the front edge of the molecular array. The rear sensor is coupled with the rear edge of the molecular array (Fig. 17).

The model allowed the edge molecules to be excited with a probability proportional to the values on their corresponding macrosensors. Orientation of the global vector is transformed to the rotation angles of the robot (via spin speed of the motors) in a straightforward way. The following algorithm of robot behavior is implemented: evolve molecular array, calculate local vectors, calculate global vector, rotate robot, move robot at fixed distance, if light source is not reached go to first step, otherwise stop experiment.

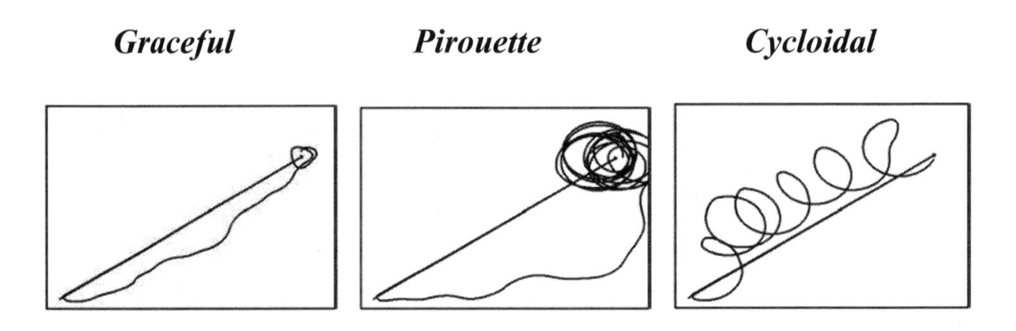

Figure 15: Example of robot trajectories for three main groups of robot locomotive behavior. The robot starts its journey at the left bottom corner of the rectangular arena. The starting point is connected to the destination point (source of light) by a straight-line segment. The robot trajectories are curve lines.

Our robot performed inside a huge arena, which has an area 1,760 times more than that of the robot. We placed a line of lights outside the arena. The behavior of the robot is recorded via a video camera, which is mounted 6 m above the arena. At the beginning of every trial we put the robot near the edge of the arena opposite the light source. The robot is turned off when it reaches the light source. The highlighted trajectory of the robot is shown in Fig. 17.

In the figure we noted that even light from a nearby corridor (two light spots at the top part of the arena) contributed to the noisy environment of the real-world experiments. As we can see from the series of snapshots in Fig. 17(b), the robot did not choose the right directions immediately. At the beginning the robot moved toward the incidental light spot. Then it "realized" that this spot is not a source of light with maximal intensity. So, it implemented a U-turn and headed toward the light target. While approaching the target the robot changed its trajectory because of the second spot of light. However, it recovered from this "mistake" quite soon and eventually hit the light target.

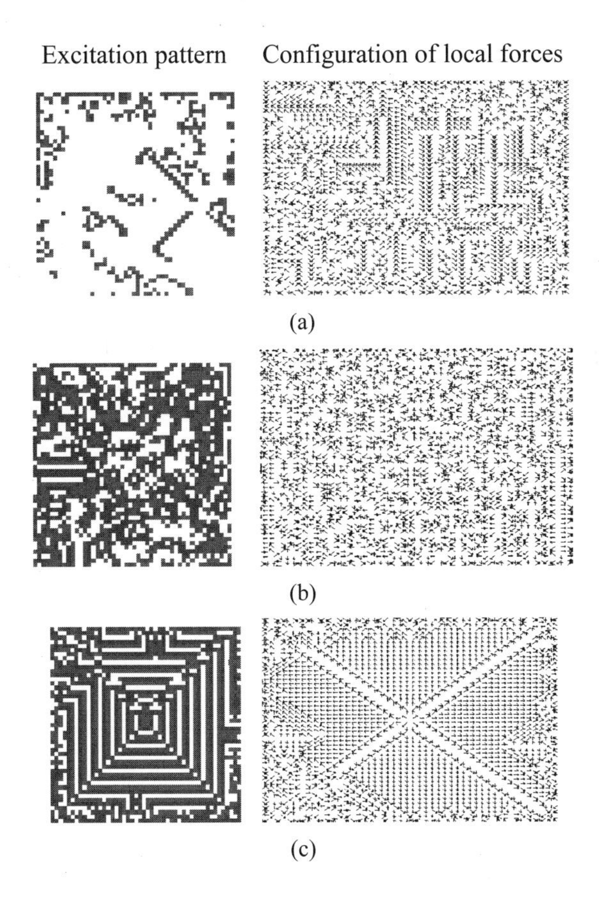

Figure 16: Excitation dynamics responsible for graceful (a), pirouette (b) and cycloidal (c) motions.

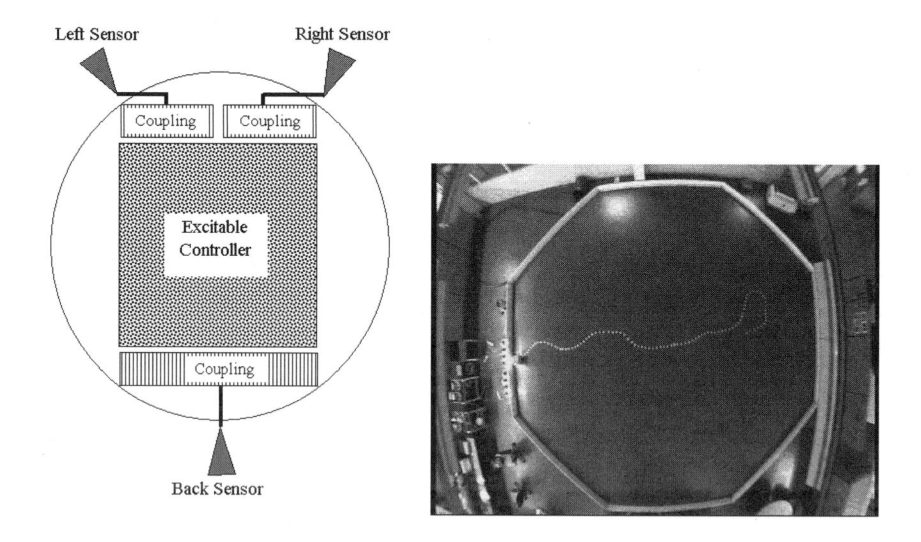

Figure 17: Conventional robot with ciliate-based controller. (a) Coupling of light sensors with simulated excitable lattice in real robot. (b) Trajectory of the robot, which searches for the source of light. The robot starts its journey at the far right part of the arena. The light source is positioned on the left.

Chemicals, as well as light, may serve as both a navigation driver and an actuator stimulus. A good example is the behavior of a species of *Amoeba*. Amoebae move due to the effective separation of its cytoplasm into a sol and gel during the formation of its pseudopodium. When some parts of amoeba cytoplasm are transformed into sol, endoplasm starts to flow to this area. Resultantly, a membrane expands and the pseudopodium is extended forward. When sollike cytoplasm reaches the end of the pseudopodium it is transformed into gel. This recurrent converting of sol into gel and back allows the organism to move purposefully. Two mechanisms, in general, may determine the amoeba's motility: changes in intracellular pressure [Yanai et al., 1995] and microtubule-dependent development of pseudopodia [Ueda and Ogihara, 1994]. The intracellular pressure is possibly generated by contracting actomyosin residing in the cortical layer causing formation. Microtubules seem to be responsible for directional stabilization of pseudopodia, which enables amoebae to undertake reorientation steps [Ueda and Ogihara, 1994].

As mentioned, the sensing-actuating of amoebae is usually associated with chemotactic behavior. When an amoeba—let us talk about *Physarum polycephalum* at this stage—moves it usually has one linear pseudopodium, which is expanding during organism motion. If some attractive chemicals are presented in the substratum, then the linear pseudopodium is split into several lateral pseudopodia, which explore a space around the main linear tip. One of the lateral pseudopodia, usually positioned at the site of a substrate with a relatively higher concentration of chemicals, is stabilized and becomes responsible for linear motion. Thus, a selection of concentration maximum is implemented by real amoeboids. An example is shown in the Fig. 18. Quite similar ideas are already employed in the algorithm for shortest path computation by *Physarum polycephalum* in [Nakagaki, Yamada, and Tóth, 2000]. However, in this installment, selection of leading pseudopodia is achieved not by its relative position at the site with a higher concentration of chemicals but by the distance of the tip of the pseudopodium from the amoeba body's "center."

Perhaps the most obvious example of "directed" behavior in amoeba is that of chemotaxis. However, it is interesting to enquire if we could also control the behavior of amoeboids by electroactivation. Quite possibly, it seems. There are examples of electrical sensitivity of both real [Korohoda et al., 2000] and artificial amoeboids [Ishida et al., 2000; Hirai et al., 2000]. It is reported that *Amoeba proteus* shows strong positive galvanotaxis. When put on the substrate influenced by a direct current field [Korohoda et al., 2000] the amoeba moves toward the cathode.

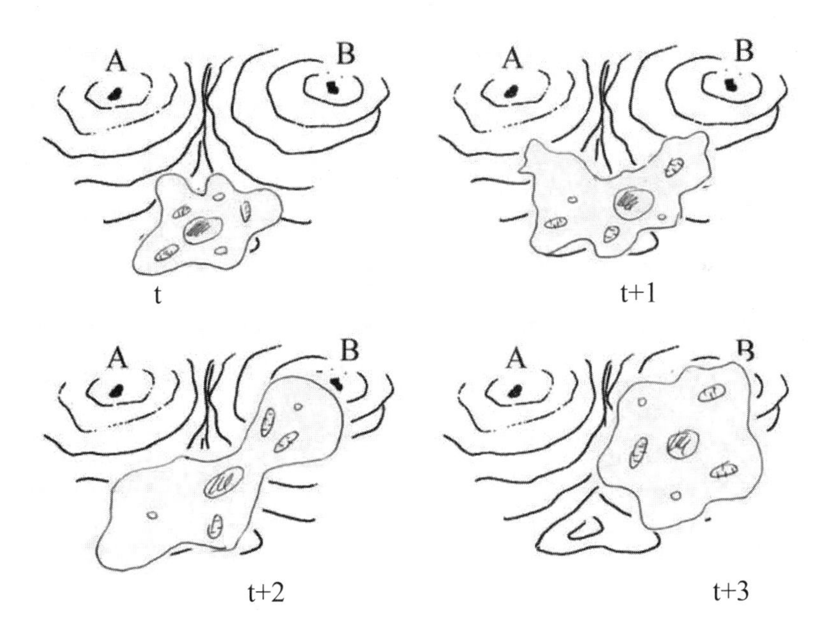

Figure 18: Chemotactic solution of a decision problem by amoebae. Concentration of attracting chemical, e.g., glucose, is higher at the site B than at the site A.

A beautiful example of artificial amoeboid like creature is offered in Ishida et al. [2000]. The liquid mobile robot, with no structure at all and easily changeable shapes, is realized there based on the paradigm of artificial amoebae developed earlier by Yokoi and Kakazu [1992]. In the experiments, the liquid metal robot is represented by mercury drops while its environment is a substrate with an array of electrodes, connected to an external controller (Fig. 19).

Figure 19: Liquid mobile robot with kind permission of authors [Ishida et al., 2000] (Reprinted with the permission of ISAB. Copyright 2000, ISAB).

It is demonstrated in the experiments that by varying the potential pattern of the electrode array, one can easily guide the mercury drop, causing splitting of the drop into several daughter drops, their fusion and different types of motion (Fig. 19). Galvanomotile polymers [Osada et al., 1992] and crystals [Hirai et al., 2000], may be promising nonmetal substrates for artificial amoeboids.

17.7.3 Micro-EAP "Integrated" Robot: A Case Study of a Phototactic System

So, how might we go about creating microrobots using EAP hybrid materials, combined with some of the ideas of excitable media discussed above? It's interesting and fun to speculate how EAP could be combined with other smart materials to create microrobot devices capable of phototaxis. Let's give our imagination some exercise.

First, let us look at the construction of a swimming "pad" that could exploit an undulating motion of a plane polymer sheet. Suppose that a polymer material is made that has the characteristics of being able to propagate some wave of excitation—perhaps some ionic concentration. Figures 20(a) and (b) illustrate the idea. Light is allowed to enter the polymer at the end of the strip only (perhaps the rest of the polymer surface is covered in some light-reflecting covering). The photonic energy alters the molecular state of the polymer substrate, which, in turn alters the state of its neighboring molecules. In this way, given some appropriate refractory period for the molecules in the altered energy state, a wave of excitation could pass along the substrate. If the excitation layer could then be bonded to an actuation layer then it might be possible for an actuation wave to travel through the actuation layer [see Figure 20(c)]. In this way the bonded laminate could ripple and execute some form of swimming behavior since the sheet of EAP is bent under the influence of an electrical current spreading along the conductive sheet. Repetitive generation of excitation waves travelling along the pad cause undulating movement of the pad and therefore the pad swims.

If the above arrangement in Fig. 20(c) was duplicated, and the two units bonded back to back, we would have the potential for differential locomotion. Like the "solarbot" mentioned above, light on one side of the robot will excite its actuator layer more on its side than the other, which biases motion toward the light.

Perhaps we could take this one stage further and bond artificial cilli, constructed from laminar EAP, to an excitation layer. Figure 21 shows this principle. The excitation layer is bonded to artificial cilli constructed from laminar EAP tubules. When the excitation wave traverses an actuator it deforms appropriately. This mechanism could provide the basis of a crawling robot.

Let us speculate further. If we put together slabs of the units shown above in Fig. 21 we could create a tubular construction as shown in Fig. 22. The diagram illustrates one slab being activated more than the others. Waves of deformed EAP tubules are seen to be travelling backwards along the activated slab causing the robot to alter its direction. When the robot is lined up with the light source all slabs become active. In this way such a tubular robot could therefore demonstrate phototactic behavior.

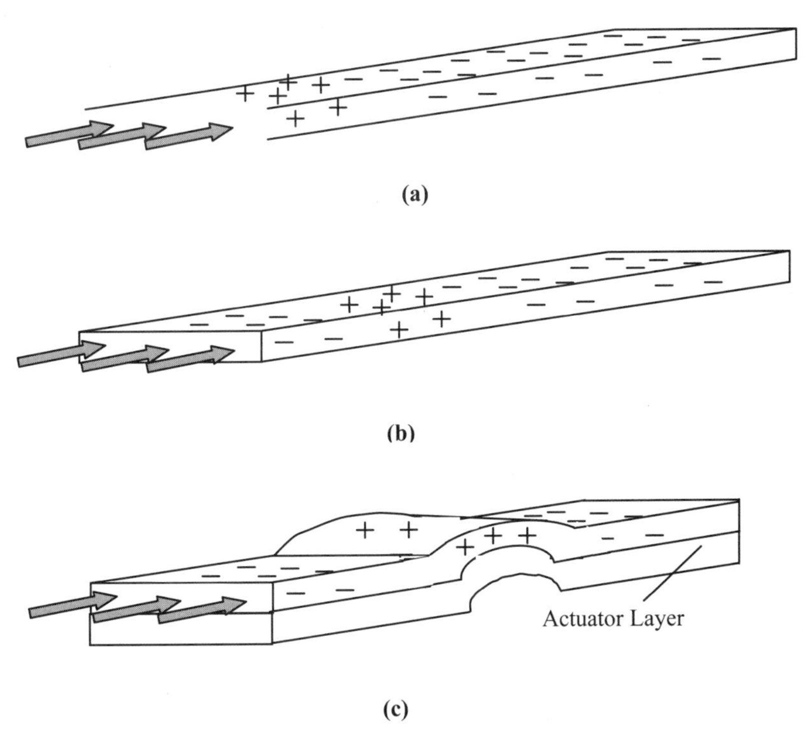

Figure 20: Integrated system with sensing, computation, and actuation. (a) Excitable media layer excited by light, (b) excitable media layer propagating excitation wave, (c) excitable media layer propagating actuation wave.

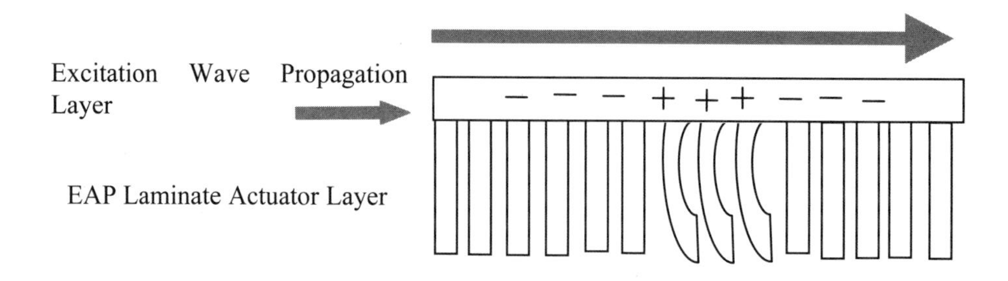

Figure 21: Traveling wave induced in the actuator layer of laminated cilia.

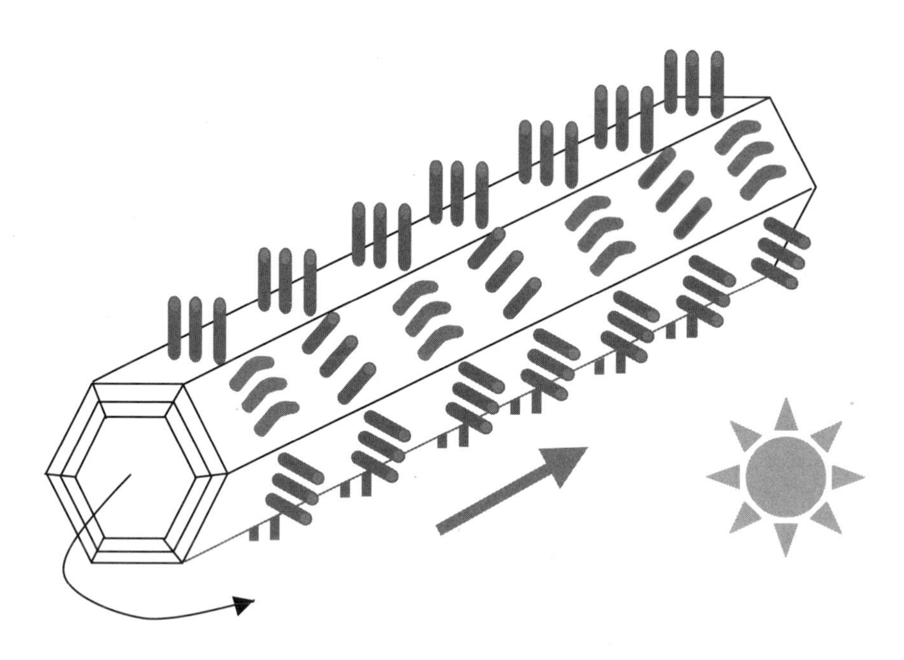

Figure 22: Travelling wave of actuators induced on one side producing phototactic behavior.

17.8 Solving the Power Problem—Toward Energetic Autonomy

Of course, robots need energy to move and to execute their behaviors. Conventional mobile robots usually employ on-board batteries, which either need replacing by humans or recharging stations (powered by generating devices). Very few, if indeed any, could be regarded as being energetically autonomous. Some researchers have recently started to look at this problem of how a robot can extract its energy from the environment. Researchers at the IAS Laboratory at the University of the West of England [Kelly et al., 2000] are building a robot capable of detecting and capturing slugs. It is envisaged that groups of robots could transport the captured biomass to a central digester, which could convert the biomass into methane thus providing the fuel for a fuel cell thus powering the robot. This collective idea is loosely modeled on the leaf cutter ant strategy. The University of South Florida [Wilkinson, 2000] has employed a microbial fuel cell to convert sugar into the energy required to power a model train. The idea of a microbial fuel cell is also interesting. It is within the realms of possibility to employ a semipermeable polymer substrate on which bacterial bio-film could exist. Perhaps, in the long term, one could imagine a tubular EAP-based robot, existing in an aqueous environment, incorporating such a biofilm as illustrated in Fig. 23(a). Nutrients could be allowed into the tube either passively or forced by tubular actuators at the mouth. The mouth itself could be constructed from EAP actuator rings behaving in a manner analogous to a sphincter. Once the nutrient was inside the "gut" it would be processed by the bio-film to produce energy in, say, the form of an energized molecule, or charge, which could diffuse through the inner membrane and power the excitation and actuator layers. The actuator layer could be comprised of a series of annular actuators. Waves of actuation along these "muscle rings" could provide the peristaltic action [see Fig. 23(b)] required by the nutrition system and possibly contribute toward locomotion as well.

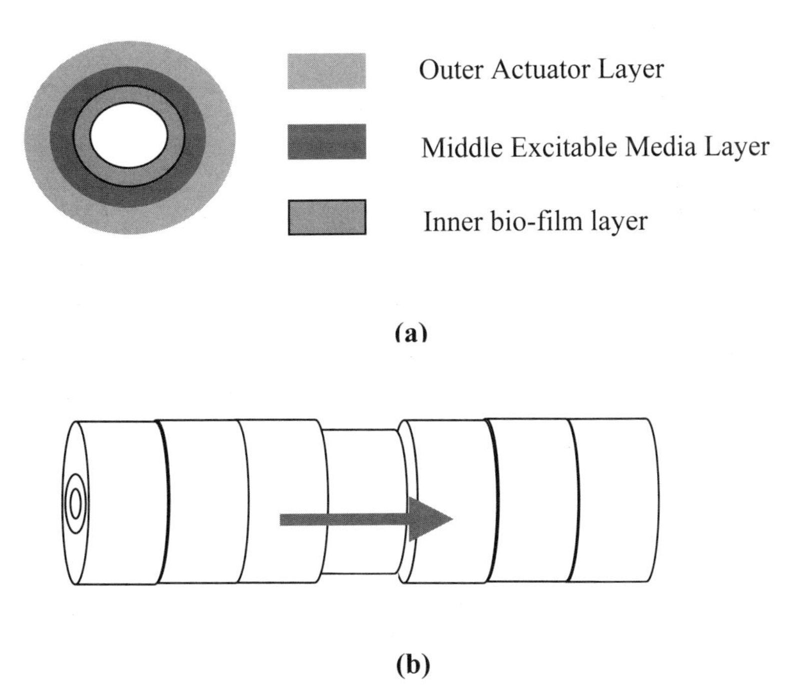

Figure 23: Robotic worm with artificial gut. (a) cross-section of concentric structure of "worm" robot, (b) peristaltic wave travelling along body axis.

17.9 The Future of Active Polymer Actuators and Robots

Clearly the robotics field can benefit from actuator technologies associated with active polymers. While not applicable to all areas, even in advanced forms, artificial muscles may make robotic platforms possible that have up to now been the stuff of science fiction. At the most basic level, active polymers may be substituted for the existing electromagnetic actuators in otherwise classic system architectures, providing benefits in the physical layout, mass, and control of the robots. In particular, platforms that have been rough approximations of animals may reach new levels of realism and functionality. Perhaps most exciting, however, is the impact of active polymers on the way robots are designed and controlled when considering a "from-scratch" design mentality. Taking advantage of the "soft" characteristics of active polymers, entirely new robotic paradigms may be created. Through the use of genetic algorithms and rapid prototyping, the design cycle may be dramatically shortened and the effectiveness of the eventual product dramatically increased.

Moreover, with the advent of new micromachining technologies including microsensing, microactuation, microelectronics, and microcomputation, it is reasonable to assume that very small mobile robots will be built in the future. Promising advances in EAP indicate that this material could be an enabling technology in the creation of small robots. EAP could be bonded with other polymers to create smart composites whose behavior can be "programmed" by judicious mixtures of different layers. MEMS material might also be able to be bonded on or encased in EAP laminates, which gives the prospect for smart hybrid systems. The intriguing futuristic prospect of integrating biofilms within EAP to generate energy from "food" in the environment has also been touched on. The phototactic tube robot represents speculation on how EAP might be employed in the construction and control of individual microrobots as well as the material characteristics of EAP that would be required. Mass production of robots of the same scale of size and complexity would also tend to follow. Current research, inspired by social insects, indicates how such machines might be controlled and how their activities might be coordinated. These systems could prove to be robust, inexpensive, and highly tailorable.

Despite the initial fears of the public and pundits concerning robots that aren't directly designed by humans, that exist in multitudes, and that are too small to see, the optimistic roboticist may look forward to a world in which mankind can manufacture polymeric systems to our benefit rather than detriment; and, moreover, we have the wisdom to differentiate. These robots can be simple and self-sufficient, perhaps even self-perpetuating (with the proper safeguards, of course). Suitable environments may range from the carpets of our homes to the pills from the pharmacy to the surface of other planets. As has always been true with robotics, we are only limited by our imaginations.

17.10 References

Adamatzky A. (2000) "Reaction-diffusion and excitable processors: a sense of the unconventional," *Parallel and Distributed Computing Practices* 3(2).

Adamatzky A. and Melhuish C. (2000) "Parallel controllers for decentralised robots: toward nanodesign, kybernetes," *Kybernetes: The International Journal of Systems & Cybernetics* 29(5), pp. 733-745.

Adamatzky A. and Melhuish C. (2000) "Phototaxis of mobile excitable lattice," *Chaos, Solitons and Fractals* 13(1), pp. 171-184.

Aron S., Deneubourg J-L., Goss S., and Pasteels J.M. (1990) "Functional self-organization illustrated by inter-nest traffic in ants: the case of the Argentine ant," in W. Alt and G. Hoffman, Eds. *Biological Motion, Lecture Notes in Biomathematics 89*, Springer-Verlag: pp. 533-547.

Bailey, S. A., Cham J.G., Cutkosky, M. R., and Full, R. J. (1999) "Biomimetic Robotic Mechanisms via Shape Deposition Manufacturing" *9th International Symposium of Robotics Research*, Snowbird, Utah, October 9-12, pp. 321-327.

BEAM (1999) http:\\www.solarbotics.com.

Binnard, M. (1999) *Design by Composition for Rapid Prototyping,* Kluwer Academic Publishers.

Braitenberg V. (1984) *Vehicles*, MIT: Cambridge.

Brooks R. (1986) "A robust layered control system for mobile robot," *IEEE Journal Robotics and Automation* 2, pp. 14-23.

Conrad J. M. and Mills J. W. (1997) *Stiquito: Advanced Experiments with A Simple and Inexpensive Robot*, IEEE Computer Science Press.

Conrad, M. (1990) "Molecular computing," *Advances in Computers* 31, pp. 235-324.

Deneubourg J.L. and Goss S. (1989) "Collective patterns and decision making." *Proc. Int. Conf. on Ethology Ecology & Evolution Vol. 1.*, Florence, Italy: 295-311.

Dutton, K., Thompson, Barraclough (1997) *The Art of Control Engineering.* Addison-Wesley.

Ekeberg O. (1993) "A combined neuronal and mechanical model of fish swimming," *Biological Cybernetics* 69, pp. 363-374.

Hasegawa Y., Ito Y., and Fukuda T. (1999) "Behaviour adaptation on behaviour-based controller for brachiation robot," *Lecture Notes in Artificial Intelligence* 1674, pp. 295-303.

Hasslacher B. and Tilden M.W. (1995) "Living Machines," *Robotics and Autonomous Systems* 15, pp. 143-169.

Hickey, G. and Kennedy, B., (2000) "Six Legged Experimental Robot," *NASA Tech Brief* NPO-20897.

Hirai H., Hasegawa M., Yagi T., Yamamoto Y., Nagashima K., Sakashita M., Aoki K., and Kikegawa Y. (2000) "Methane hydrate, amoeba or a sponge made of water molecules," *Chemical Physics Letters,* 35, pp. 490-498.

Holland O. and Melhuish C. (1996a) "Getting the most from the least: lessons for the nanoscale from minimal mobile agents," *Artificial Life V*, Nara, Japan: 59-66.
Holland O. and Melhuish C. (1996b) "Some adaptive movements of animats with single symmetrical sensors," *4th Conference on Simulation of Adaptive Behaviour*, Cape Cod: 55-64.
Holland O. and Melhuish C. (1997a) "Chorusing and Controlled Clustering for Minimal Mobile Agents," *European Conference on Artificial Life*, Brighton, UK: 539-548.
Holland O. and Melhuish C. (1997b) "An interactive method for controlling group size in multiple mobile robot systems," *International Conference on Advanced Robotics*, Monterey, USA: 201-206.
Ishida T., Iida T., Yokoi H., and Kakazu Y. (2000) "Development of a liquid metal mobile object field," *SAB 2000 Proceedings Supplement*, 6, pp. 13-19.
Kelly I., Melhuish C., and Holland O. (2000) "The Development and Energetics of SlugBot, a Robot Predator," *EUREL: European Advanced Robotics Systems Masterclass and Conference*, The University of Salford, Manchester, Vol 2.
Kennedy, B., (2000) "Three-fingered Hand with Self-adjusting Grip." *NASA Tech Brief* NPO-20907.
Komocinski, (2000) Framsticks official website, http://www.frams.alife.pl/.
Korohoda W., Mycielska M., Janda E., and Madeja Z. (1991) "Immediate and long-term galvanotactic responses of Amoeba proteus to dc electric fields," *Cell Motility and the Cytoskeleton* 45, (1), 2000, pp. 10-26.
Leger, C. (2000) *Darwin2K: an Evolutionary Approach to Automated Design for Robotic,* Kluwer Academic Publishers.
Lipson, H. and Pollack J. B. (2000) "Automatic design and manufacture of robotic lifeforms," *Nature* 406, pp. 974-978.
Mavroidis C., DeLaurentis K., Won J., and Alam M (2001) "Fabrication of Non-Assembly Mechanisms and Robotic Systems Using Rapid Prototyping," *Proceedings of the 2001 NSF Design, Manufacturing and Industrial Innovation Research Conference*, Tampa, FL, January 7-10.
Melhuish C. (1999a) "Controlling and coordinating mobile micro-robots: lessons from nature" *Proceedings of the International Mechanical Engineering Congress and Exposition, IMECE'99*, Nashville Tennessee.
Melhuish C. (1999b) "Employing secondary swarming with small scale robots: a biologically inspired collective approach," *Proceeding of 2nd International Conference on Climbing and Walking Robots – CLAWAR*, Portsmouth U.K., Professional Engineering Publishing Ltd.
Melhuish C., Holland O., and Hoddell S. (1998a) "Using chorusing for the formation of travelling groups of minimal agents," *5th International Conference on Intelligent Autonomous Systems*, Sapporo, Japan.
Melhuish C., Holland O. and Hoddell S. (1999) "Convoying: using chorusing to form travelling groups of minimal agents" *Journal of Robotics and Autonomous Systems* Vol. 28, pp. 207-216.

Melkonian M., Anderson R.A., and Schnepf E. (1991) *The Cytoskeleton of Flagellate and Ciliate Protists,* Springer Verlag.

Menzel, P. and D'Aluisio F., (2000) *Robo sapiens: Evolution of a New Species,* The MIT Press.

Nakagaki T., Yamada H., and Tóth Á. (2000) "Intelligence: Maze-solving by an amoeboid organism," *Nature* 407.

Osada Y., Okuzaki H., and Hori H. (1992) "A polymer gel with electrically driven motility," *Nature* 355 pp. 242-244.

Slotine, J. J. E., and Li W., (1991) *Applied Nonlinear Control.* Prentice Hall.

Theraulaz G., Goss S., Gervet J., and Deneubourg J.L. (1990) "Task differentiation in polistes wasp colonies: a model for self-organizing groups of robots." *1st International Conference on the Simulation of Adaptive Behavior, From Animals to Animats*, MIT Press.

Ueda M. and Ogihara S. (1994) "Microtubules are required in amoeba chemotaxis for preferential stabilization of appropriate pseudopods" *Journal of Cell Science,* 107, pp. 2071-2079.

Usami Y., Saburo H., Inaba S. and Kitaoka M. (1998) "Reconstruction of extinct animals in the computer," in: *Artificial Life VI: Proc. 6th Intern. Conf. Artificial Life,* a Bradford Book, The MIT Press, pp. 173-186.

Ventrella J. (1999) http://www.ventrella.com.

Vogel, S. (1998) *Cat's Paws and Catapult,* W.W. Norton & Company.

Vogel, S. (1988) *Life's Devices.* Princeton University Press.

Whittaker W., Urmson C., Staritz P., Kennedy B., and Ambrose R. (2000) "Robotics for assembly, inspection, and maintenance of space macrofacilities" *AIAA Space 2000*, Long Beach, CA, AIAA-2000-5288.

Wilkinson S. (2000) "Gastrobots–benefits and challenges of microbial fuel cells in food powered robot applications," *Journal of Autonomous Robots* 9(2), pp. 99-111.

Yanai M., Kenyon C.M., Butler J.P., Macklem P.T., and Kelly S.M. (1995) "Intracellular pressure is a motive force for cell motion" in *Amoeba proteus, Cell Motility and the Cytoskeleton* 33, pp. 22-29.

Yokoi H. and Kakazu Y. (1992) "Theories and applications of autonomic machines based on the vibrating potential method," in *Proc. Int. Symp. Distributed Autonomous Robotics Systems* pp. 31-38.

Ziegler J, Dittrich P, and Banzhaf W. (1997) "Towards a metabolic robot controller," *Information Processing in Cells and Tissues,* Plenum Press: New York, pp. 305-318.

CHAPTER 18

Applications of EAP to the Entertainment Industry

David Hanson
University of Texas at Dallas
and
Human Emulation Robotics, LLC

18.1 Introduction

Biologically inspired EAP actuation is facilitating myriad new mechanisms, as portrayed in Topic 7 of this book and Bar-Cohen and Breazeal (2003). While many applications are at this time visionary, the entertainment industry may reap short-term benefits from EAP. Portraying animals is essential to entertainment, and for some entertainment uses, long-term performance is not critical. Abundant movies use robotics and/or digital rendering to simulate and emulate organisms [Schraft and Schrmierer, 2000]; examples include *A Bug's Life*, *Deep Blue Sea*, *The Matrix*, *Mighty Joe Young*, and many others. EAP could significantly enhance traditional character-simulation technologies, adding biological accuracy to animatronics' actuation.

In lab tests, EAP has already shown that it can match critical properties of biological muscle, as discussed in Chapter 3. EAP also promises to be effective in configurations homologous to animal musculature—a strong advantage over other forms of actuation. Although EAP has yet to demonstrate these capacities in character animation, recent materials breakthroughs bring the possibility tantalizingly close. To make it so, collaboration between academia and industry will be indispensable. If EAP proves itself in entertainment, it will find ample resources for further research and development. This chapter seeks to elucidate how EAP might infiltrate and ultimately revolutionize entertainment, with a suggested practical approach to doing so. As an example, an existing EAP-based character animation project is described. This humanoid robotic head combines artificial intelligence (AI), computer vision, and EAP actuation, in an attempt to emulate the gestalt of the human countenance.

18.1.1 EAP's Potential to Radically Improve Entertainment

EAP materials have characteristics attractive to the entertainment industry, offering the potential to more effectively model living creatures at significantly lower cost. Viscoelastic EAP materials could provide more lifelike aesthetics, vibration and shock dampening, and more flexible actuator configurations. Moreover, multiple studies have shown that the viscoelastic properties of animal tissue are key to locomotion and general stability [Chapter 3; Full, 2000; Full and Koditschek, 1999; Dickinson et al., 2000; Full et al., 1998; Bar-Cohen and Breazeal, 2003]. In these regards, EAP prominently earns the moniker "artificial muscle."

EAP animatronics could be married with other exciting contemporary technologies. Conveying the gamut of emotion, ambulating with dynamic stability [Chapter 3; Full, 2000], governed by AI, EAP-actuated figures will be capable in ways impossible with conventional engineering alone. Ultimately, EAP-actuated animatronics may give birth to the darling of science fiction: the android (or anthroid, to be gender neutral). Such sentient beings designed to specification would be ideal for entertainment. EAP-based anthroids could star in movies, roam around theme parks, and serve as telerobots or interface devices for the web. Given the likely moderate cost of EAP actuators, EAP anthroids could quickly pervade our world as servants or even peers. An illustration of such a figure enacting a drama is shown in Fig. 1. Possibly, even props such as an anthroid's clothing could be animated with EAP.

With an estimated two billion dollars per year spent on character animation, the entertainment industry may be the most salient vehicle for the development of EAP-based humanoid robotics. Key to tapping these resources, however, is a first-generation, functional product. Devising inexpensive and robust EAP-based animation that outperforms rival entertainment technology is the task at hand.

Figure 1: Illustration of an EAP robotic actor gesturing a dramatic scene, based on Bernini's "Longina." Note that ambient props such as the actor's toga could also be animated with EAP.

18.1.2 Recent EAP Advances

For many years so-called "artificial muscles" have been reported in various forums and news media. Even though these materials showed promise, they have fallen short for use in themed entertainment. Recent years' progress, as reported in Chapter 1 and Topic 3, has added exciting new options to the palette of EAP materials [Bar-Cohen, 1999 and 2000]. These include dielectric-elastomers at SRI International, the Relaxor Ferroelectric P(VDF-TrFE) at Penn State University, the carbon nanotubes at Honeywell and University of Texas at Dallas, and the conductive polymers at JPL, MIT, San Sebastian University (Spain), Pisa University (Italy), Risoe National Laboratory (Denmark), and SFST (Santa Fe, New Mexico).

Of these EAP materials, two stand out as appearing to be closest to implementation in terms of technological readiness for character animation. First, ion exchange polymer/metal composite (IPMC) actuators have shown large strains and significant robustness in experiments at JPL, EAMEX, Japan, University of New Mexico, and University of Washington. These actuators have been demonstrated in spring 2003 in what may be the first EAP-actuated entertainment product. Developed by Eamex, Japan, and manufactured by Daiichi Kogei, Japan, this animated fish swims about an aquarium (see Fig. 1, Chapter 1) propelled by IPMC actuators energized using electromagnetic induction from coils located above and below the aquarium. Specifically an entertainment product, the fish robot confirms the idea that entertainment is a "low-hanging branch" for practical application of EAP actuators. The requirements for these fish are far different than those for a humanoid robot, so much work remains. Nevertheless, these robotic fish are an extremely promising step in the evolution of EAP-powered entertainment robots.

Second, the dielectric EAP actuators developed at SRI International [Chapter 16; Pelrine et al., 2000] have shown strain, response speed, and stored mechanical energy density in the range of biological muscles [Chapter 3; Full and Meijer, 1999]. However, the challenges to these actuators are severe; they require up to 5000 V to operate, they expand when stimulated (especially tough for facial-expression animation), and they appear delicate in construction. SRI's recent tubular actuators expand the options of these dielectric actuators, operating as gimbals that may be controlled to bend in any direction.

Many other EAP actuator technologies may pull ahead to satisfy the needs of entertainment robots. Although not fully mature, the relaxor ferroelectric [Chapter 4; Zhang et al., 1998] shows high strain, fast response, and relatively high stored mechanical energy density, likely comparable to biological muscle. This EAP material also requires lower voltages, but is difficult to mass produce. Carbon nanotubes [Chapter 8; Baughman et al., 2000] offer immensely high power density and astonishing strength at low voltages, but they induce low strain at the level of 1% and they are a long way from practicable application. On the other hand, conductive polymers [Chapter 7; Olazábal et al., 2001] offer a relatively high actuation force of several hundred times the ratio of carrying force to their own weight, and require single-digit voltage. However, they are sensitive to fatigue and cannot be strained for extensive cycles due to the material's strength constraints. The above new materials may lead to the first generation of effective EAP actuators for entertainment applications, if their shortcomings can be addressed.

In addition to synthetic polymers, biologically derived electro-active polymers may be useful as actuators. The most promising class of such biopolymers would be cytoskeletal motor proteins such as myosins, kinesins, and dyneins. These protein polymers, being the motors of most cellular and subcellular biosystems) are inherently biologically compatible, and bioextracted muscle tissues have actually been used via electroactivation to drive swimming robots [Dennis et al., 1999]. Proteomics, the science of proteins and their dynamical systems, has been exploding with advancements recently, largely thanks to improving imaging technology and supercomputing simulations. Preliminary successes in engineering these proteins as nanoactuators are encouraging, though they do not indicate any near-term applicability [Wada, et al, 2003; Kull and Endow, 2002]. Beyond the engineering breakthroughs required to implement protein actuators on a nanoscopic scale, these mechanisms will further need to be assembled into large macro-arrays in order to be capable of effecting motion in an entertainment robot.

Although EAP as artificial muscles for motor-actuation is the topic here, other electroactive polymer capabilities may be useful to entertainment and so also merit mention. Recently, substantial progress has been made in flexible polymer display screens, such as Universal Display Corporation's OLED displays, and light-emitting polymers (LEPs) are finding active use in industry at the present. One can imagine that, in time, a character's skin could light up and display text or images as desired. Perhaps more exciting still is electroactive polymer computation, which has shown rapid progress in recent years [Higgins et al., 2003]. While not as inherently fast as silicon computation, computing polymers are materially flexible have other useful characteristics, such as the low cost of mass production. Additionally, electroactive-polymer-sensing arrays have recently been successfully tested in humanlike simulated skin materials at the University of Illinois at Urbana-Champaign [Engel et al., 2003].

Many such useful capabilities of EAP can be achieved within one single polymer material, a multiplicity of functions that potentially decreases the requirements of space, material use, and cost. One exemplary multifunction material would be carbon nanotubes, which promise to simultaneously serve as mechanical structure, transistorized computer, light emitter, and actuator.

While EAP materials are progressing in their capabilities, robotics research may complimentarily facilitate the use of EAP actuators in entertainment, particularly by reducing the force required to actuate entertainment automatons. This idea motivated research conducted by the author in the spring of 2002, wherein a novel class of polymers dubbed "F'rubber" (a contraction of "face" and "rubber") was invented that requires less than one-tenth the force required by conventional elastomers to be deformed into a simulated facial expression. These materials are described in detail in Sec. 18.6.2 of this chapter.

It can be anticipated that the capabilities of EAP actuators and the requirements of robotic actuators will soon converge; yet many issues remain to be addressed.

18.1.3 Herculean Tasks Remain

Certainly, if the materials science of EAP is fledgling, the implementation of EAP into biomorphic entertainment products is presently zygotic. Utility of EAP in the entertainment industry requires a total rethinking of mechanical engineering, which is conventionally based on principles of rigid structures—gears, U-joints, and the like [See Chapter 17]. Control and feedback of these inherently compliant materials also pose a significant challenge. Most importantly, EAP itself must mature considerably to be competitive in entertainment applications. Generally, useful characteristics of EAP actuators include low voltage, high strain, high power density, robustness and durability, and the support of fully developed adjunct technology (fasteners, drivers etc.). For practical use in character animation, the technology must also be reliable, affordable, and easy to use. Detailed requirements for entertainment application of EAP will be discussed in Sec. 18.2 of this chapter.

Formal collaborations between entertainment and research institutions remain crucial to tune EAP development toward the needs of the entertainment industry. EAP will not find widespread use in entertainment until EAP actuation and the supporting discipline of viscoelastic mechanical engineering are sufficiently refined. However, one EAP actuated product that sells in high volumes will pry open the door.

18.2 Entertainment and Its Shifting Significance

Webster's defines entertainment as "something diverting or engaging." Usually, "entertainment" refers to commercially marketed diversionary arts: music, movies, TV, and such, and the industry related to these arts. The contemporary explosion of digital and communications technologies is catalyzing evolution and convergence among many classes of human endeavor, and entertainment is playing an increasingly expansive (and ever fuzzier) role. Many new uses for entertainment can be cited. Computers, for instance, are often used as entertainment, as when playing video games, and toy robots such as Sony's AIBO dog are enlivened by the substantial computers contained within them.

Additionally, many nonentertainment applications are turning to the techniques of entertainment to be made more engaging. The word processor used to write this paper will occasionally interrupt with a friendly cartoon that comedically suggests helpful tips.

But even pure entertainment can be considered purposeful in multiple ways; for instance, it can offer escape, decoration, commerce, and, of course, artistic expression, which serves as a powerful agent in the evolution of culture. Some purists in the world of fine art feel that multiplicative functions dilute what they perceive as the most powerful function of art—the expression of the individual artist. Yet combined-function applications do not necessarily rob the artist of the opportunity to create art that is more singular in function, or even ambiguous in function. As Picasso knew, good art can be both profitable and engaging. In a similar way, entertainment craft and technology can enhance many other applications, such as psychological therapy, medicine, education, and human-computer interface designs.

Classically, the functions of art and entertainment have overlapped with entertainment serving as a route for the artist to communicate ideas and visual studies and by influencing public paradigms.

This capacity of entertainment—to embody a message in a very captivating way—is co-opted extensively for nonartistic aims (e.g., TV commercials, educational TV, propaganda). The tools of engagement, so powerful in entertainment, can be foreseen to progressively pervade our lives for such utilitarian purposes as the web and other new technologies propagating in our culture. But use of these tools for "pure" creative self-expression will also spread and mutate with the flowering of new technologies.

At the heart of entertainment's power to engage is the narrative structure, which lies deep-seated in the human brain [Lamb, 1999]. In this structure, the focus always revolves around the "character," derived from the human self-image and imaging of others. In this way, human nature predetermines that new ways to animate characters will remain prominent, perpetually under high market demand. As interactivity develops, AI will likely unleash "narrative engines." For interaction to be maximally effective, it seems likely that people will desire face-to-face interaction with characters. When narrative-driven anthroids arise to fill this niche, our stories will spill off the screen, perfusing our lives with living entities. Our characters will become people, our dreams reality, all to converge along a fuzzy edge between the "real" and the "virtual" [Kurzweil, 1999]. These actions will cause a gargantuan hiccup in humanity's sense of identity, but will also cause much delight and wonder. Of course, the meaning of "entertainment" will transmute inexorably along the way.

18.3 Technical Background to Entertainment Application of EAP

Manifold technologies and sciences relate the application of EAP to the emulation of organisms. The contemporary explosion of robotics (including biomimetic mechanics and AI), materials sciences, and manufacturing technologies will be integral to EAP character animation. These are discussed below. Other disciplines could be valuable as well; for example: biotech for artificial tissue engineering, medical physiology for understanding the apparatus of expressions in humans, and anthropology and psychology for insight into paralinguistic facial expressions.

18.3.1 Biologically Inspired Engineering

In the field of robotics, new biologically derived engineering techniques are causing something of a revolution of their own. Although inventors have long turned to marvels of nature for inspiration, the practice has become considerably more formalized with the modern systems approach to biology, and with the various emerging interdisciplinary design practices that are proliferating as fast as their many names: "biologically inspired engineering," "biomimetics," "biomorphic design," and "biomorphic design." Variations in name aside, these practices generally reverse-engineer structures and principles from living organisms into technological systems, resulting in machines that run like cockroaches (Stanford's Sprawlita and McGill's Rhex), hover like a fly (Berkeley's Micro-Mechanical Flying Insect), and see and track faces like people (Cristoph van der Malsberg's and Nevengineering's Facetracker system), among many other significant capabilities.

Many areas of biologically inspired engineering exist, but only those that may improve entertainment robots warrant discussion here. These include biomimetic components engineering (which includes locomotion, intelligence, and materials other than EAPs), biologically inspired intelligent robotics (which integrates many biomimetic technologies into larger systems that mimic whole organisms), and aesthetic biomimesis, which mimics human communications.

18.3.1.1 Biologically Inspired Components: Locomotion, Materials, and Intelligence

Living systems amalgamate many levels of function: locomotion, perception, advanced materials, etc. In one approach to biomimesis, the designer reverse-engineers only one area of utility of a living system, isolated from the system as a whole. This approach has resulted in a profusion of breakthrough technologies, including face recognition, natural language processing, biped locomotion and EAP actuation. Such technologies are worth mentioning mainly because of their imminent potential of being woven into cohesive, synthetic organisms.

Biomimetic locomotion would deliver more than just heightened expressivity to a sociable entertainment robot, but may additionally increase a robot's capability to contend with the real world by means of the ability to move about and materially interact with its environment. Indeed, recent research has shown that biologically inspired, dynamic legged locomotion can substantially outperform vehicles that use wheels or tank treads alone [Steeves et al., 2002].

Humanlike bipedal locomotion in particular will be appealing in entertainment applications. Honda, Sony, AIST, Fujitsu, ERATO, and Wowwee Toys have recently demonstrated robots that can walk with robust stability while being configured in a humanlike form, some of which already being marketed as products. The Honda humanoid can be rented, the Fujitsu biped can be purchased, and Sony has expressed the intention to market its SDR-4X humanoid soon. The Wowwee "Robosapien" biped, engineered by Mark Tilden, will be on the shelves of toy stores for the 2004 holidays, and has shown even more profound capabilities thanks largely to engineered resonant dynamic-stability. Rather than static or quasi-static stability exhibited by most humanoid biped robots (i.e., the "controlled fall" that preserves the zero moment point), the Robosapien prototype employs dynamic stability, and hence is more capable over diverse terrain and is more highly energy efficient. Figure 2 shows three agile biped robots.

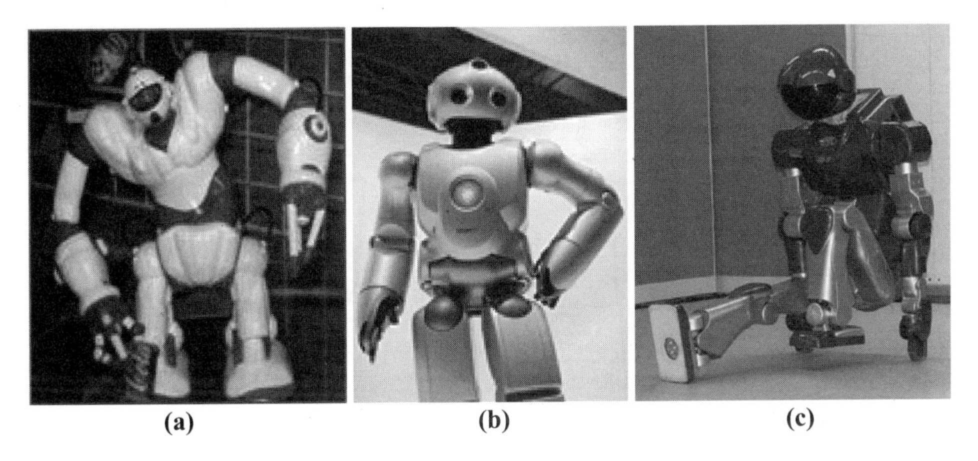

(a) **(b)** **(c)**

Figure 2: Contemporary bipedal robots: (a) Wowwee's RoboSapien (courtesy of Mark Tilden), (b) SDR-4X (courtesy Sony, copyright 2003), and (c) HRP-2P (courtesy of AIST).

Much emphasis in academic biologically inspired locomotion engineering is placed on the importance of viscoelastic properties to the dynamic stability of animal locomotion [Raibert, 1984]. The gist here is that animals cycle energy through the spring-mass action of elastomeric musculature. EAP-based mobility should tap the intellectual resources of this movement for guidance.

Scientists such as Full [Chapter 3] are studying animal movement with the perspective of a wide range of disciplines to address engineering problems for robots. This is allowing the construction of biologically inspired robots, such as those of MIT's Leg Lab, Case Western Reserve University, the Delft Biped Lab, and Stanford's Biomimetic Robot Lab (to cite a small fraction of laboratories pursuing this work). Figure 3 shows an early dynamically stable robot built by Marc Raibert and others at the MIT Leg Lab.

Much evidence presently suggests that the proper viscoelastic properties are required in a robot's materials to achieve robust dynamic stability in locomotion [Chapter 3]. In addition to EAP actuators, other biologically inspired materials may be useful, such as synthetic spider silk [Highfield, 2002], synthetic abalone shell, or synthetic ligatures. Many groups around the world are doing active research that may be applied toward robotics. Recent novel thermoplastic elastomers (TPEs) advances have shown characteristics very close to biological tissue as well. The material options for designing a biologically inspired synthetic organism are unparalleled if contrasted with even that of a decade ago. Engineering EAP actuators to utilize some of these options will return rewards in terms of performance.

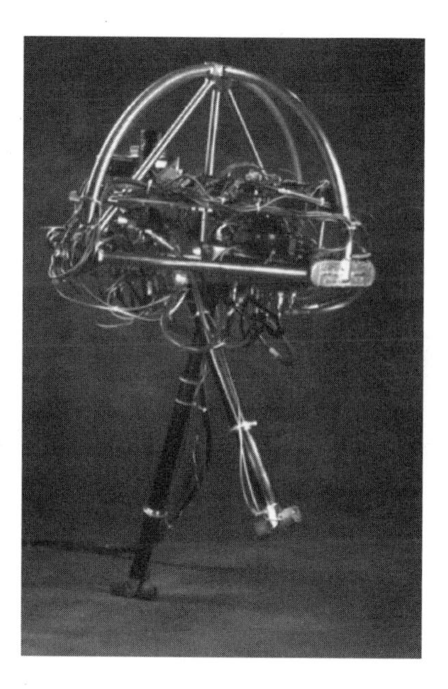

Figure 3: Early dynamic stability in action. Image provided courtesy of Marc Raibert and The MIT Leg Laboratory (copyright 1990).

Perhaps the most complex and profound undertaking of biologically inspired engineering research is that of intelligent systems, an undertaking described by many an alias: artificial intelligence, cognitive systems, computational neuroscience, and synthetic intelligence among others. All these fields commonly strive to emulate various perceptual and problem-solving capacities of the nervous system, and the results to date are dizzyingly impressive. Machines can presently track a human face and perceive facial expressions, converse with people through speech, and learn low-level semantic knowledge through naturalistic interactions, while other systems process audio data in imitation of the human auditory processing nucleus. These examples represent only a minute fraction of the huge volume of notable progress made recently in machine intelligence.

The most prominent challenge in the undertaking of biologically inspired intelligence engineering may be similar to that of biologically inspired engineering at large: the integration of many discrete elements into a biologically inspired super system that forms a "society of mind" [Minsky, 1986] and a society of body as well. This integration is largely the directed goal of biologically inspired robotics.

18.3.1.2 Biologically Inspired Intelligent Robotics

Truly intelligent robots will be woven together from many technologies—locomotion, perception, and general intelligence among them—to form elaborate machines that act as synthetic organisms. This integration of constituent technologies (some biologically inspired and others not) characterizes the field of "biologically inspired intelligent robotics," as described in the book of the same title also published by SPIE Press [Bar-Cohen and Breazeal, 2003]. The book delivers an effective snapshot of the budding field, both in terms of the component technologies and their assimilation into robotic entities. Such robots will be extremely useful for entertainment, and will be the stage upon which EAP actuators will deliver their most grand theatrical performance.

In particular, generalizing the intelligence of these machines will be elusive without integrating multiple perceptual and processing elements. The multifunctional architecture of the human mind serves as the most obvious template in this task. In humans, the emergence of intelligence is highly dependant on social nurturing, care taking, and personal interaction. If the same holds true for robots, then sociable robots may be the means (at least partially) by which machines become truly smart [Breazeal, 1997]. This sociability will also help such intelligence emerge with social values and a fondness for people; it will also make people more comfortable and able to bond with the robots. Figure 4 shows examples of sociable robots.

Figure 4: Sociable robots; left: Leonardo (courtesy of Cynthia Breazeal, MIT and Stan Winston Studios) and right: Nursebot Pearl (courtesy of Carnegie Mellon University).

The field of sociable robotics is particularly representative of biologically inspired intelligent robotics as it integrates mobility, visual perception, automated-speech interaction, and (frequently) biologically inspired facial expressions for interfacing with people. This sector is of great pertinence for entertainment robotics because they share the goal of engaging human communication and attention. This leads us to consider the aesthetic factors by which an entertaining object engages a person, factors that include static aesthetics, dynamic aesthetics, and cognitive aesthetics (social intelligence and interactivity).

18.3.1.3 Aesthetic Biomimesis

Of all the biologically inspired engineering domains, aesthetic biomimesis would be the most vital for entertainment applications. Communicative engagement of an audience is the essential purpose of entertainment. While this purpose can be greatly improved by advanced technology, such as bipedal locomotion and artificial intelligence, ultimately it is lifelike aesthetics and communicability that make entertainment effective.

Since prehistory, the fine arts have mimicked aesthetic and communicative aspects of organisms. Examples abound: Michelangelo, Da Vinci, and Disney are all best known for the portrayal of living creatures. While artists tend to formalize their powerful methods of biomimesis, merely in part at best, in the late 20th century cognitive psychologists began to scientifically consider such questions, mostly in the field dubbed "paralinguistics," with promising preliminary results [Ekman, 1996 and O'Toole, 2002].

One major effort to codify the semiotics and semantics of human facial communication has resulted in a system called the facial action coding system (FACS) [Ekman and Friesen, 1971]. Body language, also well studied [Birdwhistle 1970], will further enable robotics' sociable applications. This and future work that relate physiological expressions to increasingly subtle cognitive states will be extremely useful when automated into sociable, interactive entertainment robots.

The practical application of science of human communications and visual cognition can be considered to be "aesthetics engineering"; more specifics of this concept and use of FACS for this engineering are discussed in Sec. 18.4.2. In spite of such engineering potential, for the foreseeable future, less robustly formalized aesthetics can be expected to remain useful to entertainment robotics, mostly in the form of art. The art and technique of aesthetic biomimesis using robots is discussed further in Secs. 18.4.1 and 18.5, with a particular focus on animation with EAP actuators.

18.3.2 Advanced Manufacturing Technologies

Not only does humanity seem to be in the midst of the greatest flowering of advanced materials sciences (including recent advances in EAP); we are also experiencing an explosion in advanced manufacturing techniques that may effectively dovetail with EAP actuation [Chapter 14]. Rapid prototyping and digital design tools are enabling complex concepts to be turned into physical objects in very short cycles. Advanced silicon manufacturing techniques, largely innovated for manufacturing microprocessors, have resulted in burgeoning techniques of micro-electrico-mechanical systems (MEMS) in several interesting robot projects [Dickenson et al., 2002]. A comprehensive list of techniques that may be pertinent to manufacturing and prototyping entertainment robots with EAP actuators would be dazzling, and prohibitively long to include here.

Fusing several rapid prototyping technologies with mold making and robotics, shape-deposition manufacturing (SDM) has been described as particularly interesting for use with EAP actuators in Full [2000]. SDM uses various CAM (computer-aided manufacturing) technologies to layer and refine materials into arbitrarily complex configurations, bonded without the use of mechanical fasteners. Actuators and sensors can be imbedded directly into the "flesh" of a device, composed of materials of varied elasticity and rigidity, layered and bonded in situ. This process can achieve fully functional rapid prototypes, as well as highly efficient manufacturing procedures [Amon et al., 1996]. Images of the SDM process are shown in Fig. 5.

Figure 5: Stages in shape deposition manufacture of a small robot at Stanford's Rapid Prototyping and Biomimetic Robotics Labs. (Courtesy of Mark Cutkosky, Stanford University.)

SDM is ideal for MEMS, micro-, and nano-devices, but many principles profiled in the SDM process would be useful for engineering rubbery macro-scale animatronics, simplified by the absence of mechanical fasteners. As an example, the 15-cm (6-in.) legged robot "Sprawlita" at Stanford's SDM Rapid Prototyping Lab is shown in Fig. 6. This SDM robot is extremely robust, capable of receiving quite a pounding [Bailey et al., 2000]. This is of interest to entertainment—an industry where animatronics fail regularly, largely due to shock, vibration, friction, and fastener failure—factors which are reduced or absent in SDM elastomer devices.

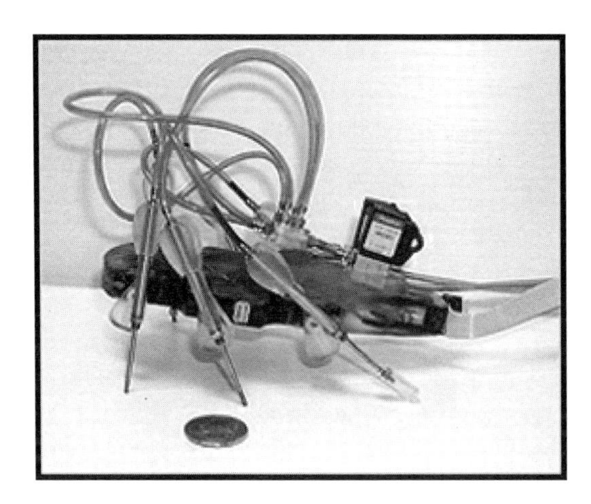

Figure 6: "Sprawlita," a shape-deposition manufactured, dynamically stable robot, built by Sean Bailey and Jonathon Clark at Stanford's Biomimetic Robot Lab. (Courtesy of Mark Cutkosky, Stanford University.)

18.3.3 Considerations of Other Actuators—EAP's Rivals

Two technologies have come to dominate the rendering of characters in Hollywood: computer-generated imagery (CGI) and mechanical animatronics. The emulation of characters has become an extremely advanced art, merging the knowledge and craft of many sources: robotics, software engineering, optics, electronics, and plastics on the technical side; sculpture, puppetry, and animation on the creative front. Some notable character-effects shops include Jim Henson's Creature Shop, Walt Disney Imagineering, Industrial Light and Magic, Edge Innovations, Stan Winston and Rick Baker Studios. Motion pictures comprise the main development bed for these crafts. Many contemporary movies will spend the majority of their budget on illusionary visual effects. Big-budget productions will spare little expense for a desired effect, which may then be used for only a few dozen takes. The lifespan of an EAP actuator in such circumstances may only be in the low thousands of cycles; still, for an EAP to be viable in this environment, it must surpass the quality of extant technology. Closer examination of this rival technology is in order.

CGI character simulation is coming to maturity, with amazing results apparent throughout cinema, TV, the web, and video games. With computer graphics, nearly any visual image becomes possible. In spite of its accomplishments, CGI still has trouble generating organic forms, especially those of humans, though this will likely change as the technology matures. A major drawback of CGI is its limitation to screen-type displays—its figures cannot be interacted within the "real" world, except via ungainly 3D displays and restrictive haptic devices. Having the "virtual" (digital images) enter our nondigital reality in the form of actual 3D entities is highly desirable, particularly for interactive entertainment. EAP could fill the gap in this regard—imagine "smart blobs" capable of adopting any shape and rheological state.

Mechanical animatronics are still the main technology of the character special effects (also known as “effects” or “FX”) industry even though they are losing ground rapidly to CGI. Animatronics use many techniques common with robotics—rigid structures and actuators, digital control systems—to simulate character expressions. This rigid understructure will be surfaced in rubber skin, hair, and clothes. Unlike robots, animatronics generally are not computer controlled but rather are operated by puppeteers. Incredibly subtle expressions and gestures are achieved. For an example refer to Disney’s *Mighty Joe Young*, which has phenomenal facial expressions, produced using remote-operator puppeteering—a form of telerobotics (Fig. 7). Several puppeteers often operate a single robot simultaneously, generating very complex gestures. Generally, the illusion of a full range of expression is quilted together from multiple shots of animatronic figures, each figure achieving a smaller range of gesture. For theme parks, the motion of an animator is captured and replayed as a digital sequence. Little, however, has been done in imbuing artificial intelligence into animatronics—a vital and lucrative path awaiting exploration.

Figure 7: Example of mechanical animatronics, used in the movie *Mighty Joe Young*, created by Cinovation. (Courtesy of Disney Enterprises, Inc. ©2001.)

Three types of actuators are prevalent in animatronics: hydraulic, pneumatic, and electric. Hydraulics are the most powerful relative to the size and weight of the actuator, but hydraulics require rather cumbersome pump and lines, and can be messy—fluids can leak or spray. Pneumatics also require a pump and lines, and can be noisy. Electrical actuation is the least messy, but conventional electric motors have the lowest energy density of the three. Moreover, traditional rigid structures present limited spatial configuration possibilities, and are further limited by the need for gearing, transmission, etc. A study evaluating EAP actuator qualities in relation to the above rivals is called for.

As stated, movie animatronics need only function a few times for a scene. For fine art, theme parks, mannequins, or toys, on the other hand, animatronics will need to withstand hundreds of thousands of cycles, with little or no maintenance. For such physically present applications, extremely robust EAP actuator products are needed. In the long run, these uses will prove more fruitful, imparting deeper immersion in the dramatic experience. Also, due to CGI's increasing dominance of screen-based media, movies of the future will nurture robotic character entertainment less than they have in the past. Prominent venues for entertainment robotics of the future will include any place a physically embodied 3D character adds value to an entertainment experience—shopping malls, theme parks, and toys may be examples.

18.3.4 The Benefits of EAP Actuation

EAP offers many improvements over traditional actuation technologies, both in function and form. Functionally, EAP promises higher power densities than any other kind of actuator [Topic 3; Baughman et al., 2000], even higher than that of the workhorse of heavy industry: hydraulic actuation. Being electric, EAP actuators would do away with bulky and fault-prone hydraulic pumps. They would eradicate the mess associated with hydraulic failure, though they would require power supplies. Given the generally moderate cost of polymers, EAP actuators are likely to be more affordable than rival actuators. EAP actuators' unitary mechanical form allows them to be packed into smaller volumes. Volume is further conserved by EAP's flexibility and potentially moldable nature. Such flexibility will also expand the options for mechanical configuration of actuators. Absence of gearing and transmission reduces points of possible failure. EAP can act as dampener, insulating a mechanism from shock and damaging resonant vibrations. Elastomeric qualities will allow for spring-mass energy cycling, improved efficiency and stability in locomotion and gesture, and a generally more lifelike appearance.

Aesthetically, EAP's affinity to biological tissue is very appealing. EAP actuation should allow more physiologically accurate animatronic design; the movement of the resulting animation would look more inherently organic. Occam's razor supports this assertion; structurally rigid mechanisms certainly represent the long route to simulating animal soft tissue. EAP's softness and elasticity can be foreseen to eliminate the distracting surface distortion caused by rigid actuator elements. Compliance and dampening characteristics of EAP closely correspond with those of biological muscle, simplifying controls and leading to more gesturally accurate animation. EAP actuators can be flat or bundled, like real muscle. Such formations could be imbedded directly into rubber skins—which are currently in widespread use for special effects. By building these artificial muscles into layers, the complex musculature of the human face could be emulated. Such a complex array of actuators will likely render much more subtle and intricate expressions than contemporary facial animatronics.

No previous actuation technology has presented so many potential benefits to animatronic character animation. If EAP actuation proves practical, it will certainly revolutionize animatronics.

18.3.5 Requirements for EAP Usefulness to Entertainment

As mentioned in Sec. 18.1.3, preferred characteristics of EAP actuators include low voltage (30-V max), high strain, high power density, robustness and durability, and the support of fully developed adjunct technology (fasteners, drivers, etc.).

Additionally, the following specifications would need to be met for EAP actuators to be reasonable for entertainment use:

- For movie use, longevity should be in the thousands of cycles; for theme park use or toys, hundreds of thousands of cycles.
- Force generation needs to be at least 15 times weight of actuator, and up to 200 grams or better.
- Response time should be in the low milliseconds. The speed of actuation needs to be at least 5-cm/sec (~2-in/sec).
- Displacement should be at least 10% of the length of the actuator.
- The actuators should preferably be configurable in sheets or bundles.
- End fasteners should be integrated into the actuators.
- The actuators should be dry, and if not, they should be thoroughly sealed.
- They should be inexpensive [less than $300 (US) for a midsize 2.5–10-cm (~1–4-inch) actuator with driver]. They also need to be easy to handle and use.
- For facial animation, the most useful actuators would be 2.5–7.5-cm actuators with 20% strain or displacement, weighing a maximum 15 grams.

These are tall demands. If they are met, however, the payoff could be colossal.

18.4 The Craft of Aesthetic Biomimesis in Entertainment

From ancient marbles to modern cinema animation, the fine arts have demonstrated the most effective biomimesis of all history. While most examples are painted, sculpted, or drawn, mechanical animation also extends back at least as far as the ancient Greeks, with the “automata” figures [Cassell, 2002]. In the twentieth century, artisans furthered this tradition of automation by combining modern materials, mechanical principles, and electronics into sophisticated robotic puppets commonly called “animatronics,” so named after Walt Disney’s “Audio-Animatronics.” Animatronic automatons now populate theme parks around the world, and can be seen in many major motion pictures as special effects.

An inherently interdisciplinary task, animatronics merges many branches of engineering technique with the classical arts. Without such a combination of skills this commercial cousin of biologically inspired robotics would not be possible. Animatronics’ collaborative spirit represents an open door for future technical innovations, including EAP actuation technology.

Mimicking the aesthetics of the human form and its dynamic action is not necessarily a simple task. Human faces are very complex with modes of expression; wrinkles lie dormant all over the face, invisible until an expression is enacted. Opposing movements evokes radically different types of folding and bunching of skin. As it is extremely soft, human skin is intimately affected by the pulling of various internal layers of tissues, each of which may have a different Young's modulus, and other physical characteristics.

Moreover, because human skin is an elaborate web of liquid-saturated cellular tissue, it is tricky to emulate with a solid elastomer. The molecules of a solid elastomer stay bonded to their nearest neighbors, and so remain geometrically constrained when distorted (unlike liquid-saturated facial flesh). This attribute causes elastomers to resist compression, and to require more force for elongation/strain than human facial skin. Alternatives to elastomers are needed to effectively simulate human tissues. While this issue is not addressed by common practices of animatronics, as described in Sec. 18.4.1 below, a solution is reached by the author's own work, described in Sec. 18.5.

18.4.1 Animatronics Practices: Methods of the Entertainment Industry

Entertainment robot applications vary widely in requirements of functionality, appearance, and durability depending on the use of the final product. For example, theme park robots must run continuously for 16 or more hours daily, 300+ days per year; whereas for a movie, an animatronic figure may need to perform only a few dozen times to achieve a certain shot. Such differences result in great variability among practices used to achieve animatronic effects. Nevertheless, certain practices are pervasive, and the differences are easily described.

18.4.1.1 Design Process

The first stage of animatronic design involves the narrative conceptualization. Several questions are asked: how does the animatronic figure fit into the story? What movements and gestures are imperative, and what does the robot need to look like? As these questions are addressed, a series of sketches are produced. For quick ideation, classical rendering tools of pencil, paints, and paper have not yet been bested. Moreover, given that people respond most immediately to complex organic forms, the skills of the classical arts to render naturalistic, real looking images are key when generating animatronic designs. Animatronic characters nearly always fall into three categories: known animal, fantastic beast or cartoon, or human.

New tools, especially 3D imaging software, have become pervasive for second phase design implementation, but the skills of the classically trained artist still remain essential, regardless of whether drawing in graphite or graphic-bytes.

In the entertainment industry, commonly used 3D modeling software packages include Maya, Studio 3D Max, and Lightwave.

Given the narrative intent of animatronics, especially for use in film, an animatronic figure's position in the story will usually be represented in the storyboards, which are sequential images that show the flow of the story in time, with the written story-script printed below corresponding images. Storyboards frequently are animated now, showing camera movements or, for themed rides, the perspective of someone in a moving ride-vehicle. These narrative constructs are linear and determinant, except in emerging media such as video games and natural language interfaces.

Once the project's creative directors decide upon the visual designs of the animatronic character and its movements, visual artists begin to collaborate with mechanical, skins, and animation artist/technician/engineers, forming a plan for technical implementation.

Often in the world of special effects, few boundaries exist between disciplines; in fact, if an animatronics' engineer is also a classically trained artist, he or she is likely to create a robot with greater lifelike anatomy, movement, and gesture. Therefore, such renaissance-style synthesis of skills and training is more often the case in entertainment special effects.

While 3D artists either sculpt or model the animatronic figure in software, the technicians will design the mechanism, usually using computer-aided design (CAD) software. Some of these now have physics engines embedded in the software, to assist with the engineering and to run fault-test simulations. Such common software includes Cobalt, Solidworks, Alias, Pro-Engineering, and Working Model 4D. One useful aspect of such digital engineering design is the gained opportunity to observe the movement of the animatronic figure, fully animated, before expending money and time on physical implementation. Another benefit of digital engineering design is that the parts designs can be sent directly to rapid-prototyping machines for quick, iterative manufacturing. Such machines range from CNC milling machines and lathes (low-end desktop CNC mills may cost as little as $1000 in 2003) to high-end stereo-lithography, ABS deposition, and metal-sputter deposition machines (which can sell for over $1,000,000 in 2003).

Concurrent to the execution of engineering designs, sculptors or digital-modelers will render the figure in three dimensions with full biomorphic detail down to individual pores, wrinkles, etc. If nondigitally sculpted, the figure will usually be produced in oil-based clay, which can then be digitized using a laser scanner for further modification in 3D software, and tight integration of the sculptural and mechanical animation designs. Oil-based clay is apt for the 3D design process: it is an infinitely high-resolution 3D display with immediate haptic feedback, very robust, and low cost. Digital haptic sculpting interfaces like FreeForm (sensable.com) are proving extremely useful too, especially for amending digital models and performing certain sculptural operations; meanwhile, imaging software engenders rapid if less sensual modeling, and permits automation of some of the biomimetic design. While some special effects shops, like George Lucas' Industrial Lights and Magic, are turning entirely to digital for execution of 3D models, by and large, oil-clay remains a useful tool for creating complex organic forms, with features not yet matched by digital design tools. The integration of high-tech and traditional design tools appears to be the best solution.

Once creative decision-makers approve a sculpture, it will be molded and cast as a rubber skin; and when the animated mechanical designs are completed, they will then be executed in engineered materials.

18.4.1.2 Skins Process

To mimic the supple flexibility of human and animal skin, animatronics studios turn to various chemical elastomers, commonly and simply called "rubber." Urethanes and silicones are the most widely used elastomers for animatronics, and each offers unique attributes and behaviors. Urethanes, showing characteristics that actually lie somewhere between classic elastomers and viscous materials [Pioggia, 2001], exhibit properties that are quite like the natural viscoelastic behavior of animal tissue. In particular, urethanes will function like both springs and dampeners, with very quick responses, even when the particular urethane is very soft. Skin-Flex urethanes, manufactured by BJB Industries, are the gold standard of special effects urethanes. Easily painted and pigmented, Skin-Flex elongates to 1000% with a shore hardness as low as 3 to 5 durometer, shore A, as soft as human skin. Another common special-effects urethane is SmoothOn's Evergreen series. The drawback of urethanes lies in their longevity; urethanes tend to degrade with time, typically over a span of months or years, and this degradation accelerated by exposure to UV radiation. Longevity of months is not an issue for movie and television effects; however, for theme park robots and other long-term, high-cycling animatronics, silicone is the preferred skins medium. For toys, thermoplastic elastomers are used; while these are very inexpensive, they are not as supple as urethanes and silicones.

Silicones are not sensitive to UV radiation and tend not to degrade significantly for years. Within critical limits, dynamic distortions of silicone can cycle millions of times without causing measurable deterioration in physical characteristics, including tensile strength and elongation. Silicones show elongation and strength characteristics similar to the best urethanes. Common brands include G.E., Dow-Corning, and Smooth-On's Dragon Skin. While silicones are easily pigmented, they are not easily painted once the silicone's cross-linking is complete. Though the addition of "plasticizers" will soften silicones to shore hardnesses near those of human skin, the softened silicone begins to exhibit kinetic induction—a delayed reaction between actuation and effective movement of the skin surface. In short, urethanes look better than silicones, but do not last as long.

To translate the sculpted animatronic figure into a rubber skin, a mold (negative impression) of the sculpture will be made. Usually this mold will have an inner surface of an elastomer (commonly silicone), surrounded by a shell, which is also called a "mother mold." This shell is composed of a hard, rigid material such as fiberglass, rigid urethane plastic, or classically, plaster reinforced with burlap. The rubber inner layer will lock into the shell with keys, or raised bumps. The shell will be composed of several interlocking parts to avoid undercuts (regions where the sculpture or cast will lock in the mold); these parts will be held together with bolts.

Usually some form of skull will reside inside the animatronic figure's head. This skull must be sculpted separately, and will frequently approximate animal physiology. To maintain anatomically correct skin thicknesses, forensics or zoological data may be consulted [Archer, 1997]; such variable skin thickness can enable lifelike dynamic skin distortions, more closely resembling natural facial expression. Generally, a cast of the skull will then be used as an inner-core to the mold so that when cast, the skin's inner and outer surfaces will conform tightly to both the exterior artistic design and the inner mechanism (Fig. 8).

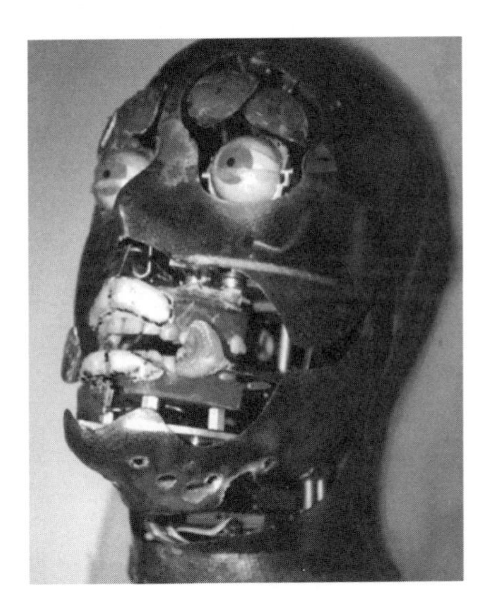

Figure 8: Animatronic mechanical system. (Courtesy of Dave Kindlon.)

Anchors in the skin provide mechanical connection-points for attaching the expressions' actuation mechanisms, while gaps in the skull permit actuators housed inside the skull to effectuate the skin movement. Anchors will frequently be composed of loose-weave fabric cast directly into the skin; such fabric spreads force over a larger area of skin, while allowing the elastomer to retain its strength by cross-linking through the voids in the fabric's weave. Because the fabric inhibits the elongation properties of the elastomer in the anchor zone, the size and placement of the anchors are critical.

Several methods exist to impart movement to the anchors. The most common method is to attach linear actuators directly to the anchors; a similar approach attaches gimbals to the anchors, thus effectuating two axes of motion (three if the gimbal sits on an extending boom). In the author's experience, a third method works better than the previous two: Kevlar cables are cast into the skin in lines that correlate to those of natural muscles. At one end, each cable attaches to an anchor; at the other, the cable slides through the skull to reach an actuator. Then, when actuated, the cable slides freely through the skin, transmitting force directly to the anchor. This method is more biologically accurate, in that it effectuates motion from the interior of the skin-mass, as do biological muscles; also, it allows layered vectors of actuation, mimicking the layers of natural facial musculature.

One notable complication results from the dissimilarity of solid elastomers and liquid-filled animal tissue. A liquid-filled cellular matrix like human skin is as easily compressible as it is elongated, whereas elastomers require considerably greater force to compress than they do to elongate. This means that either enlarged actuators are required (rendering electrical actuation prohibitively heavy and bulky), or the skin-architecture needs to be modified. Several methods for such modification exist. The first embeds liquid or air-filled pouches in the skin, but this is complicated and costly. The second method involves the strategic reduction of skin-thicknesses. The third method uses a blend of elastomer and a foaming-elastomer (which is also called "foam-rubber"). This creates a supporting internal structure inside a pure-elastomer skin surface, a structural mass that both elongates and compresses with ease. The resulting facial expressions, shown in Fig. 9, are actuated with mere hobby servos producing 19-oz in. of torque at 6 V.

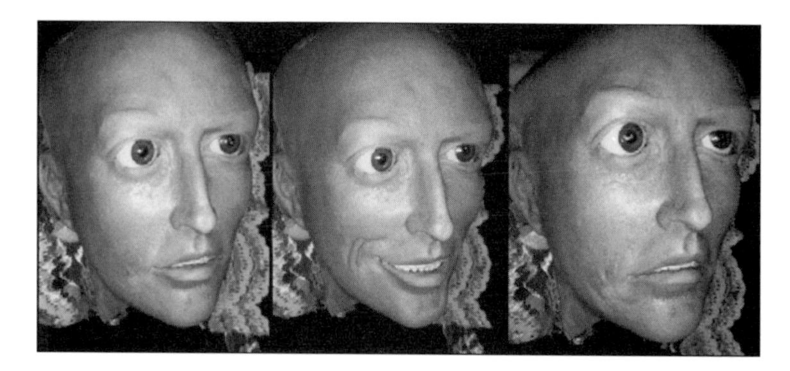

Figure 9: Facial-expression robot using elastomer-foam blend and embedded cable drives.

18.4.1.3 Mechanical Process

Animatronics technicians use proven mechanical technology, but will also often find creative, improvised mechanical solutions (provided human safety is not at stake); the rule for animatronics mechanical design is simple: if it works it is good. This approach differs from that of the engineering sciences, wherein the principles of a system are to be formalized and explained first, and then deployed. The goal of animatronics, however, is not to disseminate knowledge, but to create dramatic theatrical effects, usually on an intensively tight deadline.

Mechanically, animatronics are built using standard machining tools: milling machines, lathes, and welding equipment. Bearings, u-joints, and other common machine-components will be purchased and integrated into animatronic machine designs to reduce costs and production time. To actuate the figures, one of three actuator types will be employed: electric, pneumatic or hydraulic. Electric actuators for movie special effects will frequently consist of hobby-type servos; while for applications requiring greater longevity, more refined and durable motors such as those made by Maxon, will be employed. For greater power-to-mass density ratios, brushed, nonstepping motors are choice. Pneumatic and hydraulic actuators provide considerably greater ratios still, especially hydraulics [Hanson and Pioggia, 2001]. Neither pneumatics nor hydraulics will be appropriate for mobile animatronics, however, since these actuators require large, off-board compressors to pressurize their air or fluid. In the long run, EAPs may provide improved power densities for mobile animatronics [Hanson and Pioggia, 2001]. During the interim, the untethered entertainment robot endeavor Sony SDR-4X, shows that improved servos and controls can greatly improve the power output relative to the weight of an electric actuator.

The mechanical components an animatronic figure will attach to a frame, which functions as a skeleton of sorts. In certain cases, the mechanisms can attach to an exoskeletal shell, which gives form to the figure as well as serving as a structural frame. Form factors and spatial considerations will govern the distribution of parts inside the figure, including electronics, wiring, mechanisms, and if needed, processors and batteries.

18.4.1.4 Electronics, Controls, and Animation

At this time, automated controls in animatronics tend to be minimal and are usually devoid of artificial intelligence. For theme parks, the figures are controlled by linear, prerecorded scripts; whereas for movies, animatronics generally are controlled in real-time by human operators via telerobotics; although these remote controls (RC) are quite sophisticated in both electronics and apparatus, the animatronics industry has extended to these RC systems the simple moniker of "puppeteering." In either case, a human operator designs the motion by using principles of artistic animation, dance, or physical acting.

The rising tide of artificial intelligence will certainly enable greater autonomy in these artistic machines; it can be imagined that in time, this action will utterly change the face of narrative arts.

18.4.1.5 Final Artistry

Once the animatronics skin, mechanism, and controls are assembled, the figure will be painted and dressed in clothes, and hair (or fur). At this stage, the final animation motion will be designed, and the figure will be tested for bugs to be corrected before final use. If used for narrative arts, the final figure will then be placed in the theatrical set for final presentation, or filming.

18.4.2 The Psychology of Entertainment Robotics

The face serves as humanity's primary mode of expressing affective states [Ekman, 1971], and the human nervous system is finely attuned to understand the face's visual language [Levenson, Ekman, and Friesen, 1990]. Better mechanization of this universal visual-language will unlock a panoply of service and entertainment applications, from toys to comforting companions for the elderly.

Sylvan S. Tomkins and Carroll Izard separately concluded that there exist eight "basic" facially expressed emotions: interest, surprise, happiness, anger, fear, disgust, shame, and grief [Tomkins 1962 and Izard, 1977], and other studies indicate that such basic expressions manifest across cultures [Ekman and Friesen, 1971]. Additionally, Ekman points out that there exist distinct, large codices of subtle expressions that are culturally dependent, and have not yet been scientifically formalized.

Transcultural subtleties will be highly relevant to nonverbal communications applications, and their implementation need not necessarily await the rigor of scientific formalization. Such complex problems in aesthetics and communications biomimetics are already solved frequently by animation artists, relying on natural human talents for communication instead of scientifically formalized principles.

There is an immense body of scientific study that focuses on the expressions of humans and their associated emotions. Ekman's FACS, already mentioned above, offers some quantification to the basics in Charles Darwin's book *The Expression of the Emotions in Man and Animals* [Darwin and Ekman, 1872/1998]. Here, Darwin divides expressions into groups of opposites, theorizing that evolution eked out this system for greatest communicative clarity. Loosely following Darwin's lead, expressions can be divided to three groups: happy expressions, unhappy expressions, and expressions that can be blended.

(1) Happy expressions include joy, love, laughter, and high spirits. They are indicated by raising the cheeks, with resultant "rising sun" shape imparted to the eyes, and the smile, which consists of raising the mouth's corners, and widening and opening the mouth. In laughter, these expressions are exaggerated and tend to incorporate characteristics of grief or horror expression.

(2) Unhappy expressions include sorrow and grief, anger and contempt, dejection, guilt, fear and horror. All of these involve the downward arc of a frown, and all tend to contort the forehead.

(3) Expressions that can be married with either happy or unhappy expressions include surprise (open, relaxed mouth, wide eyes), reflection, and embarrassment.

Ideally, these basic emotional expressions can be automated and controlled by software "instincts," to be circumstantially invoked by autonomous perception or animated into a sequence by a human animator. A graphic view of a basic face with root muscular structure is shown in Fig. 10.

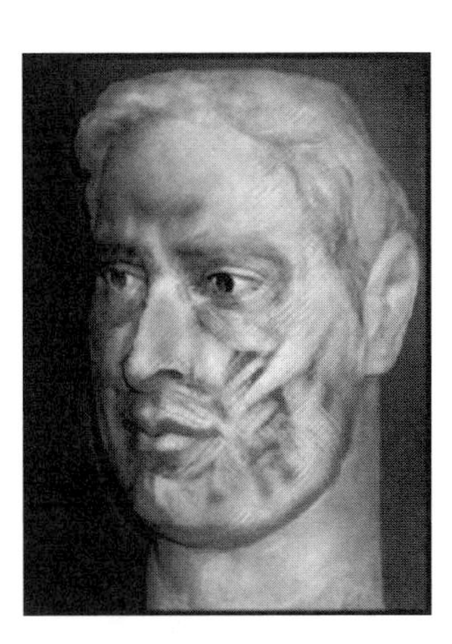

Figure 10: Basic face with root muscular structure; sculpture and illustration by David Hanson.

18.4.2.1 Body Expression and Gestures

From a distance, body gesture is all that reads of a character's movement (Fig. 11). Our visual intuition is keenly sensitive to the appearance of weight and balance; we just know if something is off. EAP applied with principles of dynamic stability could make all the difference. Our natural gestures include compliance (springiness), yet our movements are precise. One reason for this is that we have legion actuators and sensors, and distributed controls.

If EAP actuators become robust, inexpensive, small, and mass-produced, then perhaps arrays of thousands of sensors and actuators could be deposited in a robot's simulated flesh. In the meantime, mathematical principles derived from biology could allow us to approximate balance, for much improved visual effect and heightened precision [Chapter 3]. One goal could be to compile a database of gestures and expressions, with dynamic mechanical analysis, physiological analysis (part and parcel of the latter), and psychological analysis. Creating an automated codex for inputting and analyzing expressions, gestures, and various coordinated movements would be useful for this database.

Figure 11: Dynamic gestural figure with muscles exposed.

18.4.3 Verisimilitude vs. Abstraction

In the mid 1970's, Japanese roboticist Masahiro Mori speculated that if a robot were made to appear too realistic, but not perfectly so, people would reject and loathe the robots; the Robots would fall in the "Uncanny Valley." Many robotics researchers agree, and have chosen instead cartoon, mechanized, or animallike identities for their robots. The author has chosen the alternative path of verisimilitude, believing that the reason realism is so difficult is that it activates in humans expectations of a full-bandwidth nonverbal exchange. The author contends that by exploring this "Uncanny Valley," we can better understand ourselves, our aesthetics, and ultimately create more powerful art and entertainment.

18.5 A Recipe for Using EAP in Entertainment

Introducing new technology to an existing field (in this case entertainment) requires addressing technical challenges while providing results unobtainable by conventional means. In doing so, the technology needs to be reliable and affordable. Should EAP succeed in entertainment, it will find many uses aside from obvious narrative applications. EAP may also unleash humanoid GUI devices, telepresence tourism, and athletics-enhancement devices (e.g., flapping Icarus wings or gargantuan Hercules arms). A more immediate use of EAP could be in theme-park ride seats, leading to a more gripping sensory experience.

As the craze over Sony's AIBO demonstrates, there is a huge demand for ambulating character products. Walking and running figures require the same basic characteristics as for gestures—dynamic stability is paramount. While the science behind dynamic ambulation as a whole is still emerging, the dynamics of sprawled animal locomotion (lizards, cockroaches, etc.) are becoming more completely understood [Full, 1990]. Sprawled posture would be the most approachable form for EAP mobility in entertainment. Although no autonomous, dynamic biped yet exists, much headway is being made into engineering bipedal dynamic locomotion [Pratt, 2000]. Bipeds would be the more ambitious arena for EAP actuator exploration, but ultimately more rewarding for character animatronics. Figure 2 shows some existing biped projects.

Development of an anthroid face using EAP actuators is, arguably, the most important step in the introducing EAP technology to character animatronics. Currently the author is emulating facial anatomy with a robotic humanoid head, designed with dielectric elastomer EAP muscles in mind. A detailed description of the anthroid head identity emulation (IE) is provided in the next section.

18.6 Facial Expression Anthroid—Practical Test Bed for EAP

Facial expression anthroids are particularly interesting as a starting place for EAP actuation, as the force and performance requirements are smaller than those for locomotion and other tasks. While many tasks remain to be overcome for making EAP facial expression robust and otherwise feasible, one such anthroid head has been built by the author and is deployed for testing EAP artificial muscles (Fig. 12) at the Jet Propulsion Laboratory's Non-Destructive Evaluations and Advanced Actuators (NDEAA) Lab in the foothills of Pasadena [Bar-Cohen, 2002]. Such advanced materials with properties more like biological tissue may lead to tremendous leaps in biomimetic technology for entertainment [Helmer, 1998]. Considering the engineering challenges of natural facial expressions, such robots may prove useful by serving as the high-bar of achievement in robotics and EAP research.

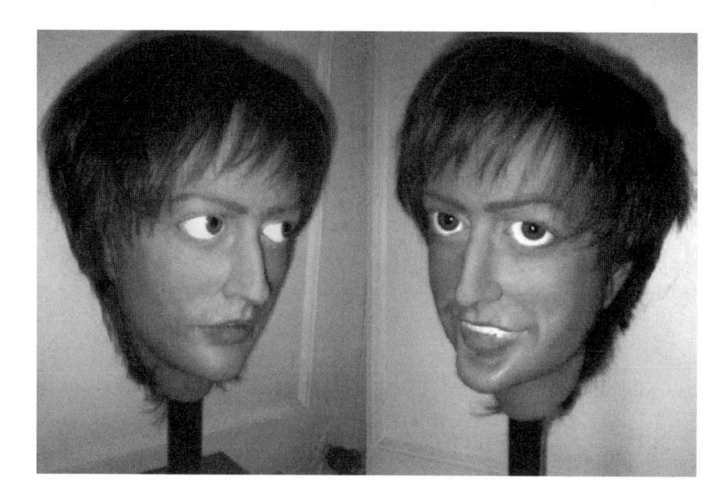

Figure 12: EAP actuator test platform built by author for JPL.

18.6.1 Android Head Structure and EAP Actuators

Here the structures of some of the robots built by the author over the last year are described. Since the IE robot was intended to be a work of fine art in its application (for example, the self portrait of the author, illustrated in Fig. 13), the first stage of IE's design involved the narrative and design conceptualization. Several questions were asked: How should the robot look? What movements and gestures are imperative to convey the identity of the subject? In addressing these questions, a series of sketches were produced, using both classical rendering tools of pencil, paints, and paper, and digital design tools of Photoshop and Maya 3D. Because the goal was to push the threshold of the robotic technology, a realistic form was chosen. Nevertheless, the author David Hanson rendered the

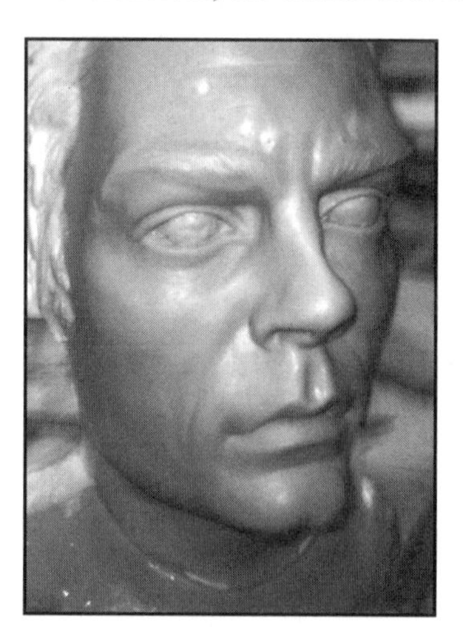

Figure 13: Robot self-portrait sculpted and instrumented by the author.

form by eye and with his hands; therefore, signature expressive distortions added interpretive personality to the work.

When building an expressive robot, one must have a strong visual and conceptual design as a foundation. The tools of classical art can be highly useful for this task. Consider the visual appearance of Michelangelo's *Pieta*: enormously full of life, almost superhuman in its expression; now imagine it furthered by effective robotic animation. Such a piece would be overwhelming in its expressive presence. The traditions of classical art and its training remain alive and indispensable throughout the world of design, including the animation and animatronics (themed robotics) industry.

During the design phase, one must ponder and devise the story of the piece, the purpose and function of the robot's interaction with people, and the complex cloud of implications of its visual and narrative concept. These considerations should also shape the design of the intelligence systems (including the "personality" of the robot), in accord with the visual and mechanical design. Given the complexity of this task, an iterative visual design process is useful if not indispensable. The process of mechanical, electronic, and control systems design was conducted in tandem with the visual and conceptual designs. For now, next considered will be the physical execution of the external visual design, starting with the sculpture.

To realize the design as a sculpted 3D form, the maximally subtle sculpting medium is desirable. At present, physical sculpting is faster and more responsive than any haptic sculpting interface system. Virtual sculpting is reaching maturity as a medium, but for now clay can be used more quickly, cheaply, and simply. Oil-based clays are the best choice, since they will not dry out and can be worked, reworked, and recycled. Before sculpting a large complex form, an armature of wood, steel, or other rigid material will be built to support the weight of the clay. The smaller IE faces shown in this article were sculpted sans armature. The 150% self-portrait, though, was rough-hewn from rigid urethane foam, a common insulation material; fine detail was then added using oil-clay. Schematically, IE is constituted of an artificial skin attached to a skulllike substrate and a muscular structure.

18.6.1.1 The Artificial Skin

To translate the sculpted animatronic figure into the biomimetic skin, a mold (negative impression) of the sculpture is made. Usually this mold will have an inner surface of an elastomer (in our case silicone), surrounded by a shell, or mother mold. This shell is composed of a hard, rigid material such as fiberglass, rigid urethane plastic, or classically, plaster reinforced with burlap. The rubber inner-layer will lock into the shell with keys, or raised bumps. The shell will be composed of several interlocking parts to avoid undercuts (regions where the sculpture or cast will lock in the mold); these parts will be held together with bolts.

For the use as artificial flesh and skin of the anthroid, numerous polymeric systems have been explored, including thermoplastic elastomers (TPEs), polysiloxanes (silicones), polyvinyl chloride, polyethylene, and polyurethane elastomers. Polymers for artificial skin require the elongation, flexibility, and appearance of the natural skin, and also must be able to maintain these features for a reasonable period of time in sunlight and other common environments. Of the synthetic elastomer options, silicones, polyurethanes, and TPEs have currently proven the most feasible for artificial skins. Given the largely liquid nature of the skin, materials that exhibit comparable ease of compression will better match the properties of skin. Polymers that rival human skin's elasticity, compressibility, and strength will require considerably less force to actuate. No solid elastomer can meet this challenge, due to adhesion of solid molecules to their nearest neighbors.

To address this challenge to the anthroids described in this chapter, in the spring of 2002 the author pursued a novel class of elastomers that would marry the highly compressible characteristics of a type of foam with the high-elongation properties of an elastomer. The resulting polymer, named "F'rubber" by the author as a contraction of "face" and "rubber," is comprised of a urethane elastomer matrix interspersed with a network of catalyzed urethane flexible foam. This material requires less than one-tenth of the force to move into expression compared to a solid elastomer skin of the same thickness and hardness. In the process of composition, a freshly catalyzed poly(urethane-urea) elastomer is introduced by mixing (while said elastomer is still liquid and not yet fully cured) into a freshly catalyzed poly(urethane-urea) foaming elastomer that is also still liquid and curing. The cured mixture expands with cells of gas distributed through the substance of the elastomer, a highly stable composition that alternates between discrete microscopic sections of nearly pure elastomer and discrete sections of foaming elastomer. In such instances, the material is easily compressed because it is permeated by gas-bubble voids, but remains highly elastic because of the action of areas or even networks of elastomer.

The "F'rubber" material elongates to 600% (as opposed to standard urethane foam's 120%), and has compression characteristics identical to those of the original foam rubber. The resulting facial expressions are actuated with hobby servos producing a mere 19 oz.-inches of torque at 6 V. This material promises to enable a powerful new class of highly expressive low-power mobile robots.

To effect the expressive movement, anchors in the skin provide mechanical connection points for attaching the expressions' actuation mechanisms, while gaps in the skull permit actuators housed inside the skull to effectuate the skin movement. Anchors were composed of loose-weave fabric cast directly into the skin. Such fabric spreads force over a larger skin area, while allowing the elastomer to retain its strength by cross-linking through the voids in the fabric's weave. Because the fabric inhibits the elongation properties of the elastomer in the anchor zone, the size and placement of the anchors was critical to allowing the skin to stretch into expression properly. The anchors are pinned into the mold, over which the skin then is poured, embedding the anchors in the skin (see Fig. 14).

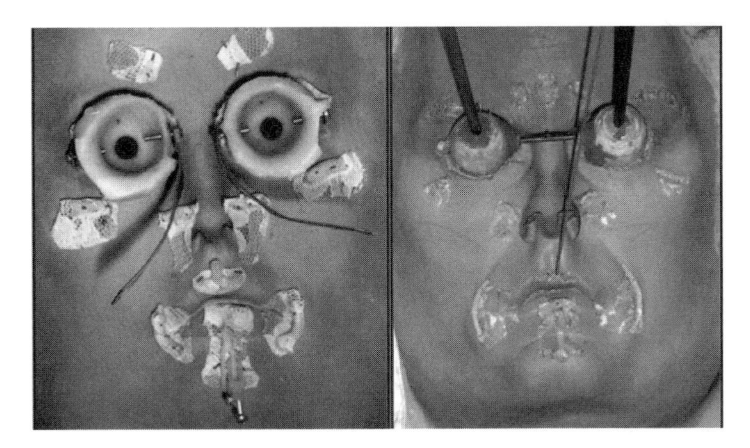

Figure 14: Force distribution anchors planted in face molds.

Several methods exist to impart movement to the anchors. The most common method is to attach linear actuators directly to the anchors; a similar approach attaches gimbals to the anchors, thus effectuating two axes of motion (three if the gimbal sits on an extending boom). In the author's experience, a third method works better than the previous two: composite fiber cables are cast into the skin, in lines that correlate to the contraction vectors of natural muscles. At one end, each cable attaches to an anchor; at the other, the cable slides through the skull to reach an actuator. Then, when actuated, the cable slides freely through the skin, transmitting force directly to the anchor. This method is more biologically accurate, in that it effectuates motion from the interior of the skin-mass, as do biological muscles; also, it allows layered vectors of actuation, there-by mimicking the layers of natural facial musculature. Although actuation is presently affected by conventional servomotors EAP actuators could replace these actuators either by being directly embedded in the skin material or by pulling on the cables from the interior of the skull.

To embed the cables and anchors, a first layer of 10-durometer shore-A rubber is painted into the face's mold in 2 to 3 layers, and is pigmented to suit the artistic design. While still tacky, the rubber actuation elements are placed upon this painted epidermis, and the force transmission elements are run through the skull's bushing and slots. At this time the skull is placed and registered, and all openings are plugged with wax to prevent leakage. Next, F'rubber is poured to fill the skin-form, and allowed to set.

Once poured, the skin is attached to a skulllike substrate. In the process of sculpting this skull, the author achieved anatomically correct proportions relative to the original portrait by using forensics data [Helmer, 1998] as reference for skin thicknesses. Notating these thicknesses with depth markers, the skull was sculpted in reverse (in negative) inside the mold of the sculpture of the face. The resulting variable skin thickness enables more lifelike dynamic skin distortions, more closely resembling natural facial expression. The skull sculpture was then itself molded, and thin composite-plastic casts were made for the final robot (Fig. 15). The skull also serves as an inner core to the mold so that when cast, the skin's inner and outer surfaces conforms tightly to both the exterior artistic design and the inner mechanism. Moreover, the skull also serves as the mechanical frame for mounting actuators and sensors.

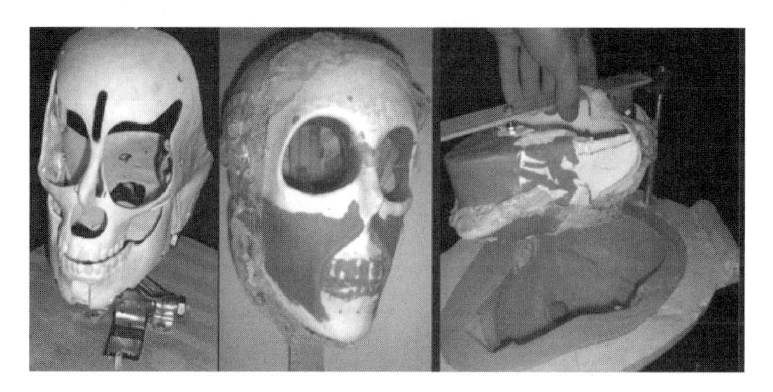

Figure 15: Mechanical frame, mold core, and mold registration system.

18.6.1.2 The IE Muscular Structure

Mechanically, the identity emulation heads are built using standard bearings, u-joints, and other common machine components. If using conventional skin materials, one would need to use pneumatic or hydraulic actuators, which provide much greater force and speed. Neither pneumatics nor hydraulics are well suited for mobile robotics, however, since these actuators require large, off-board compressors to pressurize their air or fluid. Thanks to the highly plastic characteristics of the skin materials, mere hobby servos are used to actuate the expressions. To further enable mobile paralinguistic robotic-interface devices, EAP actuators may provide the next step of improved power densities for mobile animatronics [Hanson and Pioggia, 2001]. During the interim, the IE project seeks to decrease the weight by attempting to simplify the mechanical system (Fig. 16), thinning the skin, and using lightweight composite materials. When the mechanism and skin are assembled, the figure is dressed out for final display, and the AI controls are implemented and tested.

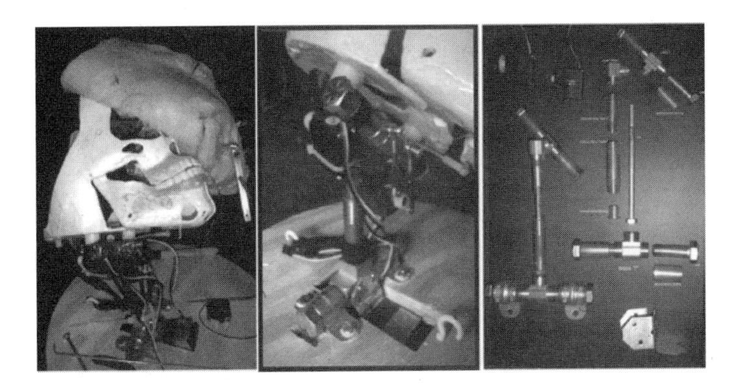

Figure 16: Robot mechanical system.

It seems that there are seven principal facial expressions that we are generally quite good at recognizing and which appear to be universal: anger, disgust, fear, happiness, interest, sadness, and surprise. Facial muscles responsible for expression can be divided in five groups: epicranius muscle, muscles of the eyelids, muscles of the lips, auricularis muscles, and muscles of the nose. In humans, all these muscles are controlled via the facial nerve. It is important to note that no muscle acts by itself. When one muscle contracts or draws together its fibers, it activates other opposing muscles, which in turn modify the action of the original contracting muscle. Most of these are small, thin or deeply embedded in fatty tissue.

Searching to replicate, with a few servos, some of these expressivities of a human face, exclusively the muscles that endow a good contribution to the expressivity were emulated while those with less influence were overlooked. In particular, a test of IE expressivity was realized by replacing the principal muscles of the lips with servo mechanisms. The muscles of the lips include the zygomaticus muscle, the buccinator muscle, the triangular muscle, the depressor labii inferioris muscle and the orbicularis oris muscle. Movements were performed by using a MINISSC II Serial Servo Controller connected to a PC. Beginning in February 2002, five prototypes were realized in 12 months (Fig. 17).

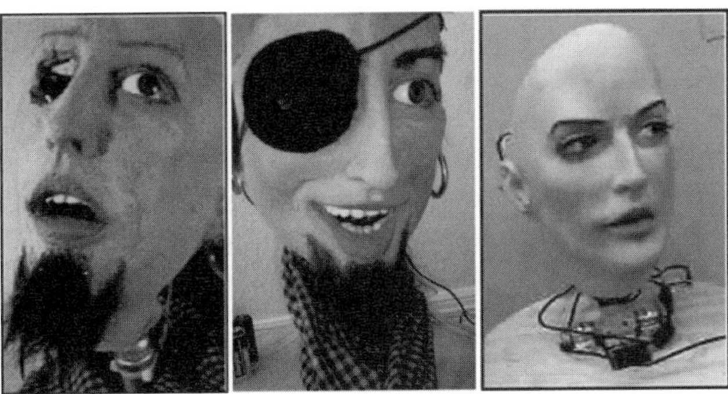

Figure 17: Robots sculpted and instrumented by author; painting and decoration done in collaboration with Kristen Nelson.

18.6.1.3 Toward AI Integration

A major goal of the IE project at the University of Texas at Dallas is the integration of a complete multimodal intelligence and perceptual control system to govern the facial robots in sociable interactions. At this point, the IE robot effectively tracks colored blobs using the CMU-cam system, and the robot can also lip-sync to prerecorded speech. We are working on implementing rudimentary natural language interaction using Carnegie Mellon's Sphinx, automatic speech recognition (ASR) software, head tracking, and facial expression recognition using the Eyematic SDK. The robot servos are currently controlled using OOpic microcontrollers.

For future improvements in the emulation of nonverbal expressions, the following issues will be considered and addressed. The complex modes of human-facial expression largely arise from the microscopic structural and physical characteristics of the organic tissues that compose the skin. Conventional elastomers cannot match the extremely soft yet tough characteristics of the human skin. Nor can it match the shearing effect among layers, intimately affected by differing Young's modulii of skin and flesh's various internal tissues. So it should be clear that this elaborate web of liquid-saturated cellular tissue would be tricky at best to emulate with the elastomers that are the staple of animatronics and biologically inspired robotics.

For animatronics, speech emulation is of great interest. Speech movements are borne primarily through the muscles surrounding the lips (orbiculis oris), lateral muscles in the cheeks (risorius muscles), and movements of the jaw. The muscles around the lips produce the most expansive panoply of movements, both emotive and speech-related. Simulating these will be the most advanced aim in animating humanoid faces. The following facial formations are the movement primitives that comprise basic speech: "ah, eh, oh, uh, oo, ee, m, n, f, j, k, l, s, sh, t, th, p, and r." Mastering these movements with an EAP humanoid would be a good start. More subtle movements could be added as the field of EAP animatronics evolves. Emotional expressions exhibited during speech go beyond the basic expressions discussed in Sec. 18.4.1. These often subtle and quick gestures tend to be idiosyncratic to individuals, and are vital to expressing individuality. Such distinguishing quirks are the purview of animators but could be supported or advanced by automation. A first approach might use speech-recognition software to break down audio into phonemes, and marry these with movement primitives to be synchronized with the timing of the audio sample. This method is derived from the traditional animation method to depict speech.

Because speech occurs in every state of emotion, it must become possible to merge speech-related movements with the sundry emotional expressions. An automated system is under construction to orchestrate this blending.

18.7 Conclusion

EAP may change animatronics, but much effort will be required. In addition to developing better EAP actuation, a discipline of viscoelastic engineering will need to supplant the traditional engineering of rigid structures. Control strategies also need to be addressed. If one single effective EAP actuator product is introduced into the entertainment market, it could generate a flurry of interest in further research and development. To be certain, a symbiosis between academia and the animatronics community is crucial at this stage to tune EAP toward entertainment needs. Many other technologies also support the use of EAP in entertainment. Of particular interest are artificial intelligence to imbue the animatronics with life and shape deposition manufacturing that allows rapid prototyping of viscoelastic mechanisms with imbedded actuators.

Successful EAP animatronics will require a vigorous study of human physiology and physical systems to derive applicable engineering principles. A healthy start has been made in an EAP-actuated anthroid head; but much work must be done before this system can be useful to entertainment. Though many obstacles lie in the way of useful EAP actuation for entertainment, now is the time for action. This technology represents so many advantages over traditional actuation, it easily warrants the adjective “revolutionary.” Those in the vanguard of EAP’s development for entertainment will gain the strategic advantage of this profound new technology and its recent breakthroughs.

18.8 Acknowledgement

The author would like to thank Dr. Yoseph Bar-Cohen, this book's editor, for his guidance, patience, and tremendously helpful suggestions. Additional thanks go to Elaine Hanson, Kristen Nelson, Thomas Linehan, Jochen Triesch and Alice O’Toole.

18.9 References

Amon, C.H., J.L. Beuth, R. Merz, F.B. Prinz, and L.E. Weiss, “Shape deposition manufacturing with microcasting: processing, thermal and mechanical issues,” *Journal of Manufacturing Science and Engineering*, (1996).

Archer, K. M. “Craniofacial Reconstruction Using Hierarchical B-Spine Interpolation,” Master’s Thesis, McGill University, 1997.

Bailey, S.A., J.G. Cham, M.R. Cutkosky, and R.J. Full. “Biomimetic robotic mechanisms via shape deposition manufacturing,” Robotics Research: the Ninth International Symposium, J. Hollerbach and D. Koditschek (Eds), Springer-Verlag, London, (2000) pp. 403-410.

Bar-Cohen, Y. and C. Breazeal, Eds., *Biologically Inspired Intelligent Robots*, SPIE Press, Bellingham,WA (2003).

Bar-Cohen, Y., (Ed.), *Proceedings of the SPIE's Electroactive Polymer Actuators and Devices Conf.*, 6th Smart Structures and Materials Symposium, SPIE Proc. Vol. 3669 (1999), pp. 1-414.

Bar-Cohen, Y., (Ed.), *Proceedings of the SPIE's Electroactive Polymer Actuators and Devices Conf.*, 7th Smart Structures and Materials Symposium, SPIE Proc. Vol. 3987 (2000) pp. 1-360.

Baughman, R.H., C. Cui, A. A. Zakhidov, Z. Iqbal, J. N. Basrisci, G. M. Spinks, G. G. Wallace, A. Mazzoldi, D. de Rossi, A. G. Rinzler, O. Jaschinski, S. Roth and M. Kertesz, "Carbon nanotube actuators," *Science*, Vol. 284, (1999) pp. 1340-1344.

Breazeal, C., *Designing Sociable Robots*, MIT Press (2002).

Brooks, RA, "Intelligence without Reason." MIT AI Lab Internal Report, (1991).

Brown, I.E., Loeb, G.E. "The physiological relevance of potentiation in mammalian fast-twitch skeletal muscle," *Soc. Neurosci.*, 23:1049, (1997).

Cassell, J. "Embodied Conversational Agents," *AI Magazine*, 22:4 (2001).

Darwin, C., Ekman, P. (Ed.), *The Expression of the Emotions in Man and Animals*, Oxford University Press, New York (1998/1872).

Dennis, R.G., Kosnik, P.E., Faulkner, J.A., Kuzon, W.M. "Contractile function of in vitro tissue-engineered skeletal muscle constructs." 44th Annual Meeting, Plastic Surgery Research Council, pp. 22-25 (1999).

Dickinson, M.H. Farley, C.T., Full, R.J., Koehl, M. A. R., Kram R., and Lehman, S., "How animals move: An integrative view," *Science* 288, (2000), pp. 100-106.

Ekman and Friesen, *Basic Emotions* (1971).

Ekman,P., "The argument and evidence about universals in facial expressions of emotion," Wagner,H., Manstead, A., (Eds), Handbook of Psychophysiology, John Wiley, London (1989).

Engel, J., Chen, J., Liu, C., "Development of polyimide flexible tactile sensor skin", J. Micromech. Microeng. 13 (May 2003) 359-366.

Full, R. J., Stokes, D. R., Ahn, A. and Josephson, R. K., "Energy absorption during running by leg muscles in a cockroach," *J. Exp Bio.* 201, pp. 997-1012, (1998).

Full, R.J and Meijer. K., "Artificial muscles and natural actuators from frogs to flies," *Proceedings of SPIE*, 2987, pp. 2-9 (1999).

Full, R.J. and Koditschek, D. E., "Templates and anchors–neuromechanical hypotheses of legged locomotion on land," *J. Exp Bio.* 202, pp. 3325-3332, (1999).

Full, R.J., and Tu, M.S., "Mechanics of six-legged runners," *J. Exp. Biol.* Vol.148, pp. 129-146, (1990).

Full, R.J., "Biological inspiration: Lessons from many-legged locomotors," in "Robotics Research 9th International Symposium," J. Hollerbach and D. Koditschek, Eds., Springer-Verlag London, pp. 337-341 (2000).

Hanson, D. and Pioggia G., "Entertainment Applications for Electrically Actuated Polymer Actuators," *Electroactive Polymer (EAP) Actuators as Artificial Muscles,* SPIE Press, Bellingham WA, Ch. 18, (2001).

Hanson, D., Pioggia G., Bar-Cohen Y., De Rossi D., "Androids: application of EAP as artificial muscles to entertainment industry," *Proceedings of SPIE*, 4329, pp. 375-379 (2001).

Higgins, S.J., Eccleston, W., Sedgi, N. and Raja M, "Plastic electronics" Education in Chemistry, Royal Society of Chemistry (2003).

Kull, F.J. and Endow, S.A. (2002) Kinesin: switch I & II and the motor mechanism. *Journal of Cell Science.* In press.

Lamb, S. M., *Pathways of the Brain, The Neurocognitive Basis of Language*, Amsterdam & Philadelphia: John Benjamins Publishing Co., (1999).

Menzel, P., D'Aluisio, F. *Robo sapiens: Evolution of a New Species*, Boston, MIT Press, (2000).

O'Toole, A., Roark, D., Abdi, H., "Recognizing moving faces:A psychological and neural synthesis," The University of Texas at Dallas, March 13, 2002.

Pelrine, R., R. Kornbluh, Q. Pei, and J. Joseph, "High speed electrically actuated elastomers with strain greater than 100%," *Science*, Vol. 287, (2000) pp. 836-839.

Pioggia, G., Di Francesco F., Geraci C., Chiarelli P., De Rossi D., "Humanlike android face equipped with EAP artificial muscles to endow expressivity," *Proceedings of SPIE,* 4329 pp. 350-356, 2001.

Pratt, J. "Exploiting inherent robustness and natural dynamics in the control of bipedal walking robots," Ph.D. thesis, Massachusetts Institute of Technology, (2000).

Raibert, M. H., *Legged Robots that Balance*, MIT Press (1986).

Schraft, R.D., and G. Schrmierer, *Service robots*, A K Peters, Ltd., Natick, MA (2000), pp.145-153.

Steeves, C., Buehler, M., Penzes, S. G., "Dynamic Behaviors for a hybrid leg-wheel mobile platform," Proceedings of SPIE, 4715, pp 75-86, (2002).

Tomkins, S. S., "Determinants of Affects," *AIG-I*, pp. 248-258. New York: Springer (1962).

Wada, Y., Bell, A., Valentin, T., Neelavar, R., Freire, S., Schmidt, J., Hamasaki, T., Montemagno, C., Satir, P., "Engineering the axonemal nanomachine for nanomounting," 11th Foresight Conference on Molecular Nanotechnology, 2003.

Zhang, Q. M., V. Bharti, and X. Zhao, "Giant electrostriction and relaxor ferroelectric behavior in electron-irradiated poly(vinylidene fluoride-trifluorethylene) copolymer," *Science*, Vol. 280, pp. 2101-2104 (1998).

CHAPTER 19

Haptic Interfaces Using Electrorheological Fluids

Constantinos Mavroidis
Rutgers University, The State University of New Jersey

Yoseph Bar-Cohen
Jet Propulsion Laboratory/California Institute of Technology

Mourad Bouzit
Rutgers University, The State University of New Jersey

19.1 Introduction

Over the past 50 years, it has been known that there are liquids that respond mechanically to electrical stimulation. These liquids, which have attracted a great deal of interest from engineers and scientists, change their viscosity electroactively. These electrorheological fluids (ERF) exhibit a rapid, reversible, and tunable transition from a fluid state to a solid-like state upon the application of an external electric field [Phule and Ginder, 1998]. Some of the advantages of ERFs are their high-yield stress, low-current density, and fast response (less than 1 ms). ERFs can apply very high electrically controlled resistive forces while their size (weight and geometric parameters) can be very small. Their long life and ability to function in a wide temperature range (as much as –40°C to +200°C) allows for the possibility of their use in distant and extreme environments. ERFs are also nonabrasive, nontoxic, and nonpolluting, meeting health and safety regulations. ERFs can be combined with other actuator types such as electromagnetic, pneumatic, or electrochemical actuators so that novel, hybrid actuators are produced with high-power density and low-energy requirements. The electrically controlled rheological properties of ERFs can be beneficial to a wide range of technologies requiring damping or resistive force generation. Examples of such applications are active vibration suppression and motion control. Several commercial applications have been explored, mostly in the automotive industry, such as ERF-based engine mounts, shock absorbers, clutches and seat dampers. Other applications include variable-resistance exercise equipment, earthquake-resistant tall structures, and positioning devices [Phule and Ginder, 1998].

While ERFs have fascinated scientists, engineers, and inventors for nearly 50 years, and have given inspiration for developing ingenious machines and mechanisms, their applications in real-life problems and the commercialization of ERF-based devices has been very limited. There are several reasons for this. Due to the complexity and nonlinearities of their behavior, their closed-loop control is a difficult problem to solve. In addition, the need for high voltage to control ERF-based devices creates safety concerns for human operators, especially when ERFs are used in devices that will be in contact with humans. Their relatively high cost and the lack of a large variety of commercially available ERFs with different properties to satisfy various design specifications have made the commercialization of ERF-based devices unprofitable. However, research on ERFs continues intensively and new ERF-based devices are being proposed [Tao, 1999]. This gives rise to new technologies that can benefit from ERFs. One such new technological area that will be described in detail here is virtual reality and telepresence, enhanced with haptic (i.e., tactile and force) feedback systems, and for use in medical applications, for example.

In this chapter, we first present a review of ERF fundamentals. Then, we discuss the engineering applications of ERFs and, more specifically, their application in haptics. We describe in detail a novel ERF-based haptic system called MEMICA (remote mechanical mirroring using controlled stiffness and actuators) that was recently conceived by researchers at Rutgers University and the Jet Propulsion Laboratory [Bar-Cohen et al., 2000a]. MEMICA is intended to provide human operators an intuitive and interactive feeling of the stiffness and forces in remote or virtual sites in support of space, medical, underwater, virtual reality, military, and field robots performing dexterous manipulation operations. MEMICA is currently being used in a system to perform virtual telesurgeries as shown in Fig. 1 [Bar-Cohen et al., 2000b]. The key aspects of the MEMICA system are miniature electrically controlled stiffness (ECS) elements and electrically controlled force and stiffness (ECFS) actuators that mirror the stiffness and forces at remote/virtual sites. The ECS elements and ECFS actuators, which make use of ERFs, are integrated on an instrumented glove. Forces applied at the robot end-effector will be reflected to the user using this ERF device where a change in the system viscosity will occur proportionally to the force to be transmitted. This chapter, as a case study on the design of ERF-based devices, describes the design, analytical modeling, and experiments that are currently underway to develop such ERF-based force feedback elements and actuators.

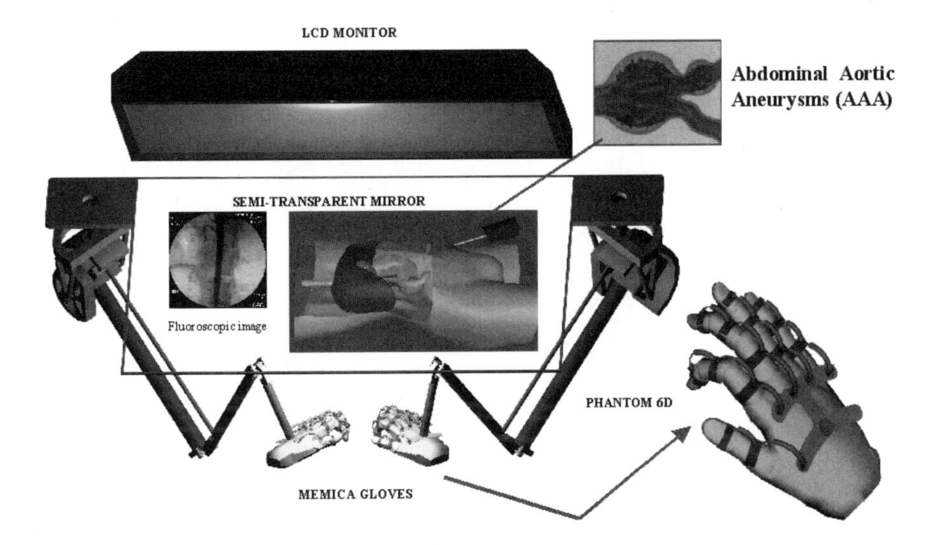

Figure 1: Performing virtual reality medical tasks via the electrorheologcal fluid based MEMICA haptic interface.

19.2 Electrorheological Fluids

Electrorheological fluids are fluids that experience dramatic changes in rheological properties, such as viscosity, in the presence of an electric field. Winslow first explained the effect in the 1940s using oil dispersions of fine powders [1949]. The fluids are made from suspensions of an insulating base fluid and particles on the order of 0.10 to 100 μm in size. The volume fraction of the particles is between 20% and 60%. The electrorheological effect, sometimes called the *Winslow* effect, is thought to arise from the difference in the dielectric constants of the fluid and particles. In the presence of an electric field, the particles, due to an induced dipole moment, will form chains along the field lines, as shown in Fig. 2. The structure induces changes to the ERF's viscosity, yield stress, and other properties, allowing the ERF to change consistency from that of a liquid to something that is viscoelastic, such as a gel, with response times to changes in electric fields on the order of milliseconds. Figure 3 shows the fluid state of an ERF without an applied electric field and the solid-like state (i.e. when an electric field is applied). Good reviews of the ERF phenomenon and the theoretical basis for ERF behavior can be found in Block and Kelly [1988], Gast and Zukoski [1989], Weiss et al. [1994], and Conrad [1998].

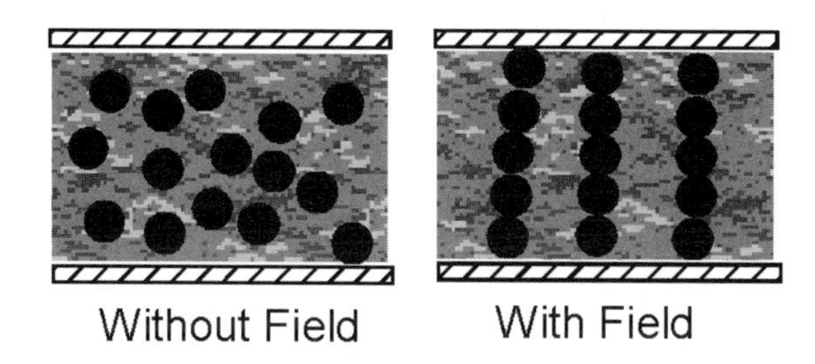

Figure 2: Particle suspension forms chains when an electric field is applied.

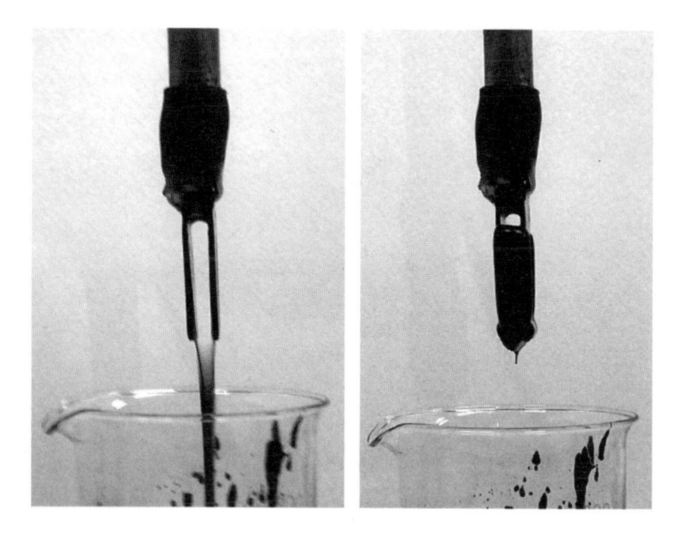

Figure 3: Electrorheological fluid at reference (left) and activated states (right). (Courtesy of Smart Technology Ltd, UK.)

Under the influence of an electric field, the ERF alters its state from a Newtonian oil to a non-Newtonian Bingham plastic. As a Bingham plastic, the ERF exhibits a linear relationship between stress and strain rate like a Newtonian fluid, only after a minimum required yield stress is exceeded. Before that point, it behaves as a solid. At stresses higher than this minimum yield stress, the fluid will flow, and the shear stress will continue to increase proportionally with the shear strain rate (see Fig. 4), so that:

$$\tau = \tau_y + \mu\dot{\gamma} \tag{1}$$

where τ is the shear stress, τ_y is the yield stress, μ is the dynamic viscosity, and γ is the shear strain. The dot over the shear strain indicates its time derivative, the shear rate. An interesting phenomenon that has been observed by many researchers is that the dynamic viscosity becomes negative at high fields (see Fig. 4). This phenomenon can be explained by assuming that fewer, or weaker, bonds are formed at higher shear rates, thus giving a smaller total yield stress and the effect of negative dynamic viscosity [ER Fluid Developments Ltd., 1998].

Yield stress τ_y and dynamic viscosity μ are two of the most important parameters that affect the design of ERF-based devices. The dynamic viscosity μ is mostly determined by the base fluid and the electric field. The field induced yield stress τ_y depends on the electric field strength. For this dependence, some theoretical models have been derived but neither one is yet able to reflect these relations properly. As a rule of thumb, one can assume that the yield stress increases quadratically with the electric field strength [Lampe, 1997].

There are two important values for the yield stress: the static yield stress $\tau_{y,s}$ and the dynamic yield stress $\tau_{y,d}$. The static yield stress is defined as the value of stress needed to initiate flow, i.e., the stress needed to change from solid to liquid. The dynamic yield stress is the value of stress needed in zero-strain rate conditions to go from a liquid to solid. Which one is larger is different from fluid to fluid. Most of the time, the static yield stress is higher than the dynamic yield stress. This phenomenon, called "stiction," is highly dependent on the particle size and shape [Weiss et al., 1994].

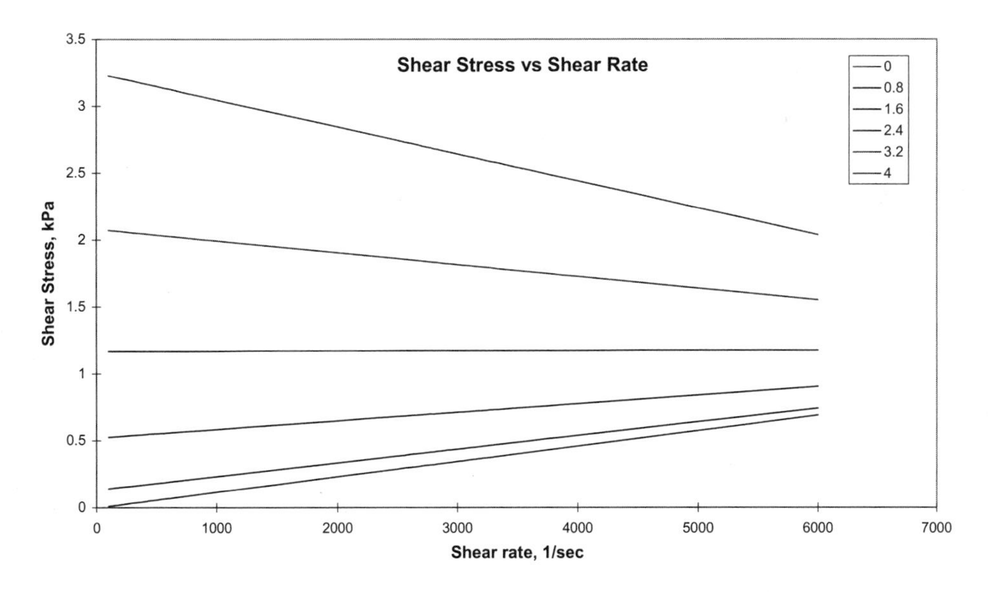

Figure 4: Relationship between shear stress and shear strain rate in an ERF. (Courtesy of Smart Technology Ltd, UK.)

Another important parameter that needs to be known for ERFs is the current density J, defined as the current per unit electrode area. This parameter is needed to estimate the power consumption of ERF-based devices. Measurement of electric current through ERF materials is believed to be the result of charge leakage between particles [Weiss et al., 1994].

ERF properties change with temperature and this can have a tremendous effect on the performance of ERF-based devices. "Good" ERFs should show constant properties over a large range of temperatures. There is no unified model describing the temperature dependence of the parameters of ERFs. This temperature dependence changes from fluid to fluid. The biggest temperature problem for ERFs results from the huge increase of current density with increasing temperatures. This increases power consumption but also increases safety concerns for human operators of ERF devices.

A good database of commercially available ERFs, including property comparison tables, can be found in Lampe [1997]. A review of the material compositions for ERF patents can be found in Weiss et al. [1994]. As an example of ERFs we present here, the electrorheological fluid LID 3354, manufactured by ER Fluid Developments Ltd. [ER Fluid Developments Ltd, 1998], is used in the work presented in the next sections.

LID 3354 is an electrorheological fluid made of a 35% volume of polymer particles in fluorosilicone base oil. It is designed for use as a general-purpose ER fluid with an optimal balance of critical properties. Its physical properties are density: 1.46×10^3 kg/m^3; viscosity: 125 mPa sec at 30°C; boiling point: > 200°C; flash point: >150°C; insoluble in water; freezing point: < –20°C. The field dependencies for this particular ERF are

$$\tau_{y,s} = C_s\left(E - E_{\text{ref}}\right), \quad \tau_{y,d} = C_d E^2, \quad \mu = \mu_o - C_v E^2, \tag{2}$$

where μ_o is the zero field viscosity; C_s, C_d, C_v, and E_{ref} are constants supplied by the manufacturer. The subscripts s and d correspond to the static and dynamic yield stresses. The formula for static yield stress is only valid for fields greater than E_{ref}. Figures 5(a), (b), and (c) are a graphical representation of Eq. (2) for the ERF LID 3354. Figure 5(d) shows the dependence of the current density at 30°C as a function of the field. Figure 5(e) shows the coefficient C_d of Eq. (2) as a function of temperature.

Control over a fluid's rheological properties offers the promise of many possibilities in engineering for actuation and control of mechanical motion. Devices that rely on hydraulics can benefit from ERF's quick response times and reduction in device complexity. Their solid-like properties in the presence of a field can be used to transmit forces over a large range, and they have a large number of other applications. A good description of the engineering applications of ERFs can be found in Duclos et al. [1992] and Coulter et al. [1994]. Devices designed to utilize ERFs include engine mounts [Sproston et al., 1994], active

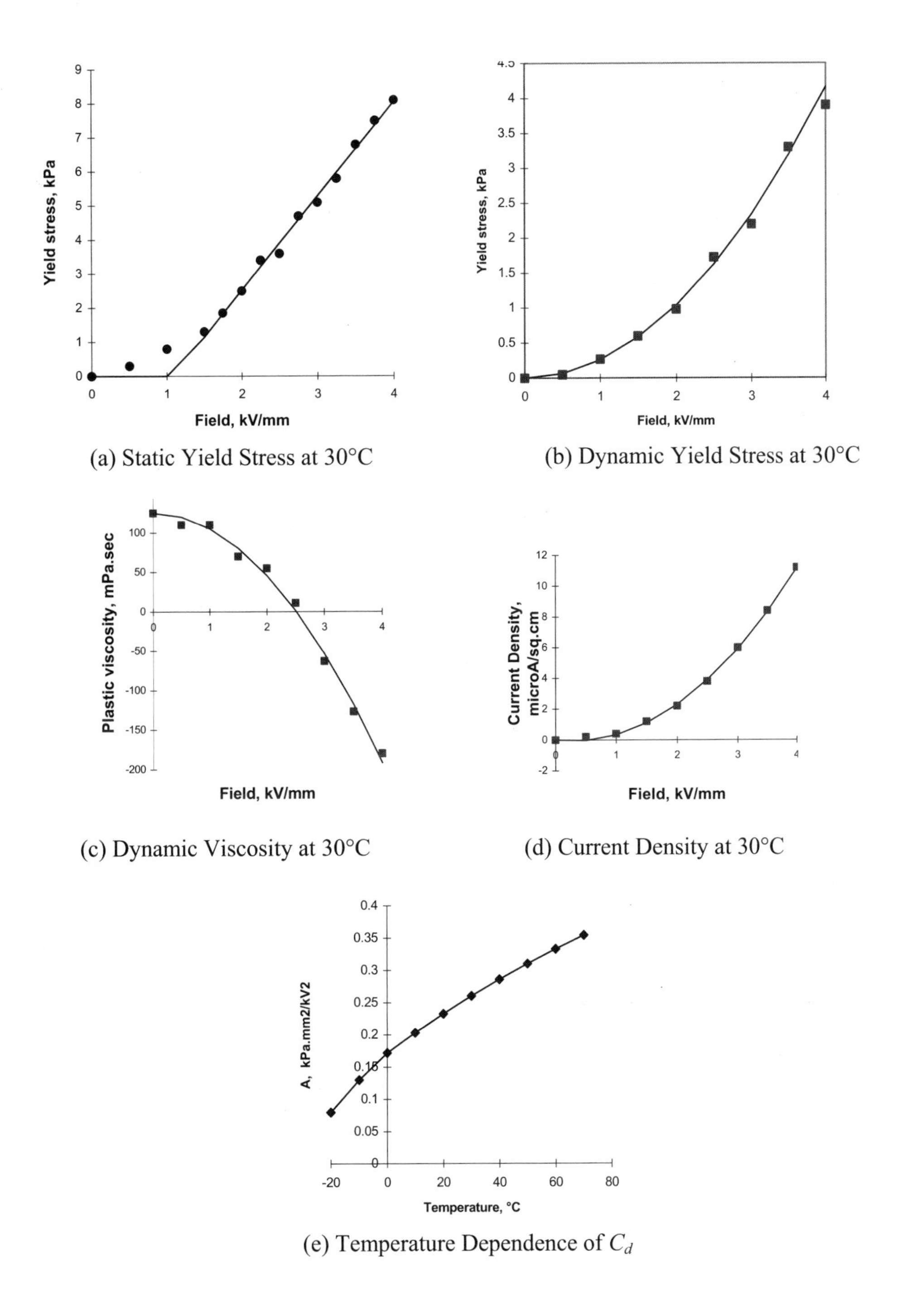

Figure 5: Technical information diagrams for the ER Fluid LID 3354 [ER Fluid Developments Ltd., 1998]. (Courtesy of Smart Technology Ltd, UK.)

dampers and vibration suppression [Choi, 1999], clutches [Bullough et al., 1993], brakes [Seed et al., 1986], and valves [Coulter et al., 1994]. An important engineering application of ERFs is vibration control; a good review of the subject can be found in Stanway et al. [1996]. The application of ERFs in robotics has been very limited. They have been used mainly as active dampers for vibration suppression [Ghandi et al., 1987; Furusho et al., 1997]. Recently, an ERF-based safety-oriented mechanism has been proposed for human-robot cooperative systems [Arai et al., 1998].

19.3 Haptic Interfaces and Electrorheological Fluids

To address the need for interacting with remote and virtual worlds, the engineering community has started developing haptic (tactile and force) feedback systems [Burdea, 1996]. At the present time, haptic feedback is a less-developed modality for interacting with remote and virtual worlds compared with visual and auditory feedback. Thus, realism especially suffers when remote and virtual tasks involve dexterous manipulation or interaction in visually occluded scenes. A very good description of the current state of the art in haptic and force feedback systems can be found in Burdea [1996] and Bar-Cohen et al. [2000c].

Tactile sensing is created by skin excitation that is usually produced by devices known as "tactile displays." These skin excitations generate the sensation of contact. Force-sensitive resistors, miniature pressure transducers, ultrasonic force sensors, piezolectric sensors, vibrotactile arrays, thermal displays, and electrorheological devices are some of the innovative technologies that have been used to generate the sensation of touch. While tactile feedback can be conveyed by the mechanical smoothness and slippage of a remote object, it cannot produce rigidity of motion. Thus, tactile feedback alone cannot convey the mechanical compliance, weight, or inertia of the virtual object being manipulated [Burdea, 1996].

Force-feedback devices are designed to apply forces or moments at specific points on the body of a human operator. The applied force or moment is equal or proportional to a force or moment generated in a remote or virtual environment. Thus, the human operator physically interacts with a computer system that emulates a virtual or remote environment. Force-feedback devices include portable and nonportable interfaces. Force-feedback joysticks, mice [Immersion Corp., 2001], and small robotic arms such as the Phantom [Sensable Technologies, 2001] are nonportable devices that allow users to feel the geometry, hardness, and/or weight of virtual objects.

Portable systems are force-feedback devices that are *grounded* to the human body. They are distinguished as *arm-exoskeletons* if they apply forces at the human arm and as *hand masters* if they apply forces at a human's wrist and/or palm. Portable hand masters are haptic interfaces that apply forces to the human hand while they are attached at the human operator forearm. In most cases, these systems look like gloves where the actuators are placed at the human forearm, and forces are transmitted to the fingers using cables, tendons and pulleys. The

CyberGrasp is an example of such a system. It is a lightweight, force-reflecting exoskeleton glove that fits over a CyberGlove and adds resistive force feedback to each finger via a network of tendons routed around an exoskeleton [Virtual Technologies, 2001]. The actuators are high-quality dc motors located in a small enclosure on the desktop. The remote reaction forces can be emulated very well; however, it is difficult to reproduce the feeling of "remote stiffness."

To date, there are no effective commercial unencumbering haptic feedback devices for the human hand. Current hand master haptic systems, while they are able to reproduce the feeling of rigid objects, present great difficulties in emulating the feeling of remote/virtual stiffness. In addition, they tend to be heavy and cumbersome with low bandwidth, and they usually allow limited operator workspace.

During the last 10 years, some researchers have proposed the use of ERFs in an effort to improve the performance of haptic interfaces. There are many properties of ERFs that can greatly improve the design of haptic devices. Their high-yield stress, combined with their small sizes, can result in miniature haptic devices that can fit easily inside the human palm without creating any obstructions to human motion. ERFs do not require any transmission elements to produce high forces, so direct-drive systems can be produced with less weight and inertia. The possibility of controlling the fluids' rheological properties gives designers of ERF-based haptic system the possibility of controlling the system compliance, and hence, they are able to accurately mirror remote or virtual compliance. Finally, ERFs respond almost instantly—in milliseconds—which can permit very high bandwidth control important for mirroring fast motions. The only concern that a designer of ERF-based haptic interfaces may have is the need for high voltages to develop the forces and compliance required. This has two consequences: (a) it increases the complexity of the electronic system needed to develop the high voltage and (b) it raises safety concerns for the human operator. Both issues can be solved easily with modern electronic circuit design techniques. Nowadays, low-power, small-size circuits can be used to generate the required high voltage using a very low current on the order of micro-amps. Consequently, the required power becomes extremely low—in the order of milliwatts—posing no hazard for human operators.

Kenaley and Cutkosky were the first to propose the use of ERFs for tactile sensing in robotic fingers [1989]. Based on that work, several researchers proposed the use of ERFs in tactile arrays to interact with virtual environments [Wood, 1998] and also as assistive devices for the blind to read the Braille system. The first to propose this application of ERFs was Monkman [1992]. Continuing this work, Taylor and his group at the University of Hull, UK, developed and tested experimentally a 5 × 5 ERF tactile array [Taylor et al., 1996]. Furusho and his group at Osaka University in Japan developed an ERF-based planar force-feedback manipulator system that interacts with a virtual environment [Sakaguchi and Furusho, 1998]. This system is actuated by low-inertia motors equipped with an ER clutch. An ERF-based force-feedback joystick has been developed at the Fraunhofer-Institute in Germany. The joystick consists of a ball and socket joint where ERF has been placed in the space between the ball and the socket. The operator feels a resistive force to his/her motion resulting from the controlled viscosity of the ERF [Böse et al., 2000].

Finally, researchers at Rutgers University and JPL are developing an ERF-based force-feedback glove called MEMICA [Bar-Cohen et al., 2000; Mavroidis et al., 2000; Pfeiffer et al., 1999]. This system is described in detail in the next section of this chapter.

19.4 MEMICA Haptic Glove

This section presents the development of a haptic interfacing mechanism that will enable a remote operator to "feel" the stiffness and forces at remote or virtual sites. These interfaces are based on novel mechanisms conceived by JPL and Rutgers University investigators in a system called MEMICA [Bar-Cohen et al., 2000; Mavroidis et al., 2000; Pfeiffer et al., 1999]. The key aspects of the MEMICA system are miniature ECS elements and ECFS actuators that mirror the forces and stiffness at remote/virtual sites. The ECS elements and ECFS actuators that make use of ERFs to achieve this feeling of remote/virtual forces are placed at selected locations on an instrumented glove to mirror the forces of resistance at the corresponding locations in the robot hand. In this section, the mechanical design of the ECS elements and of the ECFS actuators and their integration on the MEMICA glove are presented.

19.4.1 Electrically Controlled Stiffness Elements

The ECS element stiffness is modified electrically by controlling the flow of an ERF through slots on the side of a piston (Fig. 6). The ECS element consists of a piston that is designed to move inside a sealed cylinder filled with ERF. The rate of flow is controlled by electrodes facing the flowing ERF inside the channel.

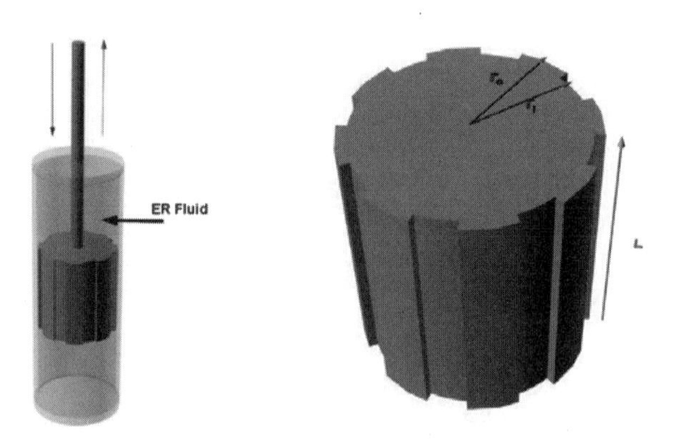

Figure 6: ECS element and its piston.

To control the "stiffness" of the ECS element, a voltage is applied between electrodes facing the slot, affecting the ability of the liquid to flow. Thus, the slot serves as a liquid valve since the increased viscosity decreases the flow rate of the ERF and varies the stiffness that is felt. To increase the stiffness bandwidth from free flow to maximum viscosity, multiple slots are made along the piston surface. To wire such a piston to a power source, the piston and its shaft are made hollow and electric wires are connected to electrode plates mounted on the side of the slots. The inside surface of the ECS cylinder surrounding the piston is made of a metallic surface and serves as the ground and opposite polarity. A sleeve covers the piston shaft to protect it from dust, jamming or obstruction. When a voltage is applied, potential is developed through the ERF along the piston channels, altering its viscosity. As a result of the increase in the ERF viscosity, the flow is slowed significantly and resistance to external axial forces increases. Section 19.5 presents the dynamic modeling equations for such an ECS element. Section 19.6 presents a parametric study of the design of the ECS element while Sec. 19.7 presents experimental results obtained from a large-scale prototype of an ECS element.

19.4.2 Electrically Controlled Force and Stiffness (ECFS) Actuator

To produce complete emulation of a mechanical "telefeeling" system, it is essential to use actuators in addition to the ECS elements in order to simulate remote reaction forces. Such a haptic mechanism needs to provide both active and resistive actuation. The active actuator can mirror the forces at the virtual/remote site by pulling the finger or other limbs backward. The main objective of this section is to present a small linear actuator that utilizes the ECS concept developed in Sec. 19.4.1. This actuator operates as an *inchworm* motor (as shown in Fig. 7) and consists of active and passive elements, i.e., two brakes and an expander, respectively. One brake locks the motor position onto a shaft and the expander advances (stretches) the motor forward. While the motor is stretched forward, the other brake clamps down on the shaft and the first brake is released. The process is repeated as necessary, inching forward (or backward) as an inchworm does in nature.

Using the controllability of the resistive aspect of the ERF, a brake can be formed to support the proposed inchworm. A schematic description of the ECFS actuator is shown in Fig. 8. The actuator consists of two pistons (brake elements) and two electromagnetic cylinders (pusher elements). Similar to ECS, each piston has several small channels with a fixed electrode plate. When an electric field is induced between the piston anode and cylinder cathode, the viscosity of the ERF increases and the flow rate of the fluid though the piston channel decreases, securing the piston to the cylinder wall.

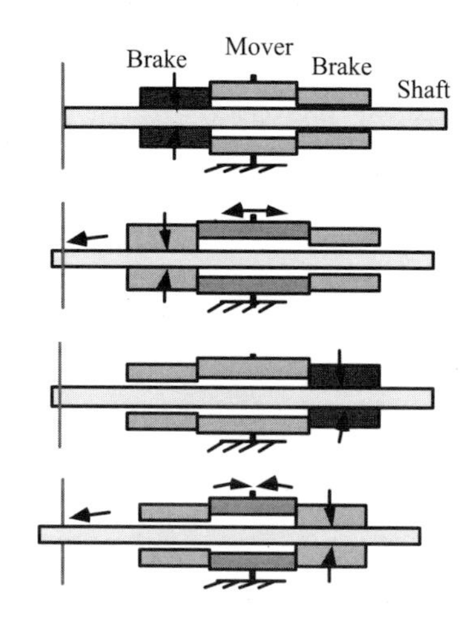

Figure 7: Concept of the inchworm motor.

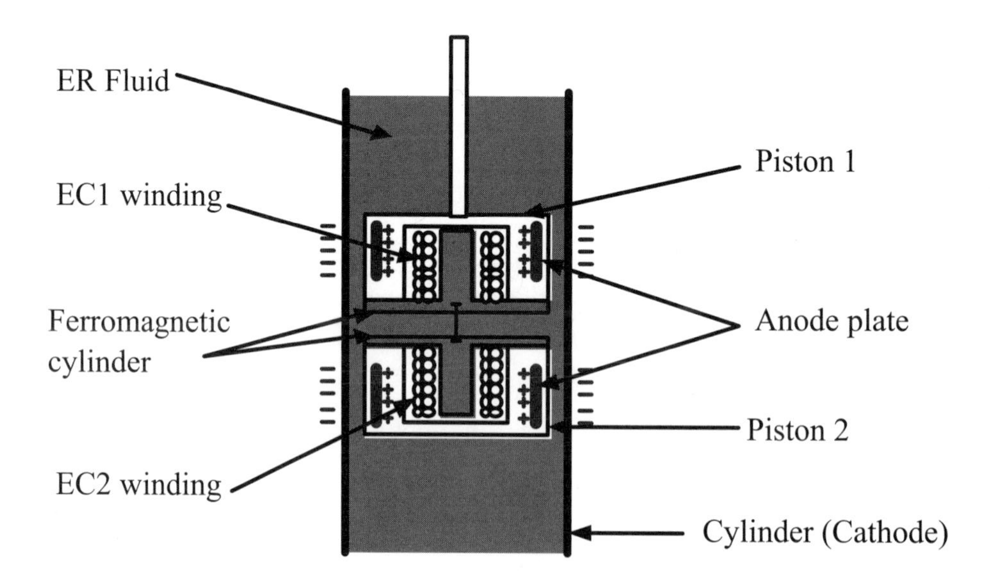

Figure 8: ECFS actuator configuration.

Each of the electromagnetic cylinders consists of a coil and a ferromagnetic core integrated within the piston. When a current impulse is passed through the winding, an electromagnetic field is induced and, depending on the current direction, the cylinder moves forward or backward. This actuation principle is shown as a set of sequence diagrams in Fig. 9. In the first step, piston P1 is fixed relative to the cylinder by activating its electrode; then, triggering the electromagnetic cylinder moves piston P2 forward. The ERF located between the two pistons is then displaced backward through the channels of piston P2. A horizontal channel is added at the surface of a ferromagnetic cylinder to increase the flow rate of the fluid. In the second step, the ERF in the channels of piston P2 is activated and P2 becomes fixed to the cylinder while P1 is disconnected from the cylinder. The current in the first winding is then reversed, changing the polarization of the magnetic cylinder, pushing P1 forward relative to P2.

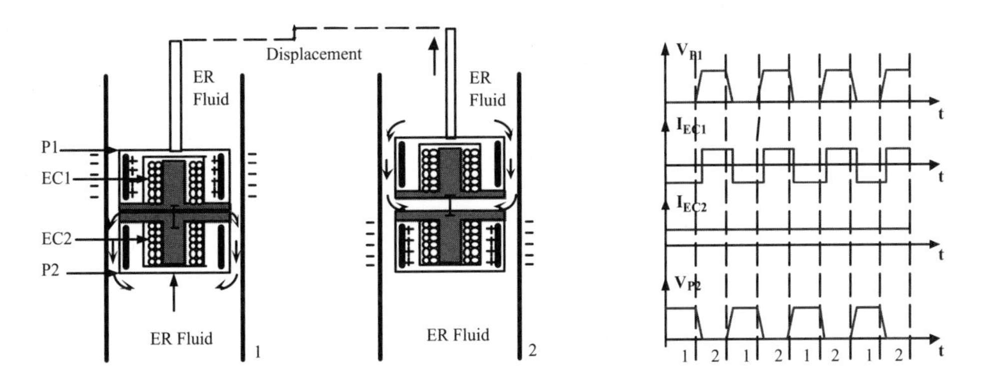

Figure 9: Sequence diagram of ECFS actuator operation.

At each cycle, the pistons move forward or backward with very small displacement (<1.5 mm). The duration of each cycle is close to a millisecond, corresponding to the response time of the ERF. The ECFS actuator can then reach a speed higher than 15 cm/s with a piston displacement equal to 0.5 mm at 3-ms cycle duration. The electromagnetic cylinder is designed to produce the same force as the resistive force of the piston inside the ERF, which is about 15 N.

19.4.3 MEMICA Haptic Glove and System

A haptic exoskeleton integrates the ECS elements and ECFS actuators at various joints. As shown in Fig. 10, the actuators are placed on the back of the fingers, out of the way of grasping motions. The natural motion of the hand is then unrestricted. Also, this configuration is capable of applying an independent force (uncoupled) on each phalange to maximize the level of stiffness/force feedback that is "felt" by the operator.

Different mounting mechanisms are evaluated and some of them are schematically represented in Fig. 11. Of these solutions, the most ergonomic seems to be the arched actuator [Fig. 11(a)], which has better fitting with the finger motion and geometry. Since ERF viscosity is higher than air, there is no need for tight tolerance for the ECFS piston and its cylinder. The second proposed solution uses a curved sliding rail, which is also suitable for a finger motion. The third solution uses a flexible tendon connected directly to the piston inside the cylinder where the tendon length can be adjustable to the user phalange length.

Once the appropriate exoskeleton mechanism is chosen, we will prototype a complete haptic glove with 16 actuators: 3 actuators for the thumb (2 for the flexion motion and 1 for the abduction/adduction motion), 4 actuators for the index (3 for the flexion motion and 1 for the abduction/adduction motion) and 3 actuators for the three other fingers (flexion only). Figure 12 contains 3D drawings of the exoskeleton glove showing the positioning of all actuators on the fingers. For precision grasping with the thumb and the index finger where abduction/adduction motion is involved, it is necessary to integrate actuators that would resist this motion.

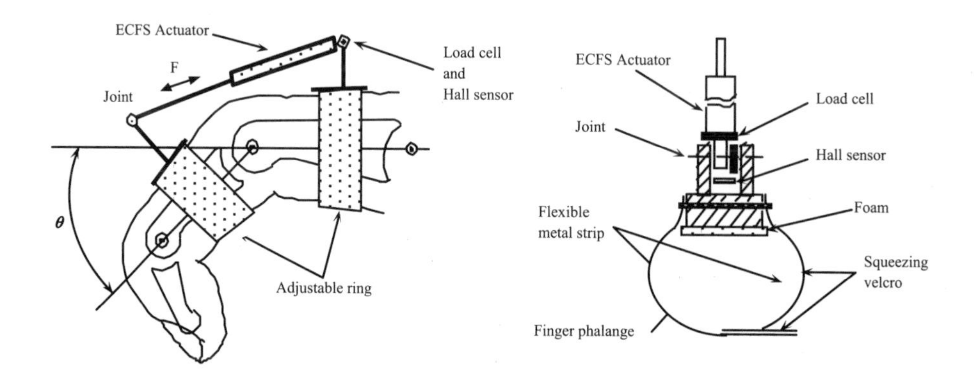

Figure 10: Mounting of an ECFS actuator on the finger phalange.

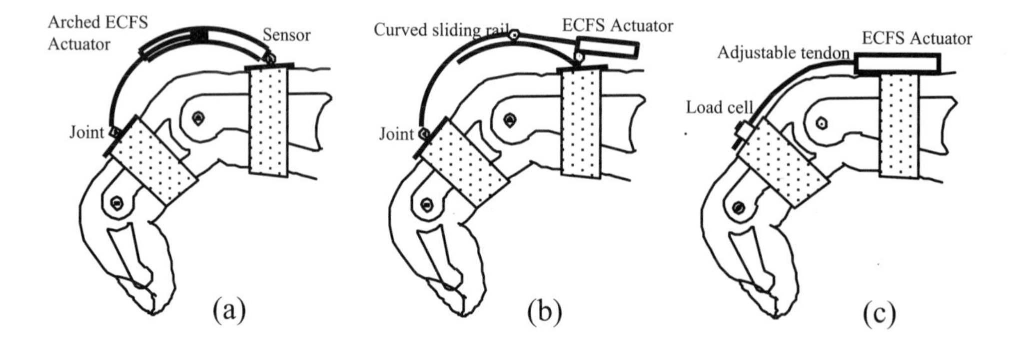

Figure 11: Different exoskeleton mechanisms.

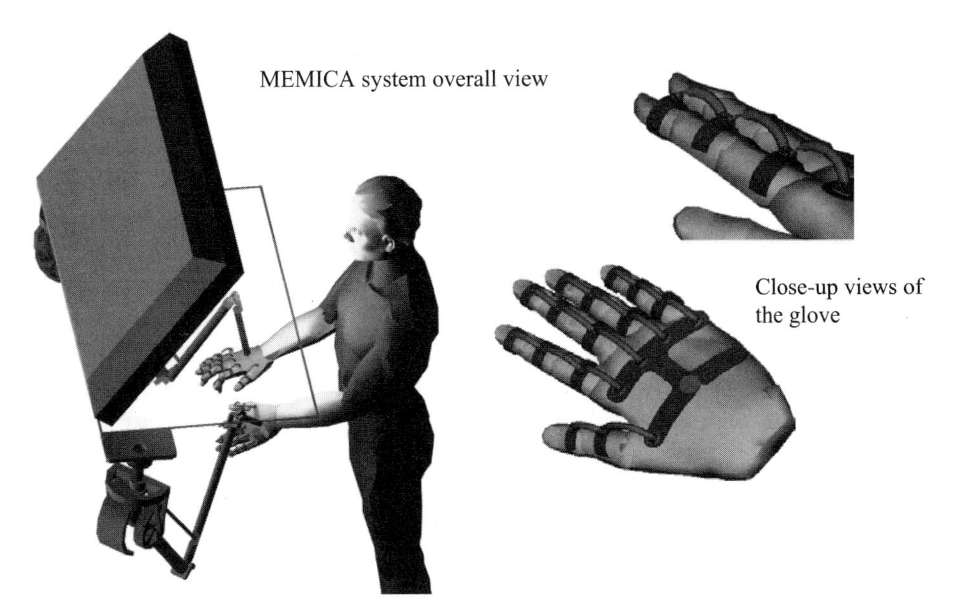

Figure 12: 3D View from the MEMICA system and close-up view of the glove.

19.5 ECS Element Model Derivation

In this section we present the dynamic modeling equations of an ECS element. These equations calculate the force that a human operator will feel when he/she is haptically interacting with an ECS element. The calculated force is the function of the applied voltage to the ERF, the geometry of the ECS element, and the characteristics of the ERF.

The field E is provided by an applied voltage V. Considering a Gaussian surface A of radius r and length l between the inner r_i, and outer r_o radii (see Fig. 9), Gauss's law is used to find the electric field. A charge of q is assigned to the charged core and the field is known to be in the radial direction, so that the dot product becomes the scalar product, and

$$\oint \vec{E} \cdot d\vec{A} = \frac{q}{\varepsilon_o} \Leftrightarrow EA = E\left(2\pi rl\right) = \frac{q}{\varepsilon_o} \Leftrightarrow \vec{E}(r) = \frac{q}{2\pi\varepsilon_o rl}\hat{r} \tag{3}$$

where ε_o is the electrical permittivity of free space.

This expression for the electric field can now be used in the definition of a difference in potential, computing the difference between the inner and outer walls, V:

$$V = -\int_{r_o}^{r_i} \left(\frac{q}{2\pi\varepsilon_o rl}\right)\hat{r} \cdot d\vec{r} = \left(\frac{q}{2\pi\varepsilon_o l}\right)\ln\left(\frac{r_o}{r_i}\right). \tag{4}$$

Equations (3) and (4) are combined to relate the electric field directly to the applied voltage and geometry:

$$E = \left[\left(\frac{q}{2\pi\varepsilon_o l}\right)\ln\left(\frac{r_o}{r_i}\right)\right]\frac{1}{\ln\left(\frac{r_o}{r_i}\right)}\frac{1}{r} = \left(\frac{V}{\ln\left(\frac{r_o}{r_i}\right)}\right)\frac{1}{r}. \tag{5}$$

The force F_{app} applied by the operator is equal to the reaction force F_R he or she will feel. This reaction force is the sum of three forces: a shear force F_τ, a pressure force F_p, and a friction force F_f. Assuming that the interface between the piston and cylinder is frictionless, the total reaction force can be computed as the sum of the remaining components.

$$F_R = F_{\text{app}} = F_\tau + F_p + F_f \approx F_\tau + F_p. \tag{6}$$

19.5.1 Static Case

Since the pressure force is a result of the flow of the ERF through the channels, there will be no pressure force term for the static force.

The shear force term is calculated by considering the entire surface area in contact with ERF, which is the surface area of all of the channels. Since shear stress for a given voltage is a function of radius, the shearing force acting on each channel will be the sum of a term calculated at the outer radius, a term calculated at the inner radius, and twice a term integrated with respect to the area of the side walls of the channel. The area of an inner or outer channel wall, i.e., $A_{rw}(r_i)$ or $A_{rw}(r_o)$ respectively, is the product of the channel length L and the arc subtended by the radius through the angular width of the channel θ: The area A_{sw} of each one of the side walls is equal to the product of the channel length L and the channel width Δr (i.e., difference in radii):

$$A_{rw}(r) = (r\theta)L, \quad A_{sw} = L\Delta r, \quad dA_{sw} = Ldr. \tag{7}$$

Since the contribution to the total reaction force from each channel is the same, the total reaction force—the product of shear stress and area—can be found by multiplying the contribution from one channel by the number of channels N:

$$F_{R,s} = N\left[\tau_s(r_o)A_{rw}(r_o) + \tau_s(r_i)A_{rw}(r_i) + 2\int_{r_i}^{r_o}\tau_s(r)\,dA_{sw}\right]. \tag{8}$$

The expressions of the shear stresses are calculated from Eqs. (1) and (2). Since the static case is being considered, the shear rate is zero in Eq. (1). Combining the information from Eqs. (1), (2), (8) and after mathematical processing, the following expression is obtained for the reactions force in static mode:

$$F_{R,s} = NC_s L\left\{\left[2+\frac{2\theta}{\ln\left(\frac{r_o}{r_i}\right)}\right]V-\left[2\Delta r+\theta\left(r_o+r_i\right)\right]E_{ref}\right\}. \tag{9}$$

19.5.2 Dynamic Case

In order to calculate the reaction force felt by the operator when the piston is moving, the dynamic shear stress and the pressure force must be considered. First, the shear force $F_{\tau,d}$ is calculated as in Eq. (8), replacing the subscript s with d:

$$F_{\tau,d} = N\left[\tau_d(r_o)A_{rw}(r_o)+\tau_d(r_i)A_{rw}(r_i)+2\int_{r_i}^{r_o}\tau_d(r)dA_{sw}\right]. \tag{10}$$

Referring back to Eq. (1), we note that the shear rate is equal to the velocity gradient given by $v/\Delta r$, with the velocity of the ERF equal in magnitude to the velocity v of the piston. Now using the expressions for dynamic yield stress and plastic viscosity from Eq. (2) in Eq. (1) and the expression for the electric field from Eq. (5). and subsequently substituting in Eq. (10), the following equation is obtained:

$$F_{\tau,d} = NL\left(C_d - C_v\frac{v}{\Delta r}\right)\left[\theta\left(\frac{1}{r_o}+\frac{1}{r_i}\right)+2\left(\frac{1}{r_i}-\frac{1}{r_o}\right)\right]\frac{V^2}{\left[\ln\left(\frac{r_o}{r_i}\right)\right]^2} + NL\mu_o\left[2+\theta\left(\frac{r_o+r_i}{\Delta r}\right)\right]v. \tag{11}$$

The next step in calculating the reaction force for the dynamic case is calculating the pressure force $F_{p,d}$. The pressure force can be determined by finding the pressure gradient in the channels, which is found through a force balance of a differential fluid element with some acceleration, a, equal in magnitude to the acceleration of the piston The differential fluid element is considered to have a differential mass dm, a length dx and an area A_f:

$$(dm)a = \frac{F_{\tau,d}}{NL}dx - A_f\left[p - \left(p + \frac{dp}{dx}dx\right)\right]. \tag{12}$$

The differential mass element *dm* can be written as a function of the density ρ:

$$dm = \rho A_f dx. \tag{13}$$

The area of the fluid is written as

$$A_f = \frac{\theta}{2\pi}\left(\pi r_o^2 - \pi r_i^2\right) = \frac{\theta}{2}\left(r_o^2 - r_i^2\right). \tag{14}$$

Simplifying Eq. (12) and solving for the pressure gradient,

$$\left(-\frac{dp}{dx}\right) = \frac{F_{\tau,d}}{NLA_f} - \rho a. \tag{15}$$

The pressure force is found by multiplying the pressure drop by the piston area A_p:

$$F_{p,d} = \Delta p A_p = \left[p - \left(p + \frac{dp}{dx}L\right)\right]A_p = -\frac{dp}{dx}LA_p, \tag{16}$$

Where the area of the piston is given by

$$A_p = \pi r_o^2 - \frac{N\theta}{2\pi}\left(\pi r_o^2 - \pi r_i^2\right) = \pi r_o^2 - \frac{N\theta}{2}\left(r_o^2 - r_i^2\right). \tag{17}$$

So,

$$F_{p,d} = F_{\tau,d}\left[\frac{\pi r_o^2}{\frac{N\theta}{2}\left(r_o^2 - r_i^2\right)} - 1\right] - \rho L\left[\pi r_o^2 - \frac{N\theta}{2}\left(r_o^2 - r_i^2\right)\right]a. \tag{18}$$

The total reaction force for the dynamic case will be the sum of these two forces. After mathematical manipulation,

$$F_{R,d} = \left[\frac{\pi r_o^2}{\frac{N\theta}{2}\left(r_o^2 - r_i^2\right)}\right] NL$$

$$\left\{\left(C_d - C_v \frac{v}{\Delta r}\right)\left(\frac{\theta}{r_o} + \frac{\theta}{r_i} + \frac{2}{r_i} - \frac{2}{r_o}\right)\frac{V^2}{\left[\ln\left(\frac{r_o}{r_i}\right)\right]^2} + \mu_o\left[2 + \theta\left(\frac{r_o + r_i}{\Delta r}\right)\right]v\right\} \tag{19}$$

$$-\rho L\left[\pi r_o^2 - \frac{N\theta}{2}\left(r_o^2 - r_i^2\right)\right]a.$$

19.6 Parametric Analysis of the Design of ECS Elements

The analytical Eqs. (9) and (19) will be used to evaluate how various geometric and input factors can influence the reaction forces felt by the user. The results from this study are very important for the design of the ECS elements.

Human studies have shown that the controllable maximum force that a human finger can exert is between 40 and 50 N [Burdea, 1996]. However, maximum exertion forces create discomfort and fatigue to the human operator. Comfortable values of exertion forces are between 15% and 25% of the controllable maximum force exerted by a human finger. Hence, the design objective is to develop an ECS element that will be able to apply a maximum force of 15 N to the operator. We are primarily interested in the dependence of the reaction forces from the ECS when the following parameters are changing: voltage applied V, motion characteristics imposed by the user such as the velocity v, acceleration a, geometric characteristics of the piston such as geometry of the channel defined by the inner and outer diameters r_i and r_o, and the angle of the channel θ. Therefore, in our study it is desired to find the ranges for these parameters that will result in the desired maximum force output of 15 N.

The parameters related to the fluid ERF LID 3354, shown in Table 1, were determined from the manufacturer's specifications [ER Fluid Developments Ltd., 1998]. The default geometric parameters of the ECS element shown in Table 2 have been determined from the dimensions of commercially available sensors and electronic equipment that will be used for measuring and actuating the device, as well as any manufacturing and machinability constraints. In the first prototype that is presented in this work (see Sec. 19.7), no effort for miniaturization was made since the goal was to prove the concept that ERFs can be used to create haptic feedback. The default values for motion characteristics were selected based on representative values of the maximum velocities and accelerations that a human finger can develop (see Table 3).

Table 1: ERF LID 3354 parameters.
(See Sec. 19.2 for parameter definition.)

C_d	0.00026
C_v	0.198 E-7
μ_0	0.125
ρ	1460 kg/m^3

Table 2: Values for the geometric parameters.
(See Section 19.5 for parameter definition.)

L	0.0254 m
r_i	0.011316 m
r_o	0.012065 m
Δr	0.000749 m
N	12
θ	0.47 rad (27°)

Table 3: Values for the motion characteristics.
(See Sec. 19.5 for parameter definition.)

a	0.01 m/s^2
v	0.1 m/s

Voltage is the principal parameter of interest in this study since it will be used for controlling the compliance of the ERF. We calculate the maximum voltage needed for achieving a reaction force of 15 N. Setting the default values in Eqs. (9) and (19) and changing the voltage the force has been calculated and is shown in Fig. 13. As expected, the relationship of force to voltage is linear in the static case and parabolic in the dynamic case. In the static case, a voltage of approximately 2 kV is needed to achieve the desired force of 15 N. In the dynamic case, the desired force output of 15 N is reached using 1 kV.

A similar parametric study revealed that the reaction force is almost independent of the velocity and acceleration imposed by the user. This is due to the fact that the velocity and acceleration contributions in the reaction force are much smaller than the voltage-related term.

The piston geometry is another important factor that affects the reaction force felt by the user. In the results presented in this section, the outer diameter of the piston changes from 0.012 m to 0.014 m while the voltage is equal to 1 kV. All other parameters take the default values shown in Tables 1, 2, and 3. The calculated reaction forces are shown in Fig. 14. It is clearly seen that, as the outer diameter increases, the reaction force decreases dramatically. On the other hand, as the outer diameter approaches the value for the channel inner diameter, the reaction forces become infinite. The thinner the piston channels are, the larger the reaction force is, hence the required voltage can be reduced.

In a similar way, the channel angle changed from 0 to 0.5 rad and the force was calculated. Under a certain minimum value of θ, the reaction force drops dramatically. Also there is a maximum limit for θ after that the reaction force is constant. Therefore, optimal values for θ are around 0.4 rad (i.e. 30 deg).

The parameters N and L affect the reaction force in a linear way. Increasing these parameters increases the force for a given voltage. However, the dimensions of the piston limit L and the values of θ limit the number of channels N.

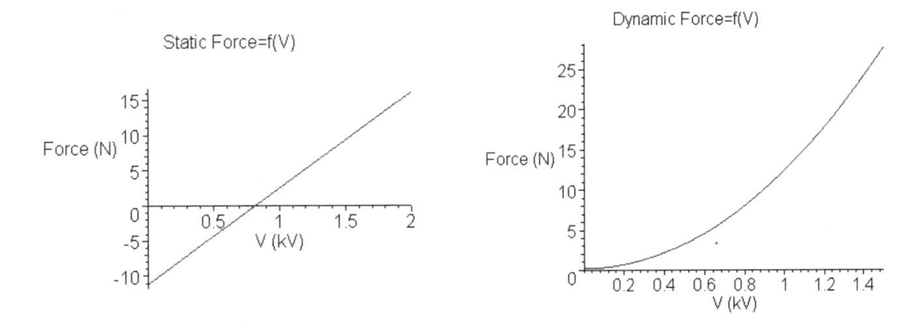

Figure 13: Force as a function of voltage.

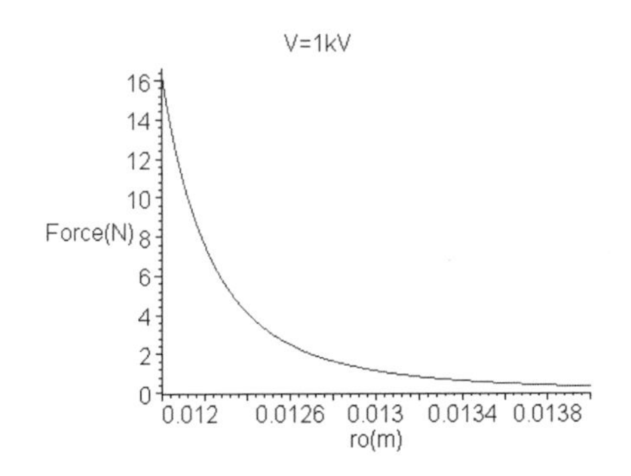

Figure 14: Force (N) vs. outer diameter r_o.

19.7 Experimental ECS System and Results

To test the concept of controlling stiffness with a miniature ECS element, a larger-scale test bed has been built at the Rutgers Robotics and Mechatronics Laboratory. This test bed, shown in Figs. 15 and 16, is equipped with temperature, pressure, force, and displacement sensors to monitor the ERF's state. The cylinder is mounted on a fixed stainless steel plate to maintain rigidity during normal force loading. The top plate is also stainless steel and serves as the base for the weight platform. Beneath the platform, around the stainless steel shaft, is a quick release collar that allows the force to be released by the operator.

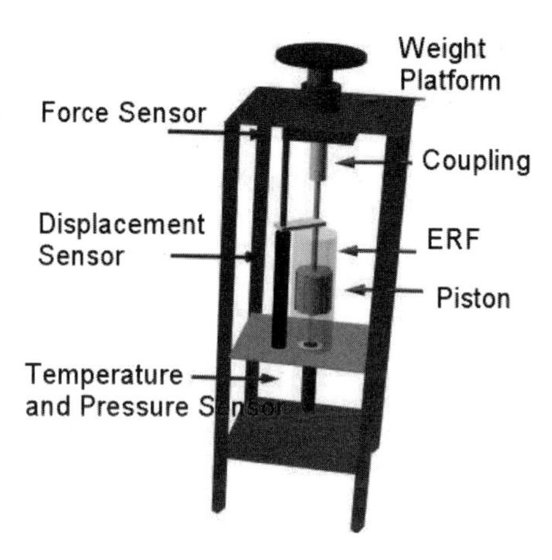

Figure 15: Experimental test bed.

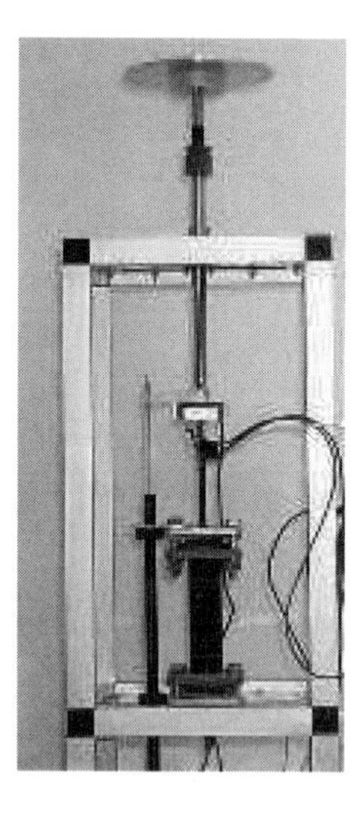

Figure 16: Actual prototype system.

The shaft, which transmits the force down into the cylinder, is restrained to only one-dimensional motion through a linear bearing mounted to the top plate. The half-inch solid shaft is reduced by an adapter to a quarter-inch aircraft steel hollow shaft. At this junction there is a load cell and flange bracket mounted for the wiper shaft of the displacement sensor. The quarter-inch shaft inserts through the ERF chamber's top plate and a small bundt cup that is needed to minimize leaking from the chamber during operation. Within the chamber, the experimental piston is attached to the shaft with e-clips secured at the top and bottom of the piston. The chamber itself is a 1-in. internal diameter beaded Pyrex piping sleeve, 6 in. in length. Pyrex allows visual observation of the ERF during actuation. In order to apply voltage to the fluid, supply wires are run down through the hollow shaft and into the piston, where the electrical connections are made to the channel plates. Threaded into the bottom plate of the chamber is the dual pressure and temperature sensor. The final sensor is mounted alongside the chamber and affixed with a flanged bracket to the chamber.

Six system parameters are measured during experimentation: voltage, current, force, displacement, pressure and temperature. All sensor signals are interfaced directly to analog-to-digital boards located in a Pentium II PC and are processed using the Rutgers WinRec v.1 real-time control and data acquisition Windows NT-based software. In addition, all sensors are connected to digital meters located inside the interface and control box. Sensor excitation voltages are supplied by 5 V from the PC or by the meter provided with the sensor itself.

Extensive experimental tests are currently underway to determine the relationship of the reaction force to the applied voltage, human motion, temperature, and pressure changes and to verify Eqs. (9) and (19). Representative results from these tests are shown in Figs. 17(a) and (b). In Fig. 17(a), no voltage is applied to the device. Four different weights equal to 2.75 lb, 5.50 lb, 8.25 lb, and 11 lb are placed individually on the weight platform. Each time the quick release collar is released, the piston displacement induced by the weight is recorded. A very fast descent of the piston is observed for all the weights. In Fig. 17(b), the same procedure is followed but this time a voltage of 2 kV is applied on the ERF. It can clearly be seen that the piston is showing a very slow descent and for the lightest weight (i.e., the 2.5 lb) no motion is observed. This experiment shows that when the electrical field is enabled, the viscosity of the ERF is such that the ECS element can resist the gravity forces from the weights.

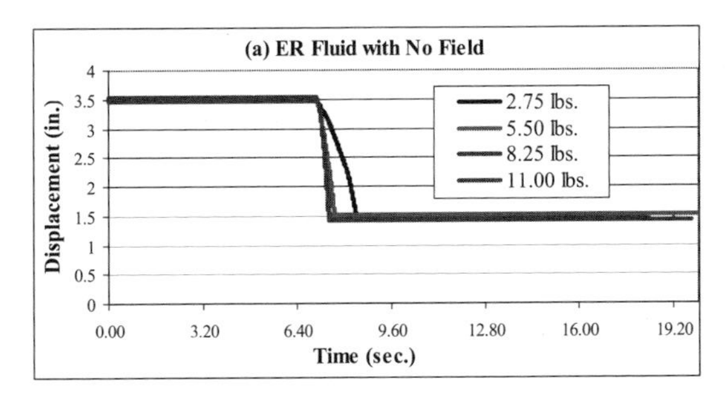

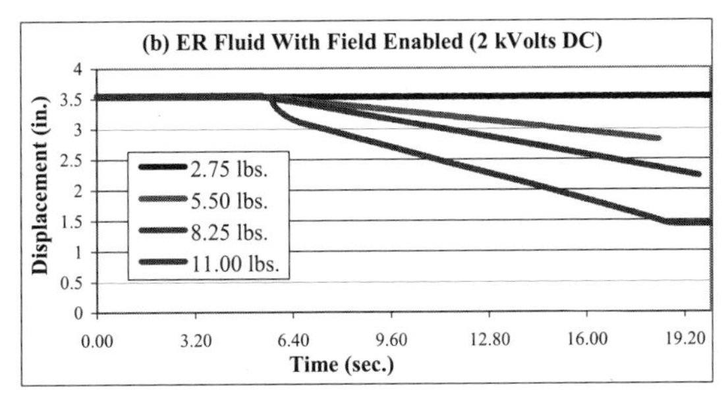

Figure 17: Piston displacement. (a) ER fluid with no field; (b) ER fluid with field enabled (2 k V dc).

19.8 Conclusions

Using EAPs as smart materials can enable the development of many interesting devices and methodologies. In this chapter, the authors presented the conceptual application of electrorheological fluids (ERFs) to address the need for haptic interfaces in such areas as automation, robotics, medicine, games, or sports. Using such EAP fluids, one may be able to construct a system that allows one to "feel" the environment compliance and reaction forces at remote or virtual robotic manipulators. The ability to have a human operator control a remote robot in the sense of telepresence addresses the realization that there are some tasks that can be best performed by a human but may be too hazardous for physical presence. Using a haptic interface as described in this chapter allows human operators to perform the tasks without the associated risks.

A new mechanism was described that can allow operators to sense the interaction of stiffness forces exerted on a robotic manipulator. An analytical model was reviewed and experimental data were described. A key to the new haptic interface is the so-called electrically controlled stiffness (ECS) element which has been demonstrated in a scaled-size experimental unit that proves the feasibility of the mechanism. A conceptually novel ERF-based haptic system called MEMICA that is based on such ECS elements was described in this chapter. MEMICA is intended for operations that support space, medical, underwater, virtual reality, military, and field robots performing dexterous manipulations. For medical applications, virtual procedures can be developed as simulators for training doctors, an exoskeleton system can be developed to augment the mobility of handicapped or ill persons, and remote surgery can be enabled.

19.9 Acknowledgments

Work at Rutgers University was supported by NASA's Jet Propulsion Laboratory (JPL) and CAIP—Center for Advanced Information Processing. Research at JPL/Caltech was carried out under a contract with the National Aeronautics Space Agency. The authors would like to thank Mr. Charles Pfeiffer, Mr. James Celectino, Mr. Alex Paljic, Ms. Jamie Lennon, Ms. Sandy Larios, and Ms. Sarah Young, Rutgers University, for providing assistance during the development of the MEMICA concept. Also, the authors would like to thank Dr. Benjamin Dolgin, Ph.D., JPL, Pasadena, CA, Dr. Deborah L. Harm, Ph.D., NASA JSC, Houston, TX, Mr. George E. Kopchok, Harbor-UCLA Medical Center, Los Angeles, CA and Rodney White, M.D., Harbor-UCLA Medical Center, Los Angeles, CA, for their helpful comments and suggestions.

19.10 References

Arai, Arai F., Akiko K., Fukuda T., Matsuura H. and Ota H., "Safety Oriented Mechanism and Control Using ER Fluid in the Joint," *Proceedings of the 1998 IEEE International Conference on Robotics and Automation,* Leuven, Belgium, May 1998, pp. 2482-2487.

Bar-Cohen, Y., Pfeiffer C., Mavroidis C., and Dolgin B., "MEMICA: a concept for reflecting remote-manipulator forces," *NASA Tech Briefs*, Vol. 24, No. 2, pp. 7a-7b, 2000a.

Bar-Cohen, Y., Mavroidis C., Bouzit M., Dolgin B., Harm D., Kopchok G., and White R., "Virtual reality robotic operation simulations using MEMICA haptic system," *SmartSystems 2000: The International Conference for Smart Systems and Robotics for Medicine and Space Applications*, Houston, Texas, September 6 to 8, 2000b.

Bar-Cohen, Y., Mavroidis, C., Pfeiffer C., Culbert C. and Magruder D., "Haptic interfaces," Chapter in *Automation, Miniature Robotics and Sensors for Non-Destructive Testing and Evaluation*, Y. Bar-Cohen, ed., The American Society for Nondestructive Testing, Inc. (ASNT), pp. 461-468, 2000c.

Bar-Cohen, Y., Mavroidis C., Bouzit M., Pfeiffer C. and Dolgin B., "Remote mechanical mirroring using controlled stiffness and actuators (MEMICA)," Rutgers Docket Number 99-0056 A US and International PCT patent application was filed by Rutgers University in September 2000d.

Block, H. and Kelly, J. P., "Electro-Rheology," *Journal of Physics, D: Applied Physics,* Vol. 21, 1988, pp. 1661.

Böse, H., Berkemeier J. and Trendler A., "Haptic system based on electrorheological fluid," *Proceedings of the ACTUATOR 2000 Conference,* 19-21 June 2000, Bremen GERMANY.

Bullough, W. A., Johnson A. R., Hosseini-Sianaki A., Makin J., and Firoozian R., "Electrorheological clutch: design, performance characterisitics and operation," *Proceedings of the Institution of Mechanical Engineers. Part I, Journal of Systems and Control Engineering,* Vol. 207, No. 2, 1993, pp. 87-95.

Burdea, G., *Force and Touch Feedback for Virtual Reality,* New York: John Wiley and Sons, 1996.

Choi, S. B., "Vibration control of a flexible structure using ER dampers," *Transactions of the ASME, Journal of Dynamic Systems, Measurement and Control,* Vol. 121, 1999, pp. 134-138.

Conrad, H., "Properties and design of electrorheological suspensions," *MRS Bulletin*, Vol. 23, No. 8, August 1998, pp. 35-42.

Coulter, J. P., Weiss K. D., and Carlson D. J., "Engineering applications of electrorheological materials," in *Advances in Intelligent Material Systems and Structures-Volume 2: Advances in Electrorheological Fluids,* Kohudic M. A., ed., Technomic Publishing Company, Inc., Lancaster, PA, 1994, pp. 64-75.

Duclos, T., Carlson, J., Chrzan, M. and Coulter, J. P., "Electrorheological fluids – materials and applications," in *Intelligent Structural Systems*, Tzou and Anderson, eds., Kluwer Academic Publishers, Netherlands, 1992, pp. 213-241.

ER Fluid Developments Ltd., "Electrorheological fluid LID 3354," *Technical Information Sheet*, United Kingdom, 1998.

Furusho, J., Zhang G. and Sakaguchi M., "Vibration suppression control of robot arms using a homogeneous-type electrorheological fluid," *Proceedings of the 1997 IEEE International Conference on Robotics and Automation,* Albuquerque, NM, 1997, pp. 3441-3448.

Gast, A. P., and Zukoski, C. F., "Electrorheological suspensions as colloidal suspensions," *Advances in Colloid and Interface Science,* Vol. 30, 1989, pp. 153.

Ghandhi, M. V., Thompson B. S., and Shakir S., "Electrorheological fluid based articulating robotic systems," in *Advances in Design Automation: Vol. 2; Robotics, Mechanisms and Machine Systems,* ASME DE Vol. 10-2, 1987, pp. 1-10.

Immersion Corp., http://www.immersion.com, 2001.

Kenaley, G. L. and Cutkosky M. R., "Electrorheological fluid-based robotic fingers with tactile sensing," *Proceedings of the 1989 IEEE International Conference on Robotics and Automation,* Scottsdale, AZ, pp. 132-136.

Lampe, D., *Materials Database on Commercially Available Electro- and Magnetorheological Fluids (ERF and MRF),* http://www.tu-dresden.de/mwilr/lampe/HAUENG.HTM, updated on 01/30/1997.

Mavroidis, C., Pfeiffer C. and Bar-Cohen Y., "Controlled compliance haptic interface using electrorheological fluids," *Proceedings of the 2000 SPIE Conference on Electro-Active Polymer Actuators and Devices (EAPAD 2000,* SPIE Proc. Vol. 3987, 2000a, pp. 300-310.

Mavroidis, C., Pfeiffer C., Celestino J. and Bar-Cohen Y., "Design and modeling of an electrorheological fluid based haptic interface," *Proceedings of the 2000 ASME Mechanisms and Robotics Conference,* Baltimore MD, September 10-13, 2000b, Paper DETC2000/MECH-14121.

Mavroidis, C., Pfeiffer C., Lennon J., Paljic A., Celestino J., and Bar-Cohen Y., "Modeling and design of an electrorheological fluid based haptic system for tele-operation of space robots," *Proceedings of the ROBOTICS 2000 Conference: The 4th International Conference and Exposition/Demonstration on Robotics for Challenging Situations and Environments*, February 27-March 2, 2000c, Albuquerque, NM, pp. 174-180.

Monkman, G. J., "Electrorheological tactile display," *Presence*, MIT Press, Vol. 1, No. 2, 1992.

Pfeiffer, C., Mavroidis C., Bar-Cohen Y., and Dolgin B., "Electrorheological fluid based force feedback device," *Proceedings of the 1999 SPIE Telemanipulator and Telepresence Technologies VI Conference,* SPIE Proc. Vol. 3840, 1999, pp. 88-99.

Phule, P. and Ginder J., "The materials science of field-responsive fluids," *MRS Bulletin*, August 1998, pp. 19-21.

Sakaguchi, M. and Furusho J., "Force display system using particle-type electrorheological fluids," *Proceedings of the 1998 IEEE International Conference on Robotics and Automation,* Leuven, Belgium, May 1998a, pp. 2586-2590.

Sakaguchi, M. and Furusho J., "Development of ER actuators and their applications to force display systems," *Proceedings of the 1998 IEEE Virtual Reality Annual International Symposium (VRAIS)*, Atlanta, GA, 1998b, pp. 66-70.

Seed, M., Hobson G. S., Tozer R. C., and Simmonds A. J., "Voltage-controlled electrorheological brake," *Proceedings of the IASTED International Symposium on Measurements, Processes and Controls,* Sicily, September 1986, pp. 280-284.

Sensable Technologies, http://www.sensable.com, 2001.

Sproston, J. L., Stanway R., Williams E. W. and Rigby S., "The electrorheological automotive engine mount," *Journal of Electrostatics*, Vol. 32, 1994, pp. 253-259.

Stanway, R., Sproston, J. L., and El-Wahed, A. K., "Applications of electrorheological fluids in vibration control: a survey," *Smart Materials and Structures*, Vol. 5, No. 4, 1996, pp. 464-482.

Tao, R., ed., *Proceedings of the Seventh International Conference on ER Fluids and MR Suspensions*, Honolulu, Hawaii, July 19-23, 1999, World Scientific Publishing Company.

Taylor, P. M., Hosseini-Sianaki A. and Varley C. J., "An electrorheological fluid-based tactile array for virtual environments," *Proceedings of the 1996 IEEE International Conference on Robotics and Automation,* Minneapolis, MN, April 1996a, pp. 18-23.

Taylor, P. M., Hosseini-Sianaki A. and Varley C. J., "Surface feedback for virtual environment systems using electrorheological fluids," *International Journal of Modern Physics B,* Vol. 10, No. 23 & 24, 1996b, pp. 3011-3018.

Virtual Technologies, http://www.virtex.com, 2001.

Weiss, K. D., Carlson D. J., and Coulter J. P., "Material aspects of electrorheological systems," in *Advances in Intelligent Material Systems and Structures-Volume 2: Advances in Electrorheological Fluids,* Kohudic M. A., ed., Technomic Publishing Company, Inc., Lancaster, PA, 1994, pp. 30-52.

Winslow, W. M., "Induced fibrillation of suspensions," *Journal of Applied Physics*, Vol. 20, 1949, pp. 1137.

Wood, D., "Editorial: tactile displays: present and future," *Displays-Technology and Applications*, Vol. 18, No. 3, 1998, pp. 125-128.

CHAPTER 20

Shape Control of Precision Gossamer Apertures

Christopher H. M. Jenkins
South Dakota School of Mines and Technology

20.1 Introduction

20.1.1 Gossamer Spacecraft

A seminal event for space-based communications occurred over four decades ago with the Echo series of satellites (Fig. 1). Echo 1 was an approximately 30.5-m diameter balloon made of 0.0127-mm thick Mylar polyester film, carrying a set of 107.9-MHz beacon transmitters for telemetry. In 1996, NASA launched from the Space Shuttle *Endeavor* the Inflatable Antenna Experiment, an approximately 14-m diameter by 30-m long antenna (Fig. 2). Today, a resurgence of interest in large, ultra-lightweight *gossamer* spacecraft is developing, due to their potential for reduced launch mass and stowed volume [Jenkins (Ed.), 2001]. Applications for gossamer technology range from planar configurations in solar sails, concentrators, and shields, to inflatable lenticulars for radar, radio, and optical uses [Chmielewski and Jenkins, 2000]. Imaging and communication applications require *precision gossamer apertures* (PGAs).

Figure 1: Echo 1 passive satellite (courtesy of NASA).

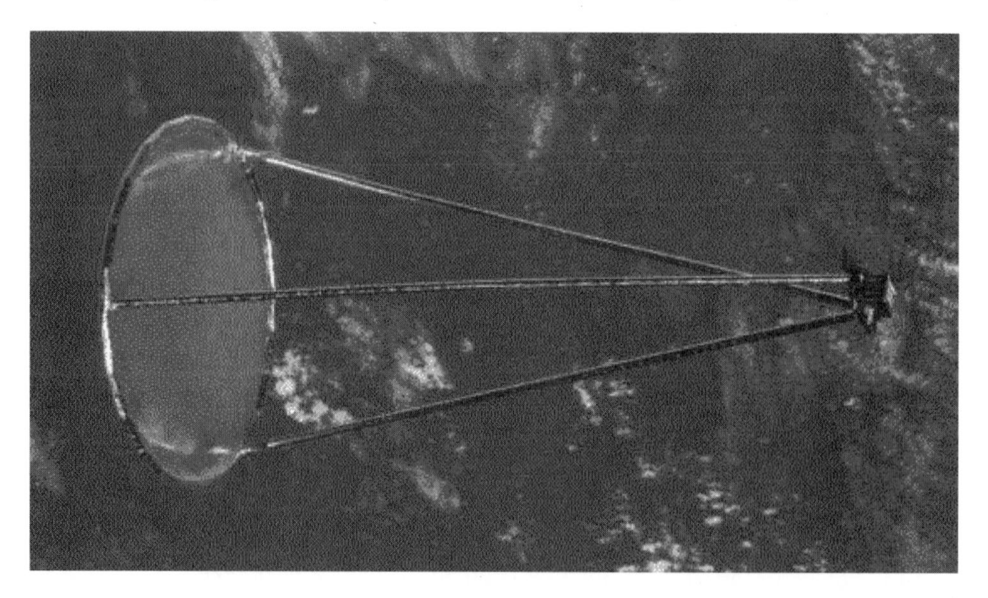

Figure 2: Inflatable Antenna Experiment on orbit. The inflatable antenna was packaged into the reusable Spartan satellite seen on the right (courtesy of NASA).

PGA technology will enable a number of space missions that have been identified by several agencies as a critical pull technology. For example, for the NASA Space and Science enterprise, precision inflatables are critical elements in meeting requirements for ultra-lightweight reflectors in the radio wavelength, for communication at high data rates, and for space-based telescopes and interferometers operating in the visible wavelength. Similarly, for NASA Earth Science, high-performance communication architectures will require precise reflector surface accuracy for both RF and optical communication. Additionally, the NASA Human Exploration and Development of Space enterprise requirements include precisely shaped reflectors for communication as well as for solar concentrators for power generation. Examples of NASA benefits from the technology development are summarized in Table 1. Additionally, DoD has a keen interest in remote sensing and global presence, while NOAA is interested in solar storm warning.

Table 1: Relevant NASA missions that benefit immediately from PGAs.

Enterprise	Missions
Earth Science	Geostorm, sun-earth connections, solar polar imager, atmospheric chemistry, climate variability
Human Exploration and Development	Solar sails, sunshields, human habitats
Space Science	Sunshields, occulters, search for origins, structure and evolution of the universe, JWST, planet imager

Two key factors are paramount for the success and user acceptance of PGA technology: deployment and on-orbit performance. In its stowed configuration, a large gossamer spacecraft may be packaged to a small fraction of its deployed volume. Autonomously deploying a highly compacted gossamer spacecraft and placing it safely into its (near) service configuration is a sophisticated operation.

Performance of gossamer apertures hinges critically on the precision of the membrane surface. The required degree of precision is highly mission dependent, and may entail one or more of the following issues: surface smoothness, deviation from desired surface profile, and slope error. Parabolic profiles are most commonly desired. The most prominent systemic error found in membranes of circular plan is a so-called "spherical aberration," often referred to as a "W-profile error," which is a measure of the deviation of the actual surface from the desired configuration.

A range of precision requirements exists. At one end are the solar sails and planar concentrators that require a moderate degree of surface flatness on membranes only a few microns thick. At the other extreme, membrane optical reflectors may require ratios of aperture diameter to figure error (rms) around 10^6 to 10^7 or more. For example, the Air Force Research Laboratory has undertaken the task of creating a large optical quality membrane telescope [Marker et al., 2001]. The membrane reflectors on these telescopes will have thickness around 10 μm. The maximum acceptable peak-to-peak figure error over the entire surface will range from 10–20 μm (this value assumes that a certain amount of secondary adaptive optics will be used to correct image errors). Passive and active means for reducing the primary figure error are under investigation [Duvvuru and Jenkins, 2003; Duvvuru et al., 2003; Jenkins et al., 1999; Kalanovic et al., 1999b].

20.1.2 Precision Gossamer Apertures

The *figure* or *shape* analysis of a precision gossamer aperture is complicated by the fact that the analysis must be performed over a range of length scales. Surface figure depends both on the global shape as well as the local *tilt* or *slope* of the surface, "local" meaning domains on the order of the wavelength of interest. The global shape enforces continuity of local slope and is required where phase or path length considerations are important. Local tilt controls the energy/surface interaction at the microscale. Thus performance depends integrally on shape at both large and very small length scales. Consider, for example, that an exact parabolic mirror with a rough surface at the microscale would have low optical performance, as would the highly polished surface of a globally undulating reflector.

Unlike classical "rigid" optics, where grinding and polishing can refine the global/local shape, PGAs require more creative means of achieving the global shape and much care in material processing to ensure local shape. Moreover, there is coupling between global and local shape perturbations. This implies a multi-scale approach, where for convenience (but somewhat arbitrarily) spatial disturbances are considered in three regimes: long wavelength, mid-wavelength, and short wavelength (all relative to the wavelength of interest). This is shown schematically in Fig. 3.

Length Scale Order: PGA radius (a) ↔ PGA thickness (h) ↔ wavelength of interest (λ)

Figure Scale Order: global figure ↔ ↔ ↔ ↔ ↔ ↔ local figure

Figure Modeling: classical mechanics ↔ wrinkling ↔ fine-scale defects

Figure 3: Multi-scale modeling of PGA surface figure.

PGAs are commonly realized physically as *membrane* structures, whose lack of bending rigidity, due to extreme thinness and/or low elastic modulus, leads to an essentially under-constrained structure that has stable equilibrium configurations only for certain loading fields [Jenkins et al., 2001; Jenkins et al., 2000b]. Under other loading conditions, large rigid body deformations can take place. In addition, these same characteristics lead to an inability to sustain compressive stress [Jenkins et al., 2000c; Kalanovic et al., 1999a]. Time-dependent and nonlinear behaviors are also common features of typical polymeric membrane materials.

20.1.3 Control Issues

System-level architecture and control is vital for optimizing PGAs, which in turn are critical to meeting weight, performance, and cost targets for mission applications. Table 2 summarizes some issues that drive control approaches for PGAs.

Table 2: Precision gossamer aperture control issues. UV = ultra-violet radiation; AO = atomic oxygen.

Generic Issues		
• Areal density: < 1 kg/m^2, < 1 g/m^2, ... • Sensor/actuator access: discrete or distributed • Sensor/actuator influence: spatial, temporal • Other issues: power limitations, redundancy		
Deployment	**Shape Achievement**	**Shape Maintenance**
• Highly compliant structures • No-compression materials (global buckling)	• Precision requirements • Plane vs. curved configuration	• Short-term perturbations: ♦ Shock/vibration ♦ Thermal loads
• High packaging ratios • Time and temperature-dependent material properties	• Inflatable vs. non-lenticular • No-compression material (local wrinkling)	• Long term perturbations: ♦ Material aging ♦ Environmental attack (UV, AO, micrometeorites,...)
• Creases and crinkles • Residual gas	• Rigidization	♦ Make up gas vs. rigidization

20.2 Shape Control of PGAs

20.2.1 Unique Challenges of PGAs

PGAs present numerous challenges, compared to their stiffer (and often smaller), more typical predecessors, in controlling their surface shape (surface figure). Glass and metallic apertures can be milled, ground, polished, fabricated, and/or supported, to provide the necessary surface precision. Access for sensors and actuators is usually good, although their influence may be limited (see further discussion below).

PGAs, on the other hand, due to their inherent high compliance (flexibility) and low mass, cannot easily be subjected to similar kinds of precision manufacturing operations. High compliance/low mass is achieved through some combination of reduced material modulus E and thickness h, and occasionally by reduced density. Reducing the mass of a PGA by reducing the thickness of plate and shell structures also reduces the bending stiffness k_b since

$$k_b \sim Eh^3.$$

Common materials for fabricating PGAs at the present time are polymer films, since these exhibit the requisite high compliance/low mass. A number of candidate polymer films exist. One of the earliest space-qualified films in use was Mylar, a polyester film usually coated with a thin layer of metal such as aluminum. Mylar was used to construct the Echo series of satellites, and is still used in high-altitude balloons and other space applications. Kapton, a polyimide, has been a commonly used membrane material for space applications, e.g., in thermal control (MLI) blankets. More recently, several polyimide derivatives (e.g., TOR, COR, CP1, and CP2) have been developed with various improved properties, such as increased resistance to atomic oxygen.

The mechanical and other material properties of special interest in the design of PGAs include:

- Coefficient of thermal expansion (CTE)
- Modulus
- Low-temperature toughness
- Low-temperature ductility
- Dimensional stability
- Environmental durability
- Aging
- Surface smoothness
- Thickness uniformity
- Bend radius (especially after coating)
- Potential to be "active" or to integrate with active materials
- Potential to be "rigidized"
- Strength.

Although strength is included in the list, only a nominal amount of strength is required; membrane/inflatables will in general be lightly loaded, with stress and strain levels at much less than their ultimate values. Low modulus values will also help to keep stress levels low, even though strain levels may be high. Failure by fracture or collapse is much more likely than tensile failure. A sample of properties for common polymer films is provided in Table 3.

Table 3: Selected properties of representative polymer films.

Film	Polymer	Density	Elastic Modulus	Strength	CTE	Ultimate Elongation
		(g/cm^3)	(GPa)	(MPa)	(ppm/°C)	(%)
CP1 and 2	Polyimide	1.4	2.6	124	47 to 51	
Kapton	Polyimide	1.42	3.0	172	20	72.0
Mylar	Polyester	1.38	3.8	172		
TOR	Polyimide	1.4	3.4	138	42	6.5
Upilex - R	Polyimide	1.39	3.7	248	28	
Upilex - S	Polyimide	1.47	8.8	393	12	42.0

Polymer films are being used with thickness ranging from a few to perhaps a hundred or so microns. For longer wavelength applications such as radar and radio, seaming together individual gores is a possible manufacturing technique. For shorter wavelength applications, notably optics, seamless monolithic apertures are highly desirable.

The technology to "grind and polish" soft, compliant apertures formed from large polymer sheets has yet to be developed, although laser ablation offers intriguing possibilities. Surface precision in PGAs is achieved first and foremost through passive means, such as

- Careful processing and fabrication techniques
- Special engineering design methods
- Sophisticated packaging and deployment procedures.

Even if the required surface precision could be achieved initially through such passive means alone (which is highly unlikely), the perturbations to the shape during service (e.g., from thermal loads, pointing dynamics, and environmental degradation) would require some means of active figure control. (The amount of control of the primary aperture can be reduced somewhat through use of adaptive secondaries or feeds, but that discussion is outside the scope of the present work.) Two important considerations for the incorporation of sensors and actuators for active structure control are *access* and *influence*.

20.2.2 Access

For PGAs, *access* for actuator placement can be severely limited. (In the following sections we refer principally to *actuators*, since much *sensing* of PGA surface precision can be done by noncontact monitoring of the aperture performance, for example, through antenna gain or optical image. However, much of what is said about access and influence of actuators can be extended to contact-type sensors as well.) Due to the extreme compliance of these structures, actuators can significantly disturb the surface figure merely by their placement on the aperture. For example, for bonded actuators the local stresses induced by the adhesive, and the increased stiffness added by the actuator and adhesive, lead to local deformation in the vicinity of the actuator, referred to as *print-through*. Contact by an actuator plunger will cause severe distortion in the neighborhood of the contact. For short wavelength applications, the local tilt deviation that can be tolerated may be on the order of a few tens of micro-radians, and local distortions of this type are unacceptable.

In general, discrete actuators contacting the surface of the PGA are undesirable, although novel solutions should not be ruled out at this time. Hence the access for actuators distributed over the aperture surface is severely limited at best. Control methods that exploit the aperture boundaries or that act without contact at all are indicated; electrostatic actuation is an example (see further discussion below).

20.2.3 Influence

The corollary issue to actuator access is the *influence* of an actuator at a particular location. In the case of PGAs, the same extreme compliance that limits access can dramatically enhance the influence of a given actuator. The influence can in principle be described by an *influence function* I that relates the spatial effectiveness of the actuator to the energy applied, time, and many other configuration and material parameters:

$$\mathsf{I} = \mathsf{I}\,(\mathbf{r}, t; \mathsf{E}, \mathbf{x}, \mathbf{D}, \sigma, \ldots)$$

where

$\mathbf{r}$ = position vector
t = time
E = applied actuation flux (energy/area)
$\mathbf{x}$ = aperture configuration
$\mathbf{D}$ = material property tensor
σ = stress tensor

In a plate or shell structure, transverse discrete loads are resisted in a relatively localized fashion by their inherent bending stiffness. Hence, the unit influence function for such structures is rather narrow, resulting from the rapid decay of the load response. On the other hand, in membrane structures like PGAs, which cannot resist loads by mustering localized bending resistance, greater global deformation of the structure is required to satisfy equilibrium. As such, the influence function of a membrane structure is relatively broader due to a much less rapid spatial decay [Hossain et al., 2003].

20.2.4 Boundary Control

As a specific example to illustrate the issues of sensor/actuator access and influence, we consider boundary control. For PGAs, there are two general locations for sensor/actuator placement: at the boundaries of the reflector or some level of distribution over its surface. The boundaries are a natural choice, due to the need to interface the membrane with the support structure, and apply to flat or curved concepts alike. The interface between the membrane reflector and the structural support system is an ideal place to add active elements, since by definition there must be some physical connection between the two (see Fig. 4).

Figure 4: Detailed view of the interconnection between aperture and support torus on a 10-m SRS Technologies reflector demonstrator. Wrinkling along aperture seams, as well as the torus, is evident.

Jenkins and coworkers have shown in numerous publications that boundary control can be an effective strategy to improve figure accuracy [Hossain et al., 2003; Ash et al., 2000; Jenkins et al., 1999; Jenkins and Marker, 1998]. Since a membrane cannot provide bending stiffness to resist boundary motions locally, the resulting *influence* of a given boundary action can be significantly greater over the surface of the membrane than for a shell structure. In fact, any locally applied couple must be enforced by global moment equilibrium, resulting in global motion of the membrane (Fig. 5).

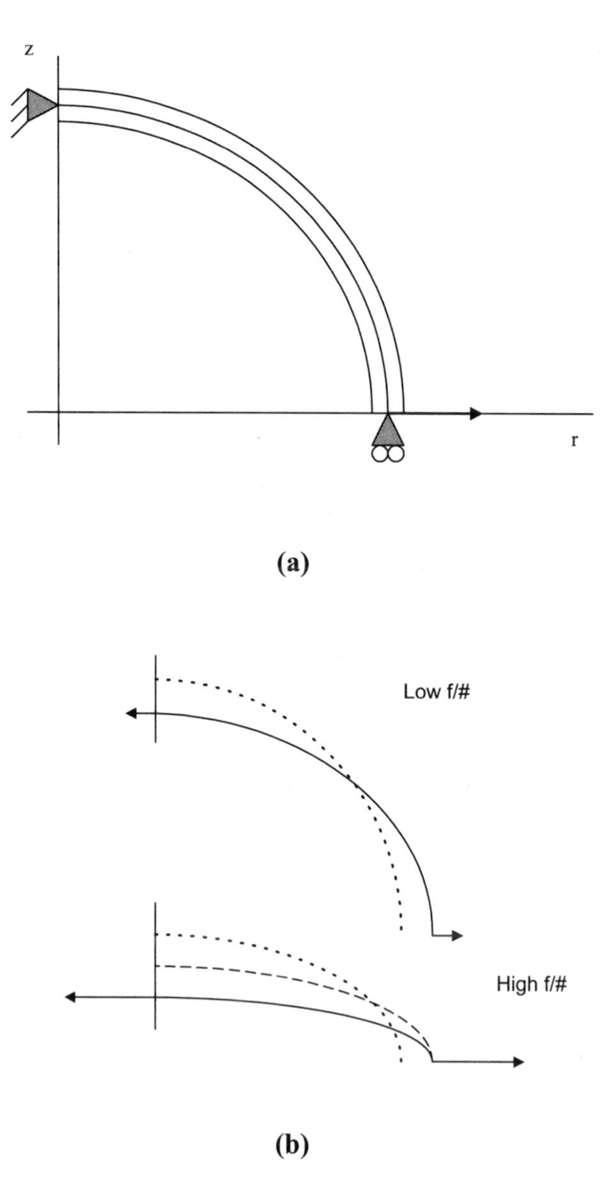

Figure 5: (a) Schematic of a curved membrane under the influence of a boundary action (b) For an equivalent boundary displacement, a reduced force is required as the *f*/# decreases, due to a lower tangential force component. Moment equilibrium requires increased apex translation as boundary force increases ("*f*/#" or "*f*-number" is the ratio of focal length to aperture diameter.)

In principle, the global motion could be quite large, but in practice it may be mitigated by such factors as pressure loads, circumferential (hoop) restraint, and some small bending rigidity.

20.2.5 Other Control Methodologies

Other concepts for active control include electrostatic and thermal actuation. Electrostatic forces can be used to deform metalized membrane reflectors. An electric potential difference between an electrode and the membrane causes an electrostatic attraction or repulsion, thus deforming the reflector surface. Active vibration damping may be a good application of electrostatic methods.

Temperature gradients applied to the reflector may be used to effect changes in the surface profile. Jenkins and coworkers have investigated such a scheme analytically for PGAs [Jenkins and Faisal, 2001]. Selective heating of the membrane surface via a laser is an intriguing possibility.

20.3 Shape Control Methodologies Involving Electroactive Polymers

20.3.1 Piezoelectric Polymers

Jenkins and coworkers have shown in simulations that manipulating the boundary (rim) of an inflatable membrane reflector reduced the figure error (deviation from a parabola) in the inflated profile [Ash et al., 2000; Jenkins et al., 1999; Jenkins and Marker, 1998]. An inflated membrane reflector was simulated using the nonlinear FEM code ABAQUS (Fig. 6). PVDF actuators [Jenkins and Vinogradov, 2000] were simulated to apply radial boundary displacements.

Figure 7 shows ABAQUS nonlinear FEM results of the RMS surface error for two cases of uniform radial boundary displacement after inflation to f/# = 1 (w_0/h = 540) from an initially plane circular plan (a/h = 4200). The radial displacements were small at 0.5% and 1.0% of the initial radius, respectively. The deformed profiles are compared to a parabola fit through the apex and boundary. It is seen that the surface error is significantly reduced (deformed profile moves toward parabolic) by the boundary manipulation. Moreover, the influence of the control action is easily seen to spread over the entire membrane surface.

In other research, Salama and coworkers [1994] investigated forming the entire aperture from PVDF. Main and coworkers [1998] have reported on use of an electron gun to provide the activation potential for PVDF strips [Main et al., 1998; Martin et al., 1998].

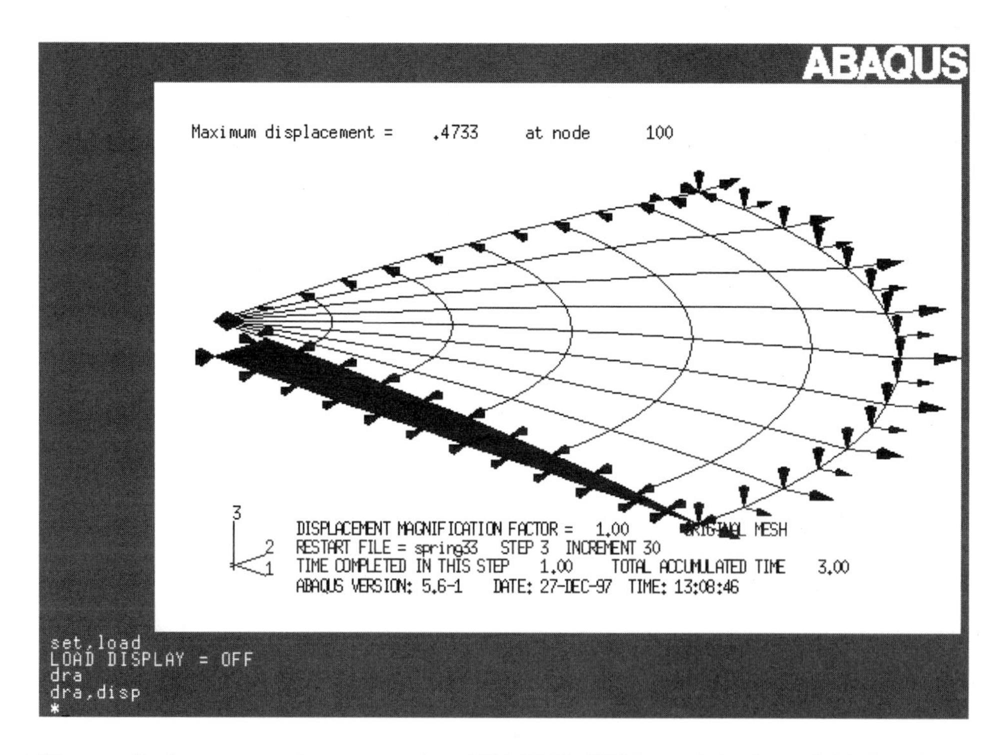

Figure 6: A one-quarter symmetry ABAQUS FEM model of an initially plane circular membrane with PVDF radial actuators.

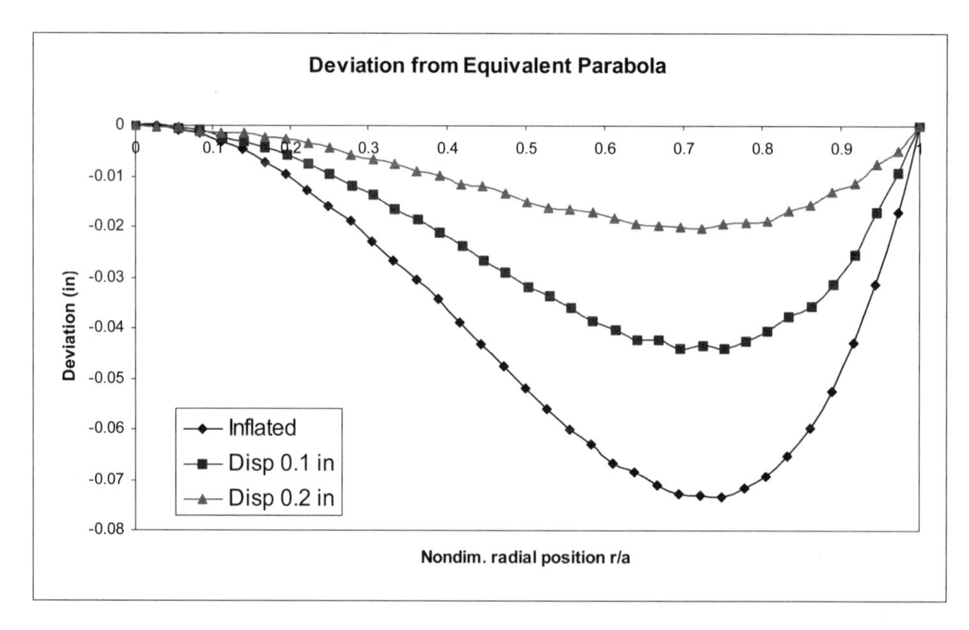

Figure 7: Improvement in W-profile error curve with uniform radial boundary displacements. r/a = 0 is the apex of the reflector, and r/a = 1 is the outer boundary where the radial displacement occurs.

Duvvuru and Jenkins [2003] showed that PVDF seams could reduce wrinkles in the neighboring gores. Jenkins and Duvvuru [Duvvuru, 2003] investigated PVDF bimorph actuators. They showed that PVDF bimorph deflections followed classical theory only up to a certain magnitude of applied voltage as shown in Fig. 8.

Jenkins and Duvvuru also considered multiplying the actuation force by placing several bimorphs together. Figure 9 shows a sketch of a double bimorph actuator connected to a load (an elastic spring in the sketch).

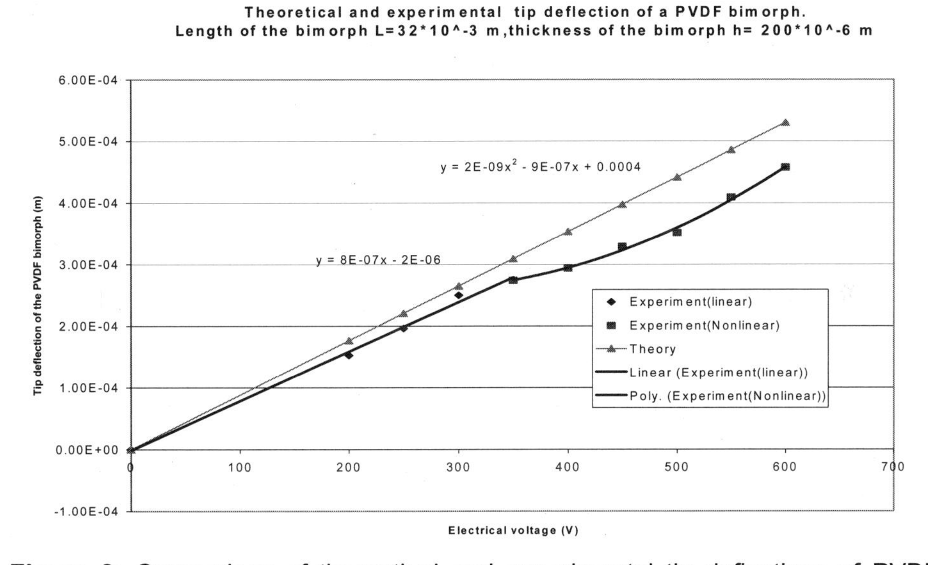

Figure 8: Comparison of theoretical and experimental tip deflections of PVDF bimorph actuator [from Duvvuru, 2003]. At about 350 V, the tip deflection of the physical test article begins to diverge from linear theory.

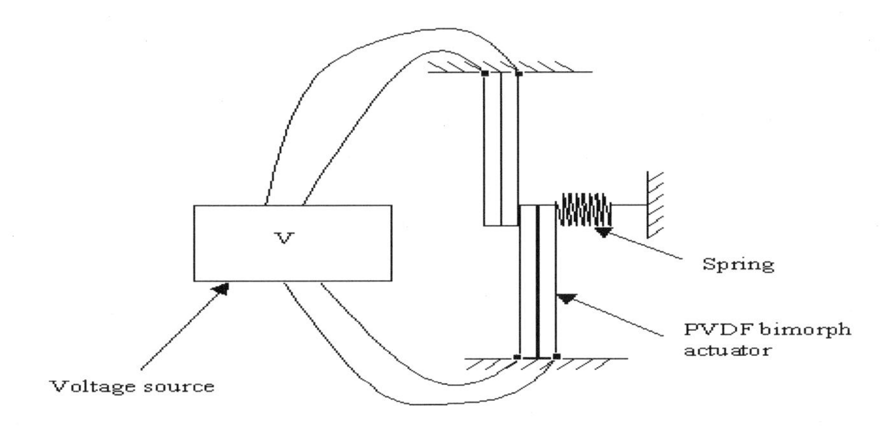

Figure 9: Schematic diagram (plan view) of the experimental set up of a spring attached to the ends of two PVDF bimorph actuators [from Duvvuru, 2003].

Figure 10 shows the displacement of the spring due to single and double bimorphs actuators. It is clear that combining multiple PVDF bimorph actuators readily increases the actuation force available.

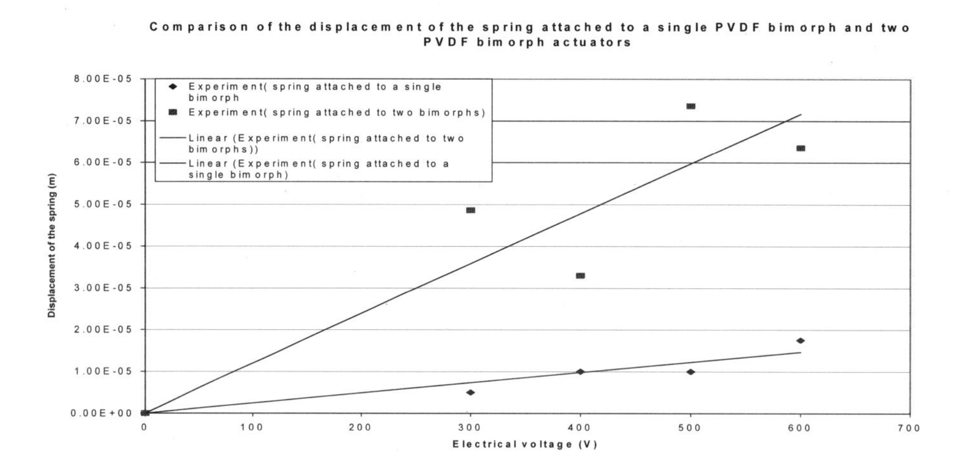

Figure 10: Comparison of the displacement of the spring attached to a single PVDF bimorph actuator and two PVDF bimorph actuators.

As a simple demonstration of the application of bimorph actuators, one might consider them placed along spines (possibly made of a shape-memory alloy running inside seam pockets) within an "inner boundary" of a PGA (Fig. 11). The bimorph actuators would be attached to the seam and when activated could afford additional shape control as shown in Figs. 12 and 13.

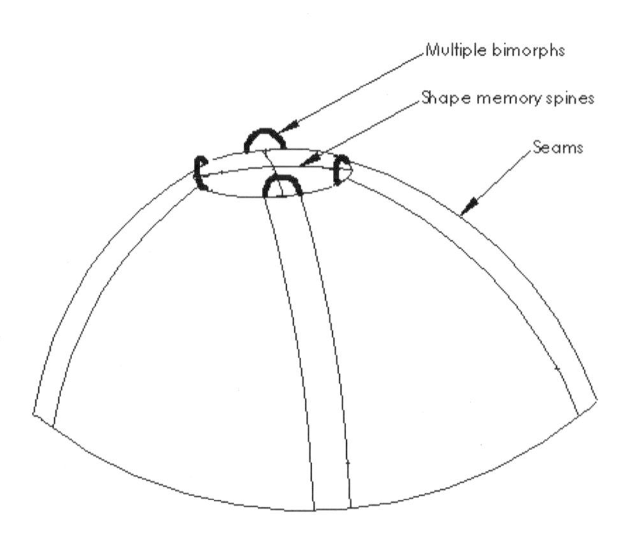

Figure 11: PGA concept using multiple PVDF bimorphs and shape memory deployment actuators.

Figure 12: Experimental set up. One end of the seam is attached to the tips of the double PVDF bimorph actuators, whose other ends are fixed in a wheel-like insulating foam rim [from Duvvuru, 2003].

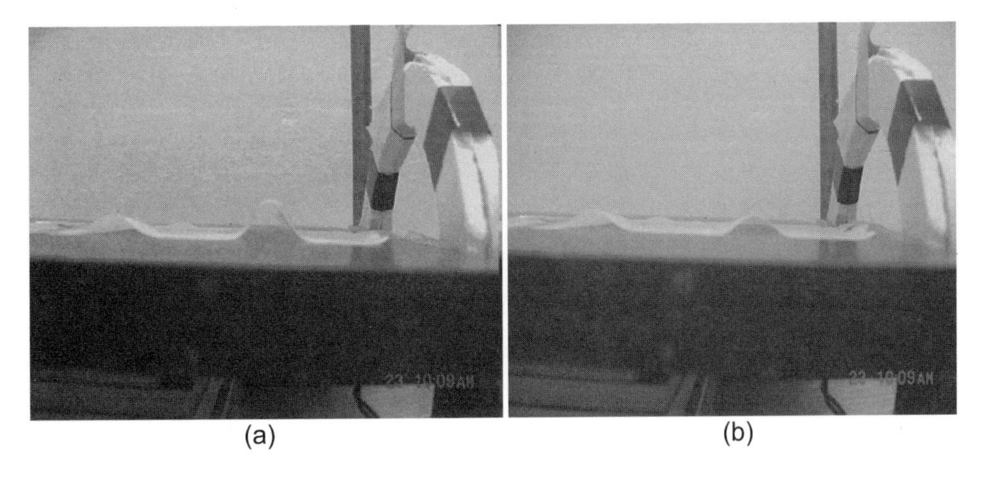

(a) (b)

Figure 13: (a) Initial position of the gore showing wrinkles. (b) Final position of the gore with wrinkle suppression after activation of the PVDF actuators [from Duvvuru, 2003].

20.3.2 Other Electroactive Polymers

In order to control membrane shape, Bar-Cohen has proposed to develop an electroactive polymer (EAP) fiber (see Fig. 14) that can be used as an imbedded actuator similar to biological muscles. The actuators will be based on a novel helically electroded composite EAP fiber (HECEF) [Bar-Cohen, et al, 1996 and 1999] that would contract under electrical excitation. Coulomb forces will induce the actuation where the thin layer (several microns) between the electrodes will induce large strains under low voltages while maintaining low power consumption.

In contrast to the piezoelectric effect, where the displacement changes linearly with the field, the electroactivity phenomenon (i.e., electrostriction and electrostatics) is related to the square of the field. The use of fiber bundles would provide redundancy, increased reliability, and larger actuation force. To make a contractile fiber actuator, the electric field would be directed along the fiber axis, which would be achieved by winding the electrodes with their surface normal aligned parallel to the fiber axis. One electrode of the pair needs to be terminated further away than the other on each end of the fiber in order to serve as a pole for the electro-activation.

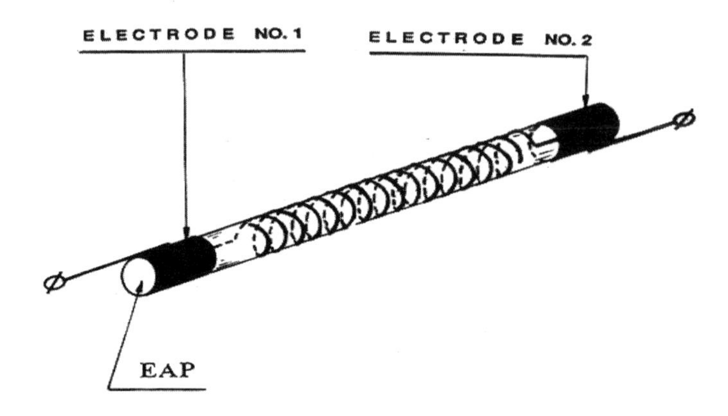

Figure 14: Helically electroded EAP fiber offers an electrically contractile actuator [Bar-Cohen et al., 1996 and 1999].

20.4 Conclusions

Two distinct, but not exclusive, trends will likely emerge in space technology development over the next several decades. Both are driven by the need to expand the possibilities of science while keeping costs down. One is the on-going trend toward miniaturization of hardware, which will keep launch mass low while still providing considerable scientific capability. The other is the trend toward larger spacecraft, because some missions will only be enabled by large collecting areas. Gossamer spacecraft technologies will play an important role in both of these trends.

Gossamer spacecraft still face the same launch mass and volume limits as any other spacecraft. Hence, one of the many challenges facing gossamer spacecraft is making them large while keeping the total mass at or below the launch threshold; in other words, reducing the areal density (mass/aperture area). At the same time, gossamer spacecraft must meet the same maximum volume constraints, and thus their packing efficiencies (deployed volume/packaged volumes) must be high. All of this implies a highly compliant structural/material system, one whose aperture combines reduced elastic modulus (at least until deployed) and thickness.

But the need to reduce mass and stiffness, which increases structural compliance, is antithetical to the need for control of structural precision and dimensional stability. Highly compliant structures are difficult to manufacture, fabricate, and assemble to the same precision tolerances as are more typical stiff structures. Some level of inherent stiffness is required to passively achieve a desired level of dimensional stability. In the case of membrane structures, for example, load-derived stiffness may be decades greater than any inherent stiffness. Simply holding out a polymer sheet from one edge will easily demonstrate its inability to support even its own weight.

The described problems and challenges underscore the need for active control of precision gossamer apertures during the mission lifetime. But what kind of actuator would one specify for a highly compliant structure? Ceramic actuators, whose own stiffness and density are incompatible with that of the aperture? Clearly, active polymer actuators, whose compliance and density more closely match that of the aperture, would be highly desirable.

This chapter has provided an introduction to precision gossamer apertures, their unique structural characteristics, and the potential applications of electroactive polymers (EAPs), including piezoelectric PVDF and others, for the control of their shape. Much still remains to be done to realize the control approaches described, but the promise of gossamer spacecraft provides strong motivation to try.

20.5 Nomenclature

a	aperture diameter
$\mathbf{D}$	material property tensor
E	elastic modulus
E	applied actuation energy/area
f/#	"f-number" or focal length to diameter ratio
h	aperture thickness
k_b	bending stiffness
$\mathbf{r}$	position vector
t	time
w_0	central deflection of the aperture
$\mathbf{x}$	aperture deformation
$\boldsymbol{\sigma}$	stress tensor

20.6 Acknowledgments

The author wishes to express his deepest appreciation to his many graduate students, in particular Mr. Hari Duvvuru, who helped demonstrate many of the author's ideas. Gratitude is also extended to Drs. Yoseph Bar-Cohen (Jet Propulsion Laboratory), Moktar Salama (Jet Propulsion Laboratory), and Aleksandra Vinogradov (Montana State University) for their contributions to this chapter.

20.7 References

Ash, J.A., Jenkins, C.H., and Marker, D.K. (2000), "Deployment of a membrane mirror with a center plunger," *41st AIAA/SDM Space Inflatables Forum*, Atlanta, GA.

Bar-Cohen Y., B. Joffe, J. Simpson and T. Xue, "Electroactive Muscle Actuators (EMA)," New Technology Report, Item No. 9649, Docket 20017, August 22, 1996.

Bar-Cohen Y., S. Leary and T. Knowles, "Helical electroded composite EAP fiber (HECEF) actuators emulating biological muscles," New Technology Report, Submitted on November 12, 1999. Docket No. 20909, Item No. 0505b, November 23, 1999.

Chmielewski, A.B. and Jenkins, C.H. (2000), "Gossamer structures: space membranes, inflatables and other expandables," in *Structures Technology for Future Aerospace Systems*, A.K. Noor, ed., AIAA Progress in Astronautics and Aeronautics Series, 201-268.

Duvvuru, H. (2003), *Active Shape Control of Gossamer Apertures*, M.S. Thesis, South Dakota School of Mines and Technology, Rapid City, SD.

Duvvuru, H. and Jenkins, C.H. (2003), "Active seam control ofgossamer apertures," *48th SPIE Int. Symp. Optical Science and Technology*, San Diego, CA.

Duvvuru, H., Hossain, A., and Jenkins, C.H. (2003), "Modeling of an active seam antenna," 4th *Gossamer Spacecraft Forum, 44th AIAA/ASME/ASCE/AHS/ASC Structures, Structural Dynamics, and Materials Conference*, Norfolk, VA.

Hossain, N.A., Jenkins, C.H., and Hill, L.R. (2003), "Analysis of a membrane-modified perimeter truss mesh antenna," *48th SPIE Int. Symp. Optical Science and Technology*, San Diego, CA.

Jenkins, C.H. (Editor) (2001), *Gossamer Spacecraft: Membrane/Inflatable Structure Technology for Space Applications*, AIAA Progress in Astronautics and Aeronautics Series, vol. 191.

Jenkins, C.H. and Faisal, S.M. (2001), "Thermal load effects on precision membranes," *J. Spacecraft Rockets*.

Jenkins, C.H. and Kalanovic, V.D. (2000), "Issues in control of space membrane/inflatable structures," *IEEE Aerospace Conference*, Big Sky, MT.

Jenkins, C.H. and Marker, D.K. (1998), "Surface precision of inflatable membrane reflectors," *J. Solar Energy Eng,* **120** (4), 298-305.

Jenkins, C.H. and Vinogradov, A. (2000), "Active polymers for space inflatables: properties and applications," *IEEE Aerospace Conference*, Big Sky, MT.

Jenkins, C.H., Ash, J.A., Wilkes, J.M., and Marker, D.K. (2000b), "Mechanics of membrane mirrors," *IASS-ICAM 2000 Computational Methods for Shell and Spatial Structures*, Crete, Greece.

Jenkins, C.H., Ash, J.T., Marker, D.K., and Wilkes, J.M. (2000a), "Near-net shape membrane mirrors using coating stress," *SPIE Opto-Southwest*, Albuquerque, NM.

Jenkins, C.H., Fitzgerald, D., and Liu, X. (2000c), "Wrinkling of an inflated membrane with thermo-elastic boundary restraint," *41st AIAA/SDM Space Inflatables Forum*, Atlanta, GA.

Jenkins, C.H., Kalanovic, V.D., Padmanabhan, K., and Faisal, S.M. (1999). "Intelligent shape control for precision membrane antennae and reflectors in space," *Smart Mat. Struct.* **8**, 1-11.

Jenkins, C.H., Schur, W.W., and Greschik, G. (2001). "Mechanics of membrane structures," in *Gossamer Spacecraft: Membrane/Inflatable Structure Technology for Space Applications,* C.H. Jenkins, ed., AIAA Progress in Astronautics and Aeronautics Series.

Kalanovic, V.D., Jenkins, C.H., and Haugen, F. (1999a). "Fuzzy control of membrane wrinkling," *Intell Automation Soft Comput.* **5**, 139-148.

Kalanovic, V.D., Padmanabhan, K., and Jenkins, C.H. (1999b), "A discrete cell model for shape control of precision membrane antennae and reflectors," *Adaptive Structures and Material Systems Symposium, Int. Mech. Engr. Conf. Expo, ASME*, Nashville, TN (invited).

Main, J.A., G. Nelson, and J. Martin (1998), "Electron gun control of smart materials," in *1998: Smart Structures and Integrated Systems*, M. Regelbrugge, ed., Proc. SPIE Vol. 3329, 688-693.

Marker, D.K., Wilkes, J.M, Carreras, R.A., Rotge, J.R., Jenkins, C.H., and Ash, J.T. (2001). "Fundamentals of membrane optics," in *Gossamer Spacecraft: Membrane/Inflatable Structure Technology for Space Applications*, C.H. Jenkins, ed., AIAA Progress in Astronautics and Aeronautics Series.

Martin, J., J. A. Main, and G. C. Nelson (1998), "Shape control of deployable membrane mirrors," *Proceedings of the ASME International Mechanical Engineering Congress and Exposition, Adaptive Structures and Material Systems*, Anaheim California, November 15-20, 1998, AD-Vol. 57, 217-223.

Salama, M., C.P. Kuo, J. Garba, B. Wada, M. Thomas (1994), "On-orbit shape correction of inflatable structures," *AIAA/ASME Adaptive Structures Forum*, paper No. 94-1771, 348-355.

TOPIC 8

Lessons Learned, Applications, and Outlook

CHAPTER 21

EAP Applications, Potential, and Challenges

Yoseph Bar-Cohen
Jet Propulsion Lab./California Institute of Technology

21.1 Introduction

One of the most attractive characteristics of electroactive polymer (EAP) materials is their actuation potential for the development of biologically inspired (so-called biomimetic) systems that are lightweight, low power, inexpensive, resilient, damage tolerant, noiseless, and agile [Bar-Cohen and Breazeal, 2003]. As described in Topic 7, various applications are currently being considered in an effort to take advantage of these unique characteristics. This chapter describes a broad prospective of the current and potential applications of EAPs, providing both a reference guide and a vision for future potentials. As described in Topic 1 and Topic 3, the number of developed polymers that exhibit significant electromechanical response has been growing steadily since the early 1990s. These materials are establishing an arsenal of choices that may provide alternatives for designers who are considering related applications. To assist potential users of EAPs in assessing the applicability of relevant EAP actuators, these materials were divided in this book into two major groups: ionic and electronic. Typical responses of example materials of these two groups are shown in Figs. 1 and 2, respectively. In Fig. 1, a starfish-shaped IPMC is shown to bend significantly. The direction of bending depends on the voltage polarity. In Fig. 2, a dielectric film is shown with a circular carbon grease electroded area that is activated by an electric field to generate expansion. The expanded elastomer film contracts to the original shape when the electric voltage is turned off. This capability to generate a large strain cannot be matched by alternative electroactive materials such as piezoceramics and shape memory alloys. However, these materials are still in an emerging phase, and further development is needed to address the challenges to their practical application.

The lesson learned by the author from studying potential planetary applications for ionomeric polymer-metal composites (IPMC) is reviewed in this chapter. Also, the application of the dielectric EAP is briefly discussed. While the identified challenges are not common to all EAP materials or even to other ionic types, this lesson shows some of the technical difficulties that need to be addressed. In addition, this chapter provides a summary of the various EAP materials and a broad review of their applications. The materials summary (see Section 21.3 and Tables 35) is given in tabulated form listing the advantages and advantages of the key EAP materials that have been reported in recent years.

Figure 1: IPMC multifinger starfish. (Courtesy of K. Oguro, Osaka National Research Institute, Osaka, Japan.)

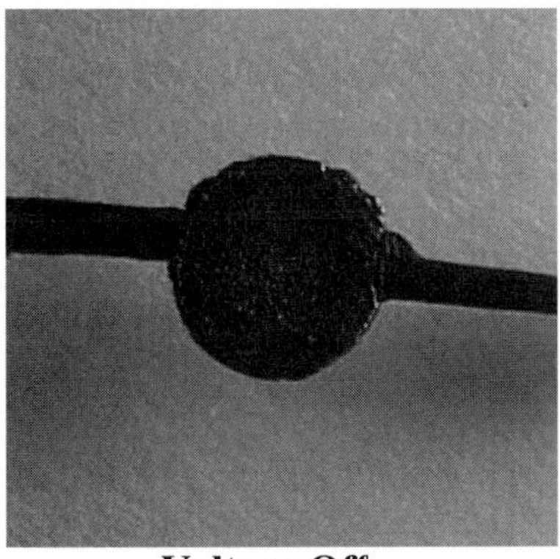

Figure 2: Dielectric actuator demonstrated to expand and relax. (Courtesy of R. Kornbluh and R. Pelrine, SRI International, copyright IASTED.)

21.2 Lesson Learned Using IPMC and Dielectric EAP

To understand the potential challenges involved with employing EAP materials as actuators, the experience that was acquired in seeking applications for IPMC and briefly for dielectric EAP are reviewed in this section. The author and his team made extensive effort to develop effective planetary applications for IPMC, and a number of issues were identified that hamper its immediate application. Detailed information about this material, its actuation mechanism, and a guide for designing related devices are given in Chapter 6 and 13. While the micro- and macro- electromechanical behavior is still not fully understood, methodic modeling and experimental studies have significantly contributed to the knowledge base [Nemat-Nasser and Li, 2000].

21.2.1 Application of IPMC as an Actuator for a Planetary Dust Wiper

Space applications are among the most demanding in terms of the harshness of the operating conditions, requiring an extremely high level of robustness and durability. For an emerging technology, the requirements and challenges associated with making hardware for space flight are very difficult to overcome. However, since such applications usually involve producing only small batches, they can provide an important avenue for introducing and experimenting with new actuators and devices. This is in contrast with commercial applications for which issues of mass production, potential consumer demand, and cost per unit can be critical to the transfer of technology to practical use.

Between 1995 and 1999, under the author's lead, a NASA study took place with the objective of improving the understanding and practicality of EAP materials and identifying planetary applications. The materials that were investigated include IPMC and dielectric EAP, which was named ESSP (Electro-Statically Stricted Polymer), and they were used as bending and longitudinal actuators, respectively. The devices that were developed include a dust wiper, gripper, robotic arm, and miniature rake. The dust wiper (Fig. 3) received the most attention and was selected as baseline in the MUSES-CN mission as a component of the Nanorover's optical/IR window. When a specific mechanism, component, or device is selected as a baseline, it is considered part of the mission hardware. The MUSES-CN mission was a joint NASA and NASDA (National Space Development Agency of Japan) mission that was scheduled for launch in January 2002, from Kagoshima, Japan, to explore the surface of a small near-Earth asteroid. The MUSES-CN mission itself was cancelled due to budget and other considerations, but it gave the field of EAP an important opportunity and brought the field to the public spotlight. The challenges to the application of IPMC and the required cost to overcome these challenges hampered the inclusion of the dust wiper in the final hardware configuration of this mission before its cancellation.

The use of IPMC was investigated jointly with NASA LaRC, Virginia Tech, Osaka National Research Institute, and Kobe University from Japan. The team used a perfluorocarboxylate-gold composite with two types of cations: tetra-n-butylammonium and lithium. An IPMC was used as an actuator to wipe the window with the aid of a novel 104-mg blade having a gold-plated fiberglass brush (Fig. 4), which was developed by ESLI (San Diego, California). When this blade is subjected to high voltage (1–2 KV), it repels dust, thus augmenting the brushing mechanism provided by the blade. A photographic view of the repelled dust and the wiper are shown in Fig. 5. Tests show that when activated in vacuum the heat losses allow the IPMC actuator to respond at temperatures as low as –100°C, where an increase in voltage compensates for the loss in performance efficiency.

Generally, developing a space-flight device requires identifying and addressing all the problems that might be encountered during its operation under the expected mission conditions. For the IPMC dust wiper, a series of issues and solutions were identified and are summarized in this section. The key issues include the critical need to protect the ionic constituents (i.e., avoid dehydration), reduce off-axis deformation, increase the actuation force, sustain the extreme temperature range and vacuum over a period of three years, and prevent electrolysis that causes the emission of hydrogen.

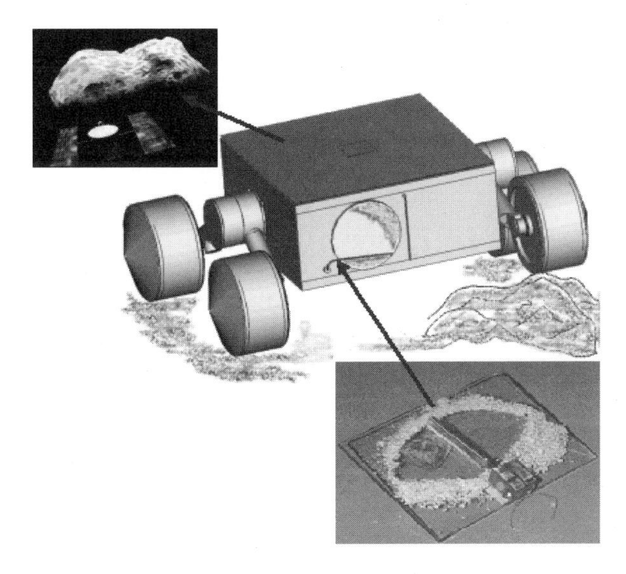

Figure 3: Graphic view of the EAP dust wiper on the MUSES-CN's mission and Nanorover (left) and a photograph of a prototype EAP dust wiper (right).

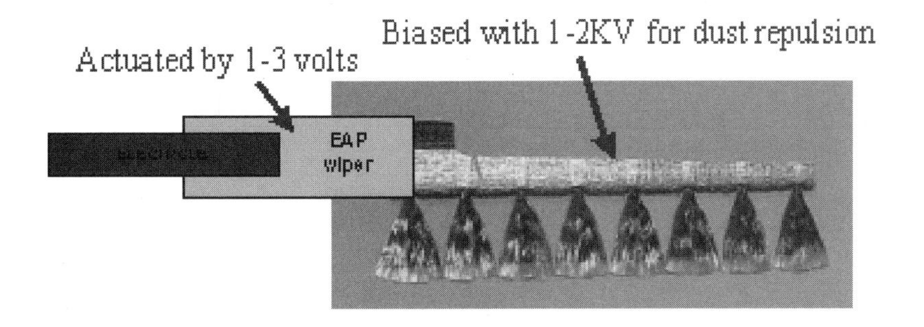

Figure 4: Combined schematic and photographic view of the EAP dust wiper.

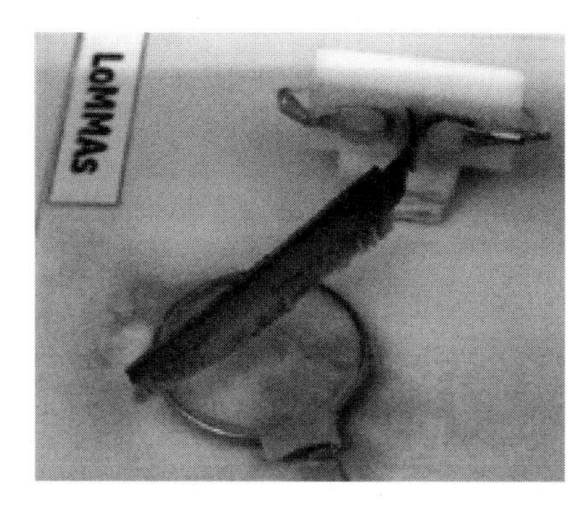

Figure 5: A view of the dust wiper activated with high voltage to repel dust.

The sensitivity of IPMC to dehydration (it is important to note that this problem is common to other ionic EAP such as carbon nanotubes and conductive polymer EAP) and the need to maintain its ionic content were addressed using a protective coating (Dow Corning 92-009). This coating was applied after the IPMC was etched to make it amenable to bonding. The application of such a protective coating emulates the role of biological skin that protects and encapsulates the blood and other life-essential body fluids. Experiments have shown that this coating allows the operation of IPMC in air, but the longest period for which a protected sample maintained response was about four months. Analysis indicates that the selected coating material is water permeable with a permeability constant of about $3 \bullet 10^3$ $cm^3 \times 10^{-9}$ per $sec/cm^2/cm$ at standard temperature and pressure (STP) over a 1 cm • Hg pressure difference (Source: Dow's data for general silicones). Assuming a 2-cm^2 electrode area and a 0.1-mm-thick coating, it is estimated that the water-loss rate is ~40–50mg/24 hrs. As observed experimentally, the actual loss is not so high due to the presence of the metallic electrode layers, which reduce the water diffusion rate. The use of a multilayered coating, possibly consisting of metallic self-assembled monolayering, may offer a better protection.

Complications arise when subjecting IPMC actuators to voltages above 1.23 V, as a result of the electrolysis that takes place. This process raises concerns since hydrogen blisters are formed under the protective coating and are expected to rupture the coating, particularly since there is an extreme vacuum on the asteroid. The electromechanical response of the ionic EAP materials can still be useful even at voltages below the electrolysis level. Further. it is important to mention that this issue of electrolysis is common to all the ionic EAP materials group, including the carbon nanotubes and conductive polymers.

Under dc activation IPMC bends relatively quickly, 0.1 to 1.0 sec (depending on the size of the cations) followed by a slow recoiling with a permanent deformation (see Chapter 6). This recoiling can be a serious issue for IPMC materials with certain cations—particularly for Na^+, for which there is a bending drift in the opposite direction even when the activating voltage is maintained at a constant level. Efforts to understand the cause of this drift have been made by various researchers including Nemat Nasser and Thomas [see Chapter 6] and Kim and Shahinpoor [Kim and Shahinpoor, 2002; Shahinpoor and Kim, 2002]. The application of IPMC is further complicated by the fact that permanent deformation is encountered after intermittent actuation, and in its current state it cannot be relaxed by electrical activation.

In order to address the requirements to survive $-155^{\circ}C < T < +125^{\circ}C$ and operate at $-125^{\circ}C < T < +60^{\circ}C$, efforts were made to develop the equivalent of automobile antifreeze fluid. Aside from the benefits of operation over a broad range of temperature, the use of such an alternative solvent offers the potential of raising the voltage level at which electrolysis occurs. Some of the liquids that were considered included N-methylformamide (NMF), which has a boiling temperature of 199°C, a melting temperature of –4°C, and a flash point of 111°C with a dielectric constant of 182.4. Unstressed strips of Nafion 117 (a polymer base of an IPMC) with an initial size of 5×37.5 mm were immersed for four days in various solvents at ambient temperature (at NASA Langley Research Center) and were tested afterward for mass and dimension changes. A summary of the weight change is listed in Table 1, where a surface area increase of 10 to 35% was observed which resulted from water absorption by the IPMC material. This water absorption was also accompanied by a loss in electromechanical response.

Table 1: Effect of various solvents on the weight of Nafion 117 (a polymer base of an IPMC).

Solvent	Change %
Ammonium hydroxide	11
Dimethylformamide (DMF)	61
Ethanol	57
Ethylene glycol	53
N-methyl-2-pyrrolidinone (NMP)	56
Water	16

In conjunction with the above-mentioned study, the use of fluids such as antifreeze was further pursued by Nemat-Nasser and Zmani [2003], who studied the use of ethylene glycol as a solvent [Chapter 6]. While this study did not address temperature issues, it focused on the material electromechanical behavior in relation to the cation composition. Through ion-exchange reaction various cations were combined in both Nafion- and Flemion-based IPMC materials. The duration of the initial bending toward the anode and the subsequent relaxation toward the cathode for Nafion-based IPMC were found to be controllable by adjusting the cation content. This result seems to promise the potential for tailorable behavior in IPMC. Other alternatives are also being considered including the use of nonvolatile electrolytes such as propylene carbonate

[Madden et al., 2002] and liquid salts (i.e., "ionic liquids"), which was demonstrated by Lu et al. [2002].

The challenges and deficiencies that affect the material performance, controllability and robustness—identified as a result of this lesson learned—and the potential solutions, are listed in Table 2. Advancing IPMC to a mature technology would necessitate overcoming these deficiencies. The challenges listed in this table are making IPMC in its current state unsuitable for the rigorous requirements of space flight hardware, particularly the mission constraints of operating on an asteroid. Future progress or alternative EAPs may provide a robust actuation capability that can be used effectively in such a planetary application.

Table 2: Challenges and identified solutions associated with the application of IPMC.

Challenge	Potential Solution
Fluorinate base—difficult to bond.	Etching the surface makes it amenable to bonding.
Extremely sensitive to dehydration.	Apply protective coating over the etched IPMC.
Off-axis bending actuation.	Constrain the free end and use a high ratio of length/width.
Remove submicron dust.	Use effective wiper-blade design for brushing and high bias voltage for repulsion.
Reverse bending drift under dc voltage.	Limit the operation to cyclic activation to minimize this effect. Also, use large cation-based IPMC.
Protective coating is permeable.	Develop alternative coating, possibly using multiple layers.
Electrolysis occurs at >1.23 V.	Use efficient IPMC that requires low actuation voltage or use solvent base that exhibits electrolysis at a higher voltage.
Residual deformation particularly after intermittent activation.	It occurs mostly after dc activation and it remains a challenge.
Difficulties to assure material reproducibility.	Still a challenge. May be overcome using mass production and protective coating.
Degradation with time due to loss of ions to the host liquid if it is immersed.	Requires electrolyte with enriched cation content of the same species as in the host liquid.

21.2.2 Dielectric EAP

Actuators that are based on EAP materials that change size or shape longitudinally offer the closest resemblance to biological muscles. By scrolling an electroded dielectric EAP [i.e., electrostatically stricted polymer (ESSP)] film to a shape of a rope, an EAP actuator was constructed that becomes longer upon activation (see detailed discussion in Chapters 4 and 16). For this purpose, films with flat flexible electrodes on both surfaces were subjected to an electric field that squeezed the film by Maxwell forces, making it wider while maintaining the material volume (Fig. 6). The produced actuator was used to lift and drop a graphite/epoxy rod serving as the equivalent of a robotic arm (see Fig. 7).

The planar configuration of such an EAP actuator with simple flat electrodes (Fig. 6) allows the production of large lateral extension when electrically activated, and the strain is proportional to the square of the field. Forming electrodes with various patterns offers the potential capability to control the film deformation and possibly leads to contraction [Chapter 12]. The merit of this approach lies in the ability to manipulate and control the shape of the electric field. After activating this actuator, the arm sustains a series of oscillations that need to be dampened to allow accurate positioning. This requires sensors and a feedback loop to support the kinematics of the system control. Several alternatives were explored, including establishment of a self-sensing capability, but more work is needed before such an arm can become practical.

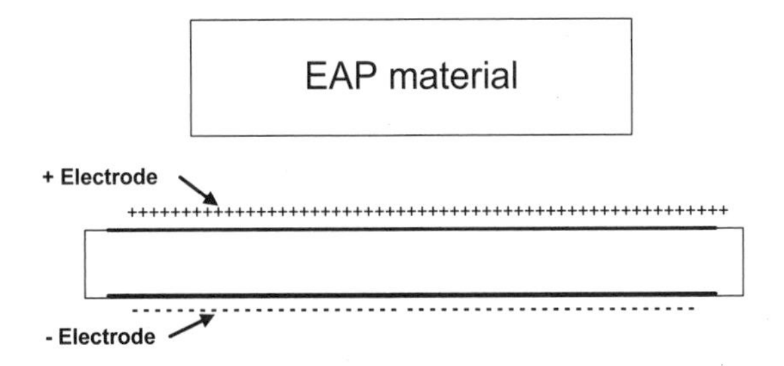

Figure 6: An electric field applied onto elastomer causes thickness contraction and lateral extension.

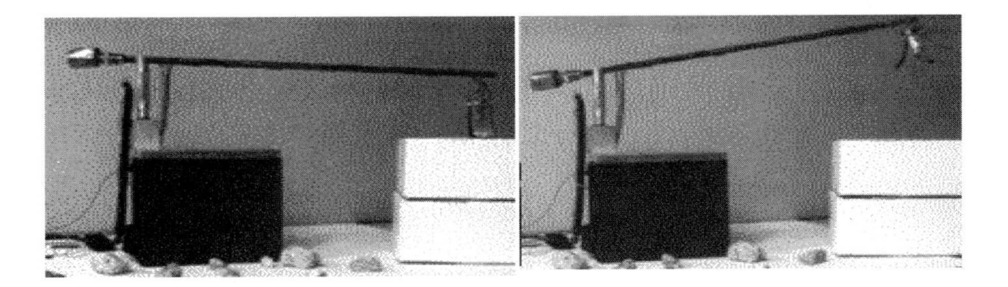

Figure 7: A dielectric EAP actuator lifts/drops a miniature robotic arm with a four-finger gripper.

21.3 Summary of Existing EAP Materials

The various EAP materials that were described in Topic 3 have advantages and disadvantages that determine both their applicability and practical use. Table 3 provides a summary of the two EAP groups and Tables 4 and 5 cover specific types of ionic and electronic EAP materials, respectively.

Table 3: A Summary of the advantages and disadvantages of the two basic EAP groups.

EAP type	Advantages	Disadvantages
Electronic EAP	• Can operate in room conditions for a long time • Rapid response (msec levels) • Can hold strain under dc activation • Induces relatively large actuation forces	• Requires high voltages (~150 MV/m). Recent development allowed for (~20 MV/m) in ferroelectric EAP. • Requires compromise between strain and stress, where >300% was demonstrated to have a relatively low actuation force. • Glass transition temperature is inadequate for low-temperature actuation tasks and, in the case of Ferroelectric EAP, high temperature applications are limited by the Curie temperature • Mostly, producing a monopolar actuation independent of the voltage polarity due to associated electrostriction effect.
Ionic EAP	• Produces large bending displacements • Requires low voltage • Natural bi-directional actuation that depends on the voltage polarity.	• Except for CPs and NTs, ionic EAPs do not hold strain under dc voltage • Slow response (fraction of a second) • Bending EAPs induce a relatively low actuation force • Except for CPs, it is difficult to produce a consistent material (particularly IPMC) • In aqueous systems the material sustains electrolysis at >1.23 V. • Need for an electrolyte and encapsulation. • Low electromechanical coupling efficiency.

21.4 Scalability Issues and Needs

Biological muscles change relatively little among species, indicating a highly optimized system (Topic 2). The main difference that distinguishes the various species is their scale factor. While EAP materials can be used to produce biologically inspired systems, scaling is expected to become an issue in EAP materials that do not exhibit linear behavior [Liu and Bar-Cohen, 1999]. Miniaturization of EAP to micro-, nano- and molecular scales is increasingly becoming a subject of research [Chapter 9; Chapter 17; Smela, et al., 1995, Smela, 1998; Smela, et al., 1999; Smela, 1999, Jager, et al., 2000]. While great emphasis is currently placed on research of miniature EAP mechanisms, making large-scale devices is still a limitation of this field and needs to be addressed as the technology evolves. This up-scaling limitation is particularly an issue in ionic EAP materials, where the low voltages and low electromechanical coupling efficiencies necessitate high electric currents [Madden et al. 2002]. One of the areas where large-scale EAP actuators are critically needed is the development of gossamer structures for space applications [Chapter 20]. Such structures are required in dimensions from many meters to possibly kilometers and may include antennas, solar sails, optical components (e.g., reflectors), and others in future NASA missions.

Conventional-scale design, as used for human-sized machines and larger, is a mature and well-established field. For small-scale mobile robots, the changes in dynamic behavior with changing size cause less than optimal the scaling down of conventional designs. For example, wheeled vehicles rely on the normal force between the wheels and ground for traction (the force is roughly proportional to the weight of the vehicle). Assuming invariance of density as a vehicle decreases in size, the weight of a geometrically similar but smaller vehicle is less by the cube of the geometric scaling. Therefore, small-wheeled vehicles have significant traction problems particularly when designing vehicles that operate in a low-gravity environment. Additionally, continuous rotation bearings exhibit proportionally more friction at small scales making the small-scale rotating mechanism less efficient than larger scale counterparts. Similar scaling issues arise when designing flying devices. Conventional-scale flight relies on Bernoulli-type (relatively inviscid) flow, which generally occurs for $Re > 10^4$. On the other hand, insect flight involves $1 < Re < 10^3$, which is far below the practical realm of typical fixed-winged flight [Fischer et al., 1999b]. Biology offers countless existence proofs of successful small-scale crawling and flying robot designs. These designs incorporate highly successful legged approaches for terrestrial locomotion and flapping approaches for flight. The primary problem in achieving winged flight or legged locomotion is developing a suitable musclelike actuator. Electromagnetic motors are used almost universally in larger scale wheeled terrestrial vehicles or propeller-driven flying vehicles, but these motors are typically not appropriate for legged or winged locomotion (also see Chapter 3). EAP actuators potentially offer the bandwidth, power-to-weight ratio, amplitude of motion, and configuration possibilities that are needed to develop useful insectlike robots.

21.5 Expected and Evolving Applications

As the technology evolves, the challenge that the editor of this book posed to the science and engineering community to develop an EAP robotic arm that can win against a human in a wrestling match is becoming closer to reality. Such a competition may occur in the next few years as scientists and engineers (such as those from SRI International) are increasingly becoming capable of developing effective EAP actuators and design capabilities. In recent years, there has been significant progress in the field of EAP toward making practical actuators, and commercial products are starting to emerge. As mentioned in Chapter 1, the first product in the form of a fish robot is now available from Eamax, Japan. A growing number of organizations are now exploring other potential applications for EAP materials, and cooperation across many disciplines are helping to overcome some of the challenges. To assist in promoting collaboration among developers and potential users of the technology, the author initiated in 1999 and

Table 4: Summary of the leading electronic EAP materials.

Principle	Advantages	Disadvantages	Reported Types
Ferroelectric Polymers			
Polymers that exhibit noncentrosymmetric sustained shape change in response to electric field. Some of these polymers have spontaneous electric polarization making them ferroelectric. Recent introduction of electron radiation in P(VDF-TrFE) copolymer with defects in their crystalline structure dramatically increased the induced strain.	• Induce relatively large strain (~5%). • Offer high mechanical energy density resulting from the relatively high elastic modulus • Permit ac switching with little generated heat • Rapid response (msec levels)	• Require high voltage (~150 MV/m). Recent development allows an order of magnitude less voltage. • Difficult to mass produce • Making thin multilayers is still a challenge and sensitive to defects. • High temperature applications are limited by the Curie temperature	• Electron-radiated P(VDF-TrFE) • P(VDF-TrFE) Terpolymers • P(VDF-TrFE-CTFE) - CTFE disrupts the order in place of the irradiation.
Dielectric EAP or ESSP			
Coulomb forces between the electrodes squeeze the material, causing it to expand in the plane of the electrodes. When the stiffness is low a thin film can be shown to stretch 200-380%.	• Large displacements reaching levels of 200–380% strain area • Rapid response (msec levels) • Inexpensive to produce	• Require high voltage (~150 MV/m) • Obtaining large displacements compromises the actuation force • Require prestrain	• Silicone • Polyurethane • Polyacrylate
Electrostrictive Graft Elastomers			
Electric field causes molecular alignment of the pendant group made of graft crystalline elastomers that are attached to the backbone.	• Strain levels of 5% • Relatively large force • Cheaper to produce • Rapid response (msec levels)	• Require high voltage (~150 MV/m)	Copolymer – poly(vinylidene-fluoride-trifluoroethylene)
Liquid Crystal Elastomers			
• Exhibit spontaneous ferroelectricity • Contracts when heated offering no-electroactive excitation	• When heated it induces large stress and strain (~ 200kPa and 45%, respec-tively) • Requires much lower field than ferroelectrics & Dielectric EAP (1.5 MV/m, 4 % strain). • Fast response (<133 Hz).	• Low electro-strictive response • Slow response • Hysteresis	• Polyacrylate • Polysiloxane

Table 5: Summary of the leading Ionic EAP materials.

Principle	Advantages	Disadvantages	Reported types
Ionic Gels (IGL)			
Application of voltage causes movement of hydrogen ions in or out of the gel. The effect is a simulation of the chemical analogue of reaction with acid and alkaline.	• Potentially capable of matching the force and energy density of biological muscles • Require low voltage	Operate very slowly —it would require very thin layers and new type of electrodes to become practical	Examples include: PAMPS, Poly(vinyl alcohol) gel with dimethyl sulfoxide, and Polyacryl-onitrile (PAN) with conductive fibers
Ionic Polymer Metal Polymers (IPMC)			
The base polymer provides channels for mobility of positive ions in a fixed network of negative ions on interconnected clusters. Electrostatic forces and mobile cation are responsible for the bending.	• Require low voltage (1–5 V) • Provide significant bending •	• Low frequency response (in the range of 1 Hz) • Extremely sensitive to dehydration • dc causes permanent deformation • Subject to hydrolysis above 1.23V • Displacement drift under dc voltage.	Base polymer: • Nafion (perfluorosulfonate made by DuPont) • Flemion (perfluorocaboxylate, made by Asahi Glass, Japan) **Cations**: tetra-n-butylammonium, Li+, and Na+ **Metal**: Pt and Gold
Conductive Polymers			
Materials that swell in response to an applied voltage as a result of oxidation or reduction, depending on the polarity causing insertion or de-insertion of (possibly solvated) ions.	• Require relatively low voltage • Induce relatively large force • Extensive body of knowledge • Biologically compatible	• Exhibit slow deterioration under cyclic actuation • Suffer fatigue after repeated activation. • Slow response (<40 Hz)	Polypyrrole, Polyethylenedioxythiophene, Poly(p-phenylene vinylene)s, Polyaniline, and Polythiophenes.
Carbon Nanotubes (CNT)			
The carbon-carbon bond of nanotubes (NT) suspended in an electrolyte changes length as a result of charge injection that affects the ionic charge balance between the NT and the electrolyte.	• Potentially provide superior work/cycle and mechanical stresses • Carbon offers high thermal stability at high temperatures <1000°C	• Expensive • Difficult to mass produce	Single- and multi-walled carbon nanotubes
Electro-rheological Fluids (ERF)			
ERFs experience dramatic viscosity change when subjected to electric field causing induced dipole moment in the suspended particles to form chains along the field lines	• Viscosity control for virtual valves • Enable haptic mechanisms with high spatial resolution	• Requires high voltage	Polymer particles in fluorosilicone base oil

is currently maintaining various forums, including a series of related websites, the semiannual WW-EAP Newsletter ,the SPIE Electroactive Polymer Actuators and Devices (EAPAD) Conference, and the EAP-in-Action Session that is held during the EAPAD Conference.

Some of the mechanisms and devices that are being considered are related to aerospace, automotive, medical, robotics, exoskeletons, articulation mechanisms, entertainment, animation, toys, clothing, haptic and tactile interfaces, noise control, transducers, power generators and smart structures. Several mechanisms and devices that are using or potentially can use EAPs were described in detail in Topic 7. This section provides a broader review of the applications with a brief description of the specific examples.

21.5.1 Human-Machine Interfaces

Interfacing between human and machine to complement or substitute our senses can enable important capabilities for possible medical applications or general use. Since 1995, a number of such interfaces, with some that employ EAP, were investigated or have been considered. Of notable significance is the ability to interface machines and the human brain. Such a capability addresses a critical element in the operation of prosthetics that may be developed using EAP actuators. A development by scientists at Duke University [Wessberg et al., 2000 and Mussa-Ivaldi, 2000] enabled this possibility where electrodes were connected to the brain of a monkey, and, using direct brain stimulation, the monkey operated a robotic arm both locally and remotely via the internet. Using such a capability to control prosthetics would require feedback to allow the human operator to "feel" the environment around the artificial limbs. Such feedback can be provided with the aid of tactile sensors, haptic devices, and other interfaces. Beside providing feedback, sensors will be needed to allow the users to monitor the prosthetics from potential damage (heat, pressure, impact, etc.) just as we are doing with biological limbs. This section describes the leading EAP interfaces that have been reported in recent years.

21.5.1.1 Haptic Interfacing — "Feeling" Virtual or Remote Stiffness and Forces

One of the shortcomings of robotic technology is the unavailability of an efficient haptic interface that allows human operators to "feel" the stiffness, forces, temperature, and vibration of an object that is being remotely or virtually manipulated. As described in Chapter 19, a system concept using electro-rheological fluids (ERF) was demonstrated that enabled such haptic capability [Bar-Cohen et al., 2000]. An operational system is envisioned consisting of myriad miniature sensors (gauging strain, force, pressure, temperature, load, etc.), a miniature novel device with electrically controlled stiffness (ECS) elements, and compact actuators that allow mirroring the stiffness and reaction forces at the mirrored site. Once miniaturized, ECS-type elements can be installed in a glove at various joints (such as those in human fingers), and they can be electrically controlled as needed to emulate feedback forces. Further, the mechanism can potentially be used to produce an exoskeleton that augments the mobility of humans as well as allowing remote or virtual operations. Potential applications of this technology include training medical staff in executing surgical procedures using virtual reality [Fisch et al. 2003]. Also, it may allow operating as remote presence by controlling such robots as the Robonaut (robotic astronaut) that was developed at the NASA Johnson Space Center. A schematic view of an ECS, its strategic placements on a glove and the Robonaut are shown in Fig. 8. A detailed description of this haptic interface technology and the related system, MEMICA (MEchanical MIrroring using Controlled stiffness and Actuators), is given in Chapter 19.

Successful development of a MEMICA simulator using virtual reality can potentially benefit medical therapy in space and at distant human habitats. The probability that a medical urgent care procedure will need to be performed in space is expected to increase with the growth in duration and distance of manned missions. A major obstacle may arise as a result of the unavailability of on-board medical staff capable of handling every possible medical procedure that may be required. To conduct emergency treatments and deal with unpredictable health problems the medical crews will need adequate tools and capability to practice the necessary procedure to minimize risk to the astronauts. With the aid of all-in-one-type surgical tools and a simulation system, astronauts with medical background would be able to practice the needed procedures and later physically perform the specific procedures. Medical staff in space may be able to sharpen their professional skills by practicing existing and downloaded new procedures.

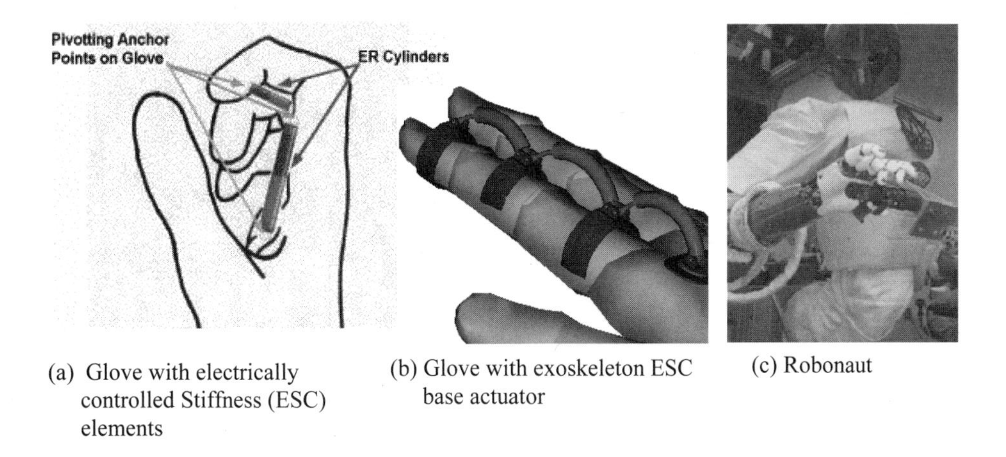

(a) Glove with electrically controlled Stiffness (ESC) elements

(b) Glove with exoskeleton ESC base actuator

(c) Robonaut

Figure 8: A schematic view of ECS elements that employ ERF making an operator “feel” the stiffness at a virtual or remote site. Such a system can potentially mirror the mechanical response of such robots as Robonaut. (Courtesy of NASA Johnson Space Center, Houston, TX.)

Generally, such a capability can also serve people who live in rural and other remote sites with no readily available full medical care capability. As an education tool employing virtual reality, training paradigms can be changed while supporting the trend in medical schools towards replacing cadaveric specimens with computerized models of human anatomy.

21.5.1.2 Tactile Interfaces

21.5.1.2.1 Simulated texture

To create complex tactile feeling (such as touching and smelling), there is a need to fuse elementary sensations that are generated by different sensing receptors [Konyo et al., 2000]. Under a joint study at Kobe University and Osaka National Research Institute (ONRI), Japan, IPMC actuators [Oguro et al., 1992] were used to produce a tactile interface to provide the human operator the required stimuli. These investigators assumed that touch sensation receptors can be generated in response to a combination of high and low frequency vibrations, whereas a sense of pressure can be produced by stimulations in the normal and shearing directions. In order to simulate a touch feeling of texture it was necessary to have a high density of distributed stimuli. A high-frequency drive can generate a vibratory stimulus that taps the skin surface, whereas a low frequency produces a large motion to stroke the skin surface in the shearing direction, projecting a sense of pressure. IPMC was chosen because it offers sufficient softness, is easy to shape, requires a relatively simple control algorithm, and it is activated by a safe, low drive voltage. A device was designed and constructed with a number of IPMC cilia, where each cilium was 2 mm wide by 5 mm long (Figs. 9 and 10).

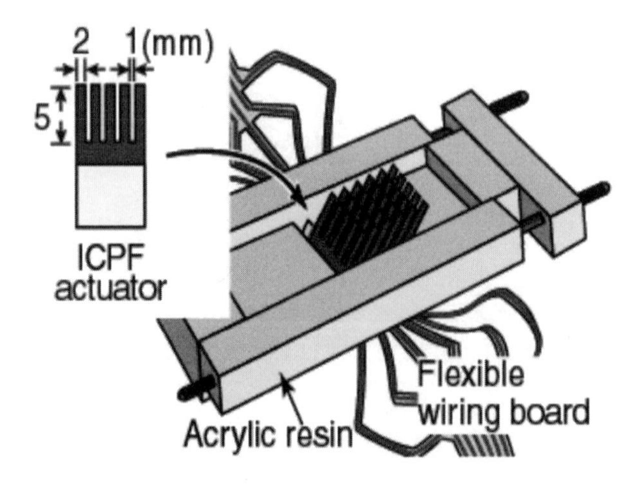

Figure 9: The structure of the ciliary device. (Courtesy of S. Tadokoro, Kobe University, Japan.)

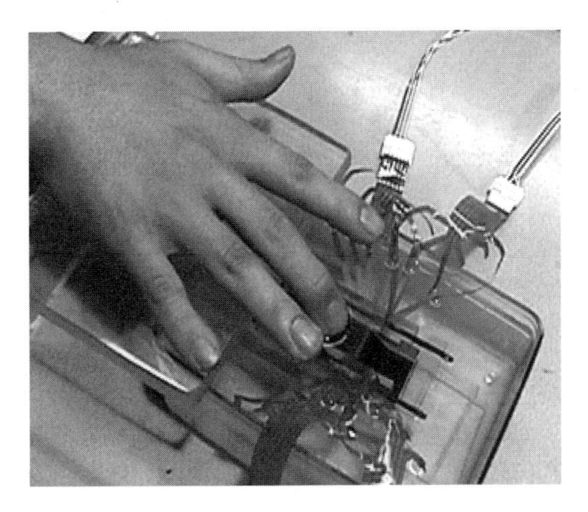

Figure 10: Experimenting finger touch of a tactile interface with controlled sensation. (Courtesy of S. Tadokoro, Kobe University, Japan.)

Experiments made with human subjects showed that over 80% of the subjects sensed a tactile feeling, with about half equating their feeling to rubbing a towel or denim. These results demonstrated that such a haptic display could project a subtle distinction of the tactile feeling of clothing. They also showed the potential of this technology to provide tactile interfacing.

21.5.1.2.2 Orientation indicator—smart flight or diving suits

The need for the capability to sense the orientation of the human body when diving, flying, or cruising in space away from earth is another area that could benefit from tactile interfacing. It is well known to divers and pilots that there are conditions in which they lose their sense of orientation making it difficult to determine which way is up or down. Such a condition can lead to critical consequences, and it is essential to assist such operators so that they rapidly regain their sense of orientation. This issue is also expected to become important to astronauts who will participate in future missions that involve significant departure from the area of earth. The earliest possibility of such a condition could occur when humans are launched to Mars. To overcome this sense of orientation loss one can develop a suit that contains mechanical stimulators at various locations along the chest and/or back. These stimulators can be actuated by EAPs and effectively tap or scratch locally the skin of the operator to create a sensation. The location of the stimulation can be synchronized in relation to gravity to provide an intuitive sense of direction. The EAP that would be used will need to be optimized to provide the necessary forces and speed of response while maintaining softness and compatibility with human factors, such as comfort and convenience.

21.5.1.2.3 Active tactile display device for reading by a blind person

EAP can be constructed in a form that can be used as an active tactile display device to present textual and graphical information to a blind person [Bar-Cohen, 1998]. The display medium can be constructed as a planar array of small cones called "reading pins" (see Fig. 11). Under computer control, reading pins would be lowered individually or in groups to produce a tactile pattern of highs and lows representing the information to be read. A person would read the pattern by scanning with the fingertips, as in reading conventional Braille print. The pins would be lowered by the use of an EAP film, such as the dielectric type that contracts significantly when subjected to an electric field. The reading pins would be mounted on an electrically insulating rubbery film on top of the EAP film, with the film made of a highly electrically resistive substrate. A selected reading pin would be lowered by applying a voltage across the thickness of the EAP film at the location directly under the pin. Electrodes to apply voltages at such locations would be formed on the top and bottom surfaces of the EAP film. The electrodes on each surface would be made of evenly spaced, parallel conductive strips, and the top and bottom electrode arrays would be crossed to obtain a square grid corresponding to the locations of pins to be lowered. Because the actuated pins would be pulled down, the information to be displayed

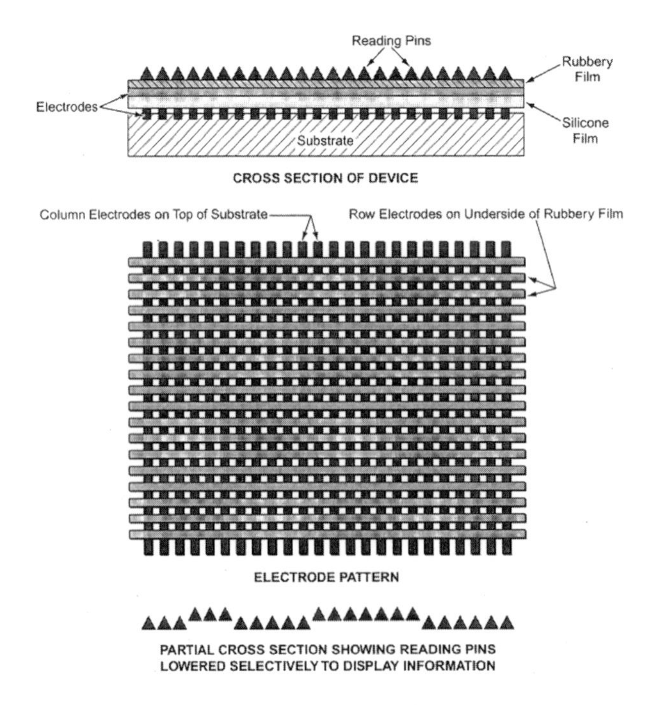

Figure 11: An EAP film at the intersection between selected row and column electrodes to which a voltage would be applied would pull down reading pins.

would have to be formatted analogously to an image on negative film; for example, pulling down pins to form valleys between the ridges would form ridges representing lines in an image. The resolution of the display could be selected by choosing the pixel widths of the letters, numbers, and other characters.

In recent years, this concept of the author [Bar-Cohen, 1998] has become a subject of several studies, and there are indications that a potential practical device may emerge as a result [Bar-Cohen, 2003; and Spinks et al. 2003]. At Wollongong University, a prototype device was constructed [Spinks et al. 2003] using four pins that can be controlled to move against a force from a spring to a distance as required and over the desired number of cycles. There are several challenges to the construction of a full screen prototype and currently these researchers are studying various related issues, including manufacturing and mass-producability of a display with as many as 8,000 pins. In parallel, researchers at Penn State University have started exploring the use of their high dielectric composite terpolymer EAP for this application [Zhang, 2003]. Using this material, which is driven by about 20 MV/m to produce 3–4% strain with 30 MPa blocking force [Chapter 4], these researchers are seeking to take advantage of the EAP capability to form a full page display with effective response.

21.5.1.2.4 Artificial nose

An electronic nose is an array of weakly specific chemical sensors, controlled and analyzed electronically, mimicking the action of the mammalian nose by recognizing patterns of response to vapors. Unlike most existing chemical sensors, which are designed to detect specific chemical compounds, the sensors in an electronic nose are not specific to any one compound, but have overlapping responses. Gases and gas mixtures can be identified by the pattern of the responses of the sensors in the array. The electronic nose or artificial nose is a concept that has been discussed since the mid 1980s. There are several such devices that have been built and tested, with some that use chemical sensors in an array [Bartlett and Gardner, 1999; and www.cyranosciences.com]. The technology is now at the level that there are commercially available electronic noses, and they have been applied to environmental monitoring and quality control in such varied fields as food processing and industrial environmental monitoring.

Chemical sensors are made from several different materials and act by several different mechanisms, including conducting polymers and insulating polymers. Conducting polymers such as polyanilines or polypyrroles can be used as the basis for a conductometric sensor, where change at the sensor is read as change in resistance. The ability of conducting polymers to detect a wide variety of compounds can be extended by mixing other polymers with the conductor [Freund and Lewis, 1995]. An electronic nose that uses polymers as the basis of the chemical sensors is under development at JPL (see Fig. 12) for such applications as event monitoring on the International Space Station.

Figure 12: A photographic view of the JPL's ENose system (left) and the chemical sensor array (right). (Courtesy of Margaret A. Ryan and NASA/JPL-Caltech.)

The polymer-based sensors used in the JPL ENose were developed at Caltech [Lonergan et al., 1996]. They are insulating polymers, which have been loaded with a conductive material such as carbon black. A thin film of the polymer/conductor composite will absorb vapor molecules into the matrix and the matrix will change shape, changing the relative orientation of the conductive particles. That change results in a change in resistance, which is used to form the pattern of response. The magnitude of the response can be related to the concentration of vapor, and mixtures of a few compounds can be deconvoluted. The library of compound patterns that the ENose contains depends on the particular space in which it is used and the hazards of that space. New compounds can be added to the library as the device is exposed to them. ENoses in different spaces can be equipped with different polymers in the array and, therefore, a different library. The polymers for an array are selected by molecular structure of the polymer and the target compounds for that array.

21.5.2 Planetary Applications

The possibility of applying EAP materials as actuators of planetary mechanisms was discussed earlier in Sec. 21.2 and in Chapter 20. Generally, the use of polymers in space has evolved to the level that flight hardware structures made of such materials are increasingly part of NASA exploration missions [Chmielewski and Jenkins, 2000]. Some of the applications include the 1997 Mars Pathfinder mission use of a balloon to cushion the landing and the IN-STEP Inflatable Antenna Experiment (shown in Fig. 13), which flew on STS-76 on May 29, 1996 (concept developed by L'Garde, Inc.). The rover that was developed at JPL using inflatable wheels is another example of space application that is under consideration and was described in Chapter 1. Inflatable space structures can be produced to have very large surfaces that can be launched in a packed form,

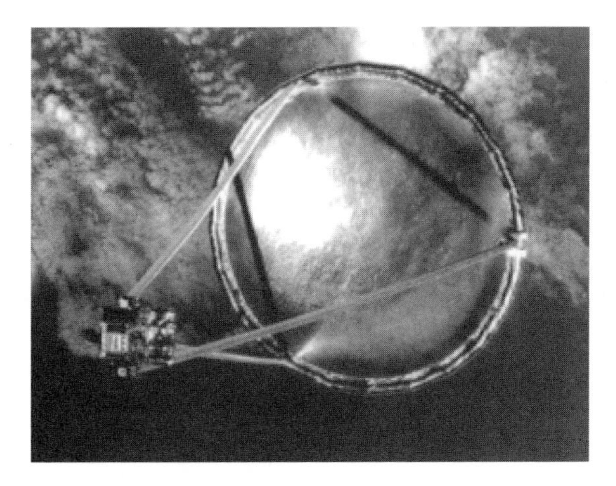

Figure 13: A Space Shuttle view of the L'Garde's Inflatable Antenna Experiment (IAE).

inflated to shape, and then rigidized, creating large structures with a very low mass. An example of inflatable and rigidized synthetic aperture radar is shown in Fig. 14. In order to obtain the maximum benefit from such structures there is a need to precisely control their shape either prior to rigidization or in real time when periodic deformation and shape control are needed. EAP materials offer the potential of providing the necessary actuation technology for such structures. Such gossamer structures are expected to enable missions that are significantly beyond the capabilities of current technologies (see Chapter 20). However, given the current limitations of the EAP technology, such capabilities can only be considered as a long-term goal.

Figure 14: Inflatable and rigidized synthetic aperture radar at the Jet Propulsion Laboratory. (Courtesy of NASA/JPL-Caltech.)

21.5.3 Controlled Weaving

21.5.3.1 Garments and Clothing

Conventional fabrics that are used to make garments and clothing have passive properties. Using EAP fibers consisting of conductive polymers would enable a new era in clothing and gossamer structures allowing controlled configuration and shape. EAP fibers can be actuated and, while maintaining flexibility, they offer the potential to adjust the thermal insulation of the clothing. Moreover, such fibers may support teleoperation and rehabilitation engineering. Sensors may be used to gauge the temperature, mechanical strain, or other properties to determine the desired action, shape, or state of the fabric weaving. At the University of Pisa, Italy, the University of Wollongong, Australia, and Santa Fe Science and Technology, Inc. (New Mexico), studies are underway to develop such fibers using a polyaniline working electrode core coated with solid electrolyte and a counter electrode [Adams et al. 2001; and de Rossi et al., 1999]. Voltage levels that are less than 2 V can be sufficient to drive these fibers, and they are expected to serve as controlled weaving. Some of the applications that are under consideration include antistatic clothing, a smart bra (at the University of Wollongong, Australia), and membranes for chemical separation.

21.5.3.2 Antigravity Suit

Sustaining high gravitational forces (G) during flight is important for military pilots who fly high-performance fighter aircraft and to astronauts during launches into space. Companies such as Celsius Aerotech (Linköping, Sweden) are exploring the possibility of using EAP to construct an antigravity (anti-G) suit to allow pilots to sustain the high forces of acceleration (positive G-forces) [Willy, 1999 and 2000]. The anti-G suit is planned to be constructed of active threads woven into the parts around legs, waist, body, and, if needed, arms. By using smart materials, the control capability can be designed into the suit without degrading the pilot's comfort or mobility. Threads are being sought that can be activated by an electric current to cause contraction using a reversible process. This technology should offer superior comfort for pilots because the suit is lighter, not tight, and has no tubes or valves. It may be possible to control the application of pressure not only as a function of acceleration, but also of time and place on the pilot, and possibly the suit could serve as a replacement of the present chest counter-pressure garment. Such suits may prevent arm pain during high-G exposure, have a short reaction time, and possibly activate a "corset effect" during an ejection process to protect the pilot from spinal injuries. Such capabilities of the G-suit may be integrated with the tactile sensing capability described in Sec. 21.5.2.

21.5.4 Robotics, Toys, and Animatronics

The potential capability of EAPs to emulate muscles may enable robotic capabilities that are in the realm of science fiction when relying on existing actuation materials [Bar-Cohen and Breazeal, 2003; Chapter 17 and 18; Kornbluh et al., 1995]. The large displacement that can be obtained using low mass, low power, and, in some of the EAPs, low voltage, makes them attractive for attempting to produce biologically inspired robots. These biomimetic robots are expected to be agile, damage tolerant, noiseless, lightweight, mass producible, and inexpensive. EAP materials are being sought as a substitute for conventional actuators, possibly eliminating the need for motors, gears, bearings, screws, etc. [Chapter 16; Bar-Cohen et al., 1999b; Fischer et al., 1999a]. Combining the bending and longitudinal strain capabilities of EAP actuators, a miniature robotic arm was designed and constructed at JPL (Sec. 21.2), and it is shown schematically in Fig. 15. The gripper (shown in Fig. 16) consists of four IPMC finger strips with hooks at the bottom emulating fingernails. Experiments

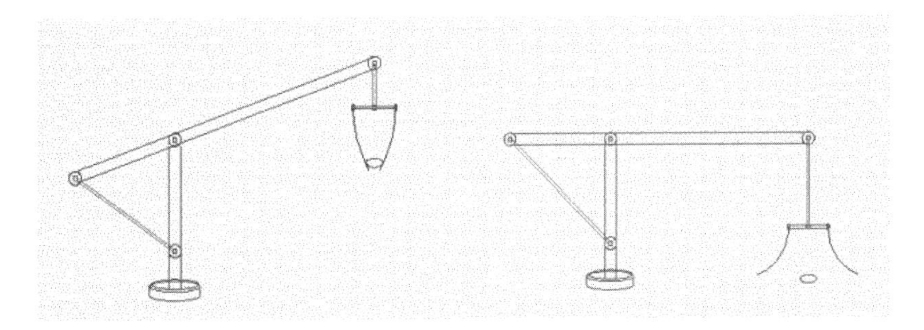

Figure: 15: Simulated view of a robotic arm taking advantage of the capability of longitudinal and bending EAP actuators.

Figure 16: Four-finger EAP gripper lifting a rock.

with this robotic arm raised concerns regarding the ability to control the kinematics of robotic components made of dielectric EAP and of IPMC since these actuators are flexible and do not provide accurate positioning. Upon activation, the arm exhibits low-frequency vibration with a relatively low damping. The inherent vibration-damping characteristics of polymers are not sufficient to suppress this low frequency vibration. Besides effective control algorithms, sensor feedback is expected to be critical to addressing this issue of precision positioning of EAP activated devices [Chapter 17].

21.5.4.1 Biologically Inspired and Insectlike Robots

In principle, musclelike EAP actuators can be used to articulate skeletal structures of biologically inspired robots [Chapter 17]. Oftentimes, nature can provide a model for equivalent mechanical locomotion [Chapter 2 and 3; Fischer et al., 1999a, b]. Developing biologically inspired robots requires: (1) application of specific circuitry; (2) power and regulators integrated into the mobility elements (legs, wings, or fins) to allow a limited degree of self-powered maneuvering (human hands free). Robot mobility requires effective actuators, sensors, and feedback algorithms, drive electronics, control system, and end-effectors (elements that are responsible for the functionality, e.g., a manipulator). The on-board electronic circuitry needs to fulfill the following functions: (1) provide regulated power; (2) control the maneuver (crawling, flying, or swimming); (3) gather mobility status and other information; and (4) communicate with teleoperators. The complete system, which includes structure, actuators, end-effectors, and circuitry, needs to be lightweight, highly functional, and low power. Umbilical systems allow the technology advancement while avoiding elements that cannot be miniaturized or made sufficiently efficient with current capability. Some of the current EAP materials can be used as potential actuator candidates for the construction of simple sea creaturelike robots. Examples of sea creatures that might be emulated are shown in Fig. 17. Bending actuators like conductive polymers, IPMC and others can function in such conditions with no threat of dehydration; however, a protective coating will still be needed to avoid chemical interaction.

One such sea creature, as a commercial product, has already emerged, having the form of a fish robot (Eamex, Japan). Example of this fish robot is shown in Fig. 1 of Chapter 1. This fish robot swims using an inductive coil for energy, thus it operates without batteries. Also, it uses IPMC actuators that simply bend upon stimulation operating as the actuator of the fin and tail, eliminating the need for complex motor, bearing, gears and other rigid elements. This fish represents a major milestone for field—it is the first reported commercial product to use electroactive polymer actuator.

Insects are incredibly fast flyers, crawlers, and swimmers. Their evolution to flying creatures allowed them to escape their predators more easily, as well as to traverse larger distances in search of food and water, to mate, and to colonize in

Figure 17: Emulating sea creatures can be a challenge for which some of the current EAP materials may provide effective actuators.

new areas. They are able to fly over large bodies of water, which are great barriers to nonflying terrestrial animals. Except for flies, all flying insects have two pairs of wings, one attached to the upper mesothorax and the other to the upper metathorax. Insect wings do not move simply by muscles pulling at the base, as one might guess. Instead, two different groups of indirect flight muscles, housed inside the thorax, work to alternately elongate and flatten the thorax in a vertical direction. The wings, wedged between the upper and lower thoracic sections, move by leveraging on a pivotal point, or fulcrum. As the vertical muscles contract, the thorax flattens and the wings move up; then the horizontal muscles contract, pulling the sides in, driving the upper and lower thorax higher, and the wings move down. Smaller direct flight muscles at the base of each wing adjust the angle of the stroke and therefore the direction of flight. The frequency of wing beats varies from species to species, from one individual to another, and even in the same individual at different times. Generally, insects such as butterflies, which have large, light bodies and large wings, need far fewer wing beats to fly than do those with small wings and relatively heavy bodies, such as a housefly or a honeybee. Maximum air speed is also highly variable, but is generally less than 30 km/hr.

Developing EAPs as muscles to flap flat wings, possibly emulating a butterfly, is a great challenge. As a model for effective flying, the monarch butterfly is an incredible migrator that flies at a speed of 120 km/day, crossing significant distances. This butterfly takes advantage of rising thermal airflows to get lift and uses these thermals to glide. Making a wing requires an EAP that can perform flapping at a bandwidth and resonance frequency that consumes minimum energy while maximizing the performance. Use of interdigital electrodes and an electronic-type EAP may allow controlling the deformation of the wings and performance of various maneuvers as needed to cause lift, drag and glide, and to allow navigation. Various wing configurations and electroding techniques can be considered for the control of the wing shape and aerodynamics.

The mobility of insects is under extensive study, and there is already a relatively large body of knowledge at such research institutes as the University of California, Berkeley [Chapter 3; Full and Tu, 1990]. Experiments studying the details of insect walking mechanisms were conducted using a windmill and surfaces with a photoelastic coating, and insects with various numbers of legs were investigated. The reduction in the size of electronic devices has reached such a level that insects can be instrumented to perform tasks that were once viewed as science fiction. In the mid-1990s, researchers at the University of Tokyo, Japan, instrumented various insects (e.g., spider, cricket, and cockroach) carrying backpacks of wireless electronics, making them like miniature "workhorses." Development of EAP actuators is expected to enable insectlike robots that can be launched into hidden areas of structures to perform various inspection and maintenance tasks. This includes operating in ruins of buildings after an earthquake in search of survivors. Application to medical tasks is described later in this section. In future years, EAP may be used to emulate the capabilities of biological creatures with integrated multidisciplinary capabilities, allowing such tasks as space exploration missions that would be performed following an innovative plot in accordance with the mission objectives. Some of the biological capabilities that may be considered include catlike soft landing, traversing distances by hopping like a grasshopper, and digging and operating cooperatively, as ants do.

21.5.4.2 Toys and Animatronics

The ability to mimic human muscles using electroactive polymers enables a variety of devices, including animatronics and toys, which can benefit the entertainment industry [See Chapter 18]. As this technology evolves, biologically inspired toys actuated by EAP materials emulating biological creatures will become a reality. One of the issues that can limit the selection of EAP actuators is the upper voltage limit that is generally being considered safe for use in toys. The current level that is being used is below 24 V, and this limit constrains the consideration of the electronic EAP actuators, which may have greater technology readiness than the ionic types. Mimicking nature would immensely expand the collection and functionality of the toys that are currently available for children. Such toys would be able to perform tasks that are impossible with existing capabilities.

The entertainment [Chapters 17 and 18] and advertising industries may also benefit from EAP. Robots that mimic the behavior of humans may be used to produce animatronics for movies, theaters, TV, and other presentation media. Humanlike or creaturelike robots operating as actors can be programmed to follow details of a script that is dictated by a computer. Moreover, the viewers of future movies or electronic shows (TV, web, etc.) may be immersed in the display/presentation medium by being connected through haptic and tactile interfaces to allow them to "feel the action" at the level of their fingers and toes rather than the current large-scale capability. Further, manikins may make welcoming gestures to customers who come to a store or show products that are being promoted. At the University of Pisa, Italy [Mazzoldi et al., 2000], and the University of Texas, Dallas [Hanson, 2003], the properties of EAP are being exploited for the development of a humanlike android, allowing replication of human facial expressions. An android system (see Fig. 18 and Chapter 18) is being developed to make facial expressions with a capability to taste and smell substances and to mimic human response to the same food tasting and odor analysis. At present, conventional electric motors are producing the required deformations related to the relevant facial expressions. However, efforts are currently under way to examine the use of electroactive polymers to serve as artificial muscles. Data is acquired, stored in a personal computer, and analyzed

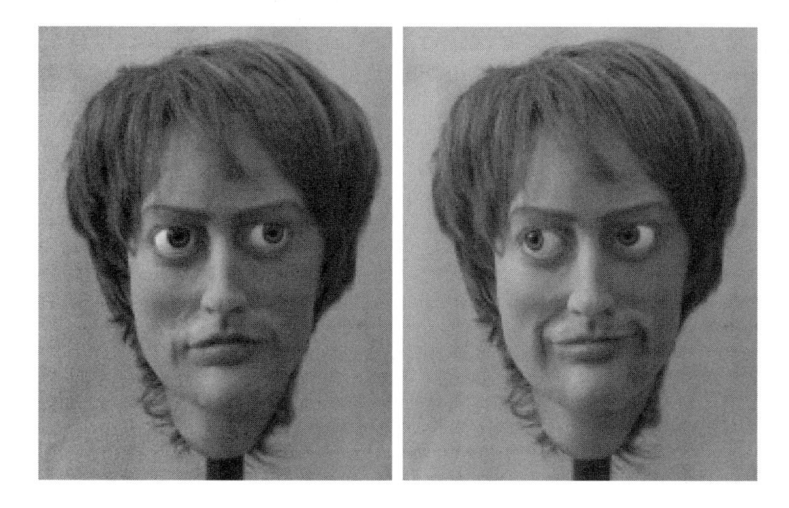

Figure 18: An android head (photographed at JPL) serves as a platform for testing EAP actuators. (The head was provided to JPL as a courtesy of D. Hanson, University of Texas, Dallas.)

through a dedicated neural network. Once effective EAP materials are chosen, they will be modeled into the control system in terms of surface shape modifications and control instructions for the creation of the desired facial expressions.

21.5.4.3 Agriculture and Ecology

Biologically inspired robots can potentially make an enormous impact on agriculture and ecology. Birds may be deterred from making their habitat next to airports, thus preventing risk to aircraft. For agriculture, miniature EAP robots may be used to deter various insects and rodents and these biologically inspired robots may be used to relocate them away from the crop to avoid the insects' damaging effect. Thus, organic produce may be more easily grown, while the environment and life cycle of many species can be better protected. Moreover, such robots may be employed in fruit and vegetable picking, particularly in the case of difficult to access or labor-intensive crops, such as berries. Concern may arise that the birds or other species might eat the EAP-driven insectlike robots, but such an issue may be properly addressed by adding repelling colors, taste, or smell.

21.5.4.4 Rapid Prototyping Miniature Robotics Using Stereolithography

As described in Chapter 15, ink-jet stereolithography and printing techniques can be used for fabrication of polymer devices. A polymer is dissolved in a volatile solvent and ejected drop by drop onto various substrates [Pede et al., 1998]. Ink-jet printing offers the potential of making complete devices driven by EAP actuators, where the devices can be produced in full 3D detail for rapid prototyping and quick mass production. Thus, polymer-made EAP-actuated insectlike robots may be fully fabricated by an ink-jet printing, potentially allowing one to rapidly implement ideas into engineering models and commercial products. Miniature conjugated polymer devices can be fabricated while avoiding chemical reactions between polymer solutions and the substrate. A deposition of conjugated polymers can be accomplished using substrate independent techniques [Chapter 7]. Patterns can be introduced in the deposition steps to make a fast and inexpensive procedure that is independent of the polymer. Further, 3D structures can be constructed by depositing different layers of polymer without damaging the underlying structures as commonly occurs in spin-coating and solvent casting [Chapter 15].

21.5.5 Medical Applications

The growing availability of EAP materials that exhibit high actuation displacements and forces is opening new avenues to bioengineering in terms of medical devices and assistance to humans in overcoming different forms of disability. Areas that are being considered include hearing aids, vocal cords, and rehabilitation robotics. For the latter, exoskeleton structures are being considered in support of rehabilitation or to augment the mobility and functionalities of patients with weak muscles.

21.5.5.1 EAP for Biological Muscle Augmentation or Replacement

The operational similarity of EAP to biological muscles in terms of flexibility, softness, and large strain makes them good candidates to operate as substitutes for human muscles. Further, EAP actuators can potentially be used as mechanisms for augmenting various body functions. A disabled person may in the future wear an exoskeleton that is driven by EAP actuators and be assisted in performing daily tasks that otherwise may be too difficult or impossible to perform. In the coming years, as the technology evolves, EAPs are expected to emerge as viable actuators offering medical options for disabled or ill persons. Toward this possibility, researchers at the Sheffield Hallam University, United Kingdom, have constructed a model of the skeleton of a hand as shown in Fig. 19 [Rust et al., 1998; Whiteley et al., 1999; Rust et al., 1999]. These researchers are seeking (in coordination with the University of Pisa, Italy, and the Jet Propulsion Laboratory, EAP actuators that can be effective in operating the various joints of the emulated skeleton. Employing such actuators offers the potential of producing prosthetics that most closely resemble the performance of the human functionality toward achieving naturally looking movements, which are vital for the self-esteem of the users.

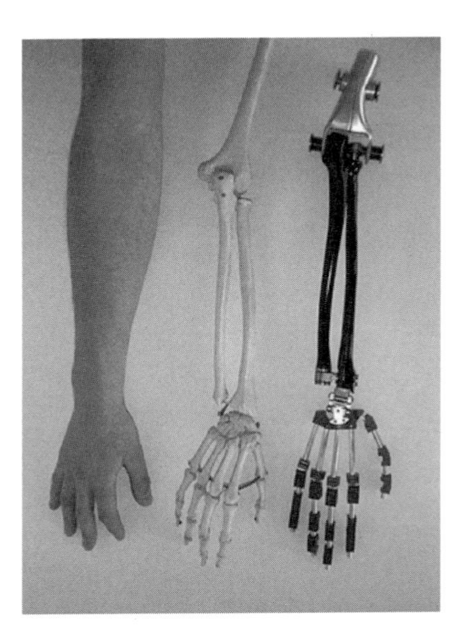

Figure 19: A photographic view of a human hand and skeleton as well as an emulated structure for which EAP actuators are being sought. (Courtesy of Graham Whiteley, Sheffield Hallam University, UK.)

21.5.5.2 Miniature In-Vivo EAP Robots for Diagnostics and Microsurgery

The idea of developing a miniature submarine that can travel through the bloodstream and conduct diagnostics and microsurgical procedures was raised in the mid 1960s. The vision of this imaginative concept was the subject of movies, but the technology in those days was far from enabling such a capability. While this capability is not feasible yet, it is becoming increasingly possible as the technology of miniaturization evolves, and EAP materials are offering an important element of this concept—the actuator. Unfortunately, travel inside the bloodstream may involve risk and will require certain forms of anchoring and control to prevent loss of the miniature vehicle or uncontrolled travel to such critical areas as the brain, where it may cause stroke or other serious consequences. As an alternative to mobility in the bloodstream, one can consider the development of a worm-type robot that travels inside tissues and muscles, much like the biological equivalent that exists in nature. A miniature wormlike robot might serve as a watchdog to detect cancer initiation as well as treat internal organs before the disease reaches a critical stage. Such a capability may help women who are concerned about breast cancer resulting from high genetic susceptibility to avoid the need to use the extreme preventive measure of mastectomy. As miniaturization and robotic technologies evolve, producing such a wormlike robot may become feasible. To take advantage of the evolving technologies, NASA, in cooperation with the National Cancer Institute (NCI), is now seeking to develop this new form of patient care using such robots. A "microscopic explorer" is being sought that would travel through the human body looking for disease, providing warning and helping mitigate medical problems. This technology may someday allow NASA to monitor astronaut health and treat conditions in space, where medical test capabilities and communication with Earth are limited. Developments in EAP would offer potential to support such challenges in terms of size, mass, and capabilities.

21.5.5.3 Catheter Steering Mechanism

As the technology of minimally invasive surgery is becoming a leading form of medical treatment, the need to guide catheters within the maze of the body's blood vessels (veins and arteries) has grown significantly. Use of a joystick-controlled catheter for in-vivo surgery to treat brain aneurysms and other conditions can offer an important tool and save many lives. Numerous mechanisms are being studied, including steering devices that are driven by such actuators as shape memory alloys using sensors to provide real-time data. To maximize the flexibility of operating inside human arteries, it is essential to use a low-stiffness steering device, and EAP actuators offer such a capability. Moreover, fiber optics and other sensors can be inserted through a tubular EAP steering mechanism and provide critical imagery for the medical staff conducting the operation. The use of EAP was investigated by the University of Pisa, Italy

[Della Santa and de Rossi, 1996], and by a joint team from Osaka National Research Institute (ONRI) and Japan Chemical Innovation Institute (JCII) [Onishi et al., 1999]. Della Santa and de Rossi's [1996] initial study involved the use of a catheter steering element (0.8-mm tube) that was made of conductive polymer based on polypyrrole and was demonstrated to bend up to 30 deg with a distal force of about 4 g. In a following study, they used a composite conductive polymer ($PANi/ClO_4$) with solid electrolyte elastomers [containing $Cu(ClO_4)_2$] to produce a catheter steerer that was 30-mm long with a 0.6- to 1.4-mm external diameter [Mazzoldi and de Rossi, 2000]. These investigators determined that articulation of the steering mechanism can be done to meet the catheter operation requirements, and the level of bending depends on the stiffness and external diameter of the EAP actuator.

A similar steering mechanism was explored by the ONRI-JCII team [Onishi et al., 1999] using an IPMC tube that was gold electroded to allow controlling the direction of bending (see Fig. 20). They divided the tube into four segments that were parallel to the steering axis. By applying opposite polarities to IPMC pairs in the plane that crosses the steerer, they were able to bend in any desired direction. This 15-mm long, 0.6-mm outer diameter steerer was demonstrated to bend over 90 deg in all directions using up to 3 V of electric stimulation between the electrode pairs.

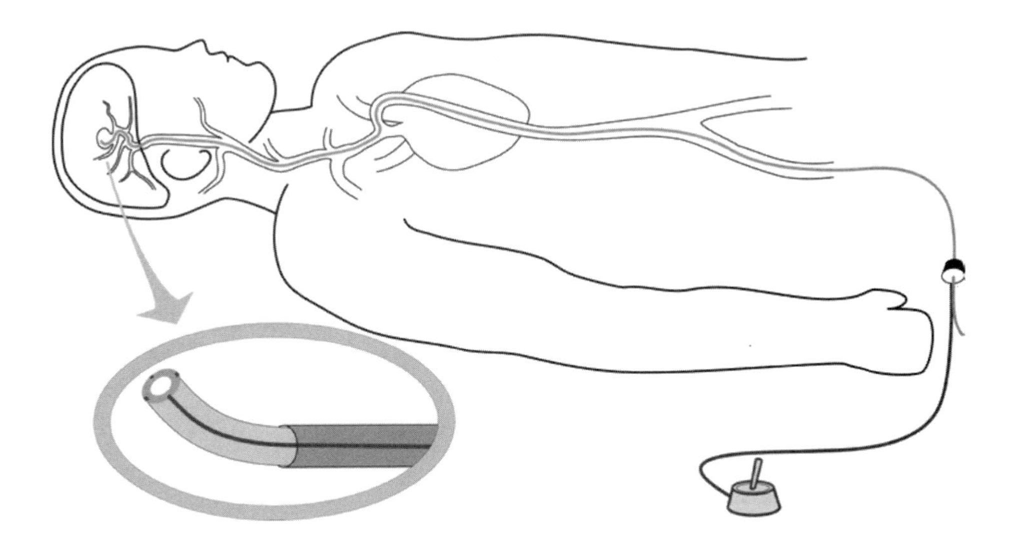

Figure 20: Active catheter guide using an IPMC type bending EAP. (Courtesy of K. Oguro, Osaka National Research Institute, Osaka, Japan.)

21.5.5.4 Tissue Growth Engineering

Recent research in tissue engineering has led to the conclusion that the material and functional properties of tissues depend on the physical and chemical environment in which they are formed [Walker, 2000]. For instance, to grow artificial arteries, the function of the cultured cells changes if the substrate upon which they are grown is strained during the growing process. Therefore, tissue engineering can be greatly enhanced by the capability to softly "move" substrates on a substrate. There are numerous challenges to such capability, including the need for cell compatibility and substrate removal. EAP materials have the potential to be formed into biological shapes and to support the cells' need for low stress and high strain, providing an advantage over other methods. Researchers at the University of Leeds, England, are currently engineering a number of tissues and seeking to perform basic experiments on deforming substrates with the possible use of EAP [Walker, 2000]. The experiments also involve testing the biocompatibility of EAP in support of the requirements and constraints of cell cultures.

21.5.5.5 Interfacing Neuron to Electronic Devices Using EAP

Studies are under way to establish an electronic interface between EAP materials and biological nerves [Otero, 1999]. If such an interface becomes successful, it would enable connecting conventional optical, acoustical, and other sensing devices to the human nervous system to ameliorate various disabilities. Increasingly, microelectronic devices are becoming smaller and significantly more capable than ever before, which, combined with the biocompatibility of some EAP materials, such as polypyrrole, may make such prosthetics realistic. For this purpose, one can use a partially oxidized polyconjugated membrane material that responds on the molecular level to ionic changes in the membrane potential [Otero et al., 1999]. These materials offer transducing from ionic to electric signals. Moreover, partially oxidized conductive polymers respond to cathodic current or polarization by expelling ions. Using transducers that allow the growth of neurons as dendrites within the polymer material may someday allow communication between artificial external electronic systems and human users by appropriate programming and training, respectively.

21.5.5.6 Active Bandage

The capability to control the level of stretching of a flexible fabric (as discussed in Sec. 21.5.1) offers the potential to develop an active bandage. Treatment in hospitals to assist blood circulation in legs is performed today by applying compressed air bladders, much like the operation of present anti-G suits. Using an active bandage activated by EAP materials could enable "smart trousers" using modern control technology. For example, the pressure application could be adapted to heart frequency and varied along the legs and the patient could perform self-treatment at home. In addition, this method might be used to widen blood vessels without surgery [Willy, 1999].

21.5.5.7 Vascular Surgery Devices—Microanastomosis Connector

In surgery of the hand, heart, brain, and spine, as well as in transplantation, it is a challenging task to reconnect two ends of divided small blood vessels. Some of the problems that may be encountered include long operation time, clotting, anastomosis patency, vascular stenosis, and foreign-body reaction. Using EAP, a connector (see Fig. 21) was developed for microanastomosis surgery that has passed the stage of proof of concept (Micromuscle, Sweden, www.micromuscle.com). The connector is electrically activated to contract using a low voltage (~1 V) and a small current. The activation time is less than 1 minute whereas the deactivation (dilatation of the connector) is passive and it starts as soon as the electric field and current are removed. The connector is used similar to a stent in angioplasty where in its contracted state it is placed inside the vessel-ends to joining them once the slow expansion is completed. The connector is designed to be nonthrombogenic and has thin walls to avoid restricting the blood flow through the vessel. Testing of the connector for biocompatibility was already initiated and it included cell cultivation and implantation into mammals. So far, it was demonstrated as noncytotoxic with minimal tissue reaction.

Figure 21: A view of an EAP-based microanastomosis connector that is currently under development is shown next to a match to illustrate its size. (Courtesy of Micromuscle, Sweden.)

21.5.6 Liquid and Gas Flow Control and Pumping

The use of EAP to articulate diaphragms and valves offers the capability to control the flow of liquids and gases. Pumps, brakes, and other devices can be developed using this capability, and the haptic interface described earlier is a novel application of the capability of EAP. There is a great need for zero mass flow (ZMF) in jets, and the Aeronautical and Maritime Research Laboratory (AMRL), Australia, is pursuing the development of a ZMF as shown in Fig. 22 21 [Giacobello, 2000]. A ZMF comprises a cavity that has a small orifice flush to the surfaces of the air vehicle at one end, and a thin diaphragm at the opposite end. If this diaphragm is made of EAP polymer, it can be driven by an alternating voltage to deform out of plane in a periodic manner. On the down stroke of the diaphragm, fluid can be drawn into the cavity, while on the upstroke fluid is forced out. Over each complete cycle of operation, the net mass injected through the orifice is zero (hence the name ZMF jet), while the net momentum out of the orifice is nonzero. This injection of momentum (or energy) near the surface of the vehicle can modify the flow detail near the surface, and in some cases can improve the aerodynamic characteristics of the vehicle. The same effect could be achieved by replacing the diaphragm with a piston or some other mechanical or pneumatic mechanism, but EAP has the advantage of simplicity and low weight, which are crucial requirements for flying vehicles.

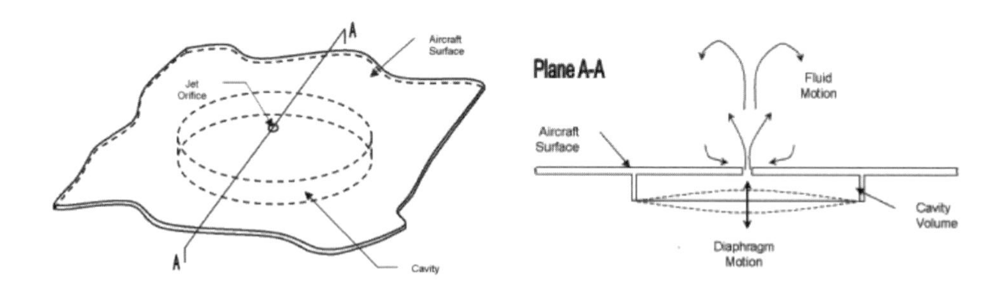

Figure 22: Schematic description of a ZMF jet. [Courtesy of M. Giacobello, DSTO (Defense Science and Technology Organization), Melbourne, Australia].

21.5.7 Noise Reduction

In recent years, concern about acoustic noise has grown significantly. Areas that need reduced noise include aircraft cabins, automobiles, and other noisy environments where comfort and hearing safety of the passengers, users, or operators may be affected. Suppressing unwanted noise can be done by active and/or passive noise control [Yang and Chen, 1999]. Active control is becoming a widespread method of reducing low-frequency noise, which is difficult to address by passive control methods (including sound absorbers). However, these techniques are less effective in damping high frequency noise. A study is under way at the Industrial Technology Research Institute, Taiwan, to develop such a noise suppression system that is activated by EAP. The use of the EAP film offers miniature, lightweight, low-cost actuators with large actuation displacement and high-power output. The smart system that is under development uses hybrid passive-active control to address noise over a wide frequency range, and it consists of an EAP film, an air gap, and a fibrous layer. The EAP film is used to minimize the reflected acoustic waves, and it operates as a loudspeaker, whereas the fibrous surface on the backside provides a feed forward least-mean-square (LMS) control.

21.5.8 Electromechanical Polymer Sensors and Transducers

The ability of polymers to convert various forms of energy from one to another makes them effective power sources, sensors, and transducers. Nafion, for example, is used in fuel cells to produce hydrogen through hydrolysis and for charge storage. As reported earlier, this polymer is also the basis of the electroactive polymer IPMC.

As an example of a polymer sensor, polyvinylidene fluoride (PVDF or PVF_2) is one of the most widely known EAP and is used as an ultrasonic transmitter and receiver for frequencies up to several tens of megahertz [Chapter 4; Bar-Cohen, 1996]. The pioneering development of effective piezoelectric polymers is attributed to Fukada and Yasuda, who discovered in the 1950s and 1960s that rolled films of polypeptides and numerous other polymers induced surface charges when stressed [Fukada and Yasuda, 1957]. A major milestone in this field was recorded in 1969 with Kawai's discovery of the strong piezoelectric effect in polyvinylidene fluoride (PVDF or PVF_2) [Kawai, 1969]. Later, other PVDF copolymers were reported, including P(VDF-TrFE) and P(VDF-TeFE) [Tasaka and Miyata, 1985].

Ferroelectric polymers are produced by a variety of techniques. In the case of PVDF, the material is mechanically drawn by extrusion and stretching, and during processing the film is subjected to a strong electrical polarization field. Without drawing, PVDF shows only weak piezoelectric behavior; the higher the molecular orientation, the stronger the response of the polarized film. After polarization, PVDF exhibits considerably stronger piezoelectric response than most other polymers [Kawai, 1969]. The discovery of the piezoelectric, and later of the pyroelectric, properties of PVDF, and the growing applications of this polymer [Tamura et al., 1975] sparked extensive research and development activities. Some of the piezoelectric polymers that are known today include: aromatic polyamides, polyparaxylene, poly-bischloromethyuloxetane (Penton), polysulfone, polyvinyl fluoride, synthetic polypeptide and cyanoethul cellulose [Wang et al., 1988].

21.5.9 Microelectromechanical Systems (MEMS)

In the past 15 years, the field of MEMS has grown tremendously to become a multi-billion dollar industry. This technology offers several distinct characteristics and advantages including mass fabrication, miniaturization, and integration of arrays. The ability of microsystems with integrated circuitry to sense and process information, compute a course of action, and control systems using actuators increases device functionality and performance. A number of existing MEMS applications have clearly demonstrated revolutionary capabilities in the areas of sensors, electronics components, guidance systems, integrated chemical analysis units, adaptive optical systems, and fluid control [Pisano, 1999; Kovacs, 1998; Fujita, 1997; Bryzek et al., 1994; Madou, 1998; Gardner, 1994; Tabib-Azar, 1994]. Progress is now being made on two major fronts:

fabrication technology and multidisciplinary applications. EAP can be a critical addition to the existing list of materials available for MEMS [Liu and Bar-Cohen, 1999; Smela et al., 1995, Smela, 1998; Smela et al., 1999; Smela, 1999, Jager et al., 2000]. Using existing microfabrication methods, large force and displacement cannot be readily achieved under practical constraint of conventional actuators in terms of power consumption, volume, or voltage. Electrostatic actuation requires significant levels of applied voltage for actuation. Magnetostatic and thermal bimetallic actuation use large power for generating magnetic fields or for providing temperature differentials, respectively. The achievable displacement of piezoelectric materials is still limited by relatively low coupling coefficients. Electroactive polymer materials with large displacement using low power are of critical importance for advancing the technology of microactuation and MEMS. Some of the applications that can be considered include: (1) pumps and valves for microfluid systems used as components of integrated chemical analysis units; (2) microactuators for active and adaptive fluid control; (3) linear microtransposition stages for microparallel assembly; and 4) position-tuning elements in micro-optical systems. In the following areas, the use of MEMS-based EAP offers significant advantages over conventional electromechanical systems and sensors [Liu and Bar-Cohen, 1999]:

(1) *Compact integration of comprehensive functionality.* Mechanical structures and active components are integrated with electronics (e.g., signal processing circuits), sensors (temperature, pH, etc.), optics, fluid components (e.g., fluid channels, micropumps, microvalves), and high-performance chemical analytical systems (e.g., electrophoresis) to realize comprehensive functional integration in "smart" sensors and actuators.
(2) *Mass fabrication and repeatability of performance.* Because integrated fabrication processes do not involve direct manual modification and assembly, device fabrication can be extremely efficient and reliable even though the size of individual devices is so small that human hands can no longer handle them. The photolithography process (described in Chapter 15) enables individual components to have extremely uniform geometry and performance, a major advantage in contrast to hand assembled instruments.
(3) *Miniaturization.* As a result of the photolithography process, MEMS systems can have extremely small features (on the order of micrometers) and have well-controlled size. The characteristic length scale of MEMS components is in the range of 1 μm to 1 cm. The small size makes it possible to insert microelectromechanical systems into a variety of applications that were previously not possible or practical. For example, pressure sensors are being integrated with automotive tires to provide on-line real-time monitoring of tire pressure [Ko et al., 1996]. Micromachined drug delivery systems are being considered for use as implantable smart drug capsules [Service, 1999]. Microinertia sensors are being used for smart projectiles to automatically adjust the trajectory for gun jump and wind factors. Micromachined digital propulsion is finding applications in

controlling the position of microsatellites [Lewis, 1999]. In these particular applications, the insertion of smart function was previously not possible with macroscopic devices.

The mechanical properties of silicon make it the primary construction material of MEMS. However, silicon has problems associated with its high cost and low fracture strain, which is on the order of 0.1%. The fact that silicon cannot tolerate large bending strain and significant impact seriously limits the possible applications. Polymers offer an attractive basis for MEMS as both a substrate and a construction material. Soft polymer materials that can be efficiently microfabricated at a low cost are highly desirable. Also, as EAP technology evolves it may become a necessity to imbed MEMS (e.g., emulate fingers with tactile sensing) into the EAP actuators to allow sensing, control, etc. Some of the characteristics that would benefit both the field of MEMS and EAP include:

- Polymers are elastic, can absorb impact energy, and can tolerate a large deformation.
- Polymers can be formed into 3D structures as well as efficient high-aspect-ratio 3D structures using micromolding processes.
- Potentially producing ionic EAPs with faster actuation can be enabled due to the short diffusion distances and smaller resistances.
- These materials are relatively inexpensive compared with silicon.

21.5.10 Other Potential Applications

Numerous other applications can be enabled or are currently being pursued using EAP, including pumps, power generators, etc. [see Topic 7]. EAP actuators and sensors can benefit ink-jet printing technology by establishing inexpensive damage-tolerant actuation and sensing mechanisms. In return, improved processing techniques can be used to fabricate miniature 3D structures, possibly made of polymers. Researchers at Penn State University, as well as Dankook, Ajou, and Korea Universities [Lee et al., 2000] are seeking to use EAP membranes and valves to produce miniature pumps. The ability of some EAP materials, such as the ferroelectric types, to operate as transducers that convert mechanical energy to electrical energy offers the potential of producing power using inexpensive devices that can convert winds or water waves to low cost electricity [Suzuki, 1993 and Ashley, 2003]. Further, as defense resources decline, the use of inexpensive and affordable technologies is becoming increasingly critical to assuring cutting-edge capability while meeting budget constraints. Using improved EAP-based hardware can significantly add to current capabilities as well as enable new ones. Devices such as insectlike robots can change the face of future battlefields, as illustrated graphically in Fig. 23.

Figure 23: Low-cost miniature EAP devices such as flying insectlike robots (top left) can change the battlefield of the future.

Butterflylike and dragonflylike robots may someday be controlled remotely to perform surveillance, information gathering, and airborne sensing. Artificial bugs that may walk on water and other biologically inspired devices may be developed with capabilities far superior to natural creatures since they are not constrained by evolution and survival needs.

21.6 EAP Characterization

Accurate information about the properties of EAP materials is critical to potential designers who are considering the construction of related mechanisms or devices. In order to assess the competitive capability of EAPs, there is a need for a performance matrix that consists of comparative performance data. Such a matrix needs to show the properties of EAP materials as compared to other classes of actuators, including piezoelectric ceramic, shape memory alloys, hydraulic actuators, and conventional motors. Studies are currently under way at JPL and other research institutes to define a unified matrix and establish effective test capabilities with emphasis on the macroscale [Chapter 15]. Key parameters are identified and test methods are being developed to allow measurements with minimum effect on the EAP. While the electromechanical properties of longitudinal electronic-type EAP materials can be addressed with some of the conventional test methods, ionic-type EAPs such as IPMC are posing technical challenges. The response of these materials suffers complexities that are associated with the mobility of the cation on the microscopic level, the strong dependence on the moisture content, and hysteretic behavior. A video camera and image processing software offer the capability of studying the deformation of IPMC strips under various mechanical loads (Fig. 24). Simultaneously, the electrical properties and the response to electrical activation are measured. Nonlinear behavior has been clearly identified in both the mechanical and electrical properties (Fig. 25). Efforts are currently under way to relate the experimental data to a nonlinear Euler-Bernoulli beam model, augmented with an eigen-strain [see Chapter 12]. Such an effort is a key to determining the properties and understanding the electromechanical behavior of EAP materials, optimizing their properties, and assuring their operational robustness.

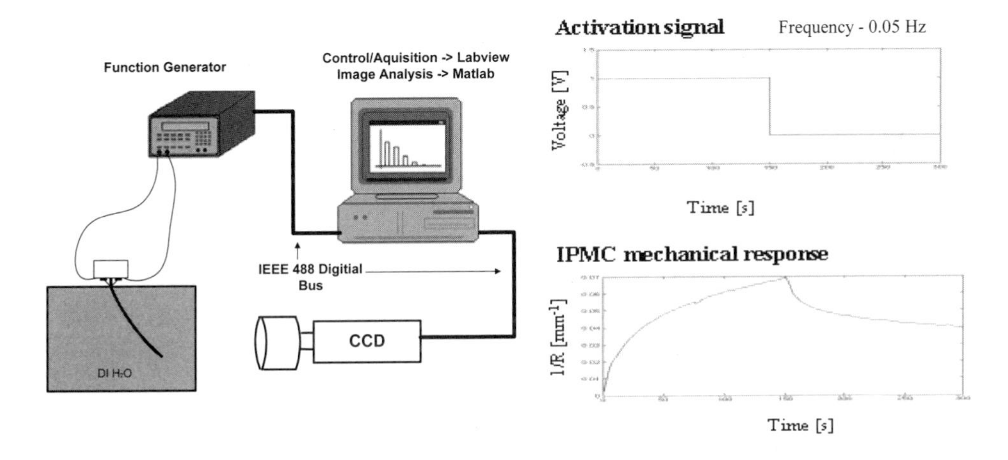

Figure 24: Mechanical test setup (left) and electromechanical response of IPMC (right). The figure on the right shows that IPMC continues to deform even under a dc field, making the measurements a challenge.

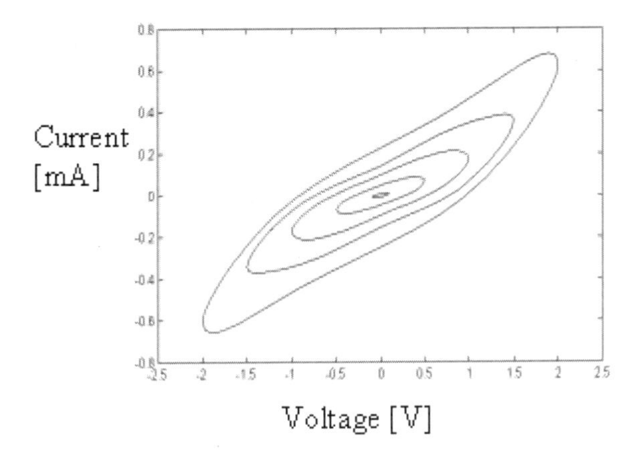

Figure 25: Nonlinear electrical response of IPMC.

21.7 Platforms for Demonstration of EAP

Developing biomimetic intelligent robots involves many disciplines and capabilities including materials, actuators, sensors, structures, functionality, control, intelligence, and autonomous operations. Mimicking nature has immensely expanded the collection and functionality of robots, allowing new classes of tasks to be performed. As technology evolves, great numbers of biologically inspired robots, emulating biological creatures are expected to emerge and find applications in our daily life. Even a simple task of staying stable while being pushed is complex for a robot to perform and will require significant progress in many technology areas.

Beside the armwrestling challenge, to promote the development of humanlike robots that are actuated by artificial muscles, two platforms were developed and made available to the author at his lab at JPL for the support of the worldwide development of EAP. These platforms include an android head (Fig. 18) that can make facial expressions and a robotic hand with joints (Fig. 26). The head can move the eyes and the lips, whereas the hand allows moving the index finger. At present, conventional electric motors are used to produce the required movement, thus leading to the desired facial expressions of the android. Once effective EAP materials are developed, they will be modeled into the control system in terms of surface-shape modifications and control instructions for the creation of controlled expressions. The robotic hand is equipped with wire-based tendons and has sensors for the operation of the various joints mimicking the human hand. This robotic hand is also driven by conventional motors allowing the up and down movement of the index finger of this hand. As with the android head, this hand is also used as a baseline for milestone comparison with the performance of emerging EAP actuators. Artificial muscles will substitute the motors when such EAP materials are developed to serve as effective actuators.

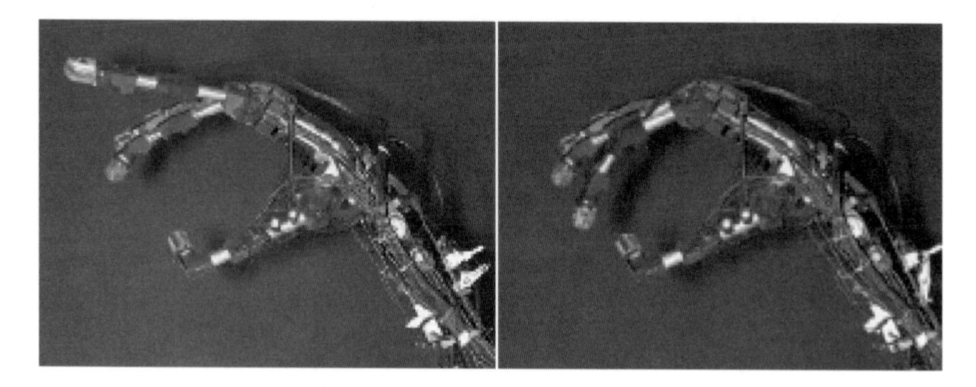

Figure 26: Robotic hand (photographed at JPL) serves as a platform for the demonstration of EAP actuators. (The arm was provided to JPL as a courtesy of G. Whiteley, Sheffield Hallam U., UK.)

Once the challenge of wrestling a human arm against a robotic arm that is driven by EAP reaches the point that the robot wins, a major milestone will be met. This accomplishment will require rapidly responding effective artificial muscles, and effective sensors and control algorithms that can counter the response of the human opponent as well. Such a success will allow the construction of robots that provide the most biologically similar capability in terms of mechanical response and performance. Thus, it will also become an important milestone for the field of biomimetics in demonstrating the capability to produce superior biologically inspired intelligent robots [Bar-Cohen and Breazeal, 2003].

21.8 Future Expectations

For many years, electroactive polymers (EAP) received relatively little attention due to their limited actuation capability and the small number of available materials. As reported in this book, there is now a relatively large arsenal of new EAP materials that exhibit large displacement in response to electrical stimulation. EAPs operational similarity to biological muscles are making them attractive as actuators, particularly their resilience, damage tolerance, and ability to induce large actuation strains (stretching, contracting, or bending). The application of these materials as actuators to drive various manipulation, mobility, and robotic devices involves multidisciplines including materials, chemistry, electromechanics, computers, and electronics. Even though the actuation force of existing EAP materials and their robustness require further improvement, there has already been a series of reported successes in the development of EAP-actuated mechanisms. Successful devices that have been reported include a fish-robot, audio speakers, catheter-steering element, miniature manipulator and robotics, miniature robotic arm, gripper, loudspeaker, active diaphragm, and dust wiper. The field of EAP has enormous potential in many application areas, and, judging from the range of inquiries that the author has received in the last five years, it seems that almost any aspect of our lives can be impacted. Some of the applications considered are still far from being practical, and it is important to tailor the requirements to the level that current materials can address. Using EAP to replace existing actuators may be a difficult challenge and therefore it is highly desirable to identify a niche application where EAP materials would not need to compete with existing technologies.

The development of niche applications is currently being sought while exploring commercial applications. Upon identifying such applications and emergence of EAP-driven products, one can expect rapid evolutionary improvements that will propel the capabilities and nurture a wide range of technology transfers. The enhancement of the actuation capability of EAP materials is studied on the fundamental level using computational chemistry, and the predictions would need to be implemented by developing new synthesis techniques [see Chapter 10]. The large strain response of EAP to electrical stimulation is nonlinear and requires adequate analytical tools for the design and control of related devices [Chapters 11–13]. Efforts are currently being made to model this nonlinear electromechanical behavior and to develop experimental techniques of material properties measurements and characterization. These efforts are leading and continue to contribute to better understanding of the origin of the electroactivity in various EAP materials, allowing improvement of their performance, and offering effective design tools for simulation of the performance of related devices. Methods of producing EAP fibers and films are being studied to effectively operate them as actuators and sensors and to improve their robustness [Chapter 14].

Generally, EAP actuators are highly agile and have the inherent ability to host embedded sensors and MEMS. Their unique characteristics offer valuable alternatives to current actuators that include electroactive ceramics and shape memory alloys. Also, inexpensive, lightweight, and low-power EAP-driven devices can be mass-produced, offering potential applications for commercial, medical, space, military, and many other fields. The making of miniature EAP-driven insectlike robots that can crawl, swim, and/or fly may become a reality as this technology evolves. Ink-jet printing techniques may potentially be employed to make complete devices that are driven by EAP actuators, where the devices can be produced with full 3D details, thereby allowing rapid prototyping and subsequent mass production possibilities. Thus, polymer-based EAP-actuated insectlike robots may be fully produced by an ink-jet printing process enabling the rapid implementation of science fiction ideas into engineering models and commercial products.

Some of the challenges that are facing the users of EAP materials and could help expand their potential applications include the availability of EAP materials with low glass transition temperature with significant response at low or high temperatures. Space applications are in great need for materials that can operate down to single digit degrees of Kelvin or high temperatures in the hundreds of Celcius, as on Venus. Another challenge to EAP is the development of large scale EAP in the form of films, fibers and others. The required dimensions can be as large as several meters or kilometers. In such dimensions, they can benefit space applications that allow development of such large gossamer structures as antennas, solar sails, optical components, and others in support of future NASA missions.

In order to exploit the highest benefit that EAP materials can offer multidisciplinary international cooperative efforts need to grow further among scientists, engineers, and other experts (e.g., medical doctors, etc.). Experts in chemistry, materials science, electro-mechanics/robotics, computer science, electronics, etc., need to work together to develop improved EAP materials, processing techniques and applications that can benefit from EAP technologies. Effective feedback sensors and control algorithms are needed to address the unique and challenging aspects of EAP actuators. If EAP-driven artificial muscles can be implanted into a human body, this technology can make a tremendously positive impact on many human lives.

Besides reviewing the state of the art and providing a tutorial and reference guide, this book addresses the various infrastructure issues related to producing

EAP actuators and devices as well as the envisioned direction of this field. Since this field is far from mature and progress is expected to change the field in future years, efforts were made in this book to cover futuristic thoughts as much as possible to help pave the path for the evolution of the field. As the technology evolves the days are nearing when the first armwrestling match between a robot that is actuated by EAP and human will be held. A robot win will make it clear that EAP performance has reached a level at which devices designed to emulate many of the physical functions that a human can perform are altogether possible.

21.9 Acknowledgments

The research at Jet Propulsion Laboratory (JPL), California Institute of Technology, was carried out under a contract with National Aeronautics Space Agency (NASA) and Defense Advanced Research Projects Agency (DARPA). The author would like also to thank everyone who contributed to the material that was used in this chapter including Dr. Margaret A. Ryan, JPL, for contributing the section on artificial nose. Special thanks to the team members of the telerobotic task LoMMAs that was funded by the NASA Code S program. The author would like to acknowledge the contribution of Dr. Sean Leary, Mark Schulman, Dr. Tianji Xu, and Andre H Yavrouian, JPL, Dr. Joycelyn Harrison, Dr. Joseph Smith, and Ji Su, NASA LaRC, Prof. Richard Claus, Virginia Tech, Prof. Constantinos Mavroidis, and Charles Pfeiffer, Rutgers University. Also, the author would like to thank Dr. Timothy Knowles, ESLI, for his assistance with the development of the EAP dust-wiper mechanism using IPMC, and Marlene Turner, Harry Mashhoudy, Brian Lucky, and Cinkiat Abidin, former graduate students of the Integrated Manufacturing Engineering (IME) Program at UCLA, for helping to construct the EAP gripper and robotic arm. The author would like to thank Chang Liu, University of Illinois at Urbana Champaign, for his contribution in the area of MEMS and Michael Goldfarb, Vanderbilt University, for his contribution in the area of biologically inspired robots. The author would like to express a special thank you to Dr. Keisuke Oguro, Osaka National Research Institute, Japan, for providing his most recent IPMC materials; and to Prof. Satoshi Tadokoro, Kobe University, Japan, for his analytical modeling effort. Prof. Mohsen Shahinpoor is acknowledged for providing IPMC samples as part of the early phase of the LoMMAs task. The author would also like to thank his NDEAA team members Dr. Xiaoqi Bao, Dr. Shyh-Shiuh Lih, and Dr. Stewart Sherrit and the former members Dr. Virginia Olazábal, Dr. José-María Sansiñena, for their help. A specific acknowledgment is made for the courteous contribution of graphics from individuals and organizations as specified next to the related graphics or tables.

21.10 References

Adams, P. N., D. Bowman, L. Brown, D. Yang and B. R. Mattes, "Molecular weight dependence of the physical properties of protonated polyanline films and fibers," Proceedings of the EAPAD 2000, Y. Bar-Cohen (Ed.), Vol. 4329 (2001) pp. 475-481.

Ashley, S., "Artificial muscles," *Scientific American* (Oct. 2003), p. 52-59.

Bar-Cohen Y. (Ed.) "Worldwide electroactive polymers newsletter," JPL's NDEAA Technologies, http://ndeaa.jpl.nasa.gov/nasa-nde/lommas/eap/WW-EAP-Newsletter.html, started in June 1999.

Bar-Cohen Y., "Active reading display for blind (ARDIB)," submitted as a New Technology Report, February 5, 1998, Item No. 0008b, Docket 20410, April 9, 1998.

Bar-Cohen, Y., (Ed.), *Proceedings of the Electroactive Polymer Actuators and Devices*, Smart Structures and Materials 1999, SPIE Proc. Vol. 3669, (1999), pp. 1-414.

Bar-Cohen Y., (Ed.), "Electro-active polymer (EAP) actuators and devices," Proceedings of the EAPAD 2000, SPIE's 7th Annual International Symposium on Smart Structures and Materials, Vol. 3987 (2000) pp. 1-360.

Bar-Cohen Y., "Electroactive polymers as artificial muscles—capabilities, potentials and challenges," Keynote Presentation, Robotics 2000 and Space 2000. Albuquerque, NM, USA, February 28 - March 2, (2000), pp. 188-196.

Bar-Cohen Y., and C. Breazeal (Eds), *Biologically Inspired Intelligent Robots*, SPIE Press, Bellingham, WA, (2003).

Bar-Cohen Y., C. Pfeiffer, C. Mavroidis, and B. Dolgin, "Remote MEchanical MIrroring using Controlled stiffness and Actuators (MEMICA)," NASA New Technology Report, Item No. 0237b, Docket 20642, January 27, 1999 and NASA Tech Briefs, Vol. 24, No. 2, (Feb. 2000), pp. 7a-7b.

Bar-Cohen Y., "Tactile Electro-Active Braille Display (TEABDI) for Presentation of Text and Graphics to Blind," NASA New Technology Report (NTR), Docket No. 40254, March 15, 2003.

Bartlett P. N., and J. W. Gardner, *Electronic Noses: Principles and Applications*, Oxford University Press (1999).

Bryzek J., K. Petersen and W. McCulley, "Micromachines on the march," *IEEE Spectrum, May* 1994.

Chmielewski A. B. and C. H. Jenkins "Flexible load-bearing structures (including inflatables)," Chapter 5, *Structures Technology for Future Aerospace Systems*, Ahmed Noor (Ed.) AIAA Progress in Astronautics and Aeronautics Series, (expected to be published in 2000)

de Rossi D., A. Della Santa and A. Mazzoldi, "Dressware: wearable hardware," *Elsevier Science, Materials Science and Engineering C*, Vol. 7 (1999) pp. 31-35.

Della S. A. and D. de Rossi, "Intervascular microcatheters steered by conductive polymer actuators," Proceedings of the 18th Annual International Conf. of the IEEE Engineering in Medicine and Biology Society, Amsterdam (1996) pp. 2203-2204.

Fisch A., C. Mavroidis, Y. Bar-Cohen, and J. Melli-Huber, "Haptic and Telepresence Robotics" Y. Bar-Cohen and C. Breazeal (Eds), *Biologically Inspired Intelligent Robots*, SPIE Press, Bellingham, WA, (2003) pp. 73-101.

Fischer G., A. G. Cox, M. Gogola, and M. K. Gordon, "Elastodynamic locomotion in mesoscale robotic insects," Bar-Cohen, Y., (Ed.), *Proceedings of the Electroactive Polymer Actuators and Devices*, Smart Structures and Materials, SPIE Proc. Vol. 3669, (1999a), pp. 362-368.

Fischer G., A.G. Cox, M. Gogola, M.K. Gordon, N. Lobontiu, D. Monopoli, E. Garcia, and M. Goldfarb, "Elastodynamic locomotion in mesoscale robotic insects," *Proceedings of the Electroactive Polymer Actuators and Devices*, Smart Structures and Materials 1999, SPIE Proc. Vol. 3669, (1999b), pp. 362-368.

Freund M. S., and N. S. Lewis, "A chemically diverse conducting polymer-based electronic nose," *Proc. Natl. Acad. Sci. U.S.A.* 1995, *92*, 2652.

Fujita H., "A decade of MEMS and its future," *Proceedings of the Tenth Annual International Workshop on MEMS (MEMS 97),* Nagoya, Japan.

Fukada, E., I. Yasuda, "On the piezoelectric effect of bone," *Jpn J. Appl. Phys.* Vol. 12 (1957), pp. 1158-1162.

Full, R.J., and Tu, M.S., "Mechanics of six-legged runners," *J. Exp. Biol.* Vol. 148, (1990) pp. 129-146.

Furukawa J., and X. Wen, "Electrostriction and piezoelectricity in ferroelectric polymers," *Japanese Journal of Applied Physics*, Vol. 23, No. 9, pp. 677-679, (1984).

Gardner J.W., *Microsensors: Principles and Applications*, Wiley, (1994).

Giacobello M., "Electro-active polymer as an actuator in flow control applications," WW-EAP Newsletter, http://ndeaa.jpl.nasa.gov/nasa-nde/lommas/eap/WW-EAP-Newsletter.html, Vol. 2, No. 1 (2000), p. 13.

Holland O., and Melhuish C. "Getting the most from the least: lessons for the nanoscale from minimal mobile agents," *Artificial Life V*, Nara, Japan (1996)

Hunter I. W., and Lafontaine, S. "A comparison of muscle with artificial actuators," *IEEE Solid-State Sensor and Actuator Workshop*, pp. 178-165, (1992).

Jager EWH., O. Inganäs and I. Lundströ̈m, "Microrobots for micrometer size objects in aqueous media: potential tools for single-cell manipulation," *Science*, Vol. 288 (2000) pp. 2335-2338.

Kawai, H., "The piezoelectricity of poly(vinylidene fluoride)," *Jpn. J. Appl. Phys.*, Vol. 8, (1969) pp. 975-976.

Kim K. J., and M. Shahinpoor, "Development of three dimensional ionic polymer-metal composites as artificial muscles," *Polymer*, Vol. 43, No. 3, pp. 797-802 (2002).

Ko W.H., Q. Qiang, and Y. Wang, "Touch mode capacitive pressure sensors for industrial applications," *Solid state sensor and actuator workshop*, (1996) p. 244.

Konyo M., S. Tadokoro, T. Takamori, K. Oguro, "Artificial tactile feel display using soft gel actuators," Proc. IEEE ICRA, (2000) pp. 3416-3421.

Kornbluh, R. Pelrine R. and Joseph, J. "Elastomeric dielectric artificial muscle actuators for small robots," *Proceeding of the 3rd IASTED International Conference*, Cancun, Mexico, June, 14-16, (1995).

Kovacs G.T.A., *Micromachined Transducers Sourcebook*, McGraw-Hill, 1998.

Lee S. K., Y. Choi, S. S. Yang, J. J. Pak, "Fabrication of electroactive polymer actuator composed of polypyrrole and solid-polymer electrolyte and its application to micropump," Proceedings of the SPIE's EAPAD 2000, 7th Annual International Symposium on Smart Structures and Materials, SPIE Proc. Vol. 3987, (2000), pp. 291-299.

Lehmann W., H. Skupin, C. Tolksdorf, E. Gebhard, R. Zentel, P. Krüger, M. Lösche and F. Kremer, "Giant lateral electrostriction in ferroelectric liquid crystalline elastomers," *Nature* 410 (6827) (Mar 22, 2001) pp. 447-450.

Lewis D.H., S.W. Janson, R.B. Cohen, E.K. Antonsson, "Digital micropropulsion," *Proc. Int. Conf. on Micro electromechanical Systems*, p. 517, 1999.

Liu C., and Y. Bar-Cohen, "Scaling laws of microactuators and potential applications of electroactive polymers in MEMS," Y. Bar-Cohen, (Ed.), *Proceedings of the Electroactive Polymer Actuators and Devices*, Smart Structures and Materials 1999, SPIE Proc. Vol. 3669, (1999), pp. 345-354.

Liu C., "Silicon micromachined sensors and actuators for fluid mechanics applications," Ph.D. Thesis, California Institute of Technology, 1996.

Lu W., A. G. Fadeev, B. Qi, E. Smela, Benjamin R. Mattes, J. Ding, G. M. Spinks, J. Mazurkiewicz, D. Zhou, G. G. Wallace, D. R. MacFarlane, S. A. Forsyth, and M. Forsyth, "Use of ionic liquids for π-conjugated polymer electrochemical devices," *Science*, Vol. 297 (July 4, 2002), pp. 983-987.

Lonergan M. C., E. J. Severin, B. J. Doleman, R. H. Grubbs and N. S. Lewis, "Array-based sensing using chemically sensitive, carbon black-polymer resistors," *Chem. Mat.* 1996, *8*, 2298.

Madden J. D.W., P. G.A. Madden, and I. W. Hunter, "Conducting polymer actuators as engineering materials", Proceedings of SPIE 9th Annual Symposium on Smart Structures and Materials: Electroactive Polymer Actuators and Devices, Y. Bar-Cohen (Ed), SPIE Press, (2002), pp. 176-190.

Madou M., *Fundamentals of Microfabrication*, CRC Press, 1998.

Mazzoldi A. and D. de Rossi, "Conductive polymer based structures for a steerable catheter," Proceedings of the SPIE's 7th Annual International Symposium on Smart Structures and Materials, SPIE Proc. Vol. 3987, (2000) pp. 237-280.

Mazzoldi A., A. Della Santa, and D. de Rossi, "Conducting polymer actuators: properties and modeling," Chapter 7, *Macromolecular System–Materials Approach*, Springer-Verlang, Berlin (2000).

Mussa-Ivaldi S., "Real brains for real robots," *Nature*, Vol. 408, (16 November 2000), pp. 305-306.

Nemat-Nasser, S., and Li, J. Y. "Electromechanical response of ionic polymer-metal composites," *J. Appl. Phys.*, v.87, no. 7, (2000), p. 3321-3331.

Nemat-Nasser, S., and S. Zamani, "Experimental study of nafion- and flemion-based ionic polymer composites (IPMCs) with ethylene glycol as solvent," Y. Bar-Cohen (Ed.), *Proceedings of the SPIE's Electroactive Polymer Actuators and Devices,* Vol. 5051, (2003), pp. 233-253

Oguro K., Y. Kawami and H. Takenake, "Bending of an ion-conducting polymer film-electrode composite by an electric stimulus at low voltage," *J. Micromachine Society*, 5, 27/30, 1992.

Onishi Z, S. Sewa, K. Asaka, N. Fujiwara and K. Oguro, "Bending response of polymer electolete actuator," Bar-Cohen, Y., (Ed.), *Proceedings of the Electroactive Polymer Actuators and Devices*, Smart Structures and Materials, SPIE Proc. Vol. 3669, (1999), pp. 121-128.

Otero T. F., I. Canero and S. Villanueva, "EAP as multifunctional and biomimetic materials," *Proceedings of the Electroactive Polymer Actuators and Devices*, Smart Structures and Materials 1999, SPIE Proc. Vol. 3669, (1999), pp. 26-34

Pede D., G. Serra and D. de Rossi, "Microfabrication of conducting polymer by ink-jet stereolithography," *Elsevier Science, Materials Science and Engineering C*, Vol. 5 (1998) pp. 289-291.

Pisano A., "MEMS 2003 and beyond: a DARPA vision of the future of MEMS," *SPIE 6th Int. Symp. on Smart Structures and Materials*, plenary presentation, 1999.

Rust C., G. Whiteley, A. Wilson, "Using practice-led design research to develop an articulated mechanical analogy of the human hand," *Journal of Medical Engineering and Technology*, Vol. 22 No. 5 (1998), pp. 226-232

Rust C., G. Whiteley, A. Wilson, "First make something - principled, creative design as a tool for multi-disciplinary research in clinical engineering," 4th Asian Design Conference, Nagaoka, Japan, (Oct 1999).

Service R., "Silicon chip find role as in vivo pharmacist," *Science*, 283 (5402), 1999.

Shahinpoor M., and K. J. Kim, "A novel physically-loaded and interlocked electrode developed for ionic polymer-metal composites (IPMCs)," *Sensors and Actuators: A. Physical*, Vol. 96, No. 2/3, pp. 125-132 (2002).

Smala E., O. Inganäs and I. Lundstrőm, "Controlled folding of micrometer size-structures," *Science*, Vol. 268, (1995), pp. 1735-1738.

Smela E., "A microfabricated movable electrochromoc "pixel" based on polypyrrole," *Advanced Materials*, Vol. 11, No. 16 (1999) pp. 1343-1345.

Smela E., "Microfabrication of PPy microactuators and other conjugated polymers devices," *J. Micromech. Microeng.*, Vol. 9 (1998), pp. 1-18.

Smela E., M. Kallenback, and J. Holdenreid, "Electrochemically driven polypyrrole bilayers for moving and possibly bulk micromachined silicon plates," *J. Microelectromech. Systems*, Vol. 8, No. 4 (1999), pp. 373-383.

Spinks G. M., G. G. Wallace, J. Ding, D. Zhou, B. Xi, and J. Gillespie, "Ionic liquids and polypyrrole helix tubes: bringing the electronc brialle screen closer to reality," Y. Bar-Cohen (Ed.), *Proceedings of the SPIE's Electroactive Polymer Actuators and Devices,* Vol. 5051, (2003), pp. 372-388.

Suzuki, M., "Mechanical power generation by solvent sensitive polymer gel," Takagi, Toshinori, ed., *Proc. Int. Conf. Intell. Mater. 1st Technomic*, Lancaster, PA (1993) pp. 283-286.

Tabib-Azar M., "Integrated optics, microstructures, and sensors," Kluwar Academic Pub., 1995, S. Sze, ed., *Semiconductor Sensors*, Wiley Interscience, 1994.

Tamura M., T. Yamguchi, T. Oyaba and T Yoshimi, "Electroacoustic transducers with piezoelectric high polymer films," *J. Audio Eng. Soc.*, Vol. 23, No. 1, (1975), pp. 21-26.

Tasaka S., and S. Miyata, "Effects of crystal-structure on piezoelectric and ferroelectric properties of copoly-(vinylidenefluoride-tetrafluoroethylene)," *J. Appl. Phys.*, Vol. 57, No. 3, (1985), pp. 906-910.

Walker P. G., "Seeking bio-compatible EAP," JPL's NDEAA Technologies, WW-EAP Newsletter http://eis.jpl.nasa.gov/ndeaa/nasa-nde/newsltr/WW-EAP_Newsletter2-1.PDF, Vol. 2, No. 1 (2000), p. 16.

Wallace, D.B., "A method of characteristics model of a DoD ink-jet device using an integral drop formation method," ASME Pub. 89-WA/FE-4, (Dec. 1989).

Wang, T. T., J. M. Herbert and A.M. Glass, (Ed.), *The Applications of Ferroelectric Polymers*, Chapman and Hall, New York (1988).

Weber, C., "Zum Zerfall eines Flussigkeitsstrables," *Z. Angew. Math. Mech.,* Vol. 11, No. 136, (1931).

Wessberg J., C. R. Stambaugh, J. D. Kralik, P. D. Beck, M. Lauback, J.C. Chapin, J. Kim, S. J. Biggs, M. A. Srinivasan and M. A. Nicolelis, "Real-time prediction of hard trajectory by ensembles of cortical neurons in primates," *Nature*, Vol. 408, (16 November 2000), pp. 361-365.

Whiteley G., A. Wilson, C. Rust, "Development of elbow and forearm joints for an anatomically analogous upper-limb prosthesis," European Medical & Biological Engineering Conference, Vienna (Nov 1999).

Willy J., "Smart materials for anti-G suit," WW-EAP Newsletter, JPL's NDEAA Technologies, Vol. 2, No. 1, (2000), p. 14.

Willy J., "Anti-G suit based on smart materials," WorldWide-ElectroActive Polymers (WW-EAP) Newsletter, JPL's NDEAA Technologies, Vol. 1, No. 2 (1999) p. 9 http://ndeaa.jpl.nasa.gov/nasa-nde/lommas/eap/WW-EAP-Newsletter.html,

Yang T. L., and W.–H. Chen, "EAP smart noise reduce system & heat dissipation device," WW-EAP Newsletter, Vol. 1, No. 2 (1999), pp. 9-10 http://ndeaa.jpl.nasa.gov/nasa-nde/lommas/eap/WW-EAP-Newsletter.html

Zhang Q.M, personal communication over e-mail, July 2003.

Zhang Q.M., T. Furukawa, Y. Bar-Cohen, and J. Scheinbeim, "Electroactive polymers (EAP)," MRS Symposium Proceedings, Vol. 600, Warrendale, PA, (1999) pp. 1-336.

Index